AQA Science Chemistry

Teacher's Book

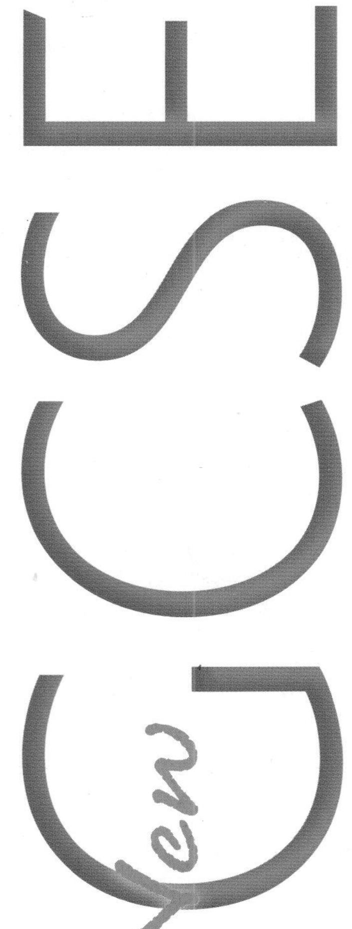

Sam Holyman

Bev Cox

John Scottow

Series Editor
Lawrie Ryan

Nelson Thornes

Published in 2011 by:
Nelson Thornes Ltd
Delta Place
27 Bath Road
CHELTENHAM
GL53 7TH
United Kingdom

12 13 14 15 / 10 9 8 7 6 5

A catalogue record for this book is available from the British Library

ISBN 978 1 4085 0830 5

Cover photograph: John Feingersh/Blend Images/Corbis

Page make-up by Tech-Set Ltd, Gateshead

Printed and bound in Spain by GraphyCems

GCSE Chemistry Contents

Welcome to AQA Chemistry!

This Chemistry Teacher's Book is written and reviewed by experienced teachers. This book is structured around the Student Book and offers guidance, advice, support for differentiation and lots of practical teaching ideas to give you what you need to teach the AQA specifications.

Learning objectives

These tell you what your students should know by the end of the lessons and relate to the learning objectives in the corresponding Student Book topic, although extra detail is provided for teachers.

Learning outcomes

These tell you what your students should be able to do to demonstrate that they have achieved against the learning objectives. These are differentiated where appropriate to provide suitable expectations for all your students. Higher Tier outcomes are labelled.

Specification link-up: Chemistry C1.1

These open every spread so you can see the AQA specification references covered in your lessons, at a glance.

Lesson structure

This provides you with guidance and ideas for tackling topics in your lessons. There are short and long starter and plenary activities so you can decide how you structure your lesson. Explicit **support** and **extension** guidance is given for some starters and plenaries.

Support

These help you to give extra support to students who need it during the main part of your lesson.

Extend

These provide ideas for how to extend the learning for higher ability students.

Further teaching suggestions

These provide you with ideas for how you might extend the lesson or offer alternative activities. These may also include extra activities or suggestions for homework.

Summary answers

All answers to questions within the Student Book are found in the Teacher's Book.

Practical support

For every practical in the Student Book you will find this corresponding feature which gives you a list of the equipment you will need to provide, safety references and additional teaching notes. There are also additional practicals given that are not found in the Student Book.

The following features are found in the Student Book, but you may find additional guidance to support them in the Teacher's Book:

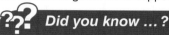

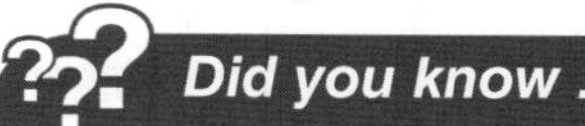

Maths skills

Activity

How Science Works

There is a chapter dedicated to 'How Science Works' in the Student Book as well as embedded throughout topics and end of chapter questions. The teacher notes within this book give you detailed guidance on how to integrate 'How Science Works' into your teaching.

End of chapter pages

At the end of each chapter you will find Summary answers and answers to the practice questions. You will also find:

Kerboodle resources

Kerboodle is our online service that holds all of the electronic resources for the series. All of the resources that support the chapter that are provided on Kerboodle are listed in these boxes.

Where you see (k) in the Student Book, you will know that there is an electronic resource on Kerboodle to support that aspect.

Just log on to www.kerboodle.com to find out more.

Practical suggestions

These list the suggested practicals from AQA that you need to be aware of. Support for these practicals can be found on Kerboodle, or are covered within the practical support section of the Teacher Book. The (k) indicates that there is a practical in Kerboodle. The (gears) indicates that there is a 'How Science Works' worksheet in Kerboodle. The (book) indicates that the practical is covered in this Teacher's Book.

Study tip

These are written to give advice on what students should remember for their exams and highlighting common errors.

H1 How does science work?

Learning objectives

Students should learn:

- that observations are often the starting point for an investigation
- that a hypothesis is a proposal intended to explain certain facts or observations
- that a prediction is an intelligent guess, based on some knowledge
- that an experiment is a way of testing your prediction
- that a conclusion is when you decide whether or not your prediction was correct.

Learning outcomes

Students should be able to:

- make first-hand observations
- distinguish between a hypothesis and a prediction
- explain the purpose of an experiment
- show how results can help to decide whether a prediction was correct.

Specification link-up: How Science Works

'How Science Works' is treated here as a separate chapter. It offers the opportunity to teach the 'thinking behind the doing' as a discrete set of procedural skills. However, it is of course an integral part of the way students will learn about science, and those skills should be nurtured throughout the course.

It is anticipated that sections of this chapter will be taught as the opportunity presents itself during the teaching programme. The chapter should also be referred back to at appropriate times when these skills are required and in preparation for the Controlled Assessment ISAs.

The thinking behind the doing

Science attempts to explain the world in which we live. It provides technologies that have had a great impact on our society and the environment. Scientists try to explain phenomena and solve problems using evidence. The data to be used as evidence must be repeatable, reproducible and valid, as only then can appropriate conclusions be made.

A scientifically literate citizen should, amongst other things, be equipped to question, and engage in debate on, the evidence used in decision-making.

The repeatability and the reproducibility of evidence refers to how much we trust the data. The validity of evidence depends on these as well as whether the research answers the question. If the data is not repeatable or reproducible the research cannot be valid.

To ensure repeatability, reproducibility and validity in evidence, scientists consider a range of ideas which relate to:

- how we observe the world
- designing investigations so that patterns and relationships between variables may be identified
- making measurements by selecting and using instruments effectively
- presenting and representing data
- identifying patterns, relationships and making suitable conclusions.

These ideas inform decisions and are central to science education. They constitute the 'thinking behind the doing' that is a necessary complement to the subject content of biology, chemistry and physics.

Lesson structure

Starters

Key words – Create a quiz looking at the meaning of key words used in this lesson: knowledge, observation, prediction and experiment. Support students by making this activity into a card sort. Extend students by asking them to put the key words into a context, i.e. the television remote control idea. *(5 minutes)*

Good science – Collect newspaper articles and news items from the television to illustrate good and poor uses of science. There are some excellent television programmes illustrating good and poor science. *(10 minutes)*

Main

- Students should begin to appreciate the 'thinking behind the doing' developed during KS3. It would be useful to illustrate this by a simple demonstration (e.g. burning a candle in an upturned gas jar) and posing questions that build into a flow diagram of the steps involved in a whole investigation. This could lead into recap questions to ascertain each individual student's progress. Emphasis should be placed on an understanding of the following terms: prediction, independent, dependent and control variables, and repeatability/hypothesis.
- It is expected that students will be familiar with:
  - the need to work safely
  - making a prediction
  - controls
  - the need for repetition of some results
  - tabulating and analysing results
  - arriving at appropriate conclusions
  - suggesting how they might improve methods.
- Revealing to the students that they use scientific thinking to solve problems during their everyday life can make their work in science more relevant. Use everyday situations to illustrate this and discussed in groups or as a class.

Support

- When carrying out the 'fireworks' activity, students could be given the headings in the form of a flow chart, with an 'either/or' option at each stage.

Extend

- Students could be allowed to use the internet to find out more about the ideal fuse-burning time for a firework. Remind students that it is illegal and dangerous to make fireworks, unless you have a licence.

For example: 'How can I clear the car windscreen when it "mists" up in the winter? It only seems to happen when we get into the car [observation]. The windscreen is probably cold, and I know we breathe out moist air [knowledge]'.

'I can use observations and knowledge to make a prediction that switching on the hot air fan next to the windscreen will clear the "mist". I can test my prediction and see what the results are. I can check again the next day to see whether I get the same results [repeatability].'

- Students should now be asked to complete the Investigating fireworks activity.

Plenaries

Misconceptions – Produce a list of statements about practical work (similar to the ones listed below), some of which are true and others false. Support students by getting them to simply write 'true' or 'false' by each statement. Extend students by asking them to write down why the false statements are untrue. *(5 minutes)*

Poor science – Using the internet, organise a competition for who can bring in the poorest example of science used to sell products – shampoo adverts are very good examples! *(10 minutes)*

Activity

Investigating fireworks

There is a great deal of information on many websites. The information includes: the history of fireworks, the chemistry of fireworks, the structure of large display fireworks and how the placement of chemicals relates to the display, how different metals are used for different colours, including their fuses, and why they continue to burn inside the firework without air; also, the electronic organisation of large displays.

Warning – misuse of fireworks can be dangerous!

H1 How does science work?

Learning objectives

- What is meant by 'How Science Works'?
- What is a prediction and why should you make one?
- What is a hypothesis?
- How can you investigate a problem scientifically?

This first chapter looks at 'How Science Works'. It is an important part of your GCSE because the ideas introduced here will crop up throughout your course. You will be expected to collect scientific **evidence** and to understand how we use evidence. These concepts will be assessed as the major part of your internal school assessment.

You will take one or more 45-minute tests. These tests are based on data you have collected previously plus data supplied for you in the test. They are called **Investigative Skills Assignments (ISA)**. The ideas in 'How Science Works' will also be assessed in your examinations.

How science works for us

Science works for us all day, every day. You do not need to know how a mobile phone works to enjoy sending text messages. But, think about how you started to use your mobile phone or your television remote control. Did you work through pages of instructions? Probably not!

You knew that pressing the buttons would change something on the screen (**knowledge**). You played around with the buttons, to see what would happen (**observation**). You had a guess based on your knowledge and observations at what you thought might be happening (**prediction**) and then tested your idea (**experiment**).

Perhaps 'How Science Works' should really be called 'How Scientists Work'.

Science moves forward by slow, steady steps. When a genius such as Einstein comes along, it takes a giant leap. Those small steps build on knowledge and experience that we already have.

The steps don't always lead in a straight line, starting with an observation and ending with a conclusion. More often than not you find yourself going round in circles, but each time you go around the loop you gain more knowledge and so can make better predictions.

links
You can find out more about your ISA by looking at H10 The ISA at the end of this chapter.

Figure 1 Albert Einstein was a genius, but he worked through scientific problems in the same way as you will in your GCSE

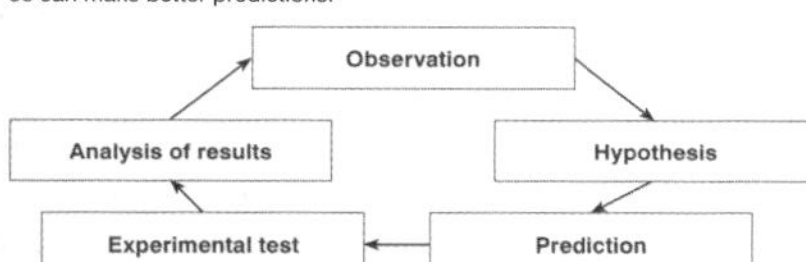

Each small step is important in its own way. It builds on the body of knowledge that we have. In 1675 a German chemist tried to extract gold from urine. He must have thought that there was a connection between the two colours. He was wrong. But after a while, with a terrible stench coming from his laboratory, the urine began to glow.

He had discovered phosphorus. Phosphorus catches fire easily. A Swedish scientist worked out how to manufacture phosphorus without the smell of urine. That is why most matches these days are manufactured in Sweden.

Activity

Investigating fireworks

Fireworks must be safe to light. Therefore you need a fuse that will last long enough to give you time to get well out of the way.

- Fuses can be made by dipping a special type of cotton into a mixture of two chemicals. One chemical (A) reacts by burning; the other (B) doesn't.
- The chemicals stick to the cotton. Once it is lit, the cotton will continue to burn, setting the firework off. The concentrations of the two chemicals will affect how quickly the fuse burns.
- In groups, discuss how you could work out the correct concentrations of the chemicals to use. You want the fuse to last long enough for you to get out of the way. However, you don't want it to burn so long that we all get bored waiting for the firework to go off!

You can use the following headings to discuss your investigation. One person should be writing your ideas down, so that you can discuss them with the rest of the class.

- What prediction can you make about the concentration of the two chemicals (A and B) and the fuse?
- What would you vary in each test? This is called the independent variable.
- What would you measure to judge the effect of varying the independent variable? This is called the dependent variable.
- What would you need to keep unchanged to make this a fair test? These are called control variables.
- Write a plan for your investigation.

Figure 2 Fireworks

Did you know ... ?

The Greeks were arguably the first true scientists. They challenged traditional myths about life. They put forward ideas that they knew would be challenged. They were keen to argue the point and come to a reasoned conclusion.

Other cultures relied on long-established myths and argument was seen as heresy.

Summary questions

1 Copy and complete this paragraph using the following words:
experiment knowledge conclusion prediction observation
You have learned before that a cup of tea loses heat if it is left standing. This is a piece of .... You make an .... that dark-coloured cups will cool faster. So you make a .... that if you have a black cup, this will cool fastest of all. You carry out an .... to get some results, and from these you make a ....

Key points

- **Observations** are often the starting point for an investigation.
- A **hypothesis** is a proposal intended to explain certain facts or observations.
- A **prediction** is an intelligent guess, based on some **knowledge**.
- An **experiment** is a way of testing your prediction.

Further teaching suggestions

Common misconceptions

Some common misconceptions that can be dealt with here and throughout the course are about:

- the purpose of controls – some students believe that it is about making accurate measurements of the independent variable.
- the purpose of preliminary work – some believe that it is the first set of results.

Some students also think:

- that the table of results is constructed after the practical work – students should be encouraged to produce the table before carrying out their work and to complete it during their work
- that anomalies are identified after the analysis – they should preferably be identified during the practical work or at the latest before any calculation of a mean
- they should automatically extrapolate the graph to its origin
- lines of best fit must be straight lines
- you repeat readings to make sure your investigation is a fair test.

Summary answers

1 knowledge, observation, prediction, experiment, conclusion

H2 Fundamental ideas about how science works

Learning objectives

Students should learn:

- to distinguish between opinion based on scientific evidence and non-scientific ideas
- the importance of continuous and categoric variables
- what is meant by valid evidence
- the difference between repeatability and reproducibility
- to look for links between the independent and dependent variables.

Learning outcomes

Students should be able to:

- identify when an opinion does not have the support of valid and reproducible science
- recognise measurements as continuous, or categoric
- suggest how an investigation might demonstrate its reliability and validity
- distinguish between repeatability and reproducibility
- state whether variables are linked, and if so, in what way.

Specification link-up: Controlled Assessment C4.3

Demonstrate an understanding of the need to acquire high quality data, by:

- appreciating that, unless certain variables are controlled, the results may not be valid *[C4.3.2 a)]*
- identifying when repeats are needed in order to improve reproducibility. *[C4.3.2 b)]*

Lesson structure

Starters

Crazy science – Show a video clip of one of the science shows that are aimed at entertainment rather than education or an advert that proclaims a scientific opinion. This should lead into a discussion of how important it is to form opinions based on sound scientific evidence. *(5 minutes)*

Types of variable – Produce a list of observable or measurable variables, e.g. colour, temperature, time, type of material. Ask students to sort them into two types: these can then be revealed as being either categoric or continuous. Support students by giving them a full definition of categoric and continuous variables. Extend students by asking them to add other examples of their own to the lists. *(10 minutes)*

Main

- Discuss some examples of adverts that make 'scientific' claims about products.
- From a light-hearted look at entertainment science, bring the thalidomide example into contrast and discuss how tragic situations can be created by forming opinions that are not supported by valid science. Search for video clips about thalidomide at www.britishpathe.com.
- Show how some metals bend more easily than others. Discuss, in small groups, the different ways in which the dependent variable could be measured, identifying these in terms of continuous and categoric measurements.
- Discuss the usefulness in terms of forming opinions of each of the proposed measurements.
- Consider that this might be a commercial proposition and the students might be advising an architect on which metal to use for a particular construction.
- Discuss how they could organise the investigation to demonstrate its validity and reproducibility to an architect.
- Discuss what sort of relationship there might be between the variables.

Plenaries

Evidence for opinions – Bring together the main features of scientific evidence that would allow sound scientific opinions to be formed from an investigation. *(5 minutes)*

Analyse conclusions – Use an example of a poorly structured investigation and allow the students to critically analyse any conclusions drawn, e.g. data from an investigation into different forms of insulation, using calorimeters and cooling curves. Support students by telling them the mistakes that had been made in the design and ask them to say why this would make the conclusion invalid. Extend students by allowing them to first identify the mistakes in the design. *(10 minutes)*

Support

- When looking at the bending of metals, ask students to suggest other properties of metals that would need to be considered when choosing a material for a particular task. Put these suggestions on 'Post-Its' on the wall.

Extend

- Discussion could range into the ethics of drug provision and the increased importance of having scientifically based opinions. It might develop into an appreciation of the limits of science in terms of ethical delivery of drugs. Decisions of this nature are not always as clear cut as scientists might want them to be. Some experts estimated the number of accidental deaths in the UK at 20 000 per year caused by prescribed medical drugs.

Practical support

Bending different metals

Equipment and materials required

Thin strips of different metals that bend to differing degrees.

Details

Some standard technique for bending the metals. If they are of different shapes, this will complicate the task.

Safety

Beware of sharp edges.

How Science Works

Fundamental ideas about how science works

H2 Fundamental ideas about how science works

Learning objectives

- How do you spot when an opinion is not based on good science?
- What is the importance of continuous and categoric variables?
- What does it mean to say that evidence is valid?
- What is the difference between a result being repeatable and a result being reproducible?
- How can two sets of data be linked?

AQA Examiner's tip

Read a newspaper article or watch the news on TV. Ask yourself whether any research presented is valid. Ask yourself whether you can trust that person's opinion and why.

Figure 1 Student recording a range of temperatures – an example of a continuous variable

Science is too important for us to get it wrong

Sometimes it is easy to spot when people try to use science poorly. Sometimes it can be funny. You might have seen adverts claiming to give your hair 'body' or sprays that give your feet 'lift'!

On the other hand, poor scientific practice can cost lives.

Some years ago a company sold the drug thalidomide to people as a sleeping pill. Research was carried out on animals to see if it was safe. The research did not include work on pregnant animals. The **opinion** of the people in charge was that the animal research showed the drug could be used safely with humans.

Then the drug was also found to help ease morning sickness in pregnant women. Unfortunately, doctors prescribed it to many women, resulting in thousands of babies being born with deformed limbs. It was far from safe.

These are very difficult decisions to make. You need to be absolutely certain of what the science is telling you.

a Why was the opinion of the people in charge of developing thalidomide based on poor science?

Deciding on what to measure: variables

Variables are physical, chemical or biological quantities or characteristics.

In an investigation, you normally choose one thing to change or vary. This is called the **independent variable**.

When you change the independent variable, it may cause something else to change. This is called the **dependent variable**.

A **control variable** is one that is kept the same and is not changed during the investigation.

You need to know about two different types of these variables:

- A **categoric variable** is one that is best described by a label (usually a word). The 'colour of eyes' is a categoric variable, e.g. blue or brown eyes.
- A **continuous variable** is one that we measure, so its value could be any number. Temperature (as measured by a thermometer or temperature sensor) is a continuous variable, e.g. 37.6 °C, 45.2 °C. Continuous variables can have values (called quantities) that can be found by making measurements (e.g. light intensity, flow rate, etc.).

b Imagine you were testing the energy given out in three different reactions (A, B and C). Would it be best to say **i** reactions A and B felt warm, but C felt hot, or **ii** reaction C got hottest, followed by A and finally B, or **iii** the rise in temperature in reaction C was 31 °C, in A it was 16 °C and in B it was 14 °C?

Making your evidence repeatable, reproducible and valid

When you are designing an investigation you must make sure that other people can get the same results as you. This makes the evidence you collect **reproducible**. This is more likely to happen if you repeat measurements and get consistent results.

A measurement is repeatable if the original experimenter repeats the investigation using the same method and equipment and obtains the same results.

A measurement is reproducible if the investigation is repeated by another person, or by using different equipment or techniques, and the same results are obtained.

You must also make sure you are measuring the actual thing you want to measure. If you don't, your data can't be used to answer your original question. This seems very obvious but it is not always quite so easy. You need to make sure that you have controlled as many other variables as you can, so that no one can say that your investigation is not **valid**. A measurement is valid if it measures what it is supposed to be measuring, with an appropriate level of performance.

c State one way in which you can show that your results are valid.

How might an independent variable be linked to a dependent variable?

Looking for a link between your independent and dependent variables is very important. The pattern of your graph or bar chart can often help you to see whether there is a link.

But beware! There may not be a link! If your results seem to show that there is no link, don't be afraid to say so. Look at Figure 2.

The points on the top graph show a clear pattern, but the bottom graph shows random scatter.

Summary questions

1 Students were asked to find the solubility of three different solids – D, E and F. Name each of the following types of dependent variable described by the students:
 a D and E were 'soluble', whereas F was 'insoluble'.
 b 0 g of F dissolved in 100 cm³ of water, 30.2 g of D dissolved in 100 cm³ of water, 25.9 g of F dissolved in 100 cm³ of water.

2 Some people believe that the artificial sweetener aspartame causes headaches and dizziness. Do you trust these opinions? What would convince you not to use aspartame?

AQA Examiner's tip

When designing your investigation you should always try to measure continuous data whenever you can. This is not always possible, so then you have to use a label (categoric variable). You might still be able to put the variables in an order so that you can spot a pattern. For example, you could describe flow rate as 'fast flowing', 'steady flow' or 'slow flowing'.

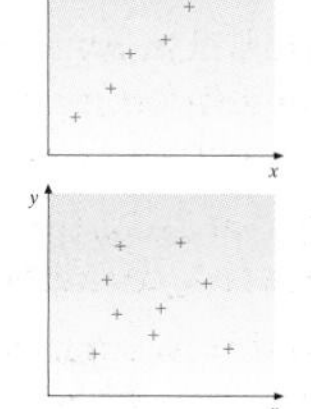

Figure 2 Which graph shows that there might be a link between *x* and *y*?

Did you know ... ?

At any time there are only about 20 atoms of francium in the entire planet. How do we know that this data is valid?

Key points

- Be on the lookout for non-scientific opinions.
- Continuous data give more information than other types of data.
- Check that evidence is repeatable, reproducible and valid.

4 5

Answers to in-text questions

a The original animal investigation did not include pregnant animals and was not carried out on human tissue, so was not valid, when the opinion was formed that it could be given to pregnant women.

b **iii** the rise in temperature in reaction C was 31 °C, in A it was 16 °C and in B it was 14 °C.

c Control all (or as many as possible) of the other variables.

Summary answers

1 **a** categoric
 b continuous

2 Believing something is not the same as knowing something. Evidence from scientific investigations would give you some facts from which you could make a reasonable conclusion.

H3 Starting an investigation

Learning objectives

Students should learn:

- how scientific knowledge can be used to observe the world around them
- how good observations can be used to make hypotheses
- how hypotheses can generate predictions that can be tested.

Learning outcomes

Students should be able to:

- state that observation can be the starting point for an investigation
- state that observation can generate hypotheses
- describe how hypotheses can generate predictions and investigations.

Specification link-up: Controlled Assessment C4.2 & C4.3

Develop hypotheses and plan practical ways to test them, by:

- being able to develop a hypothesis *[C4.1.1 a)]*
- being able to test hypotheses. *[C4.1.1 b)]*

Make observations, by:

- carrying out practical work and research, and using the data collected to develop hypotheses. *[C4.3.1]*

Lesson structure

Starters

Demo observation – Begin the lesson with a demonstration – as simple as lighting a match or more involved such as a bell ringing in a bell jar, with air gradually being withdrawn. Students should be asked, in silence, and without further prompting, to write down their observations. These should be collated and questions should be derived from those observations. *(5 minutes)*

Linking observation to knowledge – Discuss with students any unusual events they saw on the way to school. If possible, take them into the school grounds to look and listen to events. Try to link their observations to their scientific knowledge. They are more likely to notice events that they can offer some scientific explanation for. Support students by prompting with some directed questions. Extend students by getting them to start to ask questions about those observations. *(10 minutes)*

Main

- If in the laboratory, allow students to participate in a 'scientific happening' of your choice, e.g. either by blowing large and small bubbles or dropping paper cups with different masses in. Preferably use something that they have not met before, but which they will have some knowledge of.
- If students need some help at this point, they should read through the section on observation in the Student Book. Then answer in-text questions **a** and **b**.
- 'Making bubbles' practical: in groups they should discuss possible explanations for one agreed observation. Encourage a degree of lateral thinking. You might need to pose the questions for some groups, e.g. why were some bubbles larger than others? They could be told how the bubble mixture was made. Ask the group to select which of their explanations is the most likely, based on their own knowledge of science.
- Work these explanations into a hypothesis.
- Individually, each student should try in-text question **c**. Gather in ideas and hypotheses.
- Students, working in groups, can now turn this into a prediction.
- They could suggest ways in which their prediction could be tested. Identify independent, dependent and control variables and the need to make sure that they are measuring what they intend to measure.
- Go over in-text question **d** as a class.

Plenaries

Poster – Ask students to design, but not make, a poster that links 'Observation + knowledge → hypothesis → prediction → investigation'. *(5 minutes)*

Discussion – Read the following information and discuss with students the need to record observations as well as to make them.

A chemist called Karl Scheele illustrates a good example of how observation without knowledge can be a dangerous thing. He brilliantly discovered many elements including chlorine, nitrogen and oxygen, none of which he got credit for. He was keen to 'observe' as many properties of his discoveries as he could. His method included tasting them. His discovery of hydrocyanic acid probably led to his death aged 43! Priestley 'discovered' oxygen two years after Scheele. Davy 'discovered' chlorine 36 years after Scheele! *(10 minutes)*

Support

- Assist students in the 'Making bubbles' practical by giving them a list of possible explanations and asking them to choose which is the most likely.

Extend

- Extend students in the 'Making bubbles' practical by getting them to vary the amounts of detergent and glycerine added to the water and to make predictions regarding the effect this would have.

Practical support

Making bubbles

Equipment and materials required

Distilled water, 100 cm^3 plastic beaker, paper towels, washing-up liquid, glycerine, different sized loops, eye protection.

Details

Fill container three-quarters full with water and add some detergent and glycerine.

Students should make a prediction linking the size of the bubble to the size of the loop, and then use different sized loops to test their prediction.

How Science Works

H3 Starting an investigation

Learning objectives

- How can you use your scientific knowledge to observe the world around you?
- How can you use your observations to make a hypothesis?
- How can you make predictions and start to design an investigation?

Observation

As humans we are sensitive to the world around us. We can use our senses to detect what is happening. As scientists we use observations to ask questions. We can only ask useful questions if we know something about the observed event. We will not have all of the answers, but we know enough to start asking relevant questions.

If we observe that the weather has been hot today, we would not ask if it was due to global warming. If the weather was hotter than normal for several years, we could ask that question. We know that global warming takes many years to show its effect.

When you are designing an investigation you have to observe carefully which variables are likely to have an effect.

a Would it be reasonable to ask whether the iron in Figure 1 is rusting because of acid rain? Discuss your answer.

Figure 1 A rusting lock

An owner of a house noticed that the driveway up to the house had cracks in the concrete on the left side of the driveway (observation). He was concerned because the driveway had only been laid for ten weeks. The work had been done in December. Before the builder came to look at it, the owner thought of a few questions to ask the builder:

- Did the builder have the correct amount of water in the concrete?
- Did the builder use the correct amount of cement?
- Could it be the car that was causing the damage?
- Did the builder dig the foundations deep enough?
- Did the builder put the same depth of foundations on both sides?
- Could the frost have caused the damage?
- Could the bushes growing next to the drive have caused the problem?

b Discuss all of these good ideas and choose three that are the most likely.

Observations, backed up by really creative thinking and good scientific knowledge, can lead to a **hypothesis**.

Testing scientific ideas

Scientists always try to think of ways to explain how things work or why they behave in the way that they do.

After their observations, they use their understanding of science to come up with an idea that could explain what is going on. This idea is sometimes called a hypothesis. They use this idea to make a prediction. A prediction is like a guess, but it is not just a wild guess – it is based on previous understanding.

A scientist will say, 'If it works the way I think it does, I should be able to change **this** (the independent variable) and **that** will happen (the dependent variable).'

Starting an investigation

Predictions are what make science so powerful. They mean that we can work out rules that tell us what will happen in the future. For example, a weather forecaster can use knowledge and understanding to predict **wind speeds**. Knowing this, sailors and windsurfers can decide whether it would be a good day to enjoy their sport.

Knowledge of energy transfer could lead to an idea that the insides of chips cook by energy being conducted from the outside. You might predict that small, thinly sliced chips will cook faster than large, fat chips.

Figure 2 Which cook faster? Small, thinly sliced chips or larger, fat chips?

c Look at the photograph in Figure 2. How could you test your prediction about how fast chips cook?

Not all predictions are correct. If scientists find that the prediction doesn't work, it's back to the drawing board! They either amend their original idea or think of a completely new one.

Starting to design a valid investigation

observation + knowledge → hypothesis → prediction → investigation

We can test a prediction by carrying out an **investigation**. You, as the scientist, predict that there is a relationship between two variables.

The independent variable is one that is selected and changed by you, the investigator. The dependent variable is measured for each change in your independent variable. Then all other variables become control variables, kept constant so that your investigation is a fair test.

If your measurements are going to be accepted by other people, they must be valid. Part of this is making sure that you are really measuring the effect of changing your chosen variable. For example, if other variables aren't controlled properly, they might be affecting the data collected.

d Look at Figure 3. Darren was investigating the temperature change when adding anhydrous copper sulfate to water. He used a test tube for the reaction. What is wrong here?

Figure 3 Darren investigating the temperature change

Summary questions

1 Copy and complete this paragraph using the following words:
controlled dependent independent knowledge prediction hypothesis
An observation linked with scientific .... can be used to make a .... A .... links an .... variable to a .... variable. All other variables need to be ....

2 What is the difference between a prediction and a guess?

3 Imagine you were testing whether the concentration of the reactants affects the rate of reaction. The reaction might cause the solution to get hot.
 a How could you monitor the temperature?
 b What other control variables can you think of that might affect the results?

Key points

- Observation is often the starting point for an investigation.
- Testing predictions can lead to new scientific understanding.
- You must design investigations that produce valid results if you are to be believed.

6 7

Answers to in-text questions

a Yes – because you know that much of our rain is acidic and you know that some metals react with acids.

b All of the suggestions are possible, but any three of these would be the more likely:
Did he have the correct amount of water in the concrete?
Did he use the correct amount of cement?
Did he put the same depth of foundations on both sides?
Could the bushes growing next to the drive have caused the problem?

c E.g. prepare two portions of chips – one thin set and one thick set. Cook both sets and see which cooks faster. Control variables would include having the cooking oil at the same temperature for both sets, keeping the total mass of potato the same in each set.

d Darren's hand could be heating or cooling the solution. The thermometer is not in the solution. The results would not be valid.

Summary answers

1 knowledge, hypothesis, prediction, independent, dependent, controlled

2 A prediction is based on knowledge or observation, a guess is not.

3 **a** By using a thermometer to check the temperature before and after the reaction.
 b Examples include the volume of the solutions, the degree of mixing or stirring, or the type of container.

H4 Planning an investigation

Learning objectives

Students should learn:

- how to design a fair test
- how to set up a survey
- how to set up a control group or control experiment
- how to reduce risks in hazardous situations.

Learning outcomes

Students should be able to:

- identify variables that need to be controlled in an investigation
- design a survey
- design a fair test and understand the use of control variables and control groups
- identify potential hazards and take action to minimise risk.

Support

- Show students an unfair test with some obvious errors, e.g. dropping two paper 'helicopters' with different wing sizes to see which takes longer to reach the ground, but dropping them from different heights.

Extend

- Show students a complex experiment. An example might be using a bomb calorimeter. Give an example of one of the controlled variables. Ask them to list as many other variables as possible that should be controlled.

Specification link-up: Controlled Assessment C4.2 & C4.3

Assess and manage risks when carrying out practical work, by:

- identifying some possible hazards in practical situations *[C4.2.1 b)]*
- suggesting ways of managing risks. *[C4.2.1 b)]*

Demonstrate an understanding of the need to acquire high quality data, by:

- appreciating that, unless certain variables are controlled, the results may not be valid. *[C4.3.2 a)]*

Lesson structure

Starters

Risk assessment – Give students a picture sheet illustrating a situation showing a number of hazards. Ask the students to spot the hazard and write down what could be done to minimise the risk. The situation illustrated could be one in the school laboratory, or it could be outside the school environment, e.g. in the road, in a factory or on a farm. *(5 minutes)*

Head start – Start, for example, with a video clip of a 100 m race. (Search for 'marathon' or 'race' at www.video.google.com or www.bbc.co.uk). This has to be a fair test. How is this achieved? Then show the mass start of the London marathon and ask whether this is a fair test. Support students by asking why there is no official world record for a marathon. (Instead they have world best times.) This could lead to a discussion of how difficult it is to control all of the variables in the field. Extend students by going on to discuss why athletes can break the 100 m world record and this may not be recognised because of a helping wind. *(10 minutes)*

Main

- Start with a group discussion about fair testing. Highlight any misconceptions about fair testing and stress that repeat readings do not make a test fair. Effectively controlling variables is what makes a test fair.
- Challenge students with a test you set up in an 'unfair' way. You can differentiate by making some errors obvious and some more subtle. Students can observe and then generate lists of mistakes in small groups. Ask each group to give one error from their list and record what should have been done to ensure fair testing until all suggestions have been considered.
- Start group discussions on how and why we need to produce survey data. Use a topical issue here. It might be appropriate to see how it should *not* be done by using a vox pop clip from a news programme.
- Students will be familiar with the idea of a placebo, but possibly not with how it is used to set up a control group. This might need explanation.
- Consider the case of whether it is possible to tell the difference in taste if the milk is put in before or after the tea. R. A. Fisher tested this, using a double-blind taste test, and went on to devise 'Statistical Methods for Research Workers'.
- Use the school or college laboratory rules to review safety procedures.
- Ask students to carry out a risk assessment on the 'burning alcohols' experiment.

Plenaries

Key words – Using a card sort, ask students to match the definitions to the five key words introduced in this lesson. *(5 minutes)*

Survey – Ask students to imagine they have been asked to conduct a survey to find out whether or not people prefer a particular brand of toothpaste. They should produce a questionnaire that lists the questions that you could ask people on the street. Support students by supplying them with a list of questions, some of which would be relevant, others irrelevant. Ask students to tick which questions would be the most appropriate. Extend students by asking them to suggest how many people should be chosen and on what basis they should be selected. *(10 minutes)*

Practical support

Risk assessment on burning alcohols

Equipment and materials required

Small beaker, thermometer, stopwatch, spirit burner and safety equipment, eye protection.

Details

Students could use a spirit burner to heat a small quantity of water in a small beaker and measure the temperature rise over a period of time. Before doing so, they should identify any possible hazards and then write down ways in which they would minimise any risk. After carrying out the experiment, they should discuss whether or not their plans for risk reduction were sufficient.

Safety: Take care with alcohol in the spirit burner. CLEAPSS Hazcard 40A Ethanol – highly flammable/irritant. Wear eye protection.

How Science Works

H4 Planning an investigation

Learning objectives

- How do you design a fair test?
- How do you set up a survey?
- How do you set up a control group or control experiment?
- How do you reduce risks in hazardous situations?

AQA Examiner's tip

If you are asked about why it is important to keep control variables constant, you need to give a detailed explanation. Don't just answer, 'To make it a fair test.'

When you are asked to write a plan for your investigation, make sure that you give all the details. Ask yourself, 'Would someone else be able to follow my written plan and use it to do the investigation?'

Fair testing

A **fair test** is one in which only the independent variable affects the dependent variable. All other variables (called control variables) should be kept the same. If the test is not fair, the results of your investigation will not be valid.

Sometimes it is very difficult to keep control variables the same. However, at least you can **monitor** them, so that you know whether they have changed or not.

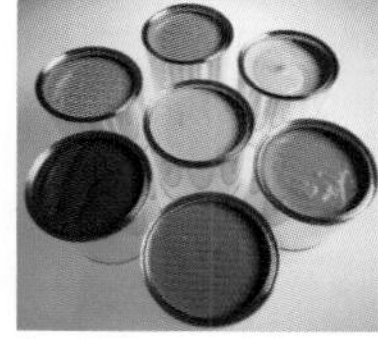

Figure 1 Different types of paint

a How would you set up an investigation to see how exposure to different amounts of sunlight affected different types of paint?

Surveys

Not all scientific investigations involve deliberately changing the independent variable.

If you were investigating the effect that using a mobile phone may have on health, you wouldn't put a group of people in a room and make them use their mobile phones to see whether they developed brain cancer!

Instead, you might conduct a **survey**. You might study the health of a large number of people who regularly use a mobile phone and compare their health with those who never use a mobile phone.

You would have to choose people of the same age and same family history to test. The larger the sample size you test, the better your results will be.

Control group

Control groups are used in investigations to try to make sure that you are measuring the variable that you intend to measure. When investigating the effects of a new drug, the control group will be given a **placebo**. This is a 'pretend' drug that actually has no effect on the patient at all. The control group think they are taking a drug but the placebo does not contain the drug. This way you can control the variable of 'thinking that the drug is working' and separate out the effect of the actual drug.

Usually neither the patient nor the doctor knows until after the trials have been completed which of the patients were given the placebo. This is known as a **double-blind trial**.

8

Planning an investigation

Risks and hazards

One of the first things you must do is to think about any potential **hazards** and then assess the **risk**.

Everything you do in life presents a hazard. What you have to do is to identify the hazard and then decide the degree of risk that it gives. If the risk is very high, you must do something to reduce it.

For example, if you decide to go out in the pouring rain, lightning is a possible hazard. However, you decide that the risk is so small that you will ignore it and go out anyway.

If you decide to cross a busy road, the cars travelling along it at high speed represent a hazard. You decide to reduce the risk by crossing at a pedestrian crossing.

Activity

Burning alcohols

Imagine you were testing alcohols to see how much energy they give out when burned.

Thermometer
Glass beaker
Water
Spirit burner
Alcohol
Tripod

- What are the **hazards** that are present?
- What could you do to reduce the **risk** from these hazards?

Figure 2 The hazard is the busy road; we reduce the risk by using a pedestrian crossing

AQA Examiner's tip

Before you start your practical work you must make sure that it is safe. What are the likely hazards? How could you reduce the risk caused by these hazards? This is known as a **risk assessment**. You may well be asked questions like this on your ISA paper.

Summary questions

1 Copy and complete this paragraph using the following words:
investigation hazards assessment risks
Before you carry out any practical .... you need to carry out a risk .... You can do this by looking for any potential .... and making sure that the .... are as small as possible.

2 Explain the difference between a control group and a control variable.

3 Briefly describe how you would go about setting up a fair test in a laboratory investigation. Give your answer as general advice.

Key points

- Care must be taken to ensure fair testing – as far as is possible.
- Control variables must be kept the same during an investigation.
- Surveys are often used when it is impossible to carry out an experiment in which the independent variable is changed.
- Control groups allow you to make a comparison.
- A risk assessment must be made when planning a practical investigation.

9

Further teaching suggestions

Which?

Look at some recent *Which?* reports on consumer goods. Discuss issues such as:

- Was the size of the sample surveyed sufficient?
- Could the people who were surveyed have been biased?
- Could the people conducting the survey have been biased?

Answers to in-text questions

a Wooden tiles (or tiles made from whatever surface the paint is intended to coat) should be painted. These should then be exposed to light of different intensities (including zero, i.e. kept in darkness) from a solar spectrum source (available commercially), and their condition recorded over time. Several tiles of the same paint should be used at each intensity. Each colour of paint should have a separate trial as the different colours will absorb different wavelengths. Temperature should be controlled to counter heating effects of the light. If necessary, it would be possible to extend the trial to include, for example, the effect of alternating light and dark.

Summary answers

1 investigation, assessment, hazards, risks

2 In an experiment is to determine the effect of changing a single variable, a **control** is often set up in which the independent variable is not changed, thus enabling a comparison to be made. If the investigation is of the survey type (q.v.), a control group is usually established to serve the same purpose.

3 Control all the variables that might affect the dependent variable, apart from the independent variable whose values you select.

H5 Designing an investigation

Learning objectives

Students should learn:

- how to choose the best values for the variables
- how to decide on a suitable range
- how to decide on a suitable interval
- how to ensure accuracy and precision.

Learning outcomes

Students should be able to:

- use trial runs to establish the best values for the variables
- use trial runs to establish a suitable range for the independent variable
- use trial runs to establish a suitable interval for the independent variable
- design a fair test that will yield valid and repeatable results.

Specification link-up: Controlled Assessment C4.1 & C4.3

Develop hypotheses and plan practical ways to test them, by:

- using appropriate technology *[C4.1.1 c)]*

Demonstrate an understanding of the need to acquire high quality data, by:

- appreciating that, unless certain variables are controlled, the results may not be valid *[C4.3.2 a)]*
- identifying when repeats are needed in order to improve reproducibility *[C4.3.2 b)]*
- recognising the value of further readings to establish repeatability and accuracy *[C4.3.2 c)]*
- considering the resolution of the measuring device *[C4.3.2 d)]*
- considering the precision of the measured data where precision is indicated by the degree of scatter from the mean *[C4.3.2 e)]*
- identifying the range of the measured data. *[C4.3.2 f)]*

Lesson structure

Starters

Interval – Give students a graph of enzyme activity that shows a peak, but where the interval on the *x*-axis is very large, so that it is difficult to judge the exact position of the peak. Ask students to suggest what other values should be tested in order to ascertain the peak more accurately. *(5 minutes)*

Demonstration – Demonstrate different ways of measuring the volume of a gas produced in a chemical reaction. These could include displacement of water in a test tube, measuring cylinder or burette. Gas syringes could also be used. Bubbles could be counted. Conclude that the most accurate method is likely to be the burette or the gas syringe. *(10 minutes)*

Main

- Preliminary work: explain that the demonstration of ways of measuring the volume of gas produced is an example of preliminary work.
- In small groups, devise a method for testing the accuracy of the two methods, e.g. using a known volume of air pumped into each device.
- Choose a person to try out this technique. Stress that we do not have a true answer. We trust the instrument and the technique that is most likely to give us the most accurate result – the one nearest the true value.
- Run a quick rate of reaction investigation.
- Demonstrate an experiment in which there is a built-in systematic error, e.g. weighing some chemicals using a filter paper without using the 'tare' function on the scales or measuring radioactivity without taking background radiation into account.
- Point out the difference between this type of systematic error and random errors. Also, how you might tell from results which type of error it is. You can still have a high degree of precision with systematic errors.

Plenaries

Prize giving! – Award a prize to the group achieving (a) the most accurate result; (b) the most precise results from a rate of reaction experiment. Let the groups try to explain their success. *(5 minutes)*

Out of range – Show students a graph of volume of gas produced plotted against time for a particular reaction. Ask students to predict from the graph what the volume would be for a time ten times longer than that plotted. Support students by asking them what the value would be if the pattern shown by the graph continued in the same way, and then ask them why this might not happen. Extend students by asking them to discuss whether the precision is greater at the start of the range, greater near the end of the range or the same throughout the range. *(10 minutes)*

Support

- Run a simple rate of reaction experiment where the students can take readings before describing what range and intervals are, using the students' data.
- Continually check on students' understanding of the key words accurate, precise, repeatable and reproducible. Rhymes or cartoon sketches might help them remember the definitions.

Extend

- Ask students to calculate an accurate rate of reaction experiment by using all of the class's data and calculate the mean. Then check this against a secondary source. Students should consider anomalous results in this calculation.

Further teaching suggestions

Repeatability

- Using the data from the rate of reaction experiment, consider the range of their repeat measurements and judge repeatability. Find the maximum range for the whole class – who got the highest reading and who got the lowest? Can we explain why?

Graphs and charts

- The 'collecting gas' experiment can be expanded – discuss whether the results should be plotted on a line graph or a bar chart.

H5 Designing an investigation

Learning objectives

- How do you make sure that you choose the best values for your variables?
- How do you decide on a suitable range?
- How do you decide on a suitable interval?
- How do you ensure accuracy and precision?

Choosing values of a variable

Trial runs will tell you a lot about how your early thoughts are going to work out.

Do you have the correct conditions?

A rate of reaction investigation may not have sufficient of one of the reactants to result in a measurable amount of product.

Have you chosen a sensible range?

Range means the maximum and minimum values of the independent or dependent variables. It is important to choose a suitable range for the independent variable, otherwise you may not be able to see any change in the dependent variable.

For example, if the results are all very similar in a rate of reaction experiment, you might not have chosen a wide enough range of concentrations.

Have you got enough readings that are close together?

The gap between the readings is known as the **interval**.

For example, you might alter the temperature to see whether this affects the rate of reaction. A set of 11 readings equally spaced over a range of 10 °C would give an interval of 1 °C, and this might not be enough to notice any change in the rate.

On the other hand, if you choose an interval of 20 °C you might miss an important change in the pattern.

Accuracy

Accurate measurements are very close to the **true value**.

Your investigation should provide data that is accurate enough to answer your original question.

However, it is not always possible to know what the true value is.

How do you get accurate data?

- You can repeat your measurements and your mean is more likely to be accurate.
- Try repeating your measurements with a different instrument and see whether you get the same readings.
- Use high-quality instruments that measure accurately.
- The more carefully you use the measuring instruments, the more accuracy you will get.

Precision, resolution, repeatability and reproducibility

A **precise** measurements is one in which there is very little spread about the mean value.

If your repeated measurements are closely grouped together, you have precision. Your measurements must be made with an instrument that has a suitable **resolution**. Resolution of a measuring instrument is the smallest change in the quantity being measured (input) that gives a perceptible change in the reading.

It's no use measuring the time for a fast reaction to finish using the seconds hand on a clock! If there are big differences within sets of repeat readings, you will not be able to make a valid conclusion. You won't be able to trust your data!

How do you get precise data?

- You have to use measuring instruments with sufficiently small scale divisions.
- You have to repeat your tests as often as necessary.
- You have to repeat your tests in exactly the same way each time.

If you repeat your investigation using the same method and equipment and obtain the same results, your results are said to be **repeatable**.

If someone else repeats your investigation in the same way, or if you repeat it by using different equipment or techniques, and the same results are obtained, it is said to be **reproducible**.

You may be asked to compare your results with those of others in your group, or with data from other scientists. Research like this is a good way of checking your results.

A word of caution!

Precision depends only on the extent of random errors – it gives no indication of how close results are to the true value. Just because your results show precision does not mean they are accurate.

a Draw a thermometer scale reading 49.5 °C, showing four results that are both accurate and precise.

Summary questions

1 Copy and complete this paragraph using the following words:
range repeat conditions readings
Trial runs give you a good idea of whether you have the correct .... to collect any data; whether you have chosen the correct .... for the independent variable; whether you have enough ....; and whether you need to do .... readings.

2 Use an example to explain how a set of repeat measurements could be accurate, but not precise.

3 Explain the difference between a set of results that are reproducible and a set of results that are repeatable.

AQA Examiner's tip

You must know the difference between accurate and precise results.

Imagine measuring the temperature after a set time when a fuel is used to heat a fixed volume of water.

Two students repeated this experiment, four times each. Their results are marked on the thermometer scales below:

- A **precise** set of repeat readings will be grouped closely together.
- An **accurate** set of repeat readings will have a mean (average) close to the true value.

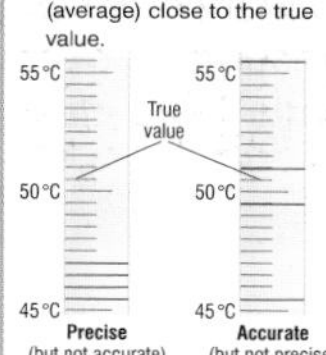

Key points

- You can use a trial run to make sure that you choose the best values for your variables.
- The range states the maximum and minimum values of a variable.
- The interval is the gap between the values of a variable.
- Careful use of the correct equipment can improve accuracy and precision.
- You should try to reproduce your results carefully.

Answers to in-text questions

a Diagram of thermometer showing the true value with four readings tightly grouped around it.

Summary answers

1 conditions, range, readings, repeat

2 Any example that demonstrates understanding of the two terms. e.g. 'I measured the yield of the reaction as 3.5 g, 4.8 g, 2.2 g, 3.8 g, 3.2 g. The average of my results is 3.5 g and the theoretical yield is 3.5 g. My results were accurate but not precise.

3 **Repeatable:** a measurement is repeatable if the original experimenter repeats the investigation using the same method and equipment and obtains the same results.

Reproducible: a measurement is reproducible if the investigation is repeated by another person, or by using different equipment or techniques, and the same results are obtained.

H6 Making measurements

Learning objectives

Students should learn:

- that they can expect results to vary
- that instruments vary in their accuracy
- that instruments vary in their resolution
- the difference between systematic errors and random errors
- that human error can affect results, and what to do with anomalies.

Learning outcomes

Students should be able to:

- distinguish between results that vary and anomalies
- explain that instruments vary in their accuracy and resolution
- explain that anomalies should be discarded or repeated before calculating a mean.

Specification link-up: Controlled Assessment C4.5

Review methodology to assess fitness for purpose, by:

- identifying causes of variation in data *[C4.5.2 a)]*
- recognising and identifying the cause of random errors. When a data set contains random errors, repeating the readings and calculating a new mean can reduce their effect *[C4.5.2 b)]*
- recognising and identifying the cause of anomalous results *[C4.5.2 c)]*
- recognising and identifying the cause of systematic errors. *[C4.5.2 d)]*

Lesson structure

Starters

Demonstration – Demonstrate different ways of measuring the width of the laboratory. Use a 30 cm rule, a metre rule, a tape and a laser/sonic measure. Discuss the relative merits of using each of these devices for different purposes. Discuss the details of the measuring instrument – its percentage accuracy, its useful range and its resolution. *(5 minutes)*

Human reaction time – Allow students to test their reaction times using a computer program (e.g. www.bbc.co.uk. Search for 'Sheep Dash'.) and then by dropping and catching a ruler, using a stopwatch. Discuss the advantages and disadvantages of each method. Support students by explaining that the human reaction time is normally about 0.2 s. Extend students on the 'dropping the ruler' method by getting them to explain whether it would be better for the same person to drop the ruler and operate the watch, or whether it would be better to use two different people. *(10 minutes)*

Main

- In small groups, plan the most accurate way to measure a person's height. Students can have any equipment they need. They will need to think about what a person's height includes, e.g. hair flat or not, shoes on or off. They might suggest a board placed horizontally on the head, using a spirit level, removing the person being measured and then using the laser/sonic measure placed on the ground.
- Stress that we do not have a true answer. We do not know the person's true height. We trust the instrument and the technique that is most likely to give us the most accurate result – the one nearest the true value.
- Demonstrate an experiment in which there is a built-in systematic error, e.g. weighing some chemicals using a filter paper without using the tare, or measuring radioactivity without taking background radiation into account.
- Point out the difference between this type of systematic error and random errors. Also, how you might tell from results which type of error it is. You can still have a high degree of precision with systematic errors.
- Complete question **a** individually.
- Encourage students to identify anomalies while carrying out the investigations so that they have an opportunity to check and replace them.

Plenaries

Human v. computer – Class discussion of data logging compared to humans when collecting data. Stress the importance of data logging in gathering data over extended or very short periods of time. *(5 minutes)*

Checklist – Ask students to draw up a checklist for an investigation so that every possible source of error is considered. Support students by giving them a list of possible sources of error that includes a mixture of relevant and irrelevant suggestions. Ask them to tick the ones that they think are relevant. Extend students by asking them to suggest what they could do to minimise the effect of any errors. *(10 minutes)*

Support

- Students will need support when interpreting data on oil and identifying evidence for systematic and random errors. For systematic errors, they should look to see whether the measured values are always larger or smaller than the calculated values. For random errors, they should look to see whether there is any scatter around the mean.

Extend

- Demonstrate a different experiment in which there is a built-in systematic error, e.g. measuring the effect of temperature on the rate of an exothermic reaction.

Further teaching suggestions

Data logging

- Data logging provides a good opportunity to exemplify changes in dependent variables.

  Use data logging to illustrate how detailed measurements taken frequently can show variation in results that would not have been seen by other methods.

  Data logging can increase the accuracy of readings that can be taken where it might not otherwise be possible to take readings accurately.

For example:

- Compare two students taking their hand temperatures – one with a thermometer, one with a logger.
- Set the logger to record room temperatures until the next lesson.
- Compare measurements using a tape measure with those of a distance sensor linked to a computer. Draw attention to the ability to measure distances as you move the sensor.

H6 Making measurements

Learning objectives

- Why do results always vary?
- How do you choose instruments that will give you accurate results?
- What do we mean by the 'resolution' of an instrument?
- What is the difference between a systematic error and a random error?
- How does human error affect results and what do you do with anomalies?

Using instruments

Try measuring the temperature of a beaker of water using a digital thermometer. Do you always get the same result? Probably not! So can we say that any measurement is absolutely correct?

In any experiment there will be doubts about actual measurements.

When you choose an instrument you need to know that it will give you the accuracy that you want. You need to be confident that it is giving a true reading.

If you have used an electric water bath, would you trust the temperature on the dial? How do you know it is the true temperature? You could use a very expensive thermometer to calibrate your water bath. The expensive thermometer is more likely to show the true temperature. But can you really be sure it is accurate?

Instruments that measure the same thing can have different sensitivities. The **resolution** of an instrument refers to the smallest change in a value that can be detected. This is one factor that determines the precision of your measurements.

Choosing the wrong scale can cause you to miss important data or make silly conclusions. We would not measure the length of a chemical bond in metres, we would use nanometres.

a Match the following weighing machines to their best use:

Used to measure	Resolution of weighing machine
Sodium chloride in a packet of cornflakes	micrograms
Cornflakes delivered to a supermarket	milligrams
Vitamin D in a packet of cornflakes	grams
Sugar added to a bowl of cornflakes	kilograms

AQA Examiner's tip

If you are asked what may have caused an error, never answer simply '**human error**' – you won't get any marks for this. You need to say what the experimenter may have done to cause the error, or give more detail, e.g. 'Human reaction time might have caused an error in the timing when using a stopwatch.'

Errors

Even when an instrument is used correctly, the results can still show differences.

Results may differ because of **random error**. This is most likely to be due to a poor measurement being made. It could be due to not carrying out the method consistently.

If you repeat your measurements several times and then calculate a mean, you will reduce the effect of random errors.

The **error** might be a **systematic error**. This means that the method was carried out consistently but an error was being repeated. A systematic error will make your readings be spread about some value other than the true value. This is because your results will differ from the true value by a consistent amount each time a measurement is made.

No number of repeats can do anything about systematic errors. If you think that you have a systematic error, you need to repeat using a different set of equipment or a different technique. Then compare your results and spot the difference!

A **zero error** is one kind of systematic error. Suppose that you were trying to measure the length of your desk with a metre rule, but you hadn't noticed that someone had sawn off half a centimetre from the end of the ruler. It wouldn't matter how many times you repeated the measurement, you would never get any nearer to the true value.

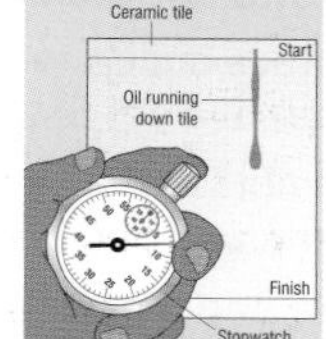

Figure 1 Matt timing the flow of oil

Check out these two sets of data that were taken from the investigation that Matt did. He tested five different oils. The bottom row is the time calculated from knowing the viscosity of the different oils:

Type of oil used	A	B	C	D	E
Time taken to flow down tile (seconds)	23.2	45.9	49.5	62.7	75.9
	24.1	36.4	48.7	61.5	76.1
Calculated time (seconds)	18.2	30.4	42.5	55.6	70.7

b Discuss whether there is any evidence of random error in these results.
c Discuss whether there is any evidence of systematic error in these results.

Anomalies

Anomalous results are clearly out of line. They are not those that are due to the natural variation you get from any measurement. These should be looked at carefully. There might be a very interesting reason why they are so different. You should always look for anomalous results and discard them before you calculate a mean, if necessary.

- If anomalies can be identified while you are doing an investigation, it is best to repeat that part of the investigation.
- If you find anomalies after you have finished collecting data for an investigation, they must be discarded.

Summary questions

1 Copy and complete this paragraph using the following words:
accurate discarded random resolution systematic use variation
There will always be some ………… in results. You should always choose the best instruments that you can in order to get the most ………… results. You must know how to ………… the instrument properly. The ………… of an instrument refers to the smallest change that can be detected. There are two types of error – ………… and ………… . Anomalies due to random error should be ………… .

2 What kind of error will most likely occur in the following situations?
  a Asking everyone in the class to measure the length of the bench.
  b Using a ruler that has a piece missing from the zero end.

Key points

- Results will nearly always vary.
- Better quality instruments give more accurate results.
- The resolution of an instrument refers to the smallest change that it can detect.
- Human error can produce random and/or systematic errors.
- We examine anomalies; they might give us some interesting ideas. If they are due to a random error, we repeat the measurements. If there is no time to repeat them, we discard them.

Answers to in-text questions

a

Used to measure	Resolution of weighing machine
Cornflakes delivered to a supermarket	kilograms
Carbohydrate in a packet of cornflakes	grams
Vitamin D in a packet of cornflakes	micrograms
Sodium chloride in a packet of cornflakes	milligrams

b First attempt for **B** is the random error.

c Average results are close to individual results, which are consistently different to the calculated time.

Summary answers

1 variation, accurate, use, resolution, random, systematic, discarded

2 **a** random
  b systematic

H7 Presenting data

Learning objectives

Students should learn:

- what is meant by the range and the mean of a set of data
- how to use tables of data
- how to display data.

Learning outcomes

Students should be able to:

- express accurately the range and mean of a set of data
- distinguish between the uses of bar charts and line graphs
- draw line graphs accurately.

Specification link-up: Controlled Assessment C4.4

Show an understanding of the value of means, by:

- appreciating when it is appropriate to calculate a mean *[C4.4.1 a)]*
- calculating the mean of a set of at least three results. *[C4.4.1 b)]*

Demonstrate an understanding of how data may be displayed, by:

- drawing tables *[C4.4.2 a)]*
- drawing charts and graphs *[C4.4.2 b)]*
- choosing the most appropriate form of presentation. *[C4.4.2 c)]*

Lesson structure

Starters

Newspapers – Choose data from the press – particularly useful are market trends where they do not use the origin (0,0). This exaggerates changes. This could relate to the use of data logging which can exaggerate normal variation into major trends. *(5 minutes)*

Spreadsheet – Prepare some data from a typical investigation that the students may have recently completed. Use all of the many ways of presenting the data in a spreadsheet program to display it. Allow students to discuss and reach conclusions as to which is the best method. Support students by presenting data as either a line graph or a simple bar chart so that they can make the link between continuous data and line graphs and between categoric data and bar charts. Extend students by showing graphs and charts that have non-linear scales or false origins. *(10 minutes)*

Main

- Choose an appropriate topic to either demonstrate or allow small groups to gather data, e.g. production of CO_2 from 0.5 M HCl and $CaCO_3$ over 1 minute; using food labels to determine saturated oil content of different foods. Choose any topic that will allow rapid gathering of data. Be aware that some data will lead to a bar chart; this might be more appropriate to groups struggling to draw line graphs.
- Students should be told what their task is and should therefore know how to construct an appropriate table. This should be done individually prior to collecting the data. Refer to the first paragraph under 'Table' in the Student Book.
- Start a group discussion on the best form of table.
- Carry out data gathering, putting data directly into the table. Refer to the second paragraph under 'Tables' in the Student Book.
- Individuals produce their own graphs. Refer to the section 'Displaying your results' in the Student Book.
- Graphs could be exchanged and marked by others in the group, using the criteria in the section mentioned above.

Plenaries

Which type of graph? – Give students different headings from a variety of tables and ask them how best to show the results graphically. This could be done as a whole class with individuals showing answers as the teacher reveals each table heading. Each student can draw a large letter 'L' (for line graph) on one side of a sheet of paper and 'B' (for bar chart) on the other, ready to show their answers. *(5 minutes)*

Key words – Students should be given key words to prepare posters for the laboratory. Key words should be taken from the summary questions in the first six sections. Support students by giving them a poster in two sections – one containing the key word, the other the definition. Students should then match the pairs together correctly. Extend students by getting them to write their own definitions. *(10 minutes)*

Support

- Some students struggle with plotting graphs. They should start with bar charts and move on to line graphs.

Extend

- Students could be asked to handle two dependent variables in the table and graph, e.g. cooling and weight loss of a beaker of water with time, with repeat readings included.
- They could also be given more difficult contexts that are more likely to produce anomalies. They could, for example, be given a context that produces both random and systematic errors.

Further teaching suggestions

ICT link-up

- Students could use a set of data within spreadsheet software to present the data as pie charts, line graphs, bar charts, etc. Allow them to decide on the most appropriate form. Care needs to be given to 'smoothing', which does not always produce a line of best fit.

How Science Works

H7 Presenting data

Learning objectives

- How do we calculate the mean from a set of data?
- How do you use tables of results?
- What is the range of the data?
- How do you display your data?

Figure 1 Student using an oxygen meter

For this section you will be working with data from this investigation:

Mel took a litre (1 dm³) of tap water. She shook it vigorously for exactly 2 minutes. She tried to get as much oxygen to dissolve in it as possible. Then she took the temperature of the water. She immediately tested the oxygen concentration, using an oxygen meter.

Tables

Tables are really good for getting your results down quickly and clearly. You should design your table **before** you start your investigation.

Your table should be constructed to fit in all the data to be collected. It should be fully labelled, including units.

You may want to have extra columns for repeats, calculations of means or calculated values.

Checking for anomalies

While filling in your table of results you should be constantly looking for anomalies.

- Check to see whether any reading in a set of repeat readings is significantly different from the others.
- Check to see whether the pattern you are getting as you change the independent variable is what you expected.

Remember, a result that looks anomalous should be checked out to see whether it really is a poor reading.

Planning your table

Mel had decided on the values for her independent variable. We always put these in the first column of a table. The dependent variable goes in the second column. Mel will find its values as she carries out the investigation.

So she could plan a table like this:

Temperature of water (°C)	Concentration of oxygen (mg/dm³)
5	
10	
16	
20	
28	

Or like this:

Temperature of water (°C)	5	10	16	20	28
Concentration of oxygen (mg/dm³)					

All she had to do in the investigation was to write the correct numbers in the second column to complete the top table.

Mel's results are shown in the alternative format in the table below:

Temperature of water (°C)	5	10	16	20	28
Concentration of oxygen (mg/dm³)	12.8	11.3	9.9	9.1	7.3

The range of the data

Pick out the maximum and the minimum values and you have the range of a variable. You should always quote these two numbers when asked for a range. For example, the range of the dependent variable is between 7.3 mg/dm³ (the lowest value) and 12.8 mg/dm³ (the highest value) – and don't forget to include the units!

a What is the range for the independent variable and for the dependent variable in Mel's set of data?

Maths skills

The mean of the data

Often you have to find the **mean** of each repeated set of measurements. The first thing you should do is to look for any anomalous results. If you find any, miss these out of the calculation. Then add together the remaining measurements and divide by how many there are.

For example:

- Mel takes four readings, 15 mg/dm³, 12 mg/dm³, 29 mg/dm³, 15 mg/dm³
- 29 mg/dm³ is an anomalous result and so is missed out. So 15 + 12 + 15 = 42
- 42 divided by three (the number of valid results) = **14 mg/dm³**

The repeat values and mean can be recorded as shown below:

Temperature of water (°C)	Concentration of oxygen (mg/dm³)			
	1st test	2nd test	3rd test	Mean
0	15	12	15	14

Displaying your results

Bar charts

If one of your variables is categoric, you should use a bar chart.

Line graphs

If you have a continuous independent and a continuous dependent variable, a line graph should be used. Plot the points as small 'plus' signs (+).

AQA Examiner's tip

When you make a table for your results, remember to include:

- headings, including the units
- a title.

When you draw a line graph or bar chart, remember to:

- use a sensible scale that is easy to work out
- use as much of the graph paper as possible; your data should occupy at least a third of each axis
- label both axes
- draw a line of best fit if it is a line graph
- label each bar if it is a bar chart.

AQA Examiner's tip

Marks are often dropped in the ISA by candidates plotting points incorrectly. Also use **a line of best fit** where appropriate – don't just join the points 'dot-to-dot'!

Summary questions

1 Copy and complete this paragraph using the following words:

categoric continuous mean range

The maximum and minimum values show the .... of the data. The sum of all the values in a set of repeat readings divided by the total number of these repeat values gives the .... Bar charts are used when you have a .... independent variable and a continuous dependent variable. Line graphs are used when you have .... independent and dependent variables.

2 Draw a graph of Mel's results from the bottom of the previous page.

Key points

- The range states the maximum and the minimum values.
- The mean is the sum of the values divided by how many values there are.
- Tables are best used during an investigation to record results.
- Bar charts are used when you have a categoric variable.
- Line graphs are used to display data that are continuous.

Answers to in-text questions

a Independent variable: 5 °C to 28 °C; dependent variable: 7.3 mg/dm³ to 12.8 mg/dm³.

Summary answers

1 range, mean, categoric, continuous

2

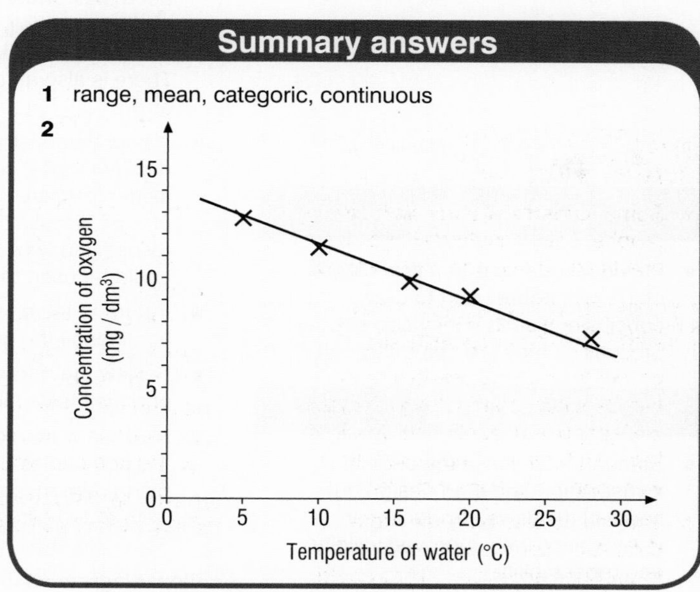

H8 Using data to draw conclusions

Learning objectives

Students should learn:

- how to use charts and graphs to identify patterns
- how to identify relationships within data
- how to draw valid conclusions from relationships
- how to evaluate the repeatability of an investigation.

Learning outcomes

Students should be able to:

- draw a line of best fit when appropriate
- identify different relationships between variables from graphs
- draw conclusions from data
- evaluate the repeatability and validity of an investigation.

Specification link-up: Controlled Assessment C4.5 & C4.6

Identify patterns in data, by:

- describing the relationship between two variables and deciding whether the relationship is causal or by association. *[C4.5.3 a)]*

Draw conclusions using scientific ideas and evidence, by:

- writing a conclusion, based on evidence that relates correctly to known facts *[C4.5.4 a)]*
- using secondary sources *[C4.5.4 b)]*
- identifying extra evidence that is required for a conclusion to be made *[C4.5.4 c)]*
- evaluating methods of data collection. *[C4.5.4 d)]*

Review hypotheses in the light of outcomes, by:

- considering whether or not any hypothesis made is supported by the evidence. *[C4.6.1a)]*

Lesson structure

Starters

Conclusions – Prepare a number of tables of results, some of which show that as *x* increases *y* increases, some that show that as *x* increases *y* decreases, and some where there is no relationship between *x* and *y*. Ask students what conclusion they can draw from each set of results. *(5 minutes)*

Starter graphs – Prepare a series of graphs that illustrate the various types of relationship in the specification. Each graph should have fully labelled axes. Students, in groups, should agree to statements that describe the patterns in the graphs. Support students by giving them graphs that illustrate simple linear relationships. Extend students by giving them more complex graphs with curved lines, and encourage them to use terms such as 'directly proportional' and 'inversely proportional'. Gather feedback from groups and discuss. *(10 minutes)*

Main

- Using the graphs from the previous lesson, students should be taught how to produce lines of best fit. Students could work individually with help from Figures 1 and 2 in the Student Book.
- They should identify the pattern in their graph.
- They now need to consider the repeatability and validity of their results. They may need their understanding of reliability and validity reinforced. Questions can be posed to reinforce their understanding of both terms. If the investigation was not carefully controlled, it is likely to be invalid, thus posing many opportunities for discussion. There is also an opportunity to reinforce other ideas such as random and systematic errors.
- A brief demonstration of a test, e.g. finding the energy transfer when burning crisps of different mass, could be used. Students should observe the teacher and make notes as the tests are carried out. They should be as critical as they can be, and in small groups discuss their individual findings. One or two students could be recording the results and two more plotting the graph, as the teacher does the tests. A spreadsheet could be used to immediately turn the results into graphs.
- Return to the original prediction. Look at the graph of the results. Ask how much confidence the group has in the results.
- Review the links that are possible between two sets of data. Ask them to decide which one their tests might support.
- Now the word 'conclusion' should be introduced and a conclusion made… if possible! It is sometimes useful to make a conclusion that is 'subject to ... e.g. the repeatability being demonstrated'.

Plenaries

Flow diagram – When pulling the lesson together, it will be important to emphasise the process involved – graph → line of best fit → pattern → question the repeatability and validity → consider the links that are possible → make a conclusion → summarise evaluation. This could be illustrated with a flow diagram generated by a directed class discussion. *(5 minutes)*

Support

- Provide students with a flow diagram of the procedure used to draw conclusions so that they can see the process as they are going through it.

Extend

- Students could take the original investigation and then design out some of the flaws, producing an investigation with improved validity and repeatability.
- Summary question 2 could be examined in some detail and the work researched on the internet.

Evaluating – Students could review the method used in the demonstration experiment of heating water. Support students by asking them to identify where errors could have been made. Extend students by asking them to suggest improvements that could be made to minimise these errors. *(10 minutes)*

Further teaching suggestions

Case studies

- Students should be able to transfer these skills to examine the work of scientists and to become critical of the work of others. Collecting scientific findings from the press and subjecting them to the same critical appraisal is an important exercise. They could be encouraged to collect these or could be given photocopies of topical issues suitable for such appraisal.

H8 Using data to draw conclusions

Learning objectives

- How do we best use charts and graphs to identify patterns?
- What are the possible relationships we can identify from charts and graphs?
- How do we draw conclusions from relationships?
- How can we decide if our results are good and our conclusions are valid?

Identifying patterns and relationships

Now that you have a bar chart or a line graph of your results you can begin to look for patterns. You must have an open mind at this point.

First, there could still be some anomalous results. You might not have picked these out earlier. How do you spot an anomaly? It must be a significant distance away from the pattern, not just within normal variation. If you do have any anomalous results plotted on your graph, circle these and ignore them when drawing the **line of best fit**.

Now look at your graph. Is there a pattern that you can see? When you have decided, draw a line of best fit that shows this pattern.

A line of best fit is a kind of visual averaging process. You should draw the line so that it leaves as many points slightly above the line as there are points below. In other words it is a line that steers a middle course through the field of points.

The vast majority of results that you get from continuous data require a line of best fit.

Remember, a line of best fit can be a straight line or it can be a curve – you have to decide from your results.

You need to consider whether your graph shows a linear **relationship**. This simply means, can you be confident about drawing a straight line of best fit on your graph? If the answer is yes – is this line positive or negative?

a Say whether graphs **i** and **ii** in Figure 1 show a positive or a negative linear relationship.

Look at the graph in Figure 2. It shows a positive linear relationship. It also goes through the origin (0,0). We call this a **directly proportional** relationship.

Your results might also show a curved line of best fit. These can be predictable, complex or very complex! Look at Figure 3 below.

i ii

Figure 1 Graphs showing linear relationships

0 0

Figure 2 Graph showing a directly proportional relationship

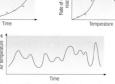

a Velocity Time

b Rate of enzyme reaction Temperature

c Air temperature Time

Figure 3 **a** Graph showing predictable results **b** Graph showing complex results **c** Graph showing very complex results

Drawing conclusions

If there is a pattern to be seen (for example as one variable gets bigger the other also gets bigger), it may be that:

- changing one has caused the other to change
- the two are related, but one is not necessarily the cause of the other.

Your conclusion must go no further than the evidence that you have.

Activity

Looking at relationships

Some people think that watching too much television can cause an increase in violence.

The table shows the number of television sets in the UK for four different years, and the number of murders committed in those years.

Year	Number of televisions (millions)	Number of murders
1970	15	310
1980	25	500
1990	42	550
2000	60	750

Plot a graph to show the relationship.

- Do you think this proves that watching television causes violence? Explain your answer.

Poor science can often happen if a wrong decision is made here. Newspapers have said that living near electricity substations can cause cancer. All that scientists would say is that there is possibly an association.

Evaluation

You will often be asked to evaluate either the method of the investigation or the conclusion that has been reached. Ask yourself: Could the method have been improved? Is the conclusion that has been made a valid one?

Summary questions

1 Copy and complete this paragraph using the following words:

anomalous complex directly negative positive

Lines of best fit can be used to identify .... results. Linear relationships can be .... or .... If a straight line goes through the origin of a graph, the relationship is .... proportional. Often a line of best fit is a curve which can be predictable or ....

2 Nasma found a newspaper article about nanoscience. Nanoparticles are used for many things, including perfumes.

There was increasing evidence that inhaled nanoparticles could cause lung inflammation. [quote from Professor Ken Donaldson]

Discuss the type of experiment and the data you would expect to see to support this conclusion.

AQA Examiner's tip

When you read scientific claims, think carefully about the evidence that should be there to back up the claim.

Key points

- Drawing lines of best fit helps us to study the relationship between variables.
- The possible relationships are linear, positive and negative, directly proportional, predictable and complex curves.
- Conclusions must go no further than the data available.
- The reproducibility of data can be checked by looking at other similar work done by others, perhaps on the internet. It can also be checked by using a different method or by others checking your method.

Answers to in-text questions

a Graph **i** – positive linear.

Graph **ii** – negative linear.

Summary answers

1 anomalous, positive, negative, directly, complex

2 Some of the following ideas: two large groups of people, one using the perfume without nanoparticles, others using nanoparticle perfume – both groups identified as not having any inflammation of the lung – each participant given questionnaire to be completed daily – questionnaire asks about clinical symptoms related to lung inflammation – [control group idea – fieldwork idea]. Data expected would include range and means (averages) of clinical factors for the two groups. Opportunity to discuss medical ethics and whether trials with animals should first be carried out.

H9 Scientific evidence and society

Learning objectives

Students should learn:

- that science must be presented in a way that takes into account the reproducibility and the validity of the evidence
- that science should be presented without bias from the experimenter
- that evidence must be checked to appreciate whether there is any political influence
- that the status of the experimenter can influence the weight attached to a scientific report.

Learning outcomes

Students should be able to:

- make judgements about the reproducibility and the validity of scientific evidence
- identify when scientific evidence might have been influenced by bias or political influence
- judge scientific evidence on its merits, taking into account the weight given to it by the status of the experimenter.

Specification link-up: Controlled Assessment C4.5

Distinguish between a fact and an opinion, by:

- recognising that an opinion might be influenced by factors other than scientific fact *[C4.5.1 a)]*
- identifying scientific evidence that supports an opinion. *[C4.5.1 b)]*

Lesson structure

Starters

Ask a scientist – It is necessary at this point to make a seamless join between work that has mostly been derived from student investigations and work generated by scientists. Students must be able to use their critical skills derived in familiar contexts and apply them to second-hand data. One way to achieve this would be to bring in newspaper cuttings on a topic of current scientific interest. They should be aware that some newspaper reporters will 'cherry-pick' sections of reports to support sensational claims that will make good headlines. Students could be supported by highlighting key words in the article. To extend students, ask them to produce a 'wish-list' of questions they would like to put to the scientists who conducted the research and to the newspaper reporter. *(5 minutes)*

Researching scientific evidence – With access to the internet, students could be given a topic to research. They should use a search engine and identify the sources of information from, say, the first six webpages. They could then discuss the relative merits of these sources in terms of potential for bias. *(10 minutes)*

Main

- The following points are best made using topics that are of immediate importance to your students. The examples used are only illustrative. Some forward planning is required to ensure that there is a plentiful supply of newspaper articles, both local and national, to support the lesson. These could be displayed and/or retained in a portfolio for reference.
- Working in pairs, students should answer in-text question **a**. They should write a few sentences about the headline and what it means to them.
- Use the next section to illustrate the possibility of bias in reporting science. Again use small group discussions, followed by whole class Plenary.
- If you have access to the internet for the whole class, it is worth pursuing the issue of mobile phone masts in relation to their political significance. Pose the question: 'What would happen to the economy of this country if it was discovered that mobile phone masts were dangerous?' Would different people come together to suppress that information? Should they be allowed to suppress scientific evidence? Stress that there is no such evidence, yet people have that fear. Why do they have that fear? Should scientists have the task of reducing that fear to proper proportions? There is much to discuss.
- Small groups can imagine that they are preparing a case against the siting of a cement works close to their village. They could be given data that relates to pollution levels from similar companies. Up-to-date data can be obtained from the internet, e.g. from the DEFRA website (www.defra.gov.uk). Students could be given the data as if it were information provided at a public enquiry for the cement works. They should be asked to prepare a case that questions, for example, the reproducibility and validity of the data. This links with work covered in C1 2.5 on limestone.

Plenaries

Contentious issues – Make a list of contentious issues on which scientists might be able to make a contribution to the debate. Examples might include the siting of wind farms or sewage works, the building of new motorways, the introduction of new drugs, etc. *(5 minutes)*

Group report – Groups should report their findings on the cement works case to the class. Support students by allowing them to present their findings by posters. Extend students by asking individuals to give a one-minute talk to the rest of the class. *(10 minutes)*

Support

- Groups could prepare posters that use scientific data to present their case for or against any of the developments discussed.

Extend

- Arrange a class debate and nominate individuals to speak for or against any of the developments discussed.

Further teaching suggestions

Role play

- Students could role-play a public enquiry. They could be given roles and asked to prepare a case for homework. The data should be available to them so that they all know the arguments before preparing their case. Possible link here with the English department. This activity could be allocated as a homework exercise.

Local visit

- Students might be able to attend a local public enquiry or even the local town council as it discusses local issues with a scientific context or considers the report of a local issue.

The limitations of science

Examples could be given of the following issues:

- We are still finding out about things and developing our scientific knowledge (e.g. the use of the hadron collider).
- There are some questions that we cannot yet answer, maybe because we do not have enough valid evidence (e.g. are mobile phones completely safe to use?).
- There are some questions that science cannot answer at all. (e.g. Why was the universe created?)

How Science Works

Scientific evidence and society

H9 Scientific evidence and society

Learning objectives

- How can science encourage people to trust its research?
- How might bias affect people's judgement of science?
- Can politics influence judgements about science?
- Do you have to be a professor to be believed?

STAR IN SCANDAL SHOCK

We Find Out What They Don't Want You To Know... And WE TELL YOU!

MOBILE PHONE TUMOUR RISK?

Swedish researchers found that the risk of developing an ear tumour increased if you used a mobile phone.

The study was of 750 people. This type of tumour affects one in 100 000 people and the risk increased four times if you used the phone for more than 10 years.

Did you know ...?

A scientist who rejected the idea of a causal link between smoking and lung cancer was later found to be being paid by a tobacco company.

AQA Examiner's tip

If you are asked about bias in scientific evidence, there are two types:

- the measuring instruments may have introduced a bias because they were not calibrated correctly
- the scientists themselves may have a biased opinion (e.g. if they are paid by a company to promote their product).

Now you have reached a conclusion about a piece of scientific research. So what is next? If it is pure research, your fellow scientists will want to look at it very carefully. If it affects the lives of ordinary people, society will also want to examine it closely.

You can help your cause by giving a balanced account of what you have found out. It is much the same as any argument you might have. If you make ridiculous claims, nobody will believe anything you have to say.

Be open and honest. If you only tell part of the story, someone will want to know why! Equally, if somebody is only telling you part of the truth, you cannot be confident about anything they say.

a An advert for a breakfast cereal claims that it has 'extra folic acid'. What information is missing? Is it important?

You must be on the lookout for people who might be biased when presenting scientific evidence. Some scientists are paid by companies to do research. When you are told that a certain product is harmless, just check out who is telling you.

b Bottles of perfume spray contain this advice: 'This finished product has not been tested on animals.' Why might you mistrust this statement?

Suppose you wanted to know about the pollution effects of burning waste in a local incinerator. Would you ask the scientist working for the incinerator company or one working in the local university?

We also have to be very careful in reaching judgements according to who is presenting scientific evidence to us. For example, if the evidence might provoke public or political problems, it might be played down.

Equally, others might want to exaggerate the findings. They might make more of the results than the evidence suggests. Take as an example the data available on animal research. Animal liberation followers may well present the *same* evidence completely differently to pharmaceutical companies wishing to develop new drugs.

c Check out some websites on limestone quarrying in the National Parks. Get the opinions of the environmentalists and those of the quarrying companies. Try to identify any political bias there might be in their opinions.

The status of the experimenter may place more weight on evidence. Suppose a quarrying company wants to convince an enquiry that it is perfectly reasonable to site a quarry in remote moorland in the UK. The company will choose the most eminent scientist in that field who is likely to support them. The small local community might not be able to afford an eminent scientist. The enquiry needs to be very careful to make a balanced judgement.

Science can often lead to the development of new materials or techniques. Sometimes these cause a problem for society where hard choices have to be made.

Scientists can give us the answers to many questions, but not to every question. Scientists have a contribution to make to a debate, but so do others such as environmentalists, economists and politicians.

The limitations of science

Science can help us in many ways but it cannot supply all the answers. We are still finding out about things and developing our scientific knowledge. For example, the Hubble telescope has helped us to revise our ideas about the beginnings of the universe.

There are some questions that we cannot answer, maybe because we do not have enough reproducible, repeatable and valid evidence. For example, research into the causes of cancer still needs much work to be done to provide data.

There are some questions that science cannot answer at all. These tend to be questions where beliefs, opinions and ethics are important. For example, science can suggest what the universe was like when it was first formed, but cannot answer the question of why it was formed.

BIODIESEL

The Fuel of the Future?

The demand for palm oil has grown tremendously in the last few years. It is used in many food products like margarine and chocolate, in cosmetics, and increasingly for making biodiesel.

Some scientists say that it is the answer to the dwindling supplies of crude oil because palm oil is a renewable resource.

Other people say that planting millions of acres of palm trees is destroying natural habitats such as rainforests and peat bogs.

Figure 1 The Hubble space telescope can look deep into space and tell us things about the universe's beginning from the formations of early galaxies

Summary questions

1 Copy and complete this paragraph using the following words:

status balanced bias political

Evidence from scientific investigations should be given in a .... way. It must be checked for any .... from the experimenter. Evidence can be given too little or too much weight if it is of .... significance. The .... of the experimenter is likely to influence people in their judgement of the evidence.

2 Collect some newspaper articles to show how scientific evidence is used. Discuss in groups whether these articles are honest and fair representations of the science. Consider whether they carry any bias.

3 Petcoke is a high carbon product from refined oil. It can be used in power stations and cement works. Owners of the Drax power station, which is running a trial use of the fuel, claim that it is cheaper than coal and can be used without harmful effects. Other groups claim that it is 'dirty fuel' and will cause environmental and health problems. Suppose you were living near Drax power station. Who would you trust to tell you whether petcoke was a safe fuel? Explain your answer.

Key points

- Scientific evidence must be presented in a balanced way that points out clearly how valid the evidence is.
- The evidence must not contain any bias from the experimenter.
- The evidence must be checked to appreciate whether there has been any political influence.
- The status of the experimenter can influence the weight placed on the evidence.

18 19

Answers to in-text questions

a E.g. what level of folic acid is required; is there a safe limit?; what evidence is there that folic acid is required?; how much do we get in a normal diet?

b It might not be safe for humans to use. The constituents of the perfume might have been tested on animals before being made into the final product.

c Identification of any political bias; this could be from companies and individuals as well as governments.

Summary answers

1 balanced, bias, political, status

2 Identification of any bias in reports.

3 Any reasonable answer, but that person should be independent, have the necessary skills as a scientist and not be capable of being influenced politically.

H10 The ISA

Learning objectives

Students should learn:

- how to write a plan
- how to make a risk assessment
- how to make a hypothesis
- how to reach a conclusion.

Learning outcomes

Students should be able to:

- structure a plan for an investigation so as to include key points such as the range and interval of the independent variable
- identify potential hazards in practical work
- show how the results of an experiment can confirm or refute a hypothesis
- reach a valid conclusion from the results of an investigation.

Specification link-up: Controlled Assessment C4.5 & C4.6

Distinguish between a fact and an opinion, by:

- recognising that an opinion might be influenced by factors other than scientific fact *[C4.5.1 a)]*
- identifying scientific evidence that supports an opinion. *[C4.5.1 b)]*

Review hypotheses in the light of outcomes, by:

- considering whether or not any hypothesis made is supported by the evidence *[C4.6.1a)]*
- developing scientific ideas as a result of observations and measurements. *[C4.6.1b)]*

Lesson structure

Starters

Structure of an investigation – Use an interactive whiteboard or sticky labels that show the different stages of an investigation and ask students to arrange them in the correct order. *(5 minutes)*

Predictions and hypotheses – Make a table containing one column of hypotheses and another column of predictions. Students should match the prediction to the correct hypothesis. *(10 minutes)*

Main

These activities may be spread over more than one lesson:

- Use a specimen ISA to guide students through the different stages that will be required.
- Start by outlining the problem that is to be investigated. Set a context for the investigation. Ask the students to develop a hypothesis and discuss.
- Research one or two (depending on the investigation) possible methods that can be used to carry out an experiment to test the hypothesis. Get them to practise making brief notes, similar to the notes they will be able to make for the real ISA.
- Review any possible hazards. Discuss how any risk associated with these hazards could be reduced.
- Discuss the control variables that should be kept constant in order to make it a fair test.
- Students should decide the range and interval of the values of the independent variable, and whether or not repeats will be needed.
- Allow the students to carry out a rough trial with the equipment in order to establish suitable values for these.
- Students should now be able to write a structured plan for the investigation.
- Ask students to design a blank table ready for the results. This should contain space to record all the measurements that will be taken during the experiment. Stress the need to include proper headings and units.
- Students carry out the investigation, recording their results.
- Draw a chart or graph of the results.
- Analyse the results and discuss any conclusion that could be reached. Make sure the students refer back to the hypothesis when making their conclusion.

Plenaries

Graph or bar chart – Give students a list of titles of different investigations and ask them to decide whether the results should be plotted on a bar chart or on a line graph. *(5 minutes)*

Comparing results – Groups should report their findings to others and compare results. Support students by making a table of pooled results. Extend students by asking individuals to give a one-minute talk to the rest of the class explaining why they think their results are or are not repeatable and reproducible. *(10 minutes)*

Support

- Groups could prepare posters that show a flow diagram for the different stages of an ISA investigation.
- Students can be provided with a plan if their plan is unworkable, unsafe or unmanageable. An example plan will be provided by the AQA. Students should not lose any marks if their plan is unworkable for a good reason (i.e. lack of equipment). However, if their plan is dangerous or unworkable this must be reflected in their mark.

Extend

- Give students a hypothesis and ask them to make a prediction based on it.

Further teaching suggestions

Writing a plan

- Give students a plan of an investigation that contains a number of errors, e.g. control variables not kept constant, or unsuitable range or interval of the independent variable, and ask them to spot and explain the mistakes.

H10 The ISA

Learning objectives

- How do you write a plan?
- How do you make a risk assessment?
- What is a hypothesis?
- How do you make a conclusion?

There are several different stages to the ISA.

Stage 1

Your teacher will tell you the problem that you are going to investigate, and you will have to develop your own hypothesis. They will also set the problem in a context – in other words, where in real life your investigation could be useful. You should have a discussion about it, and talk about different ways in which you might solve the problem. Your teacher should show you the equipment that you can use, and you should research one or two possible methods for carrying out an experiment to test the hypothesis. You should also research the context and do a risk assessment for your practical work. You will be allowed to make one side of notes on this research, which you can take into the written part of the ISA.

Figure 1 Doing practical work allows you to develop the skills needed to do well in the ISA

You should be allowed to handle the equipment and you may be allowed to carry out a preliminary experiment.

Make sure that you understand what you have to do – now is the time to ask questions if you are not sure.

AQA Examiner's tip

When you are making a blank table or drawing a graph or bar chart, make sure that you use full headings, e.g.

- the length of the leaf', **not** just 'length'
- the time taken for the reaction', **not** just 'time'
- the height from which the ball was dropped', **not** just 'height'

and don't forget to include any units.

How Science Works

Section 1 of the ISA

At the end of this stage, you will answer Section 1 of the ISA. You will need to:

- develop a hypothesis
- identify one or more variables that you need to control
- describe how you would carry out the main experiment
- identify possible hazards and say what you would do to reduce any risk
- make a blank table ready for your results.

a What features should you include in your written plan?
b What should you include in your blank table?

Stage 2

This is where you carry out the experiment and get some results. Don't worry too much about spending a long time getting fantastically accurate results – it is more important to get some results that you can analyse.

After you have got results, you will have to compare your results with those of others. You will also have to draw a graph or a bar chart.

c How do you decide whether you should draw a bar chart or a line graph?

Stage 3

This is where you answer Section 2 of the ISA. Section 2 of the ISA is all about your own results, so make sure that you look at your table and graph when you are answering this section. To get the best marks you will need to quote some data from your results.

How Science Works

Section 2 of the ISA

In this section you will need to:

- say what you were trying to find out
- compare your results with those of others, saying whether you think they are similar or different
- analyse data that is given in the paper. This data will be in the same topic area as your investigation
- use ideas from your own investigation to answer questions about this data
- write a conclusion
- compare your conclusion with the hypothesis you have tested.

You may need to change or even reject your hypothesis in response to your findings.

Summary questions

1 Copy and complete the paragraph using the words below:

control independent dependent

When writing a plan, you need to state the variable that you are deliberately going to change, called the .... variable. You also need to say what you expect will change because of this; this is called the .... variable. You must also say what .... variables you will keep constant in order to make it a fair test.

AQA Examiner's tip

When you are comparing your conclusion with the hypothesis, make sure that you also talk about the **extent** to which your results support the hypothesis. Which of these answers do you think would score the most marks?

- My results support the hypothesis.
- In my results, as *x* got bigger, *y* got bigger, as stated in the hypothesis.
- In my results, as *x* got bigger, *y* got bigger, as stated in the hypothesis, but unlike the hypothesis, *y* stopped increasing after a while.

Key points

- When you are writing the plan make sure that you include details about:
  - the range and interval of the independent variable
  - the control variables
  - the number of repeats.
- Try to put down at least two possible hazards, and say how you are going to minimise the risk from them.
- Look carefully at the hypothesis that you are given – this should give you a good clue about how to do the experiment.
- Always refer back to the hypothesis when you are writing your conclusion.

Answers to in-text questions

a Control variables, interval and range of the independent variable. Identify possible hazards and how to reduce any risk.

b Columns for quantities that are going to be measured, including complete headings and units.

c A bar chart if one of the variables is categoric, a line graph if both variables are continuous.

Summary answers

1 independent, dependent, control

Summary answers

1 Could be some differences, which would be fine, e.g. hypothesis; prediction; design; safety; controls; method; table; results; repeat; graph; conclusion; improve.

2 **a** Scientific opinion is based on repeatable, reproducible and valid evidence. An opinion might not be.
b Continuous variable because it is more powerful than an ordered or categoric variable.

3 **a** A hypothesis is an idea that fits an observation and the scientific knowledge that is available. With the knowledge that sulfur dioxide in the air forms an acid, you can make a hypothesis and then a prediction.
b Increasing the concentration of acid will increase the rate of attack.
c Rate of reaction marble chip and sulphuric acid experiment.
d Run a survey (not suitable for this prediction).

4 **a** When all variables but the one being used as the independent variable are kept constant.
b It would tell you what range of dilution to use; whether the technique used to measure pH was sensitive enough; whether repeat readings were necessary.
c By repeating tests for each given dilution and seeing how closely grouped the results are.
d The calculations would give you the true values so you could check your mean pH values against these to determine accuracy.

5 Was the clock started correctly? Was it stopped correctly? Was the oil measured out accurately? Was it put onto the correct spot? Was there a clear line at the bottom? Was the tile thoroughly washed and dried?

6 **a** Take the highest and the lowest.
b The sum of all the readings divided by the number of readings.
c When you have an ordered or categoric independent variable and a continuous dependent variable.
d When you have a continuous independent variable and a continuous dependent variable.

7 **a** Examine to see whether it is an error; if so, repeat it. If identified from the graph, it should be ignored. Be aware that it could lead to something really interesting and to a new hypothesis.
b Identify a pattern.
c That it does not go further than the data and the repeatability, reproducibility and the validity allow.
d By repeating results, by getting others to repeat your results and by checking other equivalent data.

8 **a** The science is more likely to be accepted.
b They are influenced by other factors, political bias, they are not impartial, or they are being funded by a party that have an agenda other than pure data collection and analysis.

9 **a** For many scientific developments there is a practical outcome that can be used – a technological development. Many technological developments allow further progress in science.
b Society – all of us should have an opinion.

10 **a** Increasing the load would increase the bending of the glass.

How Science Works: H1–H10

Summary questions

1 Put these words into order. They should be in the order that you might use them in an investigation.
design; prediction; conclusion; method; repeat; controls; graph; results; table; improve; safety; hypothesis

2 **a** How would you tell the difference between an opinion that was scientific and a prejudiced opinion?
b Suppose you were investigating the amount of gas produced in a reaction. Would you choose to investigate a categoric or a continuous variable? Explain why.

3 You might have seen that marble statues weather badly where there is air pollution. You want to find out why.

a You know that sulfur dioxide in the air forms an acid. How could this knowledge help you to make a prediction about the effect of sulfur dioxide on marble statues?
b Make a prediction about the effect of sulfur dioxide on marble statues.
c What experiment could you do to test your prediction?
d Suppose you are not able to carry out an experiment. How else could you test your prediction?

4 **a** What do you understand by a 'fair test'?
b Suppose you were carrying out an investigation into what effect diluting acid had on its pH. You would need to carry out a trial. Describe what a trial would tell you about how to plan your method.

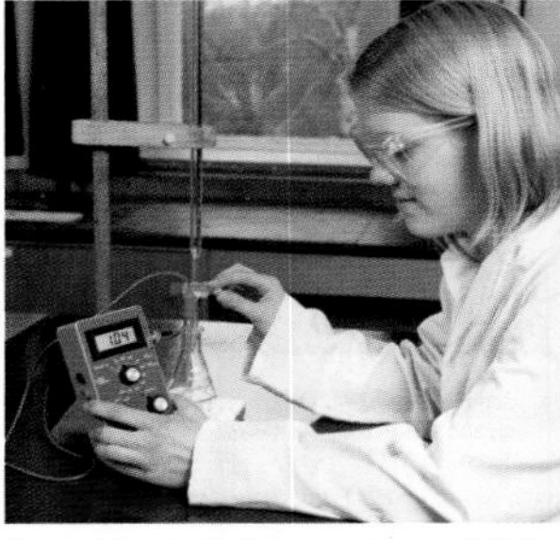

c How could you decide if your results were reliable?
d It is possible to calculate the effect of dilution on the pH of an acid. How could you use this to check on the accuracy of your results?

5 Suppose you were watching a friend carry out an investigation using the equipment shown on page 13. You have to mark your friend on how accurately he is making his measurements. Make a list of points that you would be looking for.

6 **a** How do you decide on the range of a set of data?
b How do you calculate the mean?
c When should you use a bar chart?
d When should you use a line graph?

7 **a** What should happen to anomalous results?
b What does a line of best fit allow you to do?
c When making a conclusion, what must you take into consideration?
d How can you check on the repeatability and reproducibility of your results?

22

b Line load added.
c Bending of the glass.
d E.g. same thickness of glass, line load always in the same place.
e The two readings are consistently 2 mm apart, except at 5 kN/m.
f The 5 kN/m reading could be a random error.
g Improved measuring devices.
h Yes, because there was a difference between each of the chosen line loads and no overlap between them, e.g. at 1 kN/m it bent 18 mm and at 2 kN/m it bent 37 mm.
i 95 mm.
j Graph drawn with all units and labels, the line load on the *x*-axis – it would be appropriate to plot 0, 0.
k Appropriate line of best fit.
l Directly proportional.
m Increasing the line load increases the degree of bending, up to a line load of at least 5 kN/m.
n Measure different thicknesses of polymer, using the same thicknesses of glass and test in the same way up until the glass breaks.
o Architects will be aware of the forces that the glass will be required to withstand and can choose an appropriate thickness of polymer for the task.

8 a Why is it important when reporting science to 'tell the truth, the whole truth and nothing but the truth'?

b Why might some people be tempted not to be completely fair when reporting their opinions on scientific data?

9 a 'Science can advance technology and technology can advance science.' What do you think is meant by this statement?

b Who answers the questions that start with 'Should we ... '?

10 Glass has been used for windows in buildings for a long time and is increasingly being used for structural parts as well. It is important therefore to be able to find out the strength of glass. One measure of this is the force that can be applied to glass before it breaks. Glass bends under pressure. Laminated glass is in three layers, glass on the outside sandwiching a polymer layer. This strengthens the glass.

An experiment was carried out to find out how far laminated glass would bend. The glass was supported on two wooden blocks and a load line drawn halfway between the blocks. A load was then placed on this load line and the amount of bend in the glass was measured. The load was gradually increased. Another plate of glass was then used for a second set of results.

Line load (kN/m) with 2-sided support

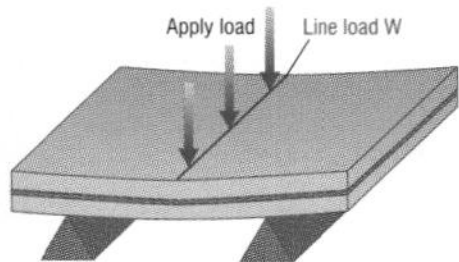

The results of the investigation are in the table.

Line load added (kN/m)	Bending Test 1 (mm)	Bending Test 2 (mm)
1	18	20
2	37	39
3	55	57
4	74	76
5	92	98

a What was the prediction for this test?

b What was the independent variable?

c What was the dependent variable?

d Suggest a control variable that should have been used.

e Is there any evidence of a systematic error in this investigation? Explain your answer.

f Is there any evidence of a random error? Explain your answer.

g How could the investigation have its accuracy improved?

h Was the precision of the bending measurement satisfactory? Provide some evidence for your answer from the data in the table.

i What is the mean for the results at a line load of 5 kN/m?

j Draw a graph of the results for the first test.

k Draw a line of best fit.

l Describe the pattern in these results.

m What conclusion can you reach?

n How might you develop this technique to show the effect of the thickness of the polymer on the breaking point of the glass?

o How might this information be used by architects wanting to protect buildings?

Examiner's comments

Changes to How Science Works

Although HSW has remained largely unchanged, there have been some additions in this specification, particularly with regard to the Controlled Assessment Unit (ISA).

These include a requirement for candidates to:

- identify potential hazards and devise a plan to minimise risk
- understand the term hypothesis
- test and/or make a prediction
- write a plan for an investigation, having been shown the basic technique to be used. Candidates should be able to decide upon issues such as the range and interval of the independent variable, the control variables and the number of repeats.

C1 1.1 Atoms, elements and compounds

Learning objectives

Students should learn:

- that elements are made of only one type of atom
- that symbols are used to represent atoms of a certain element
- the basic structure of the atom.

Learning outcomes

Most students should be able to:

- define an element
- recognise names, symbols, formulae and diagrams of elements or compounds
- label the nucleus and electrons in an atom, when the key words are given
- find symbols for elements in the periodic table.
- give examples and draw diagrams to explain the difference between elements and compounds
- draw and label an atom with the nucleus and electrons.

Specification link-up: Chemistry C1.1

- All substances are made of atoms. A substance that is made of only one sort of atom is called an element. There are about 100 different elements. Elements are shown in the periodic table. The groups contain elements with similar properties. *[C1.1.1 a)]*
- Atoms of each element are represented by a chemical symbol, e.g. O represents an atom of oxygen, and Na represents an atom of sodium. *[C1.1.1 b)]*
- Atoms have a small central nucleus, … around which there are electrons. *[C1.1.1 c)]*

Lesson structure

Starters

Anagrams – Students could try to create as many 'scientific' words as they can, using the letters from the term 'periodic table'. *(5 minutes)*

5,4,3,2,1 – Ask students to list five solid elements, four metal elements, three non-metal elements, two gaseous elements at room temperature, and one liquid element at room temperature. This task draws on the idea that most elements are solids at room temperature and most are metals. Students could be supported by being given a list of common elements or a copy of the periodic table. Students could be extended by being asked to give the symbols of the elements chosen as well as their names. *(10 minutes)*

Main

- Some students may not have had the opportunity to handle elements, apart from metals that they use in everyday life. Separate the class into groups of about five students. Give each group of students sealed samples of different elements. The students should then be instructed to sort them into different sets. Each group can feed back to the class how they classified the elements (possible sets might include: state, colour, metal/non-metal, hazard). Draw out from the students, using questions and answers, that there are a finite number of elements – about 100. Challenge the students to think about how we get the immense variety of materials in the world. This should lead on to a discussion of compounds and bonding.
- Play the elements song (this also has a flash animation): Search for 'flash animation' and 'elements' under www.privatehand.com
- Some students struggle with defining atoms, molecules, elements and compounds. A kinaesthetic, cut-and-stick activity could be used. The student should create a four-column (atom, molecule, element, compound) table in their exercise book. Different images representing atoms, molecules, elements and compounds, and the definitions of these four words, can then be given to the student to put into their table.

Plenaries

Code breaker – Ask the students to use the periodic table to decode this hidden message:

Carbon, radon, carbon, potassium, thorium, einsteinium, yttrium, M, boron, oxygen, L, sulfur. [CRaCKThEsYMBOLS]. Students could be supported by having the appropriate elements already highlighted on the periodic table, so they can find them more easily. This activity could be extended by getting the students to write their own secret message. *(5 minutes)*

Favourite – Ask students to think about their favourite element, giving reasons for their choice [e.g. copper, because it makes spiders' blood blue]. Students should then share their idea with a partner. The teacher could then ask a few pairs what their favourite elements are and why. *(10 minutes)*

Support

- Make sure that you start by showing students the elements that have a symbol which is the same as the first letter of its name. Carbon, Nitrogen and Oxygen are a good start. Stress to students that this letter is capitalised.

Extend

- Students could undertake research, using the internet, to find the origin of some symbols of elements that are unrelated to their names in English, such as Pb.
- Students could look at different representations of the periodic table, such as those which show abundance, only the naturally occurring elements, 3-D models, etc.

Answers to in-text questions

a about 100
b Because you can combine them together in millions of different ways.
c Because an element may have different names in different languages.
d metals – barium (Ba), vanadium (V), mercury (Hg)
non-metals – phosphorus (P), krypton (Kr)

Further teaching suggestions

Display work

- Each student can be assigned a different element. They are then given a piece of card to draw its chemical symbol and write five facts about the element found from secondary sources. In the following lesson, an electron diagram of one of its atoms, with the subatomic particles labelled, could be added. These cards can then be used to make a display in the classroom.

Spider diagram

- Begin to make a spider diagram using the key points on this spread. Then at the end of each lesson during this unit, or as a homework activity, the other key points from the topic can be added. By the end of the topic, a powerful revision resource will have been created.

Did you know ... ?

Most of the f-block elements were only discovered when nuclear reactors were invented. Under these conditions, these heavy elements have been synthesised. However, a number decay with very short half-lives and they are difficult to study.

For example, the element lawrencium (Lr, atomic number 103) was first discovered in 1961 in California. Lawrencium-260 has a half-life of three minutes.

Fundamental ideas

C1 1.1 Atoms, elements and compounds

Learning objectives

- What are elements made of?
- How do we represent atoms and elements?
- What is the basic structure of an atom?

Figure 1 An element contains only **one** type of atom – in this case bromine

Look at the things around you and the substances that they are made from. You will find wood, metal, plastic, glass ... the list is almost endless. Look further and the number of different substances is mind-boggling.

All substances are made of **atoms**. There are about 100 different types of atom found naturally on Earth. These can combine in a huge variety of ways. This gives us all those different substances.

Some substances are made up of only one type of atom. We call these substances elements. As there are only about 100 different types of atom, there are only about 100 different elements.

a How many different types of atom are there?
b Why can you make millions of different substances from these different types of atom?

Elements can have very different properties. Elements such as silver, copper and gold are shiny **solids**. Other elements such as oxygen, nitrogen and chlorine are **gases**.

Atoms have their own symbols

The name we use for an element depends on the language being spoken. For example, sulfur is called *schwefel* in German and *azufre* in Spanish! However, a lot of scientific work is international. So it is important that we have symbols for elements that everyone can understand. You can see these symbols in the **periodic table**.

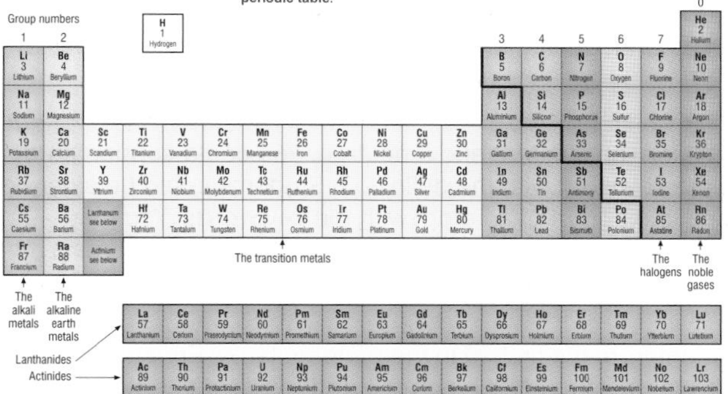

Figure 2 The periodic table shows the symbols for the elements

The symbols in the periodic table represent atoms. For example, O represents an atom of oxygen; Na represents an atom of sodium. The elements in the table are arranged in columns, called **groups**. Each group contains elements with similar chemical properties. The 'staircase' drawn in bold is the dividing line between metals and non-metals. The elements to the left of the line are metals. Those on the right of the line are non-metals.

c Why is it useful to have symbols for atoms of different elements?
d Sort these elements into metals and non-metals: phosphorus (P), barium (Ba), vanadium (V), mercury (Hg) and krypton (Kr).

Atoms, elements and compounds

Most of the substances we come across are not pure elements. They are made up of different types of atom joined together. These are called **compounds**. Chemical bonds hold the atoms tightly together in compounds. Some compounds are made from just two types of atom (e.g. water, made from hydrogen and oxygen). Other compounds consist of more different types of atom.

An atom is made up of a tiny central **nucleus** with **electrons** around it.

Electrons
Nucleus

Figure 3 Each atom consists of a small nucleus surrounded by electrons

Summary questions

1 Copy and complete using the words below:
atoms bonds molecule compounds
All elements are made up of .... When two or more atoms join together a .... is formed. The atoms in elements and .... are held tightly to each other by chemical ....
2 Explain why when we mix two elements together we can often separate them again quite easily. However, when two elements are chemically combined in a compound, they can be very difficult to separate.
3 Draw diagrams to explain the difference between an element and a compound. Use a hydrogen molecule (H_2) and a hydrogen chloride molecule (HCl) to help explain.
4 Draw a labelled diagram to show the basic structure of an atom.

Atoms, elements and compounds

links
For more information on the periodic table, see C3 1.2 The modern periodic table.

Did you know ... ?

Only 92 elements occur naturally on Earth. The other heavier elements in the periodic table have to be made artificially and might only exist for fractions of a second before they decay into other, lighter elements.

links
For more information on what is inside an atom, see C1 1.2 Atomic structure and 1.3 The arrangement of electrons in atoms.

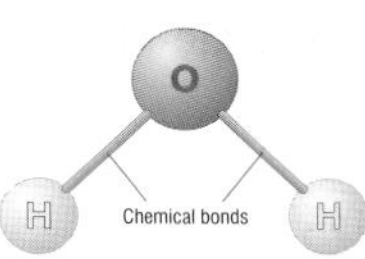

Figure 4 A grouping of two or more atoms bonded together is called a **molecule**. Chemical bonds hold the hydrogen and oxygen atoms together in the water molecule. Water is an example of a compound.

Key points

- All substances are made up of atoms.
- Elements contain only one type of atom.
- Compounds contain more than one type of atom.
- An atom has a tiny nucleus in its centre, surrounded by electrons.

24 25

Summary answers

1 atoms, molecule, compounds, bonds

2 In a mixture, the different substances are not joined to each other by chemical bonds, but in a compound the atoms are held together tightly by chemical bonds – this makes them difficult to separate.

3

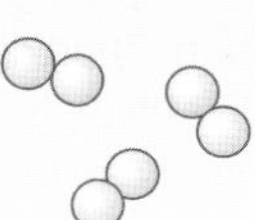

Element
e.g. hydrogen

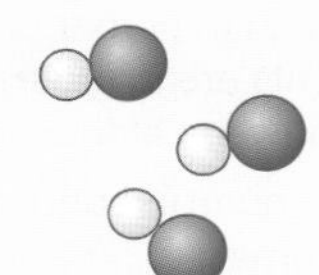

Compound
e.g. hydrogen chloride

4

Electrons
Nucleus

C1 1.2 Atomic structure

Learning objectives

Students should learn:

- the relative charges on subatomic particles
- that the number of protons equals the number of electrons in an atom
- that the number of protons in each atom of an element is its atomic number and the number of protons plus neutrons is the mass number
- that atoms are arranged in order of their atomic number in the periodic table.

Learning outcomes

Most students should be able to:

- label the subatomic particles in an atom
- state the charge of the subatomic particles in an atom
- define atomic number and mass number
- describe how the elements are ordered in the periodic table.

Some students should also be able to:

- explain why atoms are electrically neutral.

Support

- Ask students to fill in a table for a selection of elements, where they find the symbol, its atomic (proton) number from the periodic table and use this to determine the number of protons and electrons. Students often forget that the number of protons equals the number of electrons in a neutral atom.

Extend

- Ask students to write a timeline to show when, who and how the subatomic particles were discovered. Students should use secondary sources like the internet to help them. This links well to the How Science Works chapter.
- Students could research and list the general physical and chemical properties of the first 20 elements in the periodic table.

Specification link-up: Chemistry C1.1

- Atoms have a small central nucleus, which is made up of protons and neutrons and around which there are electrons. *[C1.1.1 c)]*
- The relative electrical charges are as shown. *[C1.1.1 d)]*

Name of particle	Charge
Proton	+1
Neutron	0
Electron	−1

- In an atom, the number of electrons is equal to the number of protons in the nucleus. Atoms have no overall electrical charge. *[C1.1.1 e)]*
- All atoms of a particular element have the same number of protons. Atoms of different elements have different numbers of protons. *[C1.1.1 f)]*
- The number of protons in an atom of an element is its atomic number. The sum of the protons and neutrons in an atom is its mass number. *[C1.1.1 g)]*

Lesson structure

Starters

Spot the mistake – 'When different compounds join together, atoms of the same element are made.' [Compounds do not join together; different types of atom join up to make compounds. Also, if all the atoms are the same in a material, then it is an element, not a compound.] *(5 minutes)*

Create the question – Give students three answers [atom, element, and compound]. They should create three questions that have these three answers. For example, what is the name of the smallest particle that can exist on its own? [Answer: atom.] Students can be supported by being asked to work in small groups. Students can be extended by being asked to come up with other correct answers to the same question. *(10 minutes)*

Main

- Ask students to define a proton, an electron and a neutron. You could support students by asking them to match the key word with the definition. Students could be extended by also being asked to label a diagram of an atom with the key words. You may wish to give students a partially filled table from the specification which contrasts the subatomic particles. They could then fill in the information for each of the subatomic particles using the definitions to help them.
- Give each student an element from the first 20 in the periodic table. Encourage them to research their element, define the term atomic number and represent its structure in as many different ways that they can find.
- Students could compile their diagrams into a poster, with the subatomic particles labelled. The posters could be displayed in the classroom.
- Atoms were discovered and grouped in many different ways. Supply students with cards for the first 20 elements in the periodic table. On one side of the card there should be the element's name, symbol, numbers of protons, neutrons (most common isotope) and electrons. On the reverse, its general physical and chemical properties should be listed. Students should then work in small teams to order the elements in different ways [alphabetically, number of protons, number of neutrons, physical properties, etc.]. You should tour the other teams, asking them about their grouping structure and encourage them to find new ways to group the elements.
- Then ask the students to sort their cards into the order of the periodic table. Ask the students what they notice [that the elements are in order of atomic number]. Then ask the students to summarise in one sentence how the atoms are arranged in the periodic table.
- Ensure students are familiar with 'atomic number' and 'mass number' and can use them to list the subatomic particles in atoms.

Plenary

Label – On the board draw a diagram of a helium atom. Students should copy the diagram and label the subatomic particles with their name and their charge. *(5 minutes)*

Further teaching suggestions

Elements cards

- Instead of organising cards with the elements' details on them, the elements themselves could be provided in sealed gas jars and Petri dishes. The relevant information could then be stuck onto the container with tape.

Representing the atom

- Give students lots of different examples of representing an atom, e.g. 2-D drawing, 3-D computer model, artist's impression, electron microscope. Students could look at the different ways of representing the atom and evaluate them. Students should discuss their thoughts and decide which they think is the most useful way of representing the atom and explain why.

Building mobiles

- Mobiles of atoms can be created. A 5 cm diameter circle and a ring are drawn onto a piece of card. The rings need to be drawn so that the circle fits inside and a gap is left between them [e.g. inner diameter of the ring is 13 cm and outer diameter of the ring is 15 cm]. The student cuts out the template and draws subatomic particles using colours to highlight the charges. The circle represents the nucleus and the ring the shell of electrons. They could be encouraged to find out an interesting fact and write it on another piece of card (10 cm × 5 cm). String and tape is then used to join the fact to the ring (electron shell), the ring to the circle (nucleus) and that in turn to the top part of the ring. Excess string should be available at the top to suspend the mobile. Note that to be chemically correct the students can only complete a mobile of H or He at this stage; to complete large atoms, additional rings need to be added.

Fundamental ideas

C1 1.2 Atomic structure

Learning objectives

- What is the charge on a proton, a neutron and an electron?
- What can we say about the number of protons in an atom compared with its number of electrons?
- What is the 'atomic number' and 'mass number' of an atom?
- How are atoms arranged in the periodic table?

In the middle of an atom there is a very small nucleus. This contains two types of particles, which we call **protons** and **neutrons**. A third type of particle orbits the nucleus. We call these really tiny particles electrons.

Any atom has the same number of electrons orbiting its nucleus as it has protons in its nucleus.

Protons have a positive charge. Neutrons have no charge – they are neutral. So the nucleus itself has an overall positive charge.

The electrons orbiting the nucleus are negatively charged. The relative charge on a proton is +1 and the relative charge on an electron is –1.

Because any atom contains equal numbers of protons and electrons, the positive and negative charges cancel out. So there is no overall charge on any atom. Its charge is zero. For example, a carbon atom is neutral. It has 6 protons, so we know it must have 6 electrons.

a What are the names of the three particles that make up an atom?
b An oxygen atom has 8 protons – how many electrons does it have?

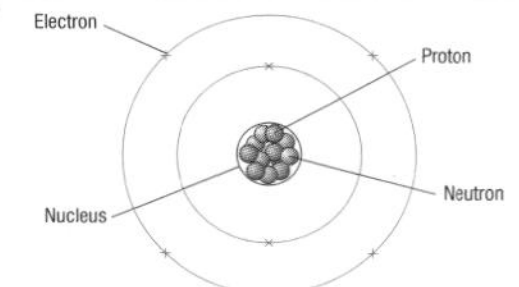

Figure 1 Understanding the structure of an atom gives us important clues to the way chemicals react together

Type of subatomic particle	Relative charge
Proton	+1
Neutron	0
Electron	–1

To help you remember the charge on the subatomic particles:

- **P**rotons are **P**ositive;
- **Neut**rons are **Neut**ral;
- so that means Electrons must be Negative!

Did you know ...?

In 1808, a chemist called John Dalton published a theory of atoms. It explained how atoms joined together to form new substances (compounds). Not everyone liked his theory though – one person wrote 'Atoms are round bits of wood invented by Mr Dalton!'

Atomic number and the periodic table

All the atoms of a particular element have the same number of protons. For example, hydrogen has 1 proton in its nucleus, carbon has 6 protons in its nucleus and sodium has 11 protons in its nucleus.

We call the **number of protons** in each atom of an element its **atomic number**.

links
For more information on the structure of atoms, see C2 3.1 The mass of atoms.

H 1

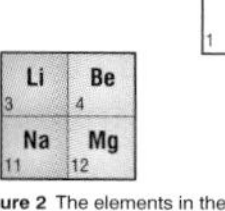

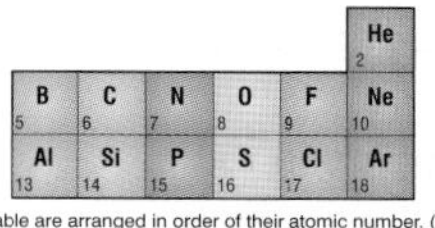

Figure 2 The elements in the periodic table are arranged in order of their atomic number. (As atoms are neutral, this is also the same order as their number of electrons.)

The elements in the periodic table are arranged in order of their atomic number (number of protons). If you are told that the atomic number of an element is 8, you can identify it using the periodic table. It will be the 8th element listed. In this case it is oxygen.

c What is the 14th element in the periodic table?

You read the periodic table from left to right, and from the top down – just like reading a page of writing.

d Look at the elements in the last group of the abbreviated periodic table in Figure 2. What pattern do you notice about the number of protons going from helium to neon to argon?

Mass number

The **number of protons plus neutrons** in the nucleus of an atom is called its **mass number**.

- So, if an atom has 4 protons and 5 neutrons, its mass number will be 4 + 5 = **9**.
- Given the atomic number and mass number, we can work out how many protons, electrons and neutrons are in an atom. For example, an argon atom has an atomic number of 18 and a mass number of 40.

  Its atomic number is 18 so it has **18 protons**. Remember that atoms have an equal number of protons and electrons. So argon also has **18 electrons**. The mass number is 40, so we know that:

  18 (the number of protons) + the number of neutrons = 40

  Therefore argon must have **22 neutrons** (as 18 + 22 = 40).

  We can summarise the last part of the calculation as:

  number of neutrons = mass number – atomic number

Summary questions

1 Copy and complete using the words below:
electrons atomic negative neutrons
In the nucleus of atoms there are protons and .... Around the nucleus there are .... which have a .... charge. In the periodic table, atoms are arranged in order of their .... number.

2 Atoms are always neutral. Explain why.

3 How many protons, electrons and neutrons do the following atoms contain?
a A nitrogen atom whose atomic number is 7 and its mass number is 14.
b A chlorine atom whose atomic number is 17 and its mass number is 35.

AQA Examiner's tip

In an atom, the number of protons is always equal to the number of electrons. You can find out the number of protons and electrons in an atom by looking up its atomic number in the periodic table.

links
For more information on the patterns in the periodic table, see C1 1.3 The arrangement of electrons in atoms.

Key points

- Atoms are made of protons, neutrons and electrons.
- Protons and electrons have equal and opposite electric charges. Protons are positively charged, and electrons are negatively charged.
- Neutrons have no electric charge. They are neutral.
- Atomic number = number of protons (= number of electrons)
  Mass number = number of protons + neutrons
- Atoms are arranged in the periodic table in order of their atomic number.

Did you know ...?

Atoms are very small. Ask students to guess how small this would be. [About 0.000 000 000 1 m in diameter.]

Summary answers

1 neutrons, electrons, negative, atomic

2 Because protons and electrons have the same amount of charge but with opposite signs, the charge on a proton is exactly cancelled out by the charge on an electron.

3 **a** 7 protons, 7 electrons and 7 neutrons.
b 17 protons, 17 electrons and 18 neutrons.

Answers to in-text questions

a protons, neutrons, electrons
b 8 electrons
c silicon
d They increase by 8 between each element.

C1 1.3 The arrangement of electrons in atoms

Learning objectives

Students should learn:

- that electrons are arranged in energy levels (shells) in an atom
- that the number of electrons in the highest energy level relates to the group number in the periodic table
- that the number of electrons in the highest energy level determines chemical properties
- that the atoms of the noble gases have stable electron arrangements.

Learning outcomes

Most students should be able to:

- state the number of electrons that can occupy the first three energy levels (shells) in an atom
- state the relationship between the number of electrons in the highest energy level and the group number
- describe how the unreactive noble gases of Group 0 have stable arrangements of electrons.

Some students should also be able to:

- explain how the number of electrons in the highest energy level relates to the chemical properties of an element.

Support

- Students could be supported by being shown diagrams of only the outer shell electrons. They could then focus on the relationship between the outer electrons and position in the periodic table.

Extend

- Extend students by asking them to reflect further on the electronic structure of the atom and the periodic table. Students should find out that the period number and the number of electron shells is the same for the first three periods.
- You could also extend students by asking them to find out about instances when noble gases do react, which noble gases they are and under what conditions they react.

Specification link-up: Chemistry C1.1

- Electrons occupy particular energy levels. Each electron in an atom is at a particular energy level (in a particular shell). The electrons in an atom occupy the lowest available energy levels (innermost available shells). Candidates may answer questions in terms of either energy levels or shells. *[C1.1.1 h)]*
- Elements in the same group in the periodic table have the same number of electrons in the highest energy levels (outer electrons) and this gives them similar chemical properties. *[C1.1.2 a)]*
- The elements in Group 0 of the periodic table are called the noble gases. They are unreactive because their atoms have stable arrangements of electrons. *[C1.1.2 b)]*

Lesson structure

Starters

Revision card sort – Give the students nine cards, each with different labels: proton, electron, neutron, nucleus (there should be two of these), shell, +1, 0, −1. Students should also be given a table on a piece of laminated card, consisting of three columns, labelled subatomic particle, charge, position, and three rows. They then sort the cards, putting them in the appropriate positions on the table. *(5 minutes)*

List – Ask students to list all the information they can remember about electrons. Then ask students to share their ideas in small groups and each group takes it in turn to write a piece of information on the board. You should address any misconceptions revealed in this activity. You could support students by giving them some key points, e.g. −1, which they have to use as a prompt to come up with pieces of information. Students could be extended by asking them to do the same for the other subatomic particles. *(10 minutes)*

Main

- Give students a diagram of an argon atom, which includes the electronic structure and the number of protons and neutrons in the nucleus. Use this diagram to explain why atoms have no charge, the positioning of electrons in shells, or energy levels. Discuss the energy levels further and the number of electrons that the shell can hold. State that the maximum number of electrons in the first three shells is 2,8,8, as shown in the argon atom. Encourage students to annotate their diagram to explain the terms 'energy level' and 'electron shell', and why an atom is neutral.
- Students should already be familiar with the periodic table from KS3. Refresh students' memories by reminding them that the columns are called 'groups' and rows are called 'periods'.
- Give students cards with the electronic structure of the first 20 elements with their names. Ask students to use the periodic table to order the cards. Students should be encouraged to reflect on the relationship between the outer electrons and the group that the element belongs in.
- Demonstrate the reactions of Group 1 metals with water and their reactions with oxygen (see 'Demonstration support'). Ensure that you relate the chemistry to the number of outer shell electrons.
- Encourage students to write word equations for each of these reactions. Students could be extended by being asked to write a general equation for the reaction.

Plenaries

Reflection – Ask students to consider a fact that they have revisited from KS3 in the lesson and a new fact that they have learned in the lesson. As they leave the classroom, ask them for their facts. You should challenge any misconceptions during the reflection plenary of the following lesson. *(5 minutes)*

Which atom? – Create flash cards to show the electronic structure of the first 20 elements in a random order. Ask the students to look at the images and work out which atom is being displayed. Students can be supported by being encouraged to use their notes and the Student Book, etc. Students could be extended by only being given the atomic number of an element and then using the periodic table to determine the number of electrons in its outer shell. *(10 minutes)*

Demonstration support

Reactions of Group 1 metals with water

Equipment and materials required

Lithium, sodium, potassium (all stored under oil), white tile, filter paper, spatula, tweezers, water trough, water, safety screen, eye protection (chemical splashproof).

Details

Half-fill the trough with water. Cut a small piece of lithium (no more than 3 mm). Wipe off the excess oil and put into the water. Repeat with the other metals.

Safety: You should wear chemical splashproof safety goggles or face visor and students should also wear safety goggles. There should be a safety screen between the water trough and the students. Be sure not to get the metal on the skin – if this does happen remove with paper or tweezers and only then wash the affected area well under cold water. CLEAPSS Hazcard 58, 88, 76 – highly flammable and corrosive. Leave equipment safe for technicians.

Reactions of Group 1 metals with oxygen

Equipment and materials required

Lithium, sodium, potassium (all stored under oil), 3 × deflagrating spoon, 3 × gas jar (with lid) of oxygen, Bunsen burner and safety equipment, safety screen, eye protection (chemical splashproof).

Details

Cut a small piece of lithium (no more than 3 mm), wipe off the excess oil and put on the deflagrating spoon. Heat the metal in the Bunsen flame until it ignites. Ask students to note the colour and quickly remove the gas jar lid and plunge the deflagrating spoon into the gas jar. Repeat with the other metals.

Safety: You should wear safety goggles or face visor and students should also wear safety goggles. There should be a safety screen between the water trough and the students. Be sure not to get the metal on the skin – if this does happen remove with paper or tweezers and only then wash the affected area well under cold water. CLEAPSS Hazcard 58, 88, 76 – highly flammable and corrosive. CLEAPSS Hazcard 69 Oxygen – oxidising. Leave equipment safe for technician.

C1 1.3

The arrangement of electrons in atoms

Learning objectives

- How are the electrons arranged inside an atom?
- How is the number of electrons in the highest energy level of an atom related to its group in the periodic table?
- How is the number of electrons in the highest energy level of an atom related to its chemical properties?
- Why are the atoms of Group 0 elements so unreactive?

One model of the atom which we use has electrons arranged around the nucleus in **shells**, rather like the layers of an onion. Each shell represents a different **energy level**. The lowest energy level is shown by the shell which is nearest to the nucleus. The electrons in an atom occupy the lowest available energy level (the shell closest to the nucleus).

a Where are the electrons in an atom?
b Which shell represents the lowest energy level in an atom?

Electron shell diagrams

We can draw diagrams to show the arrangement of electrons in an atom. A carbon atom has 6 protons, which means it has 6 electrons. Figure 1 shows how we represent an atom of carbon.

An energy level (or shell) can only hold a certain number of electrons.

- The first, and lowest, energy level holds 2 electrons.
- The second energy level can hold up to 8 electrons.
- Once there are 8 electrons in the third energy level, the fourth begins to fill up, and so on.

To save drawing atoms all the time, we can write down the numbers of electrons in each energy level. This is called the **electronic structure**. For example, the carbon atom in Figure 1 has an electronic structure of 2,4.

A silicon atom with 14 electrons has the electronic structure 2,8,4. This represents 2 electrons in the first, and lowest, energy level, then 8 in the next energy level. There are 4 in the highest energy level (its outermost shell).

The best way to understand these arrangements is to look at some examples.

C

Figure 1 A simple way of representing the arrangement of electrons in the energy levels (shells) of a carbon atom

H — 1 Hydrogen H
B — 2, 3 Boron B
O — 2,6 Oxygen O
Na — 2,8,1 Sodium Na
Ar — 2,8,8 Argon Ar
Ca — 2,8,8,2 Calcium Ca

Figure 2 Once you know the pattern, you should be able to draw the energy levels (shells) and electrons in any of the first 20 atoms (given their atomic number)

AQA Examiner's tip

Make sure that you can draw the electronic structure of the atoms for all of the first 20 elements. You will always be given their atomic number or their position in the periodic table (which tells you the number of electrons) – so you don't have to memorise these numbers.

c How many electrons can the first energy level hold?
d What is the electronic structure of sulfur (whose atoms contain 16 electrons)?

Electrons and the periodic table

Look at the elements in any one of the main groups of the periodic table. Their atoms will all have the same number of electrons in their highest energy level. These electrons are often called the outer electrons because they are in the outermost shell. Therefore, all the elements in Group 1 have one electron in their highest energy level.

Demonstration

Properties of the Group 1 elements

Your teacher will show you the Group 1 elements lithium, sodium and potassium. The elements in this group are called the **alkali metals**. Make sure you wear eye protection for all the demonstrations.

- In what ways are the elements similar?
- Watch their reactions with water and comment on the similarities.
- You might also be shown their reactions with oxygen.

LITHIUM
SODIUM
POTASSIUM

Figure 3 The Group 1 metals are all reactive metals, stored under oil

The chemical properties of an element depend on how many electrons it has. The way an element reacts is determined by the number of electrons in its highest energy level (or outermost shell). So as the elements in a particular group all have the same number of electrons in their highest energy level, they all react in a similar way.

For example:

lithium + water → lithium hydroxide + hydrogen
sodium + water → sodium hydroxide + hydrogen
potassium + water → potassium hydroxide + hydrogen

The elements in Group 0 of the periodic table are called the noble gases because they are unreactive. Their atoms have a very stable arrangement of electrons. They all have 8 electrons in their outermost shell, except for helium, which has only 2 electrons.

Summary questions

1 Copy and complete using the words below:

electrons energy group nucleus shells

The electrons in an atom are arranged around the .... in .... (energy levels). The electrons further away from the nucleus have more .... than those close to the nucleus. All elements in the same .... of the periodic table have the same number of .... in their outermost shell.

2 Using the periodic table, draw the arrangement of electrons in the following atoms and label each one with its electronic structure.

a Li **b** B **c** P **d** Ar

3 What is special about the electronic structure of neon and argon?

Key points

- The electrons in an atom are arranged in energy levels or shells.
- Atoms with the same number of electrons in their outermost shell belong in the same group of the periodic table.
- The number of electrons in the outermost shell of an element's atoms determines the way that element reacts.
- The atoms of the unreactive noble gases (in Group 0) all have very stable arrangements of electrons.

Further teaching suggestions

Periodic table

- Give students an A3 or A2 enlarged copy of the periodic table and ask them to draw the electronic structure of the first 20 directly onto the table. To extend students, ask them to find out about reactivity trends and add trend arrows on to the groups to show how reactivity changes going down/up the group.

Group 1 video

- You many wish to show videos of how all the Group 1 metals except for francium react with water. You can find suitable videos on www.teachers.tv by searching for 'Ferocious elements'.

Summary answers

1 nucleus, shells, energy, group, electrons

2 **a** Li 2, 1
b B 2, 3
c P 2, 8, 5
d Ar 2, 8, 8

3 They both have full outer shells or energy levels of electrons/very stable arrangements of electrons.

Answers to in-text questions

a Arranged around the nucleus in shells.
b The energy level closest to the nucleus.
c 2 **d** 2,8,6

C1 1.4 Forming bonds

Learning objectives

Students should learn:

- that metals and non-metals form ions when they react to make compounds
- that non-metals can bond to each other by sharing electrons, forming covalent bonds
- that a compound can be represented by its chemical formula.

Learning outcomes

Most students should be able to:

- name the two types of bonding possible in compounds
- interpret formula in terms of number and type of atoms that have joined.

Some students should also be able to:

- explain how the two types of bonding happen.

Support

- Students could be supported by being given a 'cut-and-stick' activity to contrast covalent and ionic bonding. Students should be able to sort the definitions and examples to form a table of information.

Extend

- Students could be extended by being asked to find out about compound ions, such as sulfate and nitrate ions, and how we write the formulae of their compounds. Students could also be extended by writing the formula of a compound from its name and vice versa.

Specification link-up: Chemistry C1.1

- When elements react, their atoms join with other atoms to form compounds. This involves giving, taking or sharing electrons to form ions or molecules. Compounds formed from metals and non-metals consist of ions. Compounds formed from non-metals consist of molecules. In molecules the atoms are held together by covalent bonds. *[C1.1.3 a)]*
- Chemical reactions can be represented by word equations or by symbol equations. *[C1.1.3 b)]*

Lesson structure

Starters

Fume cupboard demonstration – The formation of sodium chloride from its elements is an exciting and impressive reaction. The elements can be shown to the students in sealed containers. You could then demonstrate the formation of sodium chloride in a fume cupboard. *(5 minutes)*

Charge revision – Ask students to draw a table (similar to the one that appears in C1 1.2 Atomic structure) contrasting the charges of the subatomic particles in an atom. Students could be supported by being given the information so that they could cut and stick it into the table. Students could be extended by being asked to display this data in a labelled diagram of an atom. *(10 minutes)*

Main

- Use the step-by-step example of LiF in the Student Book to show how ionic bonds are formed.
- Students could describe the properties of these elements and their electronic structure and write it in their exercise books.
- Following this, ask students to note down the formula of the new compound and state the number of each type of atom. Using electronic structure diagrams of the outer shell electrons for lithium and fluorine, explain how the outer shell electron from lithium has been donated to fluorine. This has formed a positive lithium ion and a negative fluoride ion. Students should classify the bonding as ionic.
- Introduce the idea of covalent bonding and overlapping shells. Use H_2S, H_2O and CH_4 to illustrate covalent bonding.
- Students can form water from its elements, by completing the hydrogen pop! test (see 'Demonstration support').
- Gas jars of oxygen, boiling tubes of hydrogen and beakers of water can be used to allow the students to contrast the properties of the elements with the compound. Students should visually inspect the samples. They should be encouraged to note down the formula of water and classify the bonding as covalent.

Plenaries

Reflection – Ask the students to note down:

- What I have seen ...
- What I have heard ...
- What I have done ...

These can be collected in from the class to be used to see what the students have remembered. *(5 minutes)*

Chemical equations – Ask students to write a word equation for the reaction to make sodium chloride from its elements. Support students by giving them the chemical names and a missing word equation to complete. Extend students by asking them to write a balanced symbol equation (which could be used to introduce some of them to state symbols). *(10 minutes)*

Demonstration support

Sodium with chlorine gas

Equipment and materials required

Sodium stored under oil, gas jar of chlorine, deflagrating spoon, Bunsen burner with safety equipment, safety goggles, fume cupboard.

Details

Cut a small piece of sodium (no more than 3 mm), wipe off the excess oil and put on the deflagrating spoon. Heat the metal in the Bunsen flame until it burns, and quickly put into the gas jar. You may wish to record the demonstration using a webcam. The film can then be played back, annotated and paused on an interactive whiteboard during a discussion about the reaction.

Safety: You should wear safety goggles and the reaction should be undertaken in a fume cupboard. Be sure not to get the metal on the skin – if this does happen remove with paper or tweezers and wash the affected area well under cold water. CLEAPSS Hazcard 22 Chlorine – toxic. Warn asthmatics. CLEAPSS Hazcard 88 Sodium – highly flammable and corrosive.

Hydrogen pop! test

Equipment and materials required

Borosilicate test tube of hydrogen, splint, Bunsen burner and safety equipment, chemical splashproof safety goggles.

Details

Light the splint and remove the bung from the test tube. Quickly waft the lighted splint over the neck of the boiling tube.

More details on gas preparations in the CLEAPSS Handbook 13.2.2.

Safety: You should wear safety goggles. CLEAPSS Hazcard 48 Hydrogen – extremely flammable.

To avoid the brown fumes, a 'purer' version is suggested in CLEAPSS guide L195 'safer chemicals, safer reactions', 9.3.

Fundamental ideas

C1 1.4 Forming bonds

Learning objectives

- How do metals and non-metals bond to each other?
- How do non-metals bond to each other?
- How do we write the formula of a compound?

It is useful for us to know how atoms bond to each other in different substances. It helps us to predict and explain their properties.

How Science Works

Predicting what material to use

A team of research chemists and material scientists are working to make a new compound for the latest surfboard. Knowing about chemical bonding will make the process of designing a new compound a lot quicker.

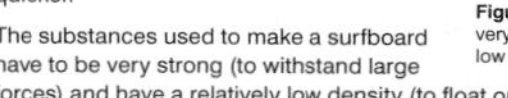

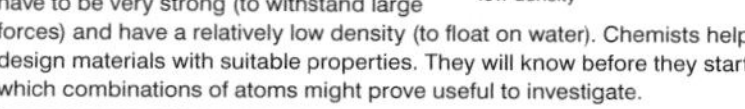

Figure 1 Surfboards have to be very strong and have a relatively low density

The substances used to make a surfboard have to be very strong (to withstand large forces) and have a relatively low density (to float on water). Chemists help design materials with suitable properties. They will know before they start which combinations of atoms might prove useful to investigate.

AQA Examiner's tip

When counting atoms, think of each symbol as a single atom and the formula of each ion as a single ion. Small numbers in a chemical formula only multiply the symbol they follow. Brackets are needed when there is more than one atom in the ion being multiplied. For example, a hydroxide ion has the formula OH^-. So calcium hydroxide, in which Ca^{2+} and OH^- combine, has the formula $Ca(OH)_2$.

Sometimes atoms react together by **transferring** electrons to form chemical bonds. This happens when metals react with non-metals. If the reacting atoms are all non-metals, then the atoms **share** electrons to form chemical bonds.

Forming ions

When a metal bonds with a non-metal, the metal atom gives one or more electrons to the non-metal atom. Both atoms become charged particles called **ions**.

- Metal atoms form positively charged ions (+).
- Non-metal atoms form negatively charged ions (–).

Opposite charges attract each other. There are strong attractions between the positive and negative ions in a compound of a metal and non-metal. These strong forces of attraction are the chemical bonds that form. They are called **ionic bonds**.

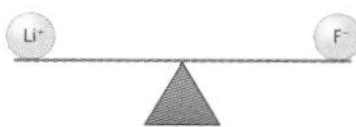

Figure 2 The positive and negative charge on the ions in a compound balance each other, making the total charge zero

To see how ions are formed we can look at an example. Lithium metal will react with the non-metal fluorine. They make the compound lithium fluoride. Lithium atoms have 3 electrons, each negatively charged. As all atoms are neutral, we know it also has 3 positive protons in its nucleus. The charges on the negative electrons are balanced by the positive protons.

When lithium reacts with fluorine it loses 1 electron. This leaves it with only 2 electrons. However, there are still 3 protons in the nucleus. Therefore the lithium ion carries a 1+ charge.	The electron lost from lithium is accepted by a fluorine atom. A fluorine atom has 9 electrons and 9 protons, making the atom neutral. However, with the extra electron from lithium, it has an extra 1– charge:
3 protons = 3+ 2 electrons = 2– Charge on ion = 1+	9 protons = 9+ 10 electrons = 10– Charge on ion = 1–
We show the formula of a lithium ion as **Li^+**.	We show the formula of a fluoride ion as **F^-**. Notice the spelling – we have a fluor**ine** atom which turns into a negatively charged fluor**ide** ion.

30

Forming bonds

In compounds between metals and non-metal, the charges on the ions always cancel each other out. This means that their compounds have no overall charge. So the formula of lithium fluoride is written as **LiF**.

a Potassium (K) is a metal. It loses one electron when it forms an ion. What is the formula of a potassium ion?

Forming molecules

Non-metal atoms bond to each other in a different way. The outermost shells of their atoms overlap and they share electrons. Each pair of shared electrons forms a chemical bond between the atoms. These are called **covalent bonds**. No ions are formed. They form molecules, such as hydrogen sulfide, H_2S, and methane, CH_4 (see Figure 3).

b What do we call the bonds between nitrogen and hydrogen atoms in an ammonia molecule, NH_3?

Hydrogen sulfide   Methane

Figure 3 There are strong covalent bonds between the non-metal atoms in each of these molecules. These are shown as lines between each atom or between the symbols of each atom in the molecule (H_2S and CH_4).

Chemical formulae

The chemical formula of an ionic compound tells us the ratio of each type of ion in the compound. We use a ratio because when ions bond together they form structures made of many millions of ions. The ratio depends on the charge on each ion. The charges must cancel each other out.

An example is magnesium chloride. Magnesium forms Mg^{2+} ions and chlorine forms Cl^- ions. So the formula of magnesium chloride is $MgCl_2$. We have 2 chloride ions for every one magnesium ion in the compound (see Figure 4).

In covalent molecules we can just count the number of each type of atom in a molecule to get its formula. Figure 3 shows two examples.

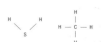

Figure 4 The 2+ positive charge on the magnesium ion balances the two 1– negative charges on the chloride ions in magnesium chloride ($MgCl_2$)

Summary questions

1 Copy and complete using the words below:

covalent lose gain ionic negative attract share positive

Metal atoms form .... ions because they .... one or more electrons when they combine with non-metals. Non-metal atoms .... electrons in the reaction, forming .... ions. The oppositely charged ions .... each other. This is called .... bonding.

When non-metals combine with each other, they form .... bonds. Their atoms .... electrons.

2 Sodium (Na) atoms lose one electron when they combine with fluorine (F). Each fluorine atom gains one electron in the reaction.

  a What is the name of the compound formed when sodium reacts with fluorine?
  b Write down the formula of a sodium ion and a fluoride ion.
  c What is the formula of the compound made when sodium reacts with fluorine.

Key points

- When atoms from different elements react together they make compounds. The formula of a compound shows the number and type of atoms that have bonded together to make that compound.
- When metals react with non-metals, charged particles called ions are formed.
- Metal atoms form positively charged ions. Non-metal atoms form negatively charged ions. These oppositely charged ions attract each other in ionic bonding.
- Atoms of non-metals bond to each other by sharing electrons. This is called covalent bonding.

31

Further teaching suggestions

Salt research

- Sodium chloride is an important compound in our diet. Ask students to research why humans need to eat salt, and what can happen if we consume too much salt.

Summary

- Students can be given secondary sources of information about ionic and covalent bonding. They should use this information and the Student Book to summarise in no more than 50 words the differences and similarities between these types of bonding. Students should be encouraged to illustrate their answer with at least one example, not yet seen in the lesson, of each type of bonding.

Answers to in-text questions

a K^+

b covalent bonds

Summary answers

1 positive, lose, gain, negative, attract, ionic, covalent, share

2 **a** Sodium fluoride
  b Na^+ and F^-
  c NaF

C1 1.5 Chemical equations

Learning objectives

Students should learn:

- how a chemical reaction can be represented
- what happens to the atoms in a chemical reaction
- that mass is conserved in a chemical reaction
- how to balance a chemical equation. **[HT only]**

Learning outcomes

Most students should be able to:

- state a definition of a chemical reaction in terms of reactants and products
- describe how atoms cannot be created or destroyed in a chemical reaction
- state that mass is conserved in a chemical reaction.

Some students should also be able to:

- explain why mass is conserved in a chemical reaction
- balance symbol equations. **[HT only]**

Support

- Some students will struggle with balancing equations. Encourage students to write out the number and type of atom in each formula under the symbol equation to complete 'atom accounting'. You may wish to relate to maths and remind students that $a + a$ is written $2a$, and this is the same in science where $2O_2$, means $2 \times O_2$.

Extend

- Extend students by introducing the idea of state symbols and getting them to use them in balanced symbol equations. You can extend students by giving them symbol equations to balance that contain compounds with several compound ions (formulae in brackets).

Specification link-up: Chemistry C1.1

- No atoms are lost or made during a chemical reaction so the mass of the products equals the mass of the reactants. *[C1.1.3 c)]*

  Controlled Assessment: C4.1 Plan practical ways to develop and test candidates' own scientific ideas. *[C4.1.1 a) b)]*; C4.4 Select and process primary and secondary data. *[C4.4.2 a)]*

Lesson structure

Starters

Sorting – Write the following word equations on the board and ask students to determine the reactants and the products of the reaction.

sodium + chlorine → sodium chloride

hydrogen + oxygen → water

Students could be supported by being given the definitions of reactants and products. Students could be extended by being given a passage describing a reaction from which they need to make the word equation. *(5 minutes)*

Card sort – Give the students separate cards with images (just use coloured circles) and others with the word 'compound', 'element', 'molecule', or 'mixture' written on them. Students should then try to match the image with the key term. *(10 minutes)*

Main

- Ask students to define the term 'chemical reaction' [a change where a new substance is made]. You may wish to encourage students to give some everyday examples of chemical reactions, e.g. rusting, burning.
- Explain to students that atoms are neither created nor destroyed in a chemical reaction, they are just rearranged. Using molecular model kits demonstrate how sulfur reacts with oxygen to make sulfur dioxide [$S + O_2 \rightarrow SO_2$]. Show students that there are the same number and type of atoms on the reactant side as on the product side, but they are rearranged. Ask them to predict how mass would therefore change in a reaction. [It remains the same.]
- Then lead students through combustion of methane using the molymod [$CH_4 + 2O_2 \rightarrow 2H_2O + CO_2$]. Give students the word equation and, through question and answer, try and get the students to give you the formulae of the chemicals. Then ask the students to use the molecular model to make one of each molecule.
- You may wish to allow students to demonstrate the conservation of mass using precipitation reactions (see 'Practical support'). Encourage students to consider how to record their results in a scientific manner. You may need to support students by giving them an appropriate results table to complete.
- Ask students to reflect on how many atoms they have on each side, and explain that you can't add just extra odd atoms, you have to add whole molecules. Students then make one extra oxygen molecule and an extra water molecule and they should see that it balances. Finally they should then write the complete symbol equation. **[HT only]**

Plenaries

Demonstration – Ignite a hydrogen balloon to get the attention of the class and generate excitement. The hydrogen will explode, as it reacts with the oxygen in the air to make water (steam). Encourage the students to write word and balanced symbol equations for this reaction. *(5 minutes)*

Balancing equations – Give the students word equations, which they then convert into balanced symbol equations.

magnesium + oxygen → magnesium oxide [$2Mg + O_2 \rightarrow 2MgO$]

ethanol + oxygen → carbon dioxide + water [$C_2H_5OH + 3O_2 \rightarrow 2CO_2 + 3H_2O$]

Students can be supported by being given the symbol equation and so they only need to balance it. Students could be extended by being given a prose description of the reaction, which they then have to use to derive the symbol equation. *(10 minutes)* **[HT only]**

Practical support

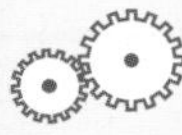

Investigating the mass of reactants and products

Equipment and materials required

0.1 mol/dm³ lead nitrate, 0.1 mol/dm³ potassium iodide, 0.1 mol/dm³ barium chloride, 0.1 mol/dm³ sodium sulfate, top-pan balance, 2 × 25 cm³ measuring cylinder, eye protection (chemical splashproof).

Details

Measure the mass of each empty measuring cylinder. In the first measuring cylinder, measure 10 cm³ of lead nitrate and record its mass. Into a second measuring cylinder, measure 10 cm³ of potassium iodide and record its mass. Carefully pour the lead nitrate into the potassium iodide and observe [bright yellow precipitate]. Record the new mass. Calculate the mass of the reactants and the mass of the products. Repeat the experiment using barium chloride and sodium sulfate [white precipitate].

Safety: Lead compounds are foetal toxins and toxic. Barium chloride is toxic. Ensure that safety googles are worn and students wash their hands after completing the experiment. CLEAPSS Hazcard 10A Barium chloride; 57A Lead nitrate; 57A Lead iodide – toxic.

Dispose of waste following CLEAPSS advice.

CLEAPSS also suggests you could conduct this experiment at 0·01 mol/dm³ lead nitrate. At this strength the lead solution is only low hazard.

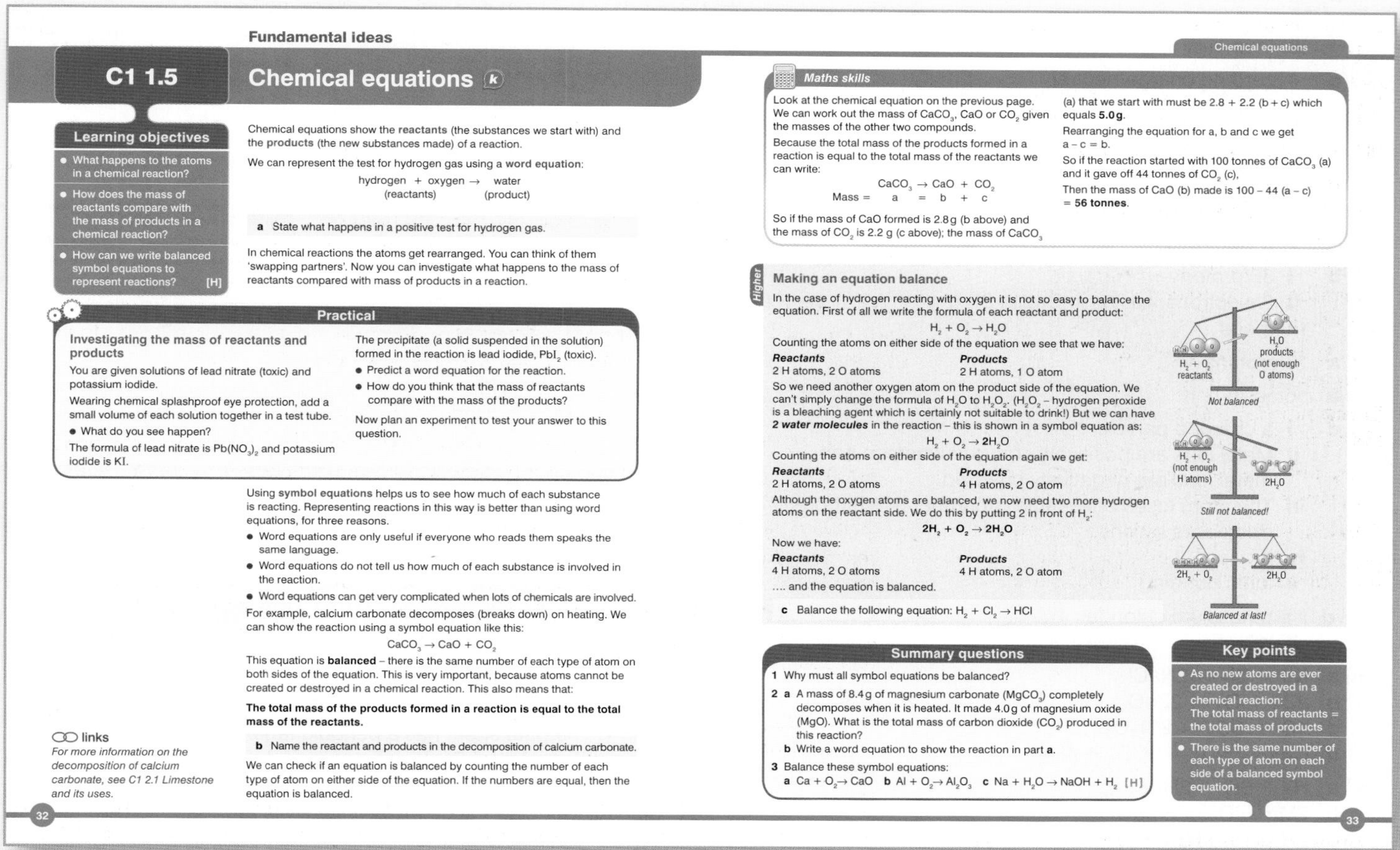

Fundamental ideas

C1 1.5 Chemical equations

Learning objectives

- What happens to the atoms in a chemical reaction?
- How does the mass of reactants compare with the mass of products in a chemical reaction?
- How can we write balanced symbol equations to represent reactions? [H]

Chemical equations show the **reactants** (the substances we start with) and the **products** (the new substances made) of a reaction.

We can represent the test for hydrogen gas using a **word equation**:

hydrogen + oxygen → water
(reactants) (product)

a State what happens in a positive test for hydrogen gas.

In chemical reactions the atoms get rearranged. You can think of them 'swapping partners'. Now you can investigate what happens to the mass of reactants compared with mass of products in a reaction.

Practical

Investigating the mass of reactants and products

You are given solutions of lead nitrate (toxic) and potassium iodide.

Wearing chemical splashproof eye protection, add a small volume of each solution together in a test tube.

- What do you see happen?

The formula of lead nitrate is $Pb(NO_3)_2$ and potassium iodide is KI.

The precipitate (a solid suspended in the solution) formed in the reaction is lead iodide, PbI_2 (toxic).

- Predict a word equation for the reaction.
- How do you think that the mass of reactants compare with the mass of the products?

Now plan an experiment to test your answer to this question.

Using **symbol equations** helps us to see how much of each substance is reacting. Representing reactions in this way is better than using word equations, for three reasons.

- Word equations are only useful if everyone who reads them speaks the same language.
- Word equations do not tell us how much of each substance is involved in the reaction.
- Word equations can get very complicated when lots of chemicals are involved.

For example, calcium carbonate decomposes (breaks down) on heating. We can show the reaction using a symbol equation like this:

$$CaCO_3 \rightarrow CaO + CO_2$$

This equation is **balanced** – there is the same number of each type of atom on both sides of the equation. This is very important, because atoms cannot be created or destroyed in a chemical reaction. This also means that:

The total mass of the products formed in a reaction is equal to the total mass of the reactants.

b Name the reactant and products in the decomposition of calcium carbonate.

We can check if an equation is balanced by counting the number of each type of atom on either side of the equation. If the numbers are equal, then the equation is balanced.

links
For more information on the decomposition of calcium carbonate, see C1 2.1 Limestone and its uses.

32

Chemical equations

Maths skills

Look at the chemical equation on the previous page. We can work out the mass of $CaCO_3$, CaO or CO_2 given the masses of the other two compounds.

Because the total mass of the products formed in a reaction is equal to the total mass of the reactants we can write:

$$CaCO_3 \rightarrow CaO + CO_2$$
$$\text{Mass} = a = b + c$$

So if the mass of CaO formed is 2.8 g (b above) and the mass of CO_2 is 2.2 g (c above); the mass of $CaCO_3$ (a) that we start with must be 2.8 + 2.2 (b + c) which equals **5.0 g**.

Rearranging the equation for a, b and c we get a − c = b.

So if the reaction started with 100 tonnes of $CaCO_3$ (a) and it gave off 44 tonnes of CO_2 (c),

Then the mass of CaO (b) made is 100 − 44 (a − c) = **56 tonnes**.

Higher

Making an equation balance

In the case of hydrogen reacting with oxygen it is not so easy to balance the equation. First of all we write the formula of each reactant and product:

$$H_2 + O_2 \rightarrow H_2O$$

Counting the atoms on either side of the equation we see that we have:

Reactants	*Products*
2 H atoms, 2 O atoms	2 H atoms, 1 O atom

So we need another oxygen atom on the product side of the equation. We can't simply change the formula of H_2O to H_2O_2. (H_2O_2 – hydrogen peroxide is a bleaching agent which is certainly not suitable to drink!) But we can have ***2 water molecules*** in the reaction – this is shown in a symbol equation as:

$$H_2 + O_2 \rightarrow 2H_2O$$

Counting the atoms on either side of the equation again we get:

Reactants	*Products*
2 H atoms, 2 O atoms	4 H atoms, 2 O atom

Although the oxygen atoms are balanced, we now need two more hydrogen atoms on the reactant side. We do this by putting 2 in front of H_2:

$$2H_2 + O_2 \rightarrow 2H_2O$$

Now we have:

Reactants	*Products*
4 H atoms, 2 O atoms	4 H atoms, 2 O atom

.... and the equation is balanced.

c Balance the following equation: $H_2 + Cl_2 \rightarrow HCl$

$H_2 + O_2$ reactants
H_2O products (not enough O atoms)
Not balanced

$H_2 + O_2$ (not enough H atoms)
$2H_2O$
Still not balanced!

$2H_2 + O_2$
$2H_2O$
Balanced at last!

Summary questions

1 Why must all symbol equations be balanced?

2 a A mass of 8.4 g of magnesium carbonate ($MgCO_3$) completely decomposes when it is heated. It made 4.0 g of magnesium oxide (MgO). What is the total mass of carbon dioxide (CO_2) produced in this reaction?

b Write a word equation to show the reaction in part **a**.

3 Balance these symbol equations:
a $Ca + O_2 \rightarrow CaO$ b $Al + O_2 \rightarrow Al_2O_3$ c $Na + H_2O \rightarrow NaOH + H_2$ [H]

Key points

- As no new atoms are ever created or destroyed in a chemical reaction: The total mass of reactants = the total mass of products
- There is the same number of each type of atom on each side of a balanced symbol equation.

33

Further teaching suggestions

Lots of balancing equations

- Students could be given a selection of the same symbol equation but balanced, unbalanced and 'balanced' by altering the formula. Ask students to select the correctly balanced symbol equation and comment on why the others are not correct.
- Give students a list of symbol equations to balance, starting off with simple equations and working up to equations which need bigger numbers.

Answers to in-text questions

a A lighted splint burns with a squeaky 'pop'.
b Reactant: calcium carbonate. Products: calcium oxide and carbon dioxide.
c $H_2 + Cl_2 \rightarrow 2HCl$

Summary answers

1 As no new atoms can be created or destroyed in a chemical reaction, the number and type of atoms in the reactants must equal the number and type of atoms in the products.

2 a 4.4 g
b magnesium carbonate → magnesium oxide + carbon dioxide

3 a $2Ca + O_2 \rightarrow 2CaO$
b $4Al + 3O_2 \rightarrow 2Al_2O_3$
c $2Na + 2H_2O \rightarrow 2NaOH + H_2$

Summary answers

1 a Elements contain only one type of atom, whereas compounds contain more than one type of atom.

b **i** 6 atoms of hydrogen
ii 3 elements
iii 9 atoms in total

2 a

Subatomic particle	Relative charge
proton	+1
neutron	0
electron	−1

b **i** in the nucleus
ii in the nucleus
iii In energy level or shells around the nucleus.

c **i** zero/no charge/neutral
ii 7 electrons

3 a **i** a non-metal
ii more metals
iii 18 protons
iv Group 0, the noble gases
v 8 electrons

b **i** 56 protons
ii 2 electrons, as it is in Group 2
iii metal

4 a 19 protons

b potassium, K

c **i** a charged particle
ii Metal ions are positively charged whereas non-metal ions are negatively charged.
iii The atom loses one electron/number of electrons decreases by one.
iv K^+
v ionic bonding

d **i** potassium chloride
ii KCl

5 a $Na + Cl_2 \rightarrow 2NaCl$
b $2Zn + O_2 \rightarrow 2ZnO$
c $4Cr + 3O_2 \rightarrow 2Cr_2O_3$
d $C_3H_8 + 5O_2 \rightarrow 3CO_2 + 4H_2O$

6 a $H_2 + Br_2 \rightarrow 2HBr$
b $2Mg + O_2 \rightarrow 2MgO$
c $2H_2O_2 \rightarrow 2H_2O + O_2$
d $2Li + 2H_2O \rightarrow 2LiOH + H_2$
e $2NaNO_3 \rightarrow 2NaNO_2 + O_2$
f $4Fe + 3O_2 \rightarrow 2Fe_2O_3$

7 a 1.6 g

b The mass of the reactants must equal the mass of the product, so we subtract the mass of iron from the mass of iron sulfide. This will give us the mass of sulfur that reacted with the 2.8 g of iron.

Fundamental ideas: C1 1.1–C1 1.5

Summary questions 

1 a What is the difference in the definitions of an element and a compound?

b The chemical formula of ethanol is written as C_2H_5OH.
i How many atoms of hydrogen are there in an ethanol molecule?
ii How many different elements are there in ethanol?
iii What is the total number of atoms in an ethanol molecule?

2 a Draw a table to show the relative charge on protons, neutrons and electrons.

b In which part of an atom do we find:
i protons
ii neutrons
iii electrons.

c i What is the overall charge on any atom?
ii A nitrogen atom has 7 protons. How many electrons does it have?

3 This question is about the periodic table of elements. You will need to use the periodic table at the back of this book to help you answer some parts of the question.

a Argon (Ar) is the 18th element in the periodic table.
i Is argon a metal or a non-metal?
ii Are there more metals or non-metals in the periodic table?
iii How many protons does an argon atom contain?
iv State the name and number of the group to which argon belongs.
v How many electrons does argon have in its highest energy level (outermost shell)?

b The element barium (Ba) has 56 electrons.
i How many protons are in the nucleus of each barium atom?
ii How many electrons does a barium atom have in its highest energy level (outermost shell)? How did you decide on your answer?
iii Is barium a metal or a non-metal?

4 The diagram below shows the arrangement of electrons in an atom.

a How many protons are in the nucleus of this atom?

b Use the periodic table at the back of this book to give the name and symbol of the element whose atom is shown here.

c This element forms ions with a 1+ charge.
i What is an ion?
ii How does the charge on the ion tell us whether the element above is a metal or non-metal?
iii Describe what happens to the number of electrons when the atom forms a 1+ ion.
iv Write the chemical formula of the ion.
v This ion can form compounds with negatively charged ions. What type of bonding will we find in these compounds?

d A compound is formed when this element reacts with chlorine gas.
i What is the name of the compound formed?
ii Chloride ions carry a 1− charge. Write the chemical formula of the compound formed.

5 What is the missing number needed to balance the following symbol equations?

a $2Na + Cl_2 \rightarrow .... NaCl$
b $2Zn + O_2 \rightarrow .... ZnO$
c $.... Cr + 3O_2 \rightarrow 2Cr_2O_3$
d $C_3H_8 + ... O_2 \rightarrow 3CO_2 + 4H_2O$ [H]

6 Balance the following symbol equations:

a $H_2 + Br_2 \rightarrow HBr$
b $Mg + O_2 \rightarrow MgO$
c $H_2O_2 \rightarrow H_2O + O_2$
d $Li + H_2O \rightarrow LiOH + H_2$
e $NaNO_3 \rightarrow NaNO_2 + O_2$
f $Fe + O_2 \rightarrow Fe_2O_3$ [H]

7 When a mixture of iron and sulfur is heated, a compound called iron sulfide is made.

In an experiment 2.8 g of iron made 4.4 g of iron sulfide.

a What mass of sulfur reacted with the 2.8 g of iron?

b Explain how you worked out your answer to part **a**.

34

Study tip

Students are expected to be familiar with the formulae of the substances named in the specification. A data sheet will be provided in the examinations for all Chemistry units, including Unit 1. The data sheet has a periodic table, the reactivity series for metals and a table of common ions, but students are not expected to work out the formulae of ionic compounds in this unit. All formulae will be given if students are asked to balance symbol equations.

Kerboodle resources 

Resources available for this chapter on Kerboodle are:

- Chapter map: Fundamental ideas)
- Support: Arranging the elements (C1 1.1)
- Extension: In orbit (C1 1.3)
- Interactive activity: Atoms and the periodic table (C1 1.4)
- Maths skills: Balancing equations (C1 1.5)
- Practical: Conservation of mass (C1 1.5)
- Revision podcast: Atom anatomy
- Test yourself: Fundamental ideas
- On your marks: Electronic structure and reactivity in groups
- Examination-style questions: Fundamental ideas
- Answers to examination-style questions: Fundamental ideas

Study tip

Bonding is covered more fully in Unit 2. In this unit, students only need to be aware of ionic and covalent bonding. They should know that compounds of metals and non-metals contain ions. They should also know that molecules are held together by covalent bonds.

AQA Examination-style questions k

1 Use numbers from the list to complete the table to show the charge on each subatomic particle.

+2 +1 0 −1 −2

Subatomic particle	Charge
electron	
neutron	
proton	

(3)

2 Use the periodic table at the back of your book to help you to answer this question.

a How many protons are in an atom of fluorine? (1)

b How many electrons are in an atom of carbon? (1)

c Complete the electronic structure of aluminium: 2,8, ... (1)

d What is the electronic structure of potassium? (1)

3 Neon is a noble gas.

a What does this tell you about its electronic structure? (1)

b Draw a diagram to show the electronic structure of neon. (2)

4 a Magnesium has the electronic structure 2,8,2. Explain, in terms of its electronic structure, why magnesium is in Group 2 of the periodic table. (1)

b Give **one** way in which the electronic structures of the atoms of Group 2 elements are:

i the same (1)

ii different. (1)

c When magnesium is heated in air it burns with a bright flame and produces magnesium oxide.

Calcium is also in Group 2. Describe what you expect to happen and what would be produced when calcium is heated in air. (2)

5 Sodium reacts with water to produce sodium hydroxide and hydrogen.

The word equation for this reaction is:

sodium + water → sodium hydroxide + hydrogen

a Name one substance in this equation that is:

i an element (1)

ii a compound (1)

iii has ionic bonds (1)

iv has covalent bonds (1)

b If 2.3 g of sodium reacted with 1.8 g of water, what would be the total mass of sodium hydroxide and hydrogen produced? Explain your answer. (2)

c Balance the symbol equation for this reaction.

.... Na + ... H_2O → .... NaOH + H_2 [H] (1)

d Lithium is in the same group of the periodic table as sodium.

i Write a word equation for the reaction of lithium with water. (1)

ii What is the formula of lithium hydroxide? (1)

iii How many atoms are shown in the formula of lithium hydroxide you have written? (1)

Practical suggestions

Practicals	AQA	k		
Modelling of atoms (using physical models or computer simulations) to illustrate chemical reactions at the atomic level.	✓		✓	
Precipitation reactions, such as lead nitrate with potassium iodide, to show conservation of mass.	✓	✓	✓	

Examination-style answers

1

Subatomic particle	Charge
electron	−1
neutron	0
proton	+1

(3 marks)

2 a 9 *(1 mark)*

b 6 *(1 mark)*

c 2,8,3 *(1 mark)*

d 2,8,8,1 *(1 mark)*

3 a It is stable or it has 8 electrons in its outer energy level/shell (not it has a full shell). *(1 mark)*

b Two electrons (dots or crosses) on inner circle, 8 electrons on outer circle. *(2 marks)*

4 a Two electrons in its highest energy level/outer shell. *(1 mark)*

b i Same number of electrons or 2 electrons in highest energy level/outer shell. (accept same number of electrons in first energy level/shell) *(1 mark)*

ii Different number of energy levels/shells. *(1 mark)*

c burns (with a bright flame)
(produces) calcium oxide *(2 marks)*

5 a i sodium or hydrogen *(1 mark)*

ii water or sodium hydroxide *(1 mark)*

iii sodium hydroxide *(1 mark)*

iv water or hydrogen *(1 mark)*

b 4.1 g
mass is conserved in reactions or mass of reactants = mass of products *(2 marks)*

c $2Na + 2H_2O \rightarrow 2NaOH + H_2$ *(1 mark)*

d i lithium + water → lithium hydroxide + hydrogen *(1 mark)*

ii LiOH *(1 mark)*

iii 3 or error carried forward from answer to (**ii**) *(1 mark)*

C1 2.1 Limestone and its uses

Learning objectives

Students should learn:

- that limestone is used to make a variety of building materials
- that calcium carbonate in limestone will undergo thermal decomposition.

Learning outcomes

Most students should be able to:

- list some uses of limestone
- write the formula of calcium carbonate
- complete a word equation for the thermal decomposition of calcium carbonate.

Some students should also be able to:

- explain the process of thermal decomposition of calcium carbonate, including the balanced symbol equation.

Support

- Students often get confused about the formula of a compound. You may wish to support students by using formula jigsaws to kinaesthetically show students how the formula is made up.

Extend

- Give students the formulae of different metal carbonates. Ask them to write out the numbers and type of each atom that they contain from the formula. You could further extend students by asking them to write the formulae of carbonates of their choice. You may wish to encourage students to use the table of common ions that are listed on their data sheet, which they would have access to in the examination.
- Students could be asked to find out why spotlights caused so many explosions in early theatres.

Specification link-up: Chemistry C1.2

- Limestone, mainly composed of the compound calcium carbonate ($CaCO_3$), is quarried and can be used as a building material. *[C1.2.1 a)]*
- Calcium carbonate can be decomposed by heating (thermal decomposition) to make calcium oxide and carbon dioxide. *[C1.2.1 b)]*

Lesson structure

Starters

Characteristics – Ask the students to consider what the following materials have in common: limestone, marble and chalk [they are all made up mainly of the same compound – calcium carbonate]. This task could be expanded to think about other groupings of these substances, e.g. chalk and limestone are sedimentary rocks, whereas marble is metamorphic. *(5 minutes)*

Classify – Students could read through the pages in this topic and classify the chemicals that are named as elements [calcium, oxygen, and carbon], compounds [calcium carbonate, calcium oxide, carbon dioxide, water] and mixtures [limestone, chalk]. Students may need to be supported by being given a reminder of the definitions of elements, compounds and mixtures. Ask them to find the symbols or the formulae of the elements and compounds. *(10 minutes)*

Main

- Using the Student Book for information, students could create a fully labelled diagram of a lime kiln. Students should include information about the raw material [limestone] and reactant [calcium carbonate], the products [carbon dioxide and calcium oxide]. Foundation Tier students should include a word equation and Higher Tier students should include a balanced symbol equation.
- Students could then create a flow chart to explain the process that occurs in a rotary lime kiln. Again students should include word and/or balanced symbol equations to detail the reactions that take place.
- Students can decompose calcium carbonate in the practical 'Thermal decomposition' in the Student Book. This reaction will be revisited later in the chapter and students should just concentrate on observing the reaction safely here and discussing the equation. Students should be encouraged to notice how brightly the calcium carbonate shines. Explain that in old theatres, acetylene and oxygen were combusted to heat limestone to produce this bright light. It was then focused using a lens and made into a spotlight.
- Often it is clear that students are confused about the scientific language, what it means and how it can be used. Students could write the key words (highlighted in the text) in their exercise book. They then need to summarise the meaning of each word in just one sentence. Note that the common names 'quicklime' and 'slaked lime' are no longer required knowledge in the current specification.

Plenaries

Fact finder – Ask students to think about one fact that they already knew before the lesson that had been reviewed and one fact that was completely new to them. *(5 minutes)*

Key-word bingo – The students should choose three of the following key words: limestone, calcium carbonate, cement, concrete, calcium oxide, thermal decomposition, calcium oxide. These can be written in the back of their exercise books, or on a pre-made bingo card that has spaces for the students to write in. Explain the word, and the student crosses it off (if they chose it). The first student to cross off all of the words could be given a reward! Students could be supported by being given the key words and a definition to refer to. Students could be extended by using the symbols, formulae or balanced symbol equations rather than the key words themselves on their bingo cards. *(10 minutes)*

Practical support

Thermal decomposition

Equipment and materials required

Gauze, tripod, Bunsen burner and safety equipment, eye protection, tongs, calcium carbonate chip.

Details

Place a piece of calcium carbonate onto a gauze mounted on a tripod. Turn the Bunsen burner to a blue flame, and direct the tip of the blue cone onto a corner of the calcium carbonate. As the Bunsen burner will need to be directed at the calcium carbonate, it will need to be picked up off the bench so extra care should to be taken. Hold the base and ensure the rubber tube is tightly secured to the gas tap and the Bunsen burner before starting the experiment. Do not allow students to overstretch the tubing. Firstly calcium carbonate will glow red/orange, then a whiter orange. Once part of the material glows white for a few minutes, turn off the Bunsen burner. Leave sufficient time for the heated product to cool down. Dispose of with tongs; do not touch the calcium oxide.

Alternatively, the calcium carbonate chip is placed on the woven steel corner of a gauze, and the tip of a roaring Bunsen flame is positioned on the chip from below, another method uses a sling/holder for the chip fashioned from thick nichrome wire, which is then clamped at the other end using a boss and clamp, and the tip of a roaring Bunsen flame is positioned on the chip from below.

Safety: CLEAPSS Hazcard 19B Calcium carbonate. CLEAPSS Hazcard 18 Calcium oxide – corrosive.

Ensure the rubber tube is tightly secured to the gas tap and the Bunsen burner before starting the experiment. Do not allow students to overstretch the tubing. Leave sufficient time for the heated product to cool down. Dispose of with tongs; do not touch the calcium oxide.

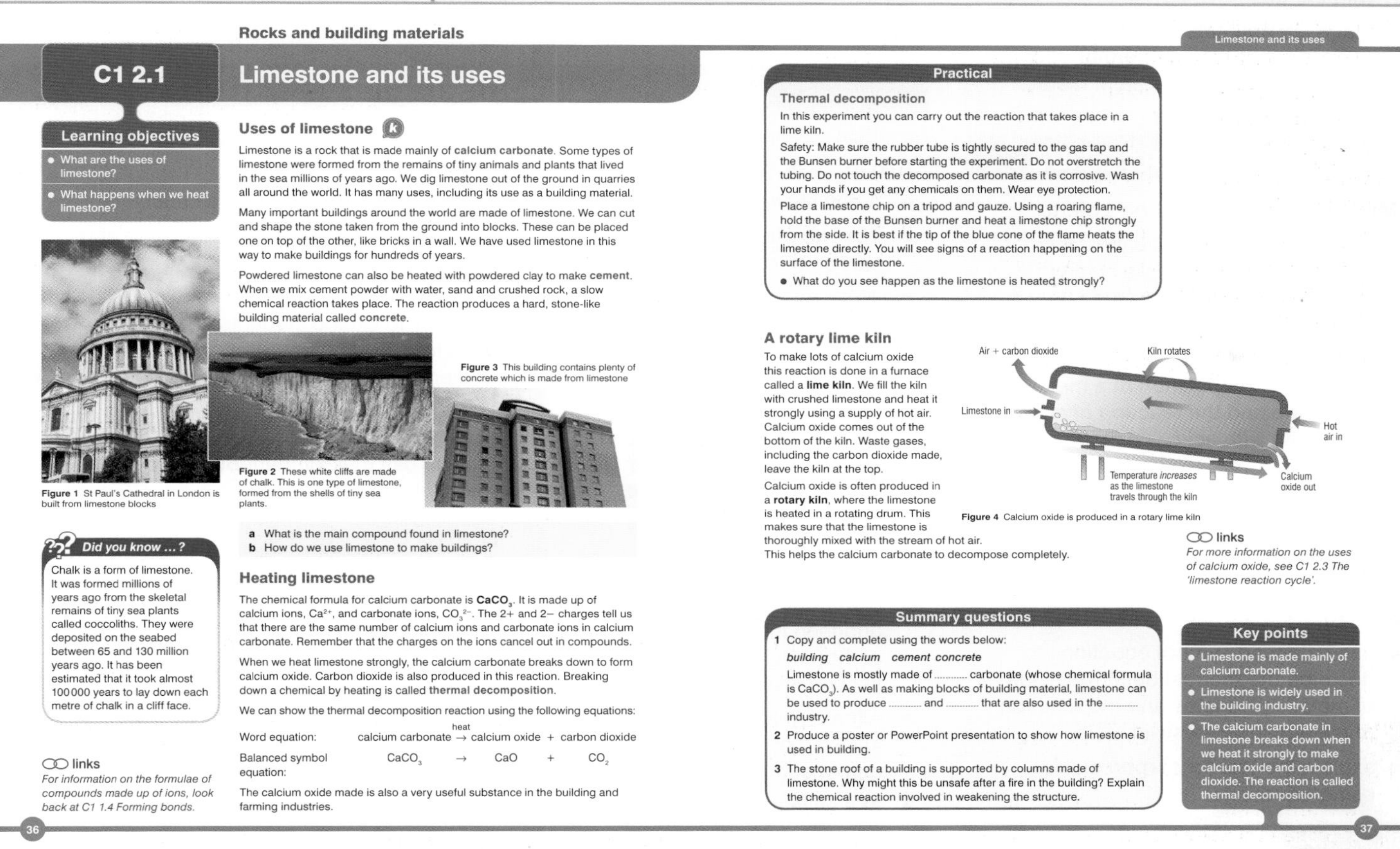

Rocks and building materials

Limestone and its uses

C1 2.1 Limestone and its uses

Learning objectives

- What are the uses of limestone?
- What happens when we heat limestone?

Uses of limestone

Limestone is a rock that is made mainly of **calcium carbonate**. Some types of limestone were formed from the remains of tiny animals and plants that lived in the sea millions of years ago. We dig limestone out of the ground in quarries all around the world. It has many uses, including its use as a building material.

Many important buildings around the world are made of limestone. We can cut and shape the stone taken from the ground into blocks. These can be placed one on top of the other, like bricks in a wall. We have used limestone in this way to make buildings for hundreds of years.

Powdered limestone can also be heated with powdered clay to make **cement**. When we mix cement powder with water, sand and crushed rock, a slow chemical reaction takes place. The reaction produces a hard, stone-like building material called **concrete**.

Figure 1 St Paul's Cathedral in London is built from limestone blocks

Figure 2 These white cliffs are made of chalk. This is one type of limestone, formed from the shells of tiny sea plants.

Figure 3 This building contains plenty of concrete which is made from limestone

a What is the main compound found in limestone?
b How do we use limestone to make buildings?

Did you know ...?

Chalk is a form of limestone. It was formed millions of years ago from the skeletal remains of tiny sea plants called coccoliths. They were deposited on the seabed between 65 and 130 million years ago. It has been estimated that it took almost 100 000 years to lay down each metre of chalk in a cliff face.

Heating limestone

The chemical formula for calcium carbonate is **$CaCO_3$**. It is made up of calcium ions, Ca^{2+}, and carbonate ions, CO_3^{2-}. The 2+ and 2− charges tell us that there are the same number of calcium ions and carbonate ions in calcium carbonate. Remember that the charges on the ions cancel out in compounds.

When we heat limestone strongly, the calcium carbonate breaks down to form calcium oxide. Carbon dioxide is also produced in this reaction. Breaking down a chemical by heating is called **thermal decomposition**.

We can show the thermal decomposition reaction using the following equations:

Word equation: calcium carbonate $\xrightarrow{\text{heat}}$ calcium oxide + carbon dioxide

Balanced symbol equation: $CaCO_3 \rightarrow CaO + CO_2$

The calcium oxide made is also a very useful substance in the building and farming industries.

links

For information on the formulae of compounds made up of ions, look back at C1 1.4 Forming bonds.

Practical

Thermal decomposition

In this experiment you can carry out the reaction that takes place in a lime kiln.

Safety: Make sure the rubber tube is tightly secured to the gas tap and the Bunsen burner before starting the experiment. Do not overstretch the tubing. Do not touch the decomposed carbonate as it is corrosive. Wash your hands if you get any chemicals on them. Wear eye protection.

Place a limestone chip on a tripod and gauze. Using a roaring flame, hold the base of the Bunsen burner and heat a limestone chip strongly from the side. It is best if the tip of the blue cone of the flame heats the limestone directly. You will see signs of a reaction happening on the surface of the limestone.

- What do you see happen as the limestone is heated strongly?

A rotary lime kiln

To make lots of calcium oxide this reaction is done in a furnace called a **lime kiln**. We fill the kiln with crushed limestone and heat it strongly using a supply of hot air. Calcium oxide comes out of the bottom of the kiln. Waste gases, including the carbon dioxide made, leave the kiln at the top.

Calcium oxide is often produced in a **rotary kiln**, where the limestone is heated in a rotating drum. This makes sure that the limestone is thoroughly mixed with the stream of hot air. This helps the calcium carbonate to decompose completely.

Figure 4 Calcium oxide is produced in a rotary lime kiln

links

For more information on the uses of calcium oxide, see C1 2.3 The 'limestone reaction cycle'.

Summary questions

1 Copy and complete using the words below:
building calcium cement concrete
Limestone is mostly made of .... carbonate (whose chemical formula is $CaCO_3$). As well as making blocks of building material, limestone can be used to produce .... and .... that are also used in the .... industry.
2 Produce a poster or PowerPoint presentation to show how limestone is used in building.
3 The stone roof of a building is supported by columns made of limestone. Why might this be unsafe after a fire in the building? Explain the chemical reaction involved in weakening the structure.

Key points

- Limestone is made mainly of calcium carbonate.
- Limestone is widely used in the building industry.
- The calcium carbonate in limestone breaks down when we heat it strongly to make calcium oxide and carbon dioxide. The reaction is called thermal decomposition.

36 37

Further teaching suggestions

Word description

- Flash cards of the key words could be created. Hold up the key word, and a student tries to describe it to you, without saying the word.

Did you know ...?

Calcium oxide was also spread over plague victims in their mass graves and over executed prisoners, as people thought it helped their bodies decompose quicker. However, it did help to kill microbes.

Answers to in-text questions

a calcium carbonate
b Limestone can be cut into blocks or processed to make materials like cement or concrete.

Summary answers

1 calcium, cement/concrete (either order), building
2 Poster or presentation.
3 The calcium carbonate in the column may have undergone thermal decomposition, making it weaker. The calcium oxide formed is much softer than the original calcium carbonate present in the columns.

C1 2.2 Reactions of carbonates

Learning objectives

Students should learn:

- that metal carbonates will undergo thermal decomposition
- that metal carbonates will react with dilute acids
- how limewater can be used as a test for carbon dioxide as it turns from colourless to cloudy when reacting with the gas
- that a fair and safe test must be used to compare the thermal decompositions of different metal carbonates.

Learning outcomes

Most students should be able to:

- list examples of metal carbonates that react similarly to the calcium carbonate in limestone when they are heated
- write word equations to describe thermal decomposition of a metal carbonate and the reaction of a metal carbonate with acid
- state the test for carbon dioxide
- plan some safety considerations and fair test ideas for the thermal decomposition of metal carbonates.

Some students should also be able to:

- use balanced symbol equations to describe the thermal decomposition of metal carbonates and the reaction of metal carbonates with acid
- explain the test for carbon dioxide using equations.

Support

- For students struggling with chemical equations, molecular model kits can be used. Explain that atoms cannot be created or destroyed, only rearranged to make the products formed.

Extend

- Students should make predictions about other metal carbonates and how they would react. You may wish to ask students to use their observations of the reaction of calcium carbonate with acid to explain how limestone buildings can be affected by rain, which is naturally weakly acidic.

Specification link-up: Chemistry C1.2

- The carbonates of magnesium, copper, zinc, calcium and sodium decompose on heating in a similar way. *[C1.2.1 c)]*
- A solution of calcium hydroxide in water (limewater) reacts with carbon dioxide to produce calcium carbonate. Limewater is used as a test for carbon dioxide. Carbon dioxide turns limewater cloudy. *[C1.2.1 e)]*
- Carbonates react with acids to produce carbon dioxide, a salt and water. Limestone is damaged by acid rain. *[C1.2.1 f)]*

Controlled Assessment: C4.2 Assess and manage risks when carrying out practical work. *[C4.2.1 a)]*; C4.3 Collect primary and secondary data. *[C4.3.2 a)]*

Lesson structure

Starters

Recap – Ask students to recall:

- A definition for thermal decomposition. [Using heat to break down a substance.]
- The two products when limestone is thermally decomposed. [Calcium oxide and carbon dioxide.]
- Write an equation for the thermal decomposition of limestone in a lime kiln. [Calcium carbonate → carbon dioxide + calcium oxide.] To extend students, you could encourage them to write a symbol equation for this reaction. Support students by reminding them that limestone is a raw material and the reactant is calcium carbonate. *(5 minutes)*

Anagram title – Tell students that today's lesson is about 'creatnois fo braconsate'? [Reactions of carbonates] Encourage students to use their Student Book to define the word carbonates and maybe even give an example detailed as an equation. *(10 minutes)*

Main

- You may wish to tackle the contents of this spread over two lessons, with all students doing both practicals. In this case, the 'Acid plus carbonates' practical can be done and then the planning for the investigation into thermal decomposition, 'Investigating carbonates', could be discussed and started, but carrying out the practical work in the second lesson.
- Alternatively, in one lesson you might do the 'Acid plus carbonates' practical, followed by 'Investigating carbonates' as a planning exercise only. You could perhaps demonstrate the decomposition of a couple of carbonates with student participation.
- At a suitable point, explain why limewater turns cloudy in the test for carbon dioxide.
- You may wish to encourage students to focus on the safety of the investigation. You could give students access to the CLEAPSS students safety sheets and ask them to write a risk assessment before completing the practical work. You may also focus on how to make the investigation a fair test by looking at identifying and classifying the different variables.
- Encourage students to attempt to write a general equation for the reactions in both practicals.

[metal carbonate → carbon dioxide + metal oxide

metal carbonate + acid → metal salt + water + carbon dioxide]

Plenaries

Symbol equations – Ask students to complete the following equations (answers in brackets):

calcium carbonate → [calcium oxide + carbon dioxide]

[magnesium carbonate] → magnesium oxide + carbon dioxide

[$CuCO_3$] → CuO + CO_2 *(5 minutes)*

Text summary – Ask the students to write down a summary of thermal decomposition and reactions of metal carbonates and acids as a text message. You may wish to give students an outline of a mobile phone to draft their text message answers into. To support students, just concentrate on thermal decomposition. To extend students, you could also ask them to include a symbol equation in their text message. *(10 minutes)*

Practical support

Acid plus carbonates

Equipment and materials required

Boiling tube, bung with delivery tube, spatula, test tube, irritant, limewater, dropping pipette, test tube rack, stand, boss, clamp, eye protection, calcium carbonate chips, 2 mol/dm^3 hydrochloric acid – irritant.

Details

Half-fill the test tube with limewater and place into the rack. Put about one spatula of a metal carbonate to be tested in a boiling tube and clamp in position. Add about 1 cm^3 of dilute acid and quickly fit the bung and delivery tube. Angle the end of the delivery tube into the test tube.

Safety: Wear eye protection. CLEAPSS Hazcard 19B Calcium carbonate; 18 Limewater – irritant; 47A Hydrochloric acid – irritant.

Further teaching suggestions

Honeycomb demonstration

- The thermal decomposition of sodium hydrogencarbonate (bicarbonate) can be illustrated by heating sugar syrup mixture to 150 °C and adding sodium bicarbonate. The carbon dioxide gas released gets trapped in the sugary mixture.

Investigating carbonates

Equipment and materials required

Boiling tube, bung with delivery tube, spatula, test tube, Bunsen burner and safety equipment, eye protection, test tube holder, stand, boss, clamp, samples of metal carbonates (e.g. calcium carbonate, sodium carbonate – irritant, potassium carbonate – irritant, magnesium carbonate, zinc carbonate, copper carbonate – harmful), limewater – irritant. An electric balance and measuring cylinders will be needed if the practical is to be carried out as an investigation into the ease of thermal decomposition.

Details

Put about one spatula of a metal carbonate to be tested in a boiling tube. Clamp into position, and fit the bung and delivery tube. Half-fill the test tube with limewater and place into the rack. Angle the end of the delivery tube into the test tube. Using the Bunsen burner, heat the carbonate and observe the limewater. If the limewater goes cloudy, then carbon dioxide has been produced and thermal decomposition has taken place. Repeat with other carbonates to compare results.

Safety: Wear eye protection throughout the practical and be aware that the boiling tube will still be hot when heating is ceased. Remove end of delivery tube from limewater before heating is stopped to prevent 'such back'. CLEAPSS Hazcard 19B Calcium carbonate; 95A Sodium/potassium carbonate – irritant; 59B Magnesium carbonate; 108B Zinc carbonate; 26 Copper carbonate – harmful; 18 Limewater – irritant. Disposal of waste as per CLEAPSS advice.

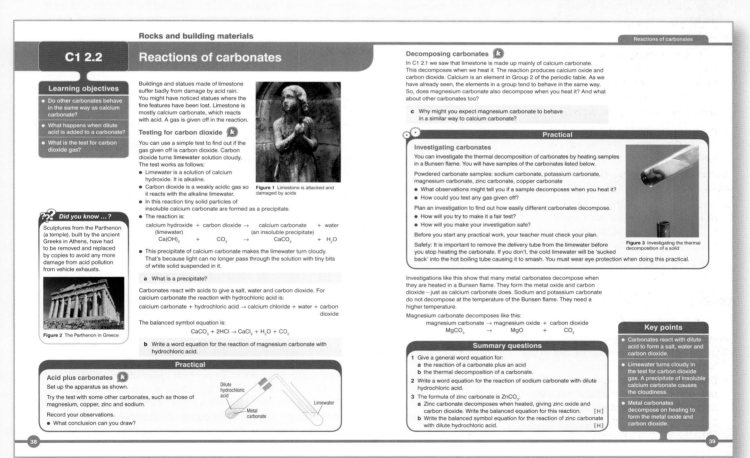

C1 2.2 Reactions of carbonates

Learning objectives

- Do other carbonates behave in the same way as calcium carbonate?
- What happens when dilute acid is added to a carbonate?
- What is the test for carbon dioxide gas?

Buildings and statues made of limestone suffer badly from damage by acid rain. You might have noticed statues where the fine features have been lost. Limestone is mostly calcium carbonate, which reacts with acid. A gas is given off in the reaction.

Figure 1 Limestone is attacked and damaged by acids

Testing for carbon dioxide

You can use a simple test to find out if the gas given off is carbon dioxide. Carbon dioxide turns **limewater** solution cloudy. The test works as follows:

- Limewater is a solution of calcium hydroxide. It is alkaline.
- Carbon dioxide is a weakly acidic gas so it reacts with the alkaline limewater.
- In this reaction tiny solid particles of insoluble calcium carbonate are formed as a precipitate.
- The reaction is:

calcium hydroxide (limewater) + carbon dioxide → calcium carbonate (an insoluble precipitate) + water

$$Ca(OH)_2 + CO_2 \rightarrow CaCO_3 + H_2O$$

- This precipitate of calcium carbonate makes the limewater turn cloudy. That's because light can no longer pass through the solution with tiny bits of white solid suspended in it.

a What is a precipitate?

Carbonates react with acids to give a salt, water and carbon dioxide. For calcium carbonate the reaction with hydrochloric acid is:

calcium carbonate + hydrochloric acid → calcium chloride + water + carbon dioxide

The balanced symbol equation is:

$$CaCO_3 + 2HCl \rightarrow CaCl_2 + H_2O + CO_2$$

b Write a word equation for the reaction of magnesium carbonate with hydrochloric acid.

Did you know ...?

Sculptures from the Parthenon (a temple), built by the ancient Greeks in Athens, have had to be removed and replaced by copies to avoid any more damage from acid pollution from vehicle exhausts.

Figure 2 The Parthenon in Greece

Practical

Acid plus carbonates

Set up the apparatus as shown.

Try the test with some other carbonates, such as those of magnesium, copper, zinc and sodium.

Record your observations.

- What conclusion can you draw?

Decomposing carbonates

In C1 2.1 we saw that limestone is made up mainly of calcium carbonate. This decomposes when we heat it. The reaction produces calcium oxide and carbon dioxide. Calcium is an element in Group 2 of the periodic table. As we have already seen, the elements in a group tend to behave in the same way. So, does magnesium carbonate also decompose when you heat it? And what about other carbonates too?

c Why might you expect magnesium carbonate to behave in a similar way to calcium carbonate?

Practical

Investigating carbonates

You can investigate the thermal decomposition of carbonates by heating samples in a Bunsen flame. You will have samples of the carbonates listed below.

Powdered carbonate samples: sodium carbonate, potassium carbonate, magnesium carbonate, zinc carbonate, copper carbonate

- What observations might tell you if a sample decomposes when you heat it?
- How could you test any gas given off?

Plan an investigation to find out how easily different carbonates decompose.

- How will you try to make it a fair test?
- How will you make your investigation safe?

Before you start any practical work, your teacher must check your plan.

Safety: It is important to remove the delivery tube from the limewater before you stop heating the carbonate. If you don't, the cold limewater will be 'sucked back' into the hot boiling tube causing it to smash. You must wear eye protection when doing this practical.

Figure 3 Investigating the thermal decomposition of a solid

Investigations like this show that many metal carbonates decompose when they are heated in a Bunsen flame. They form the metal oxide and carbon dioxide – just as calcium carbonate does. Sodium and potassium carbonate do not decompose at the temperature of the Bunsen flame. They need a higher temperature.

Magnesium carbonate decomposes like this:

magnesium carbonate → magnesium oxide + carbon dioxide

$$MgCO_3 \rightarrow MgO + CO_2$$

Summary questions

1 Give a general word equation for:
 a the reaction of a carbonate plus an acid
 b the thermal decomposition of a carbonate.
2 Write a word equation for the reaction of sodium carbonate with dilute hydrochloric acid.
3 The formula of zinc carbonate is $ZnCO_3$.
 a Zinc carbonate decomposes when heated, giving zinc oxide and carbon dioxide. Write the balanced equation for this reaction. [H]
 b Write the balanced symbol equation for the reaction of zinc carbonate with dilute hydrochloric acid. [H]

Key points

- Carbonates react with dilute acid to form a salt, water and carbon dioxide.
- Limewater turns cloudy in the test for carbon dioxide gas. A precipitate of insoluble calcium carbonate causes the cloudiness.
- Metal carbonates decompose on heating to form the metal oxide and carbon dioxide.

Answers to in-text questions

a An insoluble solid formed by the reaction of two solutions.

b magnesium carbonate + hydrochloric acid → magnesium chloride + water + carbon dioxide

c Magnesium and calcium are in the same group (group 2) of the periodic table.

Summary answers

1 **a** metal carbonate + acid → a salt + water + carbon dioxide
 b metal carbonate → metal oxide + carbon dioxide

2 sodium carbonate + hydrochloric acid → sodium chloride + water + carbon dioxide

3 **a** $ZnCO_3 \rightarrow ZnO + CO_2$
 b $ZnCO_3 + 2HCl \rightarrow ZnCl_2 + H_2O + CO_2$

C1 2.3 The 'limestone reaction cycle'

Learning objectives

Students should learn:

- how calcium oxide can be used to make calcium hydroxide
- the uses of calcium hydroxide
- the reactions in the 'limestone reaction cycle'.

Learning outcomes

Most students should be able to:

- give an example of a use of calcium hydroxide
- describe the reactions of the 'limestone reaction cycle' in word equations
- state the observations at each stage of the 'limestone reaction cycle'.

Some students should also be able to:

- write a balanced symbol equation for each reaction in the 'limestone reaction cycle'
- explain why farmers may choose to use calcium hydroxide on their fields.

Specification link-up: Chemistry C1.2

- Calcium oxide reacts with water to produce calcium hydroxide, which is an alkali that can be used in the neutralisation of acids. *[C1.2.1 d)]*
- A solution of calcium hydroxide in water (limewater) reacts with carbon dioxide to produce calcium carbonate. Limewater is used as a test for carbon dioxide. Carbon dioxide turns limewater cloudy. *[C1.2.1 e)]*

Lesson structure

Starters

Spot the mistake – Write the following sentence on the board and ask the students to spot the deliberate mistake(s):

'Limestone is a metamorphic rock and made of the pure element magnesium carbonate.'

[The mistakes are that limestone is a sedimentary rock. It is made up mainly of calcium carbonate (which is a compound, not an element) but also contains other substances, and so the rock is a mixture of substances, not an element.] *(5 minutes)*

Chemical formula – Ask students to look at the following formulae and find their chemical name. You may wish to extend students by asking them to count the number and type of atom in each formula. To support students you may wish to give the names of the compounds and they match them to the formulae:

- $CaCO_3$ [calcium carbonate]
- CaO [calcium oxide]
- $Ca(OH)_2$ [calcium hydroxide]
- CO_2 [carbon dioxide] *(10 minutes)*

Main

- Introduce the idea of a 'limestone reaction cycle'. Draw a brief outline of the cycle on the board. Explain to the students that they are going to complete a number of experiments to follow the cycle.
- Split the class into groups of two or three and ask each group to complete the practical, investigating the 'limestone reaction cycle'. Encourage students to note down any important observations and try to label the type of reaction occurring in each part of the cycle.
- Once the practical has been completed, or using the Student Book for information, a comprehensive flow chart of labelled diagrams to show the limestone reaction cycle could be made. Higher attaining students should include balanced symbol equations for each reaction.

Plenaries

Symbol equations – Ask a student to come to the front and put on eye protection, then blow through a straw into a conical flask containing a liquid (limewater). Ask the students to describe what is happening in terms of an equation. To extend students, they should be encouraged to write a balanced symbol equation. To support lower attaining students, the words of the reactants, products and an arrow could be drawn in separate text boxes in an appropriate computer program. Then using the interactive whiteboard, students could move the words and arrow to create the correct word equation. *(5 minutes)*

Key concept cards – Create a set of cards (7 cm by 3 cm), each card having a different title (calcium oxide, calcium hydroxide, calcium carbonate). Using the Student Book, students then state the chemical name, formula and how it is made in a symbol or word equation on the appropriate card. They can also be asked to add any more information that they think is important, such as uses of each of these chemicals. This is a revision technique that will help them to discriminate and select key information from text. *(10 minutes)*

Support

- Provide students with diagrams of each stage of the limestone cycle, but in the wrong order. Before students start the practical, investigating the 'limestone reaction cycle', they cut and stick these to create a pictorial method.

Extend

- Ask students to represent each stage of the practical, investigating the 'limestone reaction cycle', using balanced symbol equations. You could extend students further by asking them to generate balanced symbol equations for the neutralisation of waste acidic gases, like sulfur dioxide from the combustion of fossil fuels using calcium hydroxide.

Practical support

Investigating the 'limestone reaction cycle'

Equipment and materials required

Gauze, tripod, Bunsen burner and safety equipment, eye protection, tongs, water, 2 × boiling tube, fluted filter paper, filter funnel, glass rod, test tube rack, calcium carbonate chips, dropping pipette, straw. CLEAPSS Student Safety sheet 32.

Details

Follow the procedure detailed for the thermal decomposition of limestone in C1 2.1 Limestone and its uses.

After heating, the thermally decomposed calcium carbonate holds its heat for a long period of time. Also the calcium products are all basic and should not be touched with hands. Using the tongs, transfer the cool product to a boiling tube, add a few drops of water very slowly, one drop at a time, and observe. Then add about a third of the boiling tube of water, put a bung in the tube and shake gently. When water is added to the calcium oxide, it will often spit calcium hydroxide, so safety googles must be worn, and if the compound touches the skin it must be washed off immediately.

Filter the mixture, and keep the filtrate (limewater – irritant). Carefully take a straw and submerge it in the limewater and blow gently. The solution should turn cloudy, completing the limestone cycle.

Safety: CLEAPSS Hazcard 18 Limewater – irritant; 18 Calcium oxide – corrosive; 19B Calcium carbonate.

Ensure the rubber tube is tightly secured to the gas tap and the Bunsen burner before starting the experiment. Do not allow students to overstretch the tubing.

C1 2.3 The 'limestone reaction cycle'

Learning objectives

- How can we make calcium hydroxide from calcium oxide?
- Why is calcium hydroxide a useful substance?
- What is the 'limestone reaction cycle'?

Limestone is used very widely as a building material. We can also use it to make other materials for the construction industry.

As we saw in C1 2.1 **calcium oxide** is made when we heat limestone strongly. The calcium carbonate in the limestone undergoes thermal decomposition.

When we add water to calcium oxide it reacts to produce **calcium hydroxide**. This reaction gives out a lot of heat.

calcium oxide + water → calcium hydroxide

$$CaO + H_2O \rightarrow Ca(OH)_2$$

Although it is not very soluble, we can dissolve a little calcium hydroxide in water. After filtering, this produces a colourless solution called limewater. We can use limewater to test for carbon dioxide.

a What substance do we get when calcium oxide reacts with water?
b Describe how we can make limewater from calcium hydroxide.

Practical

Investigating the 'limestone reaction cycle'

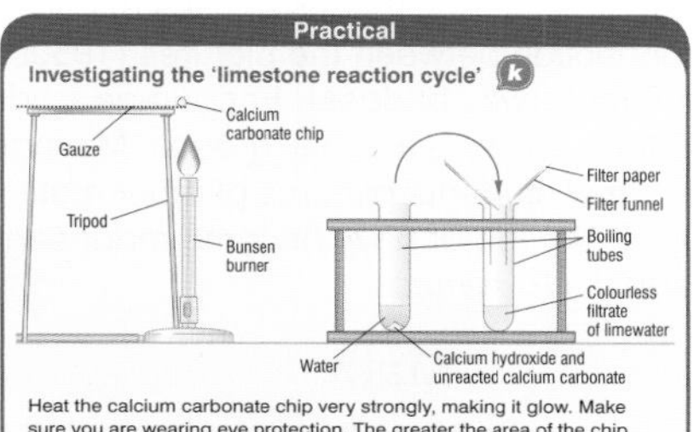

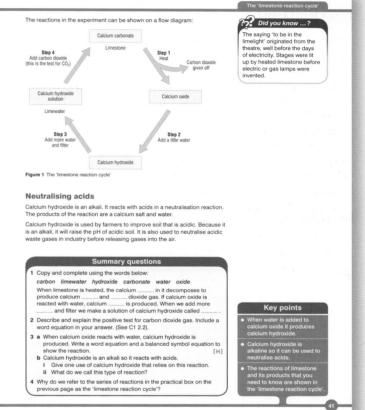

Heat the calcium carbonate chip very strongly, making it glow. Make sure you are wearing eye protection. The greater the area of the chip that glows, the better the rest of the experiment will be. This reaction produces calcium oxide (corrosive). Let the calcium oxide cool down. Then, using tongs, add it to the empty boiling tube.

Then you add a few drops of water to the calcium oxide, one drop at a time. This reaction produces calcium hydroxide.

When you dissolve this calcium hydroxide in more water and filter, it produces limewater.

Carbon dioxide bubbled through the limewater produces calcium carbonate. This turns the solution cloudy.

- The reaction between calcium oxide and water gives out a lot of energy. What do you observe during the reaction?
- Why does bubbling carbon dioxide through limewater make the solution go cloudy?

AQA Examiner's tip

Make sure that you know the limestone reaction cycle and the equations for each reaction.

links

For information on the test for carbon dioxide, look back at C1 2.2 Reactions of carbonates.

The reactions in the experiment can be shown on a flow diagram:

Calcium carbonate
Limestone
Step 1 Heat
Carbon dioxide given off
Calcium oxide
Step 2 Add a little water
Calcium hydroxide
Step 3 Add more water and filter
Calcium hydroxide solution
Limewater
Step 4 Add carbon dioxide (this is the test for CO_2)

Figure 1 The 'limestone reaction cycle'

Did you know ...?

The saying 'to be in the limelight' originated from the theatre, well before the days of electricity. Stages were lit up by heated limestone before electric or gas lamps were invented.

Neutralising acids

Calcium hydroxide is an alkali. It reacts with acids in a neutralisation reaction. The products of the reaction are a calcium salt and water.

Calcium hydroxide is used by farmers to improve soil that is acidic. Because it is an alkali, it will raise the pH of acidic soil. It is also used to neutralise acidic waste gases in industry before releasing gases into the air.

Summary questions

1 Copy and complete using the words below:
carbon limewater hydroxide carbonate water oxide
When limestone is heated, the calcium .... in it decomposes to produce calcium .... and .... dioxide gas. If calcium oxide is reacted with water, calcium .... is produced. When we add more .... and filter we make a solution of calcium hydroxide called ....

2 Describe and explain the positive test for carbon dioxide gas. Include a word equation in your answer. (See C1 2.2).

3 **a** When calcium oxide reacts with water, calcium hydroxide is produced. Write a word equation and a balanced symbol equation to show the reaction. [H]
b Calcium hydroxide is an alkali so it reacts with acids.
i Give one use of calcium hydroxide that relies on this reaction.
ii What do we call this type of reaction?

4 Why do we refer to the series of reactions in the practical box on the previous page as the 'limestone reaction cycle'?

Key points

- When water is added to calcium oxide it produces calcium hydroxide.
- Calcium hydroxide is alkaline so it can be used to neutralise acids.
- The reactions of limestone and its products that you need to know are shown in the 'limestone reaction cycle'.

Further teaching suggestions

The limestone industry

- Students could find out how parts of the limestone reaction cycle are carried out in industry, e.g. thermal decomposition of limestone occurs in a lime kiln.

Risk assessment

- You could take a Controlled Assessment focus on the practical by asking students to write a risk assessment for each stage of the 'limestone reaction cycle'.

Soil neutralisation experiment

- You could ask students to plan an experiment where they could test the effectiveness of adding calcium hydroxide to soils.

Timeline

- Students could be asked to make a timeline to show the development of the use of limestone and its products. Students should use the internet to find out key important dates when limestone and its products are first noted to be used, their properties and maybe even a sample of the material could be glued onto the timeline or a picture printed off and added. For each of the different limestone products, students should be encouraged to draw a small table listing its advantages and disadvantages.

Answers to in-text questions

a calcium hydroxide
b Dissolve it in water (and filter off any excess solid calcium hydroxide).

Summary answers

1 carbonate, oxide, carbon, hydroxide, water, limewater

2 The calcium hydroxide in limewater reacts with carbon dioxide to make a fine suspension of calcium carbonate precipitate – which makes the colourless solution turn cloudy.
calcium hydroxide + carbon dioxide → calcium carbonate + water

3 **a** calcium oxide + water → calcium hydroxide
$$CaO + H_2O \rightarrow Ca(OH)_2$$
b **i** Raising the pH of acidic soil or to neutralise acidic waste gases in industry.
ii neutralisation

4 Because the sequence of reactions starts with calcium carbonate and ends with calcium carbonate being formed again.

C1 2.4 Cement and concrete

Learning objectives

Students should learn:

- how mortar has been developed over time
- how cement is made
- what is used to make concrete
- how primary data can be improved.

Learning outcomes

Most students should be able to:

- describe how we develop materials to improve their properties
- list some uses of cement and which materials are needed to make it
- describe how concrete is made and list the properties of this material.

Some students should also be able to:

- evaluate data and suggest how their quality can be improved.

Did you know ... ?

Mortar reacts with carbon dioxide in the air. Sometimes, when old buildings are demolished, the mortar in some places still has not set. This is because the centre of the mortar was not exposed to carbon dioxide.

Support

- Students may get confused about the relationship between cement, mortar and concrete. You may wish to support students by giving them an outline of a flow chart that they can complete to explain the interaction between these materials.

Extend

- Students could be encouraged to make predictions about the effect of different amounts of raw materials on the properties of cement and mortar.
- Gypsum is added to cement. Ask the students to find out what gypsum is [calcium sulfate, $CaSO_4$] and why it is added [increases the setting time and prevents flash setting]. As the amount of gypsum increases, so does the setting time, allowing you to modify the properties of the material for a specific function.

Specification link-up: Chemistry C1.2

- Limestone is heated with clay to make cement. Cement is mixed with sand to make mortar and with sand and aggregate to make concrete. *[C1.2.1 g)]*
- Evaluate the developments in using limestone, cement and concrete as building materials, and their advantages and disadvantages over other materials. *[C1.2]*

  Controlled Assessment: C4.1 Plan practical ways to develop and test candidates' own scientific ideas. *[C4.1.1 a) b)]*; C4.2 Assess and manage risks when carrying out practical work. *[C4.2.1 a) b)]*; C4.3 Collect primary and secondary data. *[C4.3.2 a) b) c)]*; C4.5 Analyse and interpret primary and secondary data. *[C4.5.1 b)]*

Lesson structure

Starters

List – Give small groups of students a sample of cement and concrete. Ask students to list the properties of cement [opaque, hard] and concrete [opaque, hard, can be load bearing i.e. strong]. You could extend students by giving them a specific use of concrete and they have to state which of the properties they have identified makes it suitable for the job [walls in a building because cement is strong]. *(5 minutes)*

Linking pictures – Show students different images of uses of concrete through the ages. Ask them to determine the connection between the pictures [1950s prefab buildings, car parks, screed flooring, and motorway bridges]. Encourage students to think as an individual, and then compare their answers in small groups. Manage the feedback from each group. You may wish to show students pictures of other materials and examples of where they are not suitable for the job or have become redundant as new technology allows us to use concrete instead. *(10 minutes)*

Main

- By doing this they can practise their 'Controlled Assessment' planning skills. Students could be given cement, sand and different sized gravels. They could design an experiment to find out the effect of adding different proportions of each to the mix. Another investigation could be completed by comparing the effect of different gravel sizes in the mix. This can illustrate various aspects of 'Controlled Assessment', e.g. repeatability, accuracy of data collected and validity of investigational design.
- Students should consider improving the repeatability of the data collected. If they repeat a test again and get similar results the data can be described as repeatable. If they can get another person to get similar results, they could describe the results as reproducible. To further improve their results they could take a mean.
- In order to test if their results are reproducible, you could encourage students to look up strength data on the internet. They should then order the concrete data from most to least strong and see if the same pattern is borne out in their results. If their pattern is the same, the results can be described as reproducible.

Plenaries

True or false – Read out these statements and ask if they are true or false:

- Limestone is heated with clay to make cement. [True]
- Mortar is a mix of concrete and clay. [False]
- Concrete is a mixture of sand and aggregate. [True]
- Cement is a compound. [False]
- Concrete is an element. [False] *(5 minutes)*

Predict – Ask students to think about life without limestone building products. Which limestone product would they miss the most and why. You may need to support students by giving them a selection of materials made from limestone products, e.g. glass, cement, mortar, concrete, limestone blocks. To extend students further you could ask them to try and describe how their day would be affected if limestone and its products were not used. *(10 minutes)*

Practical support

Which mixture makes the strongest concrete?

Equipment and materials required

Cement, sand, gravel, spatulas, yoghurt pots, card templates to make concrete moulds, measuring cylinders (plus equipment requested by students to test strength of concrete, e.g. slotted masses, G-clamps), eye protection.

Details

Students make moulds and fill them with various concrete mixes, trying to ensure fair testing. Next lesson, they test the strength of different mixes.

Safety: See CLEAPSS document PS67-09 'Making Concrete' for suggestions and safety advice. Avoid skin contact with cement, it is an irritant. Wear eye protection when making AND testing concrete. Protect floor, feet and bench from falling weights and concrete.

Rocks and building materials

C1 2.4 Cement and concrete

Learning objectives

- How has mortar developed over time?
- How do we make cement?
- What is concrete?
- How can you improve the quality of data collected in an investigation?

Figure 1 Lime mortar is not suitable for building pools as it will not harden when in contact with water

Figure 2 The original lime mortar has flaked away from the surface of the Sphinx in Egypt, and many of the stones are now missing

How Science Works

Development of lime mortar

About 6000 years ago the Egyptians heated limestone strongly in a fire and then combined it with water. This produced a material that hardened with age. They used this material to plaster the pyramids. Nearly 4000 years later, the Romans mixed calcium hydroxide with sand and water to produce **mortar**.

Mortar holds other building materials together – for example, stone blocks or bricks. It works because the lime in the mortar reacts with carbon dioxide in the air, producing calcium carbonate again. This means that the bricks or stone blocks are effectively held together by rock.

calcium hydroxide + carbon dioxide → calcium carbonate + water

$$Ca(OH)_2 + CO_2 \rightarrow CaCO_3 + H_2O$$

The amount of sand in the mixture is very important. Too little sand and the mortar shrinks as it dries. Too much sand makes it too weak.

Even today, mortar is still used widely as a building material. However, modern mortars, made with cement in place of calcium hydroxide, can be used in a much wider range of ways than lime mortar.

Cement

Although lime mortar holds bricks and stone together very strongly, it does have some disadvantages. For example, lime mortar does not harden very quickly. It will not set at all where water prevents it from reacting with carbon dioxide.

Then people found that heating limestone with clay in a kiln produced cement. Much experimenting led to the invention of Portland cement. This is manufactured this from a mixture of limestone, clay and other minerals. They are heated and then ground up into a fine powder.

This type of cement is still in use today. The mortar used to build a modern house is made by mixing Portland cement and sand. This sets when it is mixed thoroughly with water and left for a few days.

a What does lime mortar need in order to set hard?
b Why will lime mortar not set under water?

Concrete

Sometimes builders add small stones or crushed rocks, called aggregate, to the mixture of water, cement and sand. When this sets, it forms a hard, rock-like building material called concrete.

This material is very strong. It is especially good at resisting forces which tend to squash or crush it. We can make concrete even stronger by pouring the wet mixture around steel rods or bars and then allowing it to set. This makes reinforced concrete, which is also good at resisting forces that tend to pull it apart.

Practical

Which mixture makes the strongest concrete?

Try mixing different proportions of cement, gravel and sand, then adding water, to find out how to make the strongest concrete.

- How can you test the concrete's strength?
- How could you improve the quality of the data you collect?

Summary questions

1 Copy and complete using the words below:

mortar concrete clay sand bricks

Cement is made in industry by heating limestone with ………… . It can be mixed with sand to produce …………, used to hold building materials like ………… in place. An even stronger material is made by mixing cement, ………… and aggregate to make ………… .

2 List the different ways in which limestone has been used to build your home or school.

3 Concrete and mortar are commonly used building materials. Evaluate the use of:
a concrete to make a path rather than using mortar
b mortar to bind bricks to each other rather than using concrete.

Cement and concrete

Did you know … ?

The Romans realised that they needed to add something to lime mortar to make it set in wet conditions. They found that adding brick dust or volcanic ash improved its setting. The modified mortar mixture could harden even under water. This method remained in use until the 18th century.

Figure 3 Portland cement was invented nearly 200 years ago. It is still in use all around the world today.

Key points

- Cement is made by heating limestone with clay in a kiln.
- Mortar is made by mixing cement and sand with water.
- Concrete is made by mixing crushed rocks or small stones called aggregate, cement and sand with water.

42 43

Further teaching suggestions

Museum display

- For a more creative approach to this work, students could pretend to work for Portland museum and that they have been asked to design a historical/scientific museum display. Split the students into four groups: one is responsible for the historical development of cement and concrete; the other groups detail the properties, uses and outline the basic production for cement, concrete and mortar. They must design their part of the display including visuals/scripts. If they are to use speech/videos, then storyboards need to be produced. Also they need to produce a list of artefacts that they would like on display. If you have the time, the students could actually construct their 'display' and it could be used in the classroom as an exhibit, although they may need to have images from the internet to represent the artefacts.

Contrasting cement, concrete and mortar

- Using the Student Book and everyday experiences, the students could create a table with three columns labelled cement, concrete and mortar. Each column should include the raw materials to make the product, a use and the properties which make it suitable for that particular use. This encourages students to evaluate which properties make a material fit for each purpose.

Answers to in-text questions

a Carbon dioxide must react with the lime in the mortar to produce calcium carbonate again.
b Because water prevents carbon dioxide getting to the mortar.

Summary answers

1 clay, mortar, bricks, sand, concrete

2 E.g. limestone building blocks, limestone chippings on flat roofs/roads/paths, concrete walls/paths/roads/posts, mortar to bind bricks.

3 **a** Concrete is strong in compression and can be set in large slabs. Mortar is weaker and would crack under pressure.
b Mortar is applied as a smooth, thick paste so bricks can be set level in the correct position as slight adjustments are easy to make. However, concrete has small stones in its mixture which would make it difficult to set bricks in line.

C1 2.5 Limestone issues

Learning objectives

Students should learn:

- that there are environmental, social and economic effects of quarrying limestone
- that there are advantages and disadvantages of using limestone, cement and concrete as a building material.

Learning outcomes

Most students should be able to:

- list one environmental, social and economic effect of quarrying limestone
- list at least one advantage and one disadvantage of using limestone, cement and concrete as a building material.

Some students should also be able to:

- evaluate the benefits and drawbacks of limestone quarries to the local community
- explain the advantages and disadvantages of using limestone, cement and concrete as building materials.

Specification link-up: Chemistry C1.2

- Limestone, mainly composed of the compound calcium carbonate ($CaCO_3$), is quarried and can be used as a building material. *[C1.2.1 a)]*
- Consider and evaluate the environmental, social and economic effects of exploiting limestone and producing building materials from it. *[C1.2]*
- Evaluate the developments in using limestone, cement and concrete as building materials, and their advantages and disadvantages over other materials. *[C1.2]*

Lesson structure

Starters

5,4,3,2,1 – Ask students to list five uses for limestone [making cement, making concrete, making mortar, making limewater, making building bricks]; list four names for $CaCO_3$ [limestone, chalk, marble, calcium carbonate]; list three uses of concrete [buildings, bridges, floors]; list two substances added to cement to make concrete [sand and aggregate/small stones/gravel] and list one greenhouse gas involved in the 'limestone reaction cycle' [carbon dioxide]. *(5 minutes)*

Limestone quarry – Show students a picture of a limestone quarry. Ask them to consider if they would like to live near a limestone quarry and give the value as a percentage (0% being not at all and 100% very happy). On one side of the classroom put 0% and the other 100% and ask the students to stand on the line to show how much they would like to live near a limestone quarry. Encourage some students to share their thoughts, and after these students have given their reasons, allow people to change their positions if they would like. To support students, you may wish to have some preprepared statements such as 'Claire is 90% in favour as her Dad is a coalminer and recently lost his job. She thinks that he could have the skills needed to work in the quarry.' Extend students by asking them to classify their reasons as environmental, social or economic. *(10 minutes)*

Main

- Split the class into two halves and ask them to prepare a debate on the motion: 'A new limestone quarry is a good idea in a British National Park.' Ask one half of the class to prepare an argument in favour and the other against. Students could use secondary sources of information such as textbooks and the internet. However, you may wish to give students input by providing them with fictitious people's viewpoints, such as those detailed in the Student Book, and also statistical data, such as unemployment figures and the gross income from a limestone quarry. You could support students by providing a blank table so that they can fill in the pros and cons. After students have been given adequate preparation time, you could chair the debate.
- You could supply students with a fictitious planning application for a local quarry. You could then split the class into groups and provide them with examples of people who have an interest in the planning application. Some examples might be a local unemployed teenager, local GP, member of the Green Party, Greenpeace activist, Cabinet minister, land owner, mining company executive. Ask the students to work in groups to write a biased letter from their fictitious character to the local planning officer. Students should decide if their character would be for or against and what arguments they would put in their letter to persuade the local councillors to arrive at 'the right' decision.
- Students could summarise their thoughts after the debate in a table contrasting the different building materials generated from limestone. Make a table with three headings: limestone, concrete and cement. Ask students to write statements that are advantages in green pen and disadvantages in red pen.

Support

- Students may need some help in differentiating whether statements relate to the economy, the environment or human society. It may be beneficial to provide students with a definition of each and an in-context example, so that they can refer back as necessary.

Extend

- Students should be encouraged to recognise bias and be able to generate bias and balanced arguments themselves.

Plenaries

Statements – Give students statements and ask them to order them as 'for' or 'against' a new limestone quarry and to further group them as social, economic or environmental reasons. To support students, you could give them a table to stick the statements into. Extend students by asking them to generate their own table to sort the statements. *(5 minutes)*

Revisit the limestone quarry – Show students a picture of a limestone quarry. Ask students to reconsider if they would like to live near a limestone quarry and give the value as a percentage (0% being not at all and 100% very happy). On one side of the classroom put 0% and the other 100% and ask the students to stand on the line to show how much they would like to live near a limestone quarry. Ask students to show their hands if they have changed their ideas through the lesson. Encourage some students to share their thoughts. *(10 minutes)*

How Science Works **Rocks and building materials**

C1 2.5 Limestone issues

Learning objectives

- What are the environmental, social and economic effects of quarrying limestone?
- What are the advantages and disadvantages of using limestone, cement and concrete as building materials?

Limestone is a very useful raw material, but mining for limestone can affect the local community and environment.

Limestone quarrying

Limestone is quarried from the ground. A quarry forms a huge hole in the ground. The limestone is usually blasted from a quarry by explosives. Then it is taken in giant lorries to be processed. Much of the limestone goes to cement factories which are often found near the quarry.

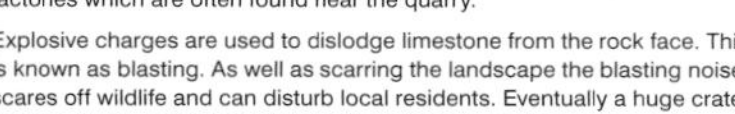

Figure 1 Limestone is often found in beautiful countryside. Quarrying the limestone scars the landscape.

Explosive charges are used to dislodge limestone from the rock face. This is known as blasting. As well as scarring the landscape the blasting noise scares off wildlife and can disturb local residents. Eventually a huge crater is formed. These can later be filled with water and can be used as a reservoir or for leisure activities. There is also the possibility of use as landfill sites for household rubbish before covering with soil and replanting.

Figure 2 Explosive charges are used to dislodge limestone from the rock face

Activity

Limestone debate

A large mining company wants to open a new limestone quarry on the edge of a National Park. Look at the views of different people affected by the planning decision to allow the quarry or not.

Take the role of one of the people shown and debate the issues involved. Assign a chairperson to make sure each person gets their say.

- Write your own response to the planning application in a letter to the mining company's managing director after your debate.

'Now we might get that by-pass we've been asking for.'

'This quarry will obviously destroy the habitats of birds and animals. A rare species of toad is found near the proposed site.'

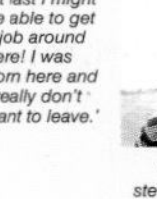

'At last I might be able to get a job around here! I was born here and I really don't want to leave.'

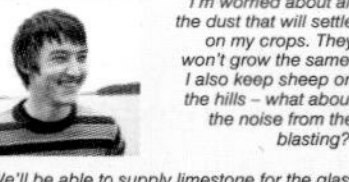

'I'm worried about all the dust that will settle on my crops. They won't grow the same. I also keep sheep on the hills – what about the noise from the blasting?'

'We'll be able to supply limestone for the glass, steel and cement industries in this region now. We predict we'll be quarrying here for 10 years – then we'll landscape the crater before moving on.'

'The lorries carrying limestone will have to go straight through our village. My daughter's primary school is on the main road.'

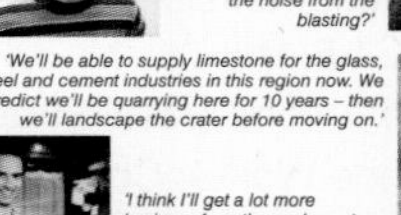

'I think I'll get a lot more business from the workers at the quarry. I might start selling sandwiches and employ someone to make them freshly each day.'

44

Limestone issues

Developments in limestone, cement and concrete

Bathroom tiles have traditionally been made from ceramics with a glazed finish to make them waterproof. They are very hard wearing. Nowadays more tiles are made from natural stone, such as travertine. These look very attractive with each tile having unique markings. However, travertine tiles are porous and can be easily scratched. They need to be sealed with a waterproof coating.

Cement is used to make mortar and concrete on building sites. Before cement mortar was invented, builders used lime mortar. However, this takes much longer to set fully than cement mortar, especially in wet conditions. The restoration of old buildings still needs lime mortar to repair brickwork. Often the old buildings have shallow, if any, foundations. Their brick walls are much more likely to move than modern buildings. With hard cement mortar this results in cracking along weak points in the walls. However, lime mortar offers more flexibility and will not crack as easily.

Figure 3 Travertine is a form of limestone. Because travertine is made up mainly of calcium carbonate, tiles and worktops can be damaged by acidic solutions.

Carbon dioxide is a greenhouse gas. The manufacture of cement contributes about 5% of the CO_2 gas produced by humans emitted into the air. About half of this comes from burning fuels used to heat the kilns that decompose limestone. The rest comes from the reaction itself:

calcium carbonate → calcium oxide + carbon dioxide

Using lime mortar would contribute less to carbon dioxide emissions as it absorbs CO_2 as it sets.

links

For information on how lime mortar reacts with CO_2 when setting, look back at C1 2.4 Cement and concrete.

Concrete is the world's most widely used building material. Concrete was first reinforced using a wire mesh to strengthen it. Nowadays we can also use:

- glass fibres
- carbon fibres
- steel rods
- poly(propene), nylon, polyesters and Kevlar.

Some of the latest research uses pulp from wood, plants and recycled paper. A little recycled paper can improve concrete's resistance to cracking, impact (making it tougher) and scratching. These reinforcing materials are shredded into small pieces before adding them to the concrete mixture.

It is much cheaper to use reinforced concrete to make a bridge than to make it from iron or steel. However, steel is much stronger (harder to snap) than concrete. Over long spans, suspension bridges can use steel's high-tensile strength in cables between concrete towers. This will support the cheap reinforced concrete sections of bridges on which cars travel. Short span bridges will always be made from reinforced concrete because of its low cost.

Figure 4 The latest high performance concretes give architects new opportunities when designing buildings

Summary questions

1 Give one effect of starting up a new limestone quarry in a National Park in each of the following:
  a an environmental effect
  b a social effect
  c an economic effect.
2 A new material has been developed called ConGlassCrete. It has large pieces of recycled glass embedded into concrete. Its surface is polished smooth which gives a very attractive finish. Give one environmental advantage and one disadvantage of using ConGlassCrete instead of slate as a building material.

Key points

- There are good and bad points about quarrying for limestone. For example, more jobs will be created but there will be a large scar on the landscape.
- Limestone, cement and concrete all have useful properties for use as building materials but the mining and processing of limestone and its products has a major effect on our environment.

45

Further teaching suggestions

Quarry study

- Students could be asked to study a real-life example of a limestone quarry and its impacts on the community. The Pennines are rich in these quarries. You may need to put together some resources to help them, or you could organise a trip to a quarry if you have one locally and see first hand what the environmental impact of mining is. Alternately the students could design a questionnaire to ask people living locally to a quarry what their opinions are of the quarry, its benefits to the community and any negative impact it has on the community.

Geological mapping

- Students could be given a geological map of the UK and then, using an OHT, they could plot the positions of the current limestone quarries. Students could then suggest where and if any other possible sites for development exist.

Summary answers

1 **a** Scar on landscape; noise pollution from explosions, crushing; dust pollution; increased traffic.
  b Less unemployment; more useful building materials being produced; improved roads.
  c More money in the area; companies in quarrying and cement industries will benefit.
2 **Advantage:** Uses recycled glass, which would not have to be remelted, using large amounts of energy, to make new objects.
  Disadvantage: Both concrete and slate require the quarrying of rock, which harm the environment but concrete needs large amounts of energy in its manufacture from limestone.

Summary answers

1 a **i** calcium carbonate → calcium oxide + carbon dioxide

ii thermal decomposition

b

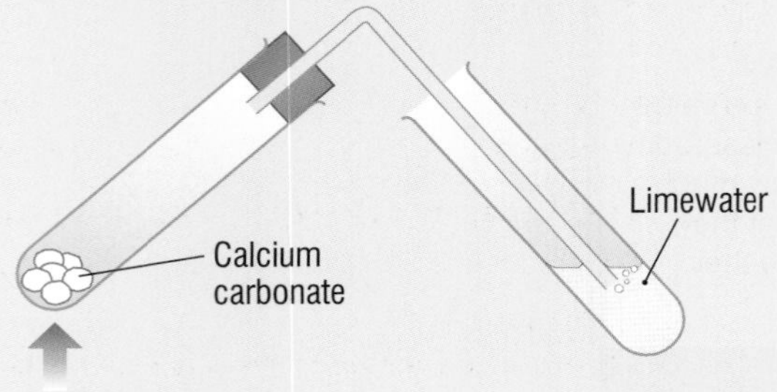

c calcium oxide + water → calcium hydroxide

2 $CaCO_3 \rightarrow CaO + CO_2$

$CaO + H_2O \rightarrow Ca(OH)_2$

3 a Limestone is crushed and roasted in a kiln with clay to make cement.

b **i** Mix with sand and water.

ii Mix with sand, aggregate/gravel/small stones/and water.

4 a potassium carbonate + hydrochloric acid → potassium chloride + water + carbon dioxide

$K_2CO_3 + 2HCl \rightarrow 2KCl + H_2O + CO_2$

b Limewater turns cloudy.

c The calcium hydroxide in limewater reacts with carbon dioxide to make a calcium carbonate precipitate, which makes the colourless solution turn cloudy.

calcium hydroxide + carbon dioxide → calcium carbonate + water

d $Ca(OH)_2 + CO_2 \rightarrow CaCO_3 + H_2O$

5 a

Ingredient	gravel	sand	cement	water
Number of buckets	4	3	1	0.5
Percentage	47	35	12	6

b Vary the composition of the mixtures, keeping dimensions of concrete moulds, volume of mixture placed in the moulds setting temperature, degree of mixing and method of testing, the same.

Suitable method for testing strength needs to be described, e.g. dropping a weight on a concrete block until it breaks, increasing the height of drop systematically.

6 a The type of metal carbonate.

b Group 1

c Group 2

d Group 2 carbonates decompose on heating in a Bunsen flame while Group 1 carbonates do not.

e No – because we haven't tested all of the carbonates of Group 1 and 2 metals.

f copper carbonate → copper oxide + carbon dioxide

g $MgCO_3 \rightarrow MgO + CO_2$

Rocks and building materials: C1 2.1–C1 2.5

Summary questions

1 In the process of manufacturing cement, calcium carbonate is broken down by heat.

a **i** Write a word equation to show the reaction that happens inside a lime kiln.

ii What do we call this type of reaction?

b Draw a diagram to show how you could test for the gas given off in the reaction described in part a.

c Write a word equation to show the reaction between calcium oxide and water.

2 Write balanced symbol equations for the reactions in Question 1 parts a and c. [H]

3 a How is limestone turned into cement?

b Given cement powder, how would you make:

i mortar

ii concrete?

4 Potassium carbonate reacts with dilute hydrochloric acid. The gas given off gives a positive test for carbon dioxide.

a Write a word equation and a balanced symbol equation to show the reaction between potassium carbonate, K_2CO_3, and dilute hydrochloric acid. [H]

b Describe what you see in a positive test for carbon dioxide.

c Explain your observations made in part **b**. Include a word equation in your answer.

d Write a balanced symbol equation for the reaction in part **c**. [H]

5 a Here is a set of instructions for making concrete:

'To make good, strong concrete, thoroughly mix together

- 4 buckets of gravel
- 3 buckets of sand
- 1 bucket of cement

When you have done this, add half a bucket of water.'

Design and fill in a table to show the percentage of each substance in the concrete mixture. Give your values to the nearest whole number.

b Describe an investigation you could use to find out which particular mixture of gravel, sand and cement makes the strongest concrete. What would you vary, what would you keep the same and how would you test the 'strength' of the concrete?

6 In an investigation into the behaviour of carbonates, a student draws the following conclusions when he heats samples of carbonates with a Bunsen burner:

Calcium carbonate	✓
Sodium carbonate	✗
Potassium carbonate	✗
Magnesium carbonate	✓
Zinc carbonate	✓
Copper carbonate	✓
(✓ = decomposes, ✗ = does not decompose	

a What was the independent variable in the investigation?

b To which group in the periodic table do sodium and potassium belong?

c To which group in the periodic table do magnesium and calcium belong?

d What do these conclusions suggest about the behaviour of the carbonates of elements in Group 1 and Group 2?

e Can you be certain about your answer to question d? Give reasons.

f Write a word equation for the thermal decomposition of copper carbonate.

g Write a balanced symbol equation for the thermal decomposition of magnesium carbonate. [H

46

Kerboodle resources

Resources available for this chapter on Kerboodle are:

- Chapter map: Rocks and building materials
- Video: Limestone: Building understanding, understanding building (C1 2.1)
- Simulation: Decomposition of carbonates (C1 2.2)
- How Science Works: Competing carbonates (C1 2.2)
- Bump up your grade: Carbonate reactions (C1 2.2)
- Support: Carbonate reactions (C1 2.2)
- Extension: Carbonate reactions (C1 2.2)
- Practical: Reactions of metal carbonates with hydrochloric acid (C1 2.2)
- Practical: The 'limestone reaction cycle' (C1 2.4)
- WebQuest: 'Modern' building materials (C1 2.4)
- Practical: Testing the strength of concrete beams (C1 2.4)
- Interactive activity: Rocks and building materials
- Revision podcast: The 'limestone reaction cycle'
- Test yourself: Rocks and building materials
- On your marks: Limestone
- Examination-style questions: Rocks and building materials
- Answers to examination-style questions: Rocks and building materials

AQA Examination-style questions (k)

1 Use words from the list to complete the sentences.

calcium carbonate calcium hydroxide calcium oxide carbon dioxide

Limestone is mainly made of the compound .... When limestone is heated strongly it decomposes producing the gas .... and solid .... When the solid reacts with water it produces .... (4)

2 Match the compounds in the list with the descriptions.

calcium carbonate copper carbonate sodium carbonate zinc carbonate

a When heated with a Bunsen burner it does not decompose. (1)
b It decomposes when heated to give zinc oxide. (1)
c It is a blue solid that produces a black solid when heated. (1)
d It can be heated with clay to make cement. (1)

3 Limestone blocks are damaged by acid rain.

Use words from the list to complete the sentences.

dissolves escapes produces reacts

Calcium carbonate in the limestone .... with acids in the rain. With sulfuric acid it .... calcium sulfate, carbon dioxide and water. The carbon dioxide .... into the air. The calcium sulfate .... in the rainwater. (4)

4 A student wanted to make calcium oxide from limestone. The student heated a piece of limestone strongly in a Bunsen burner flame.

a Complete the word equation for the reaction that happened:
calcium carbonate → calcium oxide + ... (1)

The student wanted to be sure he had made calcium oxide. He crushed the heated limestone and added water. The mixture got hot. The student cooled the mixture and filtered it. This gave a colourless solution and a white solid that was left in the filter paper.

b The student added universal indicator to the colourless solution and it turned purple.
i Name the compound in the solution that causes the indicator to turn purple. (1)
ii Explain how the student's observations show that he had made some calcium oxide by heating limestone. (2)

c The student added dilute hydrochloric acid to the white solid from the filter paper.
The mixture fizzed and produced a gas that turned limewater cloudy.
i What does this tell you about the white solid? (1)
ii Was the student successful in changing all of the limestone into calcium oxide? Explain your answer. (1)

d Write balanced equations for the three chemical reactions that the student did. [H] (3)

5 Residents living near a cement works are concerned because more children are suffering asthma attacks. Residents have also noticed that parked cars are becoming dirty because of smoke particles from the chimney.

The table shows the possible medical risk from smoke particles.

Particle size in mm	Medical effect
Larger than 0.4	No medical risks known
0.3 and smaller	Causes asthma attacks
0.2 and smaller	May cause cancer

It is also recommended that to avoid damage to health, the concentration of any particles should be no higher than 2 parts per million (ppm).

Scientists were brought in to monitor the emissions from the cement works' chimney. They positioned four sensors around the cement works to monitor airborne smoke particles.

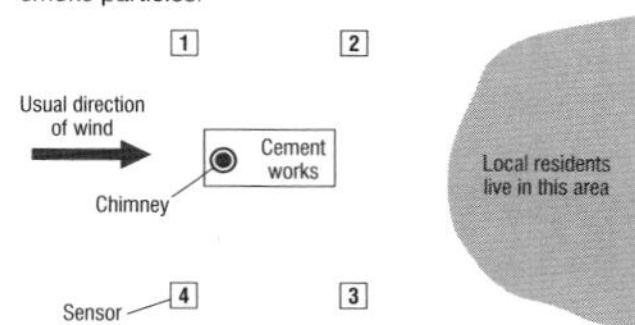

These four sensors only detect particle sizes larger than 0.5 mm and measure the concentration of particles in ppm. The scientists reported that the particle sensors showed that the average concentration of particles was 1.8 ppm. The scientists concluded that there was no risk to health.

a Suggest **two** reasons why the local residents objected to the positions of the four sensors. (2)
b What evidence did the scientists use to conclude that there was no risk to health? (1)
c The local residents were still concerned that there was a risk to health. Suggest **three** reasons why. (3)

AQA, 2009

47

Practical suggestions

Practicals	AQA	k		
Investigation of the limestone cycle: decomposition of $CaCO_3$ to give CaO, reaction with water to give $Ca(OH)_2$, addition of more water and filtering to give limewater and use of limewater to test for CO_2.	✓	✓	✓	
Thermal decomposition of $CaCO_3$ to show limelight.	✓		✓	
Honeycomb demonstration: heat sugar syrup mixture to 150 °C and add sodium bicarbonate.	✓		✓	
Making concrete blocks in moulds, investigation of variation of content and carrying out strength tests.	✓	✓		
Design and carry out an investigation of trends in the thermal decomposition of metal carbonates.	✓		✓	✓
Investigation of the reaction of carbonates with acids.	✓	✓	✓	

Examination-style answers

1 calcium carbonate, carbon dioxide, calcium oxide, calcium hydroxide *(4 marks)*

2 a sodium carbonate *(1 mark)*
b zinc carbonate *(1 mark)*
c copper carbonate *(1 mark)*
d calcium carbonate *(1 mark)*

3 reacts, produces, escapes, dissolves *(4 marks)*

4 a carbon dioxide *(1 mark)*
b i calcium hydroxide *(1 mark)*
ii Calcium hydroxide is made when calcium oxide reacts with water and this is alkaline/turns UI purple OR calcium carbonate/limestone does not react with water/is neutral and would give a green colour with UI. *(2 marks)*
c i It is or contains calcium carbonate. *(1 mark)*
ii No – because some limestone/calcium carbonate remained OR not all of the limestone/calcium carbonate decomposed on heating. *(1 mark)*
d $CaCO_3 \rightarrow CaO + CO_2$
$CaO + H_2O \rightarrow Ca(OH)_2$
$Ca(OH)_2 + 2HCl \rightarrow CaCl_2 + 2H_2O$ *(3 marks)*

5 a Any two from: not near where people/residents live, not positioned where concentration of particles likely to be highest, not positioned downwind, not between cement works and where people/residents live. *(2 marks)*
b The average/concentration was 1.8 (ppm) or the average concentration was below 2 (ppm). *(1 mark)*
c Any three from: children/people suffering asthma attacks, result was an average, readings (at some sensors) could have been higher than 2 ppm, sensors did not detect particles below 0.5 mm, small particles/particles below 0.5 mm/0.4 mm/0.3 mm/0.2 mm could (still) cause cancer/asthma. (Ignore global dimmimng, cars becoming dirty, position of sensors). *(3 marks)*

Study tip

When answering longer questions, students should be encouraged to check that they have written enough to gain all the available marks. In doing this, students often repeat the same point, so should be encouraged to check carefully what they have written. Students may use bullet points to help structure their answers to longer questions so that they are sure they have enough points.

C1 3.1 Extracting metals

Learning objectives

Students should learn:

- where metals are obtained from
- examples of how carbon can be used to extract some metals from their ores.

Learning outcomes

Most students should be able to:

- list examples of native metals and metals found in ores
- relate the use of carbon in metal extraction to the reactivity of a metal
- identify a reduction process from a description of a reaction.

Some students should also be able to:

- write a balanced equation to show the reduction of a metal oxide.

Did you know ... ?

Gold is not used just for jewellery – pure gold is used on the insides of astronauts' helmet sun-visors and in some electrical circuits. A gold compound is also used to treat arthritis sufferers.

Support

- When undertaking the 'reduction by carbon' practical, students should be asked only to reduce a copper ore, as copper is studied later on in the specification. Reducing lots of different ores may just confuse the main focus of the lesson.
- Students could use secondary sources to find out the names of copper ore [malachite], iron ore [haematite] and gold ore [trick question, it is a native metal].

Extend

- Ask students to write a balanced symbol equation for the reduction of copper oxide with carbon. You could further extend students by giving them the formula of other metal oxide compounds and asking them to write the balanced symbol equation for their reduction with carbon. Ensure that the examples are for metals that are below carbon in the reactivity series.

Specification link-up: Chemistry C1.3

- Ores contain enough metal to make it economical to extract the metal. The economics of extraction may change over time. *[C1.3.1 a)]*
- Ores are mined and may be concentrated before the metal is extracted and purified. *[C1.3.1 b)]*
- Unreactive metals such as gold are found in the Earth as the metal itself but most metals are found as compounds that require chemical reactions to extract the metal. *[C1.3.1 c)]*
- Metals that are less reactive than carbon can be extracted from their oxides by reduction with carbon, for example iron oxide is reduced in the blast furnace to make iron. *[C1.3.1 d)]*

Lesson structure

Starters

Key words – Ask students to look at each definition and to match it to its key word.

1. Removal of oxygen from a compound. [Reduction] 2. A list of elements from the most reactive to the least reactive. [Reactivity series] 3. A rock containing enough metal to make it economic to extract the metal. [Ore] 4. Unreactive metals, found as elements in nature. [Native] *(5 minutes)*

Prediction – Show students samples of ores and ask them to guess the metal that they contain. This should help students see that ores are usually mixtures of compounds and do not share the properties of the metals extracted from them. You may wish to support students by giving them the chemical name of the mineral that the ore contains. To extend students, ask them to give five things that would affect the cost of extracting the mineral from their ores. *(10 minutes)*

Main

- Show students a selection of ores and explain that the metal is locked up in a compound, often an oxide. Ask the students to suggest how the metal could be released [through a chemical reaction, some might mention reduction]. Discuss the need to 'concentrate' the metal compound from the rest of the ore in some cases e.g. copper, aluminium.
- The native metals should also be discussed. Explain that these metals are all relatively unreactive.
- Explain that carbon can be used to 'displace' the metal as long as carbon is more reactive than the metal you wish to extract.
- Ask the students to consult the reactivity series and suggest which metals could be extracted using this technique. [E.g. zinc, iron, copper, lead].
- Students carry out the 'Reduction by carbon' practical, extracting a metal from its oxide (see 'Practical support' for more details).
- You may wish to allow students to compare the reactivity of different metals (see 'Practical support' for more details). Students should be encouraged to write their observations in an appropriate results table and write word equations for all the reactions.

Plenaries

Complete the sentences – Ask the students to complete the following sentences:

- Metal ores are rocks that … [contain enough metal to make it economic to extract it].
- Gold, platinum and silver are … [native metals].
- Reduction reactions are used … [to remove oxygen from a metal oxide].

You could support students by making this a card sort. Supply both halves of the sentences and they could work in pairs to match the six cards to make the three sentences. To extend students, you could ask them to write the formula of each chemical mentioned in their sentences. *(5 minutes)*

Random questions – Create a PowerPoint presentation, with each student's name on a different slide. Set the slide show so that it is continuous. Then start the slide show; the students' names will appear one by one quickly on the screen. Press pause and one slide will hold its position, thus choosing a student. Ask this student to answer a question. Then return to the 'name generator' and press pause again. *(10 minutes)*

Practical support

Reduction by carbon

Equipment and materials required

Bunsen burner and safety equipment, eye protection, test tube/ignition tube plus test tube holders (alternatively, tripod, crucible, pipe-clay triangle), evaporating dish, spatula, carbon powder, selection of metal oxides (e.g. copper oxide – harmful, magnesium oxide, lead oxide – toxic).

Details

Mix the metal oxide thoroughly with carbon powder as a 1 : 1 ratio. Put the mixture into a test tube (or crucible secured in a pipe-clay triangle) and heat strongly in a blue Bunsen flame. Allow to cool and observe to see metal pieces. The pieces can be washed, and almost 'pan' for the pure metal. Lead oxide should be reduced in a fume cupboard.

The carbon is more reactive than copper (or lead), so it can reduce the metal oxide, leaving the metal element:

copper oxide + carbon → copper + carbon dioxide

$$2CuO + C \rightarrow 2Cu + CO_2$$

Safety: Wear eye protection, ventilate the room well, but use a fume cupboard, if it is available. Hands should be washed after the practical. CLEAPSS Hazcards 47A Hydrochloric acid; 26 Copper oxide – harmful; 56 Lead oxide – toxic. Dispose of waste correctly.

Comparing metal reactivity

Equipment and materials required

Dimple dish, 1 mol/dm^3 hydrochloric acid, 2 × dropping pipette, 3 × boiling tube, zinc chips, calcium chips, magnesium chips, copper turnings, spatula, eye protection.

Details

Put a sample of each metal into separate dimples and add a few drops of water to each metal in turn and observe. Add a new piece of each metal into clean dimples and add a few drops of hydrochloric acid to each metal in turn and observe.

Safety: Wear eye protection; ensure all sources of ignition are removed. Dispose of the waste acid and reactive metals correctly. CLEAPSS Hazcards 107 Zinc; 16 Calcium – highly flammable; 59A Magnesium – highly flammable; 26 Copper; 47A Hydrochloric acid – harmful.

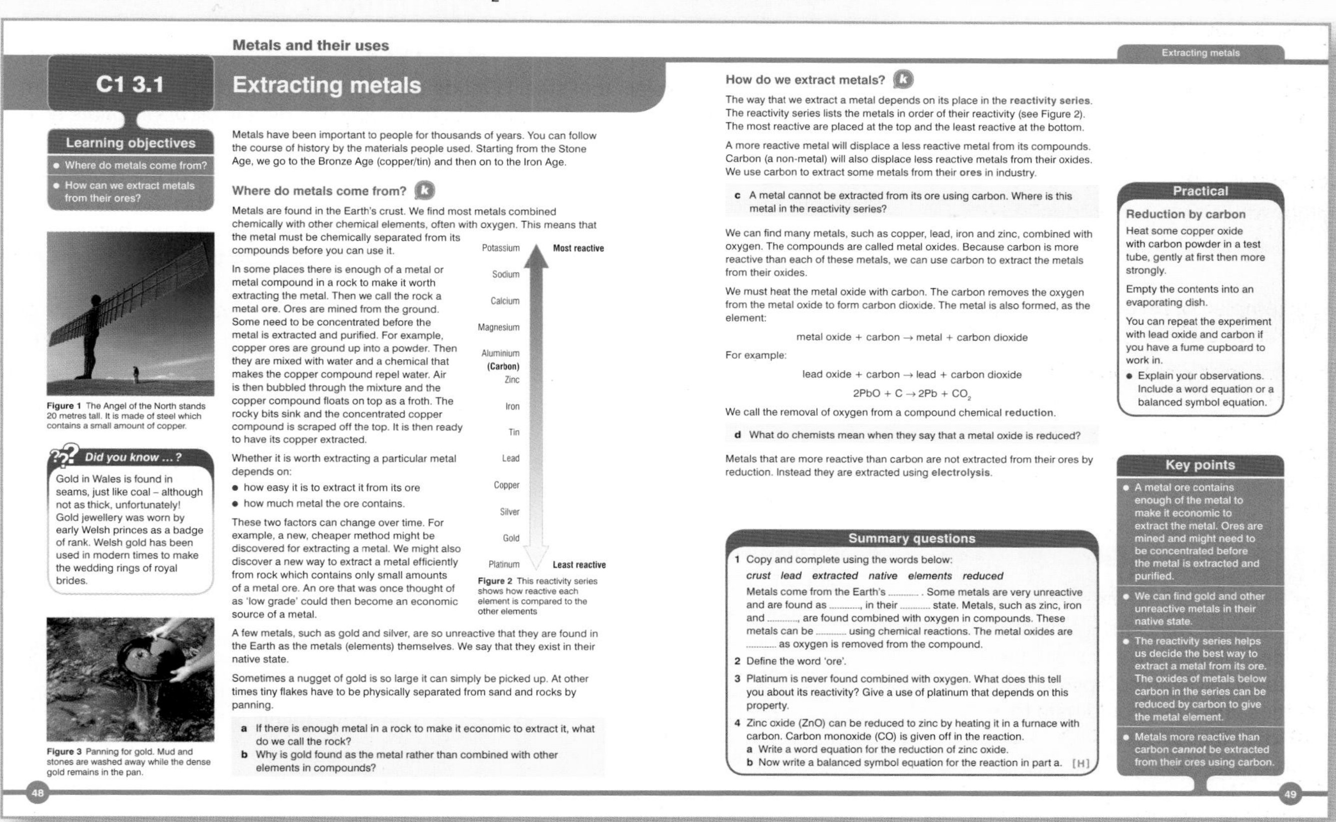

Metals and their uses

C1 3.1 Extracting metals

Learning objectives
- Where do metals come from?
- How can we extract metals from their ores?

Metals have been important to people for thousands of years. You can follow the course of history by the materials people used. Starting from the Stone Age, we go to the Bronze Age (copper/tin) and then on to the Iron Age.

Where do metals come from?

Metals are found in the Earth's crust. We find most metals combined chemically with other chemical elements, often with oxygen. This means that the metal must be chemically separated from its compounds before you can use it.

In some places there is enough of a metal or metal compound in a rock to make it worth extracting the metal. Then we call the rock a metal **ore**. Ores are mined from the ground. Some need to be concentrated before the metal is extracted and purified. For example, copper ores are ground up into a powder. Then they are mixed with water and a chemical that makes the copper compound repel water. Air is then bubbled through the mixture and the copper compound floats on top as a froth. The rocky bits sink and the concentrated copper compound is scraped off the top. It is then ready to have its copper extracted.

Whether it is worth extracting a particular metal depends on:
- how easy it is to extract it from its ore
- how much metal the ore contains.

These two factors can change over time. For example, a new, cheaper method might be discovered for extracting a metal. We might also discover a new way to extract a metal efficiently from rock which contains only small amounts of a metal ore. An ore that was once thought of as 'low grade' could then become an economic source of a metal.

Potassium — Most reactive
Sodium
Calcium
Magnesium
Aluminium
(Carbon)
Zinc
Iron
Tin
Lead
Copper
Silver
Gold
Platinum — Least reactive

Figure 2 This reactivity series shows how reactive each element is compared to the other elements

A few metals, such as gold and silver, are so unreactive that they are found in the Earth as the metals (elements) themselves. We say that they exist in their native state.

Sometimes a nugget of gold is so large it can simply be picked up. At other times tiny flakes have to be physically separated from sand and rocks by panning.

a If there is enough metal in a rock to make it economic to extract it, what do we call the rock?
b Why is gold found as the metal rather than combined with other elements in compounds?

Figure 1 The Angel of the North stands 20 metres tall. It is made of steel which contains a small amount of copper.

Did you know ...?
Gold in Wales is found in seams, just like coal – although not as thick, unfortunately! Gold jewellery was worn by early Welsh princes as a badge of rank. Welsh gold has been used in modern times to make the wedding rings of royal brides.

Figure 3 Panning for gold. Mud and stones are washed away while the dense gold remains in the pan.

48

Extracting metals

How do we extract metals?

The way that we extract a metal depends on its place in the **reactivity series**. The reactivity series lists the metals in order of their reactivity (see Figure 2). The most reactive are placed at the top and the least reactive at the bottom.

A more reactive metal will displace a less reactive metal from its compounds. Carbon (a non-metal) will also displace less reactive metals from their oxides. We use carbon to extract some metals from their **ores** in industry.

c A metal cannot be extracted from its ore using carbon. Where is this metal in the reactivity series?

We can find many metals, such as copper, lead, iron and zinc, combined with oxygen. The compounds are called metal oxides. Because carbon is more reactive than each of these metals, we can use carbon to extract the metals from their oxides.

We must heat the metal oxide with carbon. The carbon removes the oxygen from the metal oxide to form carbon dioxide. The metal is also formed, as the element:

metal oxide + carbon → metal + carbon dioxide

For example:

lead oxide + carbon → lead + carbon dioxide

$$2PbO + C \rightarrow 2Pb + CO_2$$

We call the removal of oxygen from a compound chemical **reduction**.

d What do chemists mean when they say that a metal oxide is reduced?

Metals that are more reactive than carbon are not extracted from their ores by reduction. Instead they are extracted using **electrolysis**.

Practical

Reduction by carbon

Heat some copper oxide with carbon powder in a test tube, gently at first then more strongly.

Empty the contents into an evaporating dish.

You can repeat the experiment with lead oxide and carbon if you have a fume cupboard to work in.
- Explain your observations. Include a word equation or a balanced symbol equation.

Summary questions

1 Copy and complete using the words below:
crust lead extracted native elements reduced
Metals come from the Earth's .... Some metals are very unreactive and are found as ...., in their .... state. Metals, such as zinc, iron and ...., are found combined with oxygen in compounds. These metals can be .... using chemical reactions. The metal oxides are .... as oxygen is removed from the compound.
2 Define the word 'ore'.
3 Platinum is never found combined with oxygen. What does this tell you about its reactivity? Give a use of platinum that depends on this property.
4 Zinc oxide (ZnO) can be reduced to zinc by heating it in a furnace with carbon. Carbon monoxide (CO) is given off in the reaction.
a Write a word equation for the reduction of zinc oxide.
b Now write a balanced symbol equation for the reaction in part a. [H]

Key points
- A metal ore contains enough of the metal to make it economic to extract the metal. Ores are mined and might need to be concentrated before the metal is extracted and purified.
- We can find gold and other unreactive metals in their native state.
- The reactivity series helps us decide the best way to extract a metal from its ore. The oxides of metals below carbon in the series can be reduced by carbon to give the metal element.
- Metals more reactive than carbon *cannot* be extracted from their ores using carbon.

49

Further teaching suggestions

Revision of the reactivity series

- Students should have studied the reactivity series in KS3. However, this concept underpins much of the work on metal extraction and it could be worth revising this material. For example, give each student an element from the reactivity series printed on a piece of card. On the back of the card is information about that element's reaction with water, acid and oxygen in the air. The task is for the students to line themselves up in order of reactivity using the information given.

Reactivity poster

- If you do not have access to practical equipment, ask students to make a poster detailing the reactivity series. On the poster, they could highlight which metals can be reduced using carbon and include word equations for the metal oxide being reduced. At a later date, this poster could be revisited and the metals that are extracted using eletrolysis could be added.

Answers to in-text questions

a an ore
b Because gold is very unreactive.
c above carbon
d It has had its oxygen removed, to leave the metal element.

Summary answers

1 crust, elements, native, lead, extracted, reduced
2 A metal ore is a mineral which contains enough of the metal to make it economic to extract the metal.
3 Platinum is very unreactive. It is used in jewellery and in special corrosion-resistant wires.
4 **a** zinc oxide + carbon → zinc + carbon monoxide
b $ZnO + C \rightarrow Zn + CO$

C1 3.2 Iron and steels

Learning objectives

Students should learn:

- that iron is extracted using carbon in a blast furnace
- the properties of cast iron and how this limits its usefulness
- that iron can be alloyed to make it more useful
- the properties and composition of the three main types of steel.

Learning outcomes

Most students should be able to:

- state that iron oxide is reduced to form iron in a blast furnace using carbon
- list the properties of iron produced from the blast furnace
- explain why steels are produced.

Some students should also be able to:

- give examples of different types of steels, and how their properties differ.

Support

- Some students may struggle with the properties of each type of steel. Making a simple table of the composition, properties and uses could help to consolidate these ideas.
- The properties of metals and alloys can be modelled easily by adapting the already familiar KS3 particle model. Give students 22 marbles, polystyrene balls or table tennis balls and sticky-tac. Ask them to make a model of a solid metal. Then give each group a different sized ball to add to their model by replacing one of their atoms, and explain that this is an alloy.

Extend

- To extend students, they could use secondary sources to find out the compositions of different steels and then identify a pattern in their compositions and properties. Students could also be asked to find out about the BOC (Basic Oxygen Steel) process and draw a labelled diagram of the vessel needed to make steel.

Specification link-up: Chemistry C1.3

- Metals that are less reactive than carbon can be extracted from their oxides by reduction with carbon, for example iron oxide is reduced in the blast furnace to make iron. *[C1.3.1 d)]*
- Iron from the blast furnace contains about 96% iron. The impurities make it brittle and so it has limited uses. *[C1.3.2 a)]*
- Most iron is converted into steels. Steels are alloys since they are mixtures of iron with carbon. Some steels contain other metals. Alloys can be designed to have properties for specific uses. Low-carbon steels are easily shaped, high-carbon steels are hard, and stainless steels are resistant to corrosion. *[C1.3.2 b)]*

Lesson structure

Starters

Displaying data – Explain that steel is made of a mixture of iron and carbon. Medium carbon steel is 0.59% carbon. Ask students to display this data in an appropriate scientific form. [A table of % composition and element name.]

Students could be supported by the table being supplied with some of the parts filled in. Students could be extended and asked to explain why they chose this display method [element is a categoric variable and composition is a continuous variable, but there is a big difference between these numbers making it difficult to show this data on a pie chart or bar chart]. *(5 minutes)*

List – Ask students to list as many properties of iron as they can think of, and one use of iron that draws specifically on each property they have listed [e.g. conductor of heat – saucepans; malleable – sculptures]. *(10 minutes)*

Main

- Often the scale of industrial chemistry is difficult to bring into the classroom, but videos or a site visit will help. Students could watch a video on iron production, such as *Industrial Chemistry for Schools and Colleges* (RSC).
- You may wish to show students the model blast furnace demonstration (see 'Practical support' for more details).
- Students need to appreciate that different proportions of constituents are used to make steels with a great variety of properties. Show students a recipe book, then a recipe card such as ones given free in supermarkets. Ask students to create a recipe card for making steel, including variations at the bottom of the card to make different types of steel. You need to follow a recipe to get a repeatable result.
- Ask students to create a 'lonely hearts' column for each type of steel. More artistic students could create the articles and a display could be made.
- You could develop the concept of alloys by testing the suitability of different metal wires in making springs. Students can coil the wires into springs, then test with slotted masses. Their investigations can provide data to plot extension against load graphs.

Plenaries

True or false? – Give each student a red and a green card. If the student thinks a statement is true, they hold up the green card; if they think it is false, they hold up the red card. Ask them these true/false statements:

- Steel is a mixture. [True]
- All steels contain mostly carbon. [False]
- Stainless steels are expensive. [True]
- Stainless steels rust because they contain iron. [False]
- The chemical symbol for steel is St. [False] *(5 minutes)*

List uses – For each type of steel (mild, medium and low carbon), ask the class to give an example of a use and make a note on the board. Split the class into groups and ask each group to come up with a different use for each type of steel and add the second example to their notes. Finally ask students to work as an individual to add a third use for each type of steel. *(10 minutes)*

Practical support

Investigating alloys

Equipment and materials required

Variety of wires (same gauge needed for fair testing), slotted 10 g masses, clamp stand, ruler.

Details

Students choose wires of same gauge to make springs by winding the wire around a pencil. Then they test the wires by measuring extension as successive slotted masses are added.

Model blast furnace

Equipment and materials required

Ignition tube, Bunsen burner and safety equipment, stand, boss and clamp, 0.3 g potassium manganate(VII), 0.15 g carbon powder, 0.15 g iron oxide powder, mineral wool, spatula, paper, magnet, clingfilm, eye protection.

Details

Put the potassium manganate(VII) in the bottom of an ignition tube and add a mineral wool plug. Mix together the carbon and iron oxide on a piece of paper and add on top of the mineral wool into the ignition tube. Add a second mineral wool plug. Hold the ignition tube at the neck in a clamp at an angle in a stand and boss. With the blue Bunsen flame, aim it at the carbon/iron oxide mixture for a couple of minutes and then heat the potassium manganate for a minute. Switch off the Bunsen burner. Allow the ignition tube to cool.

Using a magnet wrapped in cling film, students can take some of the iron oxide/carbon mixture and prove that it is non-magnetic. After the ignition tube has cooled, the reaction mixture can be poured on to paper and shown to be magnetic.

Safety: Wear eye protection. See CLEAPSS document PS67-8 for a detailed explanation. CLEAPSS Hazcard 81 Potassium manganate(VII) – oxidising and harmful; 55A Iron oxide. Dispose of waste safely.

Metals and their uses

C1 3.2 Iron and steels

Learning objectives

- How is iron ore reduced?
- Why is iron from a blast furnace not very useful?
- How is iron changed to make it more useful?
- What are the main types of steel?

Iron ore contains iron combined with oxygen in iron oxide. Iron is less reactive than carbon. So we can extract iron by using carbon to remove oxygen from the iron(III) oxide in the ore. We extract iron in a **blast furnace**.

Some of the iron(III) oxide reacts with carbon. The carbon reduces iron(III) oxide, forming molten iron and carbon dioxide gas. This is one of the reduction reactions which takes place in a blast furnace:

iron(III) oxide + carbon → iron + carbon dioxide

Iron straight from the blast furnace has limited uses. It contains about 96% iron and contains impurities, mainly carbon. This makes it very brittle, although it is very hard and can't be easily compressed. When molten it can be run into moulds and cast into different shapes. This **cast iron** is used to make wood-burning stoves, man-hole covers on roads, and engines.

We can treat the iron from the blast furnace to remove some of the carbon.

Removing all the carbon and other impurities from cast iron gives us pure iron. This is very soft and easily-shaped. However, it is too soft for most uses. If we want to make iron really useful we have to make sure that it contains tiny amounts of other elements. These include carbon and metals, such as nickel and chromium.

We call a metal that is mixed with other elements an **alloy**.

Steel is an alloy of iron. By adding elements in carefully controlled amounts, we can change the properties of the steel.

a Why is iron from a blast furnace very brittle?
b Why is pure iron not very useful?
c How do we control the properties of steel?

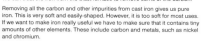

Figure 1 The iron which has just come out of a blast furnace contains about 96% iron. The main impurity is carbon.

Steels

Steel is not a single substance. Like all alloys, it is a mixture. There are lots of different types of steel. All of them are alloys of iron with carbon and/or other elements.

Carbon steels

The simplest steels are the **carbon steels**. We make these by removing most of the carbon from cast iron, just leaving small amounts of carbon (from 0.03% to 1.5%). These are the cheapest steels to make. We use them in many products, such as the bodies of cars, knives, machinery, ships, containers and structural steel for buildings.

Often these carbon steels have small amounts of other elements in them as well. High carbon steel, with a relatively high carbon content, is very strong but brittle. On the other hand, low carbon steel is soft and easily shaped. It is not as strong, but is much less likely to shatter on impact with a hard object.

Mild steel is one type of low carbon steel. It contains less than 0.1% carbon. It is very easily pressed into shape. This makes it particularly useful in mass production, such as making car bodies.

Figure 2 Low carbon steel called mild steel is easily pressed into shapes

50

Alloy steels

Low-alloy steels are more expensive than carbon steels because they contain between 1% and 5% of other metals. Each of these metals produces a steel that is well-suited for a particular use.

Figure 3 The properties of steel alloys make them ideal for use in suspension bridges

Even more expensive are the **high-alloy steels**. These contain a much higher percentage of other metals. The chromium–nickel steels are known as **stainless steels**. We use them to make cooking utensils and cutlery. They are also used to make chemical reaction vessels. That's because they combine hardness and strength with great resistance to corrosion. Unlike most other steels, they do not rust!

Figure 4 The properties of stainless steels make them ideal for making utensils and cutlery

Summary questions

1 Copy and complete the following sentences using the terms below:
carbon pure steel cast reduced
Iron(III) oxide is .... (has its oxygen removed) in a blast furnace.
Iron from the blast furnace, poured into moulds and left to solidify is called .... iron.
If all the carbon and other impurities are removed from cast iron we get .... iron.
Iron that has been alloyed with carbon and other elements is called ....
Iron that contains just a small percentage of carbon is called .... steel.

2 How does cast iron differ from pure iron?

3 a Make a table to summarise the properties and some uses of low carbon steel, high carbon steel and chromium–nickel steel.
b Why are surgical instruments made from steel containing chromium and nickel?

Iron and steels

Did you know ...?

We use nickel–steel alloys to build long-span bridges, bicycle chains and military armour-plating. That's because they are very resistant to stretching forces. Tungsten steel operates well under very hot conditions so it is used to make high-speed tools such as drill bits.

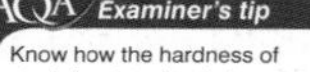

AQA Examiner's tip

Know how the hardness of steels is related to their carbon content.

Key points

- We extract iron from iron ore by reducing it using carbon in a blast furnace.
- Pure iron is too soft for it to be very useful.
- Carefully controlled quantities of carbon and other elements are added to iron to make alloys of steel with different properties.
- Important examples of steels are:
  - low carbon steels which are easily shaped,
  - high carbon steels which are very hard,
  - stainless steels which are resistant to corrosion.

51

Further teaching suggestions

Modelling steel

- A different model for iron and steel uses modelling clay. By incorporating different amounts of sand into the modelling clay, its properties can be changed. The modelling clay represents the iron and the sand the carbon. Roll the samples into a sausage shape and then pull them until they break. By comparing the fracture sites you can see the effect of making steel.

Compare and contrast

- Students could be asked to list the properties that both iron and steel share and then list the desirable properties that make steel more useful than iron. Students could also compare and contrast the reduction of iron ore and tungsten ore/zinc ore (which are also reduced using carbon).

Corus Group

- A number of resources are available from the Corus Group website (www.coruseducation.com).

The history of the blast furnace

- Students could research the historical development of the blast furnace.

Answers to in-text questions

a It contains a fairly high percentage of carbon and other impurities.
b It is too soft.
c By adding other elements in carefully controlled amounts.

Summary answers

1 reduced, cast, pure, steel, carbon

2 Cast iron is harder and more brittle than pure iron.

3 a

Type of steel	Properties
low carbon	soft, not easily shattered (malleable)
high carbon	very strong but brittle
chromium-nickel (stainless)	hard, strong, corrosion-resistant

b This is stainless steel, which is very resistant to corrosion (any rust on the instruments could harbour harmful microorganisms).

C1 3.3 Aluminium and titanium

Learning objectives

Students should learn:

- that aluminium and titanium are useful metals
- that metals more reactive than carbon are extracted by electrolysis of molten compounds
- that extraction of aluminium and titanium is expensive.

Learning outcomes

Most students should be able to:

- list some useful properties of aluminium and titanium
- state that methods of extraction used for aluminium (as an example of a metal more reactive than carbon) and titanium are expensive
- give examples of uses of aluminium and/or titanium and explain which properties makes them fit for purpose.

Some students should also be able to:

- explain why the extraction of aluminium and titanium is costly.

Specification link-up: Chemistry C1.3

- Metals that are more reactive than carbon, such as aluminium, are extracted by electrolysis of molten compounds. The use of large amounts of energy in the extraction of these metals makes them expensive. *[C1.3.1 e)]*
- Aluminium and titanium cannot be extracted from their oxides by reduction with carbon. Current methods of extraction are expensive because
  - there are many stages in the processes
  - large amounts of energy are needed. *[C1.3.1 i)]*

Lesson structure

Starters

Museum – In five different points around the room put a piece of bauxite, aluminium oxide (sealed in a Petri dish), aluminium metal, titanium metal and rutile. Each sample should be labelled with an information card (chemical's name, formula, hazards, properties, uses). Students should visit each sample and then write two sentences to summarise how the chemicals are related. [Bauxite is an aluminium ore that mainly contains aluminium oxide (Al_2O_3). Rutile is mainly titanium oxide (TiO_2), an ore of titanium.] *(5 minutes)*

Sparkler! – A demonstration to show the reactivity of aluminium: set up a Bunsen burner and sprinkle aluminium powder into the flame. The powder will combust in a twinkling effect to form aluminium oxide. Ask the students to generate a word equation. Students could be supported by being given the names of the chemicals. Extend students by asking them to give a balanced symbol equation for the reaction.
[aluminium + oxygen → aluminium oxide, $4Al + 3O_2 \rightarrow 2Al_2O_3$] *(10 minutes)*

Main

- In this lesson electrolysis is being introduced, although no details are required. Students will be familiar with charged ions from 'C1 3.1 Extracting metals' and should be able to understand that the positive metal ion is attracted to a negatively charged electrode. The use of electrolysis to extract metals is very expensive due to the amount of energy needed for this chemical reaction.
- Some students find it difficult to link specific properties of a material with its uses. Search the internet to find pictures of various items made of aluminium and titanium, e.g. a bike, a hip replacement joint, an aircraft, a saucepan, overhead cables with a pylon, a ring. Ask the students to choose which material it would be made from and give reasons for their choice.
- Explain to students that they are to make a TV commercial encouraging customers to use these metals rather than cheaper metals. Students should work in small teams to produce their one-minute commercial. Encourage students to recognise the bias needed, and to include information about why it is an expensive material.
- Students could be asked to use the Student Book to create two flow charts to explain how titanium and aluminium are extracted from their ores.

Plenaries

'I went to the shops to buy …' – This children's game can be played but the students can only give examples of items made from aluminium or titanium. The teacher could start by saying: 'I went to the shops to buy a titanium aircraft'. The first student then could say: 'I went to the shop to buy a titanium aircraft and some aluminium foil', and so on around the class. To support students, there could be a list of uses of aluminium or titanium on the board. To extend students, they could explain why that metal is chosen, e.g. I went to buy a titanium nuclear reactor, as it is unreactive and resistant to high temperatures. *(5 minutes)*

Questions and answers – Ask students to work in small teams. Complete a question and answer session: if a team gets three questions correct, then they have 'earned the right to leave'; if they are incorrect, then the question goes on to another group. A team who answers incorrectly should be given as many questions as needed to get three correct, therefore more questions than teams need to be written. *(10 minutes)*

Support

- Students could be supported by creating a 'cut-and-stick' activity or card sort activity, where the images showing the use, types of metal and property that makes it fit for that purpose are on separate rectangles or cards. Students cut them out and/or sort the information.

Extend

- Students could be extended by being given the symbol equations for each stage of the titanium extraction processes and for aluminium extraction. Students could then balance these equations and be further extended by being encouraged to use state symbols.

Practical support

Sparkler!

Equipment and materials required

Bunsen burner, safety equipment, eye protection, spatula and aluminium powder – highly flammable.

Details

Set the Bunsen burner up with a blue flame over a heatproof mat. Hold the Bunsen at an angle. Half-fill the spatula with aluminium powder, and sprinkle it into the flame. This is a very vigorous reaction.

Safety: Wear eye protection and tie back hair and loose clothing. Keep students well away from the Bunsen burner. CLEAPSS Hazcard 1 Aluminium powder – highly flammable.

Further teaching suggestions

Designing an experiment

- Students could be given samples of different metals, including aluminium and titanium. They could then design a practical to test the chemical and physical properties of the different metals. This would create an opportunity to teach aspects of Controlled Assessment, e.g. designing a fair test.

Did you know ... ?

Titanium dioxide is the white material in correction fluid and it is used in white paint.

Metals and their uses

C1 3.3 Aluminium and titanium

Learning objectives

- Why are aluminium and titanium so useful?
- What method is used to extract metals that are more reactive than carbon?
- Why does it cost so much to extract aluminium and titanium?

Although they are very strong, many metals are also very dense. This means that we cannot use them if we want to make something that has to be both strong and light. Examples are alloys for making an aeroplane or the frame of a racing bicycle.

Where we need metals which are both strong and have a low density, **aluminium** and **titanium** are often chosen. These are also metals which do not corrode.

Properties and uses of aluminium

Aluminium is a silvery, shiny metal. It is surprisingly light for a metal as it has a relatively low density. It is an excellent conductor of energy and electricity. We can also shape it and draw it into wires very easily.

Although aluminium is a relatively reactive metal, it does not corrode easily. This is because the aluminium atoms at its surface react with oxygen in air. They form a thin layer of aluminium oxide. This layer stops any further corrosion taking place.

Aluminium is not a particularly strong metal, but we can use it to form alloys. These alloys are harder, more rigid and stronger than pure aluminium.

Because of these properties, we use aluminium to make a whole range of goods. These include:

- drinks cans
- cooking foil
- saucepans
- high-voltage electricity cables
- aeroplanes and space vehicles
- bicycles.

Figure 1 We use aluminium alloys to make bicycles because of their combination of low density and strength

a Why does aluminium resist corrosion?
b How do we make aluminium stronger?

Extracting aluminium

Because aluminium is a reactive metal we cannot use carbon to displace it from its oxide. Instead we extract aluminium using electrolysis. An electric current is passed through molten aluminium oxide at high temperatures to break it down.

First we must mine the aluminium ore. This contains aluminium oxide mixed with impurities. Then the aluminium oxide is separated from the impurities. The oxide must then be melted before electrolysis can take place.

The problem with using electrolysis to extract metals is that it is a very expensive process. That's because we need to use high temperatures to melt the metal compound. Then we also need a great deal of electricity to extract the metal from its molten compound. There are also environmental issues to consider when using so much energy.

Figure 2 We use aluminium alloys to make aircraft. The alloys are strong yet have a low density so the plane can carry more passengers and cargo.

52

Aluminium and titanium

Properties and uses of titanium

Titanium is a silvery-white metal. It is very strong and very resistant to corrosion. Like aluminium it has an oxide layer on its surface that protects it. Although it is denser than aluminium, it is less dense than most other metals.

Titanium has a very high melting point – about 1660 °C – so we can use it at very high temperatures.

We use titanium for:

- the bodies of high-performance aircraft and racing bikes (because of its combination of strength and relatively low density)
- parts of jet engines (because it keeps its strength even at high temperatures)
- parts of nuclear reactors (where it can stand up to high temperatures and its tough oxide layer means that it resists corrosion)
- replacement hip joints (because of its low density, strength and resistance to corrosion).

c What properties make titanium ideal to use in jet engines and nuclear reactors?

Extracting titanium

Titanium is not particularly reactive, so we could produce it by displacing it from its oxide with carbon. But unfortunately carbon reacts with titanium metal making it very brittle. So we have to use a more reactive metal to displace titanium. We use sodium or magnesium. However, both sodium and magnesium have to be extracted by electrolysis themselves in the first place.

Before displacement of titanium can take place, the titanium ore must be processed. This involves separating the titanium oxide and converting it to a chloride. Then the chloride is distilled to purify it. Only then is it ready for the titanium to be displaced by the sodium or magnesium. Each one of these steps takes time and costs money.

d Why do we need electricity to make:
i aluminium and **ii** titanium?

Figure 3 We can use titanium inside the body as well as outside. This is an artificial hip joint, used to replace a natural joint damaged by disease or wear and tear.

links

For more information on the environmental impact of extracting metals, see C1 3.6 Metallic issues.

Summary questions

1 Copy and complete using the words below:
corrode energy expensive high low oxide reactive strong
Aluminium and titanium alloys are useful as they are .... and have a .... density. Although aluminium is reactive, it does not .... because its surface is coated with a thin, tough layer of aluminium .... Titanium does not corrode because it is not very .... and also has its oxide layer to protect it. We use large amounts of .... in the extraction of both metals from their ores which makes them .... The large number of steps involved in the extraction of the metals also contributes to their .... cost.

2 Why is titanium used to make artificial hip joints?

3 **a** Explain the different reasons why carbon cannot be used to extract: **i** aluminium, or **ii** titanium.
b Name two processes in the extraction of aluminium that require large amounts of energy.

Key points

- Aluminium and titanium are useful because they resist corrosion.
- Aluminium requires the electrolysis of molten aluminium oxide to extract it as it is too reactive to reduce using carbon.
- Aluminium and titanium are expensive because extracting them from their ores involves many stages and requires large amounts of energy.

53

Answers to in-text questions

a Because it has a layer of aluminium oxide over its surface that prevents corrosion.
b By alloying it with other elements.
c It is unreactive and very strong at high temperatures.
d **i** Because aluminium extraction requires electrolysis
ii The reactive metals (Mg or Na) used in the extraction of titanium were obtained by electrolysis.

Summary answers

1 strong, low, corrode, oxide, reactive, energy, expensive, high

2 Because of its strength and resistance to corrosion.

3 **a** **i** Carbon is not reactive enough to reduce aluminium oxide/Carbon is less reactive than aluminium so cannot displace it from its compounds.
ii Carbon forms a compound with titanium that makes the metal brittle.
b The melting of aluminium oxide and the electrolysis of aluminium oxide.

C1 3.4 Extracting copper

Learning objectives

Students should learn:

- that copper can be extracted from copper-rich ores by smelting
- that copper can be extracted by different methods from low-grade ores
- that copper can be purified by electrolysis.

Learning outcomes

Most students should be able to:

- describe how to extract copper by smelting.
- explain why low-grade ores are now being exploited
- explain two methods for extracting copper from low-grade ores

Some students should also be able to:

- evaluate the different methods of copper extraction.

Support

- Students may need support in remembering the three main methods of copper extraction. It may be worth asking students to write a list of key words as a simple explanation of the method in each case.

Extend

- To extend students, they could consider why copper extraction techniques are changing. They should also consider which method they think is best and why.

Specification link-up: Chemistry C1.3

- Copper can be extracted from copper-rich ores by heating the ores in a furnace (smelting). The copper can be purified by electrolysis. The supply of copper-rich ores is limited. *[C1.3.1 f)]*
- New ways of extracting copper from low-grade ores are being researched to limit the environmental impact of traditional mining. Copper can be extracted by phytomining, or by bioleaching. *[C1.3.1 g)]*
- Copper can be obtained from solutions of copper salts by electrolysis or by displacement using scrap iron. *[C1.3.1 h)]*

Lesson structure

Starters

Copper photos – Provide students with images of copper being used for different purposes. Ask students to suggest what they all have in common [they are all made of copper]. To extend students, ask them to list what property of copper makes them useful for each purpose [e.g. copper wiring because it is an electrical conductor, copper cooking pans because it is a thermal conductor, copper jewellery because it is shiny]. *(5 minutes)*

Crossword – Use the internet to find a simple program to make a crossword of the key words used in the lesson [copper, extraction, smelting, mining, electrolysis, phytomining, bioleaching]. To support students, you may wish to give them the clues and the key words, so that they just need to match the key word with the clue. *(10 minutes)*

Main

- Explain to students that copper is a very important metal. However, our sources of copper-rich ores are running out. Therefore scientists are developing new ways to extract copper from low-grade ores.
- Students should gain hands-on experience of smelting using the 'extracting copper from malachite' practical (see 'Practical support' for more details). Students often really enjoy this experiment and should be encouraged to detail the reaction in a word equation.
- There are three main methods of copper extraction that students need to be aware of. Split the class into three groups. Each group is going to become the 'experts' on a different part of this spread; smelting, bioleaching and phytomining.
- Each group should produce an A5 set of notes about their topic, and a puzzle to check that the class have understood their work. They could also create a small presentation on the topic, maybe using PowerPoint.
- In the following lesson, each group could deliver their presentation and their handout can be given to the class. The 'experts' are then on hand to help with the activities.
- If you have the time to spare, students could grow brassica plants that have been frequently sprayed with copper compounds, or add copper compounds to the compost. The plants will absorb the copper, which can be extracted from the biomass by electrolysis or displacement. (See C1 3.6 'Practical support' for more details.)

Plenaries

Summarise – Ask the students to write one sentence to include the following copper extraction methods: smelting, bioleaching and phytomining. Students could be supported by being given part of the sentence and they have to complete it. To extend students, they could be encouraged to explain when and why electrolysis is used to obtain pure copper. *(5 minutes)*

Consequences – Give each student a piece of A5 paper. Ask the students to write down a use of copper at the top of the paper, fold it over and pass it to the right. Then they should write their favourite extraction method and why it is their favourite, fold and pass to the right. Then write a 'boring' fact about copper extraction, fold and pass to the right, and finally a fascinating fact (something new that they have learnt in the lesson) about copper extraction. Ask the student to unfold their current piece of paper and read the comments. Ask a few students to feedback to the class. *(10 minutes)*

Practical support

Extracting copper from malachite

Equipment and materials required

Bunsen burner and safety equipment, eye protection, boiling tube, boiling tube holder, filter funnel, conical flask, spatula, copper carbonate powder (harmful), 1.0 mol/dm^3 sulfuric acid (irritant) and filter paper. **For method A:** an iron nail. **For method B:** 2 × carbon electrode, 2 × wire connectors, 2 × crocodile clip, low voltage lab pack, small beaker, copper sulfate solution low voltage (less than 1.0 mol/dm^3 would not be harmful).

Details

Fill a boiling tube with two spatulas of copper carbonate and heat on a blue Bunsen flame until the powder has change colour from green to black. Allow the mixture to cool and then add sulfuric acid until the boiling tube is half full. Fold the filter paper and rest in the filter funnel. Put the filter funnel into the neck of the conical flask and filter the mixture. Discard the filter paper and residue. **For method A:** put an iron nail in the solution and observe. **For method B:** set up a series circuit with the lab pack and electrodes. Add some extra copper sulfate solution to the filtrate in a small beaker until it is half full. Submerge the electrodes and turn on the lab pack. Observe as copper is produced at the cathode (negative electrode).

Safety: Wear eye protection, ventilate the room well, especially if method B is followed. CLEAPSS Hazcard 98A Sulfuric acid – irritant, CLEAPSS 26 Copper carbonate – harmful, 27C Copper sulfate – harmful. Hands should be washed after the practical.

Metals and their uses

C1 3.4 Extracting copper

Learning objectives

- How is copper obtained from copper-rich ores?
- What methods can be used to obtain copper from low-grade ores?
- How is copper purified?

Figure 1 Mining copper ores can leave huge scars on the landscape. This is called open-cast mining. About 90% of copper comes from open-cast mines. Our supplies of copper-rich ores are a limited resource.

Extracting copper from copper-rich ores

We extract most of our copper from **copper-rich ores**. These are a limited resource and are in danger of running out.

There are two main methods used to remove the copper from the ore.

- In one method we use sulfuric acid to produce copper sulfate solution, before extracting the copper.
- The other process is called **smelting** (roasting). We heat copper ore very strongly in a furnace with air to produce crude copper.

For example, copper can be found in an ore called chalcocite. This contains copper(I) sulfide, Cu_2S. If we heat the copper(I) sulfide in air, it decomposes to give copper metal:

copper(I) sulfide + oxygen → copper + sulfur dioxide

Care has to be taken to avoid letting sulfur dioxide gas into the air. This gas causes acid rain. So chimneys are fitted with basic 'scrubbers' to neutralise the acidic gas.

Then we use the impure copper as the positive electrode in electrolysis cells to make pure copper. About 80% of copper is still produced by smelting.

a What chemical do we use to treat copper ore in order to form copper sulfate?

Smelting and purifying copper ore uses huge amounts of heat and electricity. This costs a lot of money and will have an impact on the environment.

links

For information on the charges on metal ions, look back at C1 1.4 Forming bonds.

Practical

Extracting copper from malachite

Malachite is a copper ore containing copper carbonate. To extract the copper we first heat the copper carbonate in a boiling tube. Thermal decomposition takes place. Copper oxide is left in the tube.

- Which gas is given off?

We then add dilute sulfuric acid to the copper oxide. Stopper and shake the tube. This makes copper sulfate solution. Filter off any excess black copper oxide in the solution.

To extract the copper metal, either

1. Put an iron nail into the copper sulfate solution

- What happens to the iron nail?

Or

2. Collect some extra copper sulfate solution and place it in a small beaker. Set up the circuit as shown in Figure 2. Turn the power on until you see copper metal collecting.

- Which electrode – the positive or the negative – does the copper form on?

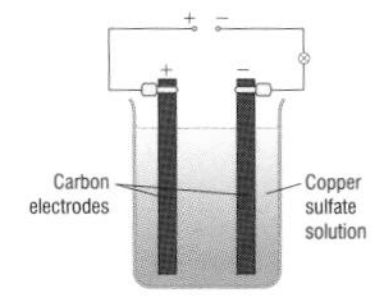

Figure 2 Extracting copper metal using electricity

54

Extracting copper

Metal ions are always positively charged. Therefore, in electrolysis they are attracted to the negative electrode. So metals are always deposited at the negative electrode. In industry the electrolysis is carried out in many cells running at once. This method gives the very pure copper needed to make electrical wiring. Electrolysis is also used to purify the impure copper extracted by smelting. In the industrial process, the electrolysis cells use copper electrodes.

The copper can also be extracted from copper sulfate solution in industry by adding scrap iron. Iron is more reactive than copper, so it can **displace** copper from its solutions:

iron + copper sulfate → iron sulfate + copper

Extracting copper from low-grade copper ores

Instead of extracting copper from our limited copper-rich ores, scientists are developing new ways to get copper from low grade ores. This would be uneconomical using traditional methods. We can now use bacteria (**bioleaching**) and even plants (**phytomining**) to help extract copper.

In phytomining, plants can absorb copper ions from low-grade copper ore as they grow. This could be on slag heaps of previously discarded waste from the processing of copper-rich ores. Then the plants are burned and the metals can be extracted from the ash. The copper ions can be 'leached' (dissolved) from the ash by adding sulfuric acid. This makes a solution of copper sulfate. Then we can use displacement by scrap iron and electrolysis to extract pure copper metal.

In bioleaching, bacteria feed on low-grade metal ores. By a combination of biological and chemical processes, we can get a solution of copper ions (called a 'leachate') from waste copper ore. Once again, we use scrap iron and electrolysis to extract the copper from the leachate.

About 20% of our copper comes from bioleaching. This is likely to increase as sources of copper-rich ores run out.

Bioleaching is a slow process so scientists are researching ways to speed it up. At present it can take years to extract 50% of the metal from a low-grade ore.

Did you know ...?

Copper metal is so unreactive that some samples of copper exist in nature as the element itself. It is found native. A huge copper boulder was discovered by a diver at the bottom of Lake Superior in North America. It was raised to the surface in 2001. It has a mass of about 15 000 kg.

links

For more information on the environmental impact of extracting metals and phytomining, see C1 3.6 Metallic issues.

Figure 3 In Australia Dr Jason Plumb looks for bacteria that can extract metals from ores. His search takes him to some exciting places – including volcanoes!

Summary questions

1 Copy and complete using the words below:

bacteria smelting electricity phytomining iron low sulfuric

Traditionally, copper can be extracted from some of its ores by heating (............). If copper ore is treated with .... acid, we get a solution of copper sulfate. We can obtain copper metal from this solution either by adding .... metal or by passing .... through the solution. Now new ways are being developed to extract copper using .... (bioleaching) or plants (............). These can extract the copper from ....-grade ores.

2 **a** Explain briefly two traditional ways of extracting copper metal.
b State an advantage of extracting copper using bacteria rather than traditional methods.
c Why can copper sometimes be found native (as the element itself)?
d When copper is purified by electrolysis, which electrode do you think that the pure copper collects at? Why?

3 Write a balanced chemical equation for the extraction of copper:
a from copper(I) sufide [H]
b from copper sulfate solution using scrap iron. [H]

Key points

- Most copper is extracted by smelting (roasting) copper-rich ores, although our limited supplies of ores are becoming more scarce.
- Copper can be extracted from copper solutions by electrolysis or by displacement using scrap iron. Electrolysis is also used to purify impure copper, e.g. from smelting.
- Scientists are developing ways to extract copper that use low-grade copper ores. Bacteria are used in bioleaching and plants in phytomining.

55

Further teaching suggestions

Classify

- Students could be asked to create a table to list the advantages and disadvantages of mining copper. Students could then be extended by classifying each statement as being economic, social or environmental. A copper mine case study can be found at www.mining-technology.com.

I'm most important

- Split the students into four groups and give each group a different key point. The groups could then create a persuasive argument why their key point is the most important and the debate could flow!

Answers to in-text questions

a sulfuric acid

Summary answers

1 smelting, sulfuric, iron, electricity, bacteria, phytomining, low

2 **a** Smelting in which a copper ore is heated to get copper. Adding sulfuric acid to get copper sulfate solution, which is electrolysed or has scrap iron added to it to displace the copper.

b We can use waste ores, previously dumped because they were too low-grade, reducing open-cast mining and scars on the landscape. Using bacteria will also conserve limited supplies of high-grade ores.

c Because it is an unreactive metal.

d It collects at the negative electrode as copper ions are positively charged (as are all metal ions).

3 **a** $Cu_2S + O_2 \rightarrow 2Cu + SO_2$

b $CuSO_4 + Fe \rightarrow FeSO_4 + Cu$

C1 3.5 Useful metals

Learning objectives

Students should learn:

- what transition metals are and their properties
- why copper is a useful metal
- that alloys are more useful than pure metals.

Learning outcomes

Most students should be able to:

- recognise transition metals and list their properties and some uses
- explain why copper is used in plumbing and for electrical wiring
- define an alloy and give an example.

Some students should also be able to:

- explain why alloys are often more useful than a pure metal.

Support

- Students may need reminding about where the metals are found in the periodic table. Give students a blank periodic table and encourage them to draw a thick black line between the metals and non-metals, colour in the non-metals and add a key. Also share with students that many metals end with -ium, e.g. calcium, magnesium, potassium.

Extend

- Students could be encouraged to find out the composition of different alloys and represent this information in a variety of appropriate formats, e.g. bar charts, tables and pie charts.
- You may wish to extend and build on the information about alloying gold. Encourage students to find out how hallmarks can be used to determine the alloy content of silver metal. Use an internet search engine and look for British hallmarks.

Specification link-up: Chemistry C1.3

- Most metals in everyday use are alloys. Pure copper, gold, iron and aluminium are too soft for many uses and so are mixed with small amounts of similar metals to make them harder for everyday use. *[C1.3.2 c)]*
- The elements in the central block of the periodic table are known as transition metals. Like other metals they are good conductors of heat and electricity and can be bent or hammered into shape. They are useful as structural materials and for making things that must allow heat or electricity to pass through them easily. *[C1.3.3 a)]*
- Copper has properties that make it useful for electrical wiring and plumbing. *[C1.3.3 b)]*
- Low density and resistance to corrosion make aluminium and titanium useful metals. *[C1.3.3 c)]*

Lesson structure

Starters

Copper poem – Ask students to write an acrostic poem with the first lines beginning with the letters from 'copper'. Students should incorporate the three different extraction techniques into their poem. *(5 minutes)*

Word search – Give students a word search containing a variety of elements from the periodic table (do not include the f-block, as students will not be given this in their GCSE examinations). Ask students to find only the transition metals. To support students, you may wish to create a word search that contains only examples of transition elements that they would be familiar with, such as gold, silver, iron. To extend students, you may wish to give students the symbols of the elements that they need to find. (There is no need to distinguish between the transition metals and d-block elements at GCSE level.) *(10 minutes)*

Main

- Show students some samples of transition metals in sealed Petri dishes. Ask them to use the periodic table to find the symbols of these elements, and where in the periodic table they are found.
- You may wish to extend students by also having some sealed samples of transition metal compounds. Students should conclude that transition metal compounds are coloured, and the same metal ion produces a similar coloured compound, no matter what the negative ion is that it is bonded to.
- Many different alloys exist, but the students need to focus on specific types. Split the class into three groups. Each group is going to become the 'experts' on a different part of this spread:
  - copper and its alloys
  - gold and its alloys
  - aluminium and its alloys.

  Ask the students to write an article for a students' science magazine about one of these topic areas. The article needs to include why the particular metal is an important material and list its properties and uses. Students should then explain why it is often alloyed, give at least one example of an alloy containing that element, its use and why the alloy is more suitable for that job than the pure metal. Students may also wish to produce their articles using a desktop publishing program.
- Then students should get into groups of three and read all of the articles, to gain an overview of the whole topic area. Students should then write a summary paragraph about transition metal alloys.

Plenaries

Key words – Place the key words, copper, alloy, transition metals, gold, aluminium, in a bag. Ask students to volunteer and pick out a key word, which they then need to explain in one sentence to the rest of the class. *(5 minutes)*

AfL (Assessment for Learning) – Give the students an examination question with a fictitious student's answer. Ask the students to work as individuals or in small groups to mark the questions. Then ask students to feed back on each question part saying what mark they would award and why. Support students by giving them an examination question from a Foundation Tier paper. Extend students by giving them a question from a Higher Tier paper. *(10 minutes)*

Further teaching suggestions

Internet research

- Show students some transition metal compounds and their solutions. Encourage them to find some uses [e.g. pottery glazes].

Which metal is best?

- Students could be given some uses of metal alloys, e.g. car bodywork, cooking pan, door handles. Students could suggest which metal alloy would be the most appropriate based on their cost and/or properties [e.g. car bodywork is made of steel as it is easily shaped, is not brittle and is strong].

Top of the class

- A variety of different answers to the same examination question on alloys and their uses could be given to groups of students. They should then order the answers from highest grade to lowest grade. Then give the students the mark scheme and ask them to reflect on their choices.

Metals and their uses

C1 3.5 Useful metals

Learning objectives

- What are transition metals and why are they so useful?
- Why is copper such a useful metal?
- Why are alloys more useful than pure metals?

Transition metals

In the centre of the periodic table there is a large block of metallic elements. They are called the **transition metals**. Many of them have similar properties. Like all metals, the transition metals are very good conductors of electricity and energy. They are strong but can also be bent or hammered into useful shapes.

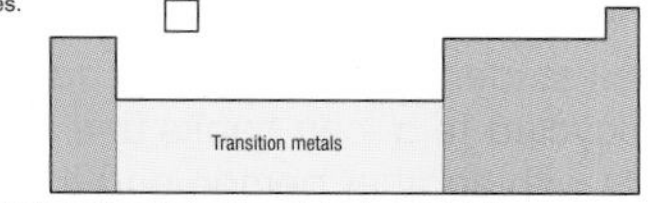

Figure 1 The position of the transition metals in the periodic table

a In which part of the periodic table do we find the transition metals?
b Name three properties of these elements.

The properties of the transition metals mean that we can use them in many different ways. You will find them in buildings and in cars, trains and other types of transport. Their strength makes them useful as building materials. We use them in heating systems and for electrical wiring because energy and electricity pass through them easily.

Figure 2 Copper is particularly useful because it is such a good conductor of electricity

Copper is a very useful transition metal. It can be bent but is still hard enough for plumbers to use as water tanks or pipes. Fortunately, it does not react with water. Copper also conducts electricity and energy very well. So it is ideal where we need:

- pipes that will carry water, or
- wires that will conduct electricity.

c What makes copper so useful for a plumber?

Figure 3 Transition metals are used in many different ways because of their useful properties

56

Useful metals

Copper alloys

Bronze was probably the first alloy made by humans, about 5500 years ago. It is usually made by mixing copper with tin. We use it to make ship's propellers because of its toughness and resistance to corrosion.

We make brass by alloying copper with zinc. Brass is much harder than copper but it is workable. It can be hammered into sheets and pressed into intricate shapes. This property is used to make musical instruments.

d Why are copper alloys more suitable for some uses than pure copper metal?

Figure 4 The Statue of Liberty in New York contains over 80 tonnes of copper

Aluminium alloys

Aluminium has a low density for a metal. It can be alloyed with a wide range of other elements. There are over 300 alloys of aluminium available. These alloys have very different properties. We can use some to build aircraft while others can be used as armour plating on tanks and other military vehicles.

Gold alloys

As with copper and iron, we can make gold and aluminium harder by adding other elements. We usually alloy gold with copper when we use it in jewellery. Pure gold wears away more easily than its alloy with copper. By varying the proportions of the two metals we also get different shades of 'gold' objects.

Figure 5 Alloying with copper makes gold more hardwearing. This is especially important in wedding rings, which many people wear most of the time.

e What property of aluminium makes it useful for making alloys in the aircraft industry?
f Apart from making gold harder, what else can alloying change?

Did you know ...?

The purity of gold is often expressed in 'carats', where 24-carat gold is almost pure gold (99.9%). If you divide the carat number by 24, you get the fraction of gold in your jewellery. So an 18-carat gold ring will contain ¾ (75%) gold.

Summary questions

1 Copy and complete using the words below:

aluminium brass aircraft bronze soft transition

The .... metals are found in the central block of the periodic table. Like pure iron, pure copper is too .... to be very useful. We can make copper harder by alloying it with tin to make ...., and with zinc to make ....

There are over 300 alloys of the low-density metal .... Many of these are used to make .... where strength is also an important property.

2 **a** Write a list of the properties of a typical transition metal.
b Why is copper metal used so much in plumbing?

3 Silver and gold are transition metals that conduct electricity even better than copper. Why do we use copper to make electric cables instead of either of these metals?

4 Why can aluminium alloys be used in so many different ways?

Key points

- The transition metals are found in the central block of elements in the periodic table.
- Transition metals have properties that make them useful for building and making things. For example, copper is used in wiring because of its high electrical conductivity.
- Copper, gold and aluminium are all alloyed with other metals to make them harder.

57

Answers to in-text questions

a In the central block (or between Groups 2 and 3).
b Three of: good conductors of heat/electricity, hard, tough, strong, can be bent/hammered into shapes, have very high melting points (except mercury).
c Used for pipes because it does not react with water and is quite easy to bend.
d They are harder or resistant to corrosion.
e Its low density.
f The colour of 'gold'.

Summary answers

1 transition, soft, bronze, brass, aluminium, aircraft

2 **a** Good conductors of energy/electricity, hard, tough, strong, can be bent/hammered into shapes, have very high melting points.
b Copper can be bent but is hard enough to be used to make pipes or tanks. It does not react with water.

3 Silver and gold are much more expensive than copper and less abundant in the Earth's crust.

4 Because there are so many different aluminium alloys, with a wide range of properties.

C1 3.6 Metallic issues

Learning objectives

Students should learn:

- the issues that arise in exploiting metal ores
- the benefits of recycling metals
- that there are advantages and disadvantages of using metals in construction.

Learning outcomes

Most students should be able to:

- list an advantage and a disadvantage of exploiting metal ores
- list reasons for the importance of recycling metals
- state a benefit and a drawback of using metals in construction.

Some students should also be able to:

- explain the benefits, in terms of social, economic and environmental, of exploiting metal ores
- explain the benefits, in terms of social, economic and environmental, of recycling metals.

Specification link-up: Chemistry C1.3

- We should recycle metals because extracting them uses limited resources and is expensive in terms of energy and effects on the environment. *[C1.3.1 j)]*
- New ways of extracting copper from low-grade ores are being researched to limit the environmental impact of traditional mining. Copper can be extracted by phytomining, or by bioleaching. *[C1.3.1 g)]*
- Copper can be obtained from solutions of copper salts by electrolysis or by displacement using scrap iron. *[C1.3.1 h)]*
- Consider and evaluate the social, economic and environmental impacts of exploiting metal ores, of using metals and of recycling metals. *[C1.3]*
- Evaluate the benefits, drawbacks and risks of using metals as structural materials. *[C1.3]*

Lesson structure

Starters

Rusting – Show students a rusty nail, or steel bar from reinforced concrete. Make a link to the construction industry. Ask students to list what needs to be present for rusting to happen [water and oxygen]. To extend students, ask them to write a balanced symbol equation for the reaction [$4Fe + 3O_2 \xrightarrow{water} 2Fe_2O_3$]. To support students, you could give them a word equation for rusting [iron + oxygen $\xrightarrow{water}$ hydrated iron oxide] and ask them to use this to help them. Note: rust is actually hydrated iron oxide or $Fe_2O_3 \cdot xH_2O$. *(5 minutes)*

Recycling think, pair square – Ask students to individually consider how they can personally get their waste metal recycled [e.g. can banks at a supermarket]. Then ask students to compare their thoughts with another person and then in small groups. Ask each group to give one new way that they can recycle metals. *(10 minutes)*

Main

- Students could work in small groups to discuss the tasks outlined in the activity 'Mining and processing metal ores – the issues'. The ideas from the group should be distilled to the most important points and two representatives should then feed back the thoughts to the rest of the group.
- Discuss the reasons that we recycle metals. Stress that extracting metals from their ores uses limited resources, uses a lot of energy and affects the environment. Stress the multi-stage process involved in extracting metals.
- Environmental issues, such as the recycling of metals, are a growing concern. Allow students to complete the activity 'Saving energy'.
- Each set of students could create a flow chart to explain the process and highlight the stages that require a lot of energy. Each student could then write a conclusion paragraph to explain how recycling metal could save energy.

Plenary

Thought experiment – Explain to students that they are going to make a prediction about the effect of rusting on steel reinforced concrete [rusting would affect the structure of the material and this affects the ability of the steel to withstand forces, so the structure would not be as strong]. Support students by giving them a writing frame to help make their prediction. Extend students by asking them to consider how the rate of rusting would be affected in a seaside structure compared to an inland structure [salt is a catalyst for rusting and so the seaside structure would rust quicker]. *(5 minutes)*

Support

- Students may find it difficult to generate some of the points to include in the 'Mining and processing metal ores – the issues' activity. Character cards could be made for the students that give a monologue from different characters containing some of the key points. These could be used to launch a discussion.

Extend

- Students could be extended by being asked to find out all of the different methods of protecting iron and steel from rust. Students could be encouraged to make a spider diagram of these methods.

Practical support

Phytomining experiment

Equipment and materials required

Compost, brassica seedlings, plant pot, scissors, string, fume cupboard, Bunsen burner and safety equipment, evaporating basin, eye protection, copper sulfate solution in a spray bottle (0.5 mol/dm^3) or copper sulfate powder (harmful), 1.0 mol/dm^3 dilute sulfuric acid (irritant), dropping pipette, 250 cm^3 conical flask, filter funnel, filter paper, stirring rod. Either magnesium powder (highly flammable) and a spatula or 2 × carbon electrodes, low-voltage power pack, 2 × connecting wire, 2 × crocodile clip, 100 cm^3 beaker.

Details

Plant the brassica seedling into a pot. Allow the plants to grow for a month. Spray the leaves of the plant with copper sulfate solution every day. An alternative to spraying is to mix copper sulfate powder into the soil used to pot the plant. Wearing gloves, harvest the leaves, and tie by their stems into a bunch and hang in the fume cupboard to dry. Wearing gloves, crush the plant leaves into an evaporating basin. Set up the Bunsen burner in the fume cupboard and use it to turn the dry leaves to ash. Allow the ash to cool and add enough acid so that the ash is easily covered and stir the mixture until the liquid turns blue. Filter the mixture into the conical flask. Then add a spatula of magnesium powder and stir. Observe as the copper is displaced from the copper sulfate. Alternatively, put the solution into a beaker, and submerge the ends of two carbon electrodes in it. Use the wires and crocodile clips to connect the electrodes in series to the power pack. Turn on the power and observe as copper is collected on the negative electrode.

Safety: Wear eye protection, ventilate room well especially if the brassica is to be sprayed. CLEAPSS Hazcard 98A Sulfuric acid – irritant. CLEAPSS Hazcard 27C Copper sulfate – harmful. Hands should be washed after the practical. Ashing of the brassica should be completed in the fume cupboard.

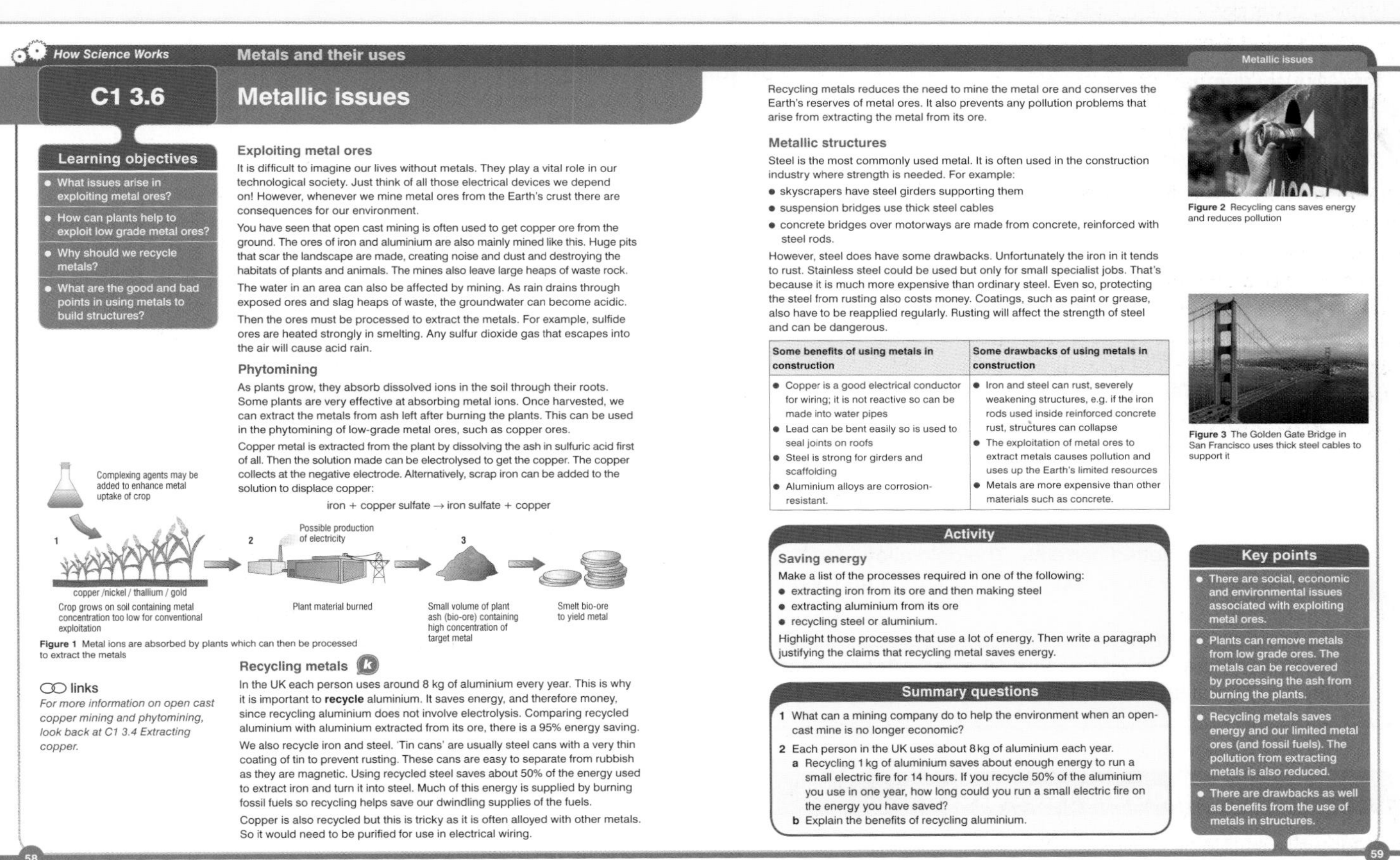

How Science Works — Metals and their uses

C1 3.6 Metallic issues

Learning objectives

- What issues arise in exploiting metal ores?
- How can plants help to exploit low grade metal ores?
- Why should we recycle metals?
- What are the good and bad points in using metals to build structures?

Exploiting metal ores

It is difficult to imagine our lives without metals. They play a vital role in our technological society. Just think of all those electrical devices we depend on! However, whenever we mine metal ores from the Earth's crust there are consequences for our environment.

You have seen that open cast mining is often used to get copper ore from the ground. The ores of iron and aluminium are also mainly mined like this. Huge pits that scar the landscape are made, creating noise and dust and destroying the habitats of plants and animals. The mines also leave large heaps of waste rock.

The water in an area can also be affected by mining. As rain drains through exposed ores and slag heaps of waste, the groundwater can become acidic.

Then the ores must be processed to extract the metals. For example, sulfide ores are heated strongly in smelting. Any sulfur dioxide gas that escapes into the air will cause acid rain.

Phytomining

As plants grow, they absorb dissolved ions in the soil through their roots. Some plants are very effective at absorbing metal ions. Once harvested, we can extract the metals from ash left after burning the plants. This can be used in the phytomining of low-grade metal ores, such as copper ores.

Copper metal is extracted from the plant by dissolving the ash in sulfuric acid first of all. Then the solution made can be electrolysed to get the copper. The copper collects at the negative electrode. Alternatively, scrap iron can be added to the solution to displace copper:

iron + copper sulfate → iron sulfate + copper

Figure 1 Metal ions are absorbed by plants which can then be processed to extract the metals

links

For more information on open cast copper mining and phytomining, look back at C1 3.4 Extracting copper.

Recycling metals

In the UK each person uses around 8 kg of aluminium every year. This is why it is important to **recycle** aluminium. It saves energy, and therefore money, since recycling aluminium does not involve electrolysis. Comparing recycled aluminium with aluminium extracted from its ore, there is a 95% energy saving.

We also recycle iron and steel. 'Tin cans' are usually steel cans with a very thin coating of tin to prevent rusting. These cans are easy to separate from rubbish as they are magnetic. Using recycled steel saves about 50% of the energy used to extract iron and turn it into steel. Much of this energy is supplied by burning fossil fuels so recycling helps save our dwindling supplies of the fuels.

Copper is also recycled but this is tricky as it is often alloyed with other metals. So it would need to be purified for use in electrical wiring.

58

Metallic issues

Recycling metals reduces the need to mine the metal ore and conserves the Earth's reserves of metal ores. It also prevents any pollution problems that arise from extracting the metal from its ore.

Metallic structures

Steel is the most commonly used metal. It is often used in the construction industry where strength is needed. For example:

- skyscrapers have steel girders supporting them
- suspension bridges use thick steel cables
- concrete bridges over motorways are made from concrete, reinforced with steel rods.

However, steel does have some drawbacks. Unfortunately the iron in it tends to rust. Stainless steel could be used but only for small specialist jobs. That's because it is much more expensive than ordinary steel. Even so, protecting the steel from rusting also costs money. Coatings, such as paint or grease, also have to be reapplied regularly. Rusting will affect the strength of steel and can be dangerous.

Some benefits of using metals in construction	Some drawbacks of using metals in construction
• Copper is a good electrical conductor for wiring; it is not reactive so can be made into water pipes • Lead can be bent easily so is used to seal joints on roofs • Steel is strong for girders and scaffolding • Aluminium alloys are corrosion-resistant.	• Iron and steel can rust, severely weakening structures, e.g. if the iron rods used inside reinforced concrete rust, structures can collapse • The exploitation of metal ores to extract metals causes pollution and uses up the Earth's limited resources • Metals are more expensive than other materials such as concrete.

Figure 2 Recycling cans saves energy and reduces pollution

Figure 3 The Golden Gate Bridge in San Francisco uses thick steel cables to support it

Activity

Saving energy

Make a list of the processes required in one of the following:

- extracting iron from its ore and then making steel
- extracting aluminium from its ore
- recycling steel or aluminium.

Highlight those processes that use a lot of energy. Then write a paragraph justifying the claims that recycling metal saves energy.

Summary questions

1 What can a mining company do to help the environment when an open-cast mine is no longer economic?

2 Each person in the UK uses about 8 kg of aluminium each year.
  a Recycling 1 kg of aluminium saves about enough energy to run a small electric fire for 14 hours. If you recycle 50% of the aluminium you use in one year, how long could you run a small electric fire on the energy you have saved?
  b Explain the benefits of recycling aluminium.

Key points

- There are social, economic and environmental issues associated with exploiting metal ores.
- Plants can remove metals from low grade ores. The metals can be recovered by processing the ash from burning the plants.
- Recycling metals saves energy and our limited metal ores (and fossil fuels). The pollution from extracting metals is also reduced.
- There are drawbacks as well as benefits from the use of metals in structures.

59

Further teaching suggestions

Charting costs

- Students could create a bar chart to compare the cost of producing 1 kg of titanium and aluminium from their ore with 1 kg of each respective metal from recycled material.

Photo of construction

- Give small groups of students a digital camera and allow them to photograph examples of uses of metals in construction around the school site. These images can then be shared with the rest of the class using a data projector.

Summary answers

1 The huge hole could be used as a landfill site for rubbish, which is then covered in soil and landscaped. Alternatively, it could be filled with water to make a reservoir/leisure activity centre.

2 a About 56 hours.
  b Less energy used in recycling, conserving the Earth's resources of fuels and ores. Less pollution from metal extraction, e.g. from open-cast mining of aluminium ore.

Summary answers

1 **a** A rock containing enough of a metal to make it economically worthwhile to extract the metal.
  b Describes a metal found in the Earth that has not combined with another element.
  c The removal of oxygen from a compound.

2 The alloy may be harder, stiffer, and it may be easier to work (shape into different objects).

3 **a** **i** bioleaching **ii** phytomining
  iii smelting **iv** electrolysis
  b bioleaching and phytomining

4 Brassica is grown on the polluted site. The plants are then harvested and burned, collecting the ash. This is processed, e.g smelted, to extract the metals.

5 E.g. extracting copper using bacteria or fungi is likely to use much less energy than conventional extraction techniques. One definite advantage is that it may be useful for extracting copper from ores that would otherwise remain uneconomic. We could also use waste ores, previously dumped because they were low grade, reducing open-cast mining and scars on the landscape.

A disadvantage is the longer length of time it takes to extract the metal. Also, the technology is at present undeveloped, so the true cost of this method of extraction is unknown.

6 **a** 869 kg **b** 23.9%
  c **i** steel/iron **ii** It is magnetic.

7 **a** copper or silver
  b Hearsay. The hardness increases up to 18 carat and only then decreases.
  c Gold alloy (carat) is the independent variable.
  d The hardness of the alloy is a continuous variable.
  e Suitable graph, which could be a line graph as both variables are continuous. However, as gold is sold in these carats, it might be more suitable to present the data as a bar chart. The axes should be fully labelled and the points plotted correctly. The gold alloy (carat) should be on the *x*-axis and the maximum hardness on the *y*-axis.
  f As the proportion of gold in the alloy increases, so the hardness increases up to about 18 carat. After this, as the alloy carat increases, the hardness decreases.

Kerboodle resources

Resources available for this chapter on Kerboodle are:

- Chapter map: Metals and their uses
- Animation: Extracting metals (C1 3.1)
- WebQuest: Find an extraction reaction – ore else! (C1 3.1)
- Practical: Extraction of metals (C1 3.1)
- Extension: Extracting zinc (C1 3.1 and C1 3.2)
- Practical: Displacement reactions of metals (C1 3.2)
- Practical: Electrolysis of copper sulfate solution (C1 3.4)
- Support: Transition metals are marvellous (C1 3.5)
- Viewpoint: Recycling metals: benefits and problems (C1 3.6)
- Interactive activity: Metals and their uses
- Revision podcast: Extracting copper
- Test yourself: Metals and their uses
- On your marks: Metals
- Examination-style questions: Metals and their uses
- Answers to examination-style questions: Metals and their uses

Metals and their uses: C1 3.1–C1 3.6

Summary questions

1 Write simple definitions for the following terms:
  a metal ore
  b native state
  c chemical reduction.

2 We can change the properties of metals by alloying them with other elements.

Write down ***three*** ways that a metal alloy may be different from the pure metal.

3 **a** What name is given to the method of extracting copper from an ore:
  i using bacteria
  ii using plants
  iii using heat
  iv using electricity?
  b Which methods in part **a** are being developed to extract copper from low-grade copper ores?

4 Describe how brassicas can be used to decontaminate 'brown-field' sites and recover the polluting metals. [H]

5 Carry out some research to find the advantages and disadvantages of using bioleaching to extract copper metal.

6 By the middle of the decade scrap car dealers are required to recover 95% of all materials used to make a car. The following table shows the metals we find in an average car:

Material	Average mass (kg)	% mass
Ferrous metal (steels)	780	68.3
Light non-ferrous metal (mainly aluminium)	72	6.3
Heavy non-ferrous metal (for example lead)	17	1.5

Other materials used include plastics, rubber and glass.
  a What is the average mass of metal in a car?
  b What percentage of a car's mass is made up of ***non-metallic*** materials?
  c **i** What is the main metal found in most cars?
  ii Which of this metal's properties allows it to be separated easily from other materials in the scrap from a car?

7 The following was overheard in a jeweller's shop:

"I would like to buy a 24-carat gold ring for my husband."

"Well madam, we would advise that you buy one which is a lower carat gold. It looks much the same but the more gold there is, the softer it is."

Is this actually the case? Let's have a look scientifically at the data.

Pure gold is said to be 24 carats. A carat is a twenty-fourth, so $24 \times \frac{1}{24} = 1$ or pure gold. So a 9-carat gold ring will have $\frac{9}{24}$ gold and $\frac{15}{24}$ of another metal, probably copper or sometimes silver. Most 'gold' sold in shops is therefore an alloy.

How hard the 'gold' is will depend on the amount of gold and on the type of metal used to make the alloy.

Here are some data on the alloys and the maximum hardness of 'gold'.

Gold alloy (carat)	Maximum hardness (BHN)
9	170
14	180
18	230
22	90
24	70

  a Which metals are used to alloy gold in jewellery?
  b The shop assistant said that 'the more gold there is, the less hard it is.' Was this based on science or was it hearsay? Explain your answer.
  c In this investigation which is the independent variable?
  d Which type of variable is 'the maximum hardness of the alloy' – continuous or categoric?
  e Plot a graph of the results.
  f What is the pattern in the results?

60

Practical suggestions

Practicals	AQA	k		
Comparing less reactive metals (gold, silver, copper) with more reactive metals, e.g. in acid.	✓		✓	
Heating metal oxides with carbon to compare reactivity, e.g. CuO, PbO, Fe_2O_3.	✓	✓		
Heating copper carbonate with charcoal to produce copper.	✓		✓	
Displacement reactions, e.g. $CuSO_4$ (aq) + Fe (using temperature sensors to investigate differences in metal reactivity).	✓	✓		
Investigation of the physical properties of metals and alloys …	✓		✓	
Electrolysis of copper sulfate solution using copper electrodes.	✓	✓		
Ignition tube demonstration of blast furnace – potassium permanganate, mineral wool plug, iron oxide mixed with carbon.	✓		✓	
Investigation of phytomining: growing brassica plants in compost with added copper sulfate or spraying brassica plants with copper sulfate solution, ashing the plants (fume cupboard), adding sulfuric acid to the ash, filtering and obtaining the metal from the solution by displacement or electrolysis.	✓		✓	

AQA Examination-style questions

1 Bicycle frames are often made from metal tubes. The metal tubes are produced using the steps in this list:
mining → concentrating → extracting → purifying → alloying → shaping
Match each of the following statements with the correct word from the list.

a The metal is produced using chemical reduction. (1)
b The metal is mixed with other metals to make it harder and stronger. (1)
c The metal ore is dug from the ground. (1)
d Waste rock is removed from the metal ore. (1)
e Other elements are removed from the metal. (1)

2 Choose the correct words from those shown to complete each sentence.

a Gold is found in the Earth as .... (1)
gold chloride gold metal gold oxide
b Iron is extracted by reacting iron oxide with .... (1)
carbon copper nitrogen
c Aluminium is extracted from aluminium oxide using .... (1)
combustion distillation electrolysis

3 Copper metal is used for electric wires. An alloy of copper, called brass, is used for pins and terminals of electric plugs.

a Copper metal is relatively soft and flexible. Give another reason why copper is used for electric wires. (1)
b Brass is an *alloy*. What is an *alloy*? (1)
c Open-cast mining of copper ore makes a very large hole.
i Suggest **one** environmental problem that is caused by open-cast mining of copper ore. (1)
ii Some copper ores contain copper sulfide, CuS. Copper sulfide in heated in air to produce copper and sulfur dioxide.
$CuS + O_2 \rightarrow Cu + SO_2$
Suggest **one** environmental problem caused by heating copper sulfide in air. (1)
d The amount of copper-rich ores is estimated to last only a few more years. New houses need several kilometres of copper wire.
i Explain why the need to use so much copper will cause a problem in the future. (1)
ii Suggest **two** ways in which society could overcome this problem. (2)

AQA, 2008

4 *In this question you will be assessed on using good English, organising information clearly and using specialist terms where appropriate.*
Most of the iron we use is converted into steels.
Describe and explain how the differences in the properties of the three main types of steel allow them to be used in different ways. (6)

5 Titanium is used in aircraft, ships and hip replacement joints. Titanium is as strong as steel but 45% lighter, and is more resistant to acids and alkalis.
Most titanium is produced from its ore, rutile (titanium oxide), by a batch process that takes up to 17 days.
Titanium oxide is reacted with chlorine to produce titanium chloride →
Titanium chloride is reacted with magnesium at 900°C in a sealed reactor for 3 days →
The reactor is allowed to cool, then opened and the titanium is separated from the magnesium chloride by hand.
Titanium reactors produce about 1 tonne of the metal per day.
Iron blast furnaces produce about 20000 tonnes of the metal per hour.

a Give **one** property of titanium that makes it more useful than steel for hip replacement joints. (1)
b Suggest **three** reasons why titanium costs more than steel. (3)

AQA, 2008

6 Phytomining uses plants to absorb metal compounds from the ground. It is often used on land that has been contaminated by normal mining. It involves these stages:
Sow seeds → grow plants → harvest plants → dry plants → burn plants → collect ash
The ash is then treated like a metal ore obtained by normal mining.

a Suggest **one** environmental advantage of phytomining compared with normal mining. (1)

The table shows information about some metals that are absorbed by plants used for phytomining.

Metal	Value of metal in £ per kg	Maximum mass of metal in plants in g per kg	Percentage (%) of metal in normal ore
Gold	25000	0.10	0.002
Nickel	17	38	2
Copper	4.9	14	0.5
Zinc	3.2	40	5
Lead	1.5	10	3

b The plants used for gold phytomining give a maximum yield of 20 tonnes of plants per hectare. Calculate the maximum value of the gold that can be recovered from 1 hectare. (2)
c One kilogram of plants used for nickel phytomining produces 150 g of ash.
What is the percentage of nickel in the ash? (2)
d Suggest reasons why phytomining has been used to produce gold, nickel and copper, but is only rarely used to produce zinc and lead. (4)

Examination-style answers

1 a extracting **b** alloying **c** mining
d concentrating **e** purifying *(5 marks)*

2 a gold metal **b** carbon **c** electrolysis *(3 marks)*

3 a good (electrical) conductor (accept low reactivity/ resistance to corrosion) *(1 mark)*

b A mixture of metals. *(1 mark)*

c i Any **one** from: eyesore, destruction of habitats, pollution of water, dust pollution, noise, traffic pollution *(1 mark)*

ii acid rain (accept sulfur dioxide is a pollutant) *(1 mark)*

d i because we are running out of copper ores or because copper ores are limited resources *(1 mark)*

ii Any **two** from: do not throw away copper/brass or put in landfill, reuse/recycle, use low grade copper ores, use other metals/materials/plastics in place of copper *(2 marks)*

4 Marks awarded for this answer will be determined by the Quality of Written Communication (QWC) as well as the standard of the scientific response.

There is a clear and detailed scientific description of three main types of steel, a difference in the properties of each, and how these are related to a different use for each type of steel. The answer shows almost faultless spelling, punctuation and grammar. It is coherent and in an organised, logical sequence. It contains a range of appropriate and relevant specialist terms used accurately. *(5–6 marks)*

There is a scientific description of two types of steel, and the difference in their properties related to their uses, or there is a description of three types of steel with either differences in properties or uses. There are some errors in spelling, punctuation and grammar. The answer has some structure and organisation. The use of specialist terms has been attempted, but not always accurately. *(3–4 marks)*

There is a brief description of at least two types of steel or two different uses of steels. The spelling, punctuation and grammar are very weak. The answer is poorly organised with almost no specialist terms and/or their use demonstrating a general lack of understanding of their meaning. *(1–2 marks)*

No relevant content. *(0 marks)*

Examples of chemistry points made in the response:

Name or description of main types of steel:

- Low-carbon steels – low percentage, less than 0.1% carbon
- Mild steels – 0.15–0.25% carbon
- High-carbon steels – up to 1.4% carbon
- Stainless steels – 10–20% of other metals, special steels, high-alloy steels, nickel-chromium steels or other named example
- Low-alloy steels – up to 5% of other metals
- Allow: carbon steels, low-alloy and high-alloy as three types.

Properties:

- Low-carbon steels – softer, more malleable, more easily shaped
- Mild steels – stronger, less easily bent
- High-carbon steels – harder, stronger, more brittle, less malleable
- Stainless steels – more resistant to corrosion, stronger
- Low-alloy steels – harder, stronger, more resistant to corrosion.

Uses:

- Low-carbon steels – wires, rivets, sheet metal
- Mild steels – general engineering purposes, reinforcing concrete
- High-carbon steels – tools, e.g. hammers, chisels, cutting tools
- Stainless steels – chemical and food industry, kitchens, cutlery
- Low-alloy steels – bridges, chains, armour plating, high-speed tools.

5 a Any **one** from: less dense/lighter, resistant to acids/alkalis/ chemicals (accept resits corrosion). *(1 mark)*

b Any **three** from: takes a long time to process, low abundance (of ore), small amount produced, batch process or blast furnace is continuous, more stages used, more energy used (per tonne of titanium, magnesium/ chlorine is expensive or produced by electrolysis, labour intensive (ignore simple references to cost/usefulness/ temperature or incorrect process). *(3 marks)*

6 a Any **one** from: No digging, less effect on landscape, can remove contamination/pollution/metals from soils. *(1 mark)*

b £50 000 (working showing 2000 g or 2 kg gold gains 1 mark) *(2 marks)*

c 25(.3)% (working showing 38/150 gains 1 mark) *(2 marks)*

d Any **four** from:
gold: high value of metal (makes it economic), low percentage of metal in normal ore
nickel: high value of metal, high percentage of nickel in ash
copper: low percentage in normal ore, normal ores running out
zinc: (relatively) low value of metal, (relatively) high percentage in normal ore
lead: (relatively) low value of metal, low mass of metal in plants, kills plants, only (cost effective when) used to clean up contaminated sites.
(Accept any other relevant reasons for specific metals or reasons correctly generalised to gold, nickel and copper or to zinc and lead). *(4 marks)*

C1 4.1 Fuels from crude oil

Learning objectives

Students should learn:

- what crude oil is
- what an alkane is
- how to represent alkanes.

Learning outcomes

Most students should be able to:

- recognise that crude oil is a mixture and state that it can be separated into fractions by distillation
- define and recognise simple alkanes
- write the correct chemical formula of an alkane represented by a structural formula.
- draw diagrams and write the formulae of simple alkanes when given named examples
- recall and use the formula C_nH_{2n+2} to give the formula of an alkane, when *n* is given.

Specification link-up: Chemistry C1.4

- Crude oil is a mixture of a very large number of compounds. *[C1.4.1 a)]*
- A mixture consists of two or more elements or compounds not chemically combined together. The chemical properties of each substance in the mixture are unchanged. It is possible to separate the substances in a mixture by physical methods including distillation. *[C1.4.1 b)]*
- Most of the compounds in crude oil consist of molecules made up of hydrogen and carbon atoms only (hydrocarbons). Most of these are saturated hydrocarbons called alkanes, which have the general formula C_nH_{2n+2}. *[C1.4.1 c)]*
- Alkane molecules can be represented in the following forms: C_2H_6. *[C1.4.2 a)]*

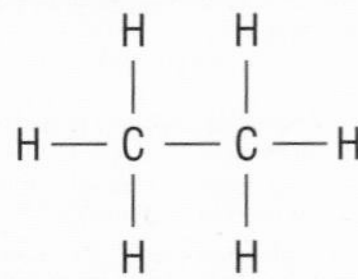

Lesson structure

Starters

What is the connection? – Show students a picture related to crude oil, e.g. a drilling rig, petroleum jelly barrier cream, petrol, a road surface or lubrication oil. Ask students to suggest the connection. [All of these products come from crude oil.] *(5 minutes)*

Oil is everywhere – It is essential for students to know that oil affects our everyday lives. Run a quick discussion about all the uses of oil, from fuel to fabrics, plastics and medicines. Ask students how they think the oil price affects all these commodities. Support students by showing them pictures of some of the products of oil to spark discussion. Extend students by giving them data on oil prices and looking to see if it correlates with the retail price index for the same period. *(10 minutes)*

Main

- Explain that crude oil is a mixture of hydrocarbons and ask the students to suggest, from their previous work in Key Stage 3, how mixtures can be separated into their components. Ask them to predict the method used to separate crude oil.
- Show the students the distillation equipment that has been set up and ask them to predict what will happen to the crude oil and why. Develop their ideas of a simple distillation into fractional distillation, using questions and answers. Demonstrate distillation. If this demonstration is done, it is important that it is not used to explain how continuous fractional distillation in a fractionating column works. In a fractionating column, the substances are being continuously evaporated and condensed on trays to achieve the separation.
- The fractions from this simple distillation could then be tested as detailed under 'Practical support' in the next lesson.
- Students often struggle with the idea that molecules are three dimensional. We often represent them in a 2-D format. Give each pair of students a molecular model kit. Then show students which atoms represent H and C, noting the size difference and the number of holes in each type of atom. Give the students the structural formulae of the first three or four alkanes and then set the students the task of making them and writing their molecular formulae. Ask them to list the similarities and differences between these molecules.

Plenaries

Observations – Draw the structural formula of butane on the board. Ask the students to write down as much information as they can about this molecule. To support students, you could give them a selection of key words to choose from. To extend students, encourage them to write the formula of the compound and consider some chemical reactions in which it could be involved, e.g. burning. *(5 minutes)*

Model a molecule – Split the class into groups. Give each group a different hydrocarbon name and some coloured sports vests or bibs. The students must use their bodies to demonstrate a hydrocarbon molecule. Those who represent carbon atoms need to sit down, so that four bonds (holding hands, feet/hands) can be created. *(10 minutes)*

Support

- You could support students by giving them a half-finished table detailing alkane names, molecular and displayed formulae. Each row should only have one missing piece of information. For very weak students this task could be a 'cut-and-stick' activity.

Extend

- You could extend students by introducing the idea that alkanes can be branched as well as straight chain molecules. Encourage students to look at different displayed formulae and generate their molecular formula. They should conclude that the branched chain saturated hydrocarbons still conform to the general formula of the alkanes and therefore are classified as alkanes.

Practical support

Fractional distillation demonstration

Equipment and materials required

A boiling tube with side arm, bung with a thermometer through it, four test tubes (as collecting tubes – ignition tubes can also be used to display small volumes of fractions), two beakers, ice/water mixture, boiling water, mineral wool, 'synthetic' crude oil (for the recipe see CLEAPSS Recipe Card 20 Crude oil or CLEAPSS 45A Hydrocarbons – aliphatic, or it can be purchased already made), Bunsen burner and safety equipment, eye protection, six watch glasses.

Details

Soak the mineral wool in the synthetic crude oil and place in the boiling tube. Fix the bung and ensure that the bulb of the thermometer is adjacent to the side arm. Put a collecting tube into an ice bath and the end of the side arm into the top of it. Gently heat the boiling tube with a Bunsen flame and notice when the temperature reading has stabilised (around 80 °C). When the temperature rises again, quickly change the current collecting tube for a new one. Repeat four times, collecting five fractions and leaving a residue in the boiling tube. During this practical, wear eye protection and complete in a well ventilated room. Each fraction can be collected at about every 50 °C up to about 300 °C. The residues will remain on the mineral wool, making the sixth fraction. The fractions can be ignited. Tip them onto mineral wool on a watch glass. Then ignite the mineral wool, taking great care.

Safety: Tie back hair and loose clothing. CLEAPSS Hazcard 45A Hydrocarbons.

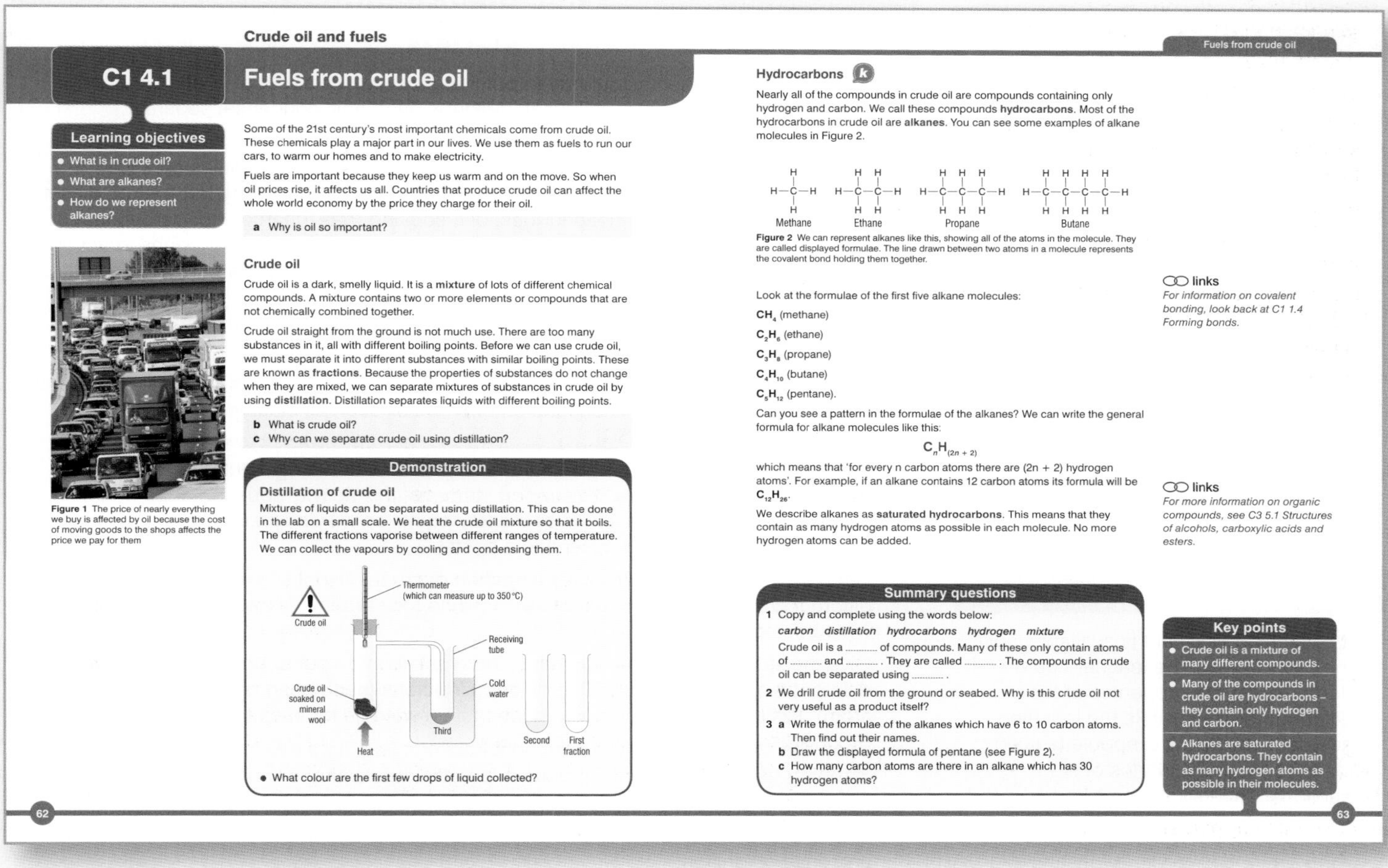

Crude oil and fuels

C1 4.1 Fuels from crude oil

Learning objectives

- What is in crude oil?
- What are alkanes?
- How do we represent alkanes?

Some of the 21st century's most important chemicals come from crude oil. These chemicals play a major part in our lives. We use them as fuels to run our cars, to warm our homes and to make electricity.

Fuels are important because they keep us warm and on the move. So when oil prices rise, it affects us all. Countries that produce crude oil can affect the whole world economy by the price they charge for their oil.

a Why is oil so important?

Figure 1 The price of nearly everything we buy is affected by oil because the cost of moving goods to the shops affects the price we pay for them

Crude oil

Crude oil is a dark, smelly liquid. It is a **mixture** of lots of different chemical compounds. A mixture contains two or more elements or compounds that are not chemically combined together.

Crude oil straight from the ground is not much use. There are too many substances in it, all with different boiling points. Before we can use crude oil, we must separate it into different substances with similar boiling points. These are known as **fractions**. Because the properties of substances do not change when they are mixed, we can separate mixtures of substances in crude oil by using **distillation**. Distillation separates liquids with different boiling points.

b What is crude oil?
c Why can we separate crude oil using distillation?

Demonstration

Distillation of crude oil

Mixtures of liquids can be separated using distillation. This can be done in the lab on a small scale. We heat the crude oil mixture so that it boils. The different fractions vaporise between different ranges of temperature. We can collect the vapours by cooling and condensing them.

Thermometer (which can measure up to 350 °C)
Crude oil
Receiving tube
Crude oil soaked on mineral wool
Cold water
Heat
Third
Second
First fraction

- What colour are the first few drops of liquid collected?

Hydrocarbons

Nearly all of the compounds in crude oil are compounds containing only hydrogen and carbon. We call these compounds **hydrocarbons**. Most of the hydrocarbons in crude oil are **alkanes**. You can see some examples of alkane molecules in Figure 2.

Methane Ethane Propane Butane

Figure 2 We can represent alkanes like this, showing all of the atoms in the molecule. They are called displayed formulae. The line drawn between two atoms in a molecule represents the covalent bond holding them together.

links *For information on covalent bonding, look back at C1 1.4 Forming bonds.*

Look at the formulae of the first five alkane molecules:

CH_4 (methane)

C_2H_6 (ethane)

C_3H_8 (propane)

C_4H_{10} (butane)

C_5H_{12} (pentane).

Can you see a pattern in the formulae of the alkanes? We can write the general formula for alkane molecules like this:

$$C_nH_{(2n+2)}$$

which means that 'for every n carbon atoms there are (2n + 2) hydrogen atoms'. For example, if an alkane contains 12 carbon atoms its formula will be $C_{12}H_{26}$.

We describe alkanes as **saturated hydrocarbons**. This means that they contain as many hydrogen atoms as possible in each molecule. No more hydrogen atoms can be added.

links *For more information on organic compounds, see C3 5.1 Structures of alcohols, carboxylic acids and esters.*

Summary questions

1 Copy and complete using the words below:
carbon distillation hydrocarbons hydrogen mixture
Crude oil is a .... of compounds. Many of these only contain atoms of .... and .... They are called .... The compounds in crude oil can be separated using ....

2 We drill crude oil from the ground or seabed. Why is this crude oil not very useful as a product itself?

3 **a** Write the formulae of the alkanes which have 6 to 10 carbon atoms. Then find out their names.
b Draw the displayed formula of pentane (see Figure 2).
c How many carbon atoms are there in an alkane which has 30 hydrogen atoms?

Key points

- Crude oil is a mixture of many different compounds.
- Many of the compounds in crude oil are hydrocarbons – they contain only hydrogen and carbon.
- Alkanes are saturated hydrocarbons. They contain as many hydrogen atoms as possible in their molecules.

Further teaching suggestions

Key words

- Ask the students to try to determine the most important word from the lesson that would help them remember all of the key points. Encourage different students to justify their choice, e.g. **alkanes** as these are saturated hydrocarbons and are contained in crude oil.

Answers to in-text questions

a Oil affects everything we do – heating and lighting our homes, transport and the goods we buy. (Also feedstock for the chemical industry.)

b Crude oil is a mixture of chemical compounds.

c Because the properties of the individual compounds in the mixture remain the same when they are mixed. Distillation relies on the different boiling points of these compounds.

Summary answers

1 mixture, carbon (hydrogen), hydrogen (carbon), hydrocarbons, distillation

2 Because there are too many substances in it.

3 **a** C_6H_{14} (hexane), C_7H_{16} (heptane), C_8H_{18} (octane), C_9H_{20} (nonane), $C_{10}H_{22}$ (decane).

b

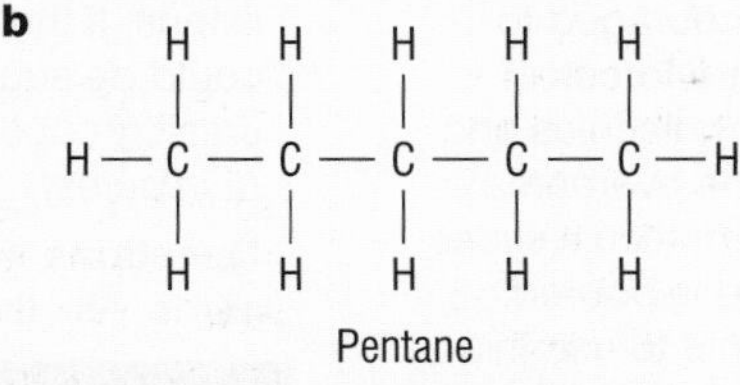

c Fourteen carbon atoms.

C1 4.2 Fractional distillation

Learning objectives

Students should learn:

- that crude oil is separated into fractions using fractional distillation
- the properties of each fraction and how they relate to the size of the molecules
- which fractions make useful fuels and why.

Learning outcomes

Most students should be able to:

- state that crude oil is separated into fractions by fractional distillation
- list how the properties change from small chain fractions to long chain fractions
- state which fractions are useful fuels.

Some students should also be able to:

- explain the key steps involved in fractional distillation
- relate the trend in properties to molecular size.

Support

- If a teaching assistant is available, split the class in two. The teacher could demonstrate the properties of the different fractions, while the teaching assistant shows the ampoule (sealed glass container) samples of the fractions. Then rotate the groups.
- Students may need some support with remembering the different displayed formulae. You could display these on the board from small to larger. Then draw an arrow from small to large molecules. Ask students to state the trend – the colour darkens, viscosity increases and ease of lighting reduces.

Extend

- Students should be encouraged to understand that there are forces of attraction between the molecules and it is these that have to be overcome and reformed as each molecule slides over others as the liquid is poured. Then encourage students to use this information to explain why larger molecules are more viscous. They could explain this model using a series of cartoons.

Specification link-up: Chemistry C1.4

- The many hydrocarbons in crude oil may be separated into fractions, each of which contains molecules with a similar number of carbon atoms, by evaporating the oil and allowing it to condense at a number of different temperatures. This process is fractional distillation. *[C1.4.2 b)]*
- Some properties of hydrocarbons depend on the size of their molecules. These properties influence how hydrocarbons are used as fuels. *[C1.4.2 c)]*

Lesson structure

Starters

Distil the order – Students could try to put these key words about distillation in order: condense, mixture, separate, boil, heat. [Mixture, heat, boil, condense, separate.] *(5 minutes)*

Fuel list – Ask students to consider what they have used today that relies on a fuel. Ask various students to give their thoughts to the class. For example, transport (petrol, diesel and more recently autogas), heating (gas, oil), cooking (gas, charcoal for barbecues) or lighting (gas, oil). Students could be supported by showing them photographs to act as a stimulus. Extend students by asking them to explain how electricity is related to fuel use. *(10 minutes)*

Main

- To contrast fractional distillation in a school lab with what happens in industry, students could watch a video on the separation of crude oil such as *Industrial Chemistry for Schools and Colleges* (RSC).
- Then the students could be given a drawing of a fractionating column, to which they would add their own notes. To support students, this activity could be adapted into a cut-and-stick exercise, which gives the key points as words and diagrams on a piece of paper and the students assemble a poster.
- Often students do not know what a fraction of crude oil looks like. Ampoules of the different crude oil fractions could be shown to the students (available from BP: www.bpes.com).
- Give the students some string beads. Then ask them to cut or break links to give different lengths to represent the different fractions (each bead represents a carbon atom). Put these onto a demonstration table – one pile for each fraction.
- Show the students samples of different fractions (e.g. light a Bunsen burner or show a camping gas bottle, sample of octane, paraffin, lubricating oil and wax).
- Ask the students to comment on the colour, viscosity and state at room temperature. Then try to ignite some of each fraction and ask the students to note the flame colour and ease of ignition.
- Ask the students to compare the properties with the length of the molecule. This task could be written up in the form of a results table.
- Link here to Controlled Assessment – relationships between variables.

Plenaries

Improvisation – Ask for volunteers to talk about a key word for 30 seconds without 'erms' or pauses. Ask the student to talk about hydrocarbons, fractions or viscosity – without any preparation! The volunteer is given a word and starts talking while the rest of the class listens. If there are any misconceptions, ask the other students to pick them out. Students could be supported by allowing them a few minutes of preparation time and working in small groups. Students could be extended by being asked to use a prop in their talk. *(5 minutes)*

Questions and answers – Give each student either a question or an answer on index cards. Ask the students to find their partners. *(10 minutes)*

Practical support

Comparing fractions (demonstration)

Equipment and materials required

Three samples of different alkanes (flammable) to represent different fractions (choose fractions with very different hydrocarbon chain lengths such as hexane, paraffin and candle wax), eye protection, four watch glasses, evaporting dish, mineral wool, dropping pipette, heatproof mat, matches, molecular model kits.

Details

Pour each alkane onto a separate watch glass, starting with the smallest carbon chain. Ask students to comment on the viscosity, then the colour. Students should conclude that the longer the hydrocarbon chain, the more viscous the liquid and the darker it is. Then, try lighting a small amount of the alkane soaked into mineral wool, in an evaporating dish. Ask students to comment on the ease of lighting and the colour of the flame. Students should conclude that the flame is 'dirtier' and the hydrocarbons are more difficult to light as the hydrocarbon's length increases. You might wish to have molecular models of each of the fractions to aid students in making the link between hydrocarbon chain length and properties.

Safety: Ensure the stock bottles of alkanes are closed before lighting matches. CLEAPSS Hazcard 45B Paraffin – harmful.

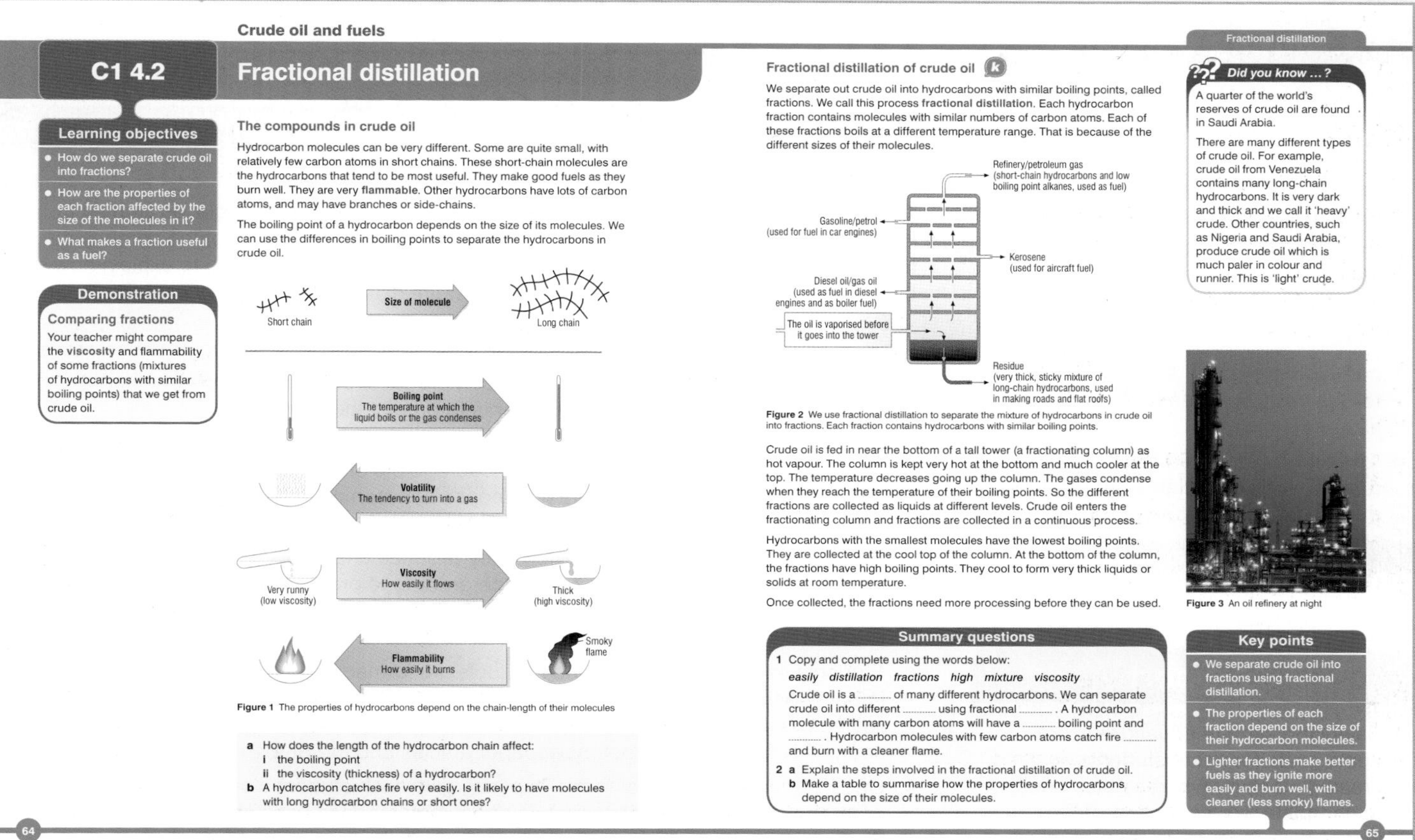

Crude oil and fuels

C1 4.2 Fractional distillation

Learning objectives
- How do we separate crude oil into fractions?
- How are the properties of each fraction affected by the size of the molecules in it?
- What makes a fraction useful as a fuel?

Demonstration

Comparing fractions

Your teacher might compare the **viscosity** and flammability of some fractions (mixtures of hydrocarbons with similar boiling points) that we get from crude oil.

The compounds in crude oil

Hydrocarbon molecules can be very different. Some are quite small, with relatively few carbon atoms in short chains. These short-chain molecules are the hydrocarbons that tend to be most useful. They make good fuels as they burn well. They are very **flammable**. Other hydrocarbons have lots of carbon atoms, and may have branches or side-chains.

The boiling point of a hydrocarbon depends on the size of its molecules. We can use the differences in boiling points to separate the hydrocarbons in crude oil.

Figure 1 The properties of hydrocarbons depend on the chain-length of their molecules

a How does the length of the hydrocarbon chain affect:
 i the boiling point
 ii the viscosity (thickness) of a hydrocarbon?
b A hydrocarbon catches fire very easily. Is it likely to have molecules with long hydrocarbon chains or short ones?

64

Fractional distillation

Fractional distillation of crude oil

We separate out crude oil into hydrocarbons with similar boiling points, called fractions. We call this process **fractional distillation**. Each hydrocarbon fraction contains molecules with similar numbers of carbon atoms. Each of these fractions boils at a different temperature range. That is because of the different sizes of their molecules.

Figure 2 We use fractional distillation to separate the mixture of hydrocarbons in crude oil into fractions. Each fraction contains hydrocarbons with similar boiling points.

Crude oil is fed in near the bottom of a tall tower (a fractionating column) as hot vapour. The column is kept very hot at the bottom and much cooler at the top. The temperature decreases going up the column. The gases condense when they reach the temperature of their boiling points. So the different fractions are collected as liquids at different levels. Crude oil enters the fractionating column and fractions are collected in a continuous process.

Hydrocarbons with the smallest molecules have the lowest boiling points. They are collected at the cool top of the column. At the bottom of the column, the fractions have high boiling points. They cool to form very thick liquids or solids at room temperature.

Once collected, the fractions need more processing before they can be used.

Did you know ...?

A quarter of the world's reserves of crude oil are found in Saudi Arabia.

There are many different types of crude oil. For example, crude oil from Venezuela contains many long-chain hydrocarbons. It is very dark and thick and we call it 'heavy' crude. Other countries, such as Nigeria and Saudi Arabia, produce crude oil which is much paler in colour and runnier. This is 'light' crude.

Figure 3 An oil refinery at night

Summary questions

1 Copy and complete using the words below:
easily distillation fractions high mixture viscosity
Crude oil is a .... of many different hydrocarbons. We can separate crude oil into different .... using fractional .... A hydrocarbon molecule with many carbon atoms will have a .... boiling point and .... Hydrocarbon molecules with few carbon atoms catch fire .... and burn with a cleaner flame.

2 **a** Explain the steps involved in the fractional distillation of crude oil.
 b Make a table to summarise how the properties of hydrocarbons depend on the size of their molecules.

Key points
- We separate crude oil into fractions using fractional distillation.
- The properties of each fraction depend on the size of their hydrocarbon molecules.
- Lighter fractions make better fuels as they ignite more easily and burn well, with cleaner (less smoky) flames.

65

Further teaching suggestions

Red diesel

- Ask students to discover the difference between red diesel and 'normal' diesel. [Brown diesel is used in cars, but red diesel is used in working vehicles, e.g. tractors. The fuel is the same but a dye is added, as red diesel is tax-free.]

Fraction names

- Ask students to find out all the different names used for each fraction, e.g. residue may also be called bitumen.

Alternate fuels

- Students could research alternatives to fossil fuels, e.g. nuclear power, hydrogen fuel cells, renewable resources.

Viscosity demonstration

- Show, by pouring them, that crude oils from different oil fields have different viscosities. Ask the students to suggest why this is so. [Different crude oils have different proportions of each type of hydrocarbon.]

Homologous series

- Students could find out the names and work out the formulae for the first 10 alkanes. Other homologous series could be considered and compared to alkanes, e.g. alkenes.

Answers to in-text questions

a **i** and **ii** Increasing hydrocarbon chain length *increases* both boiling point and viscosity of a hydrocarbon.

b short hydrocarbon chains

Summary answers

1 mixture, fractions, distillation, high, viscosity, easily

2 **a** Hot vapour enters the fractionating column which is hottest at the bottom. Vapours rise, condensing at different levels as they reach their boiling point temperature.

b Table to summarise properties in Figure 1 in C1 4.2 in Student Book.

C1 4.3 Burning fuels

Learning objectives

Students should learn:

- the combustion products formed from the complete combustion of fuels
- the pollutants produced when we burn fuels.

Learning outcomes

Most students should be able to:

- write word equations for the complete combustion of hydrocarbons
- describe differences between incomplete and complete combustion
- list pollutants formed when we burn fuels.

Some students should also be able to:

- complete balanced symbol equations for the complete combustion of simple alkanes
- explain how nitrogen oxides, sulfur dioxide and particulates are produced during the combustion process.

Support

- Balancing symbol equations is not appropriate for some students. Instead, create an activity whereby they need to complete word equations. In the first section, they must always write the word 'oxygen', in the second section of the activity, they must always write the words 'water' and 'carbon dioxide'.

Extend

- You could extend students by asking them to write balanced symbol equations for the incomplete combustion of simple alkanes. For example, they could give one equation for the formation of carbon monoxide and water, and another for carbon plus water.
- You could further extend students by asking them to find out why carbon monoxide is described as toxic. [Carbon monoxide gas binds to the haemoglobin in blood better than oxygen. When incomplete combustion happens, the carbon monoxide is inhaled and slowly reduces the oxygen-carrying ability of the blood, leading to tiredness and ultimately death.]

Specification link-up: Chemistry C1.4

- Most fuels, including coal, contain carbon and/or hydrogen and may also contain some sulfur. The gases released into the atmosphere when a fuel burns may include carbon dioxide, water (vapour), carbon monoxide, sulfur dioxide and oxides of nitrogen. Solid particles (particulates) may also be released. *[C1.4.3 a)]*
- The combustion of hydrocarbon fuels releases energy. During combustion the carbon and hydrogen in the fuels are oxidised. *[C1.4.3 b)]*
- Evaluate the impact on the environment of burning hydrocarbon fuels. *[C1.4]*

Lesson structure

Starter

Triangle – Ask students to recall and draw the combustion (fire) triangle, as studied in Key Stage 3. Students could be supported by being given an outline of the combustion triangle to fill in each side. Students could be extended by being asked to explain how a fire blanket, a CO_2 fire extinguisher and a water fire extinguisher work. *(5 minutes)*

Main

- Have the demonstration of the combustion products of hydrocarbons (methane) practical already set up. Ask the students to predict the products.
- Students probably have not considered the combustion process that takes place within Bunsen burners in any detail. Encourage the students to experiment with the Bunsen flame to observe the differences between complete and incomplete combustion. See 'Practical support' for more details.
- When hydrocarbons are combusted, they produce chemicals that are pollutants. Split the class into six groups. Give each group one product of combustion to become 'experts' in from the following list: carbon dioxide, water, carbon monoxide, particulates, sulfur dioxide and oxide of nitrogen.
- Students should find out how the products are formed, how they could have a negative impact on the environment and/or human health and how their production can be limited or stopped from entering the atmosphere. Students should use secondary sources such as the internet, newspaper articles and the Student Book to help them.
- Once students have become the 'experts', give each student a summary table. Re-order the students so each team has one expert for each product of combustion. Students should work in their new teams to complete their own summary table.

Plenaries

Equations – Ask students to complete the following equations:

Wax + oxygen → [carbon dioxide] + water

Petrol + [oxygen] → carbon dioxide + [water] + carbon + [carbon monoxide]

You could support students by giving them the missing words that they use to complete the equations. Extend students by asking them to complete balanced equations:

$CH_4 + 2O_2 \rightarrow [CO_2] + [2H_2O]$

$[6]CH_4 + [8]O_2 \rightarrow CO_2 + [12]H_2O + [2]CO + [3]C$ *(5 minutes)*

Summarise – Split the class into groups and give each team a different topic: combustion; nitrogen oxides; sulfur dioxide; particulates. They could develop a sentence that summarises what they have learned about their topic in the lesson. *(10 minutes)*

Answers to in-text questions

a carbon dioxide and water (vapour)

b methane + oxygen → carbon dioxide + water

c sulfur

d acid rain

Practical support

Products of combustion

Equipment and materials required

A candle, a small Bunsen burner and safety equipment, eye protection, a glass funnel, a boiling tube, a U-tube, limewater (irritant), ice bath, selection of delivery tubes, a water pump, two bungs with holes in (for the delivery tubes), one bung with two holes in, rubber tubing, matches, three stands, bosses and clamps, beaker, ice.

Details

Place the Bunsen burner or candle onto the heatproof mat, invert the glass funnel and clamp into position about 2 cm above the top of the candle. Using a small piece of rubber tubing, connect an 'n'-shaped delivery tube to the filter funnel. Put the other end through a bung. Mount a U-tube in a large beaker of ice and put in a few pieces of cobalt chloride paper, in one end and fit the bung connected to the funnel. Put a bung and delivery tube in the other end of the U-tube and connect it to a boiling tube of limewater (irritant). The boiling tube bung should have two delivery tubes through it. The final tube should be connected to the water pump. Turn on the water tap for the pump, and light the Bunsen burner. Water should condense and coiled in the bottom of the U-tube and the limewater should turn cloudy, indicating carbon dioxide is produced.

Safety: Eye protection should be worn. Wash hands after handling cobalt chloride paper. CLEAPSS Hazcard 18 Limewater – irritant.

Investigating combustion

Equipment and materials required

Bunsen burner and safety equipment, boiling tube, boiling tube holder, boiling tube rack, water, stop watch, 10 cm^3 measuring cylinder, eye protection.

Details

Add 10 cm^3 of tap water to a boiling tube. Hold the boiling tube just above the gas cone and time how long it takes the water to boil. Once boiling, remove the boiling tube from the flame, return the flame to the safety flame and allow the boiling tube to cool in the rack. Visually inspect the outside of the boiling tube and notice how clean it looks. Repeat the experiment with the yellow flame. This time it should take longer for the water to boil and the boiling tube will be covered in a black sooty deposit.

Safety: Eye protection should be worn. Glassware will become hot in the flames. Allow boiling tubes to cool in the rack.

Crude oil and fuels

C1 4.3 Burning fuels

Learning objectives

- What are the products of combustion when we burn fuels in a good supply of air?
- What pollutants are produced when we burn fuels?

The lighter fractions from crude oil are very useful as fuels. When hydrocarbons burn in plenty of air they release energy. The reaction produces two new substances – carbon dioxide and water.

For example, when propane burns we can write:

propane + oxygen → carbon dioxide + water

or

$$C_3H_8 + 5O_2 \rightarrow 3CO_2 + 4H_2O$$

The carbon and hydrogen in the fuel are **oxidised** completely when they burn like this. 'Oxidised' means adding oxygen in a chemical reaction in which oxides are formed.

links

For information on useful fractions from crude oil, look back at C1 4.2 Fractional distillation.

Figure 1 On a cold day we can often see the water produced when fossil fuels burn

Practical

Products of combustion

We can test the products given off when a hydrocarbon burns as shown in Figure 2.

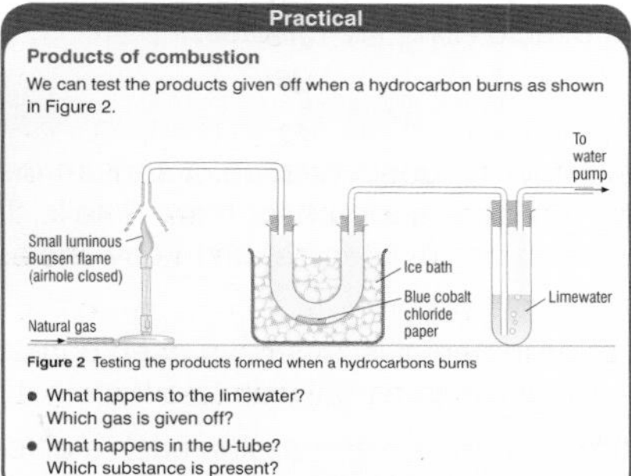

Figure 2 Testing the products formed when a hydrocarbons burns

- What happens to the limewater? Which gas is given off?
- What happens in the U-tube? Which substance is present?

a What are the names of the two substances produced when hydrocarbons burn in plenty of air?
b Methane is the main gas in natural gas. Write a word equation for methane burning in plenty of air.

Pollution from fuels

All fossil fuels – oil, coal and natural gas – produce carbon dioxide and water when they burn in plenty of air. But as well as hydrocarbons, these fuels also contain other substances. Impurities containing sulfur found in fuels cause us major problems.

All fossil fuels contain at least some sulfur. This reacts with oxygen when we burn the fuel. It forms a gas called **sulfur dioxide**. This gas is poisonous. It is also acidic. This is bad for the environment, as it is a cause of acid rain. Sulfur dioxide can also cause engine corrosion.

c When fuels burn, what element present in the impurities in a fossil fuel may produce sulfur dioxide?
d Which pollution problem does sulfur dioxide gas contribute to?

When we burn fuels in a car engine, even more pollution can be produced.

- When there is not enough oxygen inside an engine, we get **incomplete combustion**. Instead of all the carbon in the fuel turning into carbon dioxide, we also get **carbon monoxide** gas (CO) formed.

  Carbon monoxide is a poisonous gas. Your red blood cells pick up this gas and carry it around in your blood instead of oxygen. So even quite small amounts of carbon monoxide gas are very bad for you.
- The high temperature inside an engine also allows the nitrogen and oxygen in the air to react together. This reaction makes **nitrogen oxides**. These are poisonous and can trigger some people's asthma. They also cause acid rain.
- Diesel engines burn hydrocarbons with much bigger molecules than petrol engines. When these big molecules react with oxygen in an engine they do not always burn completely. Tiny solid particles containing carbon and unburnt hydrocarbons are produced. These **particulates** get carried into the air. Scientists think that they may damage the cells in our lungs and even cause cancer.

Burning fuels

Figure 3 A combination of many cars in a small area and the right weather conditions can cause smog to be formed. This is a mixture of SMoke and fOG.

Summary questions

1 Copy and complete using the words below:

monoxide carbon nitrogen oxidised particulates sulfur water

When hydrocarbons burn in a good supply of air, ... dioxide and .... are made, as the carbon and hydrogen in the fuel are .... As well as these compounds other substances such as .... dioxide may be made which causes acid rain. Other pollutants that may be formed include .... oxides, carbon .... and ....

2 Explain how **a** sulfur dioxide **b** nitrogen oxides and **c** particulates are produced when fuels burn in vehicles.

3 **a** Natural gas is mainly methane (CH_4). Write a balanced symbol equation for the complete combustion of methane. [H]
b When natural gas burns in a faulty gas heater it can produce carbon monoxide (and water). Write a balanced symbol equation to show this reaction. [H]

Key points

- When we burn hydrocarbon fuels in plenty of air the carbon and hydrogen in the fuel are completely oxidised. They produce carbon dioxide and water.
- Sulfur impurities in fuels burn to form sulfur dioxide which can cause acid rain.
- Changing the conditions in which we burn hydrocarbon fuels can change the products made.
- In insufficient oxygen, we get poisonous carbon monoxide gas formed. We can also get particulates of carbon (soot) and unburnt hydrocarbons, especially if the fuel is diesel.
- At the high temperatures in engines, nitrogen from the air reacts with oxygen to form nitrogen oxides. These cause breathing problems and can cause acid rain.

66 67

Further teaching suggestions

Further research

- Using secondary sources, ask students to find out what fuel and car manufacturing companies are doing to reduce emissions of sulfur dioxide, nitrogen oxides and particulates. Using secondary sources, ask the students to find out why carbon monoxide detectors are important and how they work.

Summary answers

1 carbon, water, oxidised, sulfur, nitrogen, monoxide, particulates

2 **a** When sulfur from impurities in fuel burns it reacts with oxygen to form sulfur dioxide.
b At the high temperatures in the engine, nitrogen gas from the air reacts with oxygen to form nitrogen oxides.
c Incomplete combustion of the fuel can produce small particles of carbon and unburnt hydrocarbons.

3 **a** $CH_4 + 2O_2 \rightarrow CO_2 + 2H_2O$
b $2CH_4 + 3O_2 \rightarrow 2CO + 4H_2O$

C1 4.4 Cleaner fuels

Learning objectives

Students should learn:

- that burning fuels has an environmental impact
- how we can reduce the pollution from burning fuels.

Learning outcomes

Most students should be able to:

- state what causes global warming, global dimming and acid rain
- list some ways of reducing pollutants released when we burn fuels
- explain how acid rain is produced, and how it can be reduced.

Some students should also be able to:

- discuss the relationship between global dimming and global warming
- explain in detail methods of reducing pollutants from fuels.

Answers to in-text questions

a We are burning much more fossil fuel than 100 years ago.

b Four from: particulates, unburnt hydrocarbons, sulfur dioxide, nitrogen oxides, carbon monoxide, carbon dioxide.

Support

- To support students with 'the pyramid' activity, some information could already be printed on the internet, e.g. the start of a diagram, a prose with missing words or just a title. Encourage students to use lots of colour as this aids their learning. They might even want to include a key – e.g. each time they mention the word acid, or an example of an acid, they might write it in red.

Extend

- Often the concepts of global warming, climate change and the greenhouse effect are confused by students. Extend students by asking them to define each of these terms in no more than 50 words each. Then ask students to explain how these terms are related, again in no more than 50 words.

Specification link-up: Chemistry C1.4

- Sulfur dioxide and oxides of nitrogen cause acid rain, carbon dioxide causes global warming, and solid particles cause global dimming. *[C1.4.3 c)]*
- Sulfur can be removed from fuels before they are burned, for example in vehicles. Sulfur dioxide can be removed from the waste gases after combustion, for example in power stations. *[C1.4.3 d)]*
- Evaluate the impact on the environment of burning hydrocarbon fuels. *[C1.4]*
- Consider and evaluate the social, economic and environmental impacts of the uses of fuels. *[C1.4]*

Lesson structure

Starters

Photographs – Show the students an image of a drought area, polar ice caps and flooding. Topical pictures can be found on the web. Ask the students to link the pictures. [Scientists believe these are all effects of global warming.] Then show the students a forest damaged by acid rain, a weathered statue and a weathered building. Again ask the students to link the images. [They are the effects of acid rain.] *(5 minutes)*

Demonstration – Ignite sulfur in oxygen to demonstrate the production of sulfur dioxide in a fume cupboard, wearing eye protection.

Ask the students to write a word equation for the reaction between sulfur and oxygen. Support students by asking them to complete the partly written word equation. Extend students by asking for a balanced symbol equation for the reaction [sulfur + oxygen → sulfur dioxide]. Then ask them how this relates to burning fuels. [Sulfur is found in impurities in fossil fuels; this reaction can occur as the fuel is burned, which leads to the production of acid rain.] *(10 minutes)*

Main

- You may wish to investigate how the concentration of sodium metabisulfite affects the growth of cress seeds. See 'Practical support' for more details. This is a model of the effect of acid rain on plants. You could extend students by encouraging them to design their own experiment.
- During KS3, a number of environmental issues have been considered; but global dimming and its interdependence with global warming have not been studied.
- Split the class into three different groups and assign each a different environmental issue: acid rain, global warming or global dimming. The groups could be given access to different research materials, e.g. the internet and library books. Each member of the group should become an expert on its environmental issue. Then make teams of three. Each team should have one specialist on each issue. Each new group should then answer the questions set out in the objectives.
- Students already have a large body of background knowledge about environmental issues gained from KS3 science, geography and citizenship. Give each student a square-based pyramid net. Each triangular face should contain information about a different environmental issue (as detailed above), using the 'cleaning up our act' section in the Student Book to ensure that the science content is correct. The base should contain information about what can be done to reduce these problems. Once the information has been drawn and written, the students could cut it out and assemble the pyramid.

Plenaries

Crossword – Give the students a crossword to complete in pairs. These can be tailor-made if a puzzle generator is used, e.g. www.discoveryschool.com. *(5 minutes)*

AfL (Assessment for Learning) – Give students an examination question with fictitious answers: a weak, an average and an excellent answer. Students could order them to show which they think is the worst or best answer and why they think this. Then feed back the group's ideas to the class through a question-and-answer session. To support students, use a question from a foundation paper and to extend students use a question from a higher paper. *(10 minutes)*

Practical support

Modelling acid rain

Equipment and materials required

Gas jar of oxygen, sulfur flowers, spatula, deflagrating spoon, Bunsen burner, dropping pipette, water, universal indicator solution, eye protection (chemical splashproof).

Details

Put half a spatula of sulfur in a deflagrating spoon. Heat it in a blue Bunsen flame until it begins to combust (a blue flame is visible). Put it into a gas jar of oxygen. When the reaction is complete, add a little water and universal indicator solution to show the acidic nature of sulfur dioxide.

Safety: Complete in a fume hood, as sulfur dioxide is harmful, wear eye protection and wash hands after use. CLEAPSS Hazcard 69 Oxygen – oxidising; 96A Sulfur; 97 Sulfur dioxide – toxic; 32 Universal indicators – highly flammable and harmful.

Modelling the effect of acid rain

Equipment and materials required

Cotton wool, two Petri dishes, cress seeds, 0.05 mol/dm^3 and 0.1 mol/dm^3 sodium metabisulfite solutions (prepared and kept in a fume cupboard), distilled water, sticky labels, sunny window sill.

Details

Open the Petri dishes, and put cotton wool in the tops or bottoms to make three open dishes for germinating the seeds. Sprinkle the same number of cress seeds on each piece of cotton wool. Then add water to one Petri dish and different concentrations of sodium metabisulfite to the others. Label your dishes and leave in a sunny position. Ensure that the seeds remain moist using the appropriate water or sodium metabisulfite solution. Monitor the growth rate over a week.

Safety: Complete in a well ventilated room, wear eye protection and wash hands after use. CLEAPSS Hazcard 92 Sodium metabisulfite.

Warn asthmatics not to breathe in sulfur dioxide fumes.

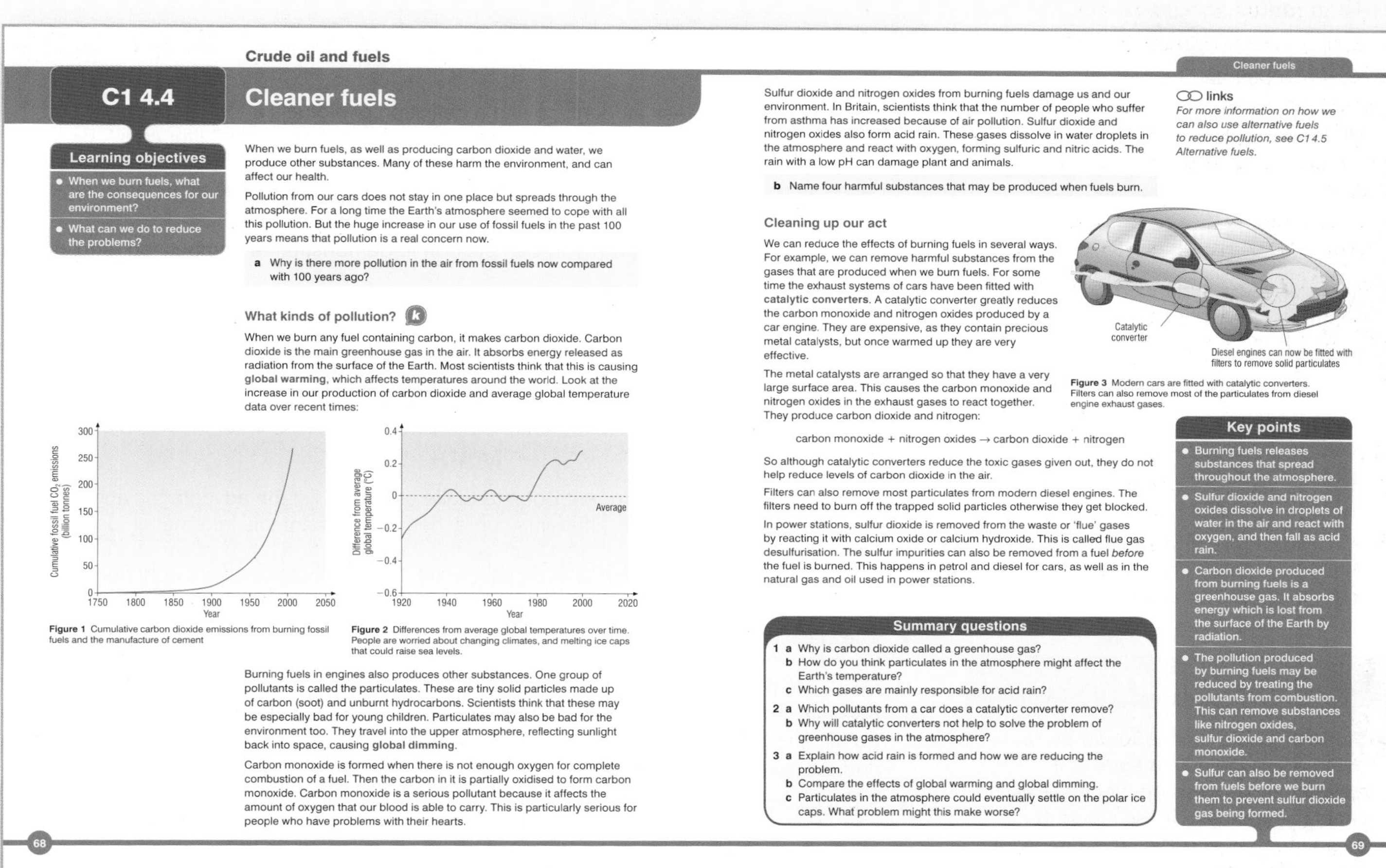

Crude oil and fuels

C1 4.4 Cleaner fuels

Learning objectives
- When we burn fuels, what are the consequences for our environment?
- What can we do to reduce the problems?

When we burn fuels, as well as producing carbon dioxide and water, we produce other substances. Many of these harm the environment, and can affect our health.

Pollution from our cars does not stay in one place but spreads through the atmosphere. For a long time the Earth's atmosphere seemed to cope with all this pollution. But the huge increase in our use of fossil fuels in the past 100 years means that pollution is a real concern now.

a Why is there more pollution in the air from fossil fuels now compared with 100 years ago?

What kinds of pollution?

When we burn any fuel containing carbon, it makes carbon dioxide. Carbon dioxide is the main greenhouse gas in the air. It absorbs energy released as radiation from the surface of the Earth. Most scientists think that this is causing **global warming**, which affects temperatures around the world. Look at the increase in our production of carbon dioxide and average global temperature data over recent times:

Figure 1 Cumulative carbon dioxide emissions from burning fossil fuels and the manufacture of cement

Figure 2 Differences from average global temperatures over time. People are worried about changing climates, and melting ice caps that could raise sea levels.

Burning fuels in engines also produces other substances. One group of pollutants is called the particulates. These are tiny solid particles made up of carbon (soot) and unburnt hydrocarbons. Scientists think that these may be especially bad for young children. Particulates may also be bad for the environment too. They travel into the upper atmosphere, reflecting sunlight back into space, causing **global dimming**.

Carbon monoxide is formed when there is not enough oxygen for complete combustion of a fuel. Then the carbon in it is partially oxidised to form carbon monoxide. Carbon monoxide is a serious pollutant because it affects the amount of oxygen that our blood is able to carry. This is particularly serious for people who have problems with their hearts.

68

Cleaner fuels

Sulfur dioxide and nitrogen oxides from burning fuels damage us and our environment. In Britain, scientists think that the number of people who suffer from asthma has increased because of air pollution. Sulfur dioxide and nitrogen oxides also form acid rain. These gases dissolve in water droplets in the atmosphere and react with oxygen, forming sulfuric and nitric acids. The rain with a low pH can damage plant and animals.

b Name four harmful substances that may be produced when fuels burn.

links

For more information on how we can also use alternative fuels to reduce pollution, see C1 4.5 Alternative fuels.

Cleaning up our act

We can reduce the effects of burning fuels in several ways. For example, we can remove harmful substances from the gases that are produced when we burn fuels. For some time the exhaust systems of cars have been fitted with **catalytic converters**. A catalytic converter greatly reduces the carbon monoxide and nitrogen oxides produced by a car engine. They are expensive, as they contain precious metal catalysts, but once warmed up they are very effective.

Figure 3 Modern cars are fitted with catalytic converters. Filters can also remove most of the particulates from diesel engine exhaust gases.

The metal catalysts are arranged so that they have a very large surface area. This causes the carbon monoxide and nitrogen oxides in the exhaust gases to react together. They produce carbon dioxide and nitrogen:

carbon monoxide + nitrogen oxides → carbon dioxide + nitrogen

So although catalytic converters reduce the toxic gases given out, they do not help reduce levels of carbon dioxide in the air.

Filters can also remove most particulates from modern diesel engines. The filters need to burn off the trapped solid particles otherwise they get blocked.

In power stations, sulfur dioxide is removed from the waste or 'flue' gases by reacting it with calcium oxide or calcium hydroxide. This is called flue gas desulfurisation. The sulfur impurities can also be removed from a fuel *before* the fuel is burned. This happens in petrol and diesel for cars, as well as in the natural gas and oil used in power stations.

Summary questions

1 **a** Why is carbon dioxide called a greenhouse gas?
  b How do you think particulates in the atmosphere might affect the Earth's temperature?
  c Which gases are mainly responsible for acid rain?

2 **a** Which pollutants from a car does a catalytic converter remove?
  b Why will catalytic converters not help to solve the problem of greenhouse gases in the atmosphere?

3 **a** Explain how acid rain is formed and how we are reducing the problem.
  b Compare the effects of global warming and global dimming.
  c Particulates in the atmosphere could eventually settle on the polar ice caps. What problem might this make worse?

Key points
- Burning fuels releases substances that spread throughout the atmosphere.
- Sulfur dioxide and nitrogen oxides dissolve in droplets of water in the air and react with oxygen, and then fall as acid rain.
- Carbon dioxide produced from burning fuels is a greenhouse gas. It absorbs energy which is lost from the surface of the Earth by radiation.
- The pollution produced by burning fuels may be reduced by treating the pollutants from combustion. This can remove substances like nitrogen oxides, sulfur dioxide and carbon monoxide.
- Sulfur can also be removed from fuels before we burn them to prevent sulfur dioxide gas being formed.

69

Summary answers

1 **a** It absorbs energy from the Earth as it cools down, preventing it escaping to space.
  b They might lower the temperature as they reflect sunlight back into space, so it doesn't get a chance to warm the Earth's surface.
  c Sulfur dioxide and nitrogen oxides.

2 **a** Carbon monoxide and nitrogen oxides (also unburnt hydrocarbons).
  b Catalytic converters do not remove carbon dioxide from exhaust gases.

3 **a** Sulfur dioxide and nitrogen oxides react with oxygen and water to form sulfuric and nitric acids, which fall to ground in rain, snow, hail or mist. Reduce by removing sulfur from fuels/removing SO_2 from gases released/removing nitrogen oxides by catalytic converters.
  b Global warming is the increase in average temperature of the Earth (due to an increase in greenhouse gases) whereas global dimming will reduce temperatures (as particulates reduce the energy reaching the Earth's surface from the Sun).
  c As they are darker in colour than the snow, they might absorb more energy from the Sun than the white snow, increasing the risk of polar ice caps melting.

C1 4.5 Alternative fuels

Learning objectives

Students should learn:

- the definition of a biofuel and some examples
- some of the advantages and disadvantages of biofuels
- that many scientists are interested in developing hydrogen as a fuel.

Learning outcomes

Most students should be able to:

- state a definition of a biofuel
- give an example of a biofuel
- recognise advantages and disadvantages of using biofuels
- explain why scientists are interested in developing hydrogen as a fuel.

Some students should also be able to:

- give a detailed, balanced argument for the use of biofuels, ethanol and hydrogen as fuels.

Support

- Students could be supported by being given the advantages and disadvantages of using biodiesel as simple sentences. These students could then cut and stick the sentences to order them into a table.

Extend

- Students could be extended by asking them to write word equations or even balanced symbol equations for the complete combustion of the different biofuels, e.g. ethanol, biogas (methane).
- Students could be encouraged to find out other forms of biofuel in common use, e.g. biogas from sewage. They could find out where these biofuels are being used and their advantages and disadvantages. Students could compile their findings into a summary table contrasting biodiesel, bioethanol, wood, biomass and biogas.

Specification link-up: Chemistry C1.4

- Biofuels, including biodiesel and ethanol, are produced from plant material. There are economic, ethical and environmental issues surrounding their use. *[C1.4.3 e)]*
- Consider and evaluate the social, economic and environmental impacts of the uses of fuels. *[C1.4]*
- Evaluate developments in the production and uses of better fuels, for example ethanol and hydrogen. *[C1.4]*
- Evaluate the benefits, drawbacks and risks of using plant materials to produce fuels. *[C1.4]*

Lesson structure

Starters

What do you think? – Ask students to find a definition of biofuels and then to consider if they agree with the following statements: 'Biofuel should be the only fuel for cars.' 'Biofuel should be mixed with fuels from crude oil for cars.' 'Biofuels should not be used in cars.' For each statement ask them to express the extent to which they agree as a percentage. *(5 minutes)*

Demonstration – Ignite a small hydrogen balloon to demonstrate a combustion reaction using hydrogen.

Ask the students to write a word equation, support students by asking questions during the demonstration and encourage students to detail the reactants and products. Students could be extended by asking for a balanced symbol equation for the reaction [hydrogen + oxygen → water/$2H_2 + O_2 \rightarrow 2H_2O$]. Ask students to use this observation to explain how hydrogen can be used as a fuel. *(10 minutes)*

Main

- Ask students to use the Student Book to make a table to list the advantages and disadvantages of biodiesel. Students could then supplement this information using secondary sources of information such as the internet to add more information. Each statement should be a summarised bullet point. This will help students construct a balanced argument for the use of biofuels.
- The class could then be split into groups of three. Each member of the group should be given a different task, as detailed in the biodiesel-fuel from plants activity box in the Student Book. Once the students have completed their task, each group should review each team member's work and discuss which biofuel they think is most likely to be used in 20 years' time. Encourage each group to feedback to the class their choice and the reasons why.
- Students could be asked to make 'top-trump' cards for different ways of powering a car. Students should consider biodiesel, bioethanol, hydrogen, petrol/diesel and autogas (methane), listing the advantages and disadvantages of each. They could then give a score out of 100 to give an arbitrary figure for their overall rating. Students could then consider their cards and make a list from best to worst fuel and construct an argument for why they think this.

Plenaries

Biofuel – Explain to students that B100 is 100 per cent biodiesel and B0 would be 100 per cent diesel (made from crude oil). Explain to students that diesel in petrol stations now, by law, must contain a percentage of biodiesel. Ask students to suggest the composition of B6 [6 per cent biodiesel, 94 per cent diesel] and B10 [10 per cent biodiesel, 90 per cent diesel]. You could support students by giving them the diesel content and so students just need to find the biodiesel content. Students could be extended by being asked to suggest why biodiesel mix is being used rather than pure biodiesel. *(5 minutes)*

Reflect: what do you think? – Ask students to consider the same three statements that they examined at the start of the lesson: biofuel should be the only fuel for cars; biofuel should be mixed with fuels from crude oil for cars; biofuels should not be used in cars. Again for each statement ask them to give a percentage of agreement. Then ask students to consider if their views have changed, and why or why not. Ask for a few volunteers to share their thoughts. *(10 minutes)*

Practical support

Demonstrate combustion of hydrogen

Equipment and materials required

Rubber party balloon filled with hydrogen gas, a splint secured on the end of a 1 m ruler with a rubber band, Bunsen burner, eye protection.

Details

Rest the hydrogen balloon on the ceiling away from anything that could combust or be damaged (such as lights or projectors). Put on eye protection and move students a suitable distance away. Light the splint and put it on the stretched part of the rubber balloon.

Safety: Wear eye protection and make sure students are a suitable distance. Do not use a large balloon and be aware that ceiling tiles can be damaged.

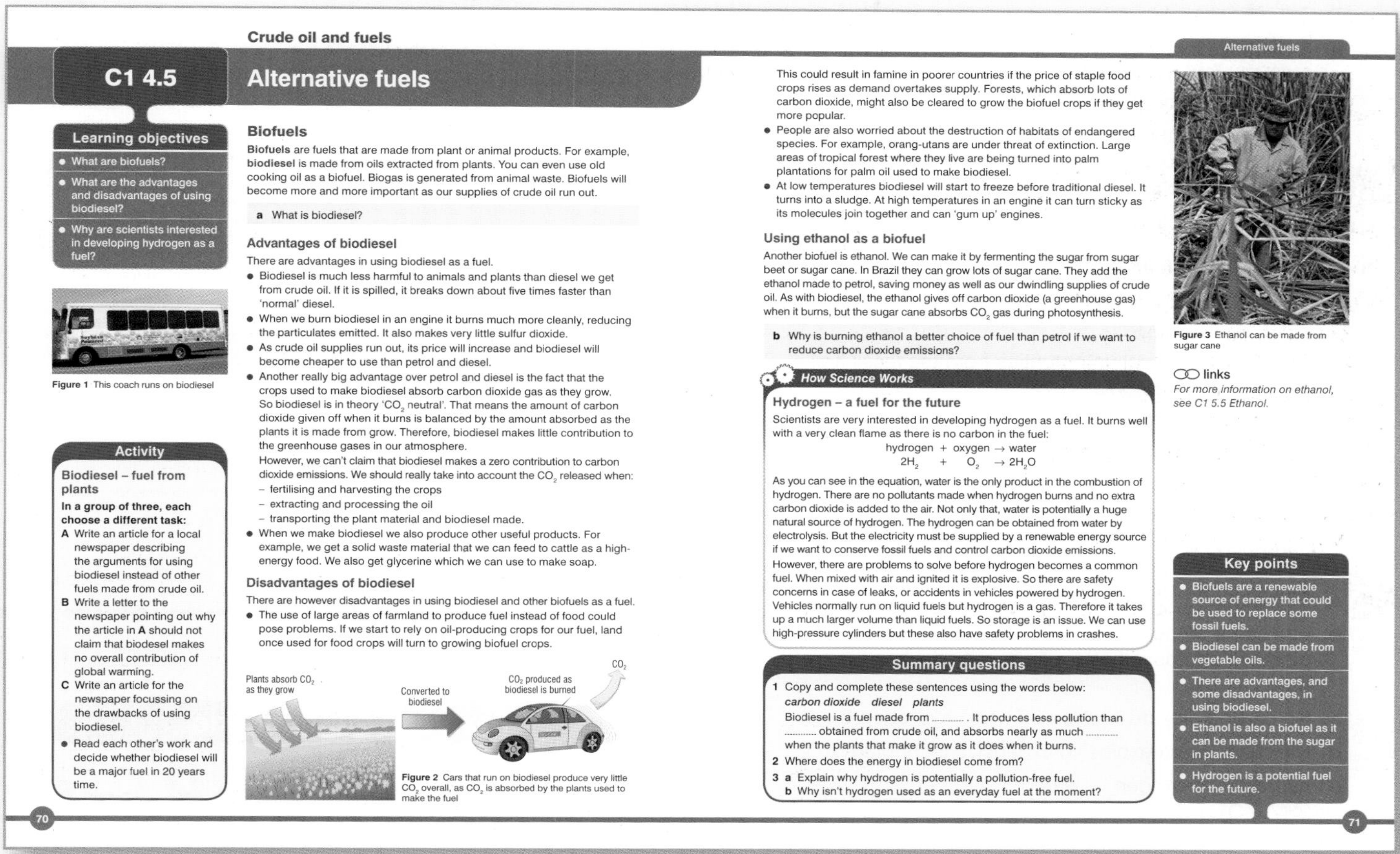

C1 4.5 Alternative fuels

Learning objectives

- What are biofuels?
- What are the advantages and disadvantages of using biodiesel?
- Why are scientists interested in developing hydrogen as a fuel?

Figure 1 This coach runs on biodiesel

Biofuels

Biofuels are fuels that are made from plant or animal products. For example, **biodiesel** is made from oils extracted from plants. You can even use old cooking oil as a biofuel. Biogas is generated from animal waste. Biofuels will become more and more important as our supplies of crude oil run out.

a What is biodiesel?

Advantages of biodiesel

There are advantages in using biodiesel as a fuel.

- Biodiesel is much less harmful to animals and plants than diesel we get from crude oil. If it is spilled, it breaks down about five times faster than 'normal' diesel.
- When we burn biodiesel in an engine it burns much more cleanly, reducing the particulates emitted. It also makes very little sulfur dioxide.
- As crude oil supplies run out, its price will increase and biodiesel will become cheaper to use than petrol and diesel.
- Another really big advantage over petrol and diesel is the fact that the crops used to make biodiesel absorb carbon dioxide gas as they grow. So biodiesel is in theory 'CO_2 neutral'. That means the amount of carbon dioxide given off when it burns is balanced by the amount absorbed as the plants it is made from grow. Therefore, biodiesel makes little contribution to the greenhouse gases in our atmosphere.
  However, we can't claim that biodiesel makes a zero contribution to carbon dioxide emissions. We should really take into account the CO_2 released when:
  - fertilising and harvesting the crops
  - extracting and processing the oil
  - transporting the plant material and biodiesel made.
- When we make biodiesel we also produce other useful products. For example, we get a solid waste material that we can feed to cattle as a high-energy food. We also get glycerine which we can use to make soap.

Activity

Biodiesel – fuel from plants

In a group of three, each choose a different task:

A Write an article for a local newspaper describing the arguments for using biodiesel instead of other fuels made from crude oil.

B Write a letter to the newspaper pointing out why the article in **A** should not claim that biodesel makes no overall contribution of global warming.

C Write an article for the newspaper focussing on the drawbacks of using biodiesel.

- Read each other's work and decide whether biodiesel will be a major fuel in 20 years time.

Disadvantages of biodiesel

There are however disadvantages in using biodiesel and other biofuels as a fuel.

- The use of large areas of farmland to produce fuel instead of food could pose problems. If we start to rely on oil-producing crops for our fuel, land once used for food crops will turn to growing biofuel crops. This could result in famine in poorer countries if the price of staple food crops rises as demand overtakes supply. Forests, which absorb lots of carbon dioxide, might also be cleared to grow the biofuel crops if they get more popular.
- People are also worried about the destruction of habitats of endangered species. For example, orang-utans are under threat of extinction. Large areas of tropical forest where they live are being turned into palm plantations for palm oil used to make biodiesel.
- At low temperatures biodiesel will start to freeze before traditional diesel. It turns into a sludge. At high temperatures in an engine it can turn sticky as its molecules join together and can 'gum up' engines.

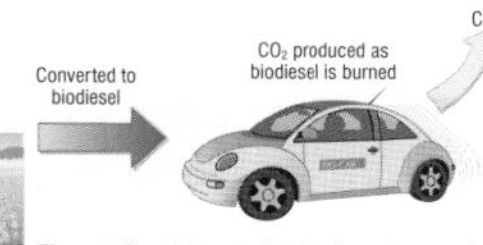

Figure 2 Cars that run on biodiesel produce very little CO_2 overall, as CO_2 is absorbed by the plants used to make the fuel

Using ethanol as a biofuel

Another biofuel is ethanol. We can make it by fermenting the sugar from sugar beet or sugar cane. In Brazil they can grow lots of sugar cane. They add the ethanol made to petrol, saving money as well as our dwindling supplies of crude oil. As with biodiesel, the ethanol gives off carbon dioxide (a greenhouse gas) when it burns, but the sugar cane absorbs CO_2 gas during photosynthesis.

b Why is burning ethanol a better choice of fuel than petrol if we want to reduce carbon dioxide emissions?

Figure 3 Ethanol can be made from sugar cane

links
For more information on ethanol, see C1 5.5 Ethanol.

How Science Works

Hydrogen – a fuel for the future

Scientists are very interested in developing hydrogen as a fuel. It burns well with a very clean flame as there is no carbon in the fuel:

$$\text{hydrogen} + \text{oxygen} \rightarrow \text{water}$$
$$2H_2 + O_2 \rightarrow 2H_2O$$

As you can see in the equation, water is the only product in the combustion of hydrogen. There are no pollutants made when hydrogen burns and no extra carbon dioxide is added to the air. Not only that, water is potentially a huge natural source of hydrogen. The hydrogen can be obtained from water by electrolysis. But the electricity must be supplied by a renewable energy source if we want to conserve fossil fuels and control carbon dioxide emissions.

However, there are problems to solve before hydrogen becomes a common fuel. When mixed with air and ignited it is explosive. So there are safety concerns in case of leaks, or accidents in vehicles powered by hydrogen. Vehicles normally run on liquid fuels but hydrogen is a gas. Therefore it takes up a much larger volume than liquid fuels. So storage is an issue. We can use high-pressure cylinders but these also have safety problems in crashes.

Summary questions

1 Copy and complete these sentences using the words below:
carbon dioxide diesel plants
Biodiesel is a fuel made from .... It produces less pollution than .... obtained from crude oil, and absorbs nearly as much .... when the plants that make it grow as it does when it burns.

2 Where does the energy in biodiesel come from?

3 **a** Explain why hydrogen is potentially a pollution-free fuel.
b Why isn't hydrogen used as an everyday fuel at the moment?

Key points

- Biofuels are a renewable source of energy that could be used to replace some fossil fuels.
- Biodiesel can be made from vegetable oils.
- There are advantages, and some disadvantages, in using biodiesel.
- Ethanol is also a biofuel as it can be made from the sugar in plants.
- Hydrogen is a potential fuel for the future.

Further teaching suggestions

Making biofuels

- You may wish to make biofuels with your students – go to www.sep.org.uk and search under 'What's New' and 'Recycling and sustainability' and 'Biofuels'.

Hydrogen as a fuel

- The University of Birmingham had the first hydrogen fuel station fitted in 2008. The station is used to fuel a fleet of cars used on the campus as part of a research project. Students could study this as a case study and consider the advantages and disadvantages of this fuel in this context. For more information go to www.newscentre.bham.ac.uk and search for hydrogen fuel.

Ethanol as a fuel

- Ethanol-to-fuel cars can be made industrially and use three different methods: hydration of ethene, fermentation with yeast, or using *E. coli*. Ask students to contrast each of these methods and suggest which is best for making ethanol for use as a fuel.

Answers to in-text questions

a The name for any fuel made from vegetable oils.

b Because the plants (e.g. sugar cane) used to make ethanol made from sugars absorbed carbon dioxide from the air as it grew, before releasing it again as ethanol burns (and in the fermentation process).

Summary answers

1 plants, diesel, carbon dioxide

2 From the Sun through photosynthesis.

3 **a** Water is its only product of combustion.
b It is difficult to store for use as a vehicle fuel; it needs a renewable source of energy to get it from water; it is highly flammable and could be explosive if there is an accident. Hydrogen fuel-cell technology is expensive.

Summary answers

1 a **i** A compound containing hydrogen and carbon only.
ii No more hydrogen can bond to its molecules.

b **i** Propane, C_3H_8.
ii Atoms of carbon and hydrogen.
iii (Covalent) bonds

c C_nH_{2n+2}

d $C_{20}H_{42}$

2 a B because it has a lower boiling point. The lower the boiling point, the higher up the fractionating column the compounds condense.

b B because its molecules will be smaller and the smaller the hydrocarbon molecule, the easier it is to ignite. It will burn with a cleaner (less sooty) flame.

c A will be more viscous than B.

3 a Carbon dioxide and water

b Global warming/climate change.

c **i** Sulfur dioxide
ii Remove the sulfur impurities from the fossil fuel before it is burned.
Absorb any sulfur dioxide using a basic substance before it escapes into the atmosphere.
iii Nitrogen oxides
iv Catalytic converters

d To remove solid particulates of carbon and unburnt hydrocarbons.

4 a Ethanol

b The same amount of carbon dioxide is absorbed by the plants in photosynthesis as is given off when the biofuel burns.

c No carbon dioxide is given off when hydrogen burns.

d Difficult to store/explosive/expensive technology

e hydrogen + oxygen → water
$2H_2 + O_2 \rightarrow 2H_2O$

5 a

Fuel	Mass burned (g)		Temperature (°C)			
			Before	After	Before	After
	1	2	1	1	2	2
Ethanol						
Propanol						
Butanol						
Pentanol						

b Variables that need to be controlled include:
distance from burner to beaker
volume of water in beaker
starting temperature of the water
insulation on beaker
time left to burn
size of flame
draughts
how the temperature was recorded.

c Stir the water before recording the temperature.
Use a well-calibrated thermometer.

d The repeats are consistent, so this would suggest that they are precise.

e As a bar graph of the mean temperature rise for each fuel.

Crude oil and fuels: C1 4.1–C1 4.5

Summary questions 

1 This question is about the alkane family of compounds.

a The alkanes are all 'saturated hydrocarbons'.
i What is a hydrocarbon?
ii What does saturated mean when describing an alkane?

b i Give the name and formula of this alkane:

```
   H  H
   |  |
H—C—C—H
   |  |
   H  H
```

ii What do the letters represent in this displayed formula?
iii What do the lines between letters represent?

c What is the general formula of the alkanes (where n = the number of carbon atoms)?

d Give the formula of the alkane with 20 carbon atoms.

2 One alkane, A, has a boiling point of 344 °C and another, B, has a boiling point of 126 °C.

a Which one will be collected nearer the top of a fractionating column in an oil refinery? Explain your choice.

b Which one will be the better fuel? Explain your choice.

c Give another difference you expect between A and B.

3 a Name the two products formed when a hydrocarbon burns in enough oxygen to ensure that complete combustion takes place.

b What problem is associated with the increased levels of carbon dioxide gas in the atmosphere?

c i What gas is given off from fossil fuel power stations that can cause acid rain?
ii Give **two** ways of stopping this gas getting into the atmosphere from power stations.
iii Name the other cause of acid rain which comes from car engines?
iv How do car makers stop the gases in part **iii** entering the air?

d Why are diesel engines now fitted with a filter for their exhaust fumes?

4 a Which one of these fuels could be termed a 'biofuel'?
Hydrogen Propane Ethanol Petrol Coal

b Biodiesel is potentially 'CO_2 neutral'. What does this mean?

c Scientists are concerned about the issue of global warming. Why is the use of hydrogen as a fuel one way to tackle the problem?

d State **two** problems with the use of hydrogen as a fuel.

e Write a word and a balanced symbol equation to show the combustion of hydrogen. [H]

5 This apparatus can be used to compare the energy given out when different fuels are burned.

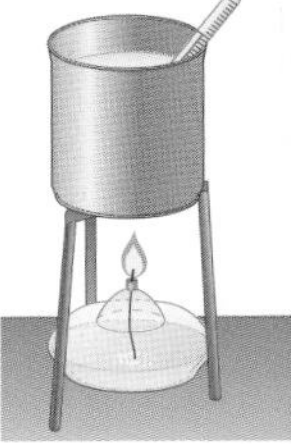

The burner is weighed before and after to determine the amount of fuel burned. The temperature of the water is taken before and after, to get the temperature rise. The investigation was repeated. From this the amount of energy released by burning a known amount of fuel can be calculated.

a Design a table that could be used to collect the data as you carry out this experiment.

A processed table of results is given below.

Fuel	Mass burned (g)	Temperature rise (°C)	
Ethanol	4.9	48	47
Propanol	5.1	56	56
Butanol	5.2	68	70
Pentanol	5.1	75	76

b List three variables that need to be controlled.

c Describe how you would take the temperature of the water to get the most accurate measurement possible.

d Do these results show precision? Explain your answer?

e How might you present these results?

72

Practical suggestions

Practicals	AQA	k	Book	Gears
Demonstration of fractional distillation of crude oil using CLEAPPS mixture (take care to avoid confusion with the continuous process in a fractionating column).	✓		✓	
Design an investigation on viscosity, ease of ignition or sootiness of flame of oils or fuels.	✓	✓	✓	
Comparison of the energy content of different fuels, for example by heating a fixed volume of water.	✓			✓
Demonstration of the production of solid particles by incomplete combustion using a Bunsen burner yellow flame or a candle flame to heat a boiling tube of cold water.	✓		✓	
Collecting and testing the products of combustion of candle wax and methane.	✓		✓	
Demonstration of burning sulfur or coal in oxygen and then testing the pH of the gas produced.	✓		✓	
Design an investigation on growing cress from seeds in various concentrations of sodium metabisulfite solution to show how acid rain affects plants.	✓		✓	

End of chapter questions

AQA Examination-style questions

The table shows some information about the first four alkanes.

Name of alkane	Formula	Boiling point in °C
Methane	CH_4	–162
	C_2H_6	–88
Propane	C_3H_8	
Butane		0

a i Name the alkane missing from the table. (1)
ii What is the formula of butane? (1)
iii Estimate the boiling point of propane. (1)

b Which one of the following is the formula of the alkane with 6 carbon atoms?

C_6H_6 C_6H_{10} C_6H_{14} C_6H_{16} (1)

c Explain why alkanes are hydrocarbons. (1)

d A molecule of methane can be represented as:

```
    H
    |
H — C — H
    |
    H
```

Draw a molecule of propane in the same way. (2)

Some crude oil was distilled in a fractionating column. The table shows the boiling ranges of three of the fractions that were collected.

Fraction	Boiling range in °C
A	60–120
B	160–230
C	240–320

a Which of these fractions is the most flammable? (1)

b Which of these fractions is the most viscous? (1)

c Which of these fractions has the smallest hydrocarbon molecules? (1)

d Why do the fractions have boiling ranges and not boiling points? (1)

3 Some landfill sites produce a gas that can be collected and burned as a fuel. The gas is mainly methane.

a Choose the word from the list to complete the sentence.

condensed distilled oxidised

During the combustion of methane the elements in the fuel are .... (1)

b Write a word equation for the complete combustion of methane, CH_4. (2)

c Under what conditions could methane burn to produce carbon monoxide? (1)

d A sample of landfill gas was burned. The waste gases contained sulfur dioxide. Explain why. (1)

4 Crude oil is a resource from which fuels can be separated.

a The name of the main fuel fractions and one of the hydrocarbons in each fraction are shown in the table.

Main fuel fraction	A hydrocarbon in this fraction	Boiling point of hydrocarbon in °C
Gases	Propane, C_3H_8	−42
Petrol	Octane, C_8H_{18}	126
Paraffin	Dodecane, $C_{12}H_{26}$	216
Diesel	Eicosane, $C_{20}H_{42}$	344

Heated crude oil

i How does the number of carbon atoms in a hydrocarbon affect its boiling point? (1)

ii Suggest the lowest temperature to which crude oil needs to be heated to vaporise all the hydrocarbons in the table.
Temperature = ... °C? (1)

iii Dodecane boils at 216 °C. At what temperature will dodecane gas condense to liquid?
Temperature = ... °C? (1)

b *In this question you will be assessed on using good English, organising information clearly and using specialist terms where appropriate.*

Describe and explain how the fractions are separated in a fractionating column. (6)

AQA, 2009

73

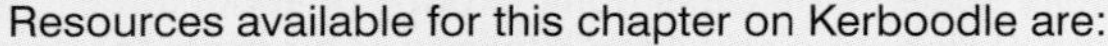

Kerboodle resources

Resources available for this chapter on Kerboodle are:

- Chapter map: Crude oil and fuels
- Support: All about alkanes (C1 4.1)
- Extension: All about alkanes (C1 4.1)
- Animation: Fractional distillation (C1 4.2)
- Practical: Investigating the fractions of crude oil (C1 4.2)
- How Science Works: Burning issues (C1 4.3)
- How Science Works: Governments get hot under the collar about global warming (C1 4.3)
- How Science Works: Don't have the foggiest about which is the foggiest! (C1 4.4)
- Interactive activity: Crude oil and fuels
- Revision podcast: Fuels and the environment
- Test yourself: Crude oil and fuels
- On your marks: Fractional distillation of crude oil
- Examination-style questions: Crude oil and fuels
- Answers to examination-style questions: Crude oil and fuels

Study tip

Some students confuse distillation with cracking, thinking that when crude oil is distilled, the molecules are broken down. Some students find it difficult to relate boiling points to the temperature gradient in the fractionating column, thinking that low boiling fractions are collected near the bottom of the column.

Examination-style answers

1 a i ethane *(1 mark)*
ii C_4H_{10} *(1 mark)*
iii any value from −50 to −30 (°C) *(1 mark)*

b C_6H_{14} *(1 mark)*

c contain only carbon and hydrogen *(1 mark)*

d Correct structure showing 3 C joined with single bonds and 8 C – H bonds. *(2 marks)*

2 a A *(1 mark)*

b C *(1 mark)*

c A *(1 mark)*

d They are mixtures or not single compounds/alkanes. *(1 mark)*

3 a oxidised *(1 mark)*

b methane + oxygen → carbon dioxide + water
reactants, products *(2 marks)*

c limited supply of air *(1 mark)*

d (land-fill gas) contains sulfur (compounds) *(1 mark)*

4 a i The greater the number (of carbon atoms) the higher its boiling point. *(1 mark)*
ii accept numbers in the range 344 to 350 (°C) *(1 mark)*
iii 216 (°C) *(1 mark)*

b There is a clear and detailed scientific description of the processes that take place during fractional distillation in a fractionating column, including evaporation, condensation and collection of fractions at different temperatures. The answer shows almost faultless spelling, punctuation and grammar. It is coherent and in an organised, logical sequence. It contains a range of appropriate or relevant specialist terms used accurately. *(5–6 marks)*

There is a scientific description that includes condensation and collection of fractions at different temperatures. There are some errors in spelling, punctuation and grammar. The answer has some structure and organisation. The use of specialist terms has been attempted, but not always accurately. *(3–4 marks)*

There is a brief description of evaporation or condensation and collection of fractions. The spelling, punctuation and grammar are very weak. The answer is poorly organised with almost no specialist terms and/or their use demonstrating a general lack of understanding of their meaning. *(1–2 marks)*

No relevant content. *(0 marks)*

Examples of chemistry points made in the response:

- crude oil is heated (to above 350 °C)
- crude oil is vaporised
- vapours/gases go into (fractionating) column
- column is hot at bottom, cool at top or correct temperature gradient indicated
- substances/vapours condense at their boiling points
- different substances/vapours condense at different levels in column
- low boiling points at the top of column or high boiling points at the bottom
- collected as liquids
- liquids/fractions contain a mixture of substances
- fractions have substances with a range of boiling points
- continuous process.

C1 5.1 Cracking hydrocarbons

Learning objectives

Students should learn:

- reasons for cracking large hydrocarbon molecules and how it is carried out
- that alkenes contain double bonds and are called unsaturated hydrocarbons.

Learning outcomes

Most students should be able to:

- state a definition of cracking
- list the general conditions needed for cracking, stating that it is a thermal decomposition reaction
- explain why cracking is carried out
- state and use the general formula to work out the formula of an alkene when *n* is given
- recognise an alkene from a structural or molecular formula
- state and carry out the test for alkenes.

Some students should also be able to:

- balance a symbol equation to represent cracking. **[HT only]**

Specification link-up: Chemistry C1.5

- Hydrocarbons can be cracked to produce smaller, more useful molecules. This process involves heating the hydrocarbons to vaporise them. The vapours are either passed over a hot catalyst or mixed with steam and heated to a very high temperature so that thermal decomposition reactions then occur. *[C1.5.1 a)]*
- The products of cracking include alkanes and unsaturated hydrocarbons called alkenes. Alkenes have the general formula C_nH_{2n}. *[C1.5.1 b)]*
- Unsaturated hydrocarbon molecules can be represented in the following forms: C_3H_6. *[C1.5.1 c)]*

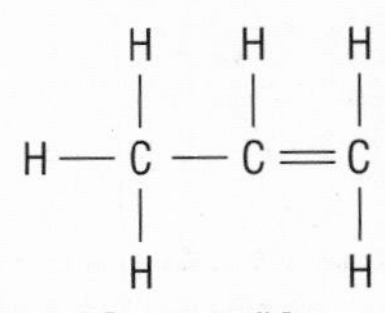

- Alkenes react with bromine water, turning it from orange to colourless. *[C1.5.1 d)]*
- Some of the products of cracking are useful as fuels. *[C1.5.1 e)]*

Lesson structure

Starters

Recap – Ask students to consider the phrase 'thermal decomposition' (recall limestone work). Ask them to define it and give an example [using heat to break down a substance; heating of calcium carbonate]. *(5 minutes)*

Comparing – On the board, draw the structural formulae of ethane and ethene. Ask the students to list the similarities and differences. [**Similarities:** same number of C atoms/H atoms all have one bond/C atoms all have four bonds/both are hydrocarbons. **Differences:** different number of H atoms/ethane is saturated, ethene is unsaturated/ ethane has C—C but ethene has C=C.] Students could be supported by being given a list of questions such as 'how many carbon atoms does each molecule have?' Students could be extended by being asked to explain what is meant by saturated and unsaturated. *(10 minutes)*

Main

- Cracking can be done on a laboratory scale in the classroom. Groups of about three are best as one student can heat the boiling tube and often it takes two students to collect the gas over water.
- Bromine water can be used to test the product and contrast this with the reactant. Ethene is flammable. This can be demonstrated by putting a lighted splint into the test tube and a flame should be seen travelling down the tube (eye protection must be worn). The flame should be smokier than the equivalent alkane (ethane) as there is a higher percentage of carbon.
- Some students can find it difficult to recall the differences between alkanes and alkenes. Encourage them to compare alkanes and alkenes in terms of their structures and how to test them. They could be given a piece of coloured paper to fold in half and make a poster comparing the two families of hydrocarbons.

Plenaries

Chemical formula – Ask the students to use the general formula for an alkene to work out the molecular and displayed formula, if $n = 2, 3, 4$ [C_2H_4, C_3H_6, C_4H_8]. Students can be supported by being given molecular model kits to help draw the display formula. Students could be extended by asking them to give alternative structures for C_4H_8. [There are two.] *(5 minutes)*

'Alkane or alkene?' practical – Give students a selection of different hydrocarbons labelled with letters and some bromine water. Their task is to identify which liquids are alkenes and which are alkanes. Any alkenes tested should decolourise the bromine water. Alkanes will not. *(10 minutes)*

Support

- The 'cracking' practical may be too difficult for students with poor manual dexterity. It could be completed as a demonstration instead. Alternatively, if a teaching assistant is available, split the class into two groups. The teaching assistant can then help small groups of students to test for alkenes while you demonstrate cracking to a small group of students.

Extend

- Students could be extended by asking them to draw a table with the name, molecular and structural formula of the first 10 alkenes. You may also wish to show students that from butane onwards, the double bond can be in different positions. You may wish to introduce the term 'isomer'.

Answers to in-text questions

a Because we can use it to make smaller, more useful hydrocarbon molecules from larger ones.

b By heating them strongly and passing them over a catalyst or by mixing them with steam and heating to a high temperature.

Practical support

Cracking

Equipment and materials required

Bunsen burner and safety equipment, S-shaped delivery tube (with bung on one end and Bunsen valve on the other), pneumatic trough, boiling tube, mineral wool, three test tubes with bungs, test-tube rack, stand, boss and clamp, a hydrocarbon (medicinal paraffin, petroleum jelly or decane), broken ceramic pot/aluminium oxide powder (catalyst), bromine water.

Details

Soak the mineral wool in paraffin and put at the bottom of the boiling tube. About 2 cm from the bung put in the catalyst (about one spatula full) and clamp in position. Heat the catalyst strongly (with a blue flame) then flash the flame towards the mineral wool, so that the hydrocarbon evaporates and reaches the catalyst. Students should collect the gas over water. The Bunsen valve will help prevent suck back within the apparatus, but tell students to remove the end of the delivery tube from the trough of water by lifting the clamp stand before they finish heating. Eye protection should be worn throughout. The collected gas can be tested with bromine water (as detailed below). The gas is highly flammable and this can be demonstrated by igniting it with a burning splint in the test tube.

Safety: medicinal paraffin, petroleum jelly or decane – flammable; CLEAPSS Hazcard 45 Bromine water – irritant and harmful.

Alkane or alkene?

Equipment and materials required

A selection of hydrocarbons (e.g. hexane, hexene, octane), dropping pipettes, test tubes, bromine water, bungs, test-tube rack, eye protection.

Details

Students should pipette about 1 cm^3 of each liquid into separate test tubes, stored in a rack. About the same amount of bromine water should be added to each test tube, and add a bung to each. Shake. Ensure that the students wear eye protection, hold the bung and base of the test tube when they shake it. Bromine vapour may be given off, therefore a well ventilated area is necessary.

Only use weak solutions of bromine water (yellow, not orange) otherwise you are unlikely to see complete discoloration (and there will be less bromine vapour released).

Safety: A selection of hydrocarbons (e.g. hexane, hexene, octane – flammable). CLEAPSS Hazcard 15B; 45 Bromine water – irritant and harmful.

Products from oil

C1 5.1 Cracking hydrocarbons

Learning objectives

- How do we make smaller, more useful molecules from larger, less useful molecules in crude oil?
- What are alkenes and how are they different from alkanes?

Some of the heavier fractions that we get by distilling crude oil are not very useful. The hydrocarbons in them are made up of large molecules. They are thick liquids or solids with high boiling points. They are difficult to vaporise and do not burn easily – so they are no good as fuels! Yet the main demand from crude oil is for fuels.

Luckily we can break down large hydrocarbon molecules in a process we call **cracking**.

The process takes place at an oil refinery in a steel vessel called a cracker.

In the cracker, a heavy fraction produced from crude oil is heated to vaporise the hydrocarbons. The vapour is then either passed over a hot catalyst or mixed with steam. It is heated to a high temperature. The hydrocarbons are cracked as thermal decomposition reactions take place. The large molecules split apart to form smaller, more useful ones.

a Why is cracking so important?
b How are large hydrocarbon molecules cracked?

Figure 2 In an oil refinery, huge crackers like this are used to break down large hydrocarbon molecules into smaller ones

Example of cracking

Decane is a medium-sized molecule with ten carbon atoms. When we heat it to 500 °C with a catalyst it breaks down. One of the molecules produced is pentane which is used in petrol.

Figure 1 Pentane (C_5H_{12}) can be used as a fuel. This is the displayed formula of pentane.

We also get propene and ethene which we can use to produce other chemicals.

$$\underset{\text{decane}}{C_{10}H_{22}} \xrightarrow{800\,°C\ +\ \text{catalyst}} \underset{\text{pentane}}{C_5H_{12}} + \underset{\text{propene}}{C_3H_6} + \underset{\text{ethene}}{C_2H_4}$$

This reaction is an example of thermal decomposition.

Notice how this cracking reaction produces different types of molecules. One of the molecules is pentane. The first part of its name tells us that it has five carbon atoms (*pent-*). The last part of its name (*-ane*) shows that it is an alkane. Like all other alkanes, pentane is a saturated hydrocarbon. Its molecules have as much hydrogen as possible in them.

The other molecules in this reaction have names that end slightly differently. They end in *-ene*. We call this type of molecule an **alkene**. The different ending tells us that these molecules are **unsaturated**. They contain a **double bond** between two of their carbon atoms. Look at Figure 3. You can see that alkenes have one double bond and have the general formula $\mathbf{C_nH_{2n}}$.

Ethene — Double bond — Propene

Figure 3 A molecule of ethene (C_2H_4) and a molecule of propene (C_3H_6). These are both alkenes – each molecule has a carbon–carbon double bond in it.

74

Cracking hydrocarbons

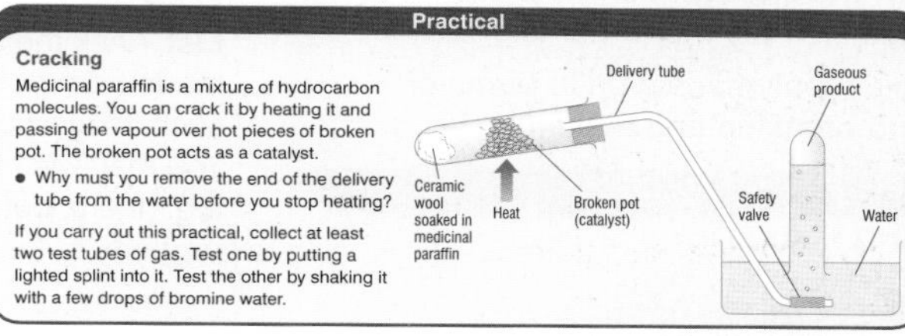

Practical

Cracking

Medicinal paraffin is a mixture of hydrocarbon molecules. You can crack it by heating it and passing the vapour over hot pieces of broken pot. The broken pot acts as a catalyst.

- Why must you remove the end of the delivery tube from the water before you stop heating?

If you carry out this practical, collect at least two test tubes of gas. Test one by putting a lighted splint into it. Test the other by shaking it with a few drops of bromine water.

A simple experiment like the one above shows that alkenes burn. They also react with bromine water (which is orange). The products of this reaction are colourless. This means that we have a good test to see if a hydrocarbon is unsaturated:

Positive test:

unsaturated hydrocarbon + bromine water (orange-yellow) → products (colourless)

Negative test:

saturated hydrocarbon + bromine water (orange) → no reaction (orange)

Did you know ...?

Ethene gas makes fruits such as bananas ripen. Bananas are picked and stored as the unripe green fruit. When they are required for display in a shop ethene gas is passed over the stored bananas to start the ripening process.

AQA Examiner's tip

Remember:
alkanes are saturated
alkenes have a double bond = (equals)

Summary questions

1 Copy and complete using the words below:
alkenes catalyst cracking double heating unsaturated
Large hydrocarbon molecules are broken down by .... them and passing them over a hot .... This is called .... Some of the molecules produced when we do this contain a .... bond. They are called .... hydrocarbons. They are examples of a group of hydrocarbons called the ....

2 Cracking a hydrocarbon makes two new hydrocarbons, A and B. When bromine water is added to A, nothing happens. Bromine water added to B turns from an orange solution to colourless.
   a Which hydrocarbon is unsaturated?
   b Which hydrocarbon is used as a fuel?
   c What type of reaction is cracking an example of?
   d Cracking can be carried out by passing large hydrocarbon molecules over a hot catalyst. State another way to crack a hydrocarbon in industry.

3 An alkene molecule with one double bond contains 7 carbon atoms. How many hydrogen atoms does it have? Write down its formula.

4 Decane (with 10 carbon atoms) is cracked into octane (with 8 carbon atoms) and ethene. Write a balanced equation for this reaction. [H]

Key points

- We can split large hydrocarbon molecules up into smaller molecules by:
  - mixing them with steam and heating them to a high temperature, or
  - by passing the vapours over a hot catalyst.
- Cracking produces saturated hydrocarbons which are used as fuels and unsaturated hydrocarbons (called alkenes).
- Alkenes react with orange bromine water, turning it colourless.

75

Further teaching suggestions

Cracking equations

- Students could be given the chemical formula of decane, and asked to generate as many balanced equations as possible, to demonstrate all the possible combinations of alkanes and alkenes that could be generated in the cracking reaction. **[HT only]**

Summary answers

1 heating, catalyst, cracking, double, unsaturated, alkenes
2 a Hydrocarbon B is unsaturated.
   b Hydrocarbon A is used as a fuel.
   c thermal decomposition
   d By mixing them with steam and heating to a high temperature.
3 Fourteen hydrogen atoms. C_7H_{14}
4 $C_{10}H_{22} \rightarrow C_8H_{18} + C_2H_4$

C1 5.2 Making polymers from alkenes

Learning objectives

Students should learn:

- that monomers join together to make polymers
- about some uses of polymers.

Learning outcomes

Most students should be able to:

- state definitions of monomers and polymers
- determine the polymer name, if the monomer name is given
- determine the monomer name, if the polymer name is given
- list some uses of polymer products.

Some students should also be able to:

- draw the structural formula of a polymer, if the monomer is given
- explain polymerisation in terms of bond breaking and making.

Specification link-up: Chemistry C1.5

- Alkenes can be used to make polymers such as poly(ethene) and poly(propene). In these reactions, many small molecules (monomers) join together to form very large molecules (polymers). For example:

$$n\ \begin{matrix} H & & H \\ | & & | \\ C & = & C \\ | & & | \\ H & & H \end{matrix} \rightarrow \left(\begin{matrix} H & & H \\ | & & | \\ -C & - & C- \\ | & & | \\ H & & H \end{matrix} \right)_n$$

ethene → poly(ethene)

[C1.5.2 a)]

- Evaluate the social and economic advantages and disadvantages of using products from crude oil as fuels or as raw materials for plastics and other chemicals. *[C1.5]*

Lesson structure

Starters

What's the connection? – On the board, have a selection of different photographs of plastics being used, e.g. drink bottles, window frames, clothes. Ask the students to suggest how all the images are linked. [They are all polymers.] *(5 minutes)*

List – Ask the students to look around the classroom and write a list of all the things that are made from polymers. Encourage the students to think about more unusual applications, such as elastic in clothes, fillings in teeth and natural polymers such as proteins and DNA. You may wish to support students by having a tray of products that are made from plastic, e.g. rulers, pens or yoghurt pots. You can extend students by asking them to state the property of the plastic that makes it useful for that application. *(10 minutes)*

Main

- You may wish to demonstrate how to make Perspex. See 'Practical support' for more details. The monomer is methyl 2-methylpropenoate and the polymer has the chemical name poly(methylpropenoate). It is better known by its brand name of Perspex. Di(dodecanoyl is used as the initiator – this starts the polymerisation reaction. Show students the displayed formula of the monomer and ask them to try to draw the displayed formula of the polymer.
- Ask students to draw a table with the following column headings: Name of monomer, displayed formula of monomer, molecular formula of monomer, name of polymer and displayed formula of polymer. Students should complete the table using secondary sources of information to help them. You may wish to support students by giving them the information for polyethene, polypropene, polychloroethene and polytetrafluoroethene. Students could then cut out the information to complete their table. Students could be extended by being encouraged to write balanced displayed formula equations for the making of the polymers that they have studied.
- A kinaesthetic approach to polymerisation would use molecular model kits to represent the polymerisation process, as described in the activity box 'Modelling polymerisation' in the Student Book.
- Students could then be encouraged to create their own model of polymerisation by using people as monomers.

Plenaries

Matching – Ask the students to generate the polymer names from these monomer names:

- Ethene → [poly(ethene)]
- Propene → [poly(propene)]
- Styrene → [poly(styrene)]
- Vinyl chloride → [poly(vinyl chloride)]
- Ethene terephthalate → [poly(ethene terephthalate)]

Students can be supported by being given the names of both the monomers and polymers, so they have to match them together. Students could be extended by being given the displayed formula of the monomer and they have to generate the displayed formula of the polymer. *(5 minutes)*

Models – Students act out their polymer models and explain them to the class. *(10 minutes)*

Support

- Students often find it difficult to focus on the C=C bond of a monomer to help them write out the polymer structure. You could draw C=C with one line coming off each carbon atom, an arrow and then the displayed structure of a polymer again with C—C, and a line from each C atom. This diagram could then be laminated. Students then use dry wipe pens and erasers to add the groups onto the carbon atoms to draw the displayed formula of one monomer and its polymer. Different examples can be considered quickly by rubbing out and starting again. Students should be able to see that the C=C is the focus and no matter how the monomer is presented, you need to focus on this area.

Extend

- Students could be extended by introducing condensation polymerisation. Students could then make nylon (this is a condensation polymer and is not on the specification). A recipe for this is in *Classic Chemistry Demonstrations* (RSC).

Practical support

Making Perspex

Equipment and materials required

Di(dodecanoyl) peroxide (0.2 g), top pan balance, 25 cm^3 plastic syringe, 25 cm^3 methyl 2-methylpropenoate, warm water bath, condenser with flat-bottom flask, stand, boss and clamp, chemical-resistant gloves, fume cupboard, eye protection. CLEAPSS Hazcard 29B – oxidising; 43B – highly flammable and irritant.

Details

Ensure that the inhibitor has been removed from the methyl 2-methylpropenoate. Complete this demonstration in a fume cupboard without any sources of ignition. Using the 25 cm^3 plastic syringe measure 25 cm^3 of methyl 2-methylpropenoate into the flat bottomed flask. Add 0.2 g of di(dodecanoyl) peroxide. Put the flat bottom flask into the warm water bath and fit a condenser in a reflux position. The liquid in the flat-bottom flask become increasingly viscous as the polymerisation occurs.

Safety: CLEAPSS Hazcard 29 Di(dodecanoyl) peroxide – oxidising; 43B Methyl 2-methylpropenoate – highly flammable and irritant. Wear chemical splashproof safety goggles.

Further teaching suggestions

Industrial video

- Students could be encouraged to look at the industrial production of polymers. Videos are available on industrial chemistry (RSC).

Paper clip model

- Students can use paper clips to model monomers, polymers and polymerisation. Give each student a few paper clips each and ask them to model the definitions. Each paper clip should represent a monomer and then the students should link some paper clips together to 'polymerise' and then the paperclip chain represents a polymer.

Plastic fact card

- Students could make fact cards about particular plastics. Each card should have the displayed formula of the monomer, polymer, its properties and uses.

Products from oil

C1 5.2 Making polymers from alkenes

Learning objectives

- What are monomers and polymers?
- How do we make polymers from alkenes?

The fractional distillation of crude oil and cracking produces a large range of hydrocarbons. These are very important to our way of life. Oil products are all around us. We simply cannot imagine life without them.

Hydrocarbons are our main fuels. We use them in our transport and at home to cook and for heating. We also use them to make electricity in oil-fired power stations.

Then there are the chemicals we make from crude oil. We use them to make things ranging from cosmetics to explosives. But one of the most important ways that we use chemicals from oil is to make plastics.

Figure 1 All of these products were manufactured using chemicals made from oil

Plastics

Plastics are made up of huge molecules made from lots of small molecules joined together. We call the small molecules **monomers**. We call the huge molecules they make **polymers**. (*Mono* means 'one' and *poly* means 'many'). We can make different types of plastic which have very different properties by using different monomers.

a List three ways that we use fuels.
b What are the small molecules that make up a polymer called?

MONOMERS → POLYMER

Ethene → etc. ●–●–●–●–●–●–● etc. Poly(ethene)

Propene → etc. ■–■–■–■–■–■–■ etc. Poly(propene)

Figure 2 Polymers are made from many smaller molecules called monomers

Ethene (C_2H_4) is the smallest unsaturated hydrocarbon molecule. We can turn it into a polymer known as poly(ethene) or polythene. Poly(ethene) is a really useful plastic. It is easy to shape, strong and transparent (unless we add colouring material to it). 'Plastic' bags, plastic drink bottles, dustbins and clingfilm are all examples of poly(ethene).

Propene (C_3H_6) is another alkene. We can also make polymers with propene as the monomer. The polymer formed is called poly(propene). It forms a very strong, tough plastic. We can use it to make many things, including carpets, milk crates and ropes.

Figure 3 Polymers produced from oil are all around us and are part of our everyday lives

c Is ethene an alkane or an alkene?
d Which polymer can we make from propene monomers?

76

Making polymers from alkenes

How do monomers join together?

When alkene molecules join together, the double bond between the carbon atoms in each molecule 'opens up'. It is replaced by single bonds as thousands of molecules join together. The reaction is called **polymerisation**.

$$H_2C{=}CH_2 + H_2C{=}CH_2 + H_2C{=}CH_2 + \text{etc.} \longrightarrow -CH_2-CH_2-CH_2-CH_2-CH_2-CH_2-\text{ etc. ...}$$

Ethene monomers → Poly(ethene)

We can also write this more simply as:

$$n\ C{=}C \longrightarrow \left(-C-C-\right)_n$$

where n is a large number

Many single ethene monomers → Long chain of poly(ethene)

Activity

Modelling polymerisation

Use a molecular model kit to show how ethene molecules polymerise to form poly(ethene).

Make sure you can see how the equation shown above represents the polymerisation reaction you have modelled.

You should also be able to describe what happens to the bonds in the reaction.

Think up a model to demonstrate the polymerisation of ethene, using people in your class as monomers.

Evaluate the ideas of other groups.

AQA Examiner's tip

The double C=C bond in ethene (an alkene) makes it much more reactive than ethane (an alkane).

Summary questions

1 Copy and complete using the words below:
polymerisation ethene monomers polymers
Plastics are made of large molecules called .... We make these by joining together lots of small, reactive molecules called .... One example of a polymer is poly(ethene), made from .... Poly(ethene) is formed as a result of a .... reaction.

2 Why is ethene the smallest possible unsaturated hydrocarbon molecule?

3 **a** Draw the displayed formula of a propene molecule, showing all its bonds.
b Draw a diagram to show how propene molecules join together to form poly(propene).
c Explain the polymerisation reaction in **b**.

Key points

- Plastics are made of polymers.
- Polymers are large molecules made when monomers (small, reactive molecules) join together. The reaction is called polymerisation.

77

Answers to in-text questions

a transport, heating, generating electricity
b monomers
c Ethene is an alkene.
d Poly(propene). (Its old name is polypropylene.)

Summary answers

1 polymers, monomers, ethene, polymerisation

2 It has two carbon atoms, the minimum number for a C=C double bond.

3 a

```
     H   H   H
     |   |   |
 H — C — C = C
     |       |
     H       H
```

b

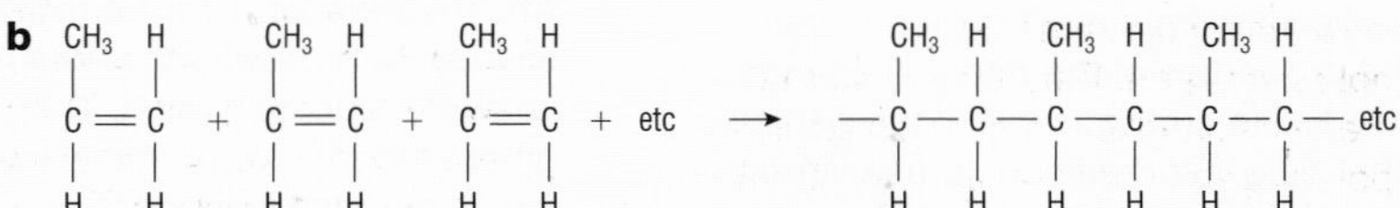

c The double bond 'opens up' in neighbouring propene molecules and forms single bonds, joining molecules together in a chain.

C1 5.3 New and useful polymers

Learning objectives

Students should learn:

- that there are new polymers being developed and being used in innovative ways
- what shape memory polymers are and what they are used for.

Learning outcomes

Most students should be able to:

- give an example of a polymer that is used because of its properties
- give an example of a polymer that has been designed for a specific job.

Some students should also be able to:

- evaluate the suitability of different polymers for particular uses.

Support

- Students often forget what a 'smart' material is. Ensure that they understand that it is a material that changes its properties as a stimulus changes. You could show students some plastic examples such as plastic drinks beakers that change colour when heated.

Extend

- Students could be extended by being asked to find out about other 'smart' materials.
- Students could be extended by being asked to find out how 'smart' materials work. The structure of 'smart' materials changes at certain thresholds for a particular stimulus. This means that the properties of the material change due to the arrangement of the particles changing. Students could use simple particle models to explain how a specific example changes.

Specification link-up: Chemistry C1.5

- Polymers have many useful applications and new uses are being developed, for example: new packaging materials, waterproof coatings for fabrics, dental polymers, wound dressings, hydrogels, smart materials (including shape memory polymers). *[C1.5.2 b)]*

  Controlled Assessment: C4.1 Plan practical ways to develop and test candidates' own scientific ideas. *[C4.1.1 a) b)]*; C4.4 Select and process primary and secondary data. *[C4.4.1 a), b)]*

Lesson structure

Starters

Monomer or polymer – Ask the students to link the following statements to the more relevant word – 'monomer' or 'polymer'. The statements could be written down or read out:

- A very long chain hydrocarbon. [polymer] • Contains a double bond. [monomer]
- Reactive molecule. [monomer] • PET. [polymer] • Ethene. [monomer]
- Plastic. [polymer] • Joins together to make a plastic. [monomer]

You could support students by giving them the definitions of a polymer and a monomer on the board. You could extend students by asking them to draw a displayed formula to represent a polymer and monomer at the end of the starter. *(5 minutes)*

Word game – Ask the students to try to make as many words as they can from: 'poly(ethyleneterephthalate)', with double points for any scientific words. *(10 minutes)*

Main

- Plastics are now often developed for a particular purpose. However, new uses for existing plastics are also being found. Set the students the task of designing a new waterproof coat. The students should list the properties of the material that they want the coat to be made out of. They should then complete the 'Evaluating plastics' practical to decide on the best material for the job and why.
- You may wish to encourage students to investigate hydrogels, a type of smart material (see 'Practical support' for more details). Encourage students to suggest how they could make their results more reliable [repeated readings and take an average]. Students could also compare their results with the information that the manufacturers provide and comment on how accurate their results were.
- The two activities above will be useful for teaching aspects of Controlled Assessment – e.g. designing investigations, planning a fair test, and societal influences on decisions made.
- Society is using polymers increasingly, but often they are used for single-use items such as ready-meal packaging. Students could be set the task to choose the best to make a drinks container. In this exercise, they should consider the size of the drinks container, aesthetics, recycling and price. This task could be completed using secondary data from the internet.
- Students need to have some understanding of the vast range in new polymers. Split the class into six groups and give each a topic: 'new packaging materials, waterproof coatings for fabrics, dental polymers, wound dressings, hydrogels, shape memory polymers'. Ask each group to come up with a PowerPoint presentation (no more than 3 minutes long) about their topic. This should include what their material is, its special properties, what it is used for and one fascinating fact. Ask the students to deliver their presentations. Have a stopwatch running, and stop the students when they reach their 3 minutes.

Plenaries

Answers – Ask the students to look back at the objectives in the Student Book and answer the 'questions' that have been posed. *(5 minutes)*

AfL (Assessment for Learning) – Split the class into pairs. Give each pair two short examination questions about new and useful polymers. Allow the students 5 minutes to answer a question each. Then ask the students to swap, and encourage them to mark the other student's work. Allow the two students to sit together 'debriefing' each other about their examination work. Support students using Foundation Tier examination questions and extend students using Higher Tier examination questions. *(10 minutes)*

Practical support

Evaluating plastics

Equipment and materials required

Swatches of different polymer fabrics (untreated nylon, treated nylon, Teflon, wool, cotton, polyester and neoprene), sandpaper, water bath, washing powder, two beakers, hairdryer, stand, boss and clamp, fabric dyes, cobalt chloride paper.

Details

Each fabric can be tested for:

Wearing/durability – Rub the fabric with sandpaper a number of times (set by the students) to see the effect on the wear of the material.

Water permeability – Carefully place on the surface of a water bath and put a piece of cobalt chloride paper onto the fabric. If water seeps through, the paper will turn from blue to pink.

Wind permeability – Set up the fabric stretched out in a stand, boss and clamp. Hold the hairdryer on cold blow at one side, and put your hand on the other, in order to feel the air flow through the fabric.

Dyeing – The coloured dyes could be applied as the instructions to determine how easy it is to change the colour of the fabric.

Wash test – Measure the fabric swatch, then wash it in washing detergent and dry. Remeasure the material to see if it has shrunk.

Safety: Cobalt chloride paper – handle as little as possible. Use freezers. Wash hands after experiment. CLEAPSS Hazcard 25 – harmful.

Investigating hydrogel

Equipment and materials required

Hydrogel (available from garden centres as a water absorbing additive for compost), 100 cm^3 measuring cylinder, water, 250 cm^3 beaker, top pan balance, dropping pipette, stirring rod.

Details

Measure 1 g of hydrogel into a 250 cm^3 beaker. Measure 100 cm^3 of tap water into a measuring cylinder. Using the dropping pipette, add water to the hydrogel and mix with the stirring rod. Keep adding water from the dropping pipette until no more water has been absorbed. Record the amount of water that the hydrogel absorbed.

Products from oil

C1 5.3 New and useful polymers

Learning objectives
- How are we using new polymers?
- What are smart polymers?

Chemists can design new polymers to make materials with special properties to do particular jobs. Medicine is one area where we are beginning to see big benefits from these 'polymers made to order'.

New polymer materials will eventually take over from fillings for teeth which contain mercury. Working with the toxic mercury every day is a potential hazard to dental workers. Other developments include:
- new softer linings for dentures (false teeth)
- new packaging material
- implants that can slowly release drugs into a patient.

a What do we mean by a 'designer polymer'?

Figure 1 A sticking plaster is often needed when we cut ourselves. Getting hurt isn't much fun – and sometimes taking the plaster off can be painful too.

Light-sensitive plasters

We all know how uncomfortable pulling a plaster off your skin can be. But for some of us taking off a plaster is really painful. Both very old and very young people have quite fragile skin. But now a group of chemists has made a plaster where the 'stickiness' can be switched off before the plaster is removed. The plaster uses a light-sensitive polymer.

1 The plaster is put on just like any normal plaster.

2 To remove the plaster, the top layer is peeled away from the lower layer which stays stuck to the skin.

3 Once the lower layer is exposed to the light, the adhesive becomes less sticky, making it easy to peel off your skin.

Figure 2 This plaster uses a light-sensitive polymer

How Science Works

Evaluating plastics

Plan an investigation to compare and evaluate the suitability of different plastics for a particular use.

For example, you might look at treated and untreated fabrics for waterproofing and 'breatheability' (gas permeability) or different types of packaging.

Hydrogels

Hydrogels are polymer chains with a few cross-linking units between chains. This makes a matrix that can trap water. These hydrogels are used as wound dressings. They let the body heal in moist, sterile conditions. This makes them useful for treating burns.

The latest 'soft' contact lenses are also made from hydrogels. To change the properties of hydrogels, scientists can vary the amount of water in their matrix structure.

78

New and useful polymers

Shape memory polymers

New polymers can also come to our rescue when we are cut badly enough to need stitches. A new 'shape memory polymer' is being developed by doctors which will make stitches that keep the sides of a cut together. When a shape memory polymer is used to stitch a wound loosely, the temperature of the body makes the thread tighten and close the wound, applying just the right amount of force.

This is an example of a '**smart polymer**', i.e. one that changes in response to changes around it. In this case a change in temperature causes the polymer to change its shape. Later, after the wound is healed, the polymer is designed to dissolve and is harmlessly absorbed by the body. So there will be no need to go back to the doctor to have the stitches out.

Figure 3 A shape memory polymer uses the temperature of the body to make the thread tighten and close the wound

New uses for old polymers

The bottles that we buy fizzy drinks in are a good example of using a plastic because of its properties. These bottles are made out of a plastic called PET.

The polymer it is made from is ideal for making drinks bottles. It produces a plastic that is very strong and tough, and which can be made transparent. The bottles made from this plastic are much lighter than glass bottles. This means that they cost less to transport and are easier for us to carry around.

Do you recycle your plastic bottles? The PET from recycled bottles is used to make polyester fibres for clothing, such as fleece jackets, and the filling for duvet covers. School uniforms and football shirts are now also made from recycled drinks bottles.

b Why is PET used to make drinks bottles?

Did you know ...?

PET is an abbreviation for poly(ethene terephthalate). It takes 5 two-litre PET lemonade bottles to make one T-shirt.

links

For more information on recycling, see C1 5.4 Plastic waste.

Summary questions

1 Copy and complete using the words below:

cold hot PET properties shape strong transparent

We choose a polymer for a job because it has certain .... For example, we make drinks bottles out of a plastic called .... because it is .... and ....
Scientists can also design 'smart' polymers, for example .... memory polymers. These change their shape when they are .... or ....

2 **a** Give one advantage of using a polymer in sticking plasters that is switched off by light making the polymer less sticky.

b Design a leaflet for a doctor to give to a patient, explaining how stitches made from smart polymers work.

Key points
- New polymers are being developed all the time. They are designed to have properties that make them specially suited for certain uses.
- Smart polymers may have their properties changed by light, temperature or by other changes in their surroundings.
- We are now recycling more plastics and finding new uses for them.

79

Further teaching suggestions

Unusual polymers
- Using secondary sources, students could research into unusual polymers, e.g. surgical glues.

Timeline
- Ask which synthetic polymers have not been around for more than a hundred years. Using secondary sources, students could produce a timeline showing the development of important (e.g. Bakelite) or interesting (e.g. superglue) polymers.

New polymer
- Students should think of a new polymer fit for a particular purpose. Students should list its properties and what specific use it is designed for.

Answers to in-text questions

a Polymers that are designed with ideal properties for a particular job.

b PET bottles are much lighter than glass bottles but are strong, tough (don't smash) and transparent.

Summary answers

1 properties, PET, strong (transparent), transparent (strong), shape, hot (cold), cold (hot)

2 **a** One advantage is needed from; makes the plaster less sticky, reduces pain when the plaster is removed, reduces the likelihood of damaging fragile skin.

b leaflet

C1 5.4 Plastic waste

Learning objectives

Students should learn:

- that problems are caused by the disposal of plastics
- what biodegradeable means
- that biodegradeable plastics can be made.

Learning outcomes

Most students should be able to:

- state a definition of biodegradeable
- give an example of how a plastic can be made biodegradeable
- explain some of the problems arising from plastic disposal.

Some students should also be able to:

- explain how a plastic can be made biodegradeable.

Specification link-up: Chemistry C1.5

- Many polymers are not biodegradable, so they are not broken down by microbes and this can lead to problems with waste disposal. *[C1.5.2 c)]*
- Plastic bags are being made from polymers and cornstarch so that they break down more easily. Biodegradable plastics made from cornstarch have been developed. *[C1.5.2 d)]*
- Evaluate the social, economic and environmental impacts of the uses, disposal and recycling of polymers. *[C1.5]*

Controlled Assessment: C4.4 Select and process primary and secondary data. *[C4.4.2 b)]*

Lesson structure

Starters

Plastic recycling – Ask students to make a list of all the ways that they can recycle plastics [recycling collection at home, plastic recycle banks e.g. at supermarkets]. *(5 minutes)*

Interpreting data – The average household bin contains 35 per cent organic waste, 30 per cent paper, 12 per cent construction materials, 9 per cent plastics, 6 per cent metal, 3 per cent glass and 5 per cent other. Ask students to display this information in a bar chart. Students could be supported by being given the labelled axes and scale. Extend students by asking them to display this information in a pie chart. *(10 minutes)*

Main

- Many plastics only degrade through physical weathering, and this process can take hundreds of years. Ask students to consider why plastics are used and the downsides of their disposal. Students could create a table listing the advantages and disadvantages of the use of common plastics.
- You could ask students to complete the activity 'sorting plastics'. Students should write a leaflet for householders to persuade them to recycle more plastic waste. You may wish to show students some examples of leaflets and adverts to help galvanise their thoughts.
- There is a wide range of different plastics, all with very different properties and uses. Some consumer goods and packaging will be made out of more than one type of plastic. Explain to students that plastics must be sorted before they can be recycled. Ask students to find out the recycling codes for plastics. They could then make an A4 or A5 label that could be put up on a recycling bin to help householders separate plastics before they are recycled.
- Plastics are being developed and modified so that they are biodegradable. Allow students to make cornstarch and test its properties. See 'Practical support' for more details.

Plenaries

Reduce, reuse, recycle – Ask students to make suggestions about their own lives that would reduce the plastic waste that they generate. Ask volunteers to share their thoughts with the whole class and manage the discussions. *(5 minutes)*

Classify – Give the students a selection of household plastic refuse and ask them to sort them into the different types of plastic (using the universal plastic identification symbol). Time the students and the first group to finish could be given a prize. Encourage students to reflect why it is expensive to recycle plastic [the plastic needs to be collected, then sorted, often manually, often before recycling can begin]. Students could be supported by being given waste plastic that is of only one type, and limiting the types of plastic to three. Students could be extended by being given things with more than one plastic, e.g. a plastic fizzy drink bottle is made from a different plastic from its cap. *(10 minutes)*

Support

- You can support students by giving them a writing frame to help them construct their leaflet for the 'sorting plastics' activity.

Extend

- You can extend students by asking them to consider why plastics are often not recycled. They could then categorise these reasons as social or economic. [Social – it is time consuming and takes effort to recycle. Economic – it is often cheaper to make new plastic than recycle, limited market for recycled plastic.]

Practical support

Investigating cornstarch

Equipment and materials required

One tablespoon of cornflour, 15 cm^3 water, three drops of cooking oil, dropping pipette, microwave oven, 100 cm^3 glass beaker, stirring rod, table spoon, food colouring (optional).

Details

In the glass beaker mix the cornflour, water, cooking oil and food colouring until an opaque paste is formed. Use a microwave to heat the mixture in the beaker until it becomes yellow gel-like material, almost transparent (takes up to 30 seconds). Remove the beaker and allow the mixture to cool. Once the mixture is cool, it can be moulded into a shape or pushed into a mould, e.g. an ice cube tray. Allow the material to fully cool and solidify (this may take up to 24 hours). The plastic's properties can then be tested, which could include dropping it in water and watching it dissolve.

Students can alter the ratios of cornflour, oil and water and experiment to determine how the properties are different.

Safety: Take care with hot materials.

Testing the properties of cornstarch

Equipment and materials required

Samples of cornflour plastic, Bunsen burner and safety equipment, tongs, water bath, mounted needle, hand lens, eye protection.

Details

Students should first look at the plastic, using the hand lens to observe the different textures. Then, using the mounted needle, they need to try to scratch the plastic to find out how hard it is and gently pull at the plastic to consider its flexibility. In a fume cupboard only, take a small piece of the plastic and hold in the Bunsen flame, using tongs, in order to find out if it is flammable and whether it is a thermosoftening or a thermosetting plastic.

Safety: Take care with burning plastics.

Products from oil

C1 5.4 Plastic waste

Learning objectives

- What are the problems caused by disposing of plastics?
- What does biodegradable mean?
- How can polymers be made biodegradable?

One of the problems with plastics is what to do with them when we've finished with them. Too much ends up as rubbish in our streets. Even the beaches in the remotest parts of the world can be polluted with plastic waste. Wildlife can get trapped in the waste or eat the plastics and die.

Not only that, just think of all the plastic packaging that goes in the bin after shopping. Most of it ends up as rubbish in landfill tips. Other rubbish in the tips rots away quite quickly. Microorganisms in the soil break it down. Many waste plastics last for hundreds of years before they are broken down completely. So they take up valuable space in our landfill sites. What was a useful property during the working life of the plastic (its lack of reactivity) becomes a disadvantage in a landfill site.

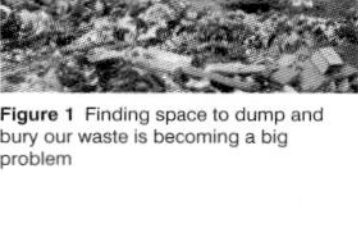

Figure 1 Finding space to dump and bury our waste is becoming a big problem

a Why are waste plastics proving to be a problem for us?

Biodegradable plastics 

Scientists are working to solve the problems of plastic waste. We are now making more plastics that do rot away in the soil when we dump them. These plastics are called **biodegradable**. They can be broken down by microorganisms.

Scientists have found different ways to speed up the decomposition. One way uses granules of cornstarch built into the plastic. The microorganisms in soil feed on the starch. This breaks the plastic up into small pieces more quickly.

Other types of plastic have been developed that are made from plant products. A plastic called PLA, poly(lactic acid), can be made from cornstarch. The plastic is totally biodegradable. It is used in food packaging. However, it cannot be put in a microwave which limits its use in ready-meal packaging.

We can also make plastic carrier bags using PLA. In carrier bags the PLA is mixed with a traditional plastic. This makes sure the bag is strong enough but will still biodegrade a lot more quickly.

Using plastics such as PLA also helps preserve our supplies of crude oil. Remember that crude oil is the raw material for many traditional plastics, such as poly(ethene).

Figure 2 The breakdown of a biodegradable plastic. PLA can be designed to break down in a few months.

Disadvantages of biodegradable plastics

However, the use of a food crop like corn to make plastics can raise the same issues as biofuels. Farmers who sell their crops to turn into fuel and plastics could cause higher food prices. The lack of basic food supplies could result in starvation in developing countries. Another problem is the destruction of tropical forests to create more farmland. This will destroy the habitats of wildlife and could affect global warming.

Other degradable plastics used for bags will break down in light. However, they will not decompose when buried in a landfill site. Probably the best solution is to reuse the same plastic carrier bags over and over again.

links

For information on the issues of using biofuels, look back at C1 4.5 Alternative fuels.

80

Plastic waste

Practical

Investigating cornstarch

Cornstarch can be fermented to make the starting material for PLA. However, cornstarch itself also has some interesting properties. You can make your own plastic material directly from cornstarch.

- How do varying the proportions of cornstarch and water affect the product?

Recycling plastics

Some plastics can be recycled. Once sorted into different types they can be melted down and made into new products. This can save energy and resources.

However, recycling plastics does tend to be more difficult than recycling paper, glass or metals. The plastic waste takes up a lot of space so is awkward to transport. Sorting out plastics into their different types adds another tricky step to the process. The energy savings are less than we get with other recycled materials. It would help recyclers if they could collect the plastics already sorted. You might have seen recycling symbols on some plastic products.

Figure 3 These symbols could help people sort out their plastic waste to help the recycling process

b How does recycling plastic waste help conserve our supplies of crude oil?

Figure 4 Recycling is becoming part of everyday life in the UK

Activity

Sorting plastics

a Imagine you are the head of your council's waste collection department. You have to write a leaflet for householders persuading them to recycle more of their plastic waste. They will be provided with extra bins to sort the plastics out before they are collected, once every two weeks.

b Write a letter back to the council from an unhappy person who is not willing to do any more recycling than they do already.

c Take a class vote on which action, **a** or **b**, you would support.

Summary questions

1 What do we mean by a biodegradable plastic?

2 **a** Why are plastics whose raw materials are plants becoming more popular?

b PLA is a biodegradable plastic. What is its monomer?

3 Non-biodegradable plastics such as poly(ethene) can be made to decompose more quickly by mixing with additives. These enable the polymer chain to be broken down by reacting with oxygen. Why might this be a waste of money if the plastic is buried and compressed under other waste in a landfill site?

Key points

- Non-biodegradable plastics cause unsightly rubbish, can harm wildlife and take up space in landfill sites.
- Biodegradable plastics are decomposed by the action of microorganisms in soil. Making plastics with starch granules in their structure help the microorganisms break down a plastic.
- We can make biodegradable plastics from plant material such as cornstarch.

81

Further teaching suggestions

Biodegradable plastics

- Ask students to find some examples of biodegradable plastics in use today, e.g. refuse sacks, nappy sacks, supermarket carrier bags, washing bags used by the NHS.

The Science Enhancement Programme

- Innovative practicals looking at recycling of plastics can be found at www.sep.org.uk. Search for 'recycling' and 'sustainability' under 'What's New'.

Local council research

- Ask students to find out how plastics are sorted and recycled by their local council. Sorting is often by hand but eddy currents, density and radiation can also be used.

Answers to in-text questions

a Plastic waste forms unsightly rubbish, which can harm wildlife. It also takes up valuable space in landfill sites.

b The raw material for making most polymers is crude oil which is a non-renewable resource. Recycling plastics means we do not have to use crude oil to make new plastics.

Summary answers

1 A plastic that is broken down by microorganisms in the soil.

2 **a** Because they do not rely on crude oil (which is a non-renewable resource) as a raw material and they are more biodegradable.

b lactic acid

3 There will be very little oxygen available to react with the polymer chains, so they will not be broken down.

C1 5.5 Ethanol

Learning objectives

Students should learn:

- that there are two main methods for making ethanol
- that there are advantages and disadvantages of each method for making ethanol.

Learning outcomes

Most students should be able to:

- describe in outline the two methods of making ethanol
- state an advantage and disadvantage of each method for making ethanol
- write the word equation and conditions for the production of ethanol from ethene
- write the word equation and conditions for the production of ethanol by fermentation.

Some students should also be able to:

- evaluate the advantages and disadvantages of using ethanol as a fuel
- write the symbol equation for the reaction of ethene with steam and for the fermentation of sugar. **[HT only]**

Answers to in-text questions

a carbon dioxide

b enzymes

c Crude oil (in the cracking of its heavy fractions).

Support

- You could support students by giving them a half-finished flow chart for each method for making ethanol industrially. Then you could give them key diagrams and statements that they cut out and add to the flow chart to finish it.

Extend

- You could extend students by asking them to find out about how genetically modified *E. coli* can be used to make ethanol.

Specification link-up: Chemistry C1.5

- Ethanol can be produced by hydration of ethene with steam in the presence of a catalyst. *[C1.5.3 a)]*
- Ethanol can also be produced by fermentation with yeast, using renewable resources. This can be represented by: sugar → carbon dioxide + ethanol. *[C1.5.3 b)]*
- Evaluate the advantages and disadvantages of making ethanol from renewable and non-renewable sources. *[C1.5]*

Lesson structure

Starters

What is the connection? – Show students a picture of bread, beer, a packet of dried yeast and a microscope image of a yeast cell and ask the students to state the connection [they all contain/are pictures of yeast]. *(5 minutes)*

Sorting – Get pictures of different fuels for cars at least A4 in size (oil, biodiesel, diesel, gas, petrol, alcohol and hydrogen). Laminate the pictures, and put sticky-tac on the back. Draw a table with two columns (renewable, non-renewable). Ask two students to explain what the two words in the table mean. Then ask seven students, in turn, to move one image into the renewable or non-renewable column. They should take the image and stick it in the appropriate column. You could support the students by giving them the definitions of renewable and non-renewable. You could extend the students by asking them to consider if electricity for powering a car is renewable or non-renewable. [If it is generated by renewable resources such as solar, it would be renewable, however it is often generated from fossil fuel power stations and is therefore non-renewable.] *(10 minutes)*

Main

- Ethanol can be made from non-renewable sources by reacting ethene and steam in a reaction called hydration. However, ethanol can also be made by fermentation. Discuss the advantages and disadvantages of each method for making ethanol.
- You may wish students to use fermentation to make their own ethanol (see 'Practical support'). You can leave the experiments set up until next lesson, then collect enough of the fermented mixture to fractionally distil. Allow students to smell the distilled ethanol. Show the students that the ethanol can be ignited on a watch glass.
- There is always development in car designs and their engines. Cars were run using petrol for many years; it has been relatively recently that other fuels have started to be investigated and used in the UK, e.g. autogas power.
- People are often slow to change, especially when innovations are based on unproven technology. So car companies will have to make some persuasive advertisements to encourage people to change their cars to run on ethanol.
- Organise students into their preferred learning styles (auditory, kinaesthetic, visual). Ask the visual group to design a billboard poster, the kinaesthetic group to make an advert for TV and the auditory group to make a radio advert. All of the marketing material should aim to persuade drivers to convert to ethanol. Students could be played TV adverts about fuel/engine developments as stimulus material for producing their own marketing material.

Plenaries

Objective answer – Ask the students to answer the objective 'questions'. You could support students by giving them a selection of answers for each question. Each answer should be correct, but some should be more scientific than others (using key terms, word equations etc.). Students should choose one of the answers to copy down and give the reason for their choice. You could extend students by asking them to write an exam-style mark scheme for these questions. Students should consider the number of marks they would award, why, alternative acceptable answers and answers that would not be good enough to be awarded a mark. *(5 minutes)*

AfL (Assessment for Learning) – Each group displays/acts out their marketing material. Question the students about any misconceptions in the adverts. Then ask students to vote for the most persuasive advert; then that group could be given a prize. *(10 minutes)*

Practical support

Fermentation

Equipment and materials required

You will need 1 g yeast, 5 g sugar, a 100 cm^3 measuring cylinder, water bath set at 40 °C, 100 cm^3 conical flask, cotton wool, a stirring rod, bung and delivery tube, test tube, limewater.

Details

Measure 50 cm^3 of warm water into the conical flask and add the sugar and stir until dissolved. Add the yeast and stir to mix. Put the cotton wool into the neck of the conical flask and leave in the water bath to ferment. Alternatively, using a bung with a delivery tube in the centre, the evolved gas could be tested. This can be achieved by half filling a test tube with limewater, then placed in a test tube rack on the outside of the water bath. Put the delivery tube into the limewater so evolved gas blows through. The distillate in the practical can be poured onto a watch glass and ignited to demonstrate its use as a fuel.

Fractional distillation of fermented ethanol mixture

Equipment and materials required

Fermented ethanol mixture, filter paper, filter funnel, round-bottom flask, heating mantle, claisen adaptor, condenser, access to cold water tap and sink, 2 × collecting flasks, watch glass, stand, boss and clamp, thermometer.

Details

Collect all of the students' fermented mixtures and filter into a round-bottom flask. Place the round-bottom flask in a heating mantle and fractionally distil. The boiling point of ethanol is 78 °C.

Safety: Ethanol is flammable, keep away from naked flames.

CLEAPSS Hazcard 40A Ethanol – highly flammable and harmful. Do not taste.

C1 5.5 Ethanol

Learning objectives

- What are the two methods used to make ethanol?
- What are the advantages and disadvantages of these two methods?

Ethanol is a member of the group of organic compounds called the alcohols. Its formula is C_2H_6O but it is more often written as C_2H_5OH. This shows the –OH group that all alcohols have in their molecules.

Making ethanol by fermentation

Ethanol is the alcohol found in alcoholic drinks. Ethanol for drinks is made by the **fermentation** of sugar from plants. Enzymes in yeast break down the sugar into ethanol and carbon dioxide gas:

$$\text{sugar (glucose)} \xrightarrow{\text{yeast}} \text{ethanol} + \text{carbon dioxide}$$

$$C_6H_{12}O_6 \longrightarrow 2C_2H_5OH + 2CO_2$$

a Which gas is given off when sugar is fermented?

b Yeast is a living thing. It is a type of fungus. What type of molecules in yeast enable it to ferment sugar?

Figure 1 Some people brew their own alcoholic drinks. The fermentation stage is often carried out by leaving the fermenting mixture in a warm place. The enzymes in yeast work best in warm conditions.

Did you know ...?

The yeast in a fermenting mixture cannot survive in concentrations of ethanol beyond about 15%. Alcoholic spirits, such as whisky or vodka, need to be distilled to increase the ethanol content to about 40% of their volume. Ethanol in high concentrations is toxic, which is why ethanol in the lab should never be drunk!

Practical

Fermentation

In this experiment you can ferment sugar solution with yeast and test the gas given off.

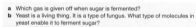

If you leave your apparatus till next lesson, your teacher can collect some fermented mixtures together and distil it to collect the ethanol formed. We use fractional distillation for the best separation as water and ethanol have similar boiling points. Ethanol boils at 78 °C. The ethanol collected will ignite and burn with a 'clean' blue flame.

links
For information on using ethanol as a fuel, look back at C1 4.5 Alternative fuels.

Ethanol is also used as a solvent. Methylated spirit is mainly ethanol. Decorators can use it to clean brushes after using an oil-based paint. It is also used to make perfume. We have already seen how ethanol can be used as a fuel. It can be mixed with petrol or just used by itself to run cars.

Making ethanol from ethene (hydration)

Ethanol for industrial use as a fuel or solvent can be made from ethene gas instead of by fermentation. Remember that ethene is made when oil companies crack hydrocarbons to make fuels. Ethene is the main by-product made in cracking. Ethene gas can react with steam to make ethanol.

$$\text{ethene} + \text{steam} \xrightarrow{\text{catalyst}} \text{ethanol}$$

$$C_2H_4 + H_2O \longrightarrow C_2H_5OH$$

links
For information on cracking, look back at C1 5.1 Cracking hydrocarbons.

This reaction is called **hydration**.

c Where do we get the ethene from to make industrial ethanol?

The reaction requires energy to heat the gases and to generate a high pressure. The reaction is reversible so ethanol can break down back into ethene and steam. So unreacted ethene and steam need to be recycled over the catalyst.

This process is continuous. It also produces no waste products. Both of these are advantages when making products in industry. When ethanol is made industrially by fermentation, the process is carried out in large vats which have to be left. This is called a batch process, which takes a lot longer than a continuous process. Carbon dioxide, a greenhouse gas, is also given off in fermentation.

Figure 2 Industrial fermentation is a slow batch process. The ethanol must be distilled off from the fermented mixture. This requires energy even though the fermentation process itself is energy efficient.

However, using ethene to make ethanol relies on crude oil which is a **non-renewable** resource. Therefore making ethanol as a biofuel, by fermenting sugars from plant material (a renewable resource), will become ever more important. The sugars are from crops such as sugar cane or sugar beet. Any cereal crop can also be used as the raw material. These need their starch to be broken down to sugars before fermentation takes place. However as we have seen before there are issues that need to be addressed when using crops for large-scale industrial processes.

links
For information on the issues of using crops for large scale industrial processes, look back at C1 4.5 Alternative fuels.

Summary questions

1 Copy and complete using the words below:
catalyst sugar yeast steam
Ethanol can be made by two processes, ethene reacting with ...., under pressure in the presence of a ...., or the fermentation of .... using enzymes in ....

2 Write a word equation to show the production of ethanol from:
a ethene
b glucose.

3 Why is a continuous process better than a batch process for making a product in industry?

4 How can people claim that the fermentation of plant materials does not contribute to the increase in carbon dioxide in the air?

Key points

- Ethanol can be made from ethene reacting with steam in the presence of a catalyst. This is called hydration.
- Ethanol is also made by fermenting sugar (glucose) using enzymes in yeast. Carbon dioxide is also made in this reaction.
- Using ethene to make ethanol needs non-renewable crude oil as its raw material whereas fermentation uses renewable plant material.

Further teaching suggestions

Magazine article

- Ask students to write a feature article in a car magazine about the developments in using ethanol as a fuel for cars.

Top Trumps

- You could ask students to make 'Top Trump'-style cards contrasting the production of ethanol by fermentation and by hydration of ethene. Students should evaluate the two methods and suggest which they think is better and why. Encourage students to try to classify their reasoning as social, environmental or economical.

Summary answers

1 steam, catalyst, sugar, yeast

2 **a** $\text{ethene} + \text{steam} \xrightarrow{\text{catalyst}} \text{ethanol}$

b $\text{glucose} \xrightarrow{\text{yeast}} \text{ethanol} + \text{carbon dioxide}$

3 The process never has to be stopped, unlike a batch process, to separate off the product and re-stock the reactants. This makes a continuous process more efficient.

4 Although fermentation gives off carbon dioxide, the gas is also taken in when the sugar was made by the plant when it made the sugar originally (during photosynthesis).

Summary answers

1 **a** A compound that contains only hydrogen and carbon.

b The use of heat and a catalyst to break down a longer chained hydrocarbon to form a shorter chain alkane and alkenes.

c A method used to separate mixtures of liquids with different boiling points.

d A molecule with only single bonds that contains only carbon and hydrogen atoms.

e A molecule with at least one C=C bond that contains only carbon and hydrogen atoms.

f Small, reactive molecules that react together to make a polymer.

g Very large molecules made from many repeating units/monomers.

h A polymer that can be broken down by microorganisms in the soil.

i The reaction in which glucose/sugar is turned into ethanol plus carbon dioxide by the action of enzymes in yeast.

2 **a** C_3H_6

b

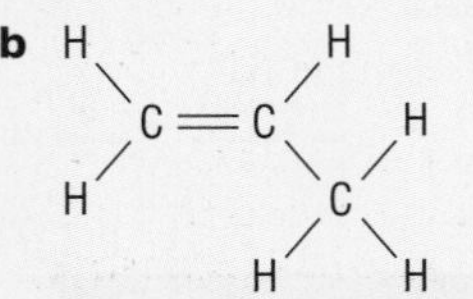

$CH_2 = CH - CH_3$

c Unsaturated, as it is an alkene and contains C=C (its molecules do not contain maximum number of hydrogen atoms).

d Put bromine water into each tube. Shake well. The tube that decolourises contains the unsaturated hydrocarbon, which is propene. The tube with propane will not change the colour of the bromine water – it will remain yellow/orange.

e **i** polymerisation

ii The polymer product is a solid at room temperature whereas the propene is a gas.

iii $nC_2H_4 \rightarrow \text{-}(CH_2 - CH_2)_n\text{-}$

3 **a** Because it is not biodegradable it takes many, many years to break down, persisting as rubbish and taking up space in landfill sites.

b They can build starch into the structure of a plastic so bacteria can feed on the starch and break the plastic into smaller pieces, which break down more quickly. Or they can use the starch to make biodegradable plastics which bacteria can use as food.

4 **a**

$$\text{ethene} + \text{steam} \xrightarrow{\text{catalyst}} \text{ethanol}$$

$$C_2H_4 + H_2O \rightarrow C_2H_5OH$$

b

$$\text{glucose} \xrightarrow{\text{yeast}} \text{ethanol} + \text{carbon dioxide}$$

$$C_6H_{12}O_6 \rightarrow 2C_2H_5OH + 2CO_2$$

5

Ethanol from ethene		Ethanol from fermenting plant sugars	
Advantage	**Disadvantage**	**Advantage**	**Disadvantage**
Continuous process	Uses up crude oil (non-renewable)	Uses plants (renewable) as its raw material	Batch process

6 Water molecules from sweat can pass through the tiny pores in the material so the wearer does not feel clammy – it is described as 'breatheable' because water vapour passes out into the air (as it does when we breathe out).

Products from oil: C1 5.1–C1 5.5

Summary questions 

1 Write simple definitions for the following words:

a hydrocarbon
b cracking
c distillation
d saturated hydrocarbon
e unsaturated hydrocarbon
f monomer
g polymer
h biodegradable polymer
i fermentation.

2 Propene is a hydrocarbon molecule containing three carbon atoms and six hydrogen atoms.

a What is the chemical formula of propene?

b Draw the display formula of propene, showing all its bonds.

c Is propene a saturated molecule or an unsaturated molecule? Explain your answer.

d You are given two unlabelled test tubes. One test tube contains propane gas, while the other test tube contains propene gas. Explain how you could test which tube contains which gas, stating clearly the results obtained in each case.

e Propene molecules will react together to form long chains.

i What do we call this type of reaction?

ii Compare the properties of the reactants to those of the product.

iii A molecule of ethene is a similar to a molecule of propene. Give an equation to show the reaction of ethene to make poly(ethene).

3 **a** Why does the disposal of much of our plastic waste cause problems?

b How can chemists help to solve the issues in part **a** using a plant material such as starch from corn?

4 **a** Write a word equation and a balanced symbol equation for the reaction between ethene and steam. [H]

b Write a word and balanced symbol equation for the fermentation of glucose. [H]

5 Draw a table showing the advantages and disadvantages of making ethanol from ethene or from sugar obtained from plant material.

6 Chemists have developed special waterproof materials made from polymers. The polymer materials have pores that are 2000 times smaller than a drop of water. However, the tiny pores are 700 times larger than a water molecule. Explain why these materials are described as 'breathable'.

7 Non-biodegradable plastic has been used for many years for growing melons. The plants are put into holes in the plastic and their shoots grow up above the plastic. The melons are protected from the soil by the plastic and grow with very few marks on them. Biodegradable plastic has been tested – to reduce the amount of non-recycled waste plastic.

In this investigation two large plots were used to grow melons. One using biodegradable plastic, the other using normal plastic. The results were as follows:

Plastic used	Total yield (kg/hectare)	Average mass of melons produced (kg)
Non-biodegradable	4829	2.4
Biodegradable	3560	2.2

a This was a field investigation. Describe how the experimenter would have chosen the two plots.

b What conclusion can you draw from this investigation?

c How could the reliability of these results be tested?

d How would you view these results if you were told that they were funded by the manufacturer of the traditional non-biodegradable plastic?

84

7 **a** The plots should have been chosen to minimise any differences in relation to, for example, soil conditions or weather conditions. Any changes in these conditions should be similar in all plots.

b The total yield is much less with the biodegradable plastic. The average melon weight is also less with the biodegradable plastic.

c To test the reliability of these results, the investigation should be repeated by other scientists. An investigation could be carried out in more controlled conditions, e.g. in a glasshouse. Increase the sample size investigated. Repeat the investigation on another soil type.

d With scepticism, because they have an interest in selling their plastic and could be using selected results. They might not be telling the whole truth.

Kerboodle resources

Resources available for this chapter on Kerboodle are:

- Chapter map: Products from oil
- Animation: Cracking crude oil (C1 5.1)
- Support: Cracking up – alkenes (C1 5.1)
- Extension: Cracking up – alkenes (C1 5.1)
- Viewpoints: In the bag (C1 5.4)
- Practical: Making a biodegradable plastic (C1 5.4)
- Interactive activity: Products from crude oil
- Revision podcast: Biodegradable polymers
- Chapter map: Products from oil
- On your marks: Using polymers
- Examination-style questions: Products from oil
- Answers to examination-style questions: Products from oil

AQA Examination-style questions

Large alkanes from crude oil are broken down to give smaller molecules.

Large alkane (e.g. $C_{12}H_{26}$) → vaporised and passed over hot catalyst → smaller alkane (e.g. C_5H_{12}) + alkene (e.g. C_2H_4)

Choose the correct word from the list to complete each sentence.

a This process is called ....
cracking distillation fermentation (1)

b The reaction is an example of thermal ....
decomposition evaporation polymerisation (1)

c The smaller alkane can be used as a ....
plastic monomer fuel (1)

d The alkene will turn bromine water ....
blue colourless orange (1)

e The general formula for an alkene is ....
C_nH_{2n-2} C_nH_{2n} C_nH_{2n+2} (1)

Ethene is used to mke the plastic poly(ethene).

a Complete the equation to show the formation of poly(ethene). (3)

$n\ CH_2{=}CH_2 \longrightarrow$

b In the equation, what does the letter *n* represent? (1)

c What name is used for the small molecules that join to make a polymer? (1)

d Name the polymer that is made from butene. (1)

e Which one of the following could be used in a similar way to make a polymer?
C_3H_6 C_3H_8 C_4H_{10} (1)

Ethanol can be used as a fuel for cars. Pure ethanol (100%) can be used in specially adapted car engines. Petrol with up to 10% ethanol can be used in ordinary car engines. To mix with petrol the ethanol must not contain any water.

Ethanol can be made from plants or from crude oil.

plants → sugars →(*fermentation*) 15% ethanol in water →(*distillation*) 96% ethanol →(*dehydration*) 100% ethanol

crude oil →(*distillation*) fractions →(*cracking*) ethene →(*catalyst + steam*) 100% ethanol

a Suggest **one** environmental advantage of making ethanol fuel from plants rather than from crude oil. (1)

b Suggest **one** economic disadvantage of producing ethanol fuel from plants rather than from crude oil. (1)

c Suggest **one** environmental disadvantage of producing ethanol fuel from plants. (1)

d 10% ethanol in petrol can be used in ordinary car engines. Suggest **one** other advantage of using 10% ethanol in petrol as a fuel rather than pure ethanol. (1)

4 Scientists develop new polymers and modify existing polymers.

a Polylactic acid (PLA) is a bioplastic that is biodegradable. It can be used to make sandwich containers, plastic cups and plastic cutlery. PLA is made from cornstarch. In the USA large amounts of maize are grown and used to make cornstarch, which has many uses. To make PLA the cornstarch is fermented with microbes to make lactic acid, which is then polymerised.

The structure of PLA is $\left(O{-}CH(CH_3){-}C({=}O) \right)_n$

i Give **one** way in which the structure of PLA is different from the structure of poly(ethene). (1)

ii Give **one** way in which the structure of PLA is similar to the structure of poly(ethene). (1)

iii Suggest what is meant by *bioplastic.* (1)

iv Suggest **two** reasons why PLA was developed. (2)

b *In this question you will be assessed on using good English, organising information clearly and using specialist terms where appropriate.*

Copper was considered to be the most suitable material to use for hot water pipes. PEX is now used as an alternative material for hot water pipes. PEX is made from poly(ethene).

Copper is extracted from its ore by a series of processes.

1 The low-grade ore is powdered and concentrated.
2 Smelting is carried out in an oxygen flash furnace. This furnace is heated to 1100 °C using a hydrocarbon fuel. The copper ore is blown into the furnace with air, producing impure, molten copper.
3 Oxygen is blown into the impure, molten copper to remove any sulfur. The copper is cast into rectangular slabs.
4 The final purification of copper is done by electrolysis.

Suggest the possible environmental advantages of using PEX instead of copper for hot water pipes. (6)

AQA, 2009

Practical suggestions

Practicals	AQA	k	📖	⚙
Demonstration of the cracking of liquid paraffin using broken pottery as the catalyst.	✓		✓	
Testing for unsaturation in the alkenes using bromine water.	✓		✓	
Making a polymer from cornstarch.	✓	✓	✓	
Demonstration of making Perspex.	✓		✓	
Molecular modelling of polymers.	✓		✓	
Design an investigation of a property of different plastics, e.g. strength, flexibility, biodegradability.	✓		✓	
Investigate the amount of water that can be absorbed by a hydrogel (e.g. those used as additives to garden composts).	✓		✓	
Testing coated fabrics for water penetration.	✓			

Examination-style answers

1 a cracking *(1 mark)*
b decomposition *(1 mark)*
c fuel *(1 mark)*
d colourless *(1 mark)*
e C_nH_{2n} *(1 mark)*

2 a single C—C, with 2H— attached to each C (1), brackets around this with – through brackets on both sides (1) subscript n outside right bracket (1) *(3 marks)*
b a large number *(1 mark)*
c monomers *(1 mark)*
d poly(butene) *(1 mark)*
e C_3H_6 *(1 mark)*

3 a One environmental advantage from: renewable (source), carbon neutral, resources conserved. *(1 mark)*
b One economic disadvantage from: more labour intensive, takes long(er) time, more steps/processing (to make pure ethanol). *(1 mark)*
c Land/crops used (that could be used for food) or deforestation. *(1 mark)*
d Can use existing supply/storage system (petrol stations/ tanks) or does not need new/different/special system. *(1 mark)*

4 a i Contains oxygen (atoms), has a double bond, is not a hydrocarbon, different monomer/repeating unit. *(1 mark)*
ii Made from monomers, has a repeating unit. *(1 mark)*
iii Plastic made from biological/plant/biomass resources or made using microbes. *(1 mark)*
iv Two from: To replace poly(ethene) or other (non-biodegradable) plastics, it is biodegradable, it uses renewable source material, cornstarch readily available, cornstarch cheap, (accept economic reasons such as to keep farmers in jobs, use up surplus corn). *(2 marks)*

b There is a clear, balanced and detailed scientific description of the environmental advantages of using PEX instead of copper for hot water pipes. The answer shows almost faultless spelling, punctuation and grammar. It is coherent and in an organised, logical sequence. It contains a range of appropriate or relevant specialist terms used accurately. *(5–6 marks)*

There is a scientific description that includes some of the environmental advantages of using PEX. There are some errors in spelling, punctuation and grammar. The answer has some structure and organization. The use of specialist terms has been attempted, but not always accurately. *(3–4 marks)*

There is a brief description of at least one advantage of using PEX. The spelling, punctuation and grammar are very weak. The answer is poorly organised with almost no specialist terms and/or their use demonstrating a general lack of understanding of their meaning. *(1–2 marks)*

No relevant content. *(0 marks)*

Examples of chemistry points made in the response:

- Less (hydrocarbon) fuel used
- Less energy used
- Less/no electricity used (allow no electrolysis)
- Carbon/carbon dioxide emissions reduced or less global warming
- Less/no pollution by sulfur dioxide/acid rain
- Conserve copper resources
- Reduces the amount of solid waste rock or less waste (allow copper ores are low-grade)
- Reduces the need to dig large hole/pits/mines.

C1 6.1 Extracting vegetable oil

Learning objectives

Students should learn:

- that oils can be extracted from plants
- that vegetable oils are important foods
- what unsaturated oils are and how we can test for them.

Learning outcomes

Most students should be able to:

- describe how oils can be extracted from plants
- recognise an unsaturated oil
- describe why plant oils are important in foods
- state the test for a compound containing a carbon–carbon double bond.

Some students should also be able to:

- detail a method for extracting and testing unsaturated oils.

Answers to in-text questions

a Pressing and (steam) distillation.

b The orange bromine water will be decolourised (turn colourless).

Support

- Some students may struggle to remember the test for saturation and to interpret the results correctly. You could give students two diagrams of two test tubes with an arrow between them. Ask them to use colours and labels to represent the test for saturation. Students can look back at their diagrams to help them interpret the results of their practical.
- Ask the students to imagine that they work for a marketing company. They should make a poster to encourage students to think about science in their everyday lives (such as the RSC posters 'scientists don't always wear white coats'). Their poster should include all the key points in this chapter and should be added to as each topic is covered.

Extend

- You could stretch students by giving them some displayed formula of different fats that we eat. Ask students to classify them as saturated, unsaturated, monounsaturated and polyunsaturated and explain why.

Specification link-up: Chemistry C1.6

- Some fruits, seeds and nuts are rich in oils that can be extracted. The plant material is crushed and the oil removed by pressing or in some cases by distillation. Water and other impurities are removed. *[C1.6.1 a)]*
- Vegetable oils are important foods and fuels as they provide a lot of energy. They also provide us with nutrients. *[C1.6.1 b)]*
- Vegetable oils that are unsaturated contain double carbon–carbon bonds. These can be detected by reacting with bromine water. *[C1.6.3 a)]*

Lesson structure

Starters

List – Ask students to make a list of oily foods. Then ask them to consider where the oil comes from, e.g. crisps – sunflower seeds, chocolate spread – nuts, olive oil – seeds. *(5 minutes)*

Sentences – Ask students to finish the sentence: 'Plants get their energy ...' Encourage students to try to finish it four times, each sentence becoming progressively more scientific, e.g.

1 Plants get their energy from the Sun.
2 Plants get their energy from sugar, that is made by using sunlight.
3 Plants get their energy using photosynthesis to make the sugar glucose.
4 Plants get their energy according to the following equation:

carbon dioxide + water → sugar (glucose) + oxygen

You may wish to support students by giving the four sentences and asking them to order the statements from the most to the least scientific. To extend students, you could ask them to write a balanced symbol equation for photosynthesis. You may wish to give students the formula of glucose [$C_6H_{12}O_6$] to aid them. Ensure that the Student Book is closed to do this extension of the starter. *(10 minutes)*

Main

- Some plants quickly release oil when they are crushed, e.g. nuts and seeds. Other oils are more difficult to extract and steam distillation needs to be used. This can be seen by crushing nuts and looking at the grease stain produced, see 'Practical support' for more detail.
- Students can complete steam distillation on a micro-scale or crush plant material to extract oil (be aware of nut allergies and eye protection must be worn) using mortar and pestle. (See if you get a translucent stain on filter paper.)
- Students should undertake the 'Testing for unsaturation' practical to reinforce the test for a saturated or unsaturated compound.
- Fats are an important part of the human diet. In the media, we often hear terms such as 'saturated, unsaturated, and polyunsaturated fats'. Students often do not realise that these terms have a scientific meaning. Ask them to consider these words and define them [saturated – no double bonds, unsaturated – contains double bonds, polyunsaturated – contains many double bonds]. They can then classify different fats as saturated or unsaturated using bromine water, see 'Practical support' for more detail.

Plenaries

Questions and answers – Ask a student to pick a number from 1 to the number of students in the class. Look at which number this corresponds to in the register. Then ask the student to generate a question about the topic studied and choose a person to answer it. The question maker then decides if the answer is correct. If misconceptions are highlighted, then you should take over the question and answer in order to correct them. *(5 minutes)*

Flow chart – Ask the students to make a flow chart to demonstrate two different ways of extracting plant oils. You may wish to support students by asking them to work in small groups and giving the stages in one flow chart as a card sort. Students then just sort the cards to put the stages in the correct order. The students in each group should then compare their flow chart with another group to review both methods. To extend students, you could ask them to evaluate the two methods and comment on which is likely to be the most expensive to operate [distillation, as this still needs to be heated and therefore uses more energy]. *(10 minutes)*

Practical support

Oils from nuts

Equipment and materials required

Paper towel, nuts (e.g. walnuts).

Details

Put a nut on a paper towel and fold it over, so that the nut is wrapped in the towel. Push onto the nut with the palm of your hand so that the nut is crushed. Remove the nut and observe the fat stains on the paper towel.

Safety: Be aware of nut allergies. Large seeds are an alternative (broad beans).

Extracting plant oil by distillation

Equipment and materials required

Orange, grater, antibumping granules, methylated spirits, silicone oil, microscale distillation apparatus (see 'Student Book').

Details

Grate the zest from part of an orange and place it in a small vial in the apparatus shown – one-quarter full. Mix with water to half fill the vial and add a few antibumping granules. Heat gently, to avoid the mixture boiling over. Collect a few drops of the 'orange oil' emulsion in a small well. Note its smell and cloudy appearance. (The apparatus is available from Edulab, Karoo Close, Bexwell, Norfolk PE38 9GA.) This process can be used to extract oils from other plant materials including lemon zest and lavender flowers.

Safety: Follow instructions for equipment. CLEAPSS Hazcard 40A Methylated spirit – highly flammable and harmful.

Testing for unsaturation

Equipment and materials required

A selection of fats (e.g. butter, margarine, dripping, lard, olive oil, vegetable oil), dropping pipettes, test tubes, bromine water, bungs, test-tube rack, water bath, ethanol, eye protection.

Details

Students should pipette about 1 cm^3 of each liquid into separate test tubes, stored in a rack. If a solid fat is to be tested, e.g. dripping, then it needs to be dissolved in ethanol. About the same amount of bromine water should be added to each test tube, and each bunged and shaken. Ensure that the students wear eye protection and hold the bung and base of the test tube when they shake it. Bromine vapour may be given off (so use pale yellow bromine water); therefore a well-ventilated area is necessary and students should wash their hands after the practical. Asthmatics may experience problems with any bromine vapour present.

Safety: CLEAPSS Hazcard 15B Bromine water. It is only harmful at concentration used. CLEAPSS Hazcard 40A Ethanol – highly flammable and harmful. No naked flames.

Plant oils

C1 6.1 Extracting vegetable oil

Learning objectives

- How do we extract oils from plants?
- Why are vegetable oils important foods?
- What are unsaturated oils and how do we detect them?

Plants use the Sun's energy to produce glucose from carbon dioxide and water during photosynthesis:

$$\text{carbon dioxide} + \text{water} \xrightarrow[\text{energy (from sunlight)}]{\text{chlorophyll}} \text{glucose} + \text{oxygen}$$

$$6CO_2 + 6H_2O \longrightarrow C_6H_{12}O_6 + 6O_2$$

Plants then turn glucose into other chemicals they need using more chemical reactions. In some cases these other chemicals can also be very useful to us. For example, the **vegetable oils** from plants, such as oilseed rape, make biofuels and foodstuffs.

We find these oils in the seeds of the rape plant. Farmers collect the seeds from the plants using a combine harvester. The seeds are then taken to a factory where they are crushed and pressed to extract their oil. The impurities are removed from the oil. It is then processed to make it into useful products.

We extract other vegetable oils using steam. For example, we can extract lavender oil from lavender plants by distillation. The plants are put into water and boiled. The oil and water evaporate together and are collected by condensing them. The water and other impurities are removed to give pure lavender oil.

Figure 1 Oilseed rape is a common sight in our countryside. As its name tells us, it is a good source of vegetable oil.

Practical

Extracting plant oil by distillation (microscale)

Take care not to let the contents of the small vial boil over.

- What does the liquid collected look and smell like?

Figure 2 Norfolk lavender oil is extracted from lavender plants by distillation

a Write down two ways we can use to extract vegetable oils from plants.

86

Extracting vegetable oil

Vegetable oils as foods and fuels

Vegetable oils are very important foods. They provide important nutrients. For example olive oil is a source of vitamin E. They also contain a great deal of energy, as the table shows. This makes them useful foods and sources of biofuels, such as biodiesel.

There are lots of different vegetable oils. Each vegetable oil contains mixtures of compounds with slightly different molecules. However, all vegetable oils have molecules which contain chains of carbon atoms with hydrogen atoms:

In some vegetable oils the hydrocarbon chains contain carbon–carbon double bonds (C=C). We call these **unsaturated oils**. We can detect the double bonds in unsaturated oils with bromine water. You know the test for double bonds from your work on alkenes.

This provides us with an important way of detecting unsaturated oils:

unsaturated oil + bromine water (orange) → colourless solution

b What will you see if you test a polyunsaturated margarine with bromine water?

Practical

Testing for unsaturation

Summary questions

1 Copy and complete using the words below:
bromine decolorised distillation energy pressing unsaturated
We can extract vegetable oils from some plants by ………… or ………… . Vegetable oils are particularly important as foods because they contain a lot of ………… . Some vegetable oils contain carbon–carbon double bonds. We call these ………… vegetable oils. They can be detected by reacting them with ………… water, which will be ………… .

2 Why might a diet containing too much vegetable oil be unhealthy?

3 A sample of vegetable oils is tested with bromine water. The solution is decolorised. Which of the following statements is true?
  a The sample contains ***only*** unsaturated oils.
  b The sample contains ***only*** saturated oils.
  c The sample may contain a mixture of saturated and unsaturated oils.
  Explain your answer.

links
For information on biofuels, look back at C1 4.5 Alternative fuels.

Energy in vegetable oil and other foods	
Food	**Energy in 100 g (kJ)**
vegetable oil	3900
sugar	1700
animal protein (meat)	1100

Figure 3 Vegetable oils have a high energy content

links
For information on the test for double bonds, look back at C1 5.1 Cracking hydrocarbons.

Did you know ...?
No more than 20% of the energy in your diet should come from fats.

Key points

- Vegetable oils can be extracted from plants by pressing or by distillation.
- Vegetable oils provide nutrients and have a high energy content. They are important foods and can be used to make biofuels.
- Unsaturated oils contain carbon–carbon double bonds (C=C). We can detect them as they decolorise bromine water.

87

Further teaching suggestions

Making perfume

- Students could extract oils from petals in order to make their own 'perfume'. Students can extract oil from fruits such as olives see www.sep.org.uk.

Summary answers

1. pressing (distillation), distillation (pressing), energy, unsaturated, bromine, decolorised
2. Because vegetable oils contain a lot of energy, which becomes stored as fat in the body if it is not used up.
3. Answer **c** – we can be certain that the sample contains unsaturated oils because it decolorises bromine water but we cannot be certain whether or not it contains saturated oils because these have no effect on bromine water.

C1 6.2 Cooking with vegetable oils

Learning objectives

Students should learn:

- reasons why people cook with vegetable oils
- what 'to harden' vegetable oils means **[HT only]**
- how to turn vegetable oils into spreads **[HT only]**

Learning outcomes

Most students should be able to:

- list advantages and a disadvantages of using vegetable oils for cooking.

Some students should also be able to:

- define the term 'harden' **[HT only]**
- explain why and how oils are turned into spreads by the addition of hydrogen to give hydrogenated oils with melting points above room temperature. **[HT only]**

Specification link-up: Chemistry C1.6

- Vegetable oils have higher boiling points than water and so can be used to cook foods at higher temperatures than by boiling. This produces quicker cooking and different flavours but increases the energy that the food releases when it is eaten. *[C1.6.1 c)]*
- Vegetable oils that are unsaturated can be hardened by reacting them with hydrogen in the presence of a nickel catalyst at about 60 °C. Hydrogen adds to the carbon–carbon double bonds. The hydrogenated oils have higher melting points so they are solids at room temperature, making them useful as spreads and in cakes and pastries. *[C1.6.3 b)]* **[HT only]**

Controlled Assessment: C4.1 Plan practical ways to develop and test candidates' own scientific ideas. *[C4.1.1 a) b) c)]*

Lesson structure

Starters

Stand-by – Put the key words: 'saturated' and 'unsaturated' onto two large pieces of paper and pin them at opposite sides of the room. Read out different examples of chemicals. Students should decide whether they are saturated or unsaturated and stand by the appropriate sign, for instance alkenes [unsaturated], alkanes [saturated], ethene [unsaturated], propane [saturated], octane [saturated]. *(5 minutes)*

Thermometer – Draw a thermometer on the board. Draw two lines to divide it approximately into thirds. Ask the students to label the states of matter, label the state change points and give the changes that occur to the organisation of particles at each state change. You could support students by using a familiar substance – water – to illustrate how the state of the substance relates to the melting point and boiling point. You could extend students by giving them the melting point and freezing point of different saturated and unsaturated fats and asking them to suggest the state at a particular temperature. *(10 minutes)*

Main

- When foods are cooked at different temperatures they have different flavours, appearances, smells and nutritional content. Students could compare potatoes that have been prepared in different ways. Encourage the students to design their own results table to record their observations. Support students by allowing them to copy a previously prepared one.
- Students taking the higher paper need to know the processes involved in creating a spread from an oil and the reasons why this is done. Split students into small groups and ask them to imagine that they are going to produce a short feature, as part of a TV programme about how food is manufactured. Give students ideas. For example, interview a chemical engineer at the oil plant. Students could then make a storyboard to show what would happen in each scene and the information that would be given. **[HT only]**
- Students could investigate the degree of saturation in a variety of oils using bromine water. 'How Science Works' concepts of fair testing and measurement can be practised. Some might try a colorimeter or a light sensor and data logger to get quantitative data on how much bromine remains after reaction with the oils. A little ethanol will help the oil and aqueous layers to mix for the test.

Plenaries

Objectives – Ask students to answer the questions posed by the objectives. You could support the students by encouraging them to work in small groups to craft their answers. You could extend them by asking them to illustrate their answers with examples. *(5 minutes)*

Question loop – Small pieces of paper are given to each student. They write a question and its answer. They then separate the question from the answer and all the papers are collected and shuffled. Each student is then given a question and an answer. The first student reads the question. The student with the correct answer reads their answer, then his or her own question and so on. *(10 minutes)*

Support

- Students could be supported during the debate by reducing the number of characters involved. Students could work in small teams to prepare just one member for an active role in the debate. They could also be provided with some statements and views that the person knows/holds and the students could use this as an input considering if they are advantages/disadvantages and whether they are facts or opinions.

Extend

- Students could be extended by being shown the displayed formula of a polyunsaturated fat and using arrow – pushing show how the hydrogen attaches across the double bond. Refer back to 'C1 1.4 Forming bonds' to remind students of the importance of electrons in bonding. This might help students visualise the process, although they do not need to know the mechanism.

Practical support

Investigating cooking

Equipment and materials required

Raw potato cores, boiled potato cores, fried potato cores, mounted needle, magnifying glass, cooker, chip pan, cooking oil, saucepan, water, stopwatch, knife.

Details

Ask the students to study the appearance of the three different samples, using the magnifying glass. They should note the colour and smell. Then, using a mounted needle, they can scratch the surface and comment on the textures.

Students should carefully cook the potato cores in cooking oil (this may be completed as a demonstration because boiling oil can be dangerous) and boiling water. They should time how long it takes for each. They could repeat the test three times to gain an average, or pool the class results in order. The subjectivity of 'the time to cook' can be discussed. Once the potatoes have been cooked, they could be tasted (but not in the laboratory), and students could comment on their different taste and texture in the mouth due to the cooking conditions. Alternatively, the potato cores can be cooked in oil and in water for equal lengths of time, allowed to cool and then tested.

Safety: Know how to extinguish a fat fire. If tasting, move to food technology room.

Comparing oils for degree of unsaturation

Equipment and materials required

Bromine water, variety of plant oils, ethanol, test tubes, bungs, dropping pipettes (possibly colorimeter, light sensor, data logging equipment), eye protection.

Details

Mix the oil and bromine water. Stopper and shake. Judge amount of decolourisation.

Safety: CLEAPSS Hazcard 15B Bromine water – irritant and harmful; 40A Ethanol – highly flammable and harmful. No naked flames.

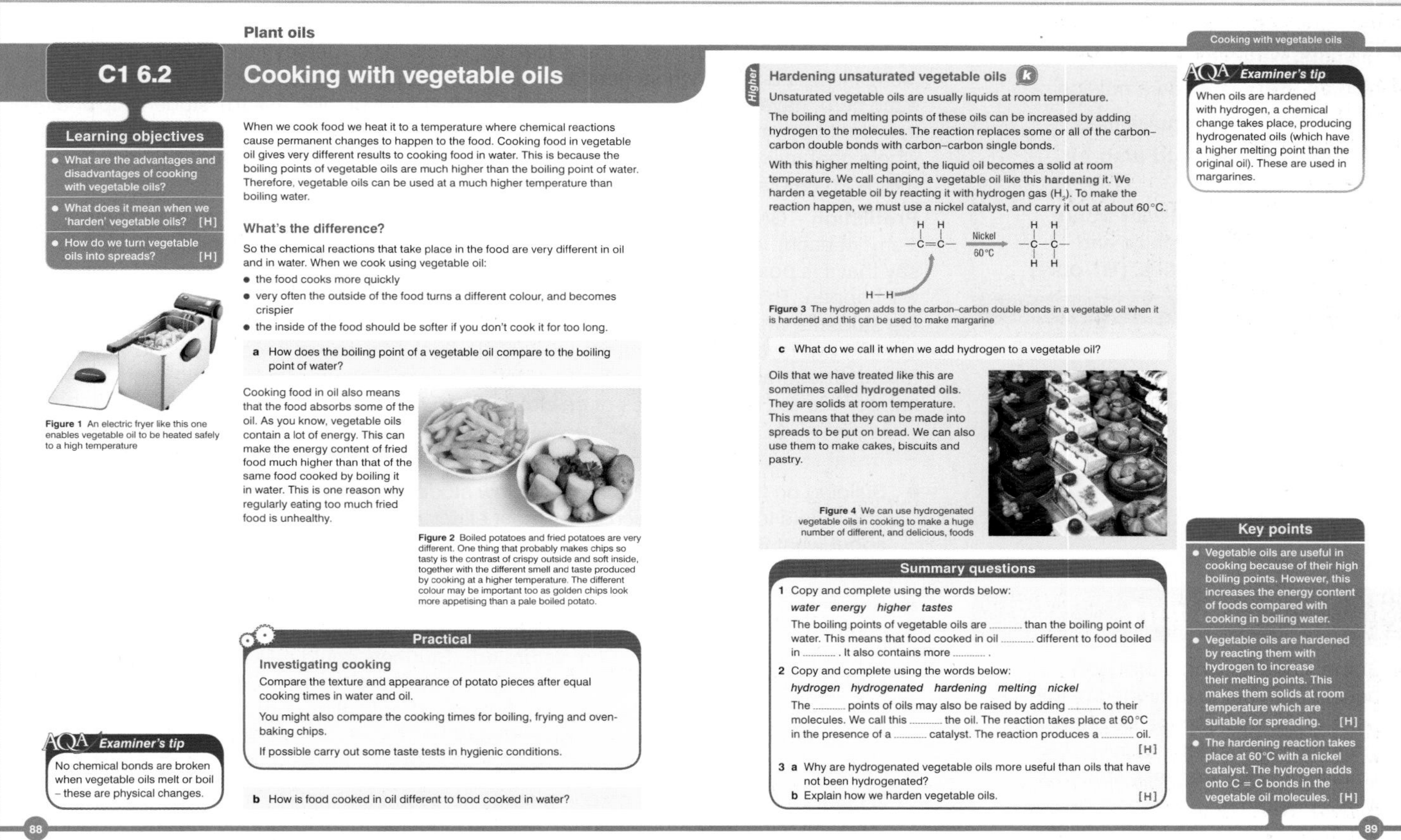

Plant oils

C1 6.2 Cooking with vegetable oils

Learning objectives

- What are the advantages and disadvantages of cooking with vegetable oils?
- What does it mean when we 'harden' vegetable oils? [H]
- How do we turn vegetable oils into spreads? [H]

When we cook food we heat it to a temperature where chemical reactions cause permanent changes to happen to the food. Cooking food in vegetable oil gives very different results to cooking food in water. This is because the boiling points of vegetable oils are much higher than the boiling point of water. Therefore, vegetable oils can be used at a much higher temperature than boiling water.

What's the difference?

So the chemical reactions that take place in the food are very different in oil and in water. When we cook using vegetable oil:

- the food cooks more quickly
- very often the outside of the food turns a different colour, and becomes crispier
- the inside of the food should be softer if you don't cook it for too long.

a How does the boiling point of a vegetable oil compare to the boiling point of water?

Figure 1 An electric fryer like this one enables vegetable oil to be heated safely to a high temperature

Cooking food in oil also means that the food absorbs some of the oil. As you know, vegetable oils contain a lot of energy. This can make the energy content of fried food much higher than that of the same food cooked by boiling it in water. This is one reason why regularly eating too much fried food is unhealthy.

Figure 2 Boiled potatoes and fried potatoes are very different. One thing that probably makes chips so tasty is the contrast of crispy outside and soft inside, together with the different smell and taste produced by cooking at a higher temperature. The different colour may be important too as golden chips look more appetising than a pale boiled potato.

Practical

Investigating cooking

Compare the texture and appearance of potato pieces after equal cooking times in water and oil.

You might also compare the cooking times for boiling, frying and oven-baking chips.

If possible carry out some taste tests in hygienic conditions.

AQA Examiner's tip

No chemical bonds are broken when vegetable oils melt or boil – these are physical changes.

b How is food cooked in oil different to food cooked in water?

88

Cooking with vegetable oils

Higher

Hardening unsaturated vegetable oils

Unsaturated vegetable oils are usually liquids at room temperature.

The boiling and melting points of these oils can be increased by adding hydrogen to the molecules. The reaction replaces some or all of the carbon–carbon double bonds with carbon–carbon single bonds.

With this higher melting point, the liquid oil becomes a solid at room temperature. We call changing a vegetable oil like this **hardening** it. We harden a vegetable oil by reacting it with hydrogen gas (H_2). To make the reaction happen, we must use a nickel catalyst, and carry it out at about 60 °C.

Figure 3 The hydrogen adds to the carbon–carbon double bonds in a vegetable oil when it is hardened and this can be used to make margarine

c What do we call it when we add hydrogen to a vegetable oil?

Oils that we have treated like this are sometimes called **hydrogenated oils**. They are solids at room temperature. This means that they can be made into spreads to be put on bread. We can also use them to make cakes, biscuits and pastry.

Figure 4 We can use hydrogenated vegetable oils in cooking to make a huge number of different, and delicious, foods

Summary questions

1 Copy and complete using the words below:

water energy higher tastes

The boiling points of vegetable oils are ………… than the boiling point of water. This means that food cooked in oil ………… different to food boiled in ………… . It also contains more ………… .

2 Copy and complete using the words below:

hydrogen hydrogenated hardening melting nickel

The ………… points of oils may also be raised by adding ………… to their molecules. We call this ………… the oil. The reaction takes place at 60 °C in the presence of a ………… catalyst. The reaction produces a ………… oil. [H]

3 a Why are hydrogenated vegetable oils more useful than oils that have not been hydrogenated?

b Explain how we harden vegetable oils. [H]

AQA Examiner's tip

When oils are hardened with hydrogen, a chemical change takes place, producing hydrogenated oils (which have a higher melting point than the original oil). These are used in margarines.

Key points

- Vegetable oils are useful in cooking because of their high boiling points. However, this increases the energy content of foods compared with cooking in boiling water.
- Vegetable oils are hardened by reacting them with hydrogen to increase their melting points. This makes them solids at room temperature which are suitable for spreading. [H]
- The hardening reaction takes place at 60 °C with a nickel catalyst. The hydrogen adds onto C = C bonds in the vegetable oil molecules. [H]

89

Further teaching suggestions

Comparing nutritional information

- Give students the nutritional information for different potato products. Ask them to draw a graph to compare their fat/calorie contents.

Displaying data

- Ask students to find out the boiling points of different fats used in cooking and to show this information in an appropriate diagram, e.g. a bar chart. You can extend students by plotting a line graph comparing molecular mass to the boiling point and determining the trend.

Flow chart

- Students could summarise the hydrogenation of fats in a flow chart format. Give students outline of the flow chart with prose, and the students could then draw a picture for each stage below the text.

Answers to in-text questions

a The boiling point of vegetable oils is much higher.

b Food cooked in oil cooks more quickly. The outside often turns a different colour and becomes crisper. The food absorbs some of the oil, which increases its energy content.

c hardening

Summary answers

1 higher, tastes, water, energy

2 melting, hydrogen, hardening, nickel, hydrogenated

3 **a** They have a higher melting point and can be used to make spreadable solids at room temperature.

b Heat the oil and hydrogen at 60 °C with a nickel catalyst.

C1 6.3 Everyday emulsions

Learning objectives

Students should learn:

- what emulsions are and how they are made
- that emulsions made from vegetable oils have many uses
- that emulsifiers keep the oil and water mixed together in an emulsion
- how an emulsifier works. **[HT only]**

Learning outcomes

Most students should be able to:

- describe what an emulsion is
- give an example of an emulsion
- describe what an emulsifier is.

Some students should also be able to:

- explain how an emulsifier works in terms of its hydrophilic and hydrophobic properties. **[HT only]**

Specification link-up: Chemistry C1.6

- Oils do not dissolve in water. They can be used to produce emulsions. Emulsions are thicker than oil or water and have many uses that depend on their special properties. They provide better texture, coating ability and appearance, for example in salad dressings, ice creams, cosmetics and paints. *[C1.6.2 a)]*
- Emulsifiers have hydrophilic and hydrophobic properties. *[C1.6.2 b)]* **[HT only]**

Controlled Assessment: C4.3 Collect primary and secondary data. *[C4.3.2 a) b)]*

Lesson structure

Starters

Order – Split the students into groups of six. Give each group a set of cards. Each card should be labelled with one of the following: 'whole milk', 'Jersey milk', 'skimmed milk', 'semi-skimmed milk', 'whole goat milk' and 'semi-skimmed goat milk'. On the back of each card, have the nutritional information for that type of milk. Ask the students to line up in order of the most to the least fat content. [The order should be: Jersey milk, whole milk, whole goat milk, semi-skimmed milk, semi-skimmed goat milk.] Introduce the idea that milk is an emulsion. *(5 minutes)*

Prediction – Give each student a bung and a test tube with water and oil in it (about 1 cm^3 of each). Ask the students how you could make the oil and water mix. Some might say that it is possible to shake it. Encourage the class to make a prediction as to what will happen. Then let the students shake their tubes and allow them time to observe. Ask the students what has happened [the oil and water mix at first but eventually separate out into two layers again]. Ask the students to think of a way of keeping oil and water mixed. Encourage them to think about doing the washing up or clothes washing, i.e. what happens to the fat from plates or on clothes? *(10 minutes)*

Main

- Building on the Starter, 'Prediction', allow students to test the effects of using different emulsifiers to find out which is the most effective. Ask the class to reach a consensus about what the control variables, independent variable and dependant variables are and how they will record their results.
- Students come into contact with many emulsions in everyday life. One of the most common is milk, which is fat suspended in water. Ask the students what type of fat is in milk [animal fat and therefore saturated]. Students could make wet slides of different types of milk to see the different emulsions using a microscope. Students could record their results in the form of a diagram, but they should also work out the magnification on the microscope (eye lens × object lens).
- Students could make another common emulsion – mayonnaise. If you can spare a lesson, a room swap can be completed into a food technology room, the students could take the mayonnaise home. Ask the students to try to identify the emulsifier in this mixture [egg yolk].
- Then students doing the Higher Tier paper can use the key words hydrophobic, hydrophilic, emulsifier and emulsion, to explain how mayonnaise is made. **[HT only]**

Plenaries

List – Ask students to list as many emulsions as they can. [For example, salad dressing, mayonnaise, salad cream, milk, ice cream, paint.] Students could be supported by being shown a selection of photographs, all of which are emulsions. Students could be extended by being asked to explain why emulsion paint is both an emulsion and a suspension. [It is a suspension of a solid (pigment) and liquid (resin). The resin is an emulsion of resin in water.] *(5 minutes)*

Label the emulsifier – Give the students a diagram of an emulsifier with its head in water and its tail in oil. Ask students to label the diagram fully using the terms 'hydrophobic' and 'hydrophilic' to explain how an emulsifier works. *(10 minutes)* **[HT only]**

Support

- Some students would probably benefit from being reminded of the parts of the microscope before using the equipment to look at everyday emulsions. Provide these students with a diagram of a microscope and ask them to label the key parts. You could turn this into a kinaesthetic activity by giving each pair of students a microscope and reading out the parts while the students point to them.

Extend

- You could extend students by telling them that emulsions are part of a group of mixtures called 'colloids'. These students could then be encouraged to define and give examples of other colloids, e.g. sol, gel, foam.
- Give students a diagram of an emulsifier and ask them to annotate it using the key words 'hydrophilic' and 'hydrophobic'. Students should then use this diagram to explain how an emulsifier works.

Practical support

Making and testing emulsions

Equipment and materials required

Boiling tube, bung, boiling tube rack, cooking oil, water, selection of liquid detergents with a dropping pipette, selection of detergents and a spatula, eye protection.

Details

Add about a quarter of a boiling tube of cooking oil and water until the boiling tube is half full. Add some detergent and put in the bung. Shake and observe. Repeat for all the different detergents.

Safety: Detergents are irritants and may be corrosive. Wear eye protection and wash hands when the practical is completed.

Making mayonnaise

Equipment and materials required

Eggs, salt, vegetable oil, vinegar, mustard powder, pepper, cooking bowl, hand whisk or fork and teaspoon.

Details

Here is a recipe for mayonnaise. Beat the eggs with a pinch of salt. Add the oil, one or two teaspoons at a time, while continuing to beat the mixture. After some of the oil has been added, add one or two teaspoons of vinegar. Continue to add oil slowly and beat the mixture. Add mustard and pepper to taste.

Safety: If tasting, move to food technology room. Be aware of food allergies and salmonella due to raw egg.

Plant oils

C1 6.3 Everyday emulsions

Learning objectives

- What are emulsions and how do we make them?
- Why are emulsions made from vegetable oils so important?
- What is an emulsifier?
- How do emulsifiers work? [H]

Emulsions in foods

The texture of food – what it feels like in your mouth – is a very important part of foods.

Some smooth foods are made from a mixture of oil and water. Everyone knows that oil and water don't mix. Just try it by pouring a little cooking oil into a glass of water. But we can get them to mix together by making the oil into very small droplets. These spread out throughout the water and produce a mixture called an **emulsion**.

A good example of this is milk. Milk is basically made up of small droplets of animal fat dispersed in water.

Figure 1 Mayonnaise is an emulsion. Smooth food has a good texture and looks as if it will taste nice – but it is not always easy to make, or to keep it smooth.

Figure 2 Milk is an emulsion made up of animal fat and water, together with some other substances

Emulsions often behave very differently to the things that we make them from. For example, mayonnaise is made from ingredients that include oil and water. Both of these are runny – but mayonnaise is not!

Another very important ingredient in mayonnaise is egg yolks. Apart from adding a nice yellow colour, egg yolks have a very important job to do in mayonnaise. They stop the oil and water from separating out into layers. Food scientists call this type of substance an **emulsifier**.

a What do we mean by 'an emulsifier'?

Emulsifiers make sure that the oil and water in an emulsion cannot separate out. This means that the emulsion stays thick and smooth. Any creamy sauce needs an emulsifier. Without it we would soon find blobs of oil or fat floating around in the sauce.

b How does an emulsifier help to make a good creamy sauce?

One very popular emulsion is ice cream. Everyday ice cream is usually made from vegetable oils, although luxury ice cream may also use animal fats.

Emulsifiers keep the oil and water mixed together in the ice cream while we freeze it. Without them, the water in the ice cream freezes separately, producing crystals of ice. That would make the ice cream crunchy rather than smooth. This happens if you allow ice cream to melt and then put it back in the freezer.

Figure 3 Ice cream contains emulsifiers

90

Everyday emulsions

Other uses of emulsions

Emulsifiers are also important in the cosmetics industry. Face creams, body lotions, lipsticks and lip gloss are all emulsions.

Emulsion paint (often just called emulsion) is a water-based paint with oil droplets dispersed throughout. It is commonly used for painting indoor surfaces such as plastered walls.

Higher

How an emulsifier works

An emulsifier is a molecule with 'a tail' that is attracted to oil and 'a head' that is attracted to water. The 'tail' is a long hydrocarbon chain. This is called the **hydrophobic** part of the emulsifier molecule. The 'head' is a group of atoms that carry a charge. This is called the **hydrophilic** part of the molecule.

The 'tails' dissolve in oil making tiny droplets. The surface of each oil droplet is charged by the 'heads' sticking out into the water. As like charges repel, the oil droplets repel each other. This keeps them spread throughout the water, stopping the oil and water separating out into two layers.

Figure 4 The structure of a typical emulsifier molecule with its water-loving (hydrophilic) head and its water-hating (hydrophobic) tail

Practical

Making and testing emulsions

Detergents act as emulsifiers.

Add a little cooking oil to some water in a boiling tube. Stopper the tube and shake. Do the same in another boiling tube but also add a drop of washing-up liquid.

- Compare the mixtures when first shaken and when left standing a while.
- You can do some tests on other types of detergent to see which is the most effective emulsifier.

Summary questions

1 Copy and complete using the words below:

emulsifier emulsion cosmetics ice mayonnaise mix separating small

Oil and water do not .... together. But if the oil droplets can be made very .... it is possible to produce a mixture of oil and water called an .... To keep the oil and water from .... we can use a chemical called an .... Important examples of food made like this include .... and .... cream. Emulsions are also important in paints and in ....

2 a Salad cream is an emulsion made from vegetable oil and water. In what ways is salad cream different from both oil and water?

b Why do we need to add an emulsifier to an emulsion like salad cream?

3 Explain how emulsifier molecules do their job. [H]

Key points

- Oils do not dissolve in water.
- Oils and water can be dispersed (spread out) in each other to produce emulsions which have special properties.
- Emulsions made from vegetable oils are used in many foods, such as salad dressings, ice creams, cosmetics and paints.
- Emulsifiers stop oil and water from separating out into layers.
- An emulsifiers works because one part of its molecule dissolves in oil (hydrophobic part) and one part dissolves in water (hydrophilic part). [H]

91

Further teaching suggestions

Milk taste test

- The different types of milk could be purchased and the students could taste them in a food technology room or outside the laboratory to see what difference fat makes to the milk (be aware of lactose-intolerant students).

Ice cream is an emulsion too!

- If there is access to an ice-cream machine, then this emulsion could also be made.

Flexicam

- If microscopes are limited, then by connecting a flexicam to the head of the microscope the image of the different milk emulsions could then be projected or shown on a TV screen.

Answers to in-text questions

a A substance that stops oil and water separating out into layers.

b It keeps the sauce thick and smooth.

Summary answers

1 mix, small, emulsion, separating, emulsifier, mayonnaise, ice, cosmetics

2 a Salad cream is thick (viscous) and is not transparent.

b To prevent the oil and water in the emulsion from separating out into layers.

3 The 'tails' of the emulsifier molecules dissolve into the oil, leaving 'heads' of the molecules lining the surface of the oil droplet. These droplets then repel each other and remain spread throughout the water.

C1 6.4 Food issues

Learning objectives

Students should learn:

- the benefits and drawbacks of using emulsifiers
- the advantages and disadvantages of using vegetable oils in our diet.

Learning outcomes

Most students should be able to:

- state one benefit and one drawback of using emulsifiers
- state one benefit and one drawback of eating vegetable oils.

Some students should also be able to:

- explain in detail the benefits and drawbacks of using emulsifiers
- explain in detail the good and bad points about vegetable oils in our diet.

Specification link-up: Chemistry C1.6

- Evaluate the effects of using vegetable oils in foods and the impacts on diet and health. *[C1.6]*
- Evaluate the use, benefits, drawbacks and risks of emulsifiers in foods. *[C1.6]*

Lesson structure

Starters

Food packaging – Give each student a piece of food packaging with an emulsifier on the ingredients label, for example custard pots. Ask the students to highlight the emulsifier and explain why it would be used in that product. You may wish to support students by giving them a selection of cards with common emulsifiers – e.g. egg yolk, lectins and E400 – and detail why they would be used in specific examples of food that match with the packaging that has been given out. You could extend students by asking them to name and give the E number (if applicable) for all the emulsifiers detailed on their food packaging. *(5 minutes)*

Dominoes – Give the students a card sort with the key words: 'food additive, emulsifier, emulsion, fat, and vegetable oil' written on it. Each card should also have a definition. Students should work in pairs to match the ends up, as in a domino game. You could support students by giving them a glossary. You could extend students by including the terms 'hydrophilic' and 'hydrophobic'. *(10 minutes)*

Main

- Split the class into two groups and give each group a different task. Ask half of the students to design and make a leaflet to be put in supermarkets to explain what emulsifiers are and why you might want to add them to your cooking.
- The remaining students should design a leaflet to be put in a pharmacy. The leaflet should explain why vegetable oils are an important part of our diet, but also the negative effects of eating too much of them.
- Pair students with the two different styles of leaflet (emulsifiers and vegetable oils). Ask each student to read his or her partner's work, then to decide if emulsifiers and vegetable fats are on balance a good or bad addition to our diet. This could be extended into a debate.
- Vegetable oils are a useful part of our diet as they provide energy and nutrients. Allow students to choose which activity from the 'Food for thought' feature they would like to complete about the advantages and disadvantages of eating vegetable oils.

Plenaries

AfL (Assessment for Learning) – Put the students leaflets, posters, scripts or articles around the room. Ask each student to 'visit' three pieces of work and use the school marking policy to assess the work. Ensure that each student also writes a constructive comment. Ask the 'owners' of the work to review what has been written and write a comment of their own. You could support students by giving them a 'level ladder' to help focus their assessment and some statements that they might want to include in their comment. You could stretch students by asking them to take their work home and action the comments and any more ideas they have to improve their work further. *(10 minutes)*

Reflection – Ask the students to think about one thing that they have learned today and one thing that they have revised. These could be facts or skills. Ask a selection of students to share their thoughts with the rest of the class. *(5 minutes)*

Support

- Students may need some help in forming arguments. You could supply facts about additives or consumption of vegetable fats and encourage them to sort the statements into pros and cons. This will help them to order their thoughts and to construct an argument. You may wish to provide the students with a writing frame to help them structure their article, poster or script.

Extend

- Students could be extended by being asked to complete the 'food for thought' activity three times. They could write an argument for the use of vegetable fats, a case against their use and a balanced piece. In this way, students will consider how some facts are stressed for persuasive purposes.

Further teaching suggestions

Food scares

- In recent years, there have been a number of food scares involving fats in our diet, e.g. trans fats made during the hardening of unsaturated fats. Students could look into these using the internet (e.g. search newspaper websites for 'food scare') and newspaper clippings kept in the school library. They could look at the economic and social effect of scares such as this and the impact on the British food industry.

Natural/unnatural

- Ask students to make a table to list some common examples of natural emulsifiers, for example egg yolk and synthetic emulsifiers, e.g. polyoxyethylene(8) stearate. Then ask students to consider the advantages and disadvantages of natural and synthetic emulsifiers. Some will be common for both. For example, an advantage of both is that they keep emulsions stable. Others will be particular to one class of emulsifiers. For instance, a disadvantage of synthetic emulsifiers is that people do not want to eat 'unnatural' chemicals.

How Science Works — Plant oils

C1 6.4 Food issues

Learning objectives

- What are the benefits and drawbacks of using emulsifiers in our food?
- What are the good and bad points about vegetable oils in our food?

Figure 1 Modern foods contain a variety of additives to improve their taste, texture or appearance, and to give them a longer shelf-life

links

For information on how an emulsifier works, look back at C1 6.3 Everyday emulsions.

Emulsifying additives

For hundreds of years we have added substances like salt or vinegar to food to help keep it longer. As our knowledge of chemistry has increased we have used other substances too, to make food look or taste better.

We call a substance that is added to food to preserve it or to improve its taste, texture or appearance a **food additive**. Additives that have been approved for use in Europe are given **E numbers**. These can be used to identify them.

a What is a food additive?

Each group of additives is given a range of E numbers. These tell us what kind of additive it is. Emulsifiers are usually given E numbers in the range 400 to 500, along with stabilisers and thickeners.

E number	Additive	What the additive does	Example
E4 _ _	emulsifiers, stabilisers and thickeners	Help to improve the texture of the food – what it feels like in your mouth. Many foods contain these additives, for example, jam and the soya proteins used in veggie burgers.	E440 – pectin

Emulsifiers stop oil and water separating out into two layers. This means that emulsifiers make it less obvious that foods are rich in oil or fat. Chocolate is a good example. The cocoa butter, which has a high energy content, is usually mixed in well, often with the help of emulsifiers. However, have you ever left a bar of chocolate past its sell-by date? Then you can see a white haze on the surface of the chocolate. This is the fatty butter starting to separate out. Then most people will throw the bar away.

So emulsifiers make oil and fat more edible in foods. They can make a mixture that is creamier and thicker in texture than either oil or water. This makes it easier and more tempting for us to eat too much fatty food.

a b

Figure 2 Which is more appetising – mayonnaise with emulsifier (**a**) or mayonnaise without emulsifier (**b**)?

92

Food issues

Vegetable oils in our diet

Everyone knows the benefits of a healthy diet. But do you know the benefits of ensuring that you eat vegetable oils as part of your diet?

Scientists have found that eating vegetable oils instead of animal fats can do wonders for the health of your heart. The saturated fats you find in things like butter and cheese can make the blood vessels of your heart become clogged up.

However, the unsaturated fats in vegetable oils (like olive oil and corn oil) are very good for you. They are a source of nutrients such as vitamin E. They also help to keep your arteries clear and reduce the chance of you having heart disease. The levels of a special fat called cholesterol in your blood give doctors an idea about your risk of heart disease. People who eat vegetable oils rather than animal fats tend to have a much lower level of 'bad' cholesterol in their blood.

b Name a vitamin that we get from olive oil.

The fats used to cook chips and other fast foods often contain certain fats that are not good for us. Scientists are concerned that eating these fats might have caused an increase in heart disease.

Changes in food labelling are very important. But many products, including fast foods, often contain high levels of potentially harmful fats from the oil they were cooked in. Yet these are exempt from labelling regulations and may be advertised as 'cholesterol-free' and 'cooked in vegetable oil'.

Figure 3 Butter contains saturated fats which raise health concerns

Activity

Food for thought

1 Write an article for a family lifestyle magazine about 'Feeding your family'. Include in this article reasons for including vegetable oils in a balanced diet and their effect on people's health.
2 Design a poster with the title 'Vegetable oils – good or bad'?
3 Write the script for a two-minute slot on local radio about the benefits and drawbacks of using emulsifiers in foods.

Figure 4 Chips have a high energy content and may contain potentially harmful fats from cooking oil

Summary questions

1 Draw a table to summarise the advantages and disadvantages of vegetable oils in our diet.
2 a Give a list of five foods that can contain emulsifiers as additives.
  b Why could it be said that emulsifiers have played a role in increasing childhood obesity rates?

Key points

- Vegetable oils are high in energy and provide nutrients. They are unsaturated and believed to be better for your health than saturated animal fats and hydrogenated vegetable oils.
- Emulsifiers improve the texture of foods enabling water and oil to mix. This makes fatty foods more palatable and tempting to eat.

93

Answers to in-text questions

a A substance that is added to food to make it keep longer or to improve its taste or appearance.

b vitamin E

Summary answers

1

Advantages of vegetable oils	Disadvantages of vegetable oils
Contain nutrients such as vitamin E. Have a high energy value to provide plenty of energy for exercise. Higher in unsaturated fats which protect against heart disease.	Have a high energy value which can turn to fat if you do not do enough exercise. Hydrogenated vegetable oils contain trans-fat, which has been associated with heart disease.

2 **a** For example, ice cream, chocolate, mayonnaise, salad cream, yoghurt.

b Emulsifiers enable oil and fats to be mixed in watery solutions without separating out to give an oily layer on top or blobs of fat. They can make foods creamier and thicker in texture which result in them being more tempting to eat.

Summary answers

1 **a** Substances produced by plants containing long hydrocarbon chains.

b Oils containing one or more C=C bonds.

c Oils containing as much hydrogen as possible in which there are no C=C bonds.

d A mixture of liquids that do not dissolve into each other.

e A substance that enables liquids that do not dissolve into each other to remain mixed without separating out into layers.

2 **a** Bromine water (which is yellow/orange) reacts with C=C bonds when the vegetable oil is unsaturated. When partial hydrogenation takes place, there are still some C=C bonds left but not as many, so it takes longer to decolourise the bromine water. Once it has been completely hardened, the oil is saturated; there are no C=C bonds, so no reaction occurs.

b Hardening plant oils increases their melting point, making them into a more solid/waxy smooth paste, which is spreadable.

c Heat the plant oil and hydrogen to 60 °C with a nickel catalyst.

3 The same-sized potato will cook quicker in boiling oil than in boiling water. Very often the outside of the food turns a different colour when cooked at the higher temperature of the boiling oil, and becomes crispier. Inside the crispy coating, the potato is soft and 'fluffy'. A boiled potato is paler in colour. Cooking food in oil also means that the food absorbs some of the oil. This makes the energy content of food cooked in oil much higher than that of the same food cooked by boiling it in water.

4 **a** The ice cream emulsion has separated, so that there are parts of the mixture which contain pure water instead of water with droplets of fat in it. When the mixture is refrozen the water freezes and forms ice that is 'crunchy' when compared to the frozen emulsion.

b To prevent the emulsion from separating out/keep it mixed.

c egg yolk

5

Water
Oil droplet
A molecule of an emulsifier
This end is attracted to water
This end is attracted to oil
Emulsifier molecule

6 **a** In its use in making biodiesel fuel.

b In its use as a method to cook foods: some of the oil is absorbed by food during cooking.

7 **a** So that the results could be compared. The instructions acted as a control. Bromine water is also hazardous.

b Controls that should have been included in the instructions:

- The amount of oil used
- How the bromine water was added
- What to look for as an end point.

c There are no anomalies. The variation in results is to be expected.

Summary questions

1 Write simple definitions for the following words:

a vegetable oils
b unsaturated oils
c saturated oils
d emulsion
e emulsifier.

2 **a** A vegetable oil removes the colour from bromine water.

It takes longer to decolourise the bromine water when the vegetable oil is partially hydrogenated.

When the vegetable oil has been completely hardened it does not react with bromine water.

Explain these observations.

b Explain why plant oils need to be hardened and the effect this has on the melting point of the oil. [H]

c Give the conditions for the reaction between a plant oil and hydrogen. [H]

3 Compare the cooking of a potato in boiling water and in vegetable oil.

4 **a** Some ice cream is left standing out on a table during a meal on a hot day. It is then put back in the freezer again. When it is taken out of the freezer a few days later, people complain that the ice cream tastes 'crunchy'. Why is this?

b A recipe for making ice cream says: 'Stir the ice cream from time to time while it is freezing.' Why must you stir ice cream when freezing homemade ice cream?

c Look at this list of ingredients for making ice cream:

8 large egg yolks
$\frac{3}{4}$ cup of sugar
$2\frac{1}{2}$ cups of whipping cream
$1\frac{1}{2}$ cups cold milk
1 vanilla pod

Which ingredient acts as an emulsifier in the mixture?

5 Draw a diagram of the structure of a typical emulsifier. [H]

6 State a use of vegetable oils where their high energy content is:

a an advantage.
b a disadvantage.

7 A teacher decided that her class should do a survey of different cooking oils to find out the degree of unsaturated oils present in them. She chose five different oils and divided them among her students. This allowed each oil to be done twice, by two different groups. They were given strict instructions as to how to do the testing.

Bromine water was added to each oil from a burette. The volume added before the mixture in the conical flask was no longer colourless was noted.

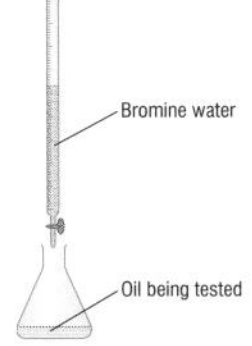

The results are in this table.

Type of oil	Amount of bromine water added (cm^3)	
	Group 1	Group 2
Ollio	24.2	23.9
Soleo	17.8	18.0
Spreo	7.9	8.1
Torneo	13.0	12.9
Margeo	17.9	17.4

a Why was it important that the teacher gave strict instructions to all of the groups on how to carry out the tests?

b List some control variables that should have been included in the instructions.

c Are there any anomalous results? How did you decide?

d What evidence is there in the results that indicate that they are reproducible?

e How might the accuracy be checked?

f How would you present these results on a graph? Explain your answer.

d The repeats are very close to each other and they have been done by two different groups.

e Accuracy could be further increased or checked by obtaining data from the manufacturers or doing a third titration.

f Results are best presented as a bar graph of mean volume of bromine water added for each oil type. This is done as the independent variable (type of oil) is categoric.

Kerboodle resources

Resources available for this chapter on Kerboodle are:

- Chapter map: Plant oils
- How Science Works: Plant oils – food for thought (C1 6.1)
- Support: Boiled or fried? (C1 6.2)
- Extension: Making margarine (C1 6.2)
- How Science Works: It's no yolk when you're a compulsive emulsive (C1 6.3)
- Practical: Making mayonnaise (C1 6.3)
- Interactive activity: Plant oils
- Revision podcast: Vegetable fat and our diet
- Test yourself: Plant oils
- On your marks: Cooking with oil
- Examination-style questions: Plant oils
- Answers to examination-style questions: Plant oils

QA Examination-style questions

1 Vegetable oils can be extracted from parts of plants that are rich in oils.

Choose the correct word from the list to complete each sentence.

a Sunflower oil is extracted from sunflower ....
leaves petals seeds (1)

b To extract olive oil the olives are crushed and ....
boiled evaporated pressed (1)

c The oil may contain small pieces of solid plant material that can be removed by ....
condensing distilling filtering (1)

d If the oil contains water it can be removed by leaving it to stand because oil and water ....
evaporate mix separate (1)

2 Lavender oil can be extracted from lavender plants by distillation.

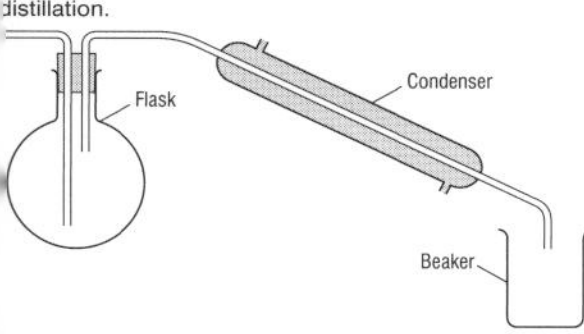

Put the following steps into the correct order, 1 to 6:
lavender plants are harvested → 1 → 2 → 3 → 4 → 5 → 6 → lavender oil is collected.

A Lavender oil and steam are condensed
B Lavender oil separates from water
C Steam is passed into the flask
D Lavender plants are put into the flask
E Lavender oil and water are collected
F Lavender oil and water evaporate (3)

3 Potatoes cooked in boiling water take about 20 minutes to cook. Potato chips can be cooked in less than 10 minutes by deep frying in hot oil. This is one reason why fast food outlets cook chips rather than potatoes.

a Explain why chips cook faster in hot oil than in boiling water. (2)

b Suggest another advantage for fast food outlets to cook chips. (1)

c Suggest a disadvantage for fast food outlets cooking chips. (1)

d Suggest an advantage for consumers who eat chips rather than boiled potatoes. (1)

e Suggest a disadvantage for consumers who eat chips rather than boiled potatoes. (1)

4 a A vegetable oil was shaken with water in flask 1 and with water and an emulsifier in flask 2. The diagrams show the results after leaving the mixtures to stand for 5 minutes.

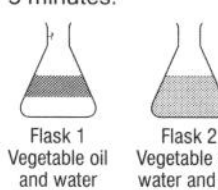

Flask 1 Vegetable oil and water

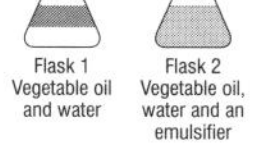

Flask 2 Vegetable oil, water and an emulsifier

a i Give a reason for the result in Flask 1. (1)
ii Explain the result in Flask 2. (2)

b Give an example of a product that contains an emulsifier and give **two** ways in which its properties are better than those of the liquids from which it is made. (3)

c Explain how an emulsifier works. Your answer should include a diagram of a simple model of an emulsifier molecule. [H] (3)

AQA, 2007

Practical suggestions

Practicals	AQA	k		
Pressing nuts (e.g. walnuts) between paper towels and studying the grease marks.	✓		✓	
Steam distillation of lavender oil, orange oil, lemon oil, olive oil, rapeseed oil or vegetable oil.	✓		✓	
Simple calorimetery investigations using small spirit burners or bottle tops to measure the energy released from various oils (weigh before and after, and measure the temperature change for a known mass of water).	✓			✓
Making emulsions, e.g. oil/water, oil/vinegar.	✓	✓	✓	
Design and carry out an investigation into the effect of emulsifiers on the stability of emulsions.	✓			✓
Using bromine water to test fats and oils for unsaturation, e.g. testing sunflower oil against butter (using colourimeter to measure level of unsaturation).	✓		✓	

Study tip

Many students are given the idea that chips are bad for you. They are an important food. Ask students to write about or discuss the positive as well as the negative aspects of foods cooked in fats and oils, when answering questions about foods cooked in fat or oil.

Examination-style answers

1 **a** seeds *(1 mark)*
b pressed *(1 mark)*
c filtering *(1 mark)*
d separate *(1 mark)*

2 In order D→C→F→A→E→B all correct *(3 marks)*
Four in correct sequence = 2 marks, three in correct sequence = 1 mark

3 **a** Oil is at higher temperature than (boiling) water, because oil has higher boiling point than water. *(2 marks)*

b Any advantage, e.g. people will buy more, people prepared to pay more (than for boiled potatoes) people prefer the taste (to boiled potatoes), easier to handle or wrap. *(1 mark)*

c Any disadvantage, e.g. waste oil for disposal (cannot go down drains), dangers of hot oil (including flammability), cost of oil (compared to water). *(1 mark)*

d Any advantage for consumer, e.g. easier to handle (can use fingers), stay hot longer, (allow taste better), if unsaturated/vegetable oils – positive impact on health or positive impact on diet e.g. provide more nutrients, provide more energy. *(1 mark)*

e Any disadvantage for consumer, e.g. negative impact on diet – more energy/calories/joules/fat/fattening, negative impact on health (if saturated oils/fats), greasy/grease on fingers/clothes. *(1 mark)*

4 **a** **i** Water and oil do not mix or are immiscible or do not dissolve into each other. *(1 mark)*
ii Any two from: forms an emulsion, does not separate, emulsifier keeps oil and water together/mixed, emulsifier stops them separating, (accept description of how an emulsifier works). *(2 marks)*

b Named example of product that is an emulsion, e.g. ice cream, salad dressing, cosmetics, paint. Any two improved properties, e.g. better coating ability, appearance, texture. *(3 marks)*

c Diagram of emulsifier molecule showing head and tail or two distinct parts.
Hydrophilic/water–attracting head and hydrophobic/oil attracting tail.
forms small globules of oil with heads on outside attracted to water or converse for water globules or diagram *(3 marks)*

Study tip

Foundation Tier students should know what emulsifiers do and Higher Tier students should be able to give a simple explanation in terms of hydrophobic and hydrophilic parts of molecules. Information about intermolecular forces is in Unit 2 and so only a simple explanation is expected in Unit 1. Students are not expected to know the formulae of fats and oils or emulsifiers. Explore the problems of unsaturation, hydrogenation and hardening of oils. Many students find these terms and ideas difficult.

Study tip

Students are expected to be familiar with methods of extracting oils. They should have seen oils extracted in the laboratory and in video clips of home-based and commercial extractions. They should understand the sequence of steam distillation but are not expected to have been taught the theory of the process other than that the oils evaporate with the steam to produce a mixture that separates when condensed.

C1 7.1 Structure of the Earth

Learning objectives

Students should learn:

- the basic structure of the Earth
- the relative size of each layer of the Earth's structure
- where minerals and other resources are found in the Earth's structure.

Learning outcomes

Most students should be able to:

- label the basic structure of the Earth
- describe the relative size of each layer of the Earth's structure
- state where minerals and other resources are found in the Earth's structure.

Some students should also be able to:

- describe how we discovered the Earth's structure.

Support

- You could support students by giving them a diagram of the Earth's structure and labels. They could then complete a 'cut-and-stick' activity to ensure that all the relevant information is correctly positioned.

Extend

- You could extend students by asking them to compare the structure of the Earth with the structure of other planets in our solar system. They could be encouraged to find out if other planets also have layers and if their compositions are similar or different.

Specification link-up: Chemistry C1.7

- The Earth consists of a core, mantle and crust and is surrounded by the atmosphere. *[C1.7.1 a)]*
- Recognise that the Earth's crust, the atmosphere and the oceans are the only source of minerals and other resources that humans need. *[C1.7]*

Lesson structure

Starters

Citizenship – Write on the board: 'How was the Earth made?' Encourage the students to list as many ideas as they can – scientific, religious and cultural. Then hold a quick discussion about all of these ideas. Explain that scientists have come up with a number of theories using the evidence they have observed although no one can be absolutely certain if any of them are correct. *(5 minutes)*

Brainstorm – Ask the students to think about geography and rocks in science. Ask them to complete a brainstorm about the structure of the Earth in the back of their books. Support students by providing them with a framework for their brainstorm. Extend students by asking them to consider the composition of the atmosphere. Ask one student to be a scribe and write on the board. They could pick students from the class to add ideas to the whole-class brainstorm. *(10 minutes)*

Main

- It is important that the students recognise a diagram of the Earth and that they can label its structure. Ask them to use secondary sources, e.g. the internet (search for 'Earth structure' at www.bbc.co.uk) and the Student Book, to make a poster about the Earth's structure.
- To ensure that students include the relevant information, state that their poster must include at least one diagram, one measurement and the key terms 'core', 'mantle', 'crust' and 'atmosphere'.
- For kinaesthetic and artistic students, a 3-D model of the Earth and its structure could be made, although this may take a couple of lessons to be completed fully.
- The students could use papier mâché to make the model of the Earth. Some students may be allergic to wallpaper paste – cellulose paste can be purchased that does not contain a fungicide. The shell should be cut in half and then decorated to show the Earth and its layers. Small pieces of coloured paper or card could be used to add information about the layers and then stuck to the model. String or thread can then be attached, so the model can be suspended, e.g. from the ceiling.

Plenaries

Label – Give the students a laminated diagram of the Earth's structure. Supply a paper towel and a washable pen. Ask the students to label as much as they can remember on their diagram. Then show them a fully labelled image (which should include a section of the Earth, depth, properties and materials) via a projector or a photocopied sheet. Ask the students to count how many of their labels were correct. *(5 minutes)*

Answers and questions – Give the students answers that are key to this topic. They should put up their hand with a question that matches the answer. Some answers could be 'Earth', 'crust', 'mantle', 'core', 'iron', 'nickel', 'atmosphere'. You could support students by asking them to work in small groups to come up with questions. You could extend them by asking them to produce a 'Foundation Tier' and a 'Higher Tier' question, both of which should have the same answer. Students should think about the choice of language in the questions. You may wish to produce more expansive answers, which would allow students to write questions that use a higher level of Bloom's taxonomy, e.g. questions involving evaluation or explanation. *(10 minutes)*

Further teaching suggestions

Interactive labelling

- An image of the Earth can be placed into interactive whiteboard software and labelled. The labels can be covered with boxes and used as a class quiz exercise.

How old is the Earth?

- Ask students to use secondary sources and find out how old scientists believe the Earth to be [about 4 600 000 000 years old]. You could extend students further by asking them to find out what evidence is used to support this idea.

Our changing planet

C1 7.1 Structure of the Earth

Learning objectives

- What is the structure of the Earth?
- What are the relative sizes of each layer of the Earth's structure?
- In which parts of the Earth's structure are our minerals and other resources found?

How big do you think the Earth is? The deepest mines go down to about 3500 m, while geologists have drilled down to more than 12 000 m in Russia. Although these figures seem large, they are tiny compared with the diameter of the Earth. The Earth's diameter is about 12 800 km. That's more than one thousand times the deepest hole ever drilled!

What's inside? 

The Earth is made up of layers that formed many millions of years ago, early in the history of our planet. Heavy materials sank towards the centre of the Earth while lighter material floated on top. This produced a structure consisting of a dense **core**, surrounded by the **mantle**. Outside the mantle there is a thin outer layer called the **crust**.

Above the Earth's crust we have a thin layer of gases called the **atmosphere**.

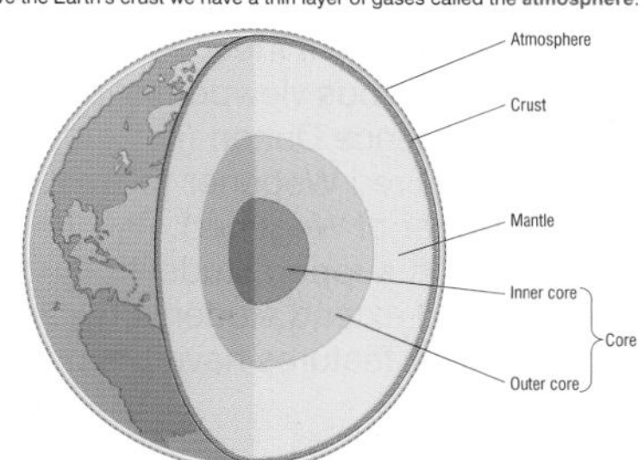

Figure 1 The structure of the Earth

The Earth's crust is a very thin layer compared to the diameter of the Earth. Its thickness can vary from as thin as 5 km under the oceans to 70 km under the continents.

Underneath the crust is the mantle. This layer is much, much thicker than the crust. It is nearly 3000 km thick. The mantle behaves like a solid, but it can flow in parts very slowly.

Finally, inside the mantle lies the Earth's core. This is about half the radius of the Earth. It is made of a mixture of the magnetic metals, nickel and iron. The core is actually two layers. The outer core is a liquid, while the inner core is solid.

a What is the solid outer layer of the Earth called?
b What is the next layer down of the Earth called?

Atmosphere	Crust	Mantle	Core
About 80% of the air in our atmosphere lies within 10 km of the surface. (Most of the rest is within 100 km but it is hard to judge exactly where our atmosphere ends and space begins).	The average thickness of the crust is about 6 km under the oceans; about 35 km under continental areas.	Starts underneath crust and continues to about 3000 km below Earth's surface. Behaves like a solid, but is able to flow very slowly.	Radius of about 3500 km. Made of nickel and iron. Outer core is liquid, inner core is solid.

Figure 2 All the resources that we depend on come from the thin crust of the Earth, its oceans and atmosphere

All the minerals and other resources that we depend on in our lives come from the thin crust of the Earth, its oceans and atmosphere. We get all the natural materials we use plus the raw materials for synthetic and processed materials from these sources. There is a limited supply of resources available to us so we should take care to conserve them for future generations.

How Science Works

Developing scientific ideas from evidence

How do we know the structure inside the Earth if nobody has ever seen it?

Scientists use evidence from earthquakes. Following an earthquake, seismic waves travel through the Earth. The waves are affected by different layers in the Earth's structure. By observing how seismic waves travel, scientists have built up our picture of the inside of the Earth.

Also, by making careful measurements, physicists have been able to measure the mass of the Earth, and to calculate its density. The density of the Earth as a whole is much greater than the density of the rocks found in the crust. This suggests that the centre of the Earth must be made from a different material to the crust. This material must have a much greater density than the material in the crust.

Did you know ...?

The temperature at the centre of the Earth is between 5000 °C and 7000 °C.

Summary questions

1 Copy and complete using the words below:
core crust mantle slowly solid thin atmosphere
The structure of the Earth consists of three layers – the .... in the centre, then the .... and the outer layer of ...., with the .... above the surface. The outer layer of the Earth is very .... compared to the Earth's diameter. The layer below this is .... but can flow in parts very ....

2 Why do some people think that the mantle is best described as a 'very thick syrupy liquid'?

3 Why should we do our best to conserve the Earth's resources?

Key points

- The Earth consists of a series of layers. Starting from the surface we have the crust, the mantle then the core in the centre. A thin layer of gases called the atmosphere surrounds the Earth.
- The Earth's limited resources come from its crust, the oceans and the atmosphere.

96 97

Did you know ...?

It is not possible to measure the temperature of the centre of the Earth directly. The scientifically accepted temperature of up to 7000 °C, is based on models of the Earth and experimentation.

Answers to in-text questions

a crust
b mantle

Summary answers

1 core, mantle, crust, atmosphere, thin, solid, slowly
2 It can flow but only very slowly.
3 Because all the raw materials we need come from the crust, oceans and atmosphere and many of them, such as mineral deposits, are non-renewable – once we have used them up they will not be replaced.

C1 7.2 The restless Earth

Learning objectives

Students should learn:

- what tectonic plates are
- why tectonic plates move
- that earthquakes and volcanic eruptions are difficult to predict.

Learning outcomes

Most students should be able to:

- state that the outer layer (crust and the upper part of the mantle) of the Earth is made up of tectonic plates
- state why the tectonic plates move
- list what happens at plate boundaries
- explain why Wegener's ideas on continental drift were not generally accepted in his time
- explain why earthquakes and volcanic activity are difficult to predict.

Some students should also be able to:

- explain in detail how convection currents cause tectonic plates to move.

Support

- To support students you may wish to provide a writing frame and a few websites for students to use for their research into plate boundaries.

Extend

- You could extend students by asking them to think about other convection currents that occur naturally on Earth [winds, ocean currents]. They could apply the same models to show convection currents in mantle to these new situations.

Specification link-up: Chemistry C1.7

- The Earth's crust and the upper part of the mantle are cracked into a number of large pieces (tectonic plates). *[C1.7.1 b)]*
- Convection currents within the Earth's mantle driven by heat released by natural radioactive processes cause the plates to move at relative speeds of a few centimetres per year. *[C1.7.1 c)]*
- The movements can be sudden and disastrous. Earthquakes and/or volcanic eruptions occur at the boundaries between tectonic plates. *[C1.7.1 d)]*
- Explain why Wegener's theory of crustal movement (continental drift) was not generally accepted for many years. *[C1.7]*
- Explain why scientists cannot accurately predict when earthquakes and volcanic eruptions will occur. *[C1.7]*

Lesson structure

Starters

Pictionary – Ask for five volunteers. They are each given a key word (crust, atmosphere, mantle, core, Earth). They should draw a picture: no noise, symbols, numbers or text are allowed; the image should explain their key word to the class. The first student in the remainder of the class to guess the word could win a prize. *(5 minutes)*

Card sort – Give the students cards with different theories about how the features on the Earth's surface were created. Start with religious viewpoints, e.g. Creationism, James Dwight Dana (The Contracting Earth), Clarence Dutton (Isostacy), Geological Society of America (Mighty Creeping Movements), Alfred Wegener (plate tectonics/continental drift) and finish with Alexander du Toit (supporter of Wegener). Ask the students to sort them into date order, and guess the dates. You could support students by giving them the century or decade the idea was published. You could extend students by asking them to think about whether scientists definitely now know how features have been created. *(10 minutes)*

Main

- The tectonic plates move on convection currents. Convection currents should have been studied in KS3. However, students may benefit from seeing the convection current demonstration.
- This can be completed as a demonstration or a class practical. Pieces from a polystyrene tile can be floated in larger beakers to model the tectonic plate movement. Discuss the limitations of this model.
- Students could be encouraged to draw a diagram of the experiment and label it (e.g. beaker, potassium manganate(VII), convection current, Bunsen burner) in blue. Then ask students to relate this to the Earth, and label these in red (e.g. mantle, magma, heat spot, tectonic plates).
- Students need to know what happens at plate boundaries and the effects of natural disasters. Show some images of plate boundaries and the destruction caused. Split the class into three groups: one group researches volcanoes, another earthquakes and the third looks at prediction and prevention of damage/loss of life.
- Each group should be supplied with video footage of natural disasters caused by earthquakes and volcanoes, for example http://video.nationalgeographic.com. Search under 'Natural Disasters'. Students then use this information, coupled with textbooks, to write a revision page for an imaginary textbook. The pages could be collected and selected ones photocopied and given to each student to add to their notes.
- Students should appreciate how theories were developed and others disproved. Ask students to imagine that they are Wegener and that they are going to attend an international conference for scientists. They are to have a 'slot' to explain his new theory and explain why previous theories were incorrect. Students should prepare a presentation that is about 3 minutes long.

Plenary

Lava lamps – Set up a lava lamp (this needs to be done at least 30 minutes before the end of the lesson). Ask the students to look at the lava lamp and relate it to the lesson. You should extend students by encouraging them to link the movement in the lava lamp with that which occurs in the mantle of the Earth. You could support students by explaining the link to plate tectonics using separate sentences and asking them to put them in order, working in pairs. *(5 minutes)*

Practical support

Convection currents

Equipment and materials required

Potassium manganate(VII) crystals (oxidising agent and harmful), tweezers, large glass beaker, cold water, Bunsen burner and safety equipment, tripod and gauze, eye protection.

Details

Fill the beaker about 75 per cent with cold water and put onto a tripod. With a set of tweezers, add one potassium manganate(VII) crystal to the bottom of the beaker. Put the Bunsen burner under the beaker at the point where the crystal is and heat on the blue flame. The convection current should become visible.

Safety: CLEAPSS Hazcard 81 Potassium manganate(VII) – oxidising and harmful. Handle carefully, as crystals will stain hands and clothing.

C1 7.2 The restless Earth

Learning objectives

- What are tectonic plates?
- Why do the plates move?
- Why is it difficult for scientists to predict when earthquakes and volcanic eruptions will occur?

Figure 1 *Mesosaurus* was a reptile that existed million years ago. Its fossils have been found in Africa and in South America.

The continents are moving

The map of the world hasn't always looked the way it does today. Look at the map shown in Figure 2. Find the western coastline of Africa and the eastern coastline of South America. Can you see how the edges of the two continents look like they could slot together?

The fossils and rock structures that we find when we look in Africa and South America are also similar. Fossils show that the same reptiles and plants once lived on both continents. Also, the layers of rock in the two continents have been built up in the same sequence.

Scientists now believe that they can explain these facts. They think that the two continents were once joined together as one land mass.

a What evidence is there that Africa and South America were once joined to each other?

Tectonic plates

Of course, the continents moved apart very, very slowly. In fact, they are still moving today, at the rate of a few centimetres each year. They move because the Earth's crust and uppermost part of the mantle is cracked into a number of huge pieces. We call these **tectonic plates**.

Deep within the Earth, radioactive atoms decay, producing vast amounts of energy. This heats up molten minerals in the mantle which expand. They become less dense and rise towards the surface. Cooler material sinks to take their place. Forces created by these **convection currents** move the tectonic plates slowly over the surface of the Earth.

Where the boundaries (edges) of the plates meet, huge stresses build up. These forces make the plates buckle and deform, and mountains may be formed. The plates may also suddenly slip past each other. These sudden movements cause earthquakes.

However, it is difficult for scientists to know exactly where and when the plates will suddenly slip like this.

b What causes an earthquake to occur?

Arctic Ocean, Greenland, Arctic Ocean, North America, Europe, Asia, Pacific Ocean, Africa, Pacific Ocean, South America, Indian Ocean, Australia, Atlantic Ocean

Key: Earthquake zones, Volcanoes, Plate boundaries

Figure 2 The distribution of volcanoes and earthquakes around the world largely follows the boundaries of the Earth's tectonic plates

How Science Works

Trying to predict the unpredictable!

Earthquakes and volcanic eruptions can be devastating. But making accurate predictions of when they will take place is difficult. Markers placed across a plate boundary or across the crater of a volcano can be monitored for movement. Scientists also monitor the angles of the slopes on volcanoes. The sides of some volcanoes start to bulge outwards before an eruption. Any abnormal readings can be used as a warning sign.

It has also been found that rocks heat up before earthquakes as a result of extreme compression. So satellites with infrared cameras can monitor the Earth's surface for unexpected rises in temperature. Our ability to predict these natural events, and evacuate people at risk, will improve as advances are made by scientists.

Figure 3 Earthquakes can be devastating to people living close by

How Science Works

Wegener's revolutionary theory

In the past, scientists thought that features like mountain ranges were caused by the crust shrinking as the early molten Earth cooled down. They thought of it rather like the skin on the surface of a bowl of custard. It tends to shrink, then wrinkle, as the custard cools down.

The idea that huge land masses once existed before the continents we know today, was put forward in the late 19th century by the geologist Edward Suess. He thought that a huge southern continent had sunk. He suggested that this left behind a land bridge (since vanished) between Africa and South America.

The idea of continental drift was put forward first by Alfred Wegener in 1915. However, his fellow scientists found Wegener's ideas hard to accept. This was mainly because he could not explain ***how*** the continents had moved. So they stuck with their existing ideas.

His theory was finally shown to be right almost 50 years later. Scientists found that the sea floor is spreading apart in some places, where molten rock is spewing out between two continents. This led to a new theory, called plate tectonics.

Did you know ...?

With the latest GPS (global positioning satellite) technology we can detect movement of tectonic plates down to 1 mm per year.

AQA Examiner's tip

The Earth's tectonic plates are made up of the crust and the upper part of the mantle (not just the crust).

Summary questions

1 Copy and complete using the words below:

convection earthquakes mantle tectonic volcanoes

The surface of the Earth is split up into a series of .... plates. These move across the Earth's surface due to .... currents in the .... Where the plates meet and rub against each other, we get .... and ....

2 a Explain how tectonic plates move.

b Why are earthquakes and volcanic eruptions difficult to predict?

3 Imagine that you are a scientist who has just heard Wegener's ideas for the first time. Write a letter to another scientist explaining what Wegener has said and why you have chosen to reject his ideas.

Key points

- The Earth's crust and upper mantle is cracked into a number of massive pieces (tectonic plates) which are constantly moving slowly.
- The motion of the tectonic plates is caused by convection currents in the mantle, due to radioactive decay.
- Earthquakes and volcanoes happen where tectonic plates meet. It is difficult to know when the plates may slip past each other. This makes it difficult to predict accurately when and where earthquakes will happen.

98 99

Further teaching suggestions

Role play

- Ask the students to think of questions or arguments that Wegener might have faced from the scientific community. A volunteer or group of volunteers could pretend to be Wegener, then other students could pose their questions/arguments for Wegener and his team to answer.

Seismologist

- Students could research the job of a seismologist and find out what their 'tools of the trade' are.

Guest speaker

- You may wish to contact your school's geography department and see if there is a geologist who could come into the science classroom as a 'guest' speaker. Allow students to prepare questions and then invite the students to ask the expert.

Did you know ...?

The earliest evidence for plate tectonics was discovered in 2002 by US/Chinese scientists. They found rocks near the Great Wall of China that show that plates were moving 2.5 billion years ago – this is about 500 million years earlier than scientists first thought.

Answers to in-text questions

a The similar shapes of their coastlines, similar rock types, similar fossils.

b Tectonic plates suddenly slipping past one another after stress builds up at their boundaries. These sudden movements cause earthquakes.

Summary answers

1 tectonic, convection, mantle, volcanoes (earthquakes), earthquakes (volcanoes)

2 a Convection currents, caused by energy from natural radioactive processes, form in the mantle beneath the tectonic plates.

b Although we know where plate boundaries lie, we cannot tell exactly when and where the forces building up will cause the sudden movement that produces an earthquake or when the magma building up in a volcano will cause an explosive eruption.

3 Letter. [Mark based on ideas from description of Wegener's ideas in the spread.]

C1 7.3 The Earth's atmosphere in the past

Learning objectives

Students should learn:

- what the Earth's atmosphere was like in the past
- how the mixture of gases in the Earth's atmosphere was produced
- how oxygen was released into the Earth's atmosphere.

Learning outcomes

Most students should be able to:

- name the gases that probably made up the Earth's early atmosphere
- list the major events that formed today's atmosphere
- state how oxygen entered the Earth's atmosphere.

Some students should also be able to:

- explain a theory how the Earth and its atmosphere were formed.

Support

- You could support students by supplying them with images and text to explain the five stages of development of the Earth's atmosphere. They could then cut and stick them into the appropriate cartoon strip boxes.

Extend

- You could extend students by asking them to discover some examples of other theories and ideas used to explain how the atmosphere developed. For example, some scientists disagree about the composition of the gases in the early atmosphere, thinking that there was a greater mix than in the currently scientifically accepted theory. Encourage students to consider what evidence has been collected to support the current main scientifically-accepted theory. Students could think about how this theory could still be refined or what evidence would need to be collected for the theory to be disproved.

Specification link-up: Chemistry C1.7

- During the first billion years of the Earth's existence, there was intense volcanic activity. This activity released the gases that formed the early atmosphere and water vapour that condensed to form the oceans. *[C1.7.2 b)]*
- There are several theories about how the atmosphere was formed. One theory suggests that during this period the Earth's atmosphere was mainly carbon dioxide and there would have been little or no oxygen gas (like the atmospheres of Mars and Venus today). There may also have been water vapour and small proportions of methane and ammonia. *[C1.7.2 c)]*
- Plants and algae produced the oxygen that is now in the atmosphere. *[C1.7.2 f)]*
- Explain and evaluate theories of the changes that have occurred and are occurring in the Earth's atmosphere. *[C1.7]*

Lesson structure

Starters

Chemical equations – Ask students to copy out the word equation for photosynthesis. Then set them the task of completing the symbol equation. Ask students to explain why this equation is so important. [This chemical reaction is believed to be responsible for putting oxygen into the Earth's atmosphere.] *(5 minutes)*

Grouping gases – Ask students to list as many gases as they can think of. Then put three titles on the board: 'element', 'compound' and 'mixture'. Encourage each student to go to the board and add a gas from their list under the correct column heading. Look at the board and, if there are any incorrect answers, tell the class how many mistakes there are. Then ask the students to see if they can pick out the errors. You could support students by giving them cards with the name of a gas and a simple particle model of the gas, rather then generating their own list. You could extend them by asking them to write the chemical formula of each gas rather than its name. *(10 minutes)*

Main

- Give students a cartoon strip with five frames, with space to draw a picture and write notes. Encourage students to use a textbook to detail five stages in the development of the atmosphere. They should draw an image and write text to explain the atmosphere's composition and how it compares to other planets in our solar system.
- Separate the class into groups. Ask the groups to imagine that they are astronauts who have landed on Earth in the different stages of atmospheric development. Ask them to design a sketch or role-play to describe the surroundings. The activity could include a predictive stage in which students consider the atmosphere in 500 years' time.
- Not all students need be involved in the acting side but all students should collect research about the development. This activity could encourage students to manage their own time and group dynamics.
- Alternatively, you could give each student a role, e.g. chair person or resource manager, and encourage the students to stick to their roles, allowing them to experience a different role in a group compared to their preferred choice.
- Groups of students could act out their sketch. The rest of the class comments on misconceptions in the sketch and then votes on the best. A prize could be awarded.

Plenaries

Ordering – Ask the students, in small groups, to arrange a series of key events in the creation of the atmosphere in chronological order. Support them by allowing them access to the Student Book to revise for one minute before they start the task. Extend students by asking them to try to remember the formula of the gases in the atmosphere during the key events. *(5 minutes)*

Pass it on – Give each group of three students a piece of paper with the same question printed on: 'Where does oxygen come from?' On the piece of paper, the first student starts the answer and is timed for 30 seconds. Then ask the students to pass the paper to their left (even if they are mid-sentence) and give the next student one minute, then pass the paper again and give the final student one-and-a-half minutes. Each student should read the answer so far, change anything that they feel is incorrect and add further information if they think they can. Then pick a few groups to read their papers. *(10 minutes)*

Further teaching suggestions

Graph work

- Students could display the composition of the Earth's atmosphere at different stages in the form of charts and graphs.

Venus

- Students could contrast the Earth's atmospheric development with that of Venus. Some scientists believe that it has had a runaway greenhouse effect that has caused its current atmosphere.

Pyramid summery

- Give students a pyramid. They should put the main idea at the top (Earth's atmosphere) and draw a line underneath it. Then they write the main points and draw a line. This continues until all the information has been summarised from C1 7.3 in the Student Book.

C1 7.3 The Earth's atmosphere in the past

Learning objectives

- What was the Earth's atmosphere like in the past?
- How were the gases in the Earth's atmosphere produced?
- How was oxygen produced?

Scientists think that the Earth was formed about 4.5 billion years ago. To begin with it was a molten ball of rock and minerals. For its first billion years it was a very violent place. The Earth's surface was covered with volcanoes belching fire and gases into the atmosphere.

Figure 1 Volcanoes moved chemicals from inside the Earth to the surface and the newly forming atmosphere

Did you know ...?

Comets could also have brought water to the Earth. As icy comets rained down on the surface of the Earth, they melted, adding to its water supplies. Even today many thousands of tonnes of water fall onto the surface of the Earth from space every year.

The Earth's early atmosphere

There are several theories about the Earth's early atmosphere. One suggests that volcanoes released carbon dioxide, water vapour and nitrogen gas and these gases formed the early atmosphere.

The water vapour in the atmosphere condensed as the Earth gradually cooled down, and fell as rain. Water collected in hollows in the crust as the rock solidified and the first oceans were formed.

As the Earth began to stabilise, the atmosphere was probably mainly carbon dioxide. There could also have been some water vapour, and traces of methane and ammonia. There would have been very little or no oxygen at that time. Some scientists believe nitrogen was another gas present at this time.

This is very like the atmospheres which we know exist today on the planets Mars and Venus.

a What was the main gas in the Earth's early atmosphere?
b How much oxygen was there in the Earth's early atmosphere?

After the initial violent years of the history of the Earth, the atmosphere remained quite stable. That is until life first appeared on Earth.

Figure 2 The surface of one of Jupiter's moons, Io, with its active volcanoes releasing gases into its sparse atmosphere. This gives us a reasonable glimpse of what our own Earth was like billions of years ago.

Oxygen in the atmosphere

There are many theories as to how life was formed on Earth billions of years ago. Scientists think that life on Earth began about 3.4 billion years ago. That is when simple organisms similar to bacteria appeared. These could make food for themselves, using the breakdown of other chemicals as a source of energy.

Later, bacteria and other simple organisms, such as algae, evolved. They could use the energy from the Sun to make their own food by photosynthesis. This produced oxygen gas as a waste product.

By two billion years ago the levels of oxygen were rising steadily as algae and bacteria thrived in the seas. More and more plants evolved. All of them were photosynthesising, removing carbon dioxide and making oxygen.

$$\text{carbon dioxide + water} \xrightarrow{\text{(energy from sunlight)}} \text{sugar + oxygen}$$

As plants evolved, they successfully colonised most of the surface of the Earth. So the atmosphere became richer and richer in oxygen. This made it possible for animals to evolve. These animals could not make their own food and needed oxygen to respire.

On the other hand, many of the earliest living microorganisms could not tolerate a high oxygen concentration (because they had evolved without it). They largely died out, as there were fewer places where they could live.

Figure 3 Some of the first photosynthesising bacteria probably lived in colonies like these stromatolites. They grew in water and released oxygen into the early atmosphere.

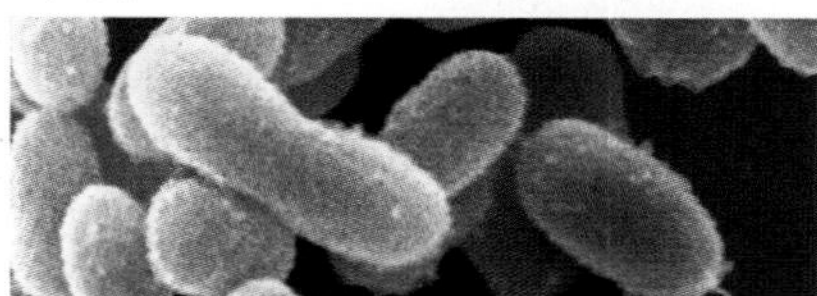

Figure 4 Not only do bacteria such as these not need oxygen – they die if they are exposed to it, but they can survive and breed in rotting tissue and other places where there is no oxygen

Did you know ...?

Scientists have reconstructed what they think the atmosphere must have been like millions of years ago based on evidence from gas bubbles trapped in ancient rocks. They also use data gathered from the atmospheres of other planets and their moons in the solar system.

Summary questions

1 Copy and complete using the words below:
dioxide methane oxygen volcanoes water
The Earth's early atmosphere probably consisted mainly of the gas carbon .... There could also have been .... vapour and nitrogen, plus small amounts of .... and ammonia. These gases were released by .... as they erupted. Plants removed carbon dioxide from the atmosphere and produced .... gas.
2 Describe how the Earth's early atmosphere was probably formed?
3 Why do scientists believe there was no life on Earth for 1.1 billion years?
4 Draw a chart that explains the early development of the Earth's atmosphere.

Key points

- The Earth's early atmosphere was formed by volcanic activity.
- It probably consisted mainly of carbon dioxide. There may also have been water vapour together with traces of methane and ammonia.
- As plants spread over the Earth, the levels of oxygen in the atmosphere increased.

Did you know ... ?

Ice cores can be drilled out from polar ice caps. The bubbles of trapped gas are laid down year by year, then released by melting the ice. This gas is tested to find out the composition of the past atmosphere many thousands of years before records could first be made.

Answers to in-text questions

a carbon dioxide
b very little or none

Summary answers

1 dioxide, water, methane, volcanoes, oxygen
2 From gases emitted by volcanoes.
3 The temperature was too high.
4 Student chart showing development from early volcanic atmosphere to the first plant-produced oxygen and the removal of carbon dioxide during photosynthesis.

C1 7.4 Life on Earth

Learning objectives

Students should learn:

- the reasons why there are many theories for how life began on Earth
- how ammonia, hydrocarbons and lightening could have been the source of life on Earth. **[HT only]**

Learning outcomes

Most students should be able to:

- explain why there are so many theories to how life started on Earth

Some students should also be able to:

- explain how ammonia, hydrocarbons and lightening could have started life on Earth. **[HT only]**
- evaluate the theories about how life started on Earth. **[HT only]**

Support

- You could support students by giving them short bulleted printouts of the main points about the different theories about how life on Earth began. They could then order them and stick them onto their timeline. Other information, such as diagrams, could be supplied, which students could also transpose onto their work.
- Students often struggle to understand how a primordial soup became a life form. Scientists are also trying to find out how this occurred and they are trying to start life from a mixture of chemicals. Theories range from lightening striking the primordial soup to dormant life forms from a meteorite starting life when they hit the surface of the early Earth. It is worth stressing to students that no one currently knows the answer. The idea of a missing part of the theory can be demonstrated by mixing a ground up vitamin tablet with protein powder and water. This has all of the chemicals needed for life but the mixture does not come into life.

Extend

- Students could be extended by being asked to suggest how scientists could try to find out if life existed elsewhere in the universe (SETI, study of meteorite, space probes).

Specification link-up: Chemistry C1.7

- There are many theories as to how life was formed billions of years ago. *[C1.7.2 d)]*
- One theory as to how life was formed involves the interaction between hydrocarbons, ammonia and lightning. *[C1.7.2 e)]* **[HT only]**
- Describe why we do not know how life was first formed. *[C1.7]* **[HT only]**

Lesson structure

Starters

Life in the universe – Ask the students to decide if they think that there is life elsewhere in the universe. Ask them to give and justify a rating out of 100, where 100 is definitely life elsewhere and 0 that there is only life on Earth. *(5 minutes)*

Have a think! – Ask students to consider how they think life could have started on Earth. Ask for volunteers to share their thoughts. Manage the class discussion. You could support students by providing them with some cards showing some theories/religious ideas. They can choose which they think is most likely. You can extend students by asking them to group the suggestions as ideas [e.g. religious views, which are not backed up by scientific evidence] and theories [ideas backed up by scientific evidence]. *(10 minutes)*

Main

- Students often have their own thoughts about how life started on Earth. However, they have rarely considered that ideas for how life has started have changed over time. Discuss with students why ideas change in science. This may be due to technological developments, research being completed and information being circulated.
- The Miller–Urey experiment was a radical experiment and its results have been re-analysed many years later. Encourage students to work in small groups to write a two-minute radio news report to explain the importance of the experiment. You may wish to record the report using a suitable piece of computer software. The best recording could be used as a podcast on the school's virtual learning platform as a revision tool. **[HT only]**
- Ask students to make a timeline, showing some of the different theories and ideas about how life started on the Earth. Their timelines should briefly outline the theory and they should bullet point any evidence used to support it or persuade people to accept it. They could also list any reasons/evidence that could be used to argue that the theory or idea is not correct. Encourage students to include the Miller–Urey experiment and the Murchison meteorite.
- You could encourage students to write an advert to try to persuade people that the generally accepted theory for how life started on Earth is correct. Students could work in their preferred learning styles and could make a billboard poster, if they are visual learners, a radio advert if they are auditory learners and they could be filmed doing a TV advert if they are kinaesthetic learners. Ask students to share their adverts with the class. They could assess each other's work. **[HT only]**

Plenaries

Life in the universe part two – Ask the students to reconsider and decide if they think that there is life elsewhere in the universe after studying this lesson. Again ask students to give a rating and reflect on their new score and original score. If it has changed, why? *(5 minutes)*

Summarise Miller–Urey – Ask the students to imagine that they are trying to advertise the Miller–Urey theory for how life began on Earth. They should try and construct an advertising strap line. Students could be supported by the whole class working together to generate the strap line. Students could be extended by being told key words that they must include. The best strap line could be given a prize. *(10 minutes)* **[HT only]**

Further teaching suggestions

Planning to create life

- Students could write the plan for the Miller–Urey experiment, detailing the aim, equipment, method and safety. **[HT only]**

Revision guide

- Students could write a page for a revision guide to try to explain how life could have started on Earth. Their page should be aimed at KS4 students, should be no larger than A4 and contain questions or activities to check that people understand the information. You could then choose the best one, photocopy it and use it as a class homework or starter.

Murchison

- Students could write a newspaper article for the day when the results from the Murchison meteorite were discovered. Students could use an appropriate desk top publishing package to write up their article. Quotations could be added, using internet search engines, from people who worked on the analysis and interpretation of results, e.g. Michael Engel, Bartholomew Nagy and Philippe Schmitt-Kopplin. **[HT only]**

Our changing planet

C1 7.4 Life on Earth

Learning objectives

- Why are there many theories about how life began on Earth?
- Why does one theory involve hydrocarbons, ammonia and lightning? [H]

Most theories of how our atmosphere developed include the arrival of living things on Earth. The oxygen in our atmosphere today is explained by photosynthesis in plants. The plants probably evolved from simple organisms like plankton and algae in the ancient oceans.

But where did the molecules that make up the cells of even the simplest living things come from? And how were they formed? Any theories to answer these questions are bound to be tentative. They will be based on assumptions. The best theories will be the ones that explain most of the widely-accepted evidence.

links

For information on theories about the Earth's early atmosphere, look back at C1 7.3 The Earth's atmosphere in the past.

Higher

Miller–Urey experiment

We know the type of molecules that make up living things. To make these we need compounds called amino acids. These amino acids make proteins.

Most amino acids contain the elements carbon, hydrogen, nitrogen and oxygen. So one way forward is to try to re-create the conditions in the early atmosphere in an experiment. Could amino acids have been made in those conditions? That is the question the scientists Miller and Urey tried to answer in 1952. Figure 1 shows a diagram of their apparatus.

They used a mixture of water (H_2O), ammonia (NH_3), methane (CH_4) and hydrogen (H_2) to model the early atmosphere. Under normal conditions, these gases do not react together. However, Miller and Urey used a high voltage to produce a spark to provide the energy needed for a reaction. This simulated lightning in a storm. The experiment ran for a week then they analysed the mixture formed. It looked like a brown soup. In it they found 11 different amino acids.

Electrodes
Electrical spark (lightning)
H_2O, CH_4, NH_3, H_2
Gases (primitive atmosphere)
Direction of water vapour circulation
To vacuum pump
Cold water in
Condenser
Sampling probe
Sampling probe
Cooled water (found to contain organic compounds)
Trap
Heat source
Water (ocean)

Figure 1 The classic Miller–Urey experiment

a Which elements make up most amino acids?

This experiment provided evidence that it was possible to make the molecules of life from gases that may have been in our early atmosphere. Miller and Urey published their findings in 1953. They froze some of the mixtures formed in their experiments and stored it. In 2008 other scientists analysed it using modern techniques. They found 22 amino acids, as well as other molecules important for life.

Theories of the composition of the early atmosphere have changed since the 1950s. For example many people think the atmosphere was mainly carbon dioxide and nitrogen before the first life on Earth. However when they carry out similar experiments to Miller and Urey, they still get similar biological molecules made.

102

Life on Earth

Higher

There are opponents of the theory that biological material can be made from non-biological material. They argue that the Miller–Urey experiment only works in the absence of oxygen. They believe that oxygen would have been present before the generally accepted time for its appearance. This would make any conclusions based on Miller–Urey or similar experimental results invalid.

Other theories

Another theory is based on analysis of meteors that crash to Earth from space. In 1969 a meteorite fell from the sky above Australia. Known as the Murchison meteorite, its mass was over 100 kg. However, more interesting were the range of organic molecules found in it.

The latest studies of fragments of the meteorite have identified about 70 different amino acids. This shows that the molecules capable of starting life on Earth might have arrived from outer space.

Figure 2 Part of the Murchison meteorite, which is rich in organic molecules – the molecules of life

b Why were scientists interested in the Murchison meteorite?

Another source of biological molecules could have been deep under the oceans. Near to volcanic vents on the seabed we get both the conditions and chemicals needed.

But just because the 'building blocks' of life might have been on Earth, it does not explain the really difficult step. How do they go on to form life?

The organic molecules, from whatever source, could have formed a 'primordial soup'. All the molecules needed to start life could have been in the seas. Then they would have had to react together to somehow make the first primitive cells. Protein molecules capable of replicating themselves might have been involved at this stage.

Others think that simple living organisms could have arrived on Earth in meteorites or comets. Their evolution had started elsewhere. This 'extraterrestrial seeding' from outer space supports the theory of life in other parts of the universe. Of course, nobody knows for sure but the search for evidence goes on.

Figure 3 Volcanic vents under the sea might have helped form a 'primordial soup' of organic molecules

Summary questions

1 Look at Figure 1.
   a Explain what Miller and Urey did in their experiment.
   b Which one of these statements best describes the outcome of their experiment:
      A It showed how life can be formed from simple molecules.
      B It showed how carbon dioxide and methane are essential parts of a living cell.
      C It showed that biological molecules can be made from substances that could have been in the early atmosphere.
      D It showed that the Earth's early atmosphere must have been made up of only carbon dioxide, ammonia, water vapour and methane. [H]
2 a What do we mean by a 'primordial soup'?
   b What role might a 'primordial soup' have played in developing life on Earth? [H]

Key points

- One theory states that the compounds needed for life on Earth came from reactions involving hydrocarbons, such as methane, and ammonia. The energy required for the reaction could have been provided by lightning. [H]
- All the theories about how life started on Earth are unproven. We can't be sure about the events that resulted in the first life-forms on Earth. [H]

103

Answers to in-text questions

a carbon, hydrogen, nitrogen and oxygen

b Because it contains organic molecules which support the theory that the molecules of life might have arrived from outer space.

Summary answers

1 a They reacted a mixture of water, ammonia, methane and hydrogen to model the early atmosphere. To produce the energy needed for reactions to take place they used a high voltage. They kept the experiment going for a week, then analysed the mixture of new compounds formed.

b C – It showed that biological molecules can be made from substance that could have been in the early atmosphere.

2 a The mixture of organic compounds brewing in prehistoric seas.

b The organic molecules could have reacted together to make biological molecules, such as self-replicating proteins, that could have gone on to make the first cells.

C1 7.5 Gases in the atmosphere

Learning objectives

Students should learn:

- the sinks for the majority of the carbon dioxide from the early atmosphere
- the main gases in the current atmosphere
- the percentage composition of the current atmosphere
- how the gases in the air can be separated. **[HT only]**

Learning outcomes

Most students should be able to:

- list the main gases in the atmosphere and the approximate proportions of gases in the atmosphere
- explain how carbon dioxide was removed from the Earth's atmosphere.

Some students should also be able to:

- explain how air can be separated by fractional distillation **[HT only]**
- state some uses of the products of the fractional distillation of air. **[HT only]**

Specification link-up: Chemistry C1.7

- For 200 million years, the proportions of different gases in the atmosphere have been much the same as they are today: about four-fifths (80%) nitrogen, about one-fifth (20% oxygen, small proportions of various other gases, including carbon dioxide, water vapour and noble gases. *[C1.7.2 a)]*
- Most of the carbon from the carbon dioxide in the air gradually became locked up in sedimentary rocks as carbonates and fossil fuels. *[C1.7.2 g)]*
- Air is a mixture of gases with different boiling points and can be fractionally distilled to provide a source of raw materials used in a variety of industrial process. *[C1.7.2 j)]* **[HT only]**
- Explain and evaluate theories of the changes that have occurred and are occurring in the Earth's atmosphere. *[C1.7]*

Controlled Assessment: C4.2 Assess and manage risks when carrying out practical work. *[C4.2.1 a) b)]*

Lesson structure

Starters

Reflection – Give the students an A4 sheet of paper and ask them to make three columns. The first should be headed with 'what I already know', then 'what I want to know' and finally 'what I know now'. Ask the students to consider the title of the topic and complete the first two columns with bullet points of information. *(5 minutes)*

Pie chart – Ask the students to estimate the percentage of each gas in the current atmosphere. This should bring out any misconceptions that oxygen is the predominant gas in the atmosphere. Then give the actual percentages so they can see how close they were. Encourage students to consider the way these data could be displayed scientifically (table, bar chart or pie chart). Support students by only asking them to complete a table of gas percentages, then represent these data in a bar chart. Extend students by asking them to present the data using a pie chart. *(10 minutes)*

Main

- After a brief explanation about the movement of carbon dioxide from the early atmosphere into rocks, students could test seashells to show that they contain carbonates. See 'Practical support' for more detail. Encourage students to reflect on the work completed in C1.2 and to use this information to help them suggest a method. You could extend students' understanding of Controlled Assessment by asking them to write a full risk assessment for their chosen method.
- The percentage composition of gases in the Earth's atmosphere has changed over time but has remained relatively constant in the last 200 million years. Ask students to imagine that they are Earth and to write a creative story about how its 'clothes' (its atmosphere) has changed and why.
- The class could be split into three groups. Students in the first write the beginning of the story, another group writes the middle and the final group writes the end. In order to complete this last suggestion, a link sentence from each section needs to be provided to the groups so they know where/how to start and/or finish their part of the story.
- Ask Higher Tier students to make a flow chart to explain how air can be separated. On their flow chart they should include details of the uses for each of the gases. The RSC Industrial Chemistry series has a short video of this process, which you could play to the students before they complete this task. You should link the fractional distillation of air to the process used to separate crude oil (C1 4.2). **[HT only]**

Plenaries

Guess what? – Ask the students to break off into pairs. Give each pair a pack of cards with separate key words per card, e.g. oxygen, nitrogen, air, gas, carbon dioxide, photosynthesis, argon, fractional distillation. The students should take it in turns to pick a card and look at the key term. They should explain it to their partner without using the key word and the other student should guess the word. You could support students by giving them the explanation to read out and then work together to match it up to the key word. You could extend them by asking them to mime the key word rather than describe it. *(5 minutes)*

Reflection part two – Ask the students to return to their A4 table. Ask them to add information to the last column, 'what I now know', that isn't included in the middle column. They should also correct any misconceptions from the middle column and ask for help if they have not found out some information that they wanted. *(10 minutes)*

Support

- You could support students by providing them with discrete sentences, which they could use to generate the story about the changing Earth's atmosphere. Flow chart boxes could be provided and information given for students to cut and stick directly onto the diagram.

Extend

- You could extend students by asking them to find out what scientists predict will happen to the composition of the Earth's atmosphere. It is important that the students list their sources. In the next lesson the information can be compared and students can consider whether bias has crept into the evidence gathered.

Practical support

Investigating sea shells

Equipment and materials required

Sea shells (e.g. cockle, oyster), pestle and mortar, spatula, hydrochloric acid (2 mol/dm^3) – irritant, 2 × dropping pipettes, boiling tube, bung with a hole in the centre fitted with a delivery tube, test tube, test-tube rack, limewater – irritant, eye protection.

Details

Students can plot their own investigation to see which type of shell contains the highest percentage of carbonate mineral. The same mass of crushed shell could be added to excess dilute hydrochloric acid. Any remaining sediment can be filtered off, washed and left to dry, then reweighed. You can show that the gas liberated is carbon dioxide using limewater. Check the students' plans, including risk assessment, before allowing any practical work to commence.

Safety: CLEAPSS Hazcard 47A Hydrochloric acid – corrosive. CLEAPSS Hazcard 18 Limewater – irritant.

C1 7.5 Gases in the atmosphere

Learning objectives

- What happened to most of the carbon dioxide in the early atmosphere?
- What are the main gases in the atmosphere today and what are their relative proportions?
- How can the gases in the air be separated? [H]

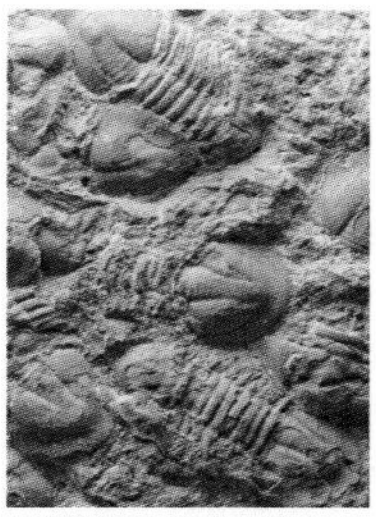

Figure 1 There is clear fossil evidence in carbonate rocks of the organisms which lived millions of years ago

We think that the early atmosphere of the Earth contained a great deal of carbon dioxide. Yet the Earth's atmosphere today only has around 0.04% of this gas. So where has it all gone? The answer is mostly into living organisms and into materials formed from living organisms.

Carbon 'locked into' rock

Carbon dioxide is taken up by plants during photosynthesis. The carbon can end up in new plant material. Then animals eat the plants and the carbon is transferred to the animal tissues, including bones, teeth and shells.

Over millions of years the dead bodies of huge numbers of these living organisms built up at the bottom of vast oceans. Eventually they formed sedimentary carbonate rocks like limestone (containing mainly calcium carbonate).

Some of these living things were crushed by movements of the Earth and heated within the crust. They formed the fossil fuels coal, crude oil and natural gas. In this way much of the carbon from carbon dioxide in the ancient atmosphere became locked up within the Earth's crust.

a Where has most of the carbon dioxide in the Earth's early atmosphere gone?

Carbon dioxide also dissolved in the oceans. It reacted and made insoluble carbonate compounds. These fell to the seabed and helped to form more carbonate rocks.

Ammonia and methane

At the same time, the ammonia and methane, from the Earth's early atmosphere, reacted with the oxygen formed by the plants.

$$CH_4 + 2O_2 \rightarrow CO_2 + 2H_2O$$
$$4NH_3 + 3O_2 \rightarrow 2N_2 + 6H_2O$$

This got rid of methane and ammonia. The nitrogen (N_2) levels in the atmosphere built up as this is a very unreactive gas.

The atmosphere today

By 200 million years ago the proportions of gases in the Earth's atmosphere had stabilised. These were much the same as they are today. Look at the percentage of gases in the atmosphere today in the pie chart in Figure 2.

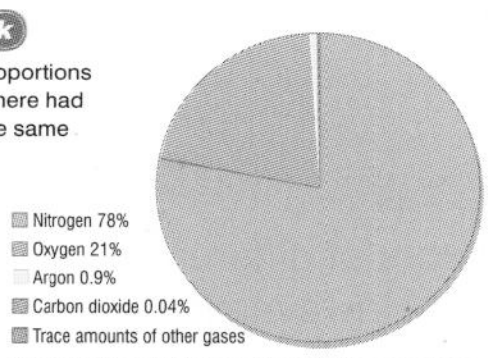

Nitrogen 78%
Oxygen 21%
Argon 0.9%
Carbon dioxide 0.04%
Trace amounts of other gases

Figure 2 The relative proportions of nitrogen, oxygen and other gases in the Earth's atmosphere

b What gas did plants produce that changed the Earth's atmosphere?

Practical

Shelly carbonates

Carry out a test to see if crushed samples of shells contain carbonates. Think of the reaction that all carbonates undergo with dilute acid. How will you test any gas given off?

- Record your findings.

Separating the gases in air (Higher)

In industry the gases are separated by the fractional distillation of liquid air.

Fractional distillation is a process in which liquids with different boiling points are separated. So first we have to get air cold enough for it to condense into a liquid. It has to be cooled to a temperature below −200 °C.

In industry they do this by compressing the air to about 150 times atmospheric pressure. This actually warms the air up. So it is cooled down to normal temperatures by passing the air over pipes carrying cold water.

But the main cooling takes place when the pressure is released. As this happens, the air is allowed to expand rapidly. This is similar to what happens in an aerosol can when pressure is released as the aerosol is sprayed. The temperature drops far enough for even the gases in air to condense to liquids. The carbon dioxide and water can be removed from the mixture as they are solids at this low temperature.

Here are the boiling points of the main substances left in the liquid air mixture: Nitrogen = −196 °C, Argon = −186 °C, Oxygen = −183 °C.

The liquid is then allowed to warm up and at −196 °C nitrogen boils off first. It is collected from the top of a tall fractionating column.

Liquid nitrogen is used to cool things down to very low temperatures. At these temperatures most things solidify. It is used to store sperm in hospitals to help in fertility treatment. Nitrogen gas is very unreactive so we use it in sealed food packaging to stop food going off. It is also used on oil tankers when the oil is pumped ashore to reduce the risk of explosion. In industry, nitrogen gas is used to make ammonia which we convert into fertilisers.

The oxygen separated off is used to help people breathe, often at the scene of an accident or in hospital. It is also used to help things react. Examples include high temperature welding and in the steel-making process.

Figure 4 Biological samples are preserved in liquid nitrogen until they are needed

links

For information on the fractionating column used to separate crude oil into fractions, look back at C1 4.2 Fractional distillation.

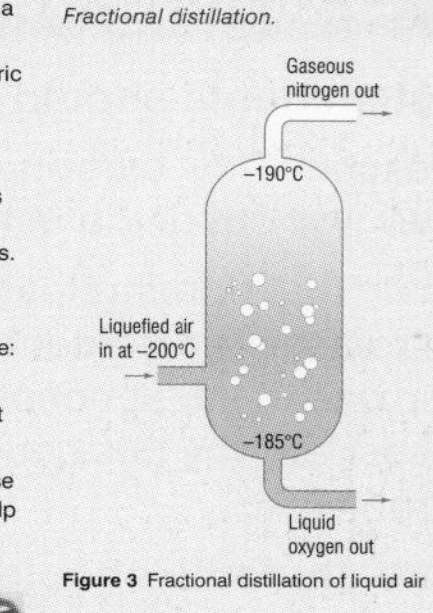

Figure 3 Fractional distillation of liquid air

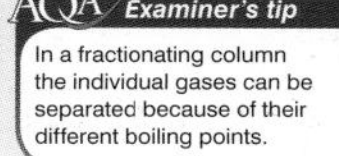

AQA Examiner's tip

In a fractionating column the individual gases can be separated because of their different boiling points.

Summary questions

1 Copy and complete the table showing the proportion of gases in the Earth's atmosphere today.

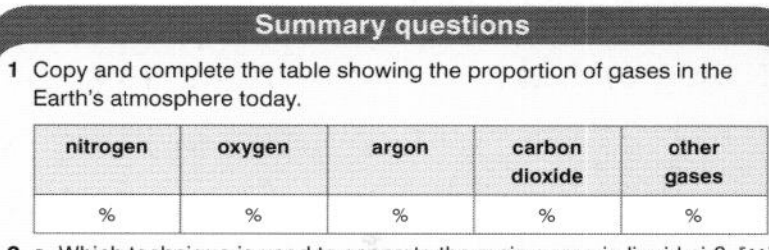

nitrogen	oxygen	argon	carbon dioxide	other gases
%	%	%	%	%

2 a Which technique is used to separate the main gases in liquid air? [H]
  b How can water and carbon dioxide be removed from the air before the gases enter the fractionating column. [H]
  c Look at the boiling points of nitrogen, argon and oxygen above:
    i Which gas boils off after nitrogen?
    ii Why is it difficult to obtain 100% pure oxygen? [H]

Key points

- The main gases in the Earth's atmosphere are oxygen and nitrogen.
- About four-fifths (80%) of the atmosphere is nitrogen, and about one-fifth (20%) is oxygen.
- The main gases in the air can be separated by fractional distillation. These gases are used in industry as useful raw materials. [H]

Further teaching suggestions

Summary mnemonic

- Students taking the Higher Tier paper could be asked to summarise fractional distillation of air into a list of words, which represent the different parts of the process. They could then create a mnemonic to help them remember the order of the stages more easily (cool the air), compress, condense, boil, collect). **[HT only]**

Key questions

- Ask students to generate questions for which each key point in the Student Book is the answer. Then cut out each question and give them to another student. Their task is to match the question with the key point. The pairs of students then feed back to each other about where their sorting was correct and about how the question was phrased.

Spider diagram

- Students taking the Higher Tier paper could make a spider diagram detailing how air is separated and the uses of each of the products. **[HT only]**

Links

- Give students a sheet of A4 paper and ask them to fold it in half. On one side they should draw a labelled diagram of fractional distillation of crude oil and a bullet point explanation as to how crude oil is separated. On the second side they should do the same but for fractional distillation of air. Hopefully students will then see that the explanation is the same but the chemical names and boiling points are different. **[HT only]**

Answers to in-text questions

a into carbonate rocks

b oxygen

Summary answers

1

nitrogen	oxygen	argon	carbon dioxide	other gases
78%	21%	0.9%	0.04%	trace

2 **a** fractional distillation
  b They solidify out by cooling the air.
  c **i** argon
    ii Because the boiling points of argon and oxygen are very close together.

C1 7.6 Carbon dioxide in the atmosphere

Learning objectives

Students should learn:

- that carbon dioxide moves in and out of the atmosphere
- that the amount of carbon dioxide is increasing in the atmosphere.

Learning outcomes

Most students should be able to:

- describe how carbon dioxide moves into and out of the atmosphere
- state that burning fossil fuels has increased the amount of carbon dioxide in the atmosphere and describe why this could cause problems.

Some students should also be able to:

- explain why there is a general trend that the amount of carbon dioxide in the air is increasing and evaluate the possible consequences.

Specification link-up: Chemistry C1.7

- The oceans also act as a reservoir for carbon dioxide but increased amounts of carbon dioxide absorbed by the oceans has an impact on the marine environment. *[C1.7.2 h)]*
- Nowadays the release of carbon dioxide by burning fossil fuels increases the level of carbon dioxide in the atmosphere. *[C1.7.2 i)]*
- Explain and evaluate the effects of human activities on the atmosphere. *[C1.7]*

Lesson structure

Starters

Demonstration – If solid carbon dioxide can be obtained, e.g. from a spare black carbon dioxide fire extinguisher, put it into water and dry ice will be created. Be careful not to handle the solid for too long or burns will be caused. Tweezers should be used to manipulate it. Students can then put their hands briefly into the water (as long as they do not touch the solid) and it will feel really cold. Ask the students to suggest what is happening. [The solid carbon dioxide is boiling in the water.] *(5 minutes)*

Describe – Supply carbon dioxide gas in a gas jar to each table. Ask the students to describe its physical appearance [colourless, transparent gas]. You could support students by giving them a list of words on the board [colourless, coloured, transparent, opaque, gas, liquid, solid] and they would have to choose from the list. You could extend students by asking them to explain the laboratory test for carbon dioxide (links to C1 2.2). Ask students to discuss their ideas in small groups and then manage whole-class feedback. *(10 minutes)*

Main

- The carbon cycle is a network of different reactions, which remove carbon from, or add it into, the atmosphere. However, in C1 students are only required to understand that carbon dioxide sinks include the oceans and sedimentary rocks and fossil fuels. Students need to know that burning fossil fuels releases carbon dioxide and that heating the oceans reduces the amount of carbon dioxide that they can absorb. They should also be aware that when carbon dioxide is absorbed, the pH of the oceans is reduced.
- Carbon dioxide is produced as a product of respiration. You may wish to measure the amount of carbon dioxide in air and contrast this to exhaled air. See 'Practical support' for more detail.
- To demonstrate the effect on the pH of water when carbon dioxide is dissolved, blow carbon dioxide through a mixture of water and universal indicator. See 'Practical support' for more detail. Ask students to suggest what affect this may have on ecosystems [it changes the species that can live in an area].
- Give students a graph that shows the percentage of carbon dioxide in the atmosphere over the last 300 years. Ask students to state the trend [carbon dioxide levels were stable until about 1850, then the levels increased, and the rate of increase is accelerating]. Students should then be encouraged to explain the trend, thinking about the use of fossil fuels.

Plenaries

Agree? – Ask for a volunteer to stand in the centre of the classroom. The volunteer should make a statement about the content of the lesson (it could be correct or deliberately incorrect), e.g. 'carbon can be found in rocks'. The rest of the class decides how much they agree with this statement. The more they agree, the closer they should stand to the person who spoke. Then ask a few students why they are positioned as they are and feedback whether the statement is correct. Ask all the students to sit down and for another volunteer to repeat the idea. *(5 minutes)*

Key reminders – Ask students to copy out the key points onto a flash card. On the other side, draw an image that will help them remember/represent the key point. Students could be supported by being supplied with the diagrams, which they should match to the key point. Students can be extended by representing the first key point with balanced symbol equations showing the hydrolysis of carbon dioxide and photosynthesis. *(10 minutes)*

Support

- You could support students by supplying key words on the board to help them follow what happens to carbon dioxide in the natural course of events. Alternatively, the exercise could be turned into a 'cut-and-stick' activity.
- Show students some posters from environmental charities such as Greenpeace and the WWF. Ask students to design a similar poster to explain how burning fossil fuels could affect marine environments.

Extend

- You could extend students by asking them to suggest or research some ways in which the amount of carbon dioxide in the atmosphere could be reduced [e.g. carbon capture schemes].

Practical support

Measuring carbon dioxide

Equipment and materials required

Carbon dioxide probe, data logger, lung volume bags.

Details

Use the carbon dioxide probe attached to a data logger to record the concentration of carbon dioxide in the air. Exhale into a lung volume bag and then put the probe into the bag and measure the concentration of carbon dioxide in exhaled air.

Acidification of water

Equipment and materials required

Carbon dioxide cylinder with regulator, rubber tube, 250 cm^3 beaker, universal indicator solution, distilled water.

Details

Half-fill the beaker with distilled water. Put a few drops of universal indictor into the water. Connect the rubber tube to the carbon dioxide cylinder and put the open end into the water. Turn the gas on, so that the flow rate is about one bubble per second. Observe the colour change of the universal indicator as the water becomes acidified.

Safety: CLEAPSS Hazcard 32 Universal indicator – highly flammable and harmful.

Our changing planet

C1 7.6 Carbon dioxide in the atmosphere

Learning objectives

- How does carbon move in and out of the atmosphere?
- Why has the amount of carbon dioxide in the atmosphere increased recently?

How Science Works

Increasing levels of carbon dioxide

Look at the data collected by scientists monitoring the proportion of carbon dioxide in the atmosphere at one location:

CO_2 concentration (ppm) — 310 to 355; Year — 1958, 64, 70, 76, 82, 88

- Why is the line not a smooth curve?
- Explain the overall trend shown by the data.

Over the past 200 million years the levels of carbon dioxide in the atmosphere have not changed much. This is due to the natural cycle of carbon in which carbon moves between the oceans, rocks and the atmosphere.

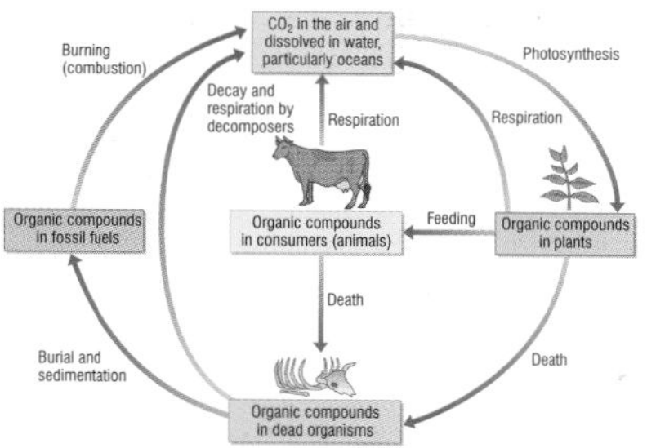

Figure 1 The level of carbon dioxide in the atmosphere has remained steady for the last 200 million years as a result of this natural cycle. However, over the past 200 years the carbon dioxide levels have risen as people started to burn more and more fossil fuels.

Left to itself, this cycle is self-regulating. The oceans act as massive reservoirs of carbon dioxide. They absorb excess CO_2 when it is produced and release it when it is in short supply. Plants also remove carbon dioxide from the atmosphere. We often call plants and oceans carbon dioxide 'sinks'.

a What has kept carbon levels roughly stable over the past 200 million years?

The changing balance

Over the recent past we have greatly increased the amount of carbon dioxide released into the atmosphere. We burn fossil fuels to make electricity, heat our homes and run our cars. This has enormously increased the amount of carbon dioxide we produce.

There is no doubt that the levels of carbon dioxide in the atmosphere are increasing.

We can record annual changes in the levels of carbon dioxide which are due to seasonal differences in the plants. The variations within each year show how important plants are for removing CO_2 from the atmosphere. But the overall trend over the recent past has been ever upwards.

The balance between the carbon dioxide produced and the carbon dioxide absorbed by 'CO_2 sinks' is very important.

links

For information about the effect humans have had on the levels of carbon dioxide in the atmosphere, look back at C1 4.4 Cleaner fuels.

106

Carbon dioxide in the atmosphere

Think about what happens when we burn fossil fuels. Carbon has been locked up for hundreds of millions of years in the fossil fuels. It is released as carbon dioxide into the atmosphere when used as fuel. For example:

$$\text{propane} + \text{oxygen} \rightarrow \text{carbon dioxide} + \text{water}$$
$$C_3H_8 + 5O_2 \rightarrow 3CO_2 + 4H_2O$$

As carbon dioxide levels in the atmosphere go up, the reactions of carbon dioxide in sea water also increase. The reactions make *insoluble* carbonates (mainly calcium carbonate). These are deposited as sediment on the bottom of the ocean. They also produce *soluble* hydrogencarbonates, mainly of calcium and magnesium. These compounds simply remain dissolved in the sea water.

In this way the seas and oceans act as a buffer, absorbing excess carbon dioxide but releasing it if necessary. However there are now signs that the seas cannot cope with all the additional carbon dioxide that we are currently producing. For example, coral reefs are dying in the more acidic conditions caused by excess dissolved carbon dioxide.

Figure 2 Most of the electricity that we use in the UK is made by burning fossil fuels

How Science Works

Thinking of solutions but at what cost?

Most of the electricity that we use in the UK is made by burning fossil fuels. This releases carbon dioxide into the atmosphere. Scientists have come up with a number of solutions. One solution would be to pump carbon dioxide produced in fossil fuel power stations deep underground to be absorbed into porous rocks. This is called 'carbon capture and storage'. It is estimated that this would increase the cost of producing electricity by about 10%.

- Give an advantage and a disadvantage of reducing carbon dioxide emissions using 'carbon capture and storage'.

Did you know ...?

Some scientists predict that global warming may mean that the Earth's average temperature could rise by as much as 5.8 °C by the year 2100!

People are worried about changing climates (including increasingly common extreme weather events) and rising sea levels as a result of melting ice caps and expansion of the warmer oceans. Low-lying land then might disappear beneath the sea.

Summary questions

1 Match up the parts of sentences:

a	Carbon dioxide levels in the Earth's atmosphere ...	A	... carbon locked up long ago is released as carbon dioxide.
b	Plants and oceans are known as ...	B	... were kept steady by the natural recycling of carbon dioxide in the environment.
c	When we burn fossil fuels ...	C	... the reactions of carbon dioxide in sea water increase.
d	As carbon dioxide levels rise ...	D	... carbon dioxide sinks

2 Draw a labelled diagram to illustrate how boiling an electric kettle may increase the amount of carbon dioxide in the Earth's atmosphere.

3 Why has the amount of carbon dioxide in the Earth's atmosphere risen so much in the recent past?

Key points

- Carbon moves into and out of the atmosphere due to plants, animals, the oceans and rocks.
- The amount of carbon dioxide in the Earth's atmosphere has risen in the recent past largely due to the amount of fossil fuels we now burn.

107

Further teaching suggestions

Investigating carbon dioxide

- All animals breathe out carbon dioxide gas, and students often talk about it, but rarely can they explain the physical and chemical properties of the gas. They could complete a series of experiments and note their observations. They could then summarise the properties of this chemical (see 'Practical support' for more details).

Model

- Ask students to make a model of a molecule of CO_2 using a molecular model kit. Describe the bonds holding the black carbon and red oxygen atoms together. You could support students by giving them the formula of carbon dioxide and the colour and number of each type of atom. You could extend students by asking them to classify the bonding in this molecule (it is covalent) (links to C1 1.4).

Answers to in-text questions

a The natural recycling of carbon dioxide between oceans, rocks and the atmosphere.

Summary answers

1 **a** B
b D
c A
d C

2 Diagram to show how an electric kettle may use energy produced by burning fossil fuels. It should include drilling, mining for fossil fuels, transport of the fossil fuels, a power station producing electricity and a kettle heating the water.

3 Mainly due to our increased burning of fossil fuels to satisfy our energy requirements in the industrial world.

Summary answers

1 a Middle layer of the Earth, made of molten rock about 3000 km thick and lying about 100 km under the surface.

b Innermost layer of the Earth made of iron and nickel.

c The relatively thin layer of gases above the Earth's surface.

d Part of the Earth's crust made of solid rock that moves and causes continental drift, earthquakes, volcanoes and mountains.

2 a The shapes of coastlines of west Africa and South America seemed like two pieces in a jig-saw. The similarity of fossils and rock types in west Africa and South America. The single land mass moved apart because the tectonic plates that make up the Earth's outer layer moved at a relative speed of a few centimetres per year. This happens because of convection currents set up in the mantle due to heat given off from radioactive atoms.

b Wegener could not adequately explain how tectonic plates could possibly move at that time (and he was not a geologist).

3 a The amount of oxygen, methane and ammonia decreased, nitrogen increased.

b Reaction of oxygen with ammonia and methane.

c i methane + oxygen → carbon dioxide + water

ii ammonia + oxygen → nitrogen + water

d Because there has been a dramatic increase in our burning fossil fuels since then, resulting in ever-increasing amounts of carbon dioxide being released.

4 a They showed that it is possible to make the organic molecules, such as amino acids, on which living things are based, from the simple molecules that could have been present in the Earth's early atmosphere.

b Although they showed that the molecules of life could be made under certain conditions, they did not show how these molecules combined to make living cells.

5 Mark poster based on facts about Earth's early atmosphere, how this has changed, the atmosphere today and how it continues to change.

6 a The graph should have the axes correctly and fully labelled. The points should be accurately plotted. The date should be on the *x*-axis and the CO_2 concentration on the *y*-axis.

b The line of best fit should be a curve.

c There should be a slow increase in the concentration of CO_2 at the start of the data, followed by more rapid increases in more recent years. Some might spot the anomalies around 1945. This might prompt a debate about the causes, which could prove very interesting. Remember these probably represent global figures and they might conclude that they would like more detailed data between 1935 and 1955 to try to sort out what might have happened. This is given in Table 1.

d That the concentration of carbon dioxide in the atmosphere has increased.

That it has increased at a faster rate in more recent years. Caution here about taking conclusions too far and resist the conclusion that this alone is evidence for the greenhouse effect or global warming.

e Yes – they are two totally different ways of deriving the data. Students might want to know of any overlap between the two sets of data and the correlation between the two. The data is given in Table 2.

Summary questions

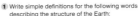

1 Write simple definitions for the following words describing the structure of the Earth:

a mantle

b core

c atmosphere

d tectonic plate.

2 Wegener suggested that all the Earth's continents were once joined in a single land mass.

a Describe the evidence for this idea, and explain how the single land mass separated into the continents we see today.

b Why were other scientists slow to accept Wegener's ideas?

3 The pie charts show the atmosphere of a planet shortly after it was formed (A) and then millions of years later (B).

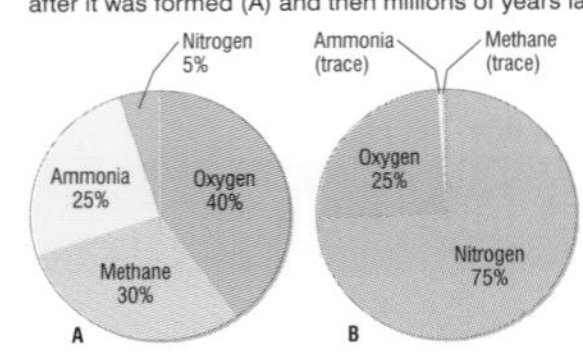

a How did the atmosphere of the planet change?

b What might have caused the change in part **a**?

c Copy and complete the word equations showing the chemical reactions that may have taken place in the atmosphere.

i methane + ... → carbon dioxide + ...

ii ammonia + ... → nitrogen + ...

d Why have levels of carbon dioxide in the Earth's atmosphere increased so dramatically over the past 200 years?

4 a Describe how the Miller–Urey experiment advanced our understanding of how life might have first formed on Earth. [H]

b Why didn't their experiment prove how life began on Earth? [H]

5 The Earth and its atmosphere are constantly changing. Design a poster to show this. It should be suitable for displaying in a classroom with children aged 10–11 years. Use diagrams and words to describe and explai ideas and to communicate them clearly to the children.

6 Core samples have been taken of the ice from Antarctica. The deeper the sample the longer it has been there. It is possible to date the ice and to take air samples from it. The air was trapped when the ice was formed. It is possible therefore to test samples of air that have been trapped in the ice for many thousands o years.

This table shows some of these results. The more rece results are from actual air samples taken from a Pacific island.

Year	CO_2 concentration (ppm)	Source
2005	379	Pacific island
1995	360	Pacific island
1985	345	Pacific island
1975	331	Pacific island
1965	320	Antarctica
1955	313	Antarctica
1945	310	Antarctica
1935	309	Antarctica
1925	305	Antarctica
1915	301	Antarctica
1905	297	Antarctica
1895	294	Antarctica
1890	294	Antarctica

a If you have access to a spreadsheet, enter the data and produce a line graph.

b Draw a line of best fit.

c What pattern can you detect?

d What conclusion can you make?

e Should the fact that the data came from two differer sources affect your conclusion? Explain why.

Table 1

Year	'35	'36	'37	'38	'39	'40
CO_2 conc. (ppm)	309.4	309.8	310.0	310.2	310.3	310.4
Year	'41	'42	'43	'44	'45	'46
CO_2 conc. (ppm)	310.4	310.3	310.2	310.1	310.1	310.1
Year	'47	'48	'49	'50	'51	'52
CO_2 conc. (ppm)	310.2	310.3	310.5	310.7	311.1	311.5
Year	'53	'54	'55			
CO_2 conc. (ppm)	311.9	312,4	313.0			

Table 2

Year	'78	'77	'76	'75	'74
Pacific island CO_2 conc. (ppm)	335.5	333.9	332.1	331.1	330.2
Antarctica CO_2 conc. (ppm)	333.7	332.6	331.5	330.3	329.2
Year	'73	'72	'71	'70	'69
Pacific island CO_2 conc. (ppm)	329.7	327.5	326.3	325.7	324.6
Antarctica CO_2 conc. (ppm)	328.0	326.9	325.8	324.8	323.8

AQA Examination-style questions

a Match a word from the list with each of the labels A, B, C and D on the diagram of the Earth.

atmosphere core crust mantle

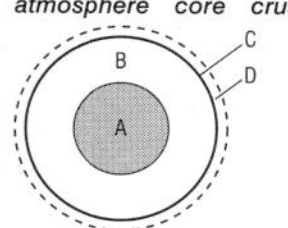

(4)

b From which parts of the Earth do we get all of our raw materials?

A atmosphere, core and crust
B atmosphere, crust and oceans
C atmosphere, core and mantle
D core, mantle and oceans (1)

a About one hundred years ago there was a scientist called Alfred Wegener. He found evidence that the continents, such as South America and Africa, had once been joined and then drifted apart.

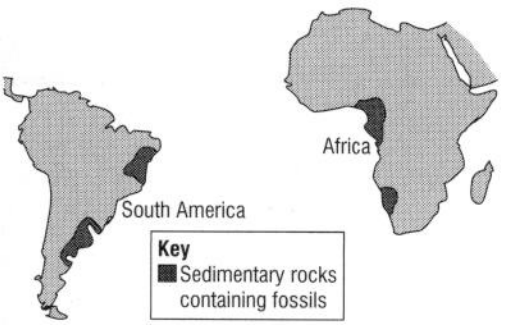

Use the diagram to suggest **two** pieces of evidence that could be used to show that the continents had once been joined. (2)

b About fifty years ago, new evidence convinced scientists that the Earth's crust is made up of tectonic plates that are moving very slowly.

Give **two** pieces of evidence that have helped to convince these scientists that the tectonic plates are moving. (2)

c Describe as fully as you can what causes the Earth's tectonic plates to move. (3)

AQA, 2009

a In the Earth's atmosphere the percentage of carbon dioxide has remained at about 0.03% for many thousands of years. The graph shows the percentage of carbon dioxide in the Earth's atmosphere over the last 50 years.

Carbon dioxide (%) 0.031–0.040; Year 1960–2010

i What was the percentage of carbon dioxide in the Earth's atmosphere in 1965? .... % (1)
ii What change has happened to the percentage of carbon dioxide in the Earth's atmosphere over the last 50 years? (1)
iii Suggest **one** reason for this change. (1)
iv Why does this change worry some people? (1)

There are different theories about the Earth's early atmosphere.

b Some scientists believe the Earth's early atmosphere was mainly carbon dioxide and water vapour. What do the scientists believe produced these gases? (1)

c In 1953 some scientists believed the Earth's early atmosphere was mainly water vapour, methane, ammonia and hydrogen. In the Miller–Urey experiment, electricity was passed through a mixture of these gases and produced amino acids, the building blocks for proteins and life. Give two reasons why the experiment does not prove that life began in this way. [H] (2)

d Most scientists agree that there was very little oxygen in the Earth's early atmosphere. Explain how the oxygen that is now in the atmosphere was produced. [H] (3)

4 The elements oxygen, nitrogen and argon can be separated from the air. Carbon dioxide and water vapour are removed from air, which is then cooled to –200°C. The liquid obtained is a mixture of oxygen, nitrogen and argon. The table shows the boiling points of these elements.

Element	Boiling point in °C
argon	–183
nitrogen	–196
oxygen	–186

Explain how these elements can be separated by fractional distillation of the liquid. [H] (3)

109

Study tip

Students find fractional distillation difficult, and even more so with negative numbers, so Question 5 gives an opportunity to check understanding of these topics.

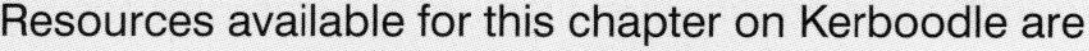

Kerboodle resources

Resources available for this chapter on Kerboodle are:

- Chapter map: Our changing planet
- Support: Where does it come from? (C1 7.1)
- Practical: How did plants alter the Earth's atmosphere? (C1 7.3)
- Maths skills: Composition of the atmosphere (C1 7.5)
- How Science Works: Analysis that's truly atmospheric (C1 7.5)
- Extension WebQuest: Capturing carbon (C1 7.6)
- Extension: What's the link? (C1 7.6)
- Interactive activity: Our changing planet
- Revision podcast: The development of the Earth's atmosphere
- Test yourself: Our changing planet
- On your marks: The early atmosphere
- Examination-style questions: Our changing planet
- Answers to examination-style questions: Our changing planet

Examination-style answers

1 a A core, B mantle, C crust, D atmosphere. *(4 marks)*
b B – atmosphere, crust and oceans *(1 mark)*

2 a Any **two** from: the shapes of the continents fit together, the same type of rocks have been found, the same fossils have been found, accept rocks match, fossils match, magma rising through a gap under the Atlantic. *(2 marks)*
b Any **two** from: earthquakes, volcanoes, idea of distance between America and Europe/Africa is increasing, oceanic ridges, formation of mountain ranges, formation of islands, nmagnetic stripes, tsunamis. *(2 marks)*
c Any **three** from: (natural) radioactivity produces heat, (causes) convection currents, in the mantle, mantle able to move or (behaves) like a liquid, create forces (to move plates). *(3 marks)*

3 a **i** 0.0317 (%) *(1 mark)*
ii increased *(1 mark)*
iii burning fossil fuels or deforestation *(1 mark)*
iv Global warming or greenhouse gas, accept causes floods or sea level rise or climate change or extreme weather. *(1 mark)*
b volcanoes or volcanic activity *(1 mark)*
c Any **two** from: cannot be sure what was in the atmosphere, only produced building blocks or amino acids are only one step towards living things, evidence is in dispute or only theory or not conclusive. *(2 marks)*
d Because plants evolved or increased or colonised the Earth, plants photosynthesise, which uses carbon dioxide or removes carbon dioxide from the atmosphere. *(3 marks)*

4 EITHER any **three** from: Liquid into fractionating column, nitrogen collected at top and oxygen collected at the bottom, argon collected from in between or part way up column, nitrogen has lowest boiling point, nitrogen collected as gas, oxygen has highest boiling point, oxygen collected as liquid, OR any **three** from: nitrogen has lowest boiling point so nitrogen boils first or is collected as gas, argon has next lowest boiling point or boils next, oxygen has highest boiling point so boils last or remains as liquid. *(3 marks)*

Practical suggestions

Practicals	AQA	k	(book icon)	(gears icon)
Investigating the composition of air by passing air over heated copper using gas syringes and measuring the percentage of oxygen. Then burning magnesium in the nitrogen to form Mg_3N_2. Add water to produce ammonia (nitrogen must have come from the air).	✓			✓
Collecting gas produced by aquatic plants and testing for oxygen (using dissolved oxygen sensor).	✓	✓		
Measuring the amount of carbon dioxide in inhaled and exhaled air (using carbon dioxide sensor).	✓		✓	
Testing the products of combustion of fuels to show that carbon dioxide is produced.	✓		✓	
Design an investigation to compare the amount of carbon dioxide released by reacting crushed shells (e.g. cockle, oyster) with dilute hydrochloric acid.	✓		✓	

Examination-style answers

1 a correctly labelled: electron, nucleus *(2 marks)*

b i oxygen, accept O_2, ignore air *(1 mark)*

ii Any one from: (water) does not pollute, (only) water is produced, no carbon dioxide (is produced), no sulfur dioxide (is produced), no nitrogen oxides (are produced), no carbon or no particles (are produced), accept no harmful gas(es), no greenhouse gas(es), no acid rain. *(1 mark)*

2 a i 8 *(1 mark)*

ii 8 *(1 mark)*

b 2,8,7 *(1 mark)*

c Because it has 7 electrons in highest energy level/outer shell, or (has structure) 2,7 or same number of outer electrons or same number (of electrons) in highest energy level/outer shell, ignore just "same number of electrons". *(1 mark)*

d Both have 8 electrons in highest energy level/outer shell or 8 outer electrons, accept both have the same number of electrons in highest energy level/outer shell, ignore "both have the same structure" their electronic structures/ arrangements are very stable. *(2 marks)*

3 a i 2 *(1 mark)*

ii 3 *(1 mark)*

iii 5 *(1 mark)*

b i zinc oxide *(1 mark)*

ii carbon dioxide *(1 mark)*

4 a Limestone is heated or thermally decomposed (to make calcium oxide), (calcium oxide is reacted with) water (to make calcium hydroxide). *(2 marks)*

b Calcium hydroxide reacts with acids, because it is an alkali. *(2 marks)*

c Limestone is/contains (mainly) calcium carbonate, carbonate(s) react with acids. *(2 marks)*

d Calcium hydroxide/alkali is caustic/corrosive or damages skin/eyes or calcium carbonate/limestone is not caustic/ corrosive or does not damage skin do not accept vague statements e.g. limestone is safer to handle. *(1 mark)*

e Any sensible suggestion, e.g. calcium hydroxide is made by heating limestone, two reactions needed (to make it), (crushed) limestone needs less energy (to make it) or only needs to be crushed. *(1 mark)*

f $CaCO_3 \rightarrow CaO + CO_2$
$CaO + H_2O \rightarrow Ca(OH)_2$ *(2 marks)*

g $Ca(OH)_2 + 2HCl \rightarrow CaCl_2 + 2H_2O$ reactants and products balancing *(2 marks)*

h $CaCO_3 + 2HCl \rightarrow CaCl_2 + CO_2 + H_2O$ reactants and products balancing *(2 marks)*

5 a Less dense or lighter (more) resistant to corrosion. *(2 marks)*

b There is a clear, logical and detailed scientific explanation of why titanium costs more than steel. The answer shows almost faultless spelling, punctuation and grammar. It is coherent and in an organised, logical sequence. It contains a range of appropriate and relevant specialist terms used accurately. *(5–6 marks)*

There is a scientific explanation of why titanium costs more than steel. There are some errors in spelling, punctuation and grammar. The answer has some structure and organization. The use of specialist terms has been attempted, but not always accurately. *(3–4 marks)*

There is a brief explanation of why titanium costs more than steel that includes at least one correct comparison of the processes or costs involved. The spelling, punctuation and grammar are very weak. The answer is poorly organised with almost no specialist terms and/or their use demonstrating a general lack of understanding of their meaning. *(1–2 marks)*

No relevant content. *(0 marks)*

Examples of chemistry points made in the response:

- Number of steps: Steel production involves fewer reactions/steps or simpler reactions/steps.
- Time and yield: titanium takes longer, smaller amounts (per day)
- Process: Iron production (allow steel or part of steel production) is continuous, titanium is batch process
- Cost of reagents: chlorine, magnesium (have to be produced by electrolysis) more expensive than carbon/coke (from coal) and/or air and/or oxygen (from air). (ignore cost of metal ores)
- Energy costs: more energy needed to produce titanium, cost of electricity (for electrolysis or electric furnace), titanium needs to be melted again
- Labour costs: Less labour needed for making iron/ steel.

6 a pressed *(1 mark)*

b (oil) does not mix (with water) *(1 mark)*

c Any **two** from: does not damage/decompose the oil, does not change its flavour, does not damage nutrients, does not use (fossil) fuel/energy for heating. *(2 marks)*

d (olive oil/it) burns (in air) produces energy *(2 marks)*

Examination-style questions

1 a The diagram shows the parts of a hydrogen atom.
Use words from the list to label the diagram.
electron group nucleus symbol

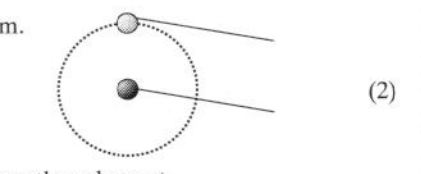

(2)

b Hydrogen can be used as a *clean fuel* for cars.

i When hydrogen burns in air, it reacts with another element.
Complete the word equation for this reaction. (1)
hydrogen + ... → water

ii Suggest **one** reason why hydrogen is called a *clean fuel*. (1)

AQA, 2008

2 Use a periodic table to help you to answer this question.
Oxygen is in Group 6 of the periodic table.

a i How many protons are in an atom of oxygen? (1)

ii How many electrons are in an atom of oxygen? (1)

b Chlorine is in Group 7 of the periodic table.
Complete the electronic structure of chlorine: 2, ... (1)

c Fluorine is also in Group 7.
Explain why in terms of electronic structure. (1)

d Neon and argon are in Group 0 of the periodic table. They are very unreactive elements. What does this tell you about their electronic structures? (2)

3 When calcium carbonate is heated it decomposes. The equation for this reaction is:
$CaCO_3 \rightarrow CaO + CO_2$

a Use numbers from the list to complete the sentences.
2 3 4 5 6

i The number of products in the equation is .... (1)

ii The formula $CaCO_3$ shows that calcium carbonate was made from .... different elements. (1)

iii The equation is balanced because there are .... atoms on both sides. (1)

b Other metal carbonates decompose in a similar way.

i Name the solid produced when zinc carbonate decomposes. (1)

ii Name the gas produced when copper carbonate decomposes. (1)

4 Farmers can use calcium hydroxide to neutralise soils that are too acidic. Limestone is mainly calcium carbonate, $CaCO_3$.
Limestone is used to make calcium hydroxide, $Ca(OH)_2$.

a What are the two reactions used to make calcium hydroxide from limestone? (2)

b Explain why calcium hydroxide neutralises soils that are too acidic. (2)

c Farmers can also use powdered limestone to neutralise soils that are too acidic. Explain why. (2)

d Suggest one reason why it may be safer for farmers to use powdered limestone instead of calcium hydroxide. (1)

e Suggest one reason why powdered limestone costs less than calcium hydroxide. (1)

f Write balanced equations for the reactions in 4(a) [H] (2)

g The formula of calcium chloride is $CaCl_2$. Write a balanced equation for the reaction of calcium hydroxide with hydrochloric acid, HCl. [H] (2)

h Write a balanced equation for the reaction of calcium carbonate with hydrochloric acid. [H] (2)

AQA Examiner's tip
When you are asked to complete a word equation for a reaction, read the information in the question carefully and you should find the names of the reactants and products.

AQA Examiner's tip
The AQA data sheet that you will have in the exam has a periodic table.

AQA Examiner's tip
Remember that each symbol represents one atom of an element and that small (subscript) numbers in a formula multiply only the atom they follow.

AQA Examiner's tip
Attempt all parts of a question. If you come to a part you cannot answer do not be put off reading the next parts.

110

5 Titanium is as strong as steel but is much more expensive. It is used to make jet engines for aircraft and to make replacement hip joints for people.

a Give two properties that make titanium better than steel for making jet engines and replacement hip joints. (2)

b *In this question you will be assessed on using good English, organising information clearly and using specialist terms where appropriate.*

Titanium is made in batches of about 10 tonnes that takes up to 15 days. The main steps to make titanium are:

- Titanium oxide is reacted with chlorine to produce titanium chloride.
- Titanium chloride is reacted with magnesium at 900 °C in a sealed reactor for three days to give a mixture of titanium and magnesium chloride.
- The reactor is cooled for 7 days, and then the mixture is removed.
- The magnesium chloride is removed from the mixture by distillation at very low pressure.
- The titanium is melted in an electric furnace and poured into moulds.

Steel is produced at about 8000 tonnes per day. The main steps to make steel are:

- Iron oxide is reacted with carbon (coke) in a blast furnace that runs continuously.
- The molten impure iron flows to the bottom of the furnace and is removed every four hours.
- Oxygen is blown into the molten iron for about 20 minutes to produce steel.
- The steel is poured into moulds.

Explain why titanium costs more than steel. (6)

AQA, 2008

AQA Examiner's tip

You may be given information about familiar or unfamiliar applications of chemistry. The information you are given should help you to answer the questions. Q5(b) requires you to organise information clearly. Think about the points in the information and decide which ones make titanium more expensive than steel. Underline or circle the points you are going to use on the question paper. Add brief notes, perhaps numbers for the order that you will use. Think about how you are going to write your answer. Rehearse it in your head before you write your answer.

6 Olives are the fruits of the olive tree. Olive oil is extracted from olives.

a Use a word from the list to complete the sentence.

condensed evaporated pressed

In the first step to extract the oil the olives are crushed and .... (1)

b This gives a mixture of liquids and solids that is left to settle.

Olive oil
Water
Solids

Why does the olive oil separate from the water? (1)

c The olive oil is removed from the water and filtered to remove any small pieces of solids.

Suggest **two** reasons why separating olive oil by this method is better than separating it by distilling. (2)

d Olive oil can be used as a fuel. Explain why. (2)

e Food can be cooked in olive oil. Give one advantage and one disadvantage of cooking food in olive oil. (2)

f Olive oil can be used with vinegar to make salad dressings. Name the type of substance that is added to salad dressings to stop them from separating. (1)

AQA Examiner's tip

Always be aware of the number of marks for a question. If it is two marks, you need to make two points in your answer. Sometimes this is obvious, as in Q6(c), but in Q6(d) you need to make sure you have not given just a single simple statement.

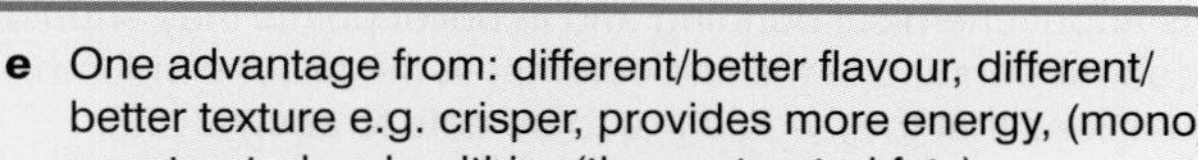

e One advantage from: different/better flavour, different/better texture e.g. crisper, provides more energy, (mono) unsaturated so healthier (than saturated fats).

One disadvantage from: provides more energy/calories/joules, costs more (than alternatives), greasy or stains clothes/fabrics. *(2 marks)*

f Emulsifier, allow a named type of emulsifier such as egg yolk. *(1 mark)*

C2 1.1 Chemical bonding

Learning objectives

Students should learn:

- how elements can form compounds
- how elements in Group 1 bond with elements in Group 7.

Learning outcomes

Most students should be able to:

- state what a compound is
- name the two types of bonding present in compounds
- represent Cl^- and Na^+ using a diagram.

Some students should also be able to:

- explain why atoms bond
- explain and work out the charge on an ion
- explain the formation of ions when a Group 1 and Group 7 element react together.

Specification link-up: Chemistry C2.1

- Compounds are substances in which atoms of two or more elements are chemically combined. *[C2.1.1 a)]*
- Chemical bonding involves either transferring or sharing electrons in the highest occupied energy levels (shells) of atoms in order to achieve the electronic structure of a noble gas. *[C2.1.1 b)]*
- When atoms form chemical bonds by transferring electrons, they form ions. Atoms that lose electrons become positively charged ions. Atoms that gain electrons become negatively charged ions. Ions have the electronic structure of a noble gas (Group 0). *[C2.1.1 c)]*
- The elements in Group 1 of the periodic table, the alkali metals, all react with non-metal elements to form ionic compounds in which the metal ion has a single positive charge. *[C2.1.1 d)]*
- The elements in Group 7 of the periodic table, the halogens, all react with the alkali metals to form ionic compounds in which the halide ions have a single negative charge. *[C2.1.1 e)]*
- Write formulae for ionic compounds from given symbols and ionic charges. *[C2.1]*
- Represent the electronic structure of the ions in sodium chloride… *[C2.1]*

Lesson structure

Starters

Spider diagram – Ask students to think back to the work they completed in C1 about the fundamental ideas in chemistry. Ask students to work in small groups to make a spider diagram of the information about bonding that they can remember. You may wish to support students by giving them a few statements or diagrams to get them started. You could extend students by asking them to look at a different group's work and give feedback about statements that they think are not quite correct. *(5 minutes)*

Reflect and plan – Ask students to draw a table. The first column is what I know, the second column, what I want to know and the third column, what I **now** know. Explain to students that this chapter is about chemical bonding and in bullet points they should list all that they already know about bonding in the first column. In the second column, students should then write questions that they would like to be able to answer after studying this topic area. The final column should remain blank. When the whole of this chapter has been covered, students should review this, completing the final column of additional information that they have gained. *(10 minutes)*

Main

- Students may have seen the formation of sodium chloride in C1 1.4. Provide students with sodium chloride and ask them to note down its properties. They should be encouraged to contrast the properties of the compound with those of its constituent elements and reflect on the fact that a compound and its elements can have completely different properties.
- Students need to be able to understand ionic bonding and represent it as dot and cross diagrams. Ask students to use the Student Book to create a poster about ionic bonding. They should include at least two diagrams of ions, and one diagram of an ionic bond.
- Stress to students that ionic bonding involves electron transfer, but the actual ionic bonds arise from the electrostatic attraction between the oppositely charged ions formed as a result of electron transfer.

Plenaries

Match – Give students the names of ionic compounds and their formulae. Students should match them up. [e.g. copper sulfate – $CuSO_4$; aluminium chloride – $AlCl_3$; iron oxide – Fe_2O_3.] *(5 minutes)*

Classify – Give students a list of compounds and ask them to classify the bonding as ionic or covalent. You may wish to support students by highlighting the definition of each type of bonding given in C1.1.4. You could extend students by asking them to write the formula of each of the compounds. You could further extend students by asking them to draw diagrams of the compounds. [Sodium chloride (ionic, NaCl), carbon dioxide (covalent, CO_2), sulfur dioxide (covalent, SO_2), potassium bromide (ionic, KBr).] *(10 minutes)*

Support

- Using A4 mini-whiteboards, encourage students to draw the electronic structure of simple atoms, including the number of protons and neutrons in the nucleus. Ask students to consider the outer shell electrons and whether the atom would lose or gain electrons. They should rub out or add in electrons in the correct places on their diagrams. Write the number of electrons gained or lost at the top right of the board. Then ask students to consider the charge: electrons add negative charge, so removing them leaves a positive charge.

Extend

- Ask students to consider molecular ions. Give students a diagram of an ionic compound, then students could label the ionic bond and the covalent bonds that it contains.

Further teaching suggestions

Models

- Students could look at and draw models of a salt crystal. They could label the ions.

Drawing and naming

- You could give students the electronic structure of different ions and then ask them to name them.

Dot and cross diagrams

- You could make a card sort of different ions with their symbol in the centre of the ionic diagram. Positive ions should be on the red card and negative ions on the blue card. The greater the charge, the darker the colour of the card. Students could then generate the dot and cross diagrams of a variety of ionic compounds, easily and quickly swapping ions to show the compounds.

Computer modelling

- You may wish to allow students to use computer modelling to simulate how electrons are shared between atoms to make molecules. You could extend students and allow them to make their own simulations using simple programs which will allow students to make their own flash animations.

Answers to in-text questions

a Mixing involves a physical change (easily reversed and no new substances are formed). Reacting involves a chemical change, which does involve making new substances.

b covalent

c ionic

Structure and bonding

C2 1.1 Chemical bonding

Learning objectives

- How do elements form compounds?
- How do the elements in Group 1 bond with the elements in Group 7?

You already know that we can mix two substances together without either of them changing. For example, we can mix sand and salt together and then separate them again. No change will have taken place. But in chemical reactions the situation is very different.

When the atoms of two or more elements react they make a compound.

A compound contains two or more elements which are chemically combined.

The compound formed is different from the elements and we cannot get the elements back again easily. We can also react compounds together to form other compounds. However, the reaction of elements is easier to understand as a starting point.

a What is the difference between mixing two substances and reacting them?

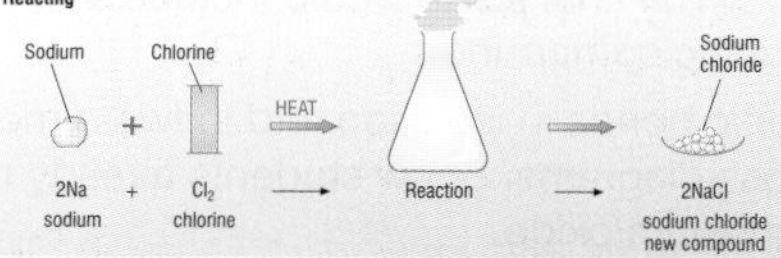

Figure 1 The difference between mixing and reacting. Separating mixtures is usually quite easy, but separating substances once they have reacted can be quite difficult. Why do atoms react?

When an atom has an arrangement of electrons like a noble gas in Group 0, it is stable and unreactive. However, most atoms do not have this electronic structure. When atoms react they take part in changes which give them a stable arrangement of electrons. They may do this by either:

- sharing electrons, which we call **covalent bonding**
- transferring electrons, which we call **ionic bonding**.

Sodium 2,8,1

Na^+ 2,8

Figure 2 A positive sodium ion (Na^+) is formed when a sodium atom loses an electron during ionic bonding

Losing electrons to form positive ions

In ionic bonding the atoms involved lose or gain electrons to form charged particles called ions. The ions have the electronic structure of a noble gas. So, for example, if sodium (2,8,1) loses one electron it is left with the stable electronic structure of neon (2,8).

112

Chemical bonding

However, it is also left with one more proton in its nucleus than there are electrons around the nucleus. The proton has a positive charge so the sodium atom has now become a positively charged ion. The sodium ion has a single positive charge. We write the formula of a sodium ion as Na^+. The electronic structure of the Na^+ ion is 2,8. This is shown in Figure 2.

Gaining electrons to form negative ions

When non-metals react with metals, the non-metal atoms gain electrons to achieve a stable noble gas structure. Chlorine, for example, has the electronic structure 2,8,7. By gaining a single electron, it gets the stable electronic structure of argon (2,8,8).

In this case there is now one more electron than there are positive protons in the nucleus. So the chlorine atom becomes a negatively charged ion. This carries a single negative charge. We write the formula of the chloride ion as Cl^-. Its electronic structure is 2,8,8. This is shown in Figure 3.

Chlorine 2,8,7

Cl^- 2,8,8

Figure 3 A negative chloride ion (Cl^-) is formed when a chlorine atom gains an electron during ionic bonding

b When atoms join together by *sharing* electrons, what type of bond is formed?

c When ions join together as a result of *gaining* or *losing* electrons, what type of bond is this?

Representing ionic bonding

Metal atoms, which need to lose electrons, react with non-metal atoms, which need to gain electrons. So when sodium reacts with chlorine, each sodium atom loses an electron and each chlorine atom gains that electron. They both form stable ions. The electrostatic attraction between the oppositely charged Na^+ ions and Cl^- ions is called ionic bonding.

2,8,1 2,8,7 → 2,8 2,8,8

NaCl (Na^+Cl^-)

Figure 4 The formation of sodium chloride (NaCl) – an example of ion formation by transferring an electron

We can show what happens in a diagram. The electrons of one atom are represented by dots, and the electrons of the other atom are represented by crosses. This is shown in Figure 4.

Summary questions

1 Copy and complete using the words below:

covalent difficult compound gaining ionic losing new noble

When two elements react together they make a .... substance called a .... It is .... to separate the elements after the reaction. Some atoms react by sharing electrons. We call this .... bonding. Other atoms react by .... or .... electrons. We call this .... bonding. When atoms react in this way they get the electronic structure of a .... gas.

2 Draw diagrams to show the ions that would be formed when the following atoms are involved in ionic bonding. For each one, state how many electrons have been lost or gained and show the charge on the ions formed.

a aluminium (Al) **b** fluorine (F) **c** potassium (K) **d** oxygen (O)

Key points

- Elements react together to form compounds by gaining or losing electrons or by sharing electrons.
- The elements in Group 1 react with the elements in Group 7. As they react, atoms of Group 1 elements can each lose one electron to gain the stable electronic structure of a noble gas. This electron can be given to an atom from Group 7, which then also achieves the stable electronic structure of a noble gas.

113

Summary answers

1 new, compound, difficult, covalent, losing (gaining), gaining (losing), ionic, noble

2 **a** Al^{3+} ion, 3 electrons lost **b** F^- ion, 1 electron gained **c** K^+ ion, 1 electron lost **d** O^{2-} ion, 2 electrons gained

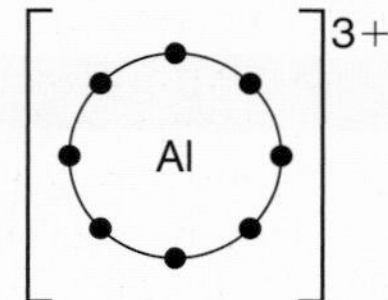

F – 2,8

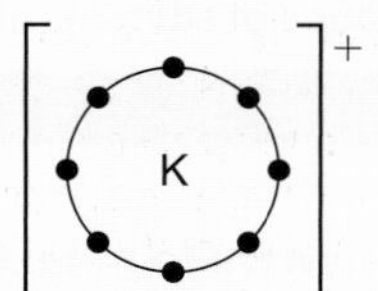

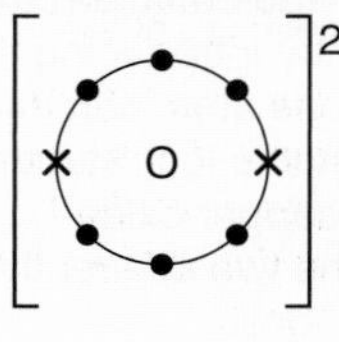

(Outer shell electrons only shown)

C2 1.2 Ionic bonding

Learning objectives

Students should learn:

- how ions are held together in a giant structure
- that elements which aren't in Groups 1 and 7 can also form ions.

Learning outcomes

Most students should be able to:

- describe ionic bonding in terms of electrostatic forces of attraction
- draw the dot and cross diagram for magnesium oxide and calcium chloride.

Some students should also be able to:

- describe in detail an ionic lattice
- apply their knowledge of ionic bonding to draw dot and cross diagrams for other ionic compounds.

Did you know ... ?

Using X-rays, scientists have discovered that ionic lattices aren't perfect. There are sections where the ions aren't perfectly arranged, and if you tap the crystal it will break along cleavage planes.

Support

- Give students a sheet with dot and cross diagrams of a sodium ion, magnesium ion, calcium ion, oxide ion and two chloride ions. Encourage them to represent sodium chloride, magnesium oxide and calcium chloride by cutting out the ions and sticking them into the correct arrangement.

Extend

- Encourage students to research and discover the charges of transition metals and to find out different examples of ionic lattices involving these elements. You could further extend students by introducing them to the idea that one atom can form more than one stable ion. Students could compare copper oxide produced with the two differently charged copper ions.

Specification link-up: Chemistry C2.1

- When atoms form chemical bonds by transferring electrons, they form ions. Atoms that lose electrons become positively charged ions. Atoms that gain electrons become negatively charged ions. Ions have the electronic structure of a noble gas (Group 0). *[C2.1.1. c)]*
- An ionic compound is a giant structure of ions. Ionic compounds are held together by strong electrostatic forces of attraction between oppositely charged ions. These forces act in all directions in the lattice and this is called ionic bonding. *[C2.1.1 f)]*
- Write formulae for ionic compounds from given symbols and ionic charges. *[C2.1]*
- Represent the electronic structure of the ions in … magnesium oxide and calcium chloride. *[C2.1]*

Lesson structure

Starters

Define – Give students the following definitions and ask them to decide the key word that relates to each:

- A charged particle formed by an atom losing or gaining electrons. [Ion]
- All the atoms have the same number of protons. [Element]
- More than one type of atom chemically bonded together. [Compound]
- A chemical bond formed by transferring electrons. [Ionic bond]
- A section of the periodic table whose elements all form 1+ ions. [Group 1/Alkali metals] *(5 minutes)*

Guess the ion – Give the students a diagram of chlorine, sodium, calcium, oxygen and calcium ions. Encourage them to use the periodic table and the textbook to name each ion. You can support students by giving them the names of the ions and asking them to match them with the diagrams. You could extend students by asking them to suggest some ionic compounds that the ions could form. *(10 minutes)*

Main

- The students could make magnesium oxide in the lab in a variety of ways. Encourage students to represent the reaction in terms of a flow chart. The first stages should detail the structure of the atoms of the elements. The middle stages could show the observations of the reaction. The final stage would include a dot and cross diagram representing the resulting ionic compound.
- Often students consider ionic bonds in isolation and in two-dimensional particle diagrams or in dot and cross diagrams. Show students already made-up structures of calcium chloride and magnesium oxide.
- Students could then compare the structures to the actual substances (crystalline powders).
- Then, give students molecular model kits and ask them to generate a sodium chloride crystal. Use questions to encourage students to understand which part of the model represents ions and bonds and what the bonds are (electrostatic forces of attraction between oppositely charged ions).

Plenaries

Question – Split the class into four groups and give each team a different key term: 'ions', 'ionic bond', 'lattice', 'charge'. Each group should write a question which matches its answer. Students then share their questions with the rest of the class. *(5 minutes)*

AfL (Assessment for Learning) – Give students an examination question with a fictitious student answer. Encourage the students to work in small groups to mark the work. Then ask the class to say how many marks they would award the student. Reveal the actual mark. The questions could be selected from a Foundation or a Higher Tier paper and the groups of students could be selected accordingly. *(10 minutes)*

Answers to in-text questions

a giant lattice
b Electrostatic forces of attraction between oppositely charged ions.

Practical support

Burning magnesium

Equipment and materials required

Magnesium ribbon, magnesium powder, spatula, tongs, Bunsen burner and safety equipment, eye protection, blue plastic/glass.

Details

Students hold a small piece (< 1 cm) of magnesium ribbon in tongs. The metal should be held at the top of the blue gas cone in the Bunsen flame. As soon as it ignites, they need to remove it from the flame. This reaction produces a bright light that can blind if looked at directly, so either encourage the students to look past the reaction or through the special blue plastic.

Then ask the students to sprinkle about half a spatula of magnesium powder directly into the blue Bunsen flame and observe. As the surface area is greater, it will combust more quickly and produces a twinkling effect. Make sure that the students hold the Bunsen at an angle or the magnesium powder may fall down the Bunsen chimney and fuse the collar to the chimney.

Safety: CLEAPSS Hazcard 59A Magnesium ribbon, magnesium powder – highly flammable. During both of these reactions, eye protection must be worn. Magnesium oxide powder will aspirate, so it is advisable to complete this in a well-ventilated area, or demonstrate.

C2 1.2 Ionic bonding

Learning objectives

- How are ionic compounds held together?
- Which elements, other than those in Groups 1 and 7, form ions?

You have seen how positive and negative ions form during some reactions. Ionic compounds are usually formed when metals react with non-metals.

The ions formed are held next to each other by very strong forces of attraction between the oppositely charged ions. This electrostatic force of attraction, which acts in all directions, is called **ionic bonding**.

The ionic bonds between the charged particles result in an arrangement of ions that we call a **giant structure** (or a **giant lattice**). If we could stand among the ions they would seem to go on in all directions forever.

The force exerted by an ion on the other ions in the lattice acts equally in all directions. This is why the ions in a giant structure are held so strongly together.

The giant structure of ionic compounds is very regular. This is because the ions all pack together neatly, like marbles in a box.

a What name do we give to the arrangement of ions in an ionic compound?
b What holds the ions together in this structure?

Figure 1 Part of the giant ionic lattice (3-D network) of sodium and chloride ions in sodium chloride

Sometimes the atoms reacting need to gain or lose two electrons to gain a stable noble gas structure. An example is when magnesium (2,8,2) reacts with oxygen (2,6). When these two elements react they form magnesium oxide (MgO). This is made up of magnesium ions with a double positive charge (Mg^{2+}) and oxide ions with a double negative charge (O^{2-}).

We can represent the atoms and ions involved in forming ionic bonds by **dot and cross diagrams**. In these diagrams we only show the electrons in the outermost shell of each atom or ion. This makes them quicker to draw than the diagrams on the previous page. Figure 2 on the next page shows an example.

Figure 2 When magnesium oxide (MgO) is formed, the reacting magnesium atoms lose two electrons and the oxygen atoms gain two electrons

Another example of an ionic compound is calcium chloride. Each calcium atom (2,8,8,2) needs to lose two electrons but each chlorine atom (2,8,7) needs to gain only one electron. This means that two chlorine atoms react with every one calcium atom to form calcium chloride. So the formula of calcium chloride is $CaCl_2$.

Figure 3 The formation of calcium chloride ($CaCl_2$)

Did you know ...?

The structure of ionic lattices is investigated by passing X-rays through them.

Summary questions

1 **a** Copy and complete the table:

Atomic number	Atom	Electronic structure of atom	Ion	Electronic structure of ion
8	O			$[2,8]^{2-}$
19		2,8,8,1	K^+	
17	Cl		Cl^-	
20		2,8,8,2		

b Explain why potassium chloride is KCl but potassium oxide is K_2O.
c Explain why calcium oxide is CaO but calcium chloride is $CaCl_2$.

2 Draw dot and cross diagrams to show how you would expect the following elements to form ions together:
a lithium and chlorine
b calcium and oxygen
c aluminium and fluorine.

Key points

- Ionic compounds are held together by strong forces of attraction between the oppositely charged ions. This is called ionic bonding.
- Besides the elements in Groups 1 and 7, other elements that can form ionic compounds include those from Groups 2 and 6.

Further teaching suggestions

Dot and cross diagrams

- Students could be encouraged to draw more complex dot and cross diagrams, e.g. for aluminium oxide. Encourage students to consider the arrangement of the ions, how the same charged ions would repel, therefore the positive and negatively charged ions alternate in all directions. Students could evaluate dot and cross diagrams as a means of representing ionic compounds, contrasting them with other methods of representing ionic compounds such as molecular model kits and computer models. Ask students to draw the dot and cross diagrams to represent sodium chloride, magnesium oxide and calcium chloride. To extend this activity, students could research and draw a diagram to represent the ionic lattice for each compound.

Iron wool and halogens

- You may wish to show students the reactions between iron wool and bromine, chlorine and iodine (see CLEAPSS 13.2.5). Follow CLEAPSS precautions.

Summary answers

1 **a**

Atomic number	Atom	Electronic structure of atom	Ion	Electronic structure of ion
8	O	2,6	O^{2-}	$[2,8]^{2-}$
19	K	2,8,8,1	K^+	$[2,8,8]^+$
17	Cl	2,8,7	Cl^-	$[2,8,8]^-$
20	Ca	2,8,8,2	Ca^{2+}	$[2,8,8]^{2+}$

b K forms K^+ and Cl forms Cl^-. Therefore one type of each ion is needed to make the compound electrically neutral. But O forms O^{2-}, so two K^+ are needed.

c Ca forms Ca^{2+} and O forms O^{2-}. Therefore one type of each ion is needed to make the compound electrically neutral. But Cl forms Cl^-, so two Cl^- are needed.

2 Students should show dot and cross diagrams for the three reactions **a**, **b** and **c**. These diagrams should use the same structure as the examples given in the Student Book. Lithium chloride should show one ion of lithium (Li^+) and one chloride ion (Cl^-). Calcium oxide should show one ion of calcium (Ca^{2+}) and one oxygen ion (O^{2-}). Aluminium chloride should show one ion of aluminium (Al^{3+}) and three chloride ions (Cl^-).

C2 1.3 Formulae of ionic compounds

Learning objectives

Students should learn:

- how to write the formula of an ionic compound when its ions are given.

Learning outcomes

Most students should be able to:

- determine the formula of simple ionic compounds when the ions are given.

Some students should also be able to:

- determine the formula of ionic compounds that include brackets in the formula, when the ions are given.

Support

- Some students may struggle with this concept. Focus only on simple ions and you may wish to introduce a method for generating the formula without understanding why the subscript numbers are different. Students should write the symbols of the elements with the compound metal first and then the non-metal. They then add the charges of each ion. Ignore the + or − and cross the number down from the metal to the subscript position of the non-metal and vice versa. If there are any 1s, rewrite the formula without them, and cancel the subscripts down if this is possible.

Extend

- It is interesting that there are many examples of negative ions that consist of a group of atoms, but very few examples of positive ions with more than one. Extend students by asking them to find out examples of ions containing more than one atom, other than those listed in the Student Book.

Specification link-up: Chemistry C2.1

- Write formulae for ionic compounds from given symbols and ionic charges. *[C2.1]*

Lesson structure

Starters

Draw that ion – Give students mini-whiteboards, pens and erasers. Ask them to draw the electronic structure of atoms or ions limited to the first 20 elements in the periodic table. Only ask them to draw one structure at a time, then the students show the answer to you and you can give instant feedback and encouragement. Support students by allowing them to look back in their notes, use the Student Book or even peek at other students' answers. You could extend students by asking them to write the shorthand notation for each of the electronic structures as well as the diagram. *(5 minutes)*

Count the atoms – Give the students a selection of different ionic compounds and ask them to count the atoms that they contain. [e.g. CuO – 2 atoms, $CuSO_4$ – 6 atoms, $(NH_4)_2SO_4$ – 15 atoms.] *(5 minutes)*

Main

- Explain to students that, when you touch an ionic compound, you do not get an electric shock. [The number of positive charges is the same as the number of negative charges.]
- Students should be quite familiar with sodium chloride and this is a good starting point. Ask students to suggest the type of bonding present [ionic] and how they know [it is a compound made of a metal and a non-metal]. Then ask for a volunteer to come to the board and write the symbol and charge of a sodium ion (Na^+). Ask a second volunteer to do the same for chlorine (Cl^-). Ask students to suggest how many chloride ions you would need for every sodium ion, reiterating that the charges have to balance. Students should recognise that there is a one-to-one ratio and therefore the formula is NaCl.
- Ask students to write the formulae of potassium chloride [KCl], sodium bromide [NaBr] and lithium iodide [LiI], as all of these compounds also have a one-to-one ratio of ions. Circulate around the classroom, making the students answer directly in their book and helping where necessary. You could extend students by giving them examples of ions also in a one-to-one relationship but with charges greater than 1, e.g. calcium sulfide [CaS].
- Use the same questioning technique to build up the formula for calcium chloride. Again ask students to generate the formulae for the following compounds: magnesium bromide [$MgBr_2$] barium iodide [BaI_2]. Extend students by asking them to generate the formula of sodium oxide [Na_2O]. Once again, circulate around the room giving instant feedback and help where necessary.
- Ask students to generate five names of ionic compounds (using a metal reacted with a non-metal). You could extend students by encouraging them to use ions that contain more than one atom as well. They can then swap their lists with their neighbours and write the formulae for the names that have been written down. Students then swap the paper back. The authors of the name consider the formulae written and award a score out of five to their partners. Students may wish to discuss their marking. Clarification from the teacher may be needed in some cases.

Plenaries

Sort it out – Give the students a selection of formulae and compound names. Students should then write the name and the correct formula. *(5 minutes)*

Objective question – Ask the students to answer the question posed in the objective. Students could be supported by being given a selection of correct answers, each more scientific than the last. They have to choose which one makes the most sense to them and copy it out. Students could be extended by being asked to include a worked example in their answer. *(10 minutes)*

Further teaching suggestions

Table

- You could ask students to write a table of ions, where they write the name of the ion, formula, electronic structure and an example of a compound that contains that ion.

Jigsaw puzzle

- You could create a jigsaw puzzle where students have to match the name of the ionic compound with its formula. Only one tessellation would give the correct shape and this makes the activity very easy to assess and allows students to have a lot of practice at recognising and generating formulae.

C2 1.3 Formulae of ionic compounds

Learning objectives

- How can we write the formula of an ionic compound, given its ions?

In this chapter we have seen how three different ionic compounds are formed. You should understand how atoms turn to ions when sodium chloride, magnesium oxide and calcium chloride are formed from their elements.

The overall charge on any ionic compound is zero. The compounds are neutral. Therefore we do not have to draw dot and cross diagrams to work out each ionic formula. As long as we know or are given the charge on the ions in a compound we can work out its formula.

a What is the overall charge on an ionic compound?

If we look at the three examples above, we can see how the charges on the ions in a compound cancel out:

Ionic compound	Ratio of ions in compound	Formula of compound
sodium chloride	$Na^+ : Cl^-$ 1 : 1	NaCl
magnesium oxide	$Mg^{2+} : O^{2-}$ 1 : 1	MgO
calcium chloride	$Ca^{2+} : Cl^-$ 1 : 2	$CaCl_2$

b What is the formula of magnesium chloride?

We can work out the formula of some ions given a copy of the periodic table. Remember that in your exams you will have a Data Sheet which includes a periodic table and a table showing the charges of common ions.

Groups of metals

- The atoms of Group 1 elements form 1+ ions, e.g. Li^+.
- The atoms of Group 2 elements form 2+ ions, e.g. Ca^{2+}.

Groups of non-metals

- The atoms of Group 7 elements form 1− ions, e.g. F^-.
- The atoms of Group 6 elements form 2− ions, e.g. S^{2-}.

The names of compounds of transition metals contain the charge on their ions in brackets in roman numerals. This is because they can form ions carrying different sizes of positive charge. For example, iron can form 2+ and 3+ ions. So the name iron(III) oxide tells us that the iron is present as Fe^{3+} ions in this compound.

c What is the formula of lithium sulfide?
d What is the formula of iron(III) oxide?

Did you know ...?

Common salt is sodium chloride. In just 58.5 g of salt there are over 600 000 000 000 000 000 000 000 ions of Na^+ and the same number of Cl^- ions.

More complicated ions

Some ions are made up of more than one element. When you studied limestone, you learned that the formula of calcium carbonate is $CaCO_3$. It contains calcium ions, Ca^{2+}, and carbonate ions, CO_3^{2-}. The carbonate ions contain carbon and oxygen. However, the rule about cancelling out charges still applies as in one-element ions. Calcium carbonate is $CaCO_3$ as the 2+ and 2− ions in the ionic compound cancel out in the ratio 1 : 1.

Two-element ions you might come across are shown in the table below:

Name of ion	Formula of ion	Example of compound
hydroxide	OH^-	calcium hydroxide, $Ca(OH)_2$
nitrate	NO_3^-	magnesium nitrate, $Mg(NO_3)_2$
carbonate	CO_3^{2-}	sodium carbonate, Na_2CO_3
sulfate	SO_4^{2-}	calcium sulfate, $CaSO_4$

Notice how the formula of a compound containing a two-element ion sometimes contains brackets. To write calcium hydroxide as $CaOH_2$ would be misleading. It would tell us the ratio of Ca : O : H ions was 1 : 1 : 2. However, as there are twice as many hydroxide ions as calcium ions, the ratio should be 1 : 2 : 2. This is why we write the formula as $Ca(OH)_2$.

e What is the formula of calcium nitrate?

Figure 1 Haematite is an ore of iron. It is mined (as here) and used as a source of iron(III) oxide for the blast furnace in the extraction of iron.

AQA Examiner's tip

When naming compounds we use the ending –ide for simple non-metal ions such as oxide or sulfide. However, we use –ate for ions that include oxygen, such as sulfate, nitrate and carbonate.

AQA Examiner's tip

You do not have to learn the charges on ions – they are on the data sheet.

Think of each symbol as a single atom and the formula of each ion as a single ion. Small numbers multiply only the symbol they follow.

Brackets are needed when there is more than one type of atom in the ion being multiplied.

Summary questions

1 Using the charges on the ions given on this spread, give the formula of:
  a calcium oxide
  b lithium oxide
  c magnesium chloride

2 Draw a table with K^+, Mg^{2+} and Fe^{3+} down the side and Br^-, OH^-, NO_3^- and SO_4^{2-} across the top. Then fill in the formula of the compound in each cell of the table.

3 a The formula of strontium nitrate is $Sr(NO_3)_2$. What is the charge on a strontium ion?
  b The formula of aluminium sulfate is $Al_2(SO_4)_3$. What is the charge on an aluminium ion?

Key points

- The charges on the ions in an ionic compound always cancel each other out.
- The formula of an ionic compound shows the ratio of ions present in the compound.
- Sometimes we need brackets to show the ratio of ions in a compound, e.g. magnesium hydroxide, $Mg(OH)_2$.

Answers to in-text questions

- **a** 0
- **b** $MgCl_2$
- **c** Li_2S
- **d** Fe_2O_3
- **e** $Ca(NO_3)_2$

Summary answers

1 **a** CaO **b** Li_2O **c** $MgCl_2$

2

	Br^-	OH^-	NO_3^-	SO_4^{2-}
K^+	KBr	KOH	KNO_3	K_2SO_4
Mg^{2+}	$MgBr_2$	$Mg(OH)_2$	$Mg(NO_3)_2$	$MgSO_4$
Fe^{3+}	$FeBr_3$	$Fe(OH)_3$	$Fe(NO_3)_3$	$Fe_2(SO_4)_3$

3 **a** 2+ **b** 3+

C2 1.4 Covalent bonding

Learning objectives

Students should learn:

- how a covalent bond is formed
- the types of substance that have covalent bonds.

Learning outcomes

Most students should be able to:

- state a simple definition for a covalent bond
- draw a dot and cross diagram for simple covalent bonds (hydrogen, chlorine, hydrogen chloride, water)
- name an element that has a giant covalent structure (carbon, diamond).

Some students should also be able to:

- explain the formation of a covalent bond
- draw dot and cross diagrams for more complex covalent substances (methane, ammonia and oxygen)
- explain the bonding in a giant covalent structure and give an example, e.g. silicon oxide.

Answers to in-text questions

a covalent
b A giant covalent structure (macromolecule).

Support

- Some students will need additional support to understand fully how to generate dot and cross diagrams. To help students understand how to draw these diagrams, give them the electronic structures of key atoms (hydrogen, chlorine) using overhead transparencies. Then ask them to arrange the images so that each atom has a noble gas structure, by overlapping one image on to another. When they have the correct arrangement of electrons, they can copy the diagram into their books. This process can be demonstrated on an overhead projector in front of the class.

Extend

- You could extend students by asking them to represent molecular ions using dot and cross diagrams.

Specification link-up: Chemistry C2.1

- When atoms share pairs of electrons, they form covalent bonds. These bonds between atoms are strong. Some covalently bonded substances consist of simple molecules such as H_2, Cl_2, O_2, HCl, H_2O, NH_3 and CH_4. Others have giant covalent structures (macromolecules), such as diamond and silicon dioxide. *[C2.1.1 g)]*
- Represent the covalent bonds in molecules such as water, ammonia, hydrogen, hydrogen chloride, methane and oxygen, and in giant structures such as diamond and silicon dioxide … . *[C2.1]*

Lesson structure

Starters

Card sort – Give the students separate cards with images (just coloured circles and dot and cross diagrams) and words on them: 'compound', 'element', 'molecule', 'mixture', 'ionic bond', 'ion'. Students should then try to match the image with the key term. *(5 minutes)*

Demonstration – Ignite a small hydrogen balloon to get the attention of the class and generate excitement. The hydrogen will explode as it reacts with the oxygen in the air to make water (steam). Encourage the students to write a word equation for this reaction. You could support students by giving them the names of the chemicals and they have to select which are the reactants and which are the products. Students could be extended by being asked to write a balanced symbol equation. *(10 minutes)*

Main

- Students need to be able to draw dot and cross diagrams for certain molecules and some students should be able to explain how and why they are formed. Ask for volunteers to draw the electronic structure of hydrogen on the board.
- Then ask the students how many electrons it needs to obtain a noble gas structure [one]. Explain how and why hydrogen atoms make diatomic molecules.
- Ask the students to draw the dot and cross diagram of a chlorine molecule and explain, in no more than two sentences, why chlorine forms a diatomic molecule.
- Demonstrate how hydrogen and chlorine atoms bond to form hydrogen chloride. Then ask students to draw the other dot and cross diagrams for the appropriate molecules.
- Most substances that form covalent bonds make discrete molecules. However, students need to be aware of some macromolecules, e.g. carbon and silicon oxide.
- Give students molecular model kits and ask them to make models of different molecules. Then focus on carbon, and ask them to make a carbon molecule using only single bonds – hopefully the students will find that the structure is never-ending.
- Explain to students that this is an example of a macromolecule and that it, too, is made of covalent bonds. This links with C2 2.3.

Plenaries

Ionic/covalent – Give the students a piece of paper. Ask them to write in large letters 'ionic bonding' and 'covalent bonding', one on each side of the paper. Then read out the following substances, and the students should hold up the side of the paper that shows which type of bonding is present:

- Methane [Covalent]
- Sodium chloride [Ionic]
- Oxygen [Covalent]
- Water [Covalent]
- Magnesium oxide [Ionic]
- Carbon [Covalent]
- Silicon oxide [Covalent]
- Ammonia [Covalent]
- Calcium chloride [Ionic]
- Hydrogen chloride [Covalent]

You could support students by putting the definitions and one example of ionic and covalent bonding on the board for them to refer to. You could extend students by asking them to generate the formula for each of these compounds. *(5 minutes)*

Model – Split the class into teams and give each team a different covalent compound to represent. The students in each team should represent atoms and their hands and feet can make up to four bonds (e.g. C in methane). Each team should 'act out' its molecule and explain it to the class. *(10 minutes)*

Practical support

Burning hydrogen

Equipment and materials required

Small rubber balloon, string, hydrogen gas cylinder (pressurised gas), metre rule, splint, tape, matches.

Details

Fill a rubber balloon with hydrogen from a gas cylinder. Tie the balloon with a piece of string onto a tap or 1 kg mass, so that it is clear of any flammable materials, students and the ceiling.

Then light a splint taped to the end of a metre rule. Hold the lighted splint onto the stretched part of the rubber.

Safety: The demonstrator should wear eye protection and be aware that hot rubber can fly from the balloon. CLEAPSS Hazcard 48 Hydrogen – flammable. Soap bubbles filled with hydrogen are an alternative here.

Further teaching suggestions

Allotropes

- Students could define the term 'allotrope' and look up other allotropes of carbon [fullerenes]. This links with C2 2.3.

Naming

- You could ask students to suggest the names of other covalent compounds and try to draw dot and cross diagrams of them.

Comparing representations

- Students could evaluate the use of dot and cross diagrams to represent covalent compounds, comparing this representation with molecular model kits, space-saving diagrams, computer models and ball-and-stick diagrams.

Gases in air

- You could ask students to use dot and cross diagrams to show the gases in the air. This reinforces the work on bonding (as argon has a full outer shell of electrons and is therefore a monatomic gas) and nitrogen is a useful extension as this molecule contains a triple bond.

Computer modelling

- You may wish to allow students to use computer modelling to simulate how electrons are shared between atoms to make molecules. You could extend students and allow them to make their own simulations using simple programmes that would allow students to make their own flash animations.

Structure and bonding

C2 1.4 Covalent bonding

Learning objectives

- How are covalent bonds formed?
- What types of substance have covalent bonds?

Reactions between metals and non-metals usually result in ionic bonding. However, many, many compounds are formed in a very different way. When non-metals react together their atoms share pairs of electrons to form molecules. We call this **covalent bonding**.

Simple molecules

The atoms of non-metals generally need to gain electrons to achieve stable outer energy levels. When they react together neither atom can give away electrons. So they get the electronic structure of a noble gas by sharing electrons. The atoms in the molecules are then held together by the shared pairs of electrons. We call these strong bonds between the atoms covalent bonds.

a What is the bond called when two atoms share a pair of electrons?

Figure 1 Most of the molecules in substances which make up living things are held together by covalent bonds between non-metal atoms

Hydrogen atoms 1 → Hydrogen molecule — A shared pair of electrons gives both atoms a stable arrangement and forms a covalent bond

Oxygen atoms 2,6 → Oxygen molecule — This is a double covalent bond (two pairs of electrons involved). Only the electrons in the highest energy level (outer shell) are shown here

Figure 2 Atoms of hydrogen and oxygen join together to form stable molecules. The atoms in H_2 and O_2 molecules are held together by strong covalent bonds.

Hydrogen chloride HCl

Water H_2O

Methane CH_4

Figure 3 The principles of covalent bonding remain the same however many atoms are involved

Sometimes in covalent bonding each atom brings the same number of electrons to share. But this is not always the case. Sometimes the atoms of one element will need several electrons, while the other element only needs one more electron for each atom to get a stable arrangement. In this case, more atoms become involved in forming the molecule.

We can represent the covalent bonds in substances such as water, ammonia and methane in a number of ways. Each way represents the same thing. The method chosen depends on what we want to show.

Water H_2O — a, b, c

Figure 4 We can represent a covalent compound by showing **a** the highest energy levels (or outer shells), **b** the outer electrons in a dot and cross diagram or **c** the number of covalent bonds

118

Covalent bonding

Giant covalent structures

Many substances containing covalent bonds consist of small molecules, for example, H_2O. However, some covalently bonded substances are very different. They have giant structures where huge numbers of atoms are held together by a network of covalent bonds. These are sometimes referred to as macromolecules.

Diamond has a giant covalent structure. In diamond, each carbon atom forms four covalent bonds with its neighbours. This results in a rigid giant covalent lattice.

Carbon atoms linked together by covalent bonds

Figure 5 Part of the giant covalent structure of diamond

Figure 6 Diamonds owe their hardness to the way the carbon atoms are arranged in a giant covalent structure

Silicon dioxide (silica) is another substance with a giant covalent structure.

b What do we call the structure of a substance held together by a network of covalent bonds?

Summary questions

1 Copy and complete using the words below:

covalent giant molecules macromolecules shared

When non-metal atoms react together they make .... bonds. The atoms in these bonds are held together by .... electrons. Most substances held together by covalent bonds consist of ...., but some have .... covalent structures, sometimes called ....

2 Draw diagrams, showing all the electrons, to represent the covalent bonding between the following atoms.
   a two hydrogen atoms
   b two chlorine atoms
   c a hydrogen atom and a fluorine atom

3 Draw dot and cross diagrams to show the covalent bonds when:
   a a nitrogen atom bonds with three hydrogen atoms
   b a carbon atom bonds with two oxygen atoms.

Key points

- Covalent bonds are formed when atoms share pairs of electrons.
- Many substances containing covalent bonds consist of simple molecules, but some have giant covalent structures.

119

Summary answers

1 covalent, shared, molecules, giant, macromolecules

2 a

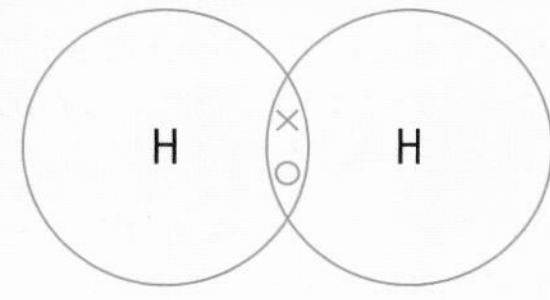

b

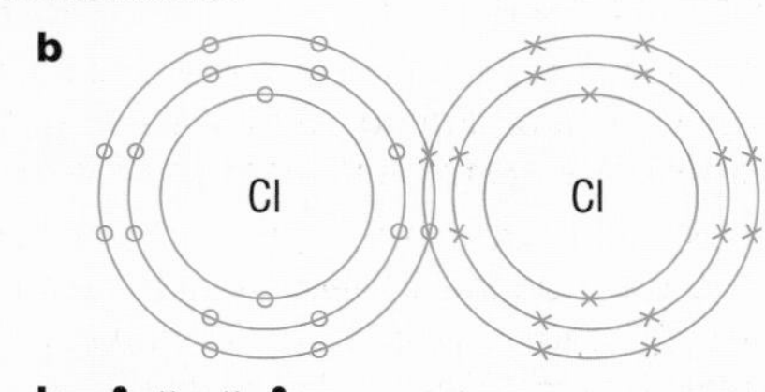

c H F

3 a H ×• N •× H, with H below N

b O C O

C2 1.5 Metals

Learning objectives

Students should learn:

- how atoms in metals are arranged
- what holds metal atoms (ions) together. **[HT only]**

Learning outcomes

Most students should be able to:

- list examples of elements that have a giant metallic structure
- describe the bonding in metals. **[HT only]**

Some students should also be able to:

- explain metallic bonding and structures in words and a labelled diagram, including delocalised electrons. **[HT only]**

Specification link-up: Chemistry C2.1

- Metals consist of giant structures of atoms arranged in a regular pattern. *[C2.1.1 h)]*
- The electrons in the highest occupied energy levels (outer shell) of metal atoms are delocalised and so free to move through the whole structure. This corresponds to a structure of positive ions with electrons between the ions holding them together by strong electrostatic attractions. *[C2.1.1 i)]* **[HT only]**
- Represent the bonding in metals … . *[C2.1]* **[HT only]**

Lesson structure

Starters

Odd one out – List the following metals: nickel, cobalt, iron and steel. Ask students to look at the list and decide which metal doesn't fit the pattern and why. [Steel is the odd one out as it is an alloy, whereas nickel, cobalt and iron are elements.] *(5 minutes)*

5, 4, 3, 2, 1 – Ask students to write a list of: five metal symbols, four metal properties, three magnetic elements, two metals used in jewellery and one metal that is a liquid at room temperature. Build up a class list on the board through questions and answers. You could support students by displaying a periodic table that has a colour-coded key to show the metals, non-metals and the state of each element at room temperature. You could extend students by asking them to define a metal. *(10 minutes)*

Main

- Metals are made up of grains, which can be seen using microscopes. Show students images of the grains and grain boundaries. The grains order themselves in metals and form crystals.
- Students can grow their own metal crystals by completing a solution displacement reaction. You could encourage students to record the metal crystal in the form of a diagram in their book.
- Visible metal crystals can be found in a variety of structures in and around the school site. Students could go out and about in small teams, armed with digital cameras to find examples. When students return to the class, they could display the photographs to the rest of the class.
- Discuss the nature of metallic bonding with Higher Tier students. **[HT only]**

Plenaries

Label – Give the students an unlabelled diagram of a metal structure (as shown in the specification). Ask the students to label the diagram as fully as possible. You could support students by making this a 'cut-and-stick' activity. You could extend them by asking them to evaluate this model and suggest its advantages and disadvantages. *(5 minutes)*

Explain and define – Ask students to explain metallic bonding to their neighbours. Then each pair should try to distil its explanation into a concise definition. Encourage a few couples to share and evaluate their definitions. *(10 minutes)* **[HT only]**

Answers to in-text questions

a The metal atoms (or ions) are arranged in a regular pattern.

b The outermost electrons.

Support

- Students could make a three-dimensional model of a metallic structure using art materials, e.g. polystyrene balls, string, paint and a shoe box.

Extend

- Ask students to consider how the charge on the metal ion affects the strength of the metallic bond. This is good preparation for C2. 2.4. Ask students to consider which metal would have the strongest bonding: sodium or aluminium, and why. [Al (aluminium), as it has more delocalised electrons.]

Practical support

Growing silver crystals

Equipment and materials required

Boiling tube, boiling tube rack, copper wire, silver nitrate solution less than 0.4 M, eye protection.

Details

- Wrap the copper wire into a spring shape and put it into a boiling tube. Wear eye protection. Add silver nitrate solution to the boiling tube so that it is about half full.
- Leave the solution to allow displacement to occur. It would be best if the experiment were left until next lesson and reviewed as a Starter.

Safety: CLEAPSS Hazcard 87 Silver nitrate solution – irritant.

Survey of metallic crystals

Equipment and materials required

Clipboards, digital camera, stationery.

Details

- Split the class into small groups. Take the students to an example of a metal crystal, e.g. a galvanised dustbin on the school site.
- Give the students a time limit (10 minutes) to find further examples of metal crystals around the school. They should record, in a table, the photograph number, the place where the crystal was found and the item on which it was found.

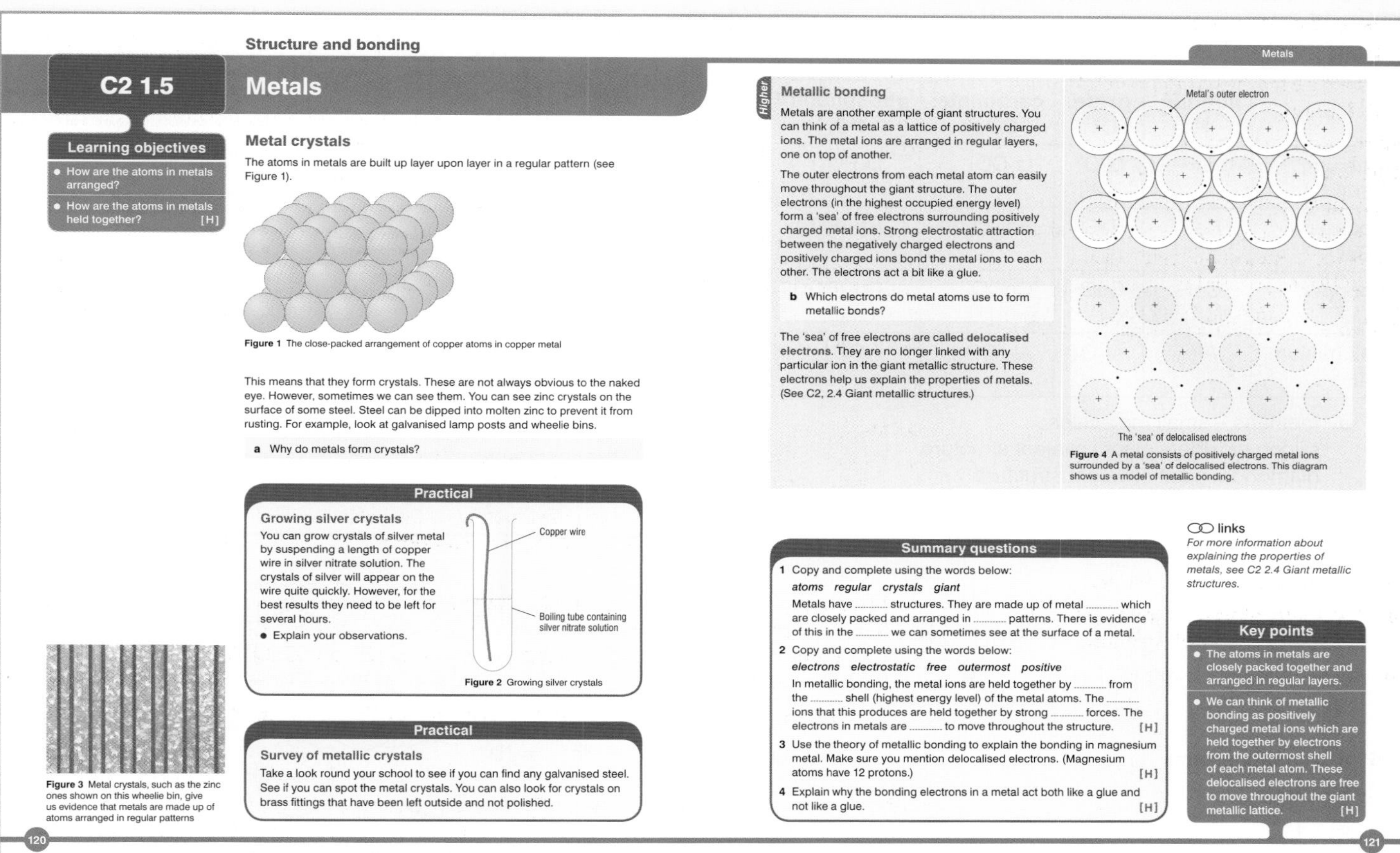

Structure and bonding

C2 1.5 Metals

Learning objectives

- How are the atoms in metals arranged?
- How are the atoms in metals held together? [H]

Metal crystals

The atoms in metals are built up layer upon layer in a regular pattern (see Figure 1).

Figure 1 The close-packed arrangement of copper atoms in copper metal

This means that they form crystals. These are not always obvious to the naked eye. However, sometimes we can see them. You can see zinc crystals on the surface of some steel. Steel can be dipped into molten zinc to prevent it from rusting. For example, look at galvanised lamp posts and wheelie bins.

a Why do metals form crystals?

Practical

Growing silver crystals

You can grow crystals of silver metal by suspending a length of copper wire in silver nitrate solution. The crystals of silver will appear on the wire quite quickly. However, for the best results they need to be left for several hours.

- Explain your observations.

Figure 2 Growing silver crystals

Figure 3 Metal crystals, such as the zinc ones shown on this wheelie bin, give us evidence that metals are made up of atoms arranged in regular patterns

Practical

Survey of metallic crystals

Take a look round your school to see if you can find any galvanised steel. See if you can spot the metal crystals. You can also look for crystals on brass fittings that have been left outside and not polished.

120

Metals

Higher

Metallic bonding

Metals are another example of giant structures. You can think of a metal as a lattice of positively charged ions. The metal ions are arranged in regular layers, one on top of another.

The outer electrons from each metal atom can easily move throughout the giant structure. The outer electrons (in the highest occupied energy level) form a 'sea' of free electrons surrounding positively charged metal ions. Strong electrostatic attraction between the negatively charged electrons and positively charged ions bond the metal ions to each other. The electrons act a bit like a glue.

b Which electrons do metal atoms use to form metallic bonds?

The 'sea' of free electrons are called **delocalised electrons**. They are no longer linked with any particular ion in the giant metallic structure. These electrons help us explain the properties of metals. (See C2, 2.4 Giant metallic structures.)

Figure 4 A metal consists of positively charged metal ions surrounded by a 'sea' of delocalised electrons. This diagram shows us a model of metallic bonding.

Summary questions

1 Copy and complete using the words below:
atoms regular crystals giant
Metals have .... structures. They are made up of metal .... which are closely packed and arranged in .... patterns. There is evidence of this in the .... we can sometimes see at the surface of a metal.

2 Copy and complete using the words below:
electrons electrostatic free outermost positive
In metallic bonding, the metal ions are held together by .... from the .... shell (highest energy level) of the metal atoms. The .... ions that this produces are held together by strong .... forces. The electrons in metals are .... to move throughout the structure. [H]

3 Use the theory of metallic bonding to explain the bonding in magnesium metal. Make sure you mention delocalised electrons. (Magnesium atoms have 12 protons.) [H]

4 Explain why the bonding electrons in a metal act both like a glue and not like a glue. [H]

links

For more information about explaining the properties of metals, see C2 2.4 Giant metallic structures.

Key points

- The atoms in metals are closely packed together and arranged in regular layers.
- We can think of metallic bonding as positively charged metal ions which are held together by electrons from the outermost shell of each metal atom. These delocalised electrons are free to move throughout the giant metallic lattice. [H]

121

Further teaching suggestions

Electrolysis

- Students could electrolyse copper sulfate solution in order to grow copper crystals or grow lead crystals from zinc foil suspended in lead nitrate solution.
  Safety: CLEAPSS Hazcard 27C Copper sulfate – harmful. CLEAPSS Hazcard 57A Lead nitrate – toxic.

Crystal models

- If the school has any examples, students could be shown different metal crystals and metal structural models.

Images

- Digital images taken by students or from the internet could be used by the students to create a PowerPoint display about metal crystals and where you can find them. They could then present them to their classmates.

Summary answers

1. giant, atoms, regular, crystals
2. electrons, outermost, positive, electrostatic, free
3. Each magnesium atom donates its two outermost electrons into the delocalised 'sea' of electrons. The electrostatic force of attraction between the negatively charged delocalised electrons and the positively charged metal ions hold the ions in position in the giant structure.
4. Like glue – because they hold the atoms together. Unlike glue – because atoms can still move/flow past each other when force is applied. The electrons can also move freely through the structure.

Summary answers

1 **Compound** – A substance made up of different types of atom/elements chemically bonded together.
Ionic bonding – The electrostatic force of attraction between oppositely charged ions.
Covalent bond – A pair of electrons shared between two atoms.

2 a iron(II) chloride, potassium oxide, lead bromide, silver nitrate.

b These compounds all contain metals and non-metals.

c covalent bonding

d i H_2S ii $FeCl_2$

e Because iron can form two/more than one ion (Fe^{2+} and Fe^{3+}).

3

	fluoride, F^-	oxide, O^{2-}	carbonate, CO_3^{2-}	phosphate(V), PO_4^{3-}
lithium, Li^+	LiF	Li_2O	Li_2CO_3	Li_3PO_4
barium, Ba^{2+}	BaF_2	BaO	$BaCO_3$	$Ba_3(PO_4)_2$
copper, Cu^{2+}	CuF_2	CuO	$CuCO_3$	$Cu_3(PO_4)_2$
aluminium, Al^{3+}	AlF_3	Al_2O_3	$Al_2(CO_3)_3$	$AlPO_4$

4 a methane – small molecules
silicon dioxide – giant covalent structure
diamond – giant covalent structure
ammonia – small molecules

b For ammonia (NH_3) And/or

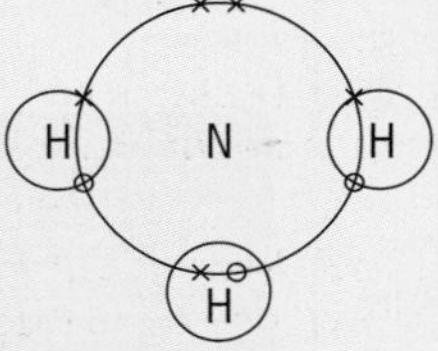

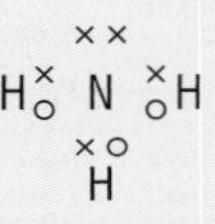

5 a A (magnesium)

b Group 7.

c i

H
H C H
H

where C (carbon) = Q

ii covalent bonding

d i

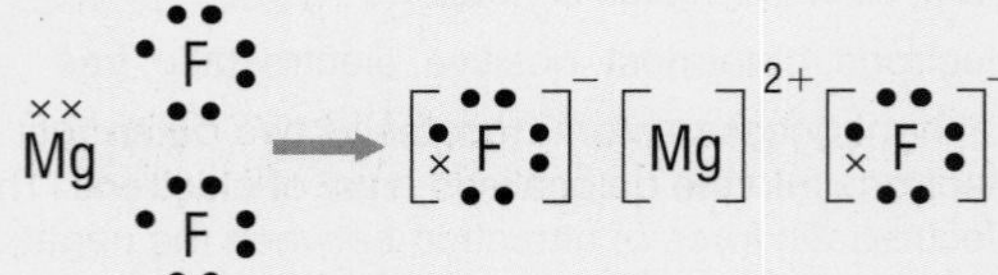

where Mg = A and F = C

ii ionic bonding

iii MgF_2

6 a Atoms held together by sharing a pair of electrons to form a covalent bond, joining two F atoms together.

F — F

Sharing electrons produces attractive force

b

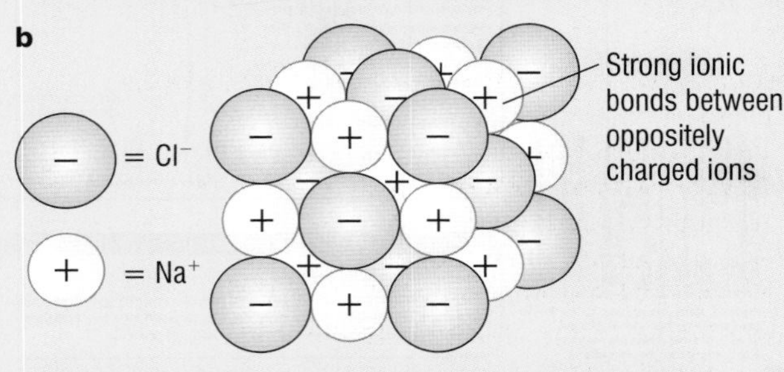

Allow a 2-D single layer of the ionic lattice showing oppositely charged ions next to each other with suitable labelling.

7 O C O

Structure and bonding: C2 1.1–C2 1.5

Summary questions

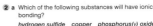

1 Define the following terms:
compound
ionic bonding
covalent bond

2 a Which of the following substances will have ionic bonding?
hydrogen sulfide copper phosphorus(V) oxide iron(II) chloride potassium oxide lead bromide silver nitrate

b Explain how you decided on your answers in part **a**.

c What type of bonding will the remaining substances in the list have?

d What is the formula of:
i hydrogen sulfide
ii iron(II) chloride.

e Why does iron(II) chloride have roman numerals in its name?

3 Copy and complete the following table with the formula of each compound formed.
(The first one is done for you).

	fluoride, F^-	oxide, O^{2-}	carbonate, CO_3^{2-}	phosphate(V), PO_4^{3-}
lithium, Li^+	LiF			
barium, Ba^{2+}				
copper, Cu^{2+}				
aluminium, Al^{3+}				

4 a Which of the following substances are made up of small molecules and which have a giant covalent structure?
methane, CH_4
silicon dioxide, SiO_2
diamond, C
ammonia, NH_3

b Draw a dot and cross diagram to show the bonding in ammonia.

5 The diagrams show the arrangement of electrons in energy levels in three atoms:
(The letters are NOT the chemical symbols.)

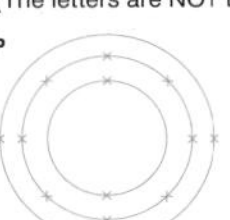

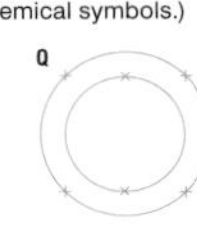

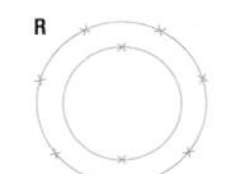

a Which atom belongs to Group 2 of the periodic table?

b To which group does atom R belong?

c i Atom Q bonds with four atoms of hydrogen. Draw a dot and cross diagram to show the compound that is formed.
ii What do we call the type of bonding between the atom of Q and the hydrogen atoms?

d i Draw dot and cross diagrams to show how atom P bonds with R atoms.
ii What do we call the type of bonding in the compound formed by P and R?
iii What is the formula of the compound formed by P and R?

6 Describe, with diagrams, how the particles are held together in the following substances:
a a molecule of fluorine (F_2)
b a salt crystal (NaCl).

7 Draw a diagram which shows how the atoms in carbon dioxide, O=C=O, bond to each other. [H

122

AQA Examination-style questions

se a periodic table and a table of charges on ions to help u to answer these questions.

Choose a word from the list to complete each sentence.

a When metals react with non-metals electrons are .... (1)

combined shared transferred

b When non-metal elements combine their atoms are held together by .... bonds. (1)

covalent ionic metallic

Choose a description from the list for each of the substances.

giant covalent giant ionic metal simple molecule

a ammonia, NH_3 c lithium, Li

b diamond, C d sodium oxide, Na_2O (4)

Choose a number from the list to complete each sentence.

0 1 2 3 4 6 7

a The elements in Group .... in the periodic table all form ions with a charge of 1+. (1)

b The elements in Group .... in the periodic table all form ions with a charge of 2–. (1)

c The elements in Group 4 in the periodic table all form .... covalent bonds. (1)

d The aluminium ion has a charge of .... + (1)

a Choose the correct formula from the list for iron(III) chloride.

FeCl Fe_3Cl $FeCl_3$ Fe_3Cl_3 (1)

b Choose the formula from the list for each of these ionic compounds.

NaS $NaSO_4$ $Na(SO_4)_2$ Na_2S NaS_2 Na_2SO_4

i sodium sulfide (1)

ii sodium sulfate (1)

Calcium hydroxide, $Ca(OH)_2$, is an ionic compound.

Which of these ions in the list are the ions in calcium hydroxide?

Ca^+ Ca^{2+} Ca^{4+} OH^- OH_2^- OH^{2-} (2)

Sodium reacts with chlorine. The reaction forms sodium chloride.

a Use words from the list to answer the questions.

compound element hydrocarbon mixture

Which word best describes:

i sodium (1)

ii sodium chloride? (1)

b When sodium reacts with chlorine the sodium atoms change into sodium ions. The diagrams represent a sodium atom and a sodium ion.

Sodium atom (Na) Sodium ion (Na^+)

Use the diagrams to help you explain how a sodium atom turns into a sodium ion. (2)

c i The diagram below represents a chlorine atom. When chlorine reacts with sodium the chlorine forms negative chloride ions.

Copy and complete the diagram below to show how the outer electrons are arranged in a chloride ion (Cl^-). (1)

ii Chloride ions are strongly attracted to sodium ions in sodium chloride.

Explain why. (1)

AQA, 2010

7 Chlorine can form compounds with ionic or covalent bonds.

a Potassium chloride, KCl, has ionic bonds. Draw dot and cross diagrams to show what happens to potassium atoms and chlorine atoms when they react to form potassium chloride. You only need to show the outer electrons in your diagrams. (4)

b Hydrogen chloride, HCl, has covalent bonds. Draw a dot and cross diagram to show the bonding in hydrogen chloride. (2)

8 Sodium metal is a giant structure of sodium atoms.

Explain how the atoms are held together in sodium metal. [H] (3)

123

Kerboodle resources

Resources available for this chapter on Kerboodle are:

- Chapter map: Structure and bonding
- Video: Chemistry at work – Medicinal chemist
- Support: Making sense of ions (C2 1.1)
- Animation: Ionic bonding (C2 1.2)
- Maths skills: Ionic formulae (C2 1.3)
- Extension: What's the formula? (C2 1.3)
- Animation: Overview of covalent bonding (C2 1.4)
- Interactive activity: Structure and bonding
- Revision podcast: Different types of bonding (C2 1.2–3, 1.5)
- Test yourself: Structure and bonding
- On your marks: Bonding
- Examination-style questions: Chemical bonding
- Answers to examination-style questions: Chemical bonding

Examination-style answers

1 **a** transferred *(1 mark)*

b covalent *(1 mark)*

2 **a** simple molecule *(1 mark)*

b giant covalent *(1 mark)*

c metal *(1 mark)*

d giant ionic *(1 mark)*

3 **a** 1 *(1 mark)*

b 6 *(1 mark)*

c 4 *(1 mark)*

d 3 *(1 mark)*

4 **a** $FeCl_3$ *(1 mark)*

b **i** Na_2S *(1 mark)*

ii Na_2SO_4 *(1 mark)*

5 Ca^{2+}, OH^- *(2 marks)*

6 **a** **i** element *(1 mark)*

ii compound *(1 mark)*

b It loses an electron. *(2 marks)*

c **i** Right-hand diagram with seven crosses and one dot in outermost shell (allow diagram with all crosses). *(1 mark)*

ii Oppositely charged ions attract each other. *(1 mark)*

7 **a** Each correctly drawn part of the diagram should be answered one mark.

- Correctly drawn potassium atom (one electron in outer shell).
- Correctly drawn chlorine atom (seven electrons in outer shell).
- Correctly drawn potassium ion: no electron in outer shell (*accept eight electrons in outer shell*) and brackets with + charge top right.
- Correctly drawn chloride ion: eight electrons in outer shell and brackets with – charge top right.

(4 marks)

b For example, H atom: circle with a single dot, Cl atom: circle with seven crosses overlapping H circle so that a dot and cross are shared or alternative with no circles. Must be one electron different from seven others and a shared pair including that electron clearly shown for two marks. One mark for a shared pair of electrons, or eight electrons all with the same symbol. *(2 marks)*

8 Electrons in the highest (occupied) energy level/outer shell are delocalised/free (to move), leaving or surrounding positive ions (in regular lattice/arrangement), electrons attract/hold together the positive ions. *(3 marks)*

Practical suggestions

Practicals	AQA	k	📖	⚙
Molecular modelling.	✓		✓	
Modelling electron transfer and electron sharing using computer simulations.	✓		✓	
Group 1 and Group 7 reactions, e.g. sodium with chlorine.	✓		✓	
The reactions of bromine, chlorine and iodine with iron wool.	✓		✓	
Growing metal crystals by displacement reactions using metals and salts.	✓		✓	
Modelling metal structures using polyspheres and bubble rafts.	✓		✓	

C2 2.1 Giant ionic structures

Learning objectives

Students should learn:

- why ionic compounds are solids at room temperature and have high melting points
- why ionic compounds conduct electricity when they are molten or are dissolved in water.

Learning outcomes

Most students should be able to:

- state that ionic compounds have high melting points and are solid at room temperature
- describe how ionic compounds can conduct electricity when they are molten or dissolved in water.

Some students should also be able to:

- explain in detail why ionic compounds have high melting points
- explain in detail why ionic compounds can conduct electricity when molten or in solution.

Answers to in-text questions

a attractive electrostatic forces
b Because of the strong attractive electrostatic forces holding the oppositely charged ions together and the large number of ionic bonds in the crystal lattice.
c Because the ions are free to move.

Support

- You may find it easier if, in the experiment, you focus only on sodium chloride. This is a chemical that students experience every day and is safe to use. You could also provide students with the results table, so they just have to fill it in.

Extend

- Students could investigate more properties of ionic compounds (see 'Practical support'). You could encourage them to explain other physical properties of ionic compounds, e.g. symmetry of crystals, cleavage and hardness, in terms of the structure of the ionic lattice.

Specification link-up: Chemistry C2.2

- Ionic compounds have regular structures (giant ionic lattices) in which there are strong electrostatic forces in all directions between oppositely charged ions. These compounds have high melting points and high boiling points because of the large amounts of energy needed to break the many strong bonds. *[C2.2.2 a)]*
- When melted or dissolved in water, ionic compounds conduct electricity because the ions are free to move and carry the current. *[C2.2.2 b)]*
- Suggest the type of structure of a substance given its properties. *[C2.2]*

Lesson structure

Starters

List – Ask students to make a list of as many ionic substances as they can. Through question and answer, build up a list of them on the board and ask the students what they all have in common, other than that they have the same bonding. [All will be compounds and most examples given will probably consist of a metal bonded to a non-metal]. *(5 minutes)*

Anagrams – Give the students the set of anagrams below on the board. They should work out the key word in each case and then define it. You could support students by giving them the key words and definitions, so that they match them up. You could extend students by asking them to give examples of each of the key words.

- noi [Ion – a charged particle made when an atom loses or gains electrons].
- ttalcei [Lattice – a 3D arrangement of particles in a giant structure].
- conii dobn [Ionic bond – the electrostatic force of attraction between oppositely charged ions]. *(10 minutes)*

Main

- There are many examples of ionic compounds in everyday life, but students rarely consider the properties of these substances. Encourage them to investigate the properties of sodium chloride and potassium chloride. Students should first design a table to record information about the appearance, hardness, melting point and conductivity in different states. As it will be near impossible to melt these substances using a Bunsen burner, encourage students to use reference material to find their melting and boiling points.
- Students can then complete the practical detailed in the Student Book, in which they use a simple series circuit to explore the conductivity of ionic compounds when solid and dissolved. Once students have completed the experiment, they can work in groups and use the model of ionic bonding to explain their observations. Ask each group to feed back its ideas to the whole class and distil out a conclusion that all students could record in their notes.
- Give the students four same-coloured index cards. On one side, they should write a key physical property [high melting point, soluble in water and conducts in a solution or a liquid]. Then, on the reverse of each card, they should explain that property using key scientific terms and at least one labelled diagram.
- The fourth card should be a title card about ionic bonding. Ask the students to punch holes in the cards and the front of their exercise books. Using a piece of string the cards could be secured to the book and removed to add other revision cards about bonding. (You could also give other sets of coloured cards to the students for them to make notes about covalent and metallic bonding. They could add these to the set about ionic bonds.)

Plenaries

Summary – Split the class into three groups. Ask each team to explain why ionic compounds have certain physical properties. [Soluble in water, high melting point, conduct electricity when molten or dissolved, etc.] *(5 minutes)*

AfL (Assessment for Learning) – Give students an examination question about the properties of ionic compounds. Time the students (about 1 minute per mark). Then ask them to swap their answers with a partner, and give out mark schemes. Encourage the students to mark the work as it is presented (no discussions about 'what they really meant'). Then swap the papers back and you could collect in the marks. To differentiate this activity, use Foundation paper or Higher Tier past paper questions, depending on the tier of entry for each group of students. *(10 minutes)*

Practical support

Testing conductivity

Equipment and materials required

Sodium chloride, potassium chloride, two carbon electrodes, lab pack, two crocodile clips, three wires, lamp, beaker, wash bottle with water, glass rod, eye protection.

Details

Students put some sodium chloride/potassium chloride crystals into a beaker (to about a depth of 1 cm). Then submerge the electrodes and connect them to a lamp and power supply. Then turn on the power and make observations. Then add water from the wash bottle (half fill the beaker) and observe. Encourage the students to swirl the beaker to help the ionic compound dissolve.

- The lamp does not light at first but does once the salt dissolves.
- The ions are stuck in position within the giant lattice. However, as the sodium chloride dissolves in water, ions become free to move, carrying a charge between the electrodes, the lamp lights up.

Safety: Please note that this electrolysis will produce a small amount of chlorine gas. It should therefore be completed in a well-ventilated area. Be aware of students with respiratory problems (e.g. asthma) as the gas can aggravate them.

Other properties of ionic compounds

Equipment and materials required

Sodium chloride, potassium chloride, hand lens, mounted needle, spatula, two boiling tubes, boiling tube rack, boiling tube holder, beaker, glass rod, Bunsen burner and safety equipment, eye protection.

Details

Students to investigate the following:

Appearance – Students to look at the crystal through a hand lens and draw a diagram of the crystal.

Hardness – Students to try to scratch the surface with a mounted needle, and view the area using the hand lens.

Melting point – Students to put about half a spatula of the compound into a boiling tube and then hold it just above the blue gas cone in a roaring Bunsen flame. The tube should be at an angle, not pointing at any faces. Eye protection should be worn. The students should keep the compound in the flame until they have decided if it has a high or low melting point – the compounds are unlikely to melt. Then they place the boiling tube on a flameproof mat to cool.

Solubility – Students to half fill a beaker with warm water from the tap, then add a few crystals and swirl or stir with a glass rod.

Structure and properties

C2 2.1 Giant ionic structures

Learning objectives

- Why do ionic compounds have high melting points?
- Why do ionic compounds conduct electricity when we melt them or dissolve them in water?

AQA Examiner's tip

Remember that every ionic compound has a giant structure. The oppositely charged ions in these structures are held together by strong electrostatic forces of attraction. These act in all directions.

We have already seen that an ionic compound consists of a giant structure of ions arranged in a lattice. The attractive electrostatic forces between the oppositely charged ions act in all directions and are very strong. This holds the ions in the lattice together very tightly.

a What type of force holds the ions together in an ionic compound?

Strong electrostatic forces of attraction called ionic bonds

Figure 1 The attractive forces between the oppositely charged ions in an ionic compound are very strong. The regular arrangement of ions in the giant lattice enables ionic compounds to form crystals.

It takes a lot of energy to break up a giant ionic lattice. There are lots of strong ionic bonds to break. To separate the ions we have to overcome all those electrostatic forces of attraction. This means that ionic compounds have high melting points and boiling points. Look at the graph in Figure 2.

b Why do ionic compounds have high melting points and boiling points?

Temperature (°C): 1800, 1600, 1400, 1200, 1000, 800, 600
Melting point
Boiling point
NaCl KCl CsCl

Figure 2 The many strong forces of attraction in a lattice of ions mean that ionic compounds have high melting points and boiling points

Once we have supplied enough energy to separate the ions from the lattice, they are free to move around. That's when the ionic solid melts and becomes a liquid. The ions are free to move anywhere in this liquid. Therefore they can carry their electrical charge through the molten liquid. A solid ionic compound cannot conduct electricity. That's because its ions are held in a fixed position in the lattice. They cannot move around. They can only vibrate 'on the spot' when solid.

Bulb lights as current flows
Moving ions carry the electrical charge through the molten potassium chloride
Molten potassium chloride

Figure 3 Because the ions are free to move, a molten ionic compound can conduct electricity

124

Giant ionic structures

Many ionic compounds will dissolve in water. When we dissolve an ionic compound in water, the lattice is split up by the water molecules. Then the ions are free to move around in the solution formed. Just as molten ionic compounds will conduct electricity, solutions of ionic compounds will also conduct electricity. The ions are able to move to an oppositely charged electrode dipped in the solution (See Figure 3).

c Why can ionic compounds conduct electricity when they are molten or dissolved in water?

Ionic solid	Molten ionic compound	Ionic compound in solution
Ions are fixed in a lattice. They vibrate but cannot move around – it does not conduct electricity.	High temperature provides enough energy to overcome the many strong attractive forces between ions. Ions are free to move around within the molten compound – it does conduct electricity.	Water molecules separate ions from the lattice. Ions are free to move around within the solution – it does conduct electricity.

Practical

Testing conductivity

Using a circuit as shown in Figure 3, dip a pair of electrodes into a 1 cm depth of sodium chloride crystals. What happens?

Now slowly add water.

- What happens to the bulb?

Repeat the experiment using potassium chloride.

- Explain your observations.

Summary questions

1 Copy and complete using the words below:

attraction conduct high lattice molten move oppositely solution

Ionic compounds have .... melting points and boiling points because of the many strong electrostatic forces of .... between .... charged ions in the giant .... Ionic compounds will .... electricity when .... or in .... because the ions are able to .... freely around in the liquids.

2 Why is seawater a better conductor of electricity than water from a freshwater lake?

Key points

- It takes a lot of energy to break the many strong ionic bonds which hold a giant ionic lattice together. So ionic compounds have high melting points. They are all solids at room temperature.
- Ionic compounds will conduct electricity when we melt them or dissolve them in water. That's because their ions can then move freely around and can carry charge through the liquid.

125

Further teaching suggestions

Explain a use

- Students could choose an ionic compound and explain why it is used for a specific purpose. For example, sodium fluoride is used in drinking water. Fluoride has been found to make teeth stronger and sodium fluoride dissolves easily into the drinking water supply. Students could then make 'top-trump' style fact cards and these could be used to make a colourful display in the classroom.

Summary answers

1 high, attraction, oppositely, lattice, conduct, molten, solution, move

2 Because it contains dissolved salt (ions).

C2 2.2 Simple molecules

Learning objectives

Students should learn:

- the properties of substances made up of simple molecules
- why simple molecular substances have low melting and boiling points **[HT only]**
- that substances made up of simple molecules do not conduct electricity.

Learning outcomes

Most students should be able to:

- recognise substances made up of simple molecules
- list examples of substances made up of simple molecules
- state the physical properties of substances made up of simple molecules
- state why substances made up of simple molecules do not conduct electricity.

Some students should also be able to:

- explain why substances made up of simple molecules have low melting and boiling points. **[HT only]**

Support

- Give students the properties of some simple molecular compounds and a few 'red herrings' that are not the properties. Students could then list the properties of simple molecular compounds in their notes.

Extend

- Students could observe the sublimation of iodine. Put some iodine crystals into a conical flask and seal with a bung. Then heat the flask with running warm water and the iodine will sublime. Then run the flask under cold water and the iodine will solidify again. Ask the students to explain what is happening in terms of intermolecular forces.
Safety: Be aware of broken glass (if the flask is dropped) and do not allow the students to touch iodine as it will stain their skin.

Specification link-up: Chemistry C2.2

- Substances that consist of simple molecules are gases, liquids or solids that have relatively low melting points and boiling points. *[C2.2.1a)]*
- Substances that consist of simple molecules have only weak forces between the molecules (intermolecular forces). It is these intermolecular forces that are overcome, not the covalent bonds, when the substance melts or boils. *[C2.2.1b)]* **[HT only]**
- Substances that consist of simple molecules do not conduct electricity because the molecules do not have an overall electric charge. *[C2.2.1c)]*
- Suggest the type of structure of a substance given its properties. *[C2.2]*

Lesson structure

Starters

Crosswords – Create a crossword in which the answers are key words that will be needed in the lesson: 'molecule', 'electron', 'covalent', 'bond', 'compound', 'element', 'dot and cross'. There are some useful free websites such as www.discoveryschool.com where bespoke puzzles can be made quickly. *(5 minutes)*

Models – Split the class into small groups. Show the dot and cross diagram of hydrogen, then instruct the groups to make a model (using a molecular model kit) of this diagram and hold it in the air so that you can easily check. Then repeat with the other simple molecules that the students should know: hydrogen chloride, water, oxygen, ammonia and methane. Support students by supplying them with cards with diagrams of the molecules to make. Students then hold up the relevant model. Extend students by asking them to generate dot and cross diagrams themselves. *(10 minutes)*

Main

- Students often find it difficult to make connections between pieces of information. A mind map can be a very useful tool to help them make links within a topic.
- Split the students into groups of about four, and give them a felt pen each, some sticky-tac, a piece of sugar paper and a pack of A6 word cards. You should make these cards in advance, including key words for this topic: 'atom', 'electron', 'bond', 'covalent', 'melting point', 'boiling point', 'intermolecular forces', 'insulator', 'molecule', 'solid', 'liquid', 'gas'.
- The students should stick two key words on to the page: one should be the start of the sentence, the other the end. They should draw an arrow in the correct direction linking the words, then on the arrow write the middle section of the sentence. There is no limit to the number of arrows that can be drawn to and from each.
- Students come into contact with many covalent compounds in everyday life, but they have probably not considered their properties. Give them a selection of covalently bonded compounds, e.g. water, ethanol, iodine, sulfur.
- Ask them to discuss, in groups, any similarities in appearance [dull]. Then they could determine the physical properties of this group of substances experimentally.
- If it is not possible to complete the practical, students could use data books to obtain melting point, boiling point and conductivity information.

Plenaries

Stand by – Ask for a volunteer to stand in the centre of the classroom. The volunteer should make a statement about simple covalent molecules (either his or her own idea or read from prepared statement cards). The rest of the class decide how much they agree with this statement. The more they agree, the closer they should stand to the person who spoke. *(5 minutes)*

Summarise – Ask the students to summarise the physical properties of simple molecular compounds and give brief reasons for them in a bullet-point format. *(10 minutes)*

Practical support

Conductivity

Equipment and materials required

Beaker, water, carbon electrodes, lamp, power pack, wires, ethanol CLEAPSS Hazcard 40A – highly/flammable/harmful, solid wax pieces.

Details

Half-fill a beaker with water and put in the carbon electrodes. Set up a simple series circuit with a lamp and power pack, to see if the liquid conducts. Repeat the experiment with ethanol and solid wax pieces.

Safety: Keep ethanol away from any naked flames.

Melting and boiling points

Equipment and materials required

Boiling tubes, boiling-tube rack, boiling-tube holder, Bunsen burner and safety equipment, thermometer, water, wax, eye protection.

Details

Ask students to predict whether covalent compounds have low or high melting/boiling points and why they think this. Get the students to prepare a results table.

They need to put a small piece of wax into a boiling tube and heat it in a blue flame, noting the temperature at which the wax begins to melt. They then put about a 1 cm depth of water into a boiling tube and heat in a blue Bunsen flame, noting the temperature of boiling.

C2 2.2 Simple molecules

Learning objectives

- Which type of substances has low melting points and boiling points?
- Why do these substances have low melting points and boiling points? [H]
- Why don't these substances conduct electricity?

When the atoms of non-metal elements react to form compounds, they share electrons in their outer shells. In this way each atom gains the electron arrangement of a noble gas. The bonds formed like this are called covalent bonds.

Figure 1 Covalent bonds hold the atoms found within molecules tightly together

a How are covalent bonds formed?

Many substances made up of covalently bonded molecules have low melting points and boiling points. Look at the graph in Figure 2.

Figure 2 Substances made of simple molecules usually have low melting points and boiling points

These substances have low melting points and boiling points. This means that many substances made up of simple molecules are liquids or gases at room temperature. Others are solids with quite low melting points, such as iodine (I_2) and sulfur (S_8).

b Do the compounds shown on the graph exist as solids, liquids or gases at 20 °C?

c You have a sample of ammonia (NH_3) at −120 °C. Describe the changes that you would see as the temperature of the ammonia rises to 20 °C (approximately room temperature).

Intermolecular forces

Higher

Covalent bonds are very strong. So the atoms within each molecule are held very tightly together. However, each molecule tends to be quite separate from its neighbouring molecules. The attraction between the individual molecules in a covalent substance is relatively small. We say that there are weak **intermolecular forces** between molecules. Overcoming these forces does not take much energy.

Look at the molecules in a sample of chlorine gas:

Figure 3 Covalent bonds and the weak forces between molecules in chlorine gas. It is the weak intermolecular forces that are overcome when substances made of simple molecules melt or boil. The covalent bonds are not broken.

d How strong are the forces between the atoms in a covalently bonded molecule?

e How strong are the forces between molecules in a covalent substance? [H]

We have seen that ionic compounds will conduct electricity when they are liquids. But although a substance that is made up of simple molecules may be a liquid at room temperature, it will not conduct electricity. Look at the demonstration below.

There is no overall charge on the simple molecules in a compound like ethanol. So their molecules cannot carry electrical charge. This makes it impossible for substances made up of simple molecules to conduct electricity.

f Why don't simple molecular substances conduct electricity?

AQA Examiner's tip

Although the covalent bonds in molecules are strong, the forces between molecules are weak. [H]

links

For information on ionic compounds conducting electricity, look back at C2 2.1 Giant ionic structures.

Demonstration

Conductivity

- What happens?

Figure 4 Compounds made of simple molecules do not conduct electricity

Summary questions

1 Copy and complete using the words below:

boiling solids covalent melting molecules strongly liquids

Non-metals react together to form ………… which are held together by ………… bonds. These hold the atoms together very ………… . If these substances are made of simple molecules, they have low ………… points and ………… points. So at room temperature they often exist as gases and ………… or as ………… which melt relatively easily.

2 A compound called sulfur hexafluoride (SF_6) is used to stop sparks forming inside electrical switches designed to control large currents. Explain why the properties of this compound make it particularly useful in electrical switches.

3 The melting point of hydrogen chloride is −115 °C, whereas sodium chloride melts at 801 °C. Explain why. [H]

Key points

- Substances made up of simple molecules have low melting points and boiling points.
- The forces between simple molecules are weak. These weak intermolecular forces explain why substances made of simple molecules have low melting points and boiling points. [H]
- Simple molecules have no overall charge, so they cannot carry electrical charge. Therefore substances made of simple molecules do not conduct electricity.

Further teaching suggestions

Displaying data

- Information from data books including databases such as the CD-ROM *RSC Data Book* could be used to get melting point/boiling point/conductivity information. The data for a particular group of covalent compounds could be collected and students could represent the data in a bar chart format.

Ice and water

- Students could find out how the intermolecular forces between water molecules make ice less dense than liquid water.

Answers to in-text questions

a By sharing pairs of electrons.

b gases

c It would be a solid to start with but at about −80 °C, it would melt into a liquid. Then at about −35 °C, it would boil to form ammonia gas. At 20 °C, it remains a gas.

d very strong

e Weak (compared with the forces between atoms held together by a covalent bond).

f The molecules have no overall electric charge to carry the electric current.

Summary answers

1 molecules, covalent, strongly, melting (boiling), boiling (melting), liquids, solids

2 It is a (simple) molecular substance so it doesn't conduct electricity.

3 In sodium chloride, to separate the oppositely charged sodium and chloride ions requires a lot of energy to overcome the strong electrostatic forces of attraction operating in every direction. In hydrogen chloride, there are only weak forces of attraction between individual HCl molecules, so they require far less energy to separate them. (No covalent bonds are broken in the process of melting.)

C2 2.3 Giant covalent structures

Learning objectives

Students should learn:

- some of the physical properties of substances with giant covalent structures
- why diamond is hard and graphite is slippery
- why graphite conducts electricity **[HT only]**
- what fullerenes are. **[HT only]**

Learning outcomes

Most students should be able to:

- list examples of substances with giant covalent structures
- recognise giant covalent structures
- state the physical properties of graphite and diamond
- explain some physical properties of diamond and graphite, such as melting point and hardness, in terms of their structures.

Some students should also be able to:

- explain in detail what a giant covalent structure is
- explain why graphite conducts electricity in terms of delocalised electrons and its softness in terms of weak intermolecular forces between layers **[HT only]**
- explain what a fullerene is. **[HT only]**

Specification link-up: Chemistry C2.2

- Atoms that share electrons can also form giant structures or macromolecules. Diamond and graphite (forms of carbon) and silicon dioxide (silica) are examples of giant covalent structures (lattices) of atoms. All the atoms in these structures are linked to other atoms by strong covalent bonds and so they have very high melting points. *[C2.2.3 a)]*
- In diamond, each carbon atom forms four covalent bonds with other carbon atoms in a giant covalent structure, so diamond is very hard. *[C2.2.3 b)]*
- In graphite, each carbon atom bonds to three others, forming layers. The layers are free to slide over each other because there are no covalent bonds between the layers and so graphite is soft and slippery. *[C2.2.3 c)]*
- In graphite, one electron from each carbon atom is delocalised. These delocalised electrons allow graphite to conduct heat and electricity. *[C2.2.3 d)]* **[HT only]**
- Carbon can also form fullerenes with different numbers of carbon atoms. Fullerenes can be used for drug delivery into the body, in lubricants, as catalysts, and in nanotubes for reinforcing materials, e.g. in tennis rackets. *[C2.2.3 e)]* **[HT only]**
- Relate the properties of substances to their uses. *[C2.2]*
- Suggest the type of structure of a substance given its properties. *[C2.2]*

Lesson structure

Starters

List – Show students different images of diamonds – in the raw state, in jewellery and on a saw or drill. Ask the students to list different uses of diamonds and then ask for a few suggestions. *(5 minutes)*

Word search – Give students a word search about covalent bonding; however do not include the list of words. Encourage students to think about the topic and use the double-page spread in the Student Book to work out which words they should be finding. Support students by using only the key words that have been written in bold in the Student Book. Extend students by asking them to write the definitions for each of the words in the word search. *(10 minutes)*

Main

- Although graphite and diamond are both carbon, they have completely different structures. Give students some graphite to handle (this may be a bit messy!) and if it is possible, samples of diamonds to study. You may wish to allow students to investigate the melting point and solubility as detailed in practical support for C2.2 Simple molecules. Brainstorm the different properties of each of these materials, but explain they are the same element.
- Foundation candidates need to recognise and describe a macromolecule limited to silicon dioxide, carbon and diamond. Higher Tier students also need to be able to explain the properties of diamond, graphite and fullerenes based on their structures and give some uses of fullerenes.
- Ask students to make a poster to contrast the properties of these two substances and explain them in terms of their structures.
- Students could use the Student Book for information to create an A-map about giant covalent structures. They should select three colours. In the centre of the page, in one colour only, they should write the key phrase 'Giant covalent structures' and draw a small image that might help them to remember this. This colour is then not used again. The second colour is then used to create four long, wavy lines. Following the contour of the line, the student should write 'formation', 'graphite', 'diamond' and 'fullerene' on separate lines, each including an image. Each idea is then added to, with a third colour, again with wavy lines. Each line again should contain key words or phrases to summarise that branch of thought, and include an image to help the student remember. Encourage the students to complete one branch before moving to the next.

Support

- Give students diagrams of the structures of diamond and graphite. On a separate sheet include the labels for the diagram and encourage the students to annotate their work.

Extend

- Ask students to find out other examples of giant covalent structures, e.g. silicon dioxide and fullerenes. They should find out and draw a diagram of the structure and annotate the diagram to explain its properties in terms of the structure.

Plenaries

Explain – Split the students into pairs, and give them a pack of cards with key terms on it: 'graphite', 'diamond', 'silicon dioxide', 'delocalised electrons' **[HT only]**, 'lattice'. The cards should be face down in front of the group. Each student in turn should take a card and try to explain the key term, but the guesser must draw a labelled diagram to match the key term. *(5 minutes)*

Models – Show the students molecular models of diamond and graphite. Ask them to list the similarities and differences. Ask for a volunteer to write them on the board. Then, through questions and answers, build up bullet points on the board. Support students by giving them a list of different statements, and they have to decide which is a similarity and which a difference. Volunteers could come to the board and write the statements into a table. Extend students by including different fullerene structures, e.g. buckminsterfullerene and nanotubes. **[HT only]** *(10 minutes)*

Answers to in-text questions

a giant covalent (lattice)

b High melting and boiling points, hard, insoluble in water.

c delocalised (free) electrons

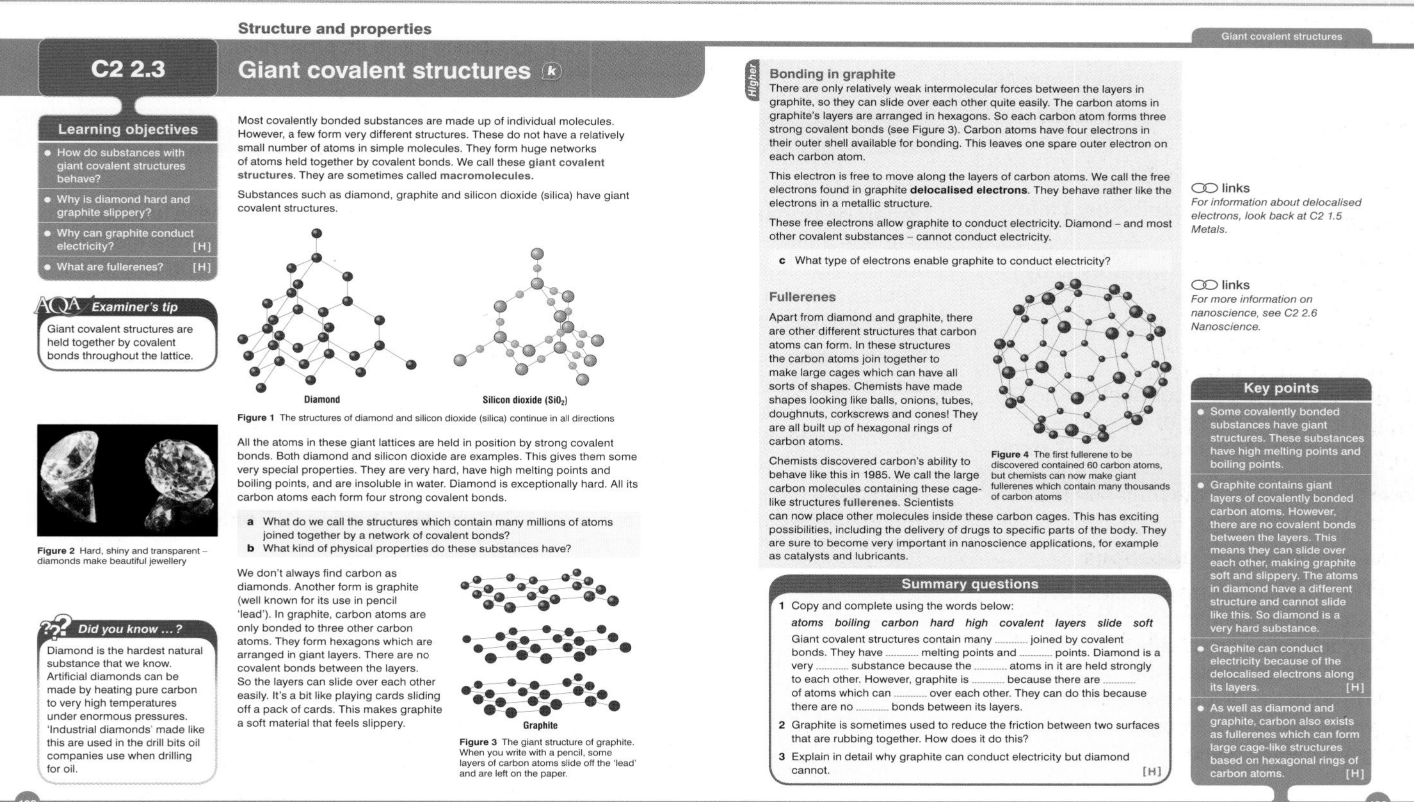

Structure and properties

Giant covalent structures

C2 2.3 Giant covalent structures

Learning objectives
- How do substances with giant covalent structures behave?
- Why is diamond hard and graphite slippery?
- Why can graphite conduct electricity? [H]
- What are fullerenes? [H]

AQA Examiner's tip
Giant covalent structures are held together by covalent bonds throughout the lattice.

Most covalently bonded substances are made up of individual molecules. However, a few form very different structures. These do not have a relatively small number of atoms in simple molecules. They form huge networks of atoms held together by covalent bonds. We call these **giant covalent structures**. They are sometimes called **macromolecules**.

Substances such as diamond, graphite and silicon dioxide (silica) have giant covalent structures.

Diamond   Silicon dioxide (SiO_2)

Figure 1 The structures of diamond and silicon dioxide (silica) continue in all directions

Figure 2 Hard, shiny and transparent – diamonds make beautiful jewellery

All the atoms in these giant lattices are held in position by strong covalent bonds. Both diamond and silicon dioxide are examples. This gives them some very special properties. They are very hard, have high melting points and boiling points, and are insoluble in water. Diamond is exceptionally hard. All its carbon atoms each form four strong covalent bonds.

a What do we call the structures which contain many millions of atoms joined together by a network of covalent bonds?
b What kind of physical properties do these substances have?

We don't always find carbon as diamonds. Another form is graphite (well known for its use in pencil 'lead'). In graphite, carbon atoms are only bonded to three other carbon atoms. They form hexagons which are arranged in giant layers. There are no covalent bonds between the layers. So the layers can slide over each other easily. It's a bit like playing cards sliding off a pack of cards. This makes graphite a soft material that feels slippery.

Graphite

Figure 3 The giant structure of graphite. When you write with a pencil, some layers of carbon atoms slide off the 'lead' and are left on the paper.

Did you know ...?
Diamond is the hardest natural substance that we know. Artificial diamonds can be made by heating pure carbon to very high temperatures under enormous pressures. 'Industrial diamonds' made like this are used in the drill bits oil companies use when drilling for oil.

Higher

Bonding in graphite

There are only relatively weak intermolecular forces between the layers in graphite, so they can slide over each other quite easily. The carbon atoms in graphite's layers are arranged in hexagons. So each carbon atom forms three strong covalent bonds (see Figure 3). Carbon atoms have four electrons in their outer shell available for bonding. This leaves one spare outer electron on each carbon atom.

This electron is free to move along the layers of carbon atoms. We call the free electrons found in graphite **delocalised electrons**. They behave rather like the electrons in a metallic structure.

These free electrons allow graphite to conduct electricity. Diamond – and most other covalent substances – cannot conduct electricity.

c What type of electrons enable graphite to conduct electricity?

Fullerenes

Apart from diamond and graphite, there are other different structures that carbon atoms can form. In these structures the carbon atoms join together to make large cages which can have all sorts of shapes. Chemists have made shapes looking like balls, onions, tubes, doughnuts, corkscrews and cones! They are all built up of hexagonal rings of carbon atoms.

Chemists discovered carbon's ability to behave like this in 1985. We call the large carbon molecules containing these cage-like structures **fullerenes**. Scientists can now place other molecules inside these carbon cages. This has exciting possibilities, including the delivery of drugs to specific parts of the body. They are sure to become very important in nanoscience applications, for example as catalysts and lubricants.

Figure 4 The first fullerene to be discovered contained 60 carbon atoms, but chemists can now make giant fullerenes which contain many thousands of carbon atoms

Summary questions

1 Copy and complete using the words below:
atoms boiling carbon hard high covalent layers slide soft
Giant covalent structures contain many .... joined by covalent bonds. They have .... melting points and .... points. Diamond is a very .... substance because the .... atoms in it are held strongly to each other. However, graphite is .... because there are .... of atoms which can .... over each other. They can do this because there are no .... bonds between its layers.

2 Graphite is sometimes used to reduce the friction between two surfaces that are rubbing together. How does it do this?

3 Explain in detail why graphite can conduct electricity but diamond cannot. [H]

links
For information about delocalised electrons, look back at C2 1.5 Metals.

links
For more information on nanoscience, see C2 2.6 Nanoscience.

Key points
- Some covalently bonded substances have giant structures. These substances have high melting points and boiling points.
- Graphite contains giant layers of covalently bonded carbon atoms. However, there are no covalent bonds between the layers. This means they can slide over each other, making graphite soft and slippery. The atoms in diamond have a different structure and cannot slide like this. So diamond is a very hard substance.
- Graphite can conduct electricity because of the delocalised electrons along its layers. [H]
- As well as diamond and graphite, carbon also exists as fullerenes which can form large cage-like structures based on hexagonal rings of carbon atoms. [H]

128   129

Further teaching suggestions

The hardness of pencils
- Graphite is used in pencil leads. Encourage students to research how the different levels of hardness (e.g. H and HB) are achieved.

Developments in carbon
- Fullerenes are a new and exciting class of carbon allotropes. Students could research into their discovery and development and create a timeline. This could include names of important scientists who work in their field of research, and diagrams of the macromolecules.
  Students could find a use for graphite and explain which property makes it suitable for this use.
  Life Gem is a company that offers to cremate pets and humans in a special way, which turns them into diamonds. Encourage students to use the internet to look up this idea. However, be sensitive to the emotional development of students in your class especially if they are recently bereaved.

Sand
- Sand is mainly silicon dioxide and has a macrostructure. You may wish to give students some samples of sand and allow them to investigate the properties using the same equipment as in 'Practical support' 'C2.2 Simple molecules'.

Summary answers

1 atoms, high, boiling, hard, carbon, soft, layers, slide, covalent

2 The graphite is used to coat the two surfaces. As they rub together, the layers of atoms in the graphite slip over each other, reducing the friction.

3 Graphite can conduct electricity because of the delocalised (free) electrons in its structure. These arise because each carbon is only bonded to three other carbon atoms. This leaves one electron to become delocalised. However, in diamond, all four outer electrons on each carbon atom are used in covalent bonding, so there are no delocalised electrons.

C2 2.4 Giant metallic structures

Learning objectives

Students should learn:

- why metals can be bent and shaped
- why alloys are harder than pure metals
- why metals can conduct electricity and energy/heat **[HT only]**
- the shape of memory alloys.

Learning outcomes

Most students should be able to:

- state the physical properties of metals
- suggest which property of a metal makes it suitable for a specific job
- state a definition of an alloy
- explain why metals can be bent into different shapes
- explain what a shape memory alloy is and give an example.

Some students should also be able to:

- explain why metals conduct electricity and energy in terms of delocalised electrons in their structures. **[HT only]**

Answers to in-text questions

- **a** The layers of atoms can slide over each other when force is applied.
- **b** By delocalised electrons from the outer energy levels of the metal atoms.
- **c** Through the delocalised electrons, which are able to move through the giant metallic lattice.

Support

- Give students a diagram of the structure of a metal crystal and statements that can be used to annotate the diagram to explain the model. Extend students by giving them information about the properties of metals, explaining these in terms of the structure. Students can then complete a 'cut-and-stick' activity to label the diagram fully.

Extend

- Ask students to find out some other models of metallic structure such as molecular models and computer modelling. Students can then evaluate the different models. They should notice that different models are more appropriate for explaining certain properties of metals.

Specification link-up: Chemistry C2.2

- Metals conduct heat and electricity because of the delocalised electrons in their structures. *[C2.2.4 a)]* **[HT only]**
- The layers of atoms in metals are able to slide over each other and so metals can be bent and shaped. *[C2.2.4 b)]*
- Alloys are usually made from two or more different metals. The different sized atoms of the metals distort the layers in the structure, making it more difficult for them to slide over each other and so make alloys harder than pure metals. *[C2.2.4 c)]*
- Shape memory alloys can return to their original shape after being deformed, e.g. Nitinol used in dental braces. *[C2.2.4 d)]*
- Relate the properties of substances to their uses. *[C2.2]*
- Suggest the type of structure of a substance given its properties. *[C2.2]*

Lesson structure

Starters

Smart demonstration – Obtain a comedy spoon from a joke shop. These are made from a shape memory alloy. When you submerge them into warm water, they change from looking like 'normal' teaspoons to twisted spoons. Demonstrate this action in a glass beaker of warm water, asking a student to stir. When the spoon has become twisted, ask the students to try to explain what has happened. *(5 minutes)*

Flash boards – Give the students A4 whiteboards (or laminated paper), a washable pen and eraser. Ask them to draw an electronic diagram of a metal of their choice and then hold their answers up to you [these should be dot and cross diagrams]. Support students by encouraging them to use their Student Books to help them or to look at other people's answers and the teacher's response. They can then draw a diagram of a metal structure in a solid (linking back to the particle model in KS3). Extend students by asking them to merge the two diagrams to form a model of metallic bonding [**HT only**] (as shown in the Student Book, C2 1.5 Metals). *(10 minutes)*

Main

- Metals are used in everyday life but often students do not consider which properties make them useful for certain jobs. Give the students adverts or catalogues to look through. Ask them to pick items that use metals and cut them out. They could then make a poster using these items, explaining which part is metal. Students could explain the useful property in terms of metallic bonding. It is important that each poster contains an example of a metal being ductile, a metal being malleable and a shape-memory alloy. Higher Tier students should also include an example of a metal conducting heat and a metal conducting electricity.
- Metal properties can be modelled in a variety of ways such as using equal-sized polystyrene balls, built up layer by layer. Show the students any pre-made molecular models that the school may have. Soap bubbles can be used to represent metal atoms and show ductility and malleability. Adding a different-sized bubble models an alloy.
- Higher Tier students should then explain their results relating the model to metallic bonding.

Plenary

'Circle of truth' – This is an interactive, self-marking exercise designed to be used on an interactive whiteboard. To create this activity, open board-specific interactive whiteboard software, or PowerPoint. Firstly, in a text box, type in the title 'Which are properties of metals?' In a small font size, and in separate text boxes, write wrong answers, e.g. 'dull', 'brittle' and 'insulator'. Then draw a circle. You may wish to add the text 'circle of truth' and group the objects. This circle should hide the previously written text, i.e. the wrong answers. Then, in separate text boxes write the correct answers: 'can be drawn out into wires', 'can be hammered into shapes', 'conductor of electricity', 'conductor of energy' and 'high melting point'. To use this activity, ask the students for volunteers to come to the board and suggest an answer to the question [Which are the properties of metals?]. They should then drag the circle to their answer. If they are correct, the answer is still visible; if it is incorrect, then the circle will cover the answer. You can make the questions easier to support students or include higher level concepts to extend students (using Bloom's taxonomy to determine the level of questioning). *(5 minutes)*

Practical support

Blowing bubbles

Equipment and materials required

Petri dish, pointed end of a dropping pipette, dropping pipette, soap solution, rubber tubing to connect to gas tap.

Details

Students should blow similar-sized bubbles into the Petri dish using the pipette. They observe and then blow different sized bubbles using a normal dropping pipette and observe.

You may wish to set up a flexicam or digital camera and connect to a digital projector or a TV. Then focus on the 'bubble raft' experiment. Using the image you can then explain how this experiment relates to the structure of metals.

Safety: Take care when using gas to make bubbles, it is flammable.

Structure and properties

C2 2.4 Giant metallic structures

Learning objectives

- Why can we bend and shape metals?
- Why are alloys harder than pure metals?
- Why do metals allow electricity and heat pass through them? [H]
- What are shape memory alloys?

We can hammer and bend metals into different shapes, and draw them out into wires. This is because the layers of atoms in a pure metal are able to slide easily over each other.

The atoms in a pure metal, such as iron, are held together in giant metallic structures. The atoms are arranged in closely-packed layers. Because of this regular arrangement, the atoms can slide over one another quite easily. This is why pure iron is soft and easily shaped.

a Why can metals be bent, shaped and pulled out into wires when forces are applied?

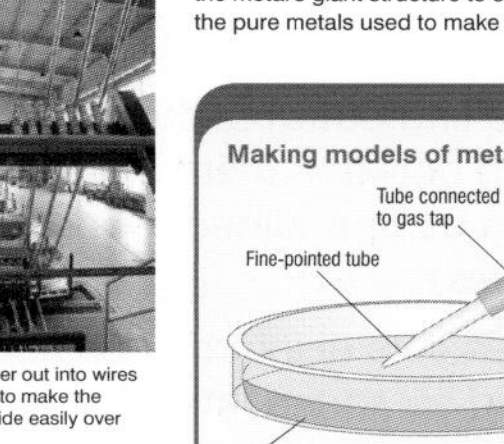

Figure 1 Drawing copper out into wires depends on being able to make the layers of metal atoms slide easily over each other

Alloys are usually mixtures of metals. However, most steels contain iron with controlled amounts of carbon, a non-metal, mixed in its structure. So there are different sizes of atoms in an alloy. This makes it more difficult for the layers in the metal's giant structure to slide over each other. So alloys are harder than the pure metals used to make them. This is shown in Figure 2.

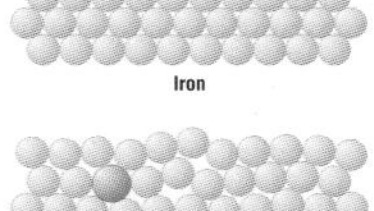

Figure 2 The atoms in pure iron are arranged in layers which can easily slide over each other. In alloys the layers cannot slide so easily because atoms of other elements change the regular structure.

Practical

Making models of metals

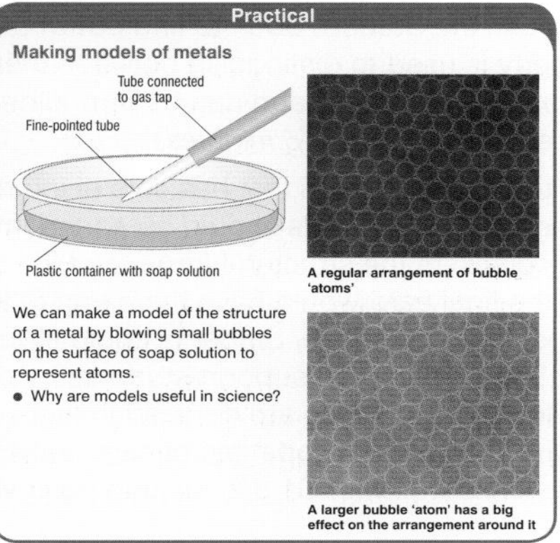

A regular arrangement of bubble 'atoms'

A larger bubble 'atom' has a big effect on the arrangement around it

We can make a model of the structure of a metal by blowing small bubbles on the surface of soap solution to represent atoms.

- Why are models useful in science?

Metal cooking utensils are used all over the world, because metals are good conductors of heat. Wherever we generate electricity, it passes through metal wires to where it is needed. That's because metals are also good conductors of electricity.

Higher

Explaining the properties of metals

The positive ions in a metal's giant structure are held together by a sea of delocalised electrons. These electrons are a bit like 'glue'. Their negative charge between the positively charged ions holds the ions in position.

However, unlike glue, the electrons are able to move throughout the whole giant lattice. Because they can move around and hold the metal ions together at the same time, the delocalised electrons enable the lattice to distort. When struck, the metal atoms can slip past one another without breaking up the metal's structure.

b How are metal atoms held together?

Metals are good conductors of heat and electricity because the delocalised electrons can flow through the giant metallic lattice. The electrical current and heat are transferred quickly through the metal by the free electrons.

c Why do metals conduct electricity and heat so well?

Figure 3 Metals are essential in our lives – the delocalised electrons mean that they are good conductors of both heat and electricity

Shape memory alloys

Some alloys have a very special property. Like all metals they can be bent (or **deformed**) into different shapes. The difference comes when you heat them up. They then return to their original shape all by themselves.

We call these metals **shape memory alloys**, which describes the way they behave. They seem to 'remember' their original shape!

We can use the properties of shape memory alloys in many ways, for example in health care. Doctors treating a badly broken bone can use alloys to hold the bones in place while they heal. They cool the alloy before it is wrapped around the broken bone. When it heats up again the alloy goes back to its original shape. This pulls the bones together and holds them while they heal.

Dentists have also made braces to pull teeth into the right position using this technique.

links

For more about the bonding in metals, look back at C2 1.5 Metals.

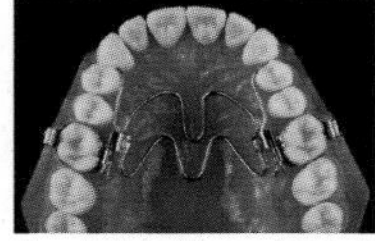

Figure 4 This dental brace pulls the teeth into the right position as it warms up. It is made of a shape memory alloy called nitinol. It is an alloy of nickel and titanium.

Summary questions

1 Copy and complete using the words below:

delocalised electricity energy heat shape slide

The positively charged .... in metals are held together by .... electrons. These also allow the layers to .... over each other so that the metal's .... can be changed. They also allow the metal to conduct .... and .... [H]

2 **a** Use your knowledge of metal structures to explain how adding larger metal atoms to a metallic lattice can make the metal harder.

b What is a shape memory alloy?

3 Explain how a dental brace made out of nitinol is more effective than a brace made out of a traditional alloy.

4 Explain why metals are good conductors of heat and electricity. [H]

Key points

- We can bend and shape metals because the layers of atoms (or ions) in a giant metallic structure can slide over each other.
- Delocalised electrons in metals enable electricity and heat to pass through the metal easily. [H]
- If a shape memory alloy is deformed, it can return to its original shape on heating.

Further teaching suggestions

Arrangement of atoms

- Metal atoms stack in different layers, e.g. ABAB or ABCABC. Encourage students to find out different atom arrangements in different metals. You may want to use polystyrene balls or computer animations to show how the atoms can slide over each other.

Using metals

- Some metals are more useful for certain jobs, but substitutes are used, e.g. silver is the best metal electrical conductor, but it degrades easily, so other metals, e.g. gold, are used in satellites. Ask students to find other interesting uses and facts about metals.

Smart materials

- Students could find out what a smart material is. They could then find out examples of shape memory alloys, and list their properties and some of their uses. Making metal crystals: You may wish to make metal crystals as detailed in C2 1.5 'Bonding in metals'.

Testing metals

- You may wish to allow students to test the properties of metals. Give them samples of metals such as metal wires (without insulation), metal rods and metal sheets. The Design and Technology department may have some off-cuts that could be used. Encourage students to look at the metal using a hand lens. Flex the material and use a simple series circuit to test conductivity. You can use data tables to find other information such as melting point, boiling point and tensile strength. Students could work in groups to make a sugar paper poster displaying their results in a variety of different forms such as chart, graphs, tables and prose.

Summary answers

1 ions, delocalised, slide, shape, energy (electricity), electricity (energy)

2 **a** This helps to stop the layers of metal atoms sliding over each other.

b A metal which (when it gets deformed) will return to its original shape on heating.

3 The brace made out of shape memory metal can be designed to fit easily, then more pressure can be applied once warmed up in the mouth. A traditional metal would need more adjustments by a dentist to have the same effect.

4 The delocalised electrons in their structures carry the charge through a metal when it conducts electricity, and the energy through it when it conducts heat.

C2 2.5 The properties of polymers

Learning objectives

Students should learn:

- that the properties of polymers depend on their monomers
- that changing reaction conditions changes the polymers that are made
- the properties of the two classes of polymers (thermosetting and thermosoftening).

Learning outcomes

Most students should be able to:

- state that the monomers used to make a polymer will affect its properties
- give an example of a polymer whose properties differ when it is formed in different conditions
- explain why a given polymer is fit for a purpose
- recall a definition of thermosetting and thermosoftening polymers and give an example of each
- explain how cross linking in thermosetting plastics stops them from melting.

Some students should also be able to:

- explain how changing the reaction conditions changes the properties of the polymer
- explain how the intermolecular forces between the polymer molecules in thermosoftening plastics affect their properties. **[HT only]**

Answers to in-text questions

a In a tangled web.
b LD – low density; HD – high density.

Support

- Give students two polymers – one thermosetting and one thermosoftening – to complete the practicals.

Extend

- Ask students how intermolecular forces affect the properties of a polymer. Students could be given different examples of polymers, showing part of their structural formulae (some branched chain, some with cross links, some with different functional groups). Then give students a property, e.g. melting point, and ask them to order the structures from low to high melting point, and use the structures to suggest why they have chosen that order.

Specification link-up: Chemistry C2.2

- The properties of polymers depend on what they are made from and the conditions under which they are made. For example, low density (LD) and high density (HD) poly(ethene) are produced using different catalysts and reaction conditions. *[C2.2.5 a)]*
- Thermosoftening polymers consist of individual, tangled polymer chains. Thermosetting polymers consist of polymer chains with cross-links between them so that they do not melt when they are heated. *[C2.2.5 b)]*
- Suggest the type of structure of a substance given its properties. *[C2.2]*

Controlled Assessment: C4.3 Collect primary and secondary data. *[C4.3.1 a)]*

Lesson structure

Starters

Define – Ask the students to use the Student Book to find out what the two groups of polymers are and what property is used to distinguish between them. [Thermosoftening polymers can be reheated to become pliable and can be remoulded, whereas thermosetting polymers will eventually char.] *(5 minutes)*

Card sort – Give the students a pack of cards that has sets of information about poly(ethene), nylon, PVC and PTFE. The cards should be sorted into sets of three for each polymer. These would consist of the structural formula of the monomer, the structural formula of the polymer, and the final card would have the name of the polymer classified into thermosoftening/thermosetting and a use. Students could then arrange the cards into a table and copy it into their book. You could support students by only giving them the cards for polyethene and nylon (both of these are covered in this double-page spread). You could extend them by giving them the properties of each polymer, and they have to suggest a use. This activity links well with C1 5.2 'Making polymers from alkenes'. *(10 minutes)*

Main

- Students could be given a sample of a thermosetting and a thermosoftening polymer and compare their properties during heating in a fume cupboard. They could also try to remould the thermosoftening polymer. Results could then be recorded as a cartoon strip, drawing images to represent different stages of the practical. Below each image, the students could be encouraged to explain the observations in terms of intermolecular forces of attraction.
- Students could experiment with the consistency of slime made with PVA glue (poly(ethenol)). The polymer strands are H-bond cross linked with borate groups. This cross linking is not permanent and most of the space in the gel is taken up with water molecules. This makes it a pliable polymer. Adding different amounts of borax changes the number of cross links. The more there are, the stiffer the polymer. Ask the students to design and conduct an experiment to find out the effect of adding different amounts of borax to PVA glue. This relates to the Controlled Assessment [AS4.3.1] in which students are required to use data to develop hypotheses.

Plenaries

Classify – Give the students a number of examples of plastic items and ask them to sort them into thermosoftening or thermosetting. For example, *thermosoftening*: a chocolate box tray, nylon clothes, student's toys, plastic beakers; *thermosetting*: epoxy resin glues, electrical equipment, pan handles. Support students by writing a definition of thermosoftening and thermosetting on the board so that students can easily refer to them. Extend students by asking them to explain why it is important that the polymer is thermosetting/thermosoftening for that particular function. *(5 minutes)*

Guess the word – Separate the class into groups of six and give out a pack of cards. Each card should be like name cards for a party with one of the following key words written on: 'thermosoftening, thermosetting, poly(ethene), LD, HD, polymer, monomer'. The students should not look at the card that they have been given, but secure it so that it faces the rest of their group. They must then ask questions, to which the others can only answer 'yes/no' in order to guess their key word. *(10 minutes)*

Practical support

Making a polymer

Equipment and materials required

A 100 cm³ measuring cylinder, 250 cm³ beaker, dropping pipettes, hot plate, stirring rod, 4 g PVA glue, 4 g borax (toxic), eye protection, (food colouring).

Details

Before the lesson, make up a solution of borax, with 100 cm³ of warm water in a beaker in a fume cupboard with the fan off. The solution may need to be warmed for it to fully dissolve. Put 100 cm³ of water into the 250 cm³ beaker and heat gently. Add 4 g of PVA slowly, while stirring. The mixture must not boil. When all the PVA has dissolved, remove it from the heat, and add a few drops of the borax solution and stir. If too much borax is added, the polymer will be brittle, if not enough is added the slime will be runny. Food colourings can be added to make slime of different colours. This should be done before the borax is added. In time, the slime will dry out. Students can handle the slime, but should wash their hands afterwards, and eye protection should be worn throughout.

Safety: Students should wear eye protection when using borax solution. CLEAPSS Hazcard 14 Borax – toxic.

Heating different plastics

Equipment and materials required

Samples of a thermosoftening and thermosetting plastic (flammable and toxic fumes may be released), tin lid, tripod, Bunsen burner and safety equipment, glass rod, eye protection.

Details

Set up the Bunsen burner in a fume cupboard. Put a sample of each plastic on the same tin lid, and position it over the Bunsen burner on a tripod. Heat gently, and observe any changes. Then, as the thermosoftening plastic becomes pliable, touch it with a glass rod and gently pull away, drawing a thread of remoulded plastic. Then heat more strongly, and the thermoset should begin to combust. (Be aware that asthmatics may be affected by any fumes that escape.)

Structure and properties

C2 2.5 The properties of polymers

Learning objectives

- Do the properties of polymers depend on the monomers we use?
- Can changing reaction conditions modify the polymers that are made?
- What are thermosetting and thermosoftening polymers?

As you know, we can make **polymers** from chemicals made from crude oil. Small molecules called monomers join together to make much bigger molecules called polymers. As the monomers join together they produce a tangled web of very long chain molecules. Poly(ethene) is an example.

The properties of a polymer depend on:

- the monomers used to make it, and
- the conditions we choose to carry out the reaction.

a How are polymer chains arranged in poly(ethene)?

Different monomers

The polymer chains in nylon are made from two different monomers. One monomer has acidic groups at each end. The other has basic groups at each end. The polymer they make is very different from the polymer chains made from hydrocarbon monomers, such as ethene. So the monomers used make a big difference to the properties of the polymer made. (See Figures 1 and 2.)

Figure 1 The forces between the molecules in poly(ethene) are relatively weak as there are no strong covalent bonds (cross links) between the molecules. This means that this plastic softens fairly easily when heated.

Figure 2 Nylon is very much stronger than poly(ethene). This climber's life depends on nylon's high-tensile strength. Nylon can withstand large forces without snapping.

Different reaction conditions

There are two types of poly(ethene). One is called high density (HD) and the other low density (LD) poly(ethene). Both are made from ethene monomers but they are formed under different conditions.

Polymer chains
LD poly(ethene) or LDPE
HD poly(ethene) or HDPE

Figure 3 The branched chains of LD poly(ethene) cannot pack as tightly together as the straighter chains in HD poly(ethene), giving them different properties

Using very high pressures and a trace of oxygen, ethene forms LD poly(ethene). The polymer chains are branched and they can't pack closely together.

Using a catalyst at 50 °C and a slightly raised pressure, ethene makes HD poly(ethene). This is made up of straighter poly(ethene) molecules. They can pack more closely together than branched chains. The HD poly(ethene) has a higher softening temperature and is stronger than LD poly(ethene).

b What do 'LD' and 'HD' stand for in the names of the two types of poly(ethene)?

132

The properties of polymers

Thermosoftening and thermosetting polymers

We can classify polymers by looking at what happens to them when they are heated. Some will soften quite easily. They will reset when they cool down. These are called **thermosoftening polymers**. They are made up of individual polymer chains that are tangled together.

Other polymers do not melt when we heat them. These are called **thermosetting polymers**. These have strong covalent bonds forming 'cross links' between their polymer chains. (See Figure 4.)

The tangled web of polymer chains are relatively easy to separate
Chains fixed together by strong covalent bonds – this is called **cross linking**
Thermosoftening polymer
Thermosetting polymer

Figure 4 Extensive cross linking by covalent bonds between polymer chains makes a thermosetting plastic that is heat-resistant and rigid

Higher

Bonding in polymers

The atoms in polymer chains are held together by very strong covalent bonds. This is true for all plastics. But the size of the forces *between* polymer molecules in different plastics can be very different.

In thermosoftening polymers the forces between the polymer chains are weak. When we heat the polymer, these weak intermolecular forces are broken. The polymer becomes soft. When the polymer cools down, the intermolecular forces bring the polymer molecules back together. Then the polymer hardens again. This type of polymer can be remoulded.

However, thermosetting polymers are different. Their monomers make covalent bonds between the polymer chains when they are first heated in order to shape them. These covalent bonds are strong, and they stop the polymer from softening. The covalent 'cross links' between chains do not allow them to separate. Even if heated strongly, the polymer will still not soften. Eventually, the polymer will char at high enough temperatures.

Summary questions

1 Copy and complete using the words below:
covalent thermosetting tangled cross links
The polymer chains in a thermosoftening polymer form a .... web. The polymer softens at relatively low temperatures. Other polymers have strong .... bonds between their chains which form .... We call these .... polymers.

2 Why do we use thermosetting polymers to make plastic kettles?

3 Polymer A starts to soften at 100 °C while polymer B softens at 50 °C. Polymer C resists heat but eventually starts to char if heated to very high temperatures.
Explain this using ideas about intermolecular forces. [H]

Practical

Modifying a polymer

Take some PVA glue . . .
. . . add a few drops of borax solution
Warm solution of PVA glue
Stir well for about 2 minutes
Slime

The glue becomes slimy because the borax makes the long polymer chains in the glue link together to form a jelly-like substance.

- How could you investigate if the properties of slime depend on how much borax you add?

Figure 5 Electrical sockets are made out of thermosetting plastics. If the plug or wires get hot, the socket will not soften.

Key points

- Monomers affect the properties of the polymers that they produce.
- Changing reaction conditions can also change the properties of the polymer that is produced.
- Thermosoftening polymers will soften or melt easily when heated. Thermosetting polymers will not soften but will eventually char if heated very strongly.

133

Further teaching suggestions

Flow chart

- You could ask students to create a flow chart to explain how poly(ethene) is made. The flow chart should have one starting point as one type of monomer is used. However, it should branch to show how HDPE and LDPE are made. You could link back to C1 4.2 Fractional distillation and C1 5.1 Cracking to show where the ethene comes from.

Nylon rope trick

- You may wish to complete the nylon rope trick, see Royal Society of Chemistry Classic Experiments.

Poly(ethene)

- Poly(ethene) can be high density (HD) or low density (LD). Each structure has different properties and so different uses. The type of poly(ethene) can be distinguished by comparing whether the sample floats or sinks in 50 : 50 water and ethanol mixture, with LD floating and HD sinking. You may wish to give students a selection of different colours of poly(ethene) pellets or small pieces of the materials and ask them to classify it using this method. It is better to use pellets or small pieces rather than sheets to avoid surface tension effects.

Summary answers

1 tangled, covalent, crosslinks, thermosetting

2 They need to be heated to 100 °C (and possibly above in the case of electrical faults) without melting.

3 Polymer A has stronger intermolecular forces than polymer B. Polymer C has stronger forces than both A and B between its polymer chains. It has covalent cross links between its polymer chains.

C2 2.6 Nanoscience

Learning objectives

Students should learn:

- what nanoscience is
- the potential benefits and risks of nanoscience.

Learning outcomes

Most students should be able to:

- recall a definition of nanoscience
- list advantages and disadvantages of nanoscience.

Some students should also be able to:

- evaluate the benefits and drawbacks of using nanotechnology.

Specification link-up: Chemistry C2.2

- Nanoscience refers to structures that are 1–100 nm in size, of the order of a few hundred atoms. Nanoparticles show different properties to the same materials in bulk and have a high surface area to volume ratio, which may lead to the development of new computers, new catalysts, new coatings, highly selective sensors, stronger and lighter construction materials, and new cosmetics such as sun tan creams and deodorants. *[C2.2.6 a)]*
- Relate the properties of substances to their uses. *[C2.2]*
- Evaluate developments and applications of new materials, e.g. nanomaterials, fullerenes and shape memory materials. *[C2.2]*

Lesson structure

Starters

Show and tell – Place various products that use nanotechnology in a tray, e.g. sun cream, some sticking plasters, self-cleaning glass. Ask students to work in small groups and suggest the connection between the products. Then reveal to the students that these are all examples of the use of nanotechnology. *(5 minutes)*

Definition – Ask students to use the textbook to find the definitions of nanoscience [the science of really tiny things], nanometer [1×10^{-9} m] and nanoparticles [a particle between 1 and 100 nm]. Support students by giving them both the key words and definitions so that they match them up. Extend students by asking them to definition the key words in prose and a labelled diagram. *(10 minutes)*

Main

- Split the class into two groups. Explain to the students that they are going to have a debate on the motion 'Nanotechnology should be banned.' Ask half of the class to prepare the 'for' debate and the other half 'against'. You may wish to give group members particular roles to help them work effectively as a team, e.g. speaker, researcher, information recorder or leader. Move around the class and help the students prepare their arguments. You may wish to have a number of resources about nanotechnology around the room open at the right page. Bring the class together and chair the debate.
- Nanotechnology is a very new science. Ask students to research the subject area using secondary sources of information and then make a radio news report. You could record the best news reports as an MP3 file and use them on the virtual learning platform as a revision file.

Plenaries

Key points – Pick famous people (either in the wider world or the school community) and put their names into a bag. Ask for volunteers to come to pick out a name. They should then read the key points in character. *(5 minutes)*

What do you think? – Ask students to raise their hands if they are 'for' or 'against' nanotechnology. Support students by asking them to work as a class to make a summary table on the board of advantages and disadvantages of nanotechnology. This will distil out the main points before students commit themselves to their decision. Extend students by reading fictitious viewpoints, e.g. Mrs Smith, who has arthritis, thinks that self-cleaning glass is a miracle as she can't easily wash her windows. Ask students to reflect on their thoughts and suggest whether these comments change their minds and why. *(10 minutes)*

Support

- Give students a table with two columns: 'Advantages' and 'Disadvantages'. They could then be given statements and they sort the statements to help them consider the two sides of the argument.

Extend

- Ask students to find out about developments in the field of nanotechnology. Ask them to consider whether this research should be publicly funded or whether it should be funded through industry only.

Further teaching suggestions

Library

- Your school's library may have cut out newspaper articles about issues such as nanotechnology. You could complete a room swap to the library and use its resources to help students research.

Reflection

- At the start of the lesson, ask students to say or write down whether they think nanotechnology should be banned and why. Then allow them to research the subject, making a list of current and predicted uses. Then ask students to reanswer the question and say whether their minds have been changed and why.

Contrast table

- Students could prepare a table of advantages and disadvantages of nanotechnology. Split the class in half, and ask half of the class to stand in a circle facing outwards. Ask the rest of the class to make a circle around the first group of students, facing inwards. So each student is in a pair. Ask the inner circle to talk for one minute positively about nanotechnology. Then ask the outer circle to talk one minute only negatively about nanotechnology. Ask the students on the outer circle to move one place to the right. Then ask the inner circle students to tell the outer students their thoughts on nanotechnology for one minute and the outer circle does the same for one minute. This activity involves students developing their listening skills and reasoning skills.

Structure and properties

C2 2.6 Nanoscience

Learning objectives

- What is nanoscience?
- What are the potential benefits and risks associated with nanoscience?

Did you know ...?

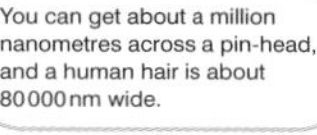

You can get about a million nanometres across a pin-head, and a human hair is about 80 000 nm wide.

Nanoscience is a new and exciting area of science. 'Nano' is a prefix like 'milli' or 'mega'. While 'milli' means 'one-thousandth', 'nano' means 'one thousand-millionth'.

1 nanometre (1 nm) = 1×10^{-9} metres (= 0.000 000 001 m or a billionth of a metre)

So nanoscience is the science of really tiny things. We are dealing with structures that are just a few hundred atoms in size or even smaller (between 1 and 100 nm in size).

We now know that materials behave very differently at a very tiny scale. Nanoparticles are so tiny that they have a huge surface area for a small volume of material. When we arrange atoms and molecules on a nanoscale, their properties can be truly remarkable.

a How many nanometres make up 1 millimetre?

Figure 1 Nanoparticles will save many people from damaged skin and cancers caused by too much UV light

Nanoscience at work

Here are some uses of nanoscience.

- Glass can be coated with titanium oxide nanoparticles. Sunshine triggers a chemical reaction which breaks down dirt which lands on the window. When it rains the water spreads evenly over the surface of the glass, washing off the broken down dirt.
- Titanium oxide and zinc oxide nanoparticles are also used in modern sun-screens. Scientists can coat nanoparticles of the metal oxide with a coating of silica. The thickness of the silica coating can be adjusted at an atomic level. These coated nanoparticles seem more effective at blocking the Sun's rays than conventional UV absorbers.
- The cosmetics industry is one of the biggest users of this new technology. The nanoparticles in face creams are absorbed deeper into the skin. They are also used in sun tan creams and deodorants.

Figure 2 Nanoparticles in cosmetic products can work deeper in the skin

The delivery of active ingredients in cosmetics can also be applied to medicines. The latest techniques being developed use nanocages of gold to deliver drugs where they need to go in the body. Researchers have found that the tiny gold particles can be injected and absorbed by tumours. Tumours have thin, leaky blood vessels with holes large enough for the gold nanoparticles to pass into. However, they can't get into healthy blood vessels.

When a laser is directed at the tumour the gold nanoparticles absorb energy and warm up. The temperature of the tumour increases enough to change the properties of its proteins but barely warms the surrounding tissue. This destroys the tumour cells without damaging healthy cells.

There is potential to use the gold nanocages to carry cancer-fighting drugs to the tumour at the same time. The carbon nanocages we met in C2 2.3 can also be used to deliver drugs in the body. Incredibly strong, yet light, nanotubes are already being used to reinforce materials (see Figure 3). The new materials are finding uses in sport, such as making very strong but light tennis racquets.

134

Nanoscience

Silver nanoparticles are antibacterial. They also act against viruses and fungi. They are used in sprays to clean operating theatres in hospitals.

Future developments?

Nanotubes are now being developed that can be used as nanowires. This will make it possible to construct incredibly small electronic circuits. Nanotubes can be used to make highly sensitive selective sensors. For example, nanotube sensors have been made that can detect tiny traces of a gas present in the breath of asthmatics before an attack. This will let patients monitor and treat their own condition without having to visit hospital to use expensive machines.

Nanowires would also help to make computers with vastly improved memory capacities and speeds.

Scientists in the US Army are developing nanotech suits – thin, or even spray-on, uniforms which are flexible and tough enough to withstand bullets and blasts. The uniforms would receive aerial views of the battlefield from satellites, transmitted directly to the soldier's brain. There would also be a built-in air conditioning system to keep the body temperature normal. Inside the suit there would be a full range of nanobiosensors that could send medical data back to a medical team.

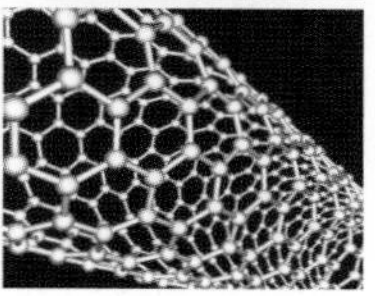

Figure 3 Nanocages can carry drugs inside them and nanotubes can reinforce materials

Possible risks

The large surface area of nanoparticles would make them very effective as catalysts. However, their large surface area also makes them dangerous. If a spark is made by accident, they may cause a violent explosion.

If nanoparticles are used more and more there is also going to be more risk of them finding their way into the air around us. Breathing in tiny particles could damage the lungs. Nanoparticles could enter the bloodstream this way, or from their use in cosmetics, with unpredictable effects. More research needs to be done to find out their effects on health and the environment.

b Why would nanoparticles make very efficient catalysts?

Activity

Whenever we are faced with a possible development in science there are two possible questions – what *can* we do? And what *should* we do? Look at the ideas about the uses of nanoscience and its future development here. Choose one idea and ask yourself 'what *can* we do and what *should* we do?' Present your answers to the rest of your group

Summary questions

1 What do we mean by 'nanoscience'?

2 In his book *Engines of Creation* K. Eric Drexler speculates that one day we may invent a nanomachine that can reproduce itself. Then the world could be overrun by so-called 'grey goo'. Some people are so worried they have called for a halt in nanoscience research. What are your views?

Key points

- Nanoscience is the study of small particles that are between 1 and 100 nanometres in size.
- Nanoparticles behave differently from the materials they are made from on a large scale.
- New developments in nanoscience are very exciting but will need more research into possible issues that might arise from their increased use.

135

Answers to in-text questions

a A million.

b They have a very large surface area.

Summary answers

1 The science of the very tiny – studying structures between 1 and 100 nm in size.

2 Students present points for and against in a balanced argument, then draw their own conclusion.

Summary answers

1 a D b A c B d C

2 a Table showing:
Giant covalent: graphite, silicon dioxide.
Giant ionic: magnesium oxide, lithium chloride.
Simple molecular: ammonia, hydrogen bromide.
Giant metallic: cobalt, manganese.

b Ammonia, hydrogen bromide.

c Graphite – it conducts electricity even though it is a giant covalent compound, because the structure contains delocalised electrons.

3 **Example of an advantage** – it can be restored to its original shape if the frame is accidently damaged.
Example of a disadvantage – it will be more expensive than a conventional frame.

4 The ionic compound is sealed in a container with two electrodes and placed in the reactor. It is connected into an electrical warning circuit. If the temperature reaches 800 °C in the reactor the compound will melt, conducting electricity and activating the alarm.

5 Metals and graphite conduct electricity because both contain delocalised (free) electrons that can flow and carry an electric current. Layers of atoms in graphite are held together only weakly by these delocalised electrons, so the layers are able to slide easily over each other, making graphite soft. Although layers of atoms can slide over each other in metals, this does not happen so easily and metals are therefore hard.

6 a i Very small or smaller (not 'small' alone) or a few atoms thick or in range 1 to 100 nm

ii Sensible idea of passing through smaller gaps, e.g. skin/pores/cells or more easily absorbed.

b Any two from: good at absorbing UV light/radiation, spread more easily, cover better, save money or use less, transparent, less chance of getting skin cancer or stops skin cancer.

c toxic (to cells or specific cells), allow harm/damage/kill cells

Structure and properties: C2 2.1–C2 2.6

Summary questions

1 Match the sentence halves together:

a	Ionic compounds have	A	.... conduct electricity when molten or in solution.
b	Ionic compounds	B	.... held together by strong electrostatic forces.
c	The oppositely charged ions in an ionic compound are	C	.... a giant lattice of ions.
d	Ionic compounds are made of	D	.... high melting points.

2 The table contains data about some different substances:

Substance	Melting point (°C)	Boiling point (°C)	Electrical conductor
cobalt	1495	2870	Good
ammonia	−78	−33	Poor
magnesium oxide	2852	3600	solid – poor liquid – good
manganese	1244	1962	Good
lithium chloride	605	1340	solid – poor liquid – good
silicon dioxide	1610	2230	Poor
hydrogen bromide	−88	−67	Poor
graphite	3652	4827	Good

a Make a table with the following headings:
Giant covalent, Giant ionic, Simple molecules,
Giant metallic.
Now write the name of each substance above in the correct column.

b Which substances are gases at 20 °C?

c One of these substances behaves in a slightly different way than its structure suggests. Why?

3 One use of shape memory alloys is to make spectacle frames. Write down **one** advantage and **one** disadvantage of using a shape memory alloy like this.

4 A certain ionic compound melts at exactly 800°C. A chemical company wants to design a device to activate a warning light and buzzer when the temperature in a chemical reactor rises above 800°C. Suggest how this ionic compound could be used in an alarm.

5 'Both graphite and metals can conduct electricity – but graphite is soft while metals are not.' Use your knowledge of the different structures of graphite and metals to explain this statement. [H]

6 Read the article about the use of nanoparticles in sun creams.

Sun creams

Many sun creams use nanoparticles. These sun creams are very good at absorbing radiation, especially ultraviolet radiation. Owing to the particle size, the sun creams spread more easily, cover better and save money because you use less. The new sun creams are also transparent, unlike traditional sun creams which are white. The use of nanoparticles is so successful that they are now used in more than 300 sun cream products.

Some sun creams contain nanoparticles of titanium oxide. Normal-sized particles of titanium oxide are safe to put on the skin.

It is thought that nanoparticles can pass through the skin and travel around the body more easily than normal-sized particles. It is also thought that nanoparticles might be toxic to some types of cell, such as skin, bone, brain and liver cells.

a i How is the size of nanoparticles different from normal-sized particles of titanium oxide?

ii Suggest how the size of nanoparticles might help them to enter the body more easily.

b Give **two** advantages of using nanoparticles in sun creams.

c Why might nanoparticles be dangerous inside the body?

AQA, 200

136

Practical suggestions

Practicals	AQA	k		
Demonstration of heating sulfur and pouring it into cold water to produce plastic sulfur.	✓		✓	
Investigating the properties of ionic compounds, e.g. NaCl: – melting point, conductivity, solubility, use of hand lens to study crystal structure.	✓	✓	✓	
Investigating the properties of covalent compounds: – simple molecules, e.g. wax, methane, hexane – macromolecules, e.g. SiO_2 (sand).	✓	✓	✓	
Investigating the properties of graphite.	✓		✓	
Demonstrations involving shape memory alloys.	✓		✓	
Investigating the properties of metals and alloys: – melting point and conductivity, hardness, tensile strength, flexibility – using models, for example using expanded polystyrene spheres or computer animations to show how layers of atoms slide – making metal crystals by displacement reactions, e.g. copper wire in silver nitrate solution.	✓		✓	
Distinguishing between LD and HD poly(ethene) using 50 : 50 ethanol : water.	✓		✓	
Making slime using different concentrations of poly(ethenol) and borax solutions.	✓		✓	
Investigating the effect of heat on polymers to find which are thermosoftening or thermosetting.	✓		✓	

AQA Examination-style questions

Match each of the substances in the table with a description from the list.

giant covalent ionic metal simple molecule

Substance	Formula	Melting point (°C)	Boiling point (°C)	Does it conduct electricity when liquid?
a	C	3550	4830	No
b	Co	1768	3142	Yes
c	CH_4	−182	−164	No
d	$CaCl_2$	1055	1873	Yes

(4)

Copper can be hammered into shape.

The structure of copper metal can be represented as shown:

a Explain why copper can be hammered into shape. (1)

b Copper can be mixed with zinc to make the alloy called brass. Brass is much harder than copper. Explain why. (2)

c Copper can be mixed with zinc and aluminium to make a shape memory alloy. What is a shape memory alloy? (2)

Choose a word from the list to complete each sentence.

different identical smart thermosoftening thermosetting

The polymers low-density poly(ethene) (LDPE) and high-density poly(ethene) (HDPE) are made from monomers that are .... The polymers are produced using catalysts and reaction conditions that are .... LDPE melts at 120°C and HDPE melts at 130°C and they have no cross links between the polymer chains so they are both .... polymers. (3)

Chloroethene, C_2H_3Cl, can be polymerised to poly(chloroethene).

a Explain in terms of its structure why chloroethene is a gas at room temperature. (2)

b Explain in terms of its structure why poly(chloroethene) is a thermosoftening polymer. (2)

[H]

5 The picture shows a copper kettle being heated on a camping stove.

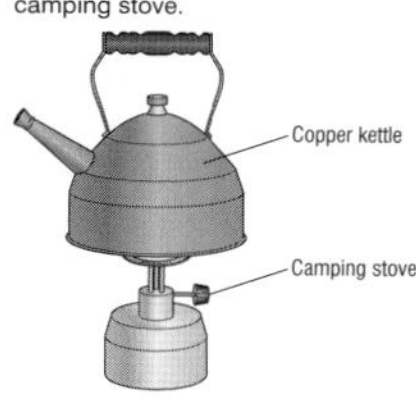

a *In this question you will be assessed on using good English, organising information clearly and using specialist terms where appropriate.*

Copper is a good material for making a kettle because it has a high melting point.

Explain why copper, like many other metals, has a high melting point.

You should describe the structure and bonding of a metal in your answer. (6)

b An aeroplane contains many miles of electrical wiring made from copper. This adds to the mass of the aeroplane.

It has been suggested that the electrical wiring made from copper could be replaced by lighter carbon nanotubes.

The diagram shows the structure of a carbon nanotube.

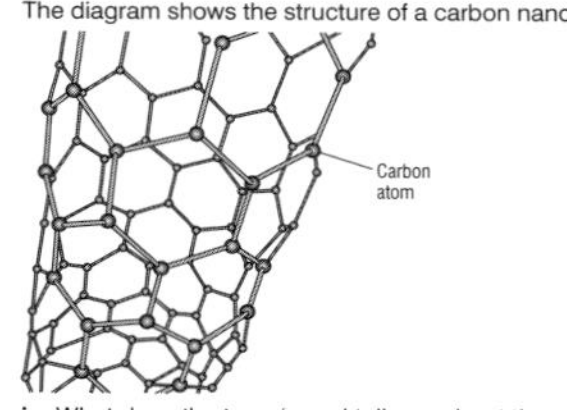

i What does the term 'nano' tell you about the carbon nanotubes? (1)

ii Like graphite, each carbon atom is joined to three other carbon atoms. Explain why the carbon nanotube can conduct electricity. (2)

AQA, 2010

Kerboodle resources

Resources available for this chapter on Kerboodle are:

- Chapter map: Structure and properties
- How Science Works: It's all about the ions (C2 2.1)
- Practical worksheet: How do the structure and bonding of a substance influence its properties? (C2 2.3)
- Support: Bonds and properties (C2 2.4)
- Extension: Bonds and properties (C2 2.4)
- How Science Works: In the slime (C2 2.5)
- Extension WebQuest: Nanotechnology (C2 2.6)
- Viewpoint: Is small better? (C2 2.6)
- Interactive activity: Structure and properties
- Revision podcast: Giant covalent structures of carbon
- Test yourself: Structure and properties
- On your marks: Structure and properties
- Examination-style questions: Structure and properties
- Answers to examination-style questions: Structure and properties

Examination-style answers

1 a giant covalent *(1 mark)*

b metal *(1 mark)*

c simple molecule *(1 mark)*

d ionic *(1 mark)*

2 a Layers of atoms can slide/move over each other. *(1 mark)*

b Zinc atoms are a different size from copper atoms they disrupt the structure, making it more difficult/less able to slide/move. *(2 marks)*

c It returns to its original shape, after being deformed or when the temperature changes or when it is heated. *(2 marks)*

3 identical, different, thermosoftening *(3 marks)*

4 a Simple molecules or small molecules, weak intermolecular forces or weak attractions between molecules. Therefore it has a low boiling point. *(2 marks)*

b There are no bonds/cross-links between polymer molecules or there are weak intermolecular forces/attractions between (polymer) molecules, these weak forces are overcome when the polymer is heated and so it softens/melts. *(2 marks)*

5 a Marks awarded for this answer will be determined by the Quality of Written Communication (QWC) as well as the standard of the scientific response.

There is a clear and detailed scientific description of why copper has a high melting point including details of the structure and bonding in a metal. The answer shows almost faultless spelling, punctuation and grammar. It is coherent and in an organised, logical sequence. It contains a range of appropriate and relevant specialist terms used accurately. *(5–6 marks)*

The answer has some structure and the use of specialist terms has been attempted, but not always accurately. There may be some errors in spelling, punctuation and grammar. There is a scientific description. There are some errors in spelling, punctuation and grammar. The answer has some structure and organisation. The use of specialist terms has been attempted, but not always accurately. *(3–4 marks)*

There is a brief description of the structure of a metal. The spelling, punctuation and grammar are very weak. The answer is poorly organised with almost no specialist terms and/or their use demonstrating a general lack of understanding of their meaning. *(1 mark)*

No relevant content. *(0 marks)*

Examples of chemistry points made in the response:

- giant structure/lattice
  atoms arranged in a regular pattern or in layers
- sea of electrons or delocalised electrons or free electrons
- awareness that outer shell/highest energy level electrons are involved
- positive ions
- (electrostatic) attractions/bonds between electrons and positive ions
- bonds/attractions (between atoms or ions and electrons) are strong
- a lot of energy/heat is needed to break these bonds/attractions.

b i The tubes are very small (not 'small' on is own), or a few atoms across, or they are 1–100 nm across. *(1 mark)*

ii Three electrons in carbon's outer shell are used in bonding, which leaves one electron free to move around or delocalise and carry current or charge. *(2 marks)*

C2 3.1 The mass of atoms

Learning objectives

Students should learn:

- what atomic number and mass number mean
- the relative masses of subatomic particles
- what isotopes are.

Learning outcomes

Most students should be able to:

- define atomic number and mass number
- use the periodic table to get atomic numbers and mass numbers for any atom and work out the number of each subatomic particle that an atom has
- state a definition of isotopes.

Some students should also be able to:

- compare the physical and chemical properties of isotopes of an element.

Specification link-up: Chemistry C2.3

- Atoms can be represented as shown in this example:
  Mass number 23
  Na
  Atomic number 11 *[C2.3.1 a)]*
- The relative masses of protons, neutrons and electrons … . *[C2.3.1 b)]*
- The total number of protons and neutrons in an atom is called its mass number. *[C2.3.1 c)]*
- Atoms of the same element can have different numbers of neutrons; these atoms are called isotopes of that element. *[C2.3.1 d)]*

Lesson structure

Starters

True or false – If students agree with these statements they should show a thumbs up sign. If they disagree, their thumbs should point downwards, and if they don't know their thumbs should be horizontal.

- Atoms are charged particles. [False]
- Atoms contain charged particles. [True]
- Electrons are found in energy levels or shells. [True]
- Electrons have a negative charge. [True]
- Protons are in the nucleus of an atom. [True]
- Neutrons are found in shells around the nucleus. [False]

Support students by putting a labelled diagram of the atom on the board for them to refer to. Extend students by asking them to reword the false statements, so that they are true. *(5 minutes)*

Definitions – Give the students the definitions on the board. Students should match each definition with a key word:

- A positive particle in an atom's nucleus. [Proton]
- A neutral particle with a relative mass of 1. [Neutron]
- The subatomic particle that is found in energy levels. [Electron]
- The number of protons in an atom. [Atomic number or Proton number]
- The number of protons and neutrons in an atom. [Mass number]

Students could then be asked to draw a diagram of an atom and label it to demonstrate all these key terms. *(10 minutes)*

Main

- The structure of the atom can be summarised in a spider diagram. Encourage the students to include information about atomic number, mass number, subatomic particles. Students first met the terms 'mass number' and 'atomic number', as well as the structure of the atom in C1 Chapter 1.
- Later in the lesson, this spider diagram could be extended to include isotopes and uses of isotopes.
- Students could be asked to include a key in their diagram. For example, use specific colours for key terms: red – proton, green – neutron, blue – electron, orange – atom, purple – isotope.
- Students need to be able to use the periodic table to work out the number of each subatomic particle in an atom. With a question and answer session, draw out how the periodic table can be used to supply information about an atom.
- Show the students how to calculate the number of each subatomic particle using mass number and proton number.
- Then ask the students to design a table to record the number of each subatomic particle in the first 20 elements. Then ask them to complete their table.
- Introduce the students to isotopes with the examples given in the Student Book.

Plenaries

Association – Split the class into pairs. Students should face their partners. All students on the right should start first, saying a word or phrase about the lesson. Then the next person says another fact/key word based on the lesson. The activity swaps between partners until all the facts are exhausted. If a student hesitates or repeats previous statements, then they have lost the association game. *(5 minutes)*

Support

- Some students will struggle with the vocabulary used in this topic area. Write a list of key words on to the board so they can use these throughout the lesson. You may wish to further support them by also having the definitions next to each word.

Extend

- Ask students to draw the structure of the two chlorine isotopes (^{35}Cl and ^{37}Cl). They could then try to explain why the relative atomic mass of chlorine is 35.5.

Think – On the board, write the symbol for carbon-12 and carbon-14 isotopes. Ask the students to use the periodic table and their knowledge of isotopes to list all the similarities between the atoms and all their differences. [Similarities – same number and arrangement of electrons, same number of protons in the nucleus, same chemical properties, same atomic number. Differences – different number of neutrons in the nucleus, different mass numbers, different physical properties.] You could support students by using hydrogen isotopes and having the structure, including the subatomic particles of the three isotopes on the board. Extend students by asking them to draw the structures of the isotopes of carbon. They could be asked to find out about the third isotope of carbon: carbon-13 and also draw this structure. *(10 minutes)*

Answers to in-text questions

a Number of protons.

b Mass of proton = mass of neutron.

c Mass of electron is much less than the mass of a proton or neutron.

d Number of neutrons = mass number minus atomic number.

e Isotopes are atoms of the same element with different numbers of neutrons (or words to that effect).

f Tritium, $^{3}_{1}H$.

Further teaching suggestions

Isotope article

- Students could use the internet to research some uses of isotopes, e.g. in medicine. Students could then use this information to make a magazine article for use in a teenage science magazine.

Heavy water

- Students could research 'heavy water', D_2O, and how its properties differ from normal water. Students should find out that isotopes have the same chemical properties because they have the same number and arrangement of electrons, but different physical properties because their masses are different.

Isotope pairs

- Students could find an example of an isotope pair. Ask them to write down the symbols, with their proton and mass numbers and state the numbers of subatomic particles in the atoms of each isotope.

How much?

C2 3.1 The mass of atoms

Learning objectives

- What is an atom's atomic number and mass number?
- What are the relative masses of protons, neutrons and electrons?
- What are isotopes?

Did you know ...?

It would take 1836 electrons to have the same mass as a single proton.

As you know, an atom consists of a nucleus containing positively charged protons, together with neutrons which have no charge. The negatively charged electrons are arranged in energy levels (shells) around the nucleus.

Every atom has the same number of electrons orbiting its nucleus as it has protons in its nucleus. The number of protons in an atom is called its atomic number.

The mass of a proton and a neutron is the same. This means that the relative mass of a neutron compared with a proton is 1. Electrons are much, much lighter than protons and neutrons. Because of this, the mass of an atom is concentrated in its nucleus. We can ignore the tiny mass of the electrons when we work out the relative mass of an atom.

Type of subatomic particle	Relative mass
Proton	1
Neutron	1
Electron	very small

a What is the atomic number of an atom?
b How does the mass of a proton compare with the mass of a neutron?
c How does the mass of an electron compare with the mass of a neutron or proton?

Mass number

Almost all of the mass of an atom is in its nucleus. This is because the mass of the electrons is so tiny. We call the total number of protons and neutrons in an atom its mass number.

We can show the atomic number and mass number of an atom like this:

Mass number → $^{12}_{6}C$ (carbon) $^{23}_{11}Na$ (sodium)
Atomic number

We can work out the number of neutrons in the nucleus of an atom by subtracting its atomic number from its mass number:

number of neutrons = mass number – atomic number

For the two examples above, carbon has 6 protons and a mass number of 12.

So the number of neutrons in a carbon atom is (12 − 6) = 6.

Sodium has an atomic number of 11 and the mass number is 23.

So a sodium atom has (23 − 11) = 12 neutrons. In its nucleus there are 11 protons and 12 neutrons.

d How do we calculate the number of neutrons in an atom?

$^{12}_{6}C$

Proton — Number of protons gives atomic number
Neutron — Number of protons plus number of neutrons gives mass number

Figure 1 An atom of carbon

138

The mass of atoms

Isotopes

Atoms of the same element always have the same number of protons. However, they can have different numbers of neutrons.

We give the name **isotopes** to atoms of the same element with different numbers of neutrons.

Isotopes always have the same atomic number but different mass numbers. For example, carbon has two common isotopes, $^{12}_{6}C$ (carbon-12) and $^{14}_{6}C$ (carbon-14). The carbon-12 isotope has 6 protons and 6 neutrons in the nucleus. The carbon-14 isotope has 6 protons and 8 neutrons.

Sometimes the extra neutrons make the nucleus unstable, so it is radioactive. However, not all isotopes are radioactive – they are simply atoms of the same element that have different masses.

e What are isotopes?

Samples of different isotopes of an element have different *physical* properties. For example, they have a different density and they may or may not be radioactive. However, they always have the same *chemical* properties. That's because their reactions depend on their electronic structure. As their atoms will have the same number of electrons, the electronic structure will be same for all isotopes of an element.

For example, hydrogen has three isotopes: hydrogen, deuterium and tritium (see Figure 2). Each has a different mass and tritium is radioactive. However, they can all react with oxygen to make water.

f Which isotope of hydrogen is heaviest?

$^{1}_{1}H$ Hydrogen
$^{2}_{1}H$ Deuterium
$^{3}_{1}H$ Tritium

Figure 2 The isotopes of hydrogen – they have identical chemical properties but different physical properties

Summary questions

1 Copy and complete using the words below:

electrons isotopes protons mass atomic one

The number of protons in an atom is called its .... number. The relative mass of a neutron compared with a proton is .... Compared with protons and neutrons .... have almost no mass. The total number of .... and neutrons in an atom is called its .... number. Atoms of an element which have different numbers of neutrons are called ....

2 State how many protons there would be in the nucleus of each of the following elements:
a **i** $^{9}_{4}Be$ **ii** $^{16}_{8}O$ **iii** $^{22}_{10}Ne$ **iv** $^{31}_{15}P$ **v** $^{79}_{35}Br$.
b State how many neutrons each atom in part **a** has.

3 **a** How do the physical properties of isotopes of the same element vary?
b Why do isotopes of the same element have identical chemical properties?

Key points

- The relative mass of protons and neutrons is 1.
- The atomic number of an atom is its number of protons (which equals its number of electrons).
- The mass number of an atom is the total number of protons and neutrons in its nucleus.
- Isotopes are atoms of the same element with different numbers of neutrons.

139

Summary answers

1 atomic, one, electrons, protons, mass, isotopes

2 **a** **i** 4 **ii** 8 **iii** 10 **iv** 15 **v** 35
b **i** 5 **ii** 8 **iii** 12 **iv** 16 **v** 44

3 **a** The atoms have a different mass/different density and they may be radioactive.
b They have identical electronic structures which governs their reactivity.

C2 3.2 Masses of atoms and moles

Learning objectives

Students should learn:

- that the masses of atoms can be compared by their relative atomic masses
- that carbon-12 is used as a standard to measure relative atomic masses **[HT only]**
- that the relative formula mass of compounds can be calculated.

Learning outcomes

Most students should be able to:

- give a definition of relative formula mass
- calculate relative formula mass if the formula and the relative atomic mass are given
- state what a mole is.

Some students should also be able to:

- give a full definition of relative atomic mass. **[HT only]**

Specification link-up: Chemistry C2.3

- The relative atomic mass of an element (A_r) compares the mass of atoms of the element with the ^{12}C isotope. It is an average value for the isotopes of the element. *[C2.3.1 e)]* **[HT only]**
- The relative formula mass (M_r) of a compound is the sum of the relative atomic masses of the atoms in the numbers shown in the formula. *[C2.3.1 f)]*
- The relative formula mass of a substance, in grams, is known as one mole of that substance. *[C2.3.1g)]*

Lesson structure

Starters

Demonstration – Have a mole of different elements premeasured in sealed containers, e.g. 12 g of carbon, 24 g of magnesium. Allow the students to handle different samples. Explain to these students that all these examples have something in common – but what? Encourage the students to use the Student Book and discuss in small groups how these samples relate to each other. They should realise that these are all examples of moles. Then ask students to raise their hands if they can tell you the mass of, e.g. 1 mole of carbon (and hold up the sample), etc. Repeat for all the samples that you have. *(5 minutes)*

Word search – Give the students a word search for the key words that they will be using in the lesson: 'relative', 'atomic', 'mass', 'formula', 'mole', 'atom'. Instead of just asking the students to find the words, give clues. Support students by giving them both the clues and the key words. They can then match them up before finding the key words in the word search. Extend students by asking them to write a definition for each of the key words. *(10 minutes)*

Main

- Higher Tier students should understand that the masses of atoms are compared to $\frac{1}{12}$ of the mass of a carbon-12 isotope. It is also worth noting that many periodic tables give the mass number of the most common isotope rather than the relative atomic mass based on natural proportions of different isotopes.
- Students need to be able to obtain A_r from the periodic table or a data book and calculate M_r. Give the students a set of cards showing the symbols of different elements (single atoms and molecules) and compound formulas. Write numbers that represent A_r or M_r on separate cards.
- Students should complete calculations and match the formula with its A_r and M_r. They should also decide if the number represents A_r or M_r. This could be made into a competition, by splitting the class into small teams.
- Students need to be able to define the key terms: 'relative atomic mass' [HT only for full definition including the use of carbon-12 as a standard], 'relative formula mass' and 'moles'.
- Give the students a template of a cube. They should write definitions of relative atomic mass, moles and relative formula mass on three faces. They should include a worked example of calculations relating to this topic on each of the remaining faces, using lots of colours. Then they cut out the template and score the lines to create sharp folds and stick the cube together.

Plenaries

Difference – Ask the students to explain the difference between the symbols A_r and Ar. Choose a volunteer to explain to the class. [Ar is the symbol for the element argon, A_r is the shorthand notation for relative atomic mass]. *(5 minutes)*

In the bag – Put the key words: 'relative atomic mass', 'relative formula mass' and 'mole' into a colourful bag. Ask for three volunteers to come to the front and remove a word in turns. After they have removed their word, they should show the class the word and explain what it means. You should interject with a question-and-answer session to help rectify any misconceptions. Support students by giving them a list of definitions on the board: the student chooses a word and then looks at the definition, picking the one they think the word matches. Extend students by listing words under the key word that they cannot use in their explanation. *(10 minutes)*

Support

- Provide students with a precut-out cube template that is also ready scored. To help further, add some information already on it, for example the titles or prose with missing words.

Extend

- Ask students to work out the formulae of compounds (either by using the Student Book or using dot and cross diagrams). Then encourage them to use the periodic table to get the A_r before working out the M_r.

Further teaching suggestions

Relative formula mass

- Give students a set of timed questions to calculate relative formula masses of different substances. You could use some past paper questions that are laminated with the mark scheme on the back. Students can then use dry-wipe pens and erasers to complete the question and mark their own answers within a time limit set by the teacher (approximately 1 minute per mark and one-third of this time to mark their answers).

Mole day

- Celebrate Mole Day: there are lots of resources on www.moleday.org.

Calculations

- Ask students to work out the mass of a mole of the following:
  - Oxygen atoms [16 g]
  - Oxygen molecules [32 g]
  - Water molecules [18 g]

  Encourage the students to show their working.

Study tip

When talking about moles of elements, be careful to specify moles of atoms or moles of molecules. For compounds, it is more complicated. Explain that a mole of a compound is related to its formula. If it is molecular, it will contain one mole of molecules. However, if it is ionic it does not consist of molecules, and it does not contain one mole of particles. For example, one mole of magnesium chloride, $MgCl_2$, contains one mole of magnesium ions and two moles of chloride ions (three moles of particles).

How much?

C2 3.2 Masses of atoms and moles

Learning objectives

- How can we compare the masses of atoms?
- What is the relative atomic mass of an element? [H]
- How can we calculate the relative formula mass of a compound from the elements it is made of?

Balanced symbol equations show us how many atoms of reactants we need to make the products. But when we actually carry out a reaction we really need to know how much to use in grams or cm^3.

For example, look at the equation:

$$Mg + 2HCl \rightarrow MgCl_2 + H_2$$

The symbol equation tells us that we need twice as many hydrogen and chlorine atoms as magnesium atoms. However, this doesn't mean that the mass of HCl will be twice the mass of Mg. This is because atoms of different elements have different masses.

To make symbol equations useful in the lab or factory we need to know more about the mass of atoms.

a Why don't symbol equations tell us directly what mass of each reactant to use in a chemical reaction?

Relative atomic masses

The mass of a single atom is so tiny that it would not be practical to use it in experiments or calculations. So instead of working with the real masses of atoms we just focus on the relative masses of different elements. We call these **relative atomic masses (A_r)**.

Relative atomic mass

Higher

We use an atom of carbon-12 ($^{12}_{6}C$) as a standard atom. We give this a 'mass' of exactly 12 units, because it has 6 protons and 6 neutrons. We then compare the masses of atoms of all the other elements with this standard carbon atom. For example, hydrogen has a relative atomic mass of 1 as most of its atoms have a mass that is one-twelfth of a $^{12}_{6}C$ atom.

The relative atomic mass of an element is usually the same as, or similar to, the mass number of its most common isotope. The A_r takes into account the proportions of any isotopes of the element found naturally. So it is an *average* mass compared with the standard carbon atom. (This is why chlorine has a relative atomic mass of 35.5, although we could never have half a proton or neutron in an atom.)

b Which atom do we use as a standard to compare relative masses of elements?

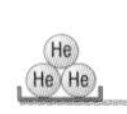

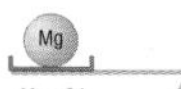

Figure 1 The relative mass of $^{12}_{6}C$ atom is 12. Compared with this, the A_r of helium is 4 and the A_r of magnesium is 24.

Relative formula masses

We can use the A_r of the various elements to work out the **relative formula mass (M_r)** of compounds. This is true whether the compounds are made up of molecules or collections of ions. A simple example is a substance like sodium chloride. We know that the A_r of sodium is 23 and the A_r of chlorine is **35.5**. So the relative formula mass of sodium chloride (NaCl) is:

$$\underset{A_r:Na}{23} + \underset{A_r:Cl}{35.5} = \underset{M_r:NaCl}{\mathbf{58.5}}$$

Another example is water. Water is made up of hydrogen and oxygen. The A_r of hydrogen is 1, and the A_r of oxygen is 16. Water has the formula H_2O. It contains two hydrogen atoms for every one oxygen, so the M_r is:

$$\underset{A_r:H \times 2}{(1 \times 2)} + \underset{A_r:O}{16} = \underset{M_r:H_2O}{\mathbf{18}}$$

c What is the relative formula mass of hydrogen sulfide, H_2S? (A_r values: H = 1, S = 32)

We can use the same approach with relatively complicated molecules like sulfuric acid, H_2SO_4. Hydrogen has a A_r of 1, the A_r of sulfur is 32 and the A_r of oxygen 16. This means that the M_r of sulfuric acid is:

$$(1 \times 2) + 32 + (16 \times 4) = 2 + 32 + 64 = \mathbf{98}$$

Moles

Saying or writing 'relative atomic mass in grams' or 'relative formula mass in grams' is rather clumsy. So chemists have a shorthand word for it: a **mole**.

They say that the relative atomic mass in grams of carbon (i.e. 12 g of carbon) is a mole of carbon atoms. One mole is simply the relative atomic mass or relative formula mass of any substance expressed in grams. A mole of any substance always contains the same number of atoms, molecules or ions. This is a huge number (6.02×10^{23}).

Summary questions

1 Copy and complete using the words below:

atom elements formula relative

The mass of an individual .... is so small that we use .... values when comparing them. We calculate the relative .... mass of a compound by adding up the relative atomic masses of its .... in the ratio given by its formula.

2 The equation for the reaction of calcium and fluorine is:

$$Ca + F_2 \rightarrow CaF_2.$$

a How many moles of fluorine molecules react with one mole of calcium atoms?

b What is the relative formula mass of CaF_2? (A_r values: Ca = 40, F = 19)

3 The relative atomic mass of helium is 4, and that of sulfur is 32. How many times heavier is a sulfur atom than a helium atom?

4 Define the term 'relative atomic mass' of an element. [H]

Did you know ...?

If you had as many soft drink cans as there are atoms in a mole they would cover the surface of the Earth to a depth of 200 miles!

AQA *Examiner's tip*

You don't have to remember the number 6.02×10^{23} or the relative atomic masses of elements. But practise calculating the mass of one mole of different substances from their formula and the relative atomic masses that you are given.

Key points

- We compare the masses of atoms by measuring them relative to atoms of carbon-12. [H]
- We work out the relative formula mass of a compound by adding up the relative atomic masses of the elements in it, in the ratio shown by its formula.
- One mole of any substance is its relative formula mass, in grams.

Answers to in-text questions

a The atoms of different elements have different masses.

b $^{12}_{6}C$/carbon-12

c 34

Summary answers

1 atom, relative, formula, elements

2 **a** 1

b 78

3 Eight times heavier.

4 The relative atomic mass of an element (A_r) is a comparison of the mass of atoms of the element with the ^{12}C isotope, taking into account the proportions of its naturally occurring isotopes.

C2 3.3 Percentages and formulae

Learning objectives

Students should learn:

- how to calculate the percentage of an element in a compound from its formula
- how to calculate the empirical formula of a compound from its percentage composition. **[HT only]**

Learning outcomes

Most students should be able to:

- calculate the percentage composition of an element in a compound.

Some students should also be able to:

- calculate the empirical formula of a compound if the percentage composition of the elements is given. **[HT only]**

Specification link-up: Chemistry C2.3

- The percentage of an element in a compound can be calculated from the relative mass of the element in the formula and the relative formula mass of the compound. *[C2.3.3 a)]*
- The empirical formula of a compound can be calculated from the masses or percentages of the elements in a compound. *[C2.3.3 b)]* **[HT only]**

Lesson structure

Starters

Measurement – Show the students different mass values on balances. Ask them to note the values shown to the nearest two decimal places. Read out the answers and ask the students to put up their hands if they got them all right, one wrong, two wrong, etc. Approach students who have been having difficulty and help them to see why they recorded the wrong answer. Support students by a worked example on the board to help them truncate data. Extend students by showing them different methods of measuring mass (bathroom scales, kitchen scales, top pan balance, accurate level balance) and using these to measure the mass of the same item. Ask students to comment on reliability and accuracy of the results and to suggest why truncation is useful. *(5 minutes)*

Concept cartoon – Show the students a concept cartoon to highlight the conservation of mass theory. Ask them to discuss the cartoon in small groups. Then feed back to the rest of the class. Address any misconceptions at this stage. *(10 minutes)*

Main

- Calculations often involve a set order of steps. Choose an example question to calculate the percentage composition of an element and write out each step onto separate cards.
- Encourage the students to order the cards and copy out the worked example correctly. Then give the students other examples to work out themselves.
- The same idea can be repeated with Higher Tier students, but for generating a formula of a compound from percentage composition data can be used instead. **[HT only]**
- Give the students two flash cards. On one side of the first card they should write how to calculate the percentage composition of an element in a compound. On the reverse, they should make up some questions of their own.
- Higher Tier students should write how to generate the formula on one side of the second card, and some questions on the reverse. **[HT only]**
- The students can then work out the answers and write them upside down on their revision card.
- In pairs, students swap their revision cards, tackle each other's questions and check their answers. They should be encouraged to feed back any problems that they are having to their partner.
- Higher Tier students could experimentally determine the formula of magnesium oxide. This requires them to be able to read a balance accurately to two decimal places. They should design their own results table and show all their working to generate the formula from their experimental data. They should calculate the mass of magnesium used (mass of initially full crucible minus mass of empty crucible). Then they should calculate the mass of oxygen in the compound (mass of crucible after heating minus mass of initially full crucible). **[HT only]**
- Once the mass of each element in the compound is known, the formula can be calculated as detailed in the Student Book. **[HT only]**

Plenaries

Calculation – Ask the students to calculate the percentage composition of each element in ammonia. They should recall the formula of ammonia from previous work. [N = 82 per cent; H = 18 per cent.] Support students by giving them the formula of ammonia and the A_r for H and N. Extend students by asking them to calculate the mass of nitrogen in 25 g of ammonia. [20.5 g.] *(5 minutes)*

On the spot – Ask for a volunteer to stand at the front of the class, and give the other students scrap paper. Read out questions to the volunteer, who should give an answer. The other students then write numbers and hold them up to show how strongly they agree with the answer. (If they strongly disagree they hold up 1; if they strongly agree they hold up 10, for example.) You could reveal the true answer. *(10 minutes)*

Support

- Some students will struggle with calculations. These students will benefit from having a worked example that they can refer to with the steps explicitly shown. Thought processes could be written in prose as reminders. Provide a writing frame that follows the worked example can also help to formalise the working.

Extend

- Give students more complex empirical formula questions to attempt, e.g. P_2O_5.

Practical support

Determining the formula of magnesium oxide

Equipment and materials required

Small strips of magnesium ribbon (less than 2 cm length), ceramic crucibles and lids, Bunsen burner and safety equipment, tongs, pipeclay triangle, tripod, accurate balance, eye protection.

Details

Students should note the mass of the crucible and lid. They should twist the magnesium ribbon into a coil shape and put it into the crucible. They should note the new mass of the crucible, put it onto the pipeclay triangle and heat it strongly in a blue Bunsen flame, lifting the lid gently and occasionally to boost the oxygen flow. They must not lift the lid up high, as some of the product will be lost as white smoke. When the lid is lifted and there is no white light, then the reaction is complete. Then they turn off the Bunsen and allow the crucible to cool, noting the mass of the crucible at the end of the reaction. **[HT only]**

Safety: Eye protection should be worn throughout the reaction and students should be warned that the crucible will retain heat for a surprising amount of time and may crack. CLEAPSS Hazcard 59 – Magnesium ribbon.

How much?

C2 3.3 Percentages and formulae

Learning objectives

- How can we calculate the percentage of an element in a compound from its formula?
- How can we calculate the empirical formula of a compound from its percentage composition? [H]

Figure 1 A small difference in the amount of metal in an ore might not seem very much. However, when millions of tonnes of ore are extracted and processed each year, it all adds up!

We can use the formula mass of a compound to calculate the percentage mass of each element in it. It's not just in GCSE Chemistry books that calculations like this are done! Mining companies decide whether to exploit mineral finds using calculations like these.

Working out the percentage of an element in a compound

Worked example 1

What percentage of the mass of magnesium oxide is actually magnesium?

Solution

We need to know the formula of magnesium oxide: MgO.

The A_r of magnesium is 24 and the A_r of oxygen is 16.

Adding these together gives us the relative formula mass is (M_r), of MgO
24 + 16 = 40

So in 40 g of magnesium oxide, 24 g is actually magnesium.

The fraction of magnesium in the MgO is:

$$\frac{\text{mass of magnesium}}{\text{total mass of compound}} = \frac{24}{40}$$

so the percentage of magnesium in the compound is:

$$\frac{24}{40} \times 100\% = \mathbf{60\%}$$

Worked example 2

A pure white powder is found at the scene of a crime. It could be strychnine, a deadly poison with the formula $C_{21}H_{22}N_2O_2$: but is it?

When a chemist analyses the powder, she finds that 83% of its mass is carbon. What is the percentage mass of carbon in strychnine? Is this the same as the white powder?

Solution

Given the A_r values: C = 12, H =1, N = 14, O = 16, the formula mass (M_r) of strychnine is:

$$(12 \times 21) + (1 \times 22) + (14 \times 2) + (16 \times 2) = 252 + 22 + 28 + 32 = 334$$

The percentage mass of carbon in strychnine is therefore:

$$\frac{252}{334} \times 100 = \mathbf{75.4\%}$$

This is ***not*** the same as the percentage mass of carbon in the white powder – so the white powder is not strychnine.

a What is the percentage mass of hydrogen in ammonia, NH_3? (A_r values: N = 14, H = 1)

Maths skills

To calculate the percentage of an element in a compound:

- Write down the formula of the compound.
- Using the A_r values from your data sheet, work out the M_r of the compound. Write down the mass of each element making up the compound as you work it out.
- Write the mass of the element you are investigating as a fraction of the M_r.
- Find the percentage by multiplying your fraction by 100.

142

Percentages and formulae

Higher

Working out the empirical formula of a compound from its percentage composition

We can find the percentage of each element in a compound by experiments. Then we can work out the simplest ratio of each type of atom in the compound. We call this simplest (whole-number) ratio its **empirical formula**.

This is sometimes the same as the actual number of atoms in one molecule (which we call the **molecular formula**) – but not always. For example, the empirical formula of water is H_2O, which is also its molecular formula. However, hydrogen peroxide has the empirical formula HO, but its molecular formula is H_2O_2.

Worked example

A hydrocarbon contains 75% carbon and 25% hydrogen by mass. What is its empirical formula? (A_r values: C = 12, H = 1)

Solution

Imagine we have 100 g of the compound. Then 75 g is carbon and 25 g hydrogen.

Work out the number of moles by dividing the mass of each element by its relative atomic mass:

For carbon: $\frac{75}{12} = 6.25$ moles of carbon atoms

For hydrogen: $\frac{25}{1} = 25$ moles of hydrogen atoms

So this tells us that 6.25 moles of carbon atoms are combined with 25 moles of hydrogen atoms.

This means that the ratio is 6.25 (C) : 25 (H).

So the simplest whole number ratio is 1 : 4 (by dividing both numbers by the smallest number in the ratio)

In other words each carbon atom is combined with 4 times as many hydrogen atoms.

So the empirical formula is $\mathbf{CH_4}$.

b A compound contains 40% sulfur and 60% oxygen. What is its empirical formula? (A_r values: S = 32, O = 16)

c 5.4 g of aluminium react exactly with 4.8 g of oxygen. What is the empirical formula of the compound formed? (A_r values: Al = 27, O = 16)

Maths skills

To work out the formula from percentage masses:

- Change the percentages given to the masses of each element in 100 g of compound.
- Change the masses to moles of atoms by dividing the masses by the A_r values. This tells you how many moles of each different element are present.
- This tells you the ratio of atoms of the different elements in the compound.
- Then the *simplest* whole-number ratio gives you the empirical formula of the compound. [H]

Summary questions

1 Copy and complete using the words below:
compound dividing hundred formula
The percentage of an element in a .... is calculated by .... the mass of the element in the compound by the relative .... mass of the compound and then multiplying the result by one ....

2 Ammonium nitrate (NH_4NO_3) is used as a fertiliser. What is the percentage mass of nitrogen in it? (A_r values: H = 1, N = 14, O = 16)

3 22.55% of the mass of a sample of phosphorus chloride is phosphorus. What is the empirical formula of this phosphorus chloride? (A_r values: P = 31, Cl = 35.5) [H]

Key points

- The relative atomic masses of the elements in a compound and its formula can be used to work out its percentage composition.
- We can calculate empirical formulae given the masses or percentage composition of elements present. [H]

143

Further teaching suggestions

Water of crystallisation

- Students could try to calculate the water of crystallisation in certain formulae, e.g. $CuSO_4{\cdot}5H_2O$.

Percentage of oxygen

- Ask students to calculate the highest proportion of oxygen in the following formulae. Encourage the students to calculate the percentage composition and to show their working.
  - CO [57 per cent O]
  - C_2H_5OH [35 per cent O]
  - CH_3COOH [53 per cent O]

Role play

- Before the lesson begins, select five famous people and write their names on a piece of paper and attach them to the underside of five chairs in the class. Then ask students to check under their chairs; if they have a famous person, they should read out the key point in the style of the famous person to the rest of the class.

Copper oxide

- You can reduce copper oxide by passing methane over it. This experiment may take up to an hour and details can be found at www.practicalchemistry.org by searching for copper(II) oxide. You can then use experimental data to determine the formula of the compound. **[HT only]**

Answers to in-text questions

a 17.6%

b SO_3

c Al_2O_3

Summary answers

1 compound, dividing, formula, hundred

2 35%

3 PCl_3

C2 3.4 Equations and calculations

Learning objectives

Students should learn:

- that balanced symbol equations show the relative numbers of molecules of reactants and products in a reaction **[HT only]**
- that balanced symbol equations can be used to calculate the masses of reactants and products. **[HT only]**

Learning outcomes

Most students should be able to:

- interpret how many moles of reactants/products are shown in a balanced symbol equation **[HT only]**
- balance symbol equations **[HT only]**
- use a balanced symbol equation to calculate the mass of reactants or products. **[HT only]**

Support

- Some students will find the calculations difficult. Give them half-finished calculations, where they need to add numbers into the working out to generate the answers. Peer mentoring could also be used, where the students are split into pairs, with Higher Tier students supporting Foundation Tier students.

Extend

- Give students more complex symbol equations to complete calculations. Include data that would encourage students to truncate answers.

Specification link-up: Chemistry C2.3

- The masses of reactants and products can be calculated from balanced symbol equations. *[C2.3.3 c)]* **[HT only]**

Lesson structure

Starters

Multiple-choice – Give each student three coloured flashcards, e.g. blue, green and red. Then create a few multiple-choice questions, one per slide on PowerPoint, with three answers, each written in a different colour to match the flash cards. This could also be achieved using Word or whiteboard software. Then show each question in turn and the students hold up the card that represents the answer they think is correct. *(5 minutes)*

Chemical equations – Ask the students to complete the following chemical equations:

1 $Mg + [O_2] \rightarrow 2MgO$
2 $CH_4 + O_2 \rightarrow [CO_2] + 2H_2O$
3 $Zn + CuSO_4 \rightarrow [Cu] + [ZnSO_4]$
4 $[NaOH] + HCl \rightarrow NaCl + H_2O$

Then students could turn these into balanced symbol equations. Support students by giving them the missing words. To extend this activity, you could ask the students to say what type of reaction each of the above represents [1 oxidation; 2 combustion/oxidation; 3 displacement; 4 neutralisation]. *(10 minutes)*

Main

- Students may have been introduced to balancing symbol equations in Key Stage 3 and will have covered it during their studies of C1. You may wish to return to the familiar equation for the oxidation of hydrogen, which is often used to demonstrate balancing equations. Advance students by explaining that the numbers that are used to balance the equation can be thought of as a ratio. Therefore two molecules of hydrogen will react with one molecule of oxygen to make two molecules of water. However, as this is a ratio, we can also say that two moles of hydrogen will react with one mole of oxygen to make two moles of water. You may then wish to work through the calculation example given in the Student Book for the reaction between hydrogen and chlorine.
- Students need to be able to work out the masses of different substances in balanced symbol equations. Create a card loop by drawing a rectangle 10 cm by 15 cm. Draw a dotted line to make a square 10 cm by 10 cm. In the square, write out questions that involve balancing equations and calculating reacting masses. Then, in the 5 cm by 10 cm rectangle, write an answer (not one that matches the question on the card). Ensure that the questions match answers on other cards so that a loop is made. Give the question loop set to small groups of students and allow them to complete the card sort. Then encourage the students to pick two of the questions and answers to copy into their book, but show their working out stage by stage (as demonstrated in the Student Book).
- Split the class into pairs, and on separate pieces of paper write enough calculation questions for one per group. Give out the questions and allow the students to start to answer for 3 minutes (timed using a stopwatch). Then ask the students to hand the paper to another group. Give the next group 3 minutes to correct the previous work and then continue with the answer. Repeat this a number of times until there has been enough time for the answers to be completed. Then return the paper back to the 'owners' where they should copy up the question and the full answer.

Plenaries

Reflection – Ask students to think about the objectives for today's lesson. Ask them to consider if they have been met, and discuss this in small groups. Ask a few groups to feed back to the rest of the class, explaining how they know that they have met the objectives. *(5 minutes)*

AfL (Assessment for Learning) – Give students some calculations but, instead of tackling the questions, encourage them to create the mark scheme. Ask the students to work in small groups discussing the questions and devising the marking points, and include alternative answers that could still be given credit, and those that definitely should not be awarded marks. This activity can be differentiated by using questions from the tier of entry students are being entered for. *(10 minutes)*

Further teaching suggestions

State symbols

- Students could include state symbols in more complex balanced symbol equations.

Scaling up

- Students may attempt industrial-sized calculations (involving tonnes). This would involve them multiplying up the masses of the different components.

Calculating key points

- Ask the students to copy out the key points, then encourage them to illustrate each of the key points by using a worked example of a calculation.

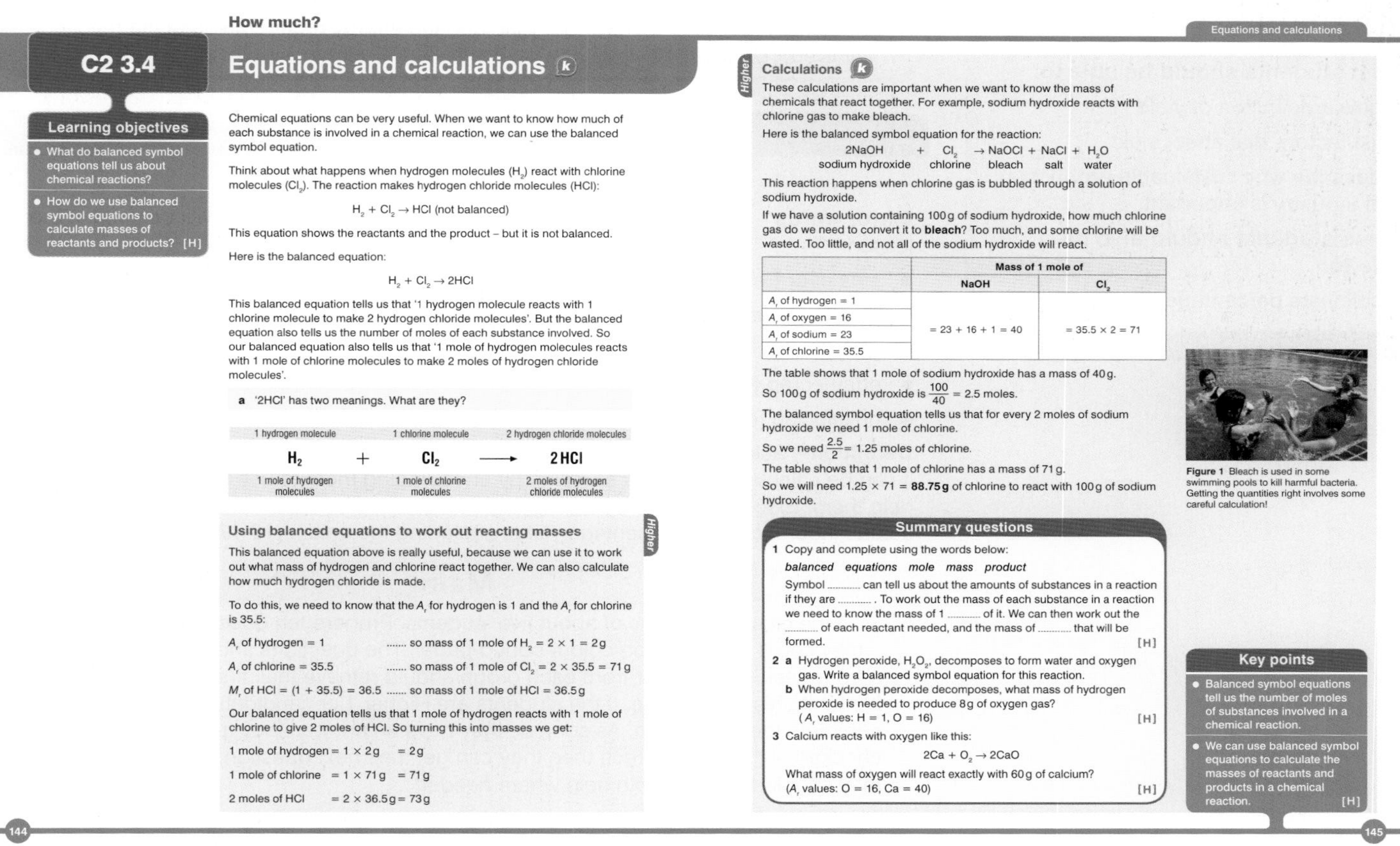

How much?

Equations and calculations

C2 3.4 Equations and calculations

Learning objectives

- What do balanced symbol equations tell us about chemical reactions?
- How do we use balanced symbol equations to calculate masses of reactants and products? [H]

Chemical equations can be very useful. When we want to know how much of each substance is involved in a chemical reaction, we can use the balanced symbol equation.

Think about what happens when hydrogen molecules (H_2) react with chlorine molecules (Cl_2). The reaction makes hydrogen chloride molecules (HCl):

$$H_2 + Cl_2 \rightarrow HCl \text{ (not balanced)}$$

This equation shows the reactants and the product – but it is not balanced.

Here is the balanced equation:

$$H_2 + Cl_2 \rightarrow 2HCl$$

This balanced equation tells us that '1 hydrogen molecule reacts with 1 chlorine molecule to make 2 hydrogen chloride molecules'. But the balanced equation also tells us the number of moles of each substance involved. So our balanced equation also tells us that '1 mole of hydrogen molecules reacts with 1 mole of chlorine molecules to make 2 moles of hydrogen chloride molecules'.

a '2HCl' has two meanings. What are they?

1 hydrogen molecule		1 chlorine molecule		2 hydrogen chloride molecules
H_2	+	Cl_2	→	2HCl
1 mole of hydrogen molecules		1 mole of chlorine molecules		2 moles of hydrogen chloride molecules

Using balanced equations to work out reacting masses (Higher)

This balanced equation above is really useful, because we can use it to work out what mass of hydrogen and chlorine react together. We can also calculate how much hydrogen chloride is made.

To do this, we need to know that the A_r for hydrogen is 1 and the A_r for chlorine is 35.5:

A_r of hydrogen = 1 .... so mass of 1 mole of H_2 = 2 × 1 = 2 g

A_r of chlorine = 35.5 .... so mass of 1 mole of Cl_2 = 2 × 35.5 = 71 g

M_r of HCl = (1 + 35.5) = 36.5 .... so mass of 1 mole of HCl = 36.5 g

Our balanced equation tells us that 1 mole of hydrogen reacts with 1 mole of chlorine to give 2 moles of HCl. So turning this into masses we get:

1 mole of hydrogen = 1 × 2 g = 2 g

1 mole of chlorine = 1 × 71 g = 71 g

2 moles of HCl = 2 × 36.5 g = 73 g

144

Calculations (Higher)

These calculations are important when we want to know the mass of chemicals that react together. For example, sodium hydroxide reacts with chlorine gas to make bleach.

Here is the balanced symbol equation for the reaction:

$$2NaOH + Cl_2 \rightarrow NaOCl + NaCl + H_2O$$

sodium hydroxide, chlorine, bleach, salt, water

This reaction happens when chlorine gas is bubbled through a solution of sodium hydroxide.

If we have a solution containing 100 g of sodium hydroxide, how much chlorine gas do we need to convert it to **bleach**? Too much, and some chlorine will be wasted. Too little, and not all of the sodium hydroxide will react.

	Mass of 1 mole of	
	NaOH	Cl_2
A_r of hydrogen = 1 A_r of oxygen = 16 A_r of sodium = 23 A_r of chlorine = 35.5	= 23 + 16 + 1 = 40	= 35.5 × 2 = 71

The table shows that 1 mole of sodium hydroxide has a mass of 40 g.

So 100 g of sodium hydroxide is $\frac{100}{40}$ = 2.5 moles.

The balanced symbol equation tells us that for every 2 moles of sodium hydroxide we need 1 mole of chlorine.

So we need $\frac{2.5}{2}$ = 1.25 moles of chlorine.

The table shows that 1 mole of chlorine has a mass of 71 g.

So we will need 1.25 × 71 = **88.75 g** of chlorine to react with 100 g of sodium hydroxide.

Figure 1 Bleach is used in some swimming pools to kill harmful bacteria. Getting the quantities right involves some careful calculation!

Summary questions

1 Copy and complete using the words below:

balanced equations mole mass product

Symbol .... can tell us about the amounts of substances in a reaction if they are .... To work out the mass of each substance in a reaction we need to know the mass of 1 .... of it. We can then work out the .... of each reactant needed, and the mass of .... that will be formed. [H]

2 a Hydrogen peroxide, H_2O_2, decomposes to form water and oxygen gas. Write a balanced symbol equation for this reaction.

b When hydrogen peroxide decomposes, what mass of hydrogen peroxide is needed to produce 8 g of oxygen gas? (A_r values: H = 1, O = 16) [H]

3 Calcium reacts with oxygen like this:

$$2Ca + O_2 \rightarrow 2CaO$$

What mass of oxygen will react exactly with 60 g of calcium? (A_r values: O = 16, Ca = 40) [H]

Key points

- Balanced symbol equations tell us the number of moles of substances involved in a chemical reaction.
- We can use balanced symbol equations to calculate the masses of reactants and products in a chemical reaction. [H]

145

Answers to in-text questions

a '2 hydrogen chloride molecules' and '2 moles of hydrogen chloride molecules'.

Summary answers

1 equations, balanced, mole, mass, product

2 a $2H_2O_2 \rightarrow 2H_2O + O_2$

b 17 g

3 24 g

C2 3.5 The yield of a chemical reaction

Learning objectives

Students should learn:

- that the amount of product made can be expressed as a yield
- how to calculate percentage yield **[HT only]**
- that it is important to maximise yield in industrial processes and reduce the waste of energy.

Learning outcomes

Most students should be able to:

- give a definition of yield
- list factors that affect yield
- describe why sustainable production in industry is important.

Some students should also be able to:

- calculate percentage yield. **[HT only]**

Specification link-up: Chemistry C2.3

- Even though no atoms are gained or lost in a chemical reaction, it is not always possible to obtain the calculated amount of a product because:
  - the reaction may not go to completion because it is reversible
  - some of the product may be lost when it is separated from the reaction mixture
  - some of the reactants may react in ways different from the expected reaction. *[C2.3.3 d)]*
- The amount of a product obtained is known as the yield. When compared with the maximum theoretical amount as a percentage, it is called the percentage yield. *[C2.3.3 e)]*
- Evaluate sustainable development issues relating the starting materials of an industrial process to the product yield and the energy requirements of the reactions involved. *[C2.3]*

Lesson structure

Starters

Mirror words – Ask the students to work out these key words, which will be used in the lesson:

- stnatcaer [reactants]
- stcudorp [products]
- dleiy [yield]
- egatnecrep [percentage]
- noitaluclac [calculation] *(5 minutes)*

Definition – Ask the students to explain what a yield is. If they are a Higher Tier, ask how it can be calculated. Support students by encouraging them to use the Student Book to help them. Extend students by asking them to look at the worked example and find the actual, theoretical and percentage yields (including the units). *(10 minutes)*

Main

- Split the class into teams of about five students. Prepare ten questions for each group – they could be written on colour-coded paper. The questions should be face-down on a front desk. A volunteer from each group should retrieve their first question and take it back to his or her table. If the students are Higher Tier candidates, the team should complete the calculation. As soon as they have an answer, they take it to you to be checked. If they are correct, then they can get their next question. If they are incorrect, they should try again, with help where needed.
- Ask Higher Tier students to look carefully at the worked example for calculating the percentage yield. Ask them to draw two flow charts to show the steps of how to complete these calculations. Then give the students a number of questions they can answer using their flow charts to help them.
- Discuss the importance of sustainable production. Students can then list the advantages.

Plenaries

Steps – Write a question on the board that involves the calculation of a yield (an example could be used from the Student Book). Put each separate number or mathematical procedure on to separate sheets of A3 paper and arrange them on the floor in the wrong order. Ask for a volunteer to stand on the starting number and move on to the next sheet physically, showing the order of the calculation. Encourage the students to describe how they would do the calculation as they stand on the different pieces of paper. Support students by asking them to work in small groups to order the steps. Extend students by having parts of the equation missing. Students should therefore not only order the steps but also complete them. *(5 minutes)* **[HT only]**

What's the question? – Give students an answer such as yield, mass lost, 100 per cent yield, reversible reaction, etc. Encourage them to work in small groups to generate questions that match the answer given. You could support students by listing questions on the board: they could then choose the question that best fits their answer. For Higher Tier students, you could include answers to calculations of percentage yield. *(10 minutes)*

Support

- Use a flow chart to show the steps that cause yield to be reduced in a chemical process. The different steps in the flowchart, such as transferring glassware etc., could be given to the students. They could then cut them out and stick them in the correct order into a predrawn flow chart outline.

Extend

- Ask students to complete yield equations that require manipulation of units, e.g. data given in tonnes or kg.

Further teaching suggestions

Theoretical and actual yields

- Students could be encouraged to reflect on the quantities of actual yield (what you make) compared with theoretical yield (maximum amount that you can make). In reality, no reaction ever produces the theoretical yield. Ask the students to think about why this is so. Note their ideas on a spider diagram in their books. Then ask them to contribute to an exhaustive diagram on the board – they should amend their own to ensure all the points are included and are correct.

Advert

- Students could make an advert for an industrial magazine, encouraging businesses to adopt more sustainable production. Students could then use the school's marking policy to assess each other's work. The author could make changes for homework based on the feedback from the AfL.

Measuring yield

- You may wish to allow students to complete the experiment of burning magnesium in a crucible as detailed in practical support 3.3 'Percentages and formulae'. Students know the mass of the magnesium that they started with. They can convert this into moles. Using the balanced symbol equation to determine that the ratio of magnesium to magnesium oxide is 1 : 1, students will then know the maximum number of moles of magnesium oxide that they could make and therefore work out the theoretical yield. Students could then measure the actual yield and Higher Tier students could calculate the percentage yield. A similar experiment can also be completed using iron wool.

C2 3.5 The yield of a chemical reaction

Learning objectives

- What do we mean by the yield of a chemical reaction and what factors affect it?
- How do we calculate the percentage yield of a chemical reaction? [H]
- Why is it important to achieve a high yield in industry and to waste as little energy as possible?

Many of the substances that we use every day have to be made from other chemicals. This may involve using complex chemical reactions. Examples include food colourings, flavourings and preservatives, the ink in your pen or printer, and the artificial fibres in your clothes. All of these are made using chemical reactions.

Imagine a reaction: A + 2B → C

If we need 1000 kg of C, we can work out how much A and B we need. All we need to know is the relative formula masses of A, B and C and the balanced symbol equation.

A + 2B → C
(reactants) (product)

a How many moles of B are needed to react with each mole of A in this reaction?
b How many moles of C will this make?

links

For information about using balanced symbol equations to predict reacting masses, look back to C2 3.4 Equations and calculations.

If we carry out the reaction, it is unlikely that we will get as much of C as we worked out. This is because our calculations assumed that all of A and B would be turned into C. We call the amount of product that a chemical reaction produces its **yield**.

It is useful to think about reactions in terms of their **percentage yield**. This compares the amount of product that the reaction *really* produces with the maximum amount that it could *possibly* produce:

$$\textbf{Percentage yield} = \frac{\textbf{amount of product produced}}{\textbf{maxmimum amount of product possible}} \times 100\%$$

Calculating percentage yield

Higher

An industrial example

Limestone is made mainly of calcium carbonate. Crushed lumps of limestone are heated in a rotating lime kiln. The calcium carbonate decomposes to make calcium oxide, and carbon dioxide gas is given off. A company processes 200 tonnes of limestone a day. It collects 98 tonnes of calcium oxide, the useful product. What is the percentage yield of the kiln, assuming limestone contains only calcium carbonate?

(A_r values: Ca = 40, C = 12, O = 16)

calcium carbonate → calcium oxide + carbon dioxide

$$CaCO_3 \rightarrow CaO + CO_2$$

Work out the relative formula masses of $CaCO_3$ and CaO.

M_r of $CaCO_3$ = 40 + 12 + (16 × 3) = 100

M_r of CaO = 40 + 16 = 56

So the balanced symbol equation tells us that:

100 tonnes of $CaCO_3$ could make 56 tonnes of CaO, assuming a 100% yield.

Higher

Therefore 200 tonnes of $CaCO_3$ could make a maximum of (56 × 2) tonnes of CaO = 112 tonnes.

So **percentage yield** $= \frac{\textbf{amount of product produced}}{\textbf{maximum amount of product possible}} \times 100\%$

$= \frac{98}{112} \times 100 = \mathbf{87.5\%}$

We can explain this yield as some of the limestone is lost as dust in the crushing process and in the rotating kiln. There will also be some other mineral compounds in the limestone. It is not 100% calcium carbonate as we assumed in our calculation.

c What is the percentage yield of a reaction?

Very few chemical reactions have a yield of 100% because:

- The reaction may be reversible (so as products form they react to re-form the reactants again).
- Some reactants may react to give unexpected products.
- Some of the product may be lost in handling or left behind in the apparatus.
- The reactants may not be completely pure.
- Some chemical reactions produce more than one product, and it may be difficult to separate the product that we want from the reaction mixture.

Sustainable production

Chemical companies use reactions to make products which they sell. Ideally, they want to use reactions with high yields (that also happen at a reasonable rate). Making a product more efficiently means making less waste. As much product as possible should be made from the reactants.

Chemical factories (or **plants**) are designed by chemical engineers. They design a plant to work as safely and economically as possible. It should waste as little energy and raw materials as possible. This helps the company to make money. It is better for the environment too as it conserves our limited resources. It also reduces the pollution we get when we use fossil fuels as sources of energy.

Summary questions

1 Copy and complete using the words below:
high maximum percentage product waste yield
The amount of .... made in a chemical reaction is called its .... The .... yield tells us the amount of product that is made compared to the .... amount that could be made. Reactions with .... yields are important because they result in less ....

2 Explain why it is good for the environment if industry finds ways to make products using high yield reactions and processes that waste as little energy as possible.

3 If the percentage yield for a reaction is 100%, 60 g of reactant A would make 80 g of product C. How much of reactant A is needed to make 80 g of product C if the percentage yield of the reaction is only 75%? [H]

Key points

- The yield of a chemical reaction describes how much product is made.
- The percentage yield of a chemical reaction tells us how much product is made compared with the maximum amount that could be made (100%).
- Factors affecting the yield of a chemical reaction include product being left behind in the apparatus and difficulty separating the products from the reaction mixture.
- It is important to maximise yield and minimise energy wasted to conserve the Earth's limited resources and reduce pollution.

Answers to in-text questions

a 2

b 1

c (amount of product made ÷ amount of product possible) × 100 per cent

Summary answers

1 product, yield, percentage, maximum, high, waste

2 It will conserve the Earth's limited resources and reduce pollution caused by waste products.

3 80 g

C2 3.6 Reversible reactions

Learning objectives

Students should learn:

- that reactions can be reversible
- how reversible reactions can be represented.

Learning outcomes

Most students should be able to:

- define a reversible reaction
- recognise a reversible reaction from its word or symbol equation.

Some students should also be able to:

- explain what a reversible reaction is, giving an example.

Specification link-up: Chemistry C2.3

- In some chemical reactions, the products of the reaction can react to produce the original reactants. Such reactions are called reversible reactions and are represented:
  $A + B \rightleftharpoons C + D$
  For example: ammonium chloride $\rightleftharpoons$ ammonia + hydrogen chloride. *[C2.3.3 f)]*

Lesson structure

Starters

Card sort – Create a card sort for the students to match the key words with their definitions. Put the cards into envelopes and give each pair of students a set to sort out on their desk. Then ask students to pick three new words and write them in their book, including the definition.

Reactants – starting substances in a chemical reaction.
Products – substances left at the end of a chemical reaction.
Reversible reaction – where the reactants make the products and the products make the reactants.
Indicators – chemicals that react with acids and alkalis to make different-coloured compounds.
Chemical reaction – a change in which a new substance is made.

Support students by encouraging them to use the Student Book to help them. Extend students by encouraging them to use the Student Book to give examples of each of the key words. *(5 minutes)*

Reflection – Ask students to draw a table with three columns headed 'What I already know', 'What I want to know', 'What I know now'. Encourage the students to look at the title of the page and the objectives. Then ask them to fill in the first column with facts they already know about the topic, presented as bullet points. Then ask them to think of questions that they think they need to have answered by the end of the lesson and note these in the second column. *(10 minutes)*

Main

- Students will be familiar with the use of indicators from their work on acids and alkalis at Key Stage 3. The colour changes produced by indicators are reversible reactions. Students could find out (experimentally or using secondary resources) the colour changes of different indicators in acidic/neutral/alkaline solutions.
- Heating ammonium chloride causes thermal decomposition to form ammonia and hydrogen chloride. This reaction can be completed experimentally in the lab by heating ammonium chloride in a boiling tube with a loosely fitted plug of mineral wool, then relating this to reversible reactions. Encourage the students to focus their attention on the cool part of the boiling tube. Ask them to explain in the form of a flow chart what is happening in the boiling tube.

Plenaries

True or false – On separate sheets of sugar paper, write the words 'true' and 'false'. Then stick them on opposite sides of the classroom. Read out the statements below. Students should stand next to the wall to represent their answer. If they don't know, they should stand in the centre of the room.

- A double-headed arrow in an equation shows that it is a non-reversible reaction. [False]
- In a reversible reaction, reactants make products and products make reactants. [True]
- All chemical reactions are reversible. [False]
- Indictors react with acids but not with alkalis. [False]
- The decomposition of ammonium chloride is a reversible reaction. [True]

Support students by giving them statements on a card and they discuss them in small groups before the movement commences. Extend students by asking them to rephrase the incorrect statements so that they are correct. *(5 minutes)*

Reflection – Ask students to review their starter table ('Reflection') and use a different colour pen/pencil to correct any misconceptions from the start of the lesson. Then ask the students to answer the questions they posed. If they can't, encourage them to talk in small groups, consult the Student Book and, if necessary, ask you. Finally, students should record, in bullet-point format, any other facts that they have picked up during the lesson, in the last column. *(10 minutes)*

Answers to in-text questions

a The reaction goes in the forward direction only.

b The reaction is reversible.

c Both the red (ALit)/(Lit^-) and blue forms of litmus are present; so mixed together in solution they look purple.

Support

- You could take the reactants and products in a reversible reaction to form a flow chart to explain the term 'reversible reaction'. This could be made into a 'cut-and-stick' activity which would generate a cycle.

Extend

- Ask students to find examples of reversible reactions that they benefit from in everyday life, e.g. oxygen binding to haemoglobin in their blood.

Practical support

Litmus

Equipment and materials required

A test tube, test tube holder, red or blue litmus solution, 1.0 mol/dm^3 hydrochloric acid, 0.04 mol/dm^3 sodium hydroxide, 2 × dropping pipettes, eye protection.

Details

Put a few drops of litmus indicator into a test tube. Add a few drops of acid and observe. Add a few drops of alkali and observe. Repeat to demonstrate the reversible reaction.

Safety: CLEAPSS Hazcard 47A Hydrochloric acid – corrosive. CLEAPSS Hazcard 31 Sodium hydroxide – corrosive. Both solutions at the concentrations used are low hazard, but eye protection should be worn throughout the experiment.

Heating ammonium chloride

Equipment and materials required

A boiling tube, boiling tube holder, Bunsen burner and safety equipment, mineral wool, spatula, ammonium chloride, eye protection.

Details

Put about half a spatula of ammonium chloride into a boiling tube and insert a mineral wool plug at the top. Gently heat in a Bunsen flame.

Safety: Eye protection should be worn at all times. As acidic hydrogen chloride gas and alkaline ammonia gas are produced, the mineral plug must be used and the reaction should be carried out in a well-ventilated room. CLEAPSS Hazcard 9A Ammonium chloride – harmful.

Heating copper sulfate

Equipment and materials required

A boiling tube, boiling tube holder, Bunsen burner and safety equipment, spatula, hydrated copper sulfate, wash bottle, eye protection.

Details

Put about half a spatula of hydrated copper sulfate into a boiling tube and heat gently on a Bunsen flame until the colour change to white is complete. Remove from the heat and allow the boiling tube to cool. Then add a few drops of water and observe.

Link

This reaction will be studied in the next chapter in terms of the energy changes involved in reversible reactions.

Safety: Be careful to allow the boiling tube to cool before adding water as the glass could crack. Eye protection should be worn throughout this experiment. CLEAPSS Hazcard 27C Copper sulfate – harmful.

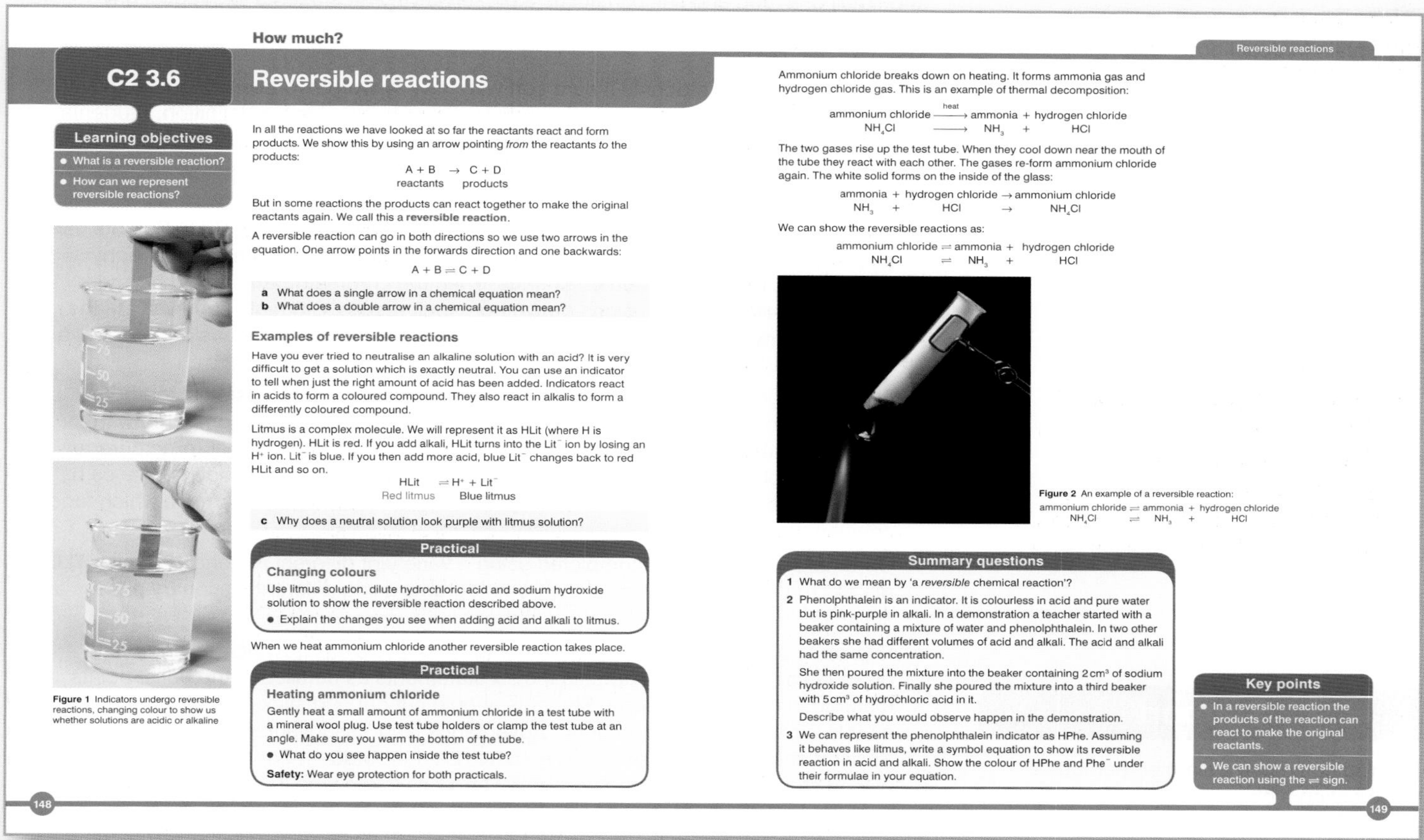

How much?

C2 3.6

Reversible reactions

Learning objectives

- What is a reversible reaction?
- How can we represent reversible reactions?

In all the reactions we have looked at so far the reactants react and form products. We show this by using an arrow pointing *from* the reactants *to* the products:

$$\underset{\text{reactants}}{A + B} \rightarrow \underset{\text{products}}{C + D}$$

But in some reactions the products can react together to make the original reactants again. We call this a **reversible reaction**.

A reversible reaction can go in both directions so we use two arrows in the equation. One arrow points in the forwards direction and one backwards:

$$A + B \rightleftharpoons C + D$$

a What does a single arrow in a chemical equation mean?
b What does a double arrow in a chemical equation mean?

Examples of reversible reactions

Have you ever tried to neutralise an alkaline solution with an acid? It is very difficult to get a solution which is exactly neutral. You can use an indicator to tell when just the right amount of acid has been added. Indicators react in acids to form a coloured compound. They also react in alkalis to form a differently coloured compound.

Litmus is a complex molecule. We will represent it as HLit (where H is hydrogen). HLit is red. If you add alkali, HLit turns into the Lit^- ion by losing an H^+ ion. Lit^- is blue. If you then add more acid, blue Lit^- changes back to red HLit and so on.

$$\underset{\text{Red litmus}}{HLit} \rightleftharpoons \underset{\text{Blue litmus}}{H^+ + Lit^-}$$

c Why does a neutral solution look purple with litmus solution?

Practical

Changing colours

Use litmus solution, dilute hydrochloric acid and sodium hydroxide solution to show the reversible reaction described above.

- Explain the changes you see when adding acid and alkali to litmus.

When we heat ammonium chloride another reversible reaction takes place.

Practical

Heating ammonium chloride

Gently heat a small amount of ammonium chloride in a test tube with a mineral wool plug. Use test tube holders or clamp the test tube at an angle. Make sure you warm the bottom of the tube.

- What do you see happen inside the test tube?

Safety: Wear eye protection for both practicals.

Figure 1 Indicators undergo reversible reactions, changing colour to show us whether solutions are acidic or alkaline

148

Reversible reactions

Ammonium chloride breaks down on heating. It forms ammonia gas and hydrogen chloride gas. This is an example of thermal decomposition:

$$\text{ammonium chloride} \xrightarrow{\text{heat}} \text{ammonia} + \text{hydrogen chloride}$$
$$NH_4Cl \longrightarrow NH_3 + HCl$$

The two gases rise up the test tube. When they cool down near the mouth of the tube they react with each other. The gases re-form ammonium chloride again. The white solid forms on the inside of the glass:

$$\text{ammonia} + \text{hydrogen chloride} \rightarrow \text{ammonium chloride}$$
$$NH_3 + HCl \rightarrow NH_4Cl$$

We can show the reversible reactions as:

$$\text{ammonium chloride} \rightleftharpoons \text{ammonia} + \text{hydrogen chloride}$$
$$NH_4Cl \rightleftharpoons NH_3 + HCl$$

Figure 2 An example of a reversible reaction:
ammonium chloride ⇌ ammonia + hydrogen chloride
$NH_4Cl \rightleftharpoons NH_3 + HCl$

Summary questions

1 What do we mean by 'a *reversible* chemical reaction'?

2 Phenolphthalein is an indicator. It is colourless in acid and pure water but is pink-purple in alkali. In a demonstration a teacher started with a beaker containing a mixture of water and phenolphthalein. In two other beakers she had different volumes of acid and alkali. The acid and alkali had the same concentration.

She then poured the mixture into the beaker containing 2 cm^3 of sodium hydroxide solution. Finally she poured the mixture into a third beaker with 5 cm^3 of hydrochloric acid in it.

Describe what you would observe happen in the demonstration.

3 We can represent the phenolphthalein indicator as HPhe. Assuming it behaves like litmus, write a symbol equation to show its reversible reaction in acid and alkali. Show the colour of HPhe and Phe^- under their formulae in your equation.

Key points

- In a reversible reaction the products of the reaction can react to make the original reactants.
- We can show a reversible reaction using the ⇌ sign.

149

Further teaching suggestions

Breathalyser

- Chromate ions undergo a reversible reaction with dichromate ions. These substances were used in the first breathalysers. Encourage students to discover the chemical reactions that occurred in these early testers.

Other reversible reactions

- You may wish to demonstrate 'blue bottle' reaction as detailed in 'RSC Classic Chemistry Experiments no. 83', or the oscillating reaction also from the 'RSC Classic Chemistry Experiments, no. 140'. You could also add alkali and acid successively to a small beaker with bromine water, or alternatively with potassium dichromate(VI) solution, to observe the reversible colour changes. Check CLEAPSS Hazcards.

Summary answers

1 The products can react to produce the original reactants.

2 The mixture turns pink-purple when added to sodium hydroxide solution and then turns back to colourless when added to the hydrochloric acid solution.

3 $\underset{\text{colourless}}{HPhe} \rightleftharpoons \underset{\text{pink-purple}}{H^+ + Phe^-}$

C2 3.7 Analysing substances

Learning objectives

Students should learn:

- how to detect and identify artificial food colourings
- that there are advantages of using instrumental analysis.

Learning outcomes

Most students should be able to:

- describe an experiment to separate coloured additives
- list advantages of modern analysis techniques.

Some students should also be able to:

- explain in detail how coloured food additives can be detected and identified using paper chromatography.

Specification link-up: Chemistry C2.3

- Chemical analysis can be used to identify additives in foods. Artificial colours can be detected and identified by paper chromatography. *[C2.3.2 b)]*
- Elements and compounds can be detected and identified using instrumental methods. Instrumental methods are accurate, sensitive and rapid and are particularly useful when the amount of a sample is very small. *[C2.3.2 a)]*

Controlled Assessment: C4.5 Analyse and interpret primary and secondary data. *[C4.5.4 c) d)]*

Lesson structure

Starters

Original additive – Show students a picture of some spices and salt fish. Ask students to suggest their connection [food additives]. Spices are used to improve the taste, colour and smell of a food and salt can act as a preservative. Ask students to suggest any examples of modern-day additives that would not have been available 200 years ago (e.g. MSG). *(5 minutes)*

Dominoes – Give the students a card sort with the key words: 'preserve, food additive, solvent, chromatography' written on. Each card should also have a definition. Students should work in pairs to match the ends up, as in a domino game. *(10 minutes)*

Main

- Food colourings are often added to food to make them more appealing. Colours can be added to sweets and savoury foods. Students can complete chromatography experiments of different food colourings. Then the chromatograms could be stuck into their book and conclusions drawn from their results. Students can evaluate the reproducibility, repeatability and validity of the experiment as a means of detecting and identifying artificial colourings.
- If chromatography is being used to identify additives, then it is essential for the conditions to be the same. This allows test chromatograms to be compared with standard ones. You may wish to allow students to make 'standard' chromatograms for food colourings, then analyse a mystery food colouring (prepared in advance by mixing some of the food colouring available to the students) to determine which food colourings are present.
- Chromatography can be completed using a variety of different materials such as TLC plates. You may be able to make chromatograms using different stationary phases and solvents.
- Explain to students that using instruments for analysis has advantages and disadvantages. Give the students the six statements as detailed in the bullet points featured in the Student Book. Students should cut out and stick each statement into a table with two columns, headed 'Advantages' and 'Disadvantages' respectively. Support students by encouraging them to use their Student Book to help them. To extend students, ask them to consider why it has taken time to develop instrumental analysis [it is dependent on technological development].

Plenaries

AfL (Assessment for Learning) – Ask students to compare chromatograms produced – which group produced the best one for analysis purposes? Take a class vote on the best chromatogram and why. *(5 minutes)*

Explain – Ask the students to explain how chromatography works. Then pick three students randomly from the register to read their explanations. Reward the best explanation. Support students by giving them statements in the wrong order to explain chromatography. Extend students by suggesting other applications for chromatography. *(10 minutes)*

Support

- You could use a simpler method of creating a chromatogram. An easier way to generate a chromatogram is to give the students a disk of filter paper. Ask them to use a paintbrush to put a sample of a food colouring into the centre. Then cut a wick (a wedge shape towards the centre). Balance the paper over a beaker with water in it. The wick must be submerged. The colours will then separate into rings.

Extend

- Encourage students to find out more about the stationary and mobile phases in paper chromatography and use them to give a more detailed explanation of chromatography.

Practical support

Detecting dyes in food colourings

Equipment and materials required

Different food colourings, capillary tubes, boiling tube, boiling tube rack, ruler, pencil, strips of chromatography paper that fits into the boiling tube.

Details

Draw a pencil line 2 cm from the bottom of the chromatography paper. Draw three pencil crosses on the base line, an equal distance apart. Dip a clean capillary tube into a food colouring. The colour will suck into the tube. Gently dot one of the crosses, trying to add only a small amount of colouring. Repeat with two further colours on the remaining crosses. Put a small amount of water in the boiling tube (about 1 cm deep). Lower the chromatography paper into the tube and place in the rack. Leave the chromatogram to develop, until the solvent line is past the last separated colour. Try not to move the chromatograms while they develop or it will not be easy to compare them.

Students will see which colourings were pure substances and which were mixtures of dyes, and can list the component colours in any mixtures.

How much?

C2 3.7 Analysing substances

Learning objectives

- What are food additives and how can we identify them?
- How can we detect artificial food colourings?
- What are the advantages of instrumental methods of analysis?

For hundreds of years we have added salt to food to preserve it. Nowadays, food technologists develop ways to improve the quality of foods. They also analyse foods to ensure they meet legal safety standards.

We call a substance that is added to food to extend its shelf life or to improve its taste or appearance a food additive. Additives that have been approved for use in Europe are given E numbers. The E numbers are like a code to identify the additives. For example, E102 is a yellow food colouring called tartrazine.

a What is a food additive?

Detecting additives

Scientists have many instruments that they can use to identify unknown compounds, including food additives. Many of these are more sensitive, automated versions of techniques we use in school labs.

One technique that is used to identify food additives is paper **chromatography**. It works because some compounds in a mixture dissolve better than others in particular solvents. Their solubility determines how far they travel across the paper.

NOODLES IN A CHICKEN & MUSHROOM FLAVOUR SAUCE MIX WITH TEXTURISED SOYA PIECES, DRIED VEGETABLES AND A SACHET OF SOY SAUCE
INGREDIENTS (after preparation)

Figure 1 Modern foods contain a variety of additives to improve their taste or appearance, and to make them keep longer

links
For more information on how chemists identify unknown substances, see C3 4.5 Chemical analysis.

Practical

Detecting dyes in food colourings

Make a chromatogram to analyse various food colourings.

- What can you deduce from your chromatogram?

Figure 2 The technique of paper chromatography that we use in schools. Techniques used to identify food additives are often based on the same principles as the simple tests we do in the school science lab.

b What happens to the food colourings when you make a paper chromatogram?

Once the compounds in a food have been separated using chromatography, they can be identified. We can compare the chromatogram with others obtained from known substances. For this we must use the same solvent at the same temperature.

Figure 3 A few years ago a batch of red food colouring was found to be contaminated with a chemical suspected of causing cancer. This dye had found its way into hundreds of processed foods. All of these had to be removed from the shelves of our supermarkets and destroyed.

150

Analysing substances

Instrumental methods

Many industries need rapid and accurate methods for analysing their products. They use modern instrumental analysis for this task.

Instrumental techniques are also important in fighting pollution. Careful monitoring of the environment using sensitive instruments is now common. This type of analysis is also used all the time in health care.

Modern instrumental methods have a number of benefits over older methods:

- they are highly accurate and sensitive
- they are quicker
- they enable very small samples to be analysed.

Against this, the main disadvantages of using instrumental methods are that the equipment:

- is usually very expensive
- takes special training to use
- gives results that can often be interpreted only by comparison with data from known substances.

c What do you think has aided the development of instrumental methods of chemical analysis?
d Why are these methods important?

Figure 4 Compared with the methods of 50 years ago, modern instrumental methods of analysis are quick, accurate and sensitive – three big advantages. They also need far fewer people to carry out the analysis than traditional laboratory analysis.

links
For more information on the instruments used by chemists to analyse substances, see C2 3.8 Instrumental analysis.

AQA Examiner's tip

Although simpler to use than bench chemistry methods, instrumental methods still need trained technicians to operate them.

Summary questions

1 Copy and complete using the words below:
additives paper analyse identify
Food scientists can .... different foods to see what .... have been used. For example, food colourings can be detected by .... chromatography. They can use results from known compounds to positively .... them.

2 **a** Carry out a survey of some processed foods. Identify some examples of food additives and explain why they have been used.
b Describe how we can separate the dyes in a food colouring and identify them.

3 What are the main advantages and disadvantages of using instrumental analysis compared with traditional practical methods?

Key points

- Additives may be added to food in order to improve its appearance, taste and how long it will keep (its shelf life).
- Food scientists can analyse foods to identify additives, e.g. by using paper chromatography.
- Modern instrumental techniques provide fast, accurate and sensitive ways of analysing chemical substances.

151

Further teaching suggestions

Sweet chromatography

- Students could investigate the dye in sweets. Different coloured sweets could be put into dimple dishes. Students wet a paintbrush to remove the dye and complete a chromatogram.

Packaging

- Ask students to choose their favourite convenience food (could be a sweet or frozen pizza, etc.). They should bring in the packaging from home, or you can use online supermarkets to pull up the ingredients lists. Encourage them to list all the additives in that food, concentrating on food colourings.

Natural additives

- A lot of people are now very concerned about food additives but do not realise that many of them are naturally occurring chemicals such as pectin (E440), which is found in fruit. Split the class into half and give each group a different task. Ask half of the students to design and make a leaflet to be put in doctors' surgeries to explain that food additives aren't necessarily harmful. The remaining students should design a leaflet warning people about additives in their food and their effects.

Answers to in-text questions

a A substance that is added to food to make it keep longer or to improve its taste or appearance.
b Each dye separates into spots of pure colour.
c The development of technologies such as electronics and computing.
d These methods are highly accurate, quick and can be used with very small quantities of materials.

Summary answers

1 analyse, additives, paper, identify
2 **a** Examples of additives in some packaged food and why they have been added.
b Description of using chromatography then comparing the chromatogram with chromatograms of known substances (or using an instrument, such as a mass spectrometer, to identify each substance separated – see C2 3.8 'Instrumental analysis').
3 **Advantages:** speed, sensitivity, ability to analyse small samples.
Disadvantages: use expensive equipment, training is needed.

C2 3.8 Instrumental analysis

Learning objectives

Students should learn:

- that gas chromatography can be used to separate compounds
- that mass spectrometers can be combined with gas chromatography to identify the components in a mixture.

Learning outcomes

Most students should be able to:

- describe the use of gas chromatography linked with mass spectrometry to identify what is in a mixture.

Some students should also be able to:

- explain in detail the technique of gas chromatography – mass spectrometry
- explain how mass spectrometry is used to determine relative molecular masses. **[HT only]**

Support

- Refer to chromatography as detailed in C2 3.7 Analysing substances as a scaffold to understanding gas chromatography (GC).

Extend

- Give students simple mass spectra and a list of compounds. They could then try to match the compound to the spectra. N.B. Students do not need to know about fragmentation patterns.

Specification link-up: Chemistry C2.3

- Elements and compounds can be detected and identified using instrumental methods. Instrumental methods are accurate, sensitive and rapid and are particularly useful when the amount of a sample is very small. *[C2.3.2 a)]*
- Gas chromatography linked to mass spectroscopy (GC-MS) is an example of an instrumental method: gas chromatography allows the separation of a mixture of compounds. The time taken for a substance to travel through the column can be used to help identify the substance. The output from the gas chromatography column can be linked to a mass spectrometer, which can be used to identify the substances leaving the end of the column … . *[C2.3.2 c)]*
- … The mass spectrometer can also give the relative molecular mass of each of the substances separated in the column. *[C2.3.2 c)]* **[HT only]**

Lesson structure

Starter

Think – Show students an image of a science lab 100 years ago and *a modern* science laboratory. Ask students to think about why instrumental analysis has only recently developed. Allow students to discuss their ideas as a class. *(5 minutes)*

Main

- Instrumental analysis is very precise. It gives similar readings every time the same sample is analysed.
- Gas chromatography is often used with mass spectrometry. A sample made of a mixture of chemicals is injected into a gas chromatography machine in order to separate the mixture. The separate samples then go into a mass spectrometer for identification.
- Ask students to imagine that they have been commissioned to write an instruction leaflet to be in the packaging of a GC-MS machine. You could show some example leaflets such as those that are supplied with toasters or TVs. Students should include a schematic diagram of how the machine works. They could create a flow chart to detail the stages involved in separation and identification.
- Ask students to work in pairs to compare their leaflets. Students should comment on the science content, choice of language, activity and presentation, giving a mark out of 10 for each of these aspects. Then they can award an overall mark out of 40. They could make amendments as homework.

Plenaries

Persuade – Ask students to work in small groups, and imagine that they are research chemists who have made a new fertiliser. Each group should make a persuasive argument for a board of directors on why they should finance the use of instrumental analysis technique rather than using traditional laboratory techniques. Support students by giving them some statements that could provoke discussion. Extend students by asking some of them to role play the board of directors and encourage them to be biased against using the techniques. *(5 minutes)*

Paper round – Give each student a piece of A4 paper. Students should work in groups of six. Ask them to write an example of instrumental analysis at the top of the paper, then fold the paper over and pass to the student at the right. Then ask the next person to write, on the folded paper from another student, one new fact that they have learned. Again fold over the paper and pass it to the right. Ask the students to continue to fold over and pass the paper after writing a revised fact, an advantage of instrumental analysis, a disadvantage of instrumental analysis and one use of mass spectrometry. After the sixth paper move, ask each student to unfold the paper and read what has been written. If time allows, ask a few students to share the work from their paper. Examination technique can be highlighted from this exercise, as some students should have answered in sentences, whereas others may have used key words that might be ambiguous. *(10 minutes)*

Further teaching suggestions

Mass spectra

- Students could be shown a spectra from mass spectrometry.

Lab visit

- A trip to a local university might be organised to see spectrometers in action.

Newspaper report

- Students could imagine that they are reporters in a newspaper in the past. They should write a news article documenting the discovery of mass spectroscopy and its uses. Students could use the internet to find out the date for such an article and people that may have given relevant quotes.

Web search

- Information about modern instrumental analysis can be found in www.sep.org.uk. Search for 'forensic chemistry'.

Videos

- The Royal Society of Chemistry has videos detailing different modern instrumental analysis techniques including mass spectrometry, as part of their Spectroscopy for Schools and Colleges.

Posters

- Show students chemical posters such as ones given free to schools from the RSC. Ask students to create a poster, which includes the key points.

Food scares

- In recent years, there have been a number of food scares involving additives, e.g. Sudan 1 dye. Students could look into these using the internet (e.g. search newspaper websites for 'food scare') and newspaper clippings kept in the school library. They could look at the economic and social effect of scares such as this and the impact on the British food industry.

How much?

C2 3.8 Instrumental analysis

Learning objectives

- How can we use gas chromatography to separate compounds in a sample mixture?
- How can we use a mass spectrometer to identify the compounds in the sample?

Analysing mixtures

Samples to be analysed are often mixtures of different compounds. So the first step is to separate the compounds. Then they can be identified using one of the many instrumental techniques available. Chemists have developed a technique called gas chromatography–mass spectrometry (GC–MS) to do this task.

- Firstly, they use **gas chromatography** to separate compounds that are easily vaporised.
- Then the separated compounds pass into another instrument – the **mass spectrometer**, which can identify each of them. The mass spectrometer is useful for identifying both elements and compounds. The pattern of peaks it produces identifies the sample.

Gas chromatography

This separation technique is similar to paper chromatography. However, instead of a solvent moving over paper, it has a gas moving through a column packed with a solid.

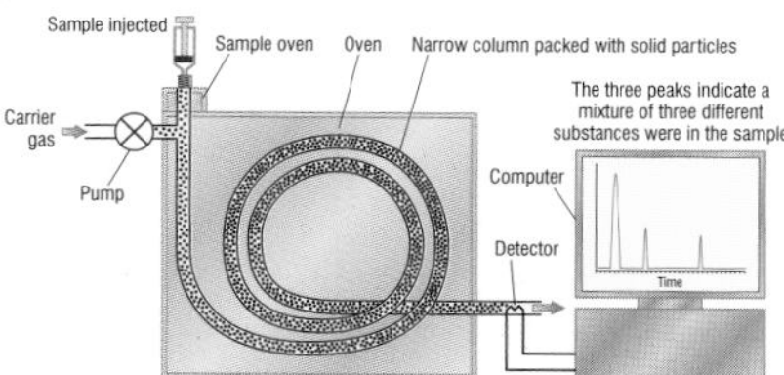

Figure 1 This is the apparatus used in gas chromatography. The solid in the column can be coated in a liquid and is sometimes then known as gas–liquid chromatography.

- First of all, the sample mixture is vaporised.
- A 'carrier' gas moves the vapour through the coiled column.
- The compounds in the sample have different attractions to the material in the column. The compounds with stronger attractions will take longer to get through the column. We say that they have a longer **retention time**.
- The compounds with weak attractions to the material in the column leave it first. They have shorter retention times.

The separated compounds can be recorded on a chart as they leave the column. Look at Figure 2 to see a gas chromatograph.

We can identify the unknown substances in the sample by comparing the chromatograph with the results for known substances. The analysis must have taken place in exactly the same conditions to compare retention times.

Substance A came out of the column first, followed by B and finally C

Figure 2 This is a gas chromatograph of a mixture of three different substances. There was more of substance A than B or C in the sample mixture.

152

Instrumental analysis

Mass spectrometry

To ensure that we identify the unknown substances the gas chromatography apparatus can be attached directly to a **mass spectrometer.** This identifies substances very quickly and accurately and can detect very small quantities in the sample.

Higher

Measuring relative molecular masses

A mass spectrometer also provides an accurate way of measuring the relative molecular (formula) mass of a compound. The peak with the largest mass corresponds to an ion with just one electron removed. As you know, the mass of an electron is so small that it can be ignored when we look at the mass of atoms. This peak is called the **molecular ion peak**. It is always found as the last peak on the right as you look at a mass spectrum. The molecular ion peak of the substance analysed in Figure 3 is at 45. So the substance has a relative molecular mass of 45.

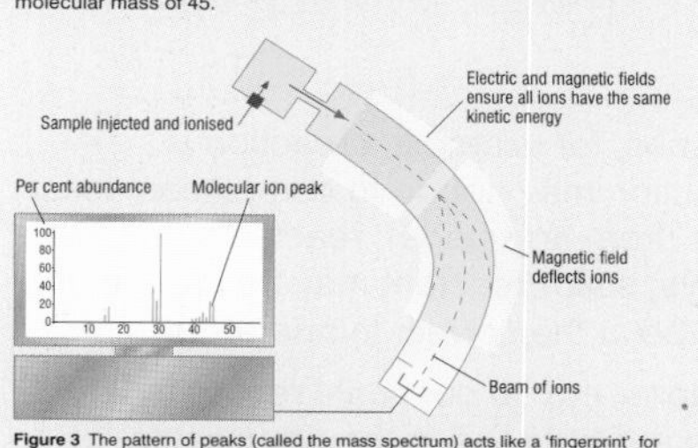

Figure 3 The pattern of peaks (called the mass spectrum) acts like a 'fingerprint' for unknown compounds. The pattern is quickly matched against a database of known compounds stored on computer.
NB You don't need to remember the details of how a mass spectrometer works.

Summary questions

1 Copy and complete using the words below:
chromatography database mass mixture fingerprint

Separating a .... of compounds can be carried out by gas .... Identifying compounds once they have been separated then uses techniques like .... spectrometry. The pattern of peaks is like a .... for each unknown compound. It is matched against known compounds on a computer ....

2 Describe how a mass spectrometer can be used to find the relative molecular mass of a compound. [H]

Key points

- Compounds in a mixture can be separated using gas chromatography.
- Once separated, compounds can be identified using a mass spectrometer.
- The mass spectrometer can be used to find the relative molecular mass of a compound from its molecular ion peak. [H]

153

Summary answers

1 mixture, chromatography, mass, fingerprint, database

2 Look to the far right of the mass spectrum to find the molecular ion peak (the peak with the highest relative mass). This gives you the relative molecular mass of the substance under analysis.

Summary answers

1 a B   b A   c D   d C

2 a 18 g
  b 16 g
  c 87 g
  d 102 g
  e 138 g
  f 158 g
  g 89 g

3 a 1   c 0.25   e 0.01
  b 3   d 0.2   f 0.05

4 a 75%
  b 8 g

5 $AlBr_3$

6 95%

7 a A reaction that can go in either direction, i.e. products → reactants as well as reactants → products.
  b An 'ordinary' reaction is normally regarded as going in one direction only, i.e. the 'forward' direction.
  c $C_2H_4 + H_2O \rightleftharpoons C_2H_5OH$

8 a 1
  b 98 kg
  c 96%
  d Any two valid reasons, for example a reaction is reversible or a reaction may not go to completion; loss/escape of product (from any stage); reactants may react in unexpected ways; sulfur/reactant may be impure; the catalyst might not be at the correct temperature.
  e Any two valid reasons, e.g. to conserve resources/sulfur; to conserve energy used in the process; to reduce energy needed for transporting sulfur; to reduce pollution/acid rain/loss of sulfur dioxide/trioxide.

How much?: C2 3.1–C2 3.8

Summary questions

1 Match up the parts of the sentences:

a	Neutrons have a relative mass of …	A	… negligible mass compared to protons and neutrons.
b	Electrons have …	B	… 1 compared to protons.
c	Protons have a relative mass of …	C	… found in its nucleus.
d	Nearly all of an atom's mass is …	D	… 1 compared to neutrons.

2 Calculate the mass of 1 mole of each of the following compounds:

a H_2O
b CH_4
c MnO_2
d Al_2O_3
e K_2CO_3
f $KMnO_4$
g $Mn(OH)_2$

(A_r values: C = 12, O = 16, Al = 27, H = 1, Ca = 40, K = 39, Mn = 55)

3 How many moles of:

a Ag atoms are there in 108 g of silver,
b P atoms are there in 93 g of phosphorus,
c Ag atoms are there in 27 g of silver,
d P atoms are there in 6.2 g of phosphorus,
e Fe atoms are there in 0.56 g of iron,
f P_4 molecules are there in 6.2 g of phosphorus?

(A_r values: Ag = 108, P = 31, Fe = 56)

4 a The chemical formula of methane is CH_4. Use the relative atomic masses in question 2 to work out the percentage by mass of carbon in methane.
  b In 32 g of methane, work out the mass of hydrogen present in the compound.

5 When aluminium reacts with bromine, 4.05 g of aluminium reacts with 36.0 g of bromine. What is the empirical formula of aluminium bromide?
(A_r values: Al = 27, Br = 80) [H]

6 In a lime kiln, calcium carbonate is decomposed to calcium oxide:

$CaCO_3 \rightarrow CaO + CO_2$

50.0 tonnes of calcium carbonate gave 26.6 tonnes of calcium oxide. Calculate the percentage yield for the process.
(A_r values: Ca = 40, O = 16, C = 12) [H]

7 a What is a reversible reaction?
  b How does a reversible reaction differ from an 'ordinary' reaction?
  c Ethene (C_2H_4) reacting with steam (H_2O) to form ethanol (C_2H_5OH) is a reversible reaction. Write the balanced symbol equation for this reaction.

8 Sulfur is mined in Poland and is brought to Britain in ships. The sulfur is used to make sulfuric acid. Sulfur is burned in air to produce sulfur dioxide. Sulfur dioxide and air are passed over a heated catalyst to produce sulfur trioxide. Water is added to sulfur trioxide to produce sulfuric acid. The reactions are:

$S + O_2 \rightleftharpoons SO_2$

$2SO_2 + O_2 \rightleftharpoons 2SO_3$

$SO_3 + H_2O \rightarrow H_2SO_4$

Relative atomic masses: H = 1; O = 16; S = 32

a How many moles of sulfuric acid are produced from one mole of sulfur?
b Calculate the maximum mass of sulfuric acid that can be produced from 32 kg of sulfur.
c In an industrial process the mass of sulfuric acid that was produced from 32 kg of sulfur was 94.08 kg. Use your answer to part b to calculate the percentage yield of this process.
d Suggest two reasons why the yield of the industrial process was less than the maximum yield.
e Give two reasons why the industrial process should produce a yield that is as close to the maximum yield as possible. [H]

154

Practical suggestions

Practicals	AQA	k	Book	Gears
Investigating food colours using paper chromatography.	✓		✓	
Working out the empirical formulae of copper oxide and magnesium oxide.	✓		✓	
Calculating yields, for example magnesium burning to produce magnesium oxide or wire wool burning to produce iron oxide.	✓	✓	✓	
There are opportunities in this section to build in the idea of instrumentation precision, e.g. for the collection of gases, the use of boiling tubes, gas jars or gas syringes.	✓		✓	
Copper sulfate – hydration/dehydration.	✓		✓	
Heating ammonium chloride in a test tube.	✓		✓	
Adding alkali and acid alternately to bromine water or to potassium chromate solution	✓		✓	
'Blue bottle' reaction (RSC Classic Chemistry Experiments no. 83).	✓		✓	
Oscillating reaction (RSC Classic Chemistry Experiments no.140).	✓		✓	

QA Examination-style questions k

1 a An atom of phosphorus can be represented as:

$^{31}_{15}P$

i What is the number of protons in this atom of phosphorus? (1)

ii What is the number of neutrons in this atom of phosphorus? (1)

iii What are the number of electrons in this atom of phosphorus? (1)

b A different atom of phosphorus can be represented as:

$^{32}_{15}P$

i What are these two atoms of phosphorus known as? (1)

ii Give one way in which these two atoms of phosphorus are different. (1)

2 Toothpastes often contain fluoride ions to help protect teeth from attack by bacteria.

Some toothpastes contain tin(II) fluoride.

This compound has the formula SnF_2.

a Calculate the relative formula mass (M_r) of SnF_2.

(Relative atomic masses: F = 19; Sn = 119) (2)

b Calculate the percentage by mass of fluorine in SnF_2. (2)

c A tube of toothpaste contains 1.2 g of SnF_2. Calculate the mass of fluorine in this tube of toothpaste. (1)

AQA, 2008

3 The diagram shows what happens when ammonium chloride is heated.

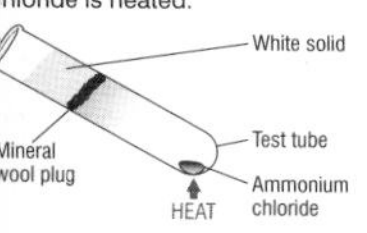

The reaction that takes place is:

$NH_4Cl(s) \rightleftharpoons NH_3(g) + HCl(g)$

a What does $\rightleftharpoons$ in the equation mean? (1)

b Explain why the white solid appears near the top of the test tube. (2)

4 The diagram shows the main parts of an instrumental method called gas chromatography linked to mass spectroscopy (GC–MS).

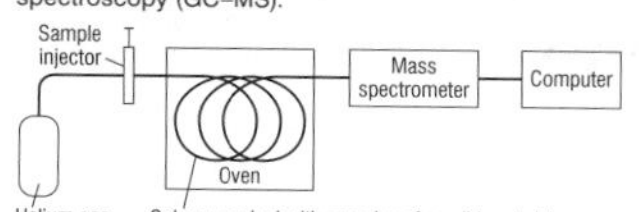

This method separates a mixture of compounds and then helps to identify each of the compounds in the mixture.

a In which part of the apparatus:

i is the mixture separated? (1)

ii is the relative molecular mass of each of the compounds in the mixture measured? (1)

iii are the results of the experiment recorded? (1)

b i Athletes sometimes take drugs because the drugs improve their performance. One of these drugs is ephedrine.

Ephedrine has the formula:

$C_{10}H_{15}NO$

What relative molecular mass (M_r) would be recorded by GC–MS if ephedrine was present in a blood sample taken from an athlete?

Show clearly how you work out your answer.

(Relative atomic masses: H = 1; C = 12; N = 14; O = 16.) (2)

ii Another drug is amphetamine, which has the formula: $C_9H_{13}N$

The relative molecular mass (M_r) of amphetamine is 135.

Calculate the percentage by mass of nitrogen in amphetamine. (Relative atomic mass: N = 14.) (2)

c Athletes are regularly tested for drugs at international athletics events. An instrumental method such as GC–MS is better than methods such as titration. Suggest why. (2)

AQA, 2010

5 A chemist thought a liquid hydrocarbon was hexane, C_6H_{14}.

Relative atomic masses: H = 1; C = 12

a Calculate the percentage of carbon in hexane. (2)

b The chemist analysed the liquid hydrocarbon and found that it contained 85.7% carbon. Calculate the empirical formula of the hydrocarbon based on this result. You must show your working to gain full marks. (4)

c Was the liquid hydrocarbon hexane? Explain your answer. [H] (1)

Kerboodle resources

Resources available for this chapter on Kerboodle are:

- Chapter map: How much?
- Support: All about atoms (C2 3.1)
- How Science Works: You need the right formula (C2 3.3)
- Maths skills: Equations and calculations (C2 3.4)
- Extension: How much? (C2 3.4)
- Viewpoints: Making chemistry better (C2 3.5)
- Practical: Calculating the percentage yield of a reaction (C2 3.5)
- Interactive activity: How much?
- Revision podcast: Mass and atoms
- Test yourself: How much?
- On your marks: How much?
- Examination-style questions: How much?
- Answers to examination-style questions: How much?

Examination-style answers

1 a i 15 *(1 mark)*

ii 16 *(1 mark)*

iii 15 *(1 mark)*

b i isotopes *(1 mark)*

ii One from: different mass (numbers), (number of) neutrons, total number of (subatomic) particles, (physical) properties *(but do not allow chemical properties)*. *(1 mark)*

2 a Give two marks for 157. If answer is incorrect either 2 × 19 + 119 or 119 + 19 = 138 gains one mark. *(2 marks)*

b Give two marks for 24.2 (accept answers in the range 24 to 24.2038). If the answer is incorrect, 25 or 38/157 × 100 or 19/157 × 100 = 12 to 12.1 or 19/138 × 100 gains one mark (allow error carried forward from part **a** so 38/**a** × 100 gains two marks if calculated correctly). *(2 marks)*

c 0.29 (accept answers in the range 0.28 to 0.3 and allow error carried forward from part **b**) *(1 mark)*

3 a Reversible (reaction). *(Accept reaction goes both ways)*. *(1 mark)*

b Heating: causes (thermal) decomposition or forward reaction and so produces gases (ammonia gas *and* hydrogen chloride gas or NH_3 and HCl). The gases are cooled near the top of the test tube, which causes the reverse reaction and so solid ammonium chloride is formed. *(2 marks)*

4 a i the column *(1 mark)*

ii the mass spectrometer *(1 mark)*

iii the computer *(1 mark)*

b i Give two marks for 165. If the answer is incorrect then evidence of correct working gains one mark, e.g 10 × 12 + 15 + 14 + 16. *(2 marks)*

ii Give two marks for 10.37%. If the answer is incorrect, then evidence of correct working gains one mark. $\frac{14}{135} \times 100\% = 10.37\%$ *(2 marks)*

c Two from: faster, more accurate, detects smaller amounts. *(2 marks)*

5 a Give two marks for 83.7 (per cent). (*Accept answers in the range 83.72 to 84.*) If the answer is incorrect, evidence of correct working gains one mark, e.g. 6 × 12/(6 × 12 + 14) × 100. *(2 marks)*

b Give one mark for each correct step: one mark for mass of hydrogen – (100 g of hydrocarbon contains 85.7 g C and 14.3 g H); one mark for dividing masses by correct A_r – 85.7/12 and 14.3/1; one mark for correct proportions – 7.14… and 14.3 or simplified ratio 1:2; 1 mark for correct empirical formula CH_2. *(4 marks)*

c Accept 'no' with a valid explanation – e.g. because the empirical formula is different/that of an alkene or it gives C_6H_{12} for six carbon atoms or it is not an alkane or because the percentage of carbon is different and compounds have fixed proportions (of elements). *Accept the answer that one cannot be sure if there is a valid explanation*, e.g. because of experimental error or only one result or one needs to confirm the result or to repeat the test. *Do not accept yes.* *(1 mark)*

C2 4.1 How fast?

Learning objectives

Students should learn:

- that chemical reactions can happen at different rates
- how the rate of a chemical reaction can be calculated.

Learning outcomes

Most students should be able to:

- state a definition of the rate of reaction
- list ways of finding the rate
- suggest a method for finding the rate in a specified reaction.

Some students should also be able to:

- explain in detail why a particular method of finding the rate is suitable for a specified reaction.

Answers to in-text questions

a How fast the reactants are turned into products.

b In order to understand (and control) how fast chemicals are made. Economics require speedy reactions to cut production and energy costs.

Summary answers

1 reactants, products, rate, slope, time

2 a Graphs for **i** and **ii** are shaped like those in the second and first practicals respectively in the Student Book.

b The steeper the graph, the faster the rate of the reaction.

Support

- Some students find graph paper confusing. Therefore supply these students with squared paper, with the scales already drawn on. Once the students have plotted their graph, you could supply these statements (in the wrong order) for students to copy on to their graph.
  - Start of reaction.
  - Fast rate of reaction.
  - Slow rate of reaction.
  - End of reaction.

Extend

- Ask students to calculate the gradient at different points of the graph and actually give a value for the rate of reaction.

Specification link-up: Chemistry C2.4

- The rate of a chemical reaction can be found by measuring the amount of a reactant used or the amount of product formed over time:

$$\text{Rate of reaction} = \frac{\text{amount of reactant used}}{\text{time}}$$

$$\text{Rate of reaction} = \frac{\text{amount of product formed}}{\text{time}}$$

[C2.4.1 a)]

- Interpret graphs showing the amount of product formed (or reactant used up) with time, in terms of the rate of the reaction. *[C2.4]*

  Controlled Assessment: C4.3 Collect primary and secondary data. *[C4.3.2 b) c)]*; C4.5 Analyse and interpret primary and secondary data. *[C4.5.2 b) d)]*

Lesson structure

Starters

Fast or slow – Give each student a piece of paper with 'fast' on one side and 'slow' on the other. Then show them different images of chemical reactions, e.g. rusting, baking a cake, cooking an egg, magnesium reacting with acid and neutralisation. They should look at each image, decide if the rate is fast or slow and hold up the card to demonstrate their answer. Images could be shown using a data projector or PowerPoint slides. *(5 minutes)*

Cut and stick – Explain to students what 'rate of reaction' is. Then give the students different magazines and catalogues. Their task is to cut out an example of a reaction with a fast rate and a slow rate. They then compare their findings with a small group of students, and they choose the best from their selection. Students could then make a class montage on sugar paper. You could support students by giving them some examples of fast and slow reactions that they could find in the magazines. You could extend students by encouraging them to represent the reactions as a word or symbol equation. *(10 minutes)*

Main

- In many chemical reactions that we use to study rate, a gas is made. Students should produce a graph so that they can interpret information about a reaction. Using the reaction between magnesium and acid, ask students to plot a graph to show the production of hydrogen over two minutes. Then ask them to annotate their graph to explain the shape. They should include information that explains why the graph starts at 0 cm^3 of gas and they should explain the shape of the graph.
- The reaction between hydrochloric acid and sodium thiosulfate is the classic example used to highlight how light can be used to determine the rate of reaction. Students could complete this experiment themselves and reflect on the technique. Alternatively, this could be completed as a class demonstration. Draw the black cross on an overhead transparency and put it onto an overhead projector. Choose a student to be in charge of the stopwatch. Then put the reaction vessel on to the cross and ask students to raise their hands when they think the cross has gone. When most of the hands are raised, stop the watch and note the time.
- Compare this with using a light sensor and data logger. A plot of the light intensity could be made. Use this as an opportunity to teach about the 'Controlled Assessment' concepts of reliability, accuracy, and experimental error. The disappearing cross experiment relies on a person deciding when the cross disappears and this introduces errors, which data logging can help to address.
- There are also opportunities here to show students how to make measurements using sensors (e.g. carbon dioxide, oxygen, pH, gas pressure and temperature) to investigate reaction rates.
- After the students have watched the demonstration of the reaction between magnesium and acid, allow them to experiment with the different methods for measuring rate of reaction detailed in 'Practical support'. They will need a supply of the hydrochloric acid and magnesium strips (and eye protection). The aim of this activity is for the students to decide the most informative method. [They should realise that the balances available are not sensitive enough (need higher resolution), there is no precipitate formed, so the disappearing cross is useless, but collection of gas is the most accurate if a gas syringe is used.] You may wish to use a pH probe to monitor the reactions. You may wish to extend students further by giving them a selection of equipment to measure the volume of gas such as a gas syringe, a water trough and a measuring cylinder or burette. Again, encourage students to evaluate the different methods and decide which would be the most accurate [gas syringe].

Plenaries

Graph interpretation – Show students a graph of the rate of a particular reaction. Ask them to interpret the graph and explain the shape of the curve. Support students by giving them graphs on squared paper and with simple numbers. Extend students by using more complex scales and data points that involve decimals. *(5 minutes)*

Card sort – Make a set of eight cards: four with diagrams showing how the rate of reaction can be found (mass change, gas collection by displacement in a measuring cylinder, gas collection in a gas syringe, disappearing cross) and four with examples of reactions that can be measured using these techniques (magnesium + acid, calcium carbonate + acid, sodium thiosulfate + acid, hydrogen peroxide + manganese dioxide). Ask the students to match the methods with reactions. Encourage them to discuss their work in small groups, then feed back to the class in a question-and-answer session. [Note: Only the sodium thiosulfate reaction can be measured using the disappearing cross. *(10 minutes)*

Further teaching suggestions

Spider diagram

- Collision theory is a fundamental concept that underpins the whole of this chapter. Students could start a spider diagram using the key points from this spread. At the end of each lesson during this chapter, encourage students to add the new key points to their diagram.

Practical support

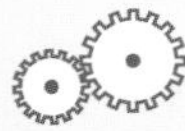

Measuring the mass of a reaction mixture

Equipment and materials required

Marble chips, 1 mol/dm³ hydrochloric acid (CLEAPSS Hazcard 47A), 250 cm³ conical flask, top-pan balance, cotton wool, stopwatch, measuring cylinder, eye protection.

Measuring the volume of gas given off

Equipment and materials required

Marble chips, 1 mol/dm³ hydrochloric acid (CLEAPSS Hazcard 47A), 250 cm³ conical flask, bung fitted with delivery tube, about 50 cm length of rubber tubing, 100 cm³ gas syringe (ensure syringe plunger is free moving), gas syringe holder, boss, stand, stopwatch, measuring cylinder, eye protection.

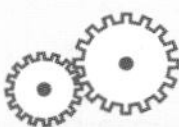

Measuring the light transmitted through a solution

Equipment and materials required

Two measuring cylinders, stopwatch, paper with large cross in the centre, conical flask, beaker, 0.2 mol/dm³ sodium thiosulfate, 0.2 mol/dm³ hydrochloric acid (irritant), eye protection (chemical splashproof goggles).

Safety: During this experiment, sulfur dioxide is produced which is toxic and can trigger asthmatic attacks. Therefore this should be completed in a well-ventilated room. Once the reaction is complete, the mixture should be disposed of following CLEAPSS guidelines in Hazcard 95C. CLEAPSS Hazcard 97 Sulfur dioxide – toxic.

Demonstration of the reaction between magnesium and acid

Equipment and materials required

Hydrochloric acid (1 mol/dm³) (CLEAPSS Hazcard 47A), strips of 1 cm magnesium ribbon (CLEAPSS Hazcard 59A), test tube, eye protection.

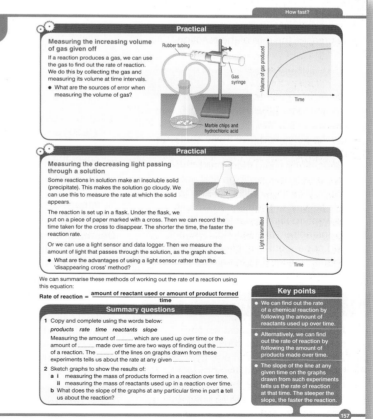

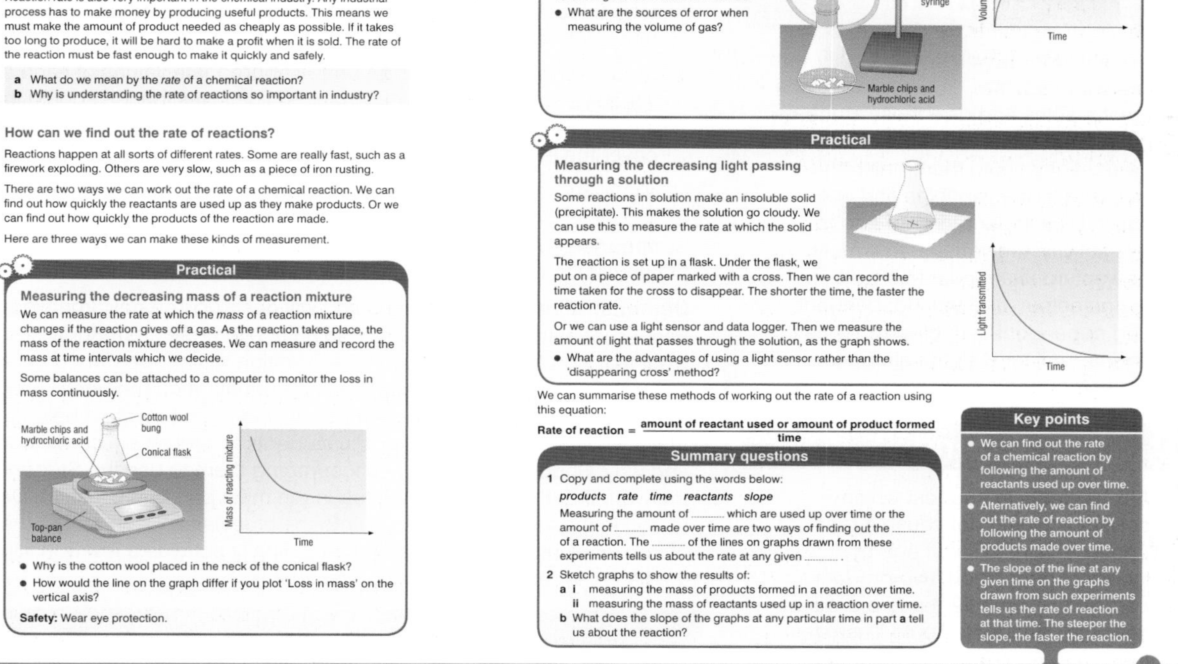

Rates and energy

How fast?

C2 4.1 How fast?

Learning objectives

- What do we mean by the rate of a chemical reaction?
- How can we find out the rate of a chemical reaction?

Figure 1 All living things depend on very precise control of the many chemical reactions happening inside their cells

The rate of a chemical reaction tells us how fast reactants turn into products. In your body, there are lots of reactions taking place all the time. They happen at rates which supply your cells with what they need, whenever required.

Reaction rate is also very important in the chemical industry. Any industrial process has to make money by producing useful products. This means we must make the amount of product needed as cheaply as possible. If it takes too long to produce, it will be hard to make a profit when it is sold. The rate of the reaction must be fast enough to make it quickly and safely.

a What do we mean by the *rate* of a chemical reaction?
b Why is understanding the rate of reactions so important in industry?

How can we find out the rate of reactions?

Reactions happen at all sorts of different rates. Some are really fast, such as a firework exploding. Others are very slow, such as a piece of iron rusting.

There are two ways we can work out the rate of a chemical reaction. We can find out how quickly the reactants are used up as they make products. Or we can find out how quickly the products of the reaction are made.

Here are three ways we can make these kinds of measurement.

Practical

Measuring the decreasing mass of a reaction mixture

We can measure the rate at which the *mass* of a reaction mixture changes if the reaction gives off a gas. As the reaction takes place, the mass of the reaction mixture decreases. We can measure and record the mass at time intervals which we decide.

Some balances can be attached to a computer to monitor the loss in mass continuously.

Marble chips and hydrochloric acid
Cotton wool bung
Conical flask
Top-pan balance
Mass of reacting mixture
Time

- Why is the cotton wool placed in the neck of the conical flask?
- How would the line on the graph differ if you plot 'Loss in mass' on the vertical axis?

Safety: Wear eye protection.

AQA Examiner's tip

The steeper the line on the graph, the faster the reaction rate.

Practical

Measuring the increasing volume of gas given off

If a reaction produces a gas, we can use the gas to find out the rate of reaction. We do this by collecting the gas and measuring its volume at time intervals.

- What are the sources of error when measuring the volume of gas?

Rubber tubing
Gas syringe
Marble chips and hydrochloric acid
Volume of gas produced
Time

Practical

Measuring the decreasing light passing through a solution

Some reactions in solution make an insoluble solid (precipitate). This makes the solution go cloudy. We can use this to measure the rate at which the solid appears.

The reaction is set up in a flask. Under the flask, we put on a piece of paper marked with a cross. Then we can record the time taken for the cross to disappear. The shorter the time, the faster the reaction rate.

Or we can use a light sensor and data logger. Then we measure the amount of light that passes through the solution, as the graph shows.

- What are the advantages of using a light sensor rather than the 'disappearing cross' method?

Light transmitted
Time

We can summarise these methods of working out the rate of a reaction using this equation:

$$\text{Rate of reaction} = \frac{\text{amount of reactant used or amount of product formed}}{\text{time}}$$

Summary questions

1 Copy and complete using the words below:
products rate time reactants slope
Measuring the amount of .... which are used up over time or the amount of .... made over time are two ways of finding out the .... of a reaction. The .... of the lines on graphs drawn from these experiments tells us about the rate at any given ....

2 Sketch graphs to show the results of:
a i measuring the mass of products formed in a reaction over time.
ii measuring the mass of reactants used up in a reaction over time.
b What does the slope of the graphs at any particular time in part **a** tell us about the reaction?

Key points

- We can find out the rate of a chemical reaction by following the amount of reactants used up over time.
- Alternatively, we can find out the rate of reaction by following the amount of products made over time.
- The slope of the line at any given time on the graphs drawn from such experiments tells us the rate of reaction at that time. The steeper the slope, the faster the reaction.

156 157

C2 4.2 Collision theory and surface area

Learning objectives

Students should learn:

- that different factors affect the rate of reaction
- what collision theory is
- how collision theory can be used to explain the effect of surface area on the rate of reaction.

Learning outcomes

Most students should be able to:

- list the factors that affect the rate of reaction
- recall a definition of collision theory
- describe how surface area affects the rate of reaction.

Some students should also be able to:

- explain in detail collision theory
- apply collision theory to explain in detail how surface area affects the rate of reaction.

Answers to in-text questions

- **a** The particles must collide with sufficient energy for a reaction to occur.
- **b** The activation energy.
- **c** The same mass of small pieces of wood.
- **d** Increasing the surface area increases the rate of the reaction.

Support

- Some students will have difficulty with isolating variables. You could provide the variables and give these students suggested measurements on separate cards. They could then match the values with the variables and use this in their experiment. To add to this activity, the variable that must be kept constant could be written in one colour, and the independent and dependent variables could be in another colour. Ask the students what the colour code means.

Extend

- Ask students to suggest which collisions would not cause a reaction (those with insufficient energy, incorrect orientation, two product particles or a reactant and a product colliding). Students could illustrate these factors as particle diagrams.

Specification link-up: Chemistry C2.4

- Chemical reactions can only occur when reacting particles collide with each other and with sufficient energy. The minimum amount of energy particles must have to react is called the activation energy. *[C2.4.1 b)]*
- Increasing the surface area of solid reactants increases the frequency of collisions and so increases the rate of reaction. *[C2.4.1 f)]*
- Interpret graphs showing the amount of product formed (or reactant used up) with time, in terms of the rate of the reaction.

  Controlled Assessment: C4.2 Assess and manage risks when carrying out practical work. *[C4.2.1 a) b)]*; C4.3 Collect primary and secondary data. *[C4.3.2 a)]*; C4.4 Select and process primary and secondary data. *[C4.4.2 c)]*

Lesson structure

Starters

Labelling – Give students a diagram of the equipment used to monitor the rate of the reaction of marble (calcium carbonate) with acid, when mass loss is being measured. Ask them to label the equipment. Support students by giving them the labels and completing the activity as a 'cut-and-stick' exercise. Extend students by explaining how this equipment could be used to monitor rate of reaction. *(5 minutes)*

Thinking – Ask students to work in pairs and list ways of monitoring the rate of reaction between an acid and a metal [mass loss, gas production]. Then ask students for ways in which this reaction could be speeded up [heat the acid, increase the surface area of the metal, increase the concentration of the acid]. *(10 minutes)*

Main

- Magnesium is often found as ribbon or in powdered form in schools. Show the students samples of each. Ask them to predict which would have the faster rate of reaction and why (using collision theory to explain). In their prediction, they should include a word equation (balanced symbol equation for Higher Tier students) to represent the reaction and what observations they would expect. They could then complete a risk assessment for the experiment, carry it out (once checked) and see if their prediction was correct.
- Marble chips come in a variety of sizes but each size is within a range. Show the students samples of different marble chips. Explain that they are going to investigate the mass lost during this experiment in order to decide how the surface area affects the rate. Encourage students to consider what the variables are in the experiment [time, temperature, concentration and volume of acid, mass of marble].
- Students should then decide on the appropriate values of each and detail which variables should be control variables. The investigation should then be completed and a conclusion written using collision theory. This provides an excellent opportunity to cover the investigative aspects of the Controlled Assessment.

Plenaries

Summarise – Ask the students to make a bullet-point list of facts about collision theory. *(5 minutes)*

Demonstration – For this demo you will need an iron nail, iron wool, iron filings, tongs, a heat-proof mat, a Bunsen burner, spatula, eye protection. Show students the iron nail, iron wool and iron filings. Ask them what reaction will happen when the iron is put in the flame and ask for a volunteer to write the word equation on the board [iron + oxygen → iron oxide]. You could extend students by asking them to write a balanced symbol equation [4]Fe + [3]O_2 → [2]Fe_2O_3. Ask the students to predict which will combust most quickly and why. Support students by encouraging them to use the Student Book and think about the work that they have completed in the main part of the lesson. Demonstrate each type of iron in the flame: hold the nail into the blue flame using tongs; then hold a small piece of iron wool into the flame using tongs; finally sprinkle a few iron filings from a spatula into the flame. Wear eye protection. *(10 minutes)*

Practical support

Which burns faster – ribbon or powder?

Equipment and materials required

Bunsen burner and safety equipment, 2 cm length of magnesium ribbon (CLEAPSS Hazcard 59A), magnesium powder (highly flammable), spatula, tongs, stopwatch, eye protection.

Details

Hold the end of the magnesium ribbon with tongs. Put the tip of the ribbon in the top of the blue gas cone. As soon as the ribbon ignites, remove it from the flame and observe. Then sprinkle about half a spatula of magnesium powder directly into the blue flame (held at an angle).

Safety: Eye protection should be worn throughout this practical. Magnesium oxide powder will be made and may enter the air; this could irritate airways and therefore the reaction should be completed in a well-ventilated room. When magnesium ribbon burns, a very bright white light is produced. This can blind if people look at it directly. Therefore encourage students to look past the reaction or alternatively use special blue glass/plastic to mute the light.

Investigating surface area

Equipment and materials required

Conical flask, 1 mol/dm³ dilute hydrochloric acid (CLEAPSS Hazcard 47A), cotton wool, top-pan balance, marble chips of different sizes, measuring cylinder, stopwatch, eye protection.

Details

Wash and dry the marble chips (to remove the powder from the surface). Measure out about 1 g of marble chips of a certain size into a conical flask. Eye protection should be worn at this point. Add 25 cm³ of acid and put in the cotton wool plug. Put the reaction vessel on to the top-pan balance and observe the mass change over time. Repeat the experiment with different sized marble chips.

Control variables are: concentration of acid, volume of acid, mass of marble, temperature.

Safety: Eye protection should be worn.

Rates and energy

C2 4.2 Collision theory and surface area

Learning objectives

- What affects the rate of a chemical reaction?
- What is collision theory?
- How does collision theory explain the effect of surface area on reaction rate?

Figure 1 There is no doubt that the chemicals in these fireworks have reacted. But how can we explain what happens in a chemical reaction?

Figure 2 Cooking – an excellent example of controlling reaction rates!

In everyday life we control the rates of chemical reactions. People often do it without knowing! For example, cooking cakes in an oven or revving up a car engine. In chemistry we need to know what affects the rate of reactions. We also need to explain why each factor affects the rate of a reaction.

There are four main factors which affect the rate of chemical reactions:

- temperature
- surface area
- concentration of solutions or pressure of gases
- presence of a catalyst.

Reactions can only take place when the particles (atoms, ions or molecules) of reactants come together. But the reacting particles don't just have to bump into each other. They also need enough energy to react when they collide. This is known as **collision theory**.

The smallest amount of energy that particles must have before they can react is called the **activation energy**.

So reactions are more likely to happen between reactant particles if we:

- increase the chance of reacting particles colliding with each other
- increase the energy that they have when they collide.

If we increase the chance of particles reacting, we will also increase the rate of reaction.

a What must happen before two particles have a chance of reacting?
b Particles must have a minimum amount of energy to be able to react. What is this energy called?

Surface area and reaction rate

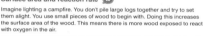

Imagine lighting a campfire. You don't pile large logs together and try to set them alight. You use small pieces of wood to begin with. Doing this increases the surface area of the wood. This means there is more wood exposed to react with oxygen in the air.

When a solid reacts in a solution, the size of the pieces of solid affects the rate of the reaction. The particles inside a large lump of solid are not in contact with the solution, so they can't react. The particles inside the solid have to wait for the particles on the surface to react first.

In smaller lumps, or in a powder, each tiny piece of solid is surrounded by solution. More particles are exposed to attack. This means that reactions can take place much more quickly.

c Which has the larger surface area – a log or the same mass of small pieces of wood?
d How does the surface area of a solid affect its rate of reaction?

158

Collision theory and surface area

Practical

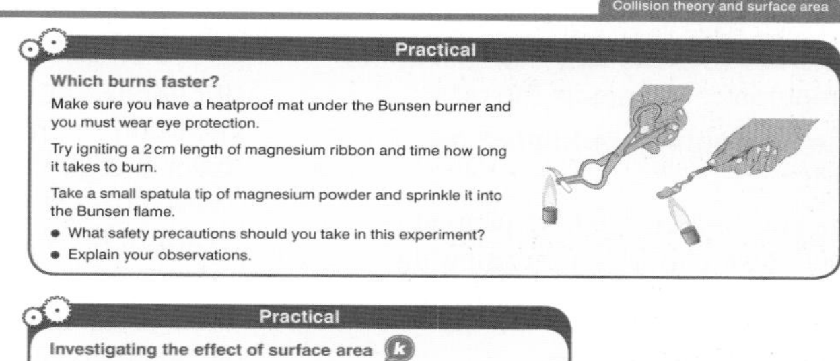

Which burns faster?

Make sure you have a heatproof mat under the Bunsen burner and you must wear eye protection.

Try igniting a 2 cm length of magnesium ribbon and time how long it takes to burn.

Take a small spatula tip of magnesium powder and sprinkle it into the Bunsen flame.

- What safety precautions should you take in this experiment?
- Explain your observations.

Practical

Investigating the effect of surface area

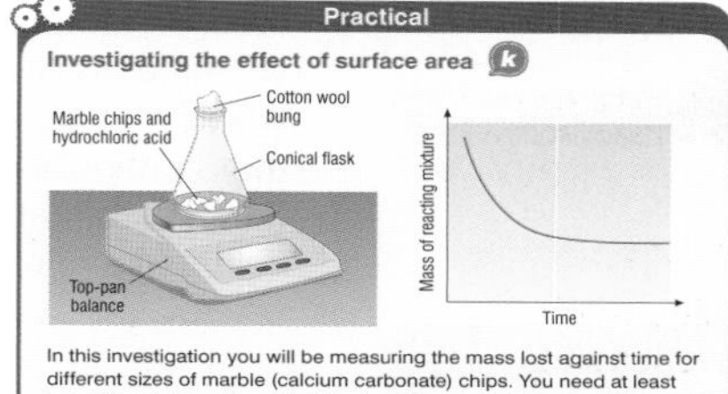

In this investigation you will be measuring the mass lost against time for different sizes of marble (calcium carbonate) chips. You need at least two different sizes of marble chips in order to vary the surface area.

- What variables should you control to make this a fair test?
- Why does this method of finding out the rate of reaction work?
- Use the data collected to draw a graph. Explain what the graph shows. (A data logger would help to plot a graph of the results.)

Safety: Wear eye protection.

AQA Examiner's tip

Particles collide all the time, but only some collisions lead to reactions.
Increasing the number of collisions in a certain time and the energy of collisions produces faster rates.
Larger surface area does not result in collisions with more energy but does increase the frequency of collisions.

Summary questions

1 Copy and complete using the words below:
energy activation collide frequently minimum
Particles can react with each other only when they .... with sufficient .... Reaction rates increase when collisions are more energetic and/or happen more .... The .... amount of energy needed for particles to react is known as the .... energy.

2 Draw a diagram to explain why it is easier to light a fire using small pieces of wood rather than large logs.

3 Why do you digest your food more quickly if you chew it well before you swallow it?

Key points

- Particles must collide, with a certain amount of energy, before they can react.
- The minimum amount of energy that particles must have in order to react is called the activation energy.
- The rate of a chemical reaction increases if the surface area of any solid reactants is increased. This increases the frequency of collisions between reacting particles.

159

Further teaching suggestions

Evaluation

- Students could complete an evaluation about the calcium carbonate and acid reaction. For example, the problems that they might have found are: 'The same mass of marble in each experiment could not be achieved because of natural variation in the marble', 'Mass was lost before the reaction vessel was put on the balance', 'The balances aren't sensitive enough to get enough information to draw graphs.' Students could then consider ways to change the method, to try to reduce these problems and ultimately the errors in the experiment.

Bath bomb

- A bath bomb is mainly a metal carbonate and an acid. When they are put into water a neutralisation reaction happens. As the two reactants dissolve, they can collide and react. Ask students to explain why a bath bomb takes a long time to react with water, but if it is crumbled up, it reacts more quickly. You could even make a bath bomb in the lesson and investigate how the surface area of the bath bomb affects the rate at which it reacts.

Summary answers

1 collide, energy, frequently, minimum, activation

2 Students should draw a diagram clearly explaining/showing that cutting a solid into smaller pieces increases its surface area.

3 Because you have increased the food's surface area for reaction with digesting chemicals.

C2 4.3 The effect of temperature

Learning objectives

Students should learn:

- that increasing the temperature affects the rate of reactions
- how collision theory can be used to explain how temperature affects rate of reaction.

Learning outcomes

Most students should be able to:

- describe how increasing the temperature affects the rate of reactions
- state two reasons why increasing the temperature increases the rate.

Some students should also be able to:

- use collision theory to explain in detail how and why increasing the temperature changes the rate of reaction.

Support

- Some students will have difficulty in understanding the link between rate of reaction and temperature. Using a familiar situation can help students to understand abstract concepts more easily. Link temperature and rate of reaction with dissolving sugar into tea. Ask students to predict which cup of tea will dissolve sugar the quickest. This physical change could be demonstrated or carried out by the students. Each group from the class could put its results from the thiosulfate experiment into spreadsheet. This could then be used to calculate a class mean quickly and plot a graph of these results.

Extend

- Students could consider the effect of temperature on a reaction where the particles are in a solid state. As the particles are only vibrating, not moving, there will be no increase or decrease in frequency of collisions as temperature is changed. However, the solids would need to be mixed for the reactant particles to come into contact with each other. They would then need an input of energy to overcome the activation energy.

Specification link-up: Chemistry C2.4

- Increasing the temperature increases the speed of the reacting particles so that they collide more frequently and more energetically. This increases the rate of reaction. *[C2.4.1 c)]*
- Interpret graphs showing the amount of product formed (or reactant used up) with time, in terms of the rate of the reaction. *[C2.4]*

Controlled Assessment: C4.1 Plan practical ways to develop and test candidates' own scientific ideas. *[C4.1.1 c)]*; C4.3 Collect primary and secondary data. *[C4.3.2 b) c)]*; C4.5 Analyse and interpret primary and secondary data. *[C4.5.2 b) d)]*

Lesson structure

Starters

Pictionary – Create packs of cards with these statements on: 'increase temperature', 'collision', 'particle', 'reactant', 'product', 'rate', 'chemical reaction'. Split the students into pairs and give them a pack of cards. Ask them to take it in turns to pick up the card and draw a picture (with no text, symbols or numbers) to get their partner to say the statement or word. *(5 minutes)*

What is in the bag? – In a bag, put in key words on separate pieces of paper: 'collision theory', 'temperature', 'particles', 'collision', 'rate', 'activation energy'. Then ask for a volunteer to come to the front and remove a key word. They can then explain what that words means. Allow students who have difficulty to go back to their seat and look through the Student Book, then talk to neighbours; invite them back at the end of the Starter. Extend students by asking them to use the word correctly in a sentence. In total, ask five volunteers to explain each of the words in turn. *(10 minutes)*

Main

- Students could experimentally determine the effect of temperature on rate of reaction by using the sodium thiosulfate reaction. As a class, decide on five temperatures that will be used – the maximum temperature should be 50 °C to minimise any sulfur dioxide liberated. In small groups each temperature is completed once, but then three groups (or the whole class) could pool their results and take a mean, making the results more reliable.
- Students could plot a graph, drawing a line of best fit. They should then use their graph and their knowledge of collision theory to draw a conclusion. This provides an opportunity to cover the investigative aspects of the Controlled Assessment.
- Ask students to draw a cartoon to show the effect of heating up a reaction in terms of its rate. Students could be encouraged to personify the particles and make a fun depiction of the reaction. To help them, the cartoon framework could be given with statements to include in a box below their images, to explain what is happening.
- Ensure that students understand the two reasons why increasing the temperature increases the rate of reaction. Some students will quote the increased frequency of collisions but fail to mention the greater proportion of collisions with energy greater than the activation energy in examination questions.

Plenaries

Spot the mistake – Ask students to spot the error in the following sentence: 'When temperature is dropped, particles have more energy but move around less and so the rate of reaction stays the same.' [When temperature is dropped, the particles have less energy and the rate of reaction will decrease.] *(5 minutes)*

Sentences – In small groups ask the students to finish the following sentence:

'As you heat up a reaction ...'

After a few minutes ask the groups to read out their finished sentences. Then choose the most scientific sentence and give that group a prize. Support students by giving them a selection of endings for the sentence – some correct, some incorrect and some more scientific than others. Encourage students to select which ending they want to use from the list. Extend students by asking them to draw a labelled diagram to illustrate their sentence. *(10 minutes)*

Practical support

Reacting magnesium and hydrochloric acid at different temperatures

Equipment and materials required

Ice bath, test tube rack, hot water bath, 1 mol/dm^3 hydrochloric acid (CLEAPSS Hazcard 47A), 1 cm magnesium strips (CLEAPSS Hazcard 59A), calcium carbonate, two measuring cylinders, six test tubes, thermometer, eye protection.

Details

Wearing eye protection, measure out 2 cm^3 of acid into each test tube, put two test tubes in an ice bath, two in a test tube rack and two in a hot-water bath. Allow the test tubes to rest in the water/ice baths for about 5 minutes to reach the appropriate temperature. Students can check the temperature of the acid using the thermometer. Add one strip of magnesium to each of the different temperatures of acid. Allow the students to observe the reaction and comment on the rate at different temperatures, encouraging them to decide how they are determining the rate [amount of bubbles produced]. Keep the lab well ventilated.

Repeat the reaction with one marble chip (calcium carbonate) in each test tube.

The effect of temperature on rate of reaction

Equipment and materials required

Two measuring cylinders, stopwatch, paper with large cross in the centre, conical flask, 0.2 mol/dm^3 sodium thiosulfate, 0.2 mol/dm^3 hydrochloric acid, ice bath, water bath/hot plate, thermometer, eye protection (chemical splashproof goggles).

Details

Measure 10 cm^3 of acid (eye protection should be worn) and 10 cm^3 of sodium thiosulfate in separate clean measuring cylinders and reduce or increase the temperatures of these solutions using ice baths/water baths/hot plates. Choose temperatures that are easy to attain, e.g. 10 °C, 20 °C, 30 °C, 40 °C, 50 °C (maximum). At least five experiments need to be completed to draw a line graph. Place the conical flask on the centre of the large cross; first add the sodium thiosulfate to the beaker. Then add the acid and start the stopwatch and swirl to mix the solutions. Stop the clock when the cross disappears and note the time. You may wish to use a light sensor.

Safety: This experiment produces sulfur dioxide, which is toxic and can trigger asthmatic attacks. Therefore this practical should be completed in a well-ventilated room. Once the reaction is complete, follow CLEAPSS guidelines in Hazcard 95C for correct disposal.

Rates and energy

C2 4.3 The effect of temperature

Learning objectives

- How does increasing the temperature affect the rate of reactions?
- How does collision theory explain this effect?

When we increase the temperature, it always increases the rate of reaction. We can use fridges and freezers to reduce the temperature and slow down the rate of reactions. When food goes off it is because of chemical reactions. Reducing the temperature slows down these reactions.

Collision theory tells us why raising the temperature increases the rate of a reaction. There are two reasons:

- particles collide more often
- particles collide with more energy.

Particles collide more often

When we heat up a substance, energy is transferred to its particles. In solutions and in gases, this means that the particles move around faster. And when particles move faster they collide more often. Imagine a lot of people walking around in the school playground blindfolded. They may bump into each other occasionally. However, if they start running around, they will bump into each other much more often.

When particles collide more frequently, there are more chances for them to react. This increases the rate of reaction.

Figure 1 Lowering the temperature will slow down the reactions that make foods go off

Particles collide with more energy

Particles that are moving around more quickly have more energy. This means that any collisions they have are much more energetic. It's like two people colliding when they're running rather than when they are walking.

When we increase the temperature of a reacting mixture, a higher proportion of the collisions will result in a reaction taking place. This is because a higher proportion of particles have energy greater than the activation energy. This second factor has a greater effect on rate than the increased frequency of collisions.

Around room temperature, if we increase the temperature of a reaction by 10 °C the rate of the reaction will roughly double.

Figure 2 Moving faster means it's more likely that you'll bump into someone else – and the collision will be harder too!

a Why does increasing the temperature increase the rate of a reaction?
b How much does a 10 °C rise in temperature increase reaction rate at room temperature?

Cold – slow movement, less frequent collisions, little energy
Hot – fast movement, more frequent collisions, more energy

Figure 3 More frequent collisions, with more energy – both of these factors increase the rate of a chemical reaction caused by increasing the temperature

Practical

The effect of temperature on rate of reaction

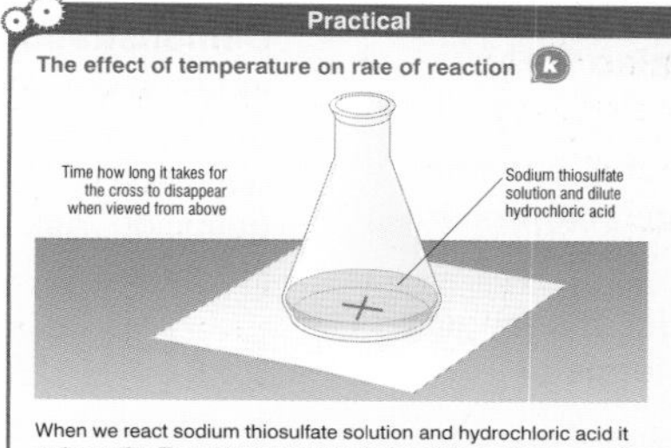

When we react sodium thiosulfate solution and hydrochloric acid it makes sulfur. The sulfur is insoluble in water. This makes the solution go cloudy. We can record the length of time it takes for the solution to go cloudy at different temperatures.

- Which variables do you have to control to make this a fair test?
- Why is it difficult to get accurate timings by eye in this investigation?
- How can you improve the **precision** of the data you collect?

Safety: Wear eye protection. Take care if you are an asthmatic.

Maths skills

As one goes up, the other comes down

In the experiment opposite we can measure the time for an X to disappear as a precipitate forms. This means that the longer the time, the slower the rate of reaction. There is an inverse relationship between time and rate. So as time increases, rate decreases. We say the rate is proportional to 1/time (also written as time^{-1}). Therefore, we can plot a graph of temperature against 1/time to investigate the effect of temperature on rate of reaction. [H]

The results of an investigation like this can be plotted on a graph (see opposite).

The graph shows how the time for the solution to go cloudy changes with temperature.

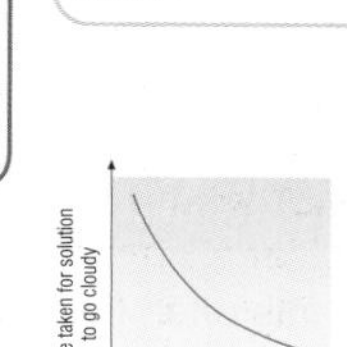

c What happens to the time it takes the solution to go cloudy as the temperature increases?

Summary questions

1 Copy and complete using the words below:
chemical collide decreases doubles energy off quickly rate reducing rise
When we increase the temperature of a reacting mixture, we increase its .... of reaction. The higher temperature makes the particles move more .... so they .... more often and the collisions have more .... At room temperature, a temperature .... of about 10°C roughly .... the reaction rate. This explains why we use fridges and freezers. .... the temperature .... the rate of the .... reactions which make food go ....

2 Water in a pressure cooker boils at a much higher temperature than water in a saucepan because it is under pressure. Why does food take longer to cook in a pan than it does in a pressure cooker?

Key points

- Reactions happen more quickly as the temperature increases.
- Increasing the temperature increases the rate of reaction because particles collide more frequently and more energetically. More of the collisions result in a reaction because a higher proportion of particles have energy greater than the activation energy.

Answers to in-text questions

a The particles collide more often and, when they do, they have more energy. So in a given time there are more collisions that have at least the activation energy required for the reaction.

b The rate roughly doubles.

c The time decreases.

Summary answers

1 rate, quickly, collide, energy, rise, doubles, reducing, decreases, chemical, off

2 It is at a higher temperature in the pressure cooker. This speeds up the chemical reactions that make food cook. The steam particles move around more quickly with greater energy when under pressure. This increases the collision frequency with food substances.

C2 4.4 The effect of concentration or pressure

Learning objectives

Students should learn:

- that increasing the concentration of reactants in solutions increases the rate of reaction
- that changing the pressure of reacting gases changes the rate of reaction.

Learning outcomes

Most students should be able to:

- state what effect increasing the concentration of reactants in solutions has on the rate of reaction
- describe what we mean by gas pressure
- state what effect increasing the pressure of reacting gases has on the rate of reaction.

Some students should also be able to:

- explain in detail the effect of changing concentration on the rate in terms of collision theory
- explain in detail the effect of changing pressure on the rate in terms of collision theory.

Support

- Students often find it difficult to understand what concentration is. Get a bottle of coloured squash and explain that there are lots of coloured particles mixed in the water. Ask a student to draw a particle diagram of this using blue pen for water particles and red pen for squash particles. Then measure 50 cm^3 of squash into a beaker and add 50 cm^3 of water. Ask a student to draw the new particle model, and explain that the concentration of squash has been halved.

Extend

- Ask students to explain the effects of changing pressure on the rate of reaction for a chemical change with gaseous reactants [an increase in pressure increases the amount of gaseous reactants in a given volume/space, making collisions more likely and rate of reaction increases]. You could extend students further to explain the effects of changing pressure on a chemical change that has no reactants in the gaseous state [no change].

Specification link-up: Chemistry C2.4

- Increasing the pressure of reacting gases increases the frequency of collisions and so increases the rate of reaction. *[C2.4.1 d)]*
- Increasing the concentration of reactants in solutions increases the frequency of collisions and so increases the rate of reaction. *[C2.4.1 e)]*
- Interpret graphs showing the amount of product formed (or reactant used up) with time, in terms of the rate of the reaction.

Controlled Assessment: C4.3 Collect primary and secondary data. *[C4.3.2 a)]*; C4.5 Analyse and interpret primary and secondary data. *[C4.5.3 a), C4.5.4 a)]*

Lesson structure

Starters

Graph – Give students an unfinished concentration–time graph with two curved lines to show reactants and products. Ask students to complete the axis labels (including units) and briefly explain the shape of the two curves. Support students by giving them the missing labels to add to the graph. Extend students by asking them to plot a few additional data points and draw a line of best fit. *(5 minutes)*

Demonstration – Show the students a bottle of undiluted squash. Then put half into a large beaker and add water. Ask the students which container has the most concentrated drink and how they know. Then ask the students to work in pairs to come up with a definition of concentration. Ask each pair to come to the board and write down their definition. Ask the whole class to consider the definitions and decide on the best. *(10 minutes)*

Main

- Students can experimentally determine the effect of concentration on rate by observing the reaction between marble chips and acid. At this point, moles have been introduced to students but not calculations in terms of moles per unit volume. Therefore, to change the concentration, the volume of the acid should be diluted with water but the volume of the mixture should remain constant, so that the experiment is fair.
- Encourage the students to plot all the curves on the same axis. Then ask students to explain their results using collision theory. (This offers another excellent opportunity to cover any investigative aspect of the Controlled Assessment.)
- Give students an A4 sheet of paper and ask them to split in it half. On one side they should explain how concentration affects rate and on the other how pressure affects rate. In each section they should define the key word (concentration/pressure) and include one labelled diagram. Ensure that students do not put their name on the front of the poster. (See Plenary on 'Exhibition'.)

Plenaries

Exhibition – Get all the students' posters from the main part of the lesson. Lay them out on the side benches and give each a number. Arrange students in small groups and ask them to rate each poster out of 10 in terms of presentation, accuracy of science and ease of understanding. *(5 minutes)*

Demonstration – For this demonstration, iron wool, tongs, deflagration spoon, a gas jar of oxygen and a Bunsen burner and safety equipment is needed. Using a safety screen between the class and the demonstration (plus eye protection), hold some iron wool into a blue Bunsen flame using tongs. Then put some iron wool on a deflagration spoon and heat it until it is glowing in the top of a blue gas cone. Then quickly put the wool into a gas jar of oxygen. Ask students, in small groups, to explain which reaction was more vigorous and why. Then choose a few students to feedback into the class. *(10 minutes)*

Answers to in-text questions

a There are more particles in the same volume, making collisions more likely.

b The higher acid concentration (green line) shows the fastest reaction because the line is steepest initially (or it finished reacting first).

Practical support

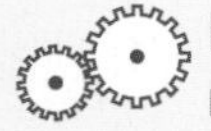

Investigating the effect of concentration on rate of reaction

Equipment and materials required

Marble chips, 1 mol/dm³ hydrochloric acid (CLEAPSS Hazcard 47A), 250 cm³ conical flask, top-pan balance, cotton wool, stopwatch, measuring cylinder, eye protection.

Details

Put about five marble chips into the bottom of a conical flask. Measure out 25 cm³ of acid (wear eye protection), and put it into the conical flask. Put a piece of cotton wool in the neck. Quickly place it on the balance and take a reading, start the stopwatch. Measure the mass of the conical flask every 10 s for 2 min. You may wish to monitor this reaction by sealing the container and using a carbon dioxide probe or gas pressure probe and data logger. Repeat for different concentrations.

Safety: Use a loose seal to prevent a build-up of pressure in the glass vessel.

Rates and energy

C2 4.4 The effect of concentration or pressure

Learning objectives

- How does increasing the concentration of reactants in solutions affect the rate of reaction?
- How does increasing the pressure of reacting gases affect the rate of reaction?

Some of our most beautiful buildings are made of limestone or marble. These buildings have stood for centuries. However, they are now crumbling away at a greater rate than before. This is because both limestone and marble are mainly calcium carbonate. This reacts with acids, leaving the stone soft and crumbly. The rate of this reaction has speeded up because the concentration of acids in rainwater has been steadily increasing.

Increasing the concentration of reactants in a solution increases the rate of reaction. That's because there are more particles of the reactants moving around in the same volume of solution. The more 'crowded' together the reactant particles are, the more likely it is that they will collide. So the more frequent collisions result in a faster reaction.

Increasing the pressure of reacting gases has the same effect. It squashes the gas particles more closely together. We have more particles of gas in a given space. This increases the chance that they will collide and react. So increasing the pressure speeds up the rate of the reaction.

a Why does increasing concentration or pressure increase reaction rate?

Figure 1 Limestone statues are damaged by acid rain. This damage happens more quickly as the concentration of the acids in rainwater increases.

Figure 2 Increasing concentration and pressure both mean that particles are closer together. This increases the frequency of collisions between particles, so the reaction rate increases.

162

The effect of concentration or pressure

Practical

Investigating the effect of concentration on rate of reaction

We can investigate the effect of changing concentration by reacting marble chips with different concentrations of hydrochloric acid:

$$CaCO_3 + 2HCl \rightarrow CaCl_2 + CO_2 + H_2O$$

We can find the rate of reaction by plotting the mass of the reaction mixture over time. The mass will decrease as carbon dioxide gas is given off in the reaction.

- How do you make this a fair test?
- What conclusion can you draw from your results?

Safety: Wear eye protection.

AQA Examiner's tip

Increasing concentration or pressure does not increase the energy with which the particles collide. However, it does increase the frequency of collisions.

If we plot the results of an investigation like the one above on a graph they look like the graph opposite:

The graph shows how the mass of the reaction mixture decreases over time at three different concentrations.

b Which line on the graph shows the fastest reaction? How can you tell?

Summary questions

1 Copy and complete using the words below:

collisions concentration faster frequency number pressure rate volume

The .... of a reaction is affected by the .... of reactants in solutions and by the .... if the reactants are gases. Both of these tell us the .... of particles that there are in a certain .... of the reaction mixture. Increasing these will increase the .... of .... between reacting particles, making .... reactions.

2 Acidic cleaners are designed to remove limescale when they are used neat. They do not work as well when they are diluted. Using your knowledge of collision theory, explain why this is.

3 You could also follow the reaction in the Practical box above by measuring the volume of gas given off over time. Sketch a graph of volume of gas against time for three different concentrations. Label the three lines as high, medium and low concentration.

Key points

- Increasing the concentration of reactants in solutions increases the frequency of collisions between particles, and so increases the rate of reaction.
- Increasing the pressure of reacting gases also increases the frequency of collisions and so increases the rate of reaction.

163

Further teaching suggestions

Why only gases?

- Ask students to explain why changing the pressure of a reaction mixture only affects a reaction with one or more reactants in the gaseous phase.

Calculate the gradient

- Once the graph of the production of gas has been drawn, students could be shown how to calculate a numerical value for rate. The graph should be a curve. They choose a particular time and draw a tangent at this point on the curve. Then they work out the gradient (change in vertical value/change in horizontal value) and this is the rate measured in units that refer to volume of gas/time.

Moles per decimetre cubed

- Introduce the idea that concentration is measured in moles per decimetre cubed (mol/dm³). Ask about the relative concentrations of a 0.25 and a 2.0 mol/dm³ solution and ask students to explain the difference in terms of particles.

Sodium thiosulfate reaction

- You may wish to investigate how the concentration of sodium thiosulfate affects its reaction with hydrochloric acid. See www.chemistryteachers.org and search for 'sodium thiosulfate'.

Summary answers

1 rate, concentration, pressure, number, volume, frequency, collisions, faster

2 There are more particles in the acid to collide with limescale particles when the cleaner is more concentrated, so increasing the rate of reaction.

3 Graph with vertical axis labelled 'Volume of gas' and horizontal axis labelled 'Time'. Three curved lines rising at different gradients, gradually levelling off. The steepest curve labelled 'high concentration' and the shallowest labelled 'low concentration'.

C2 4.5 The effect of catalysts

Learning objectives

Students should learn:

- what a catalyst is
- how catalysts affect the rate of reactions.

Learning outcomes

Most students should be able to:

- give a definition of a catalyst
- give an example of an industrial process that uses a catalyst
- list the reasons why a catalyst may be used in an industrial process.

Some students should also be able to:

- explain in detail why a catalyst would be used in an industrial process.

Specification link-up: Chemistry C2.4

- Catalysts change the rate of chemical reactions but are not used up during the reaction. Different reactions need different catalysts. *[C2.4.1 g)]*
- Catalysts are important in increasing the rates of chemical reactions used in industrial processes to reduce costs. *[C2.4.1 h)]*

  Controlled Assessment: C4.4 Select and process primary and secondary data. *[C4.4.2 c)]*

Lesson structure

Starters

True or false – Give each student a statement about catalysts. Then they must walk around the room and ask three people if they think the statement is true or false. Based on these answers, the student should decide whether the statement is true or false. You may wish to ask each student to read his or her statement to the class and say if they think it is true or false. Then you give feedback. *(5 minutes)*

Foam of death – Write the formula of hydrogen peroxide (H_2O_2) on the board and ask the students to predict what the gaseous product of the decomposition reaction could be and how it could be tested [oxygen, which is tested with a glowing splint; they will often incorrectly suggest hydrogen, which 'pops' with a lighted splint]. See 'Practical support'. You could support students by using a periodic table and asking for volunteers to suggest the elements that the compound contains. As an extension, when the foam has dried slightly, allow students to come up to the foam and relight a glowing splint. Explain that the manganese(IV) oxide was not used up in the reaction, but increased the rate and this is a catalyst. *(10 minutes)*

Main

- Hydrogen peroxide is unstable in sunlight and will decompose into oxygen and water. This process is relatively slow, but a number of catalysts can be used to speed up this reaction: chopped raw potato; chopped fresh liver; manganese(IV) oxide. Encourage students to investigate gas production using the different catalysts to decide the best. Students should consider the dependent, independent and control variables. They should take a set of results for each catalyst and draw all the lines of best fit on the same graph, giving more coverage of the investigative aspects of the Controlled Assessment.
- Encourage the students to create an eight-line poem, where the first letters of each line spell 'catalyst'. Then choose some students to read their poem to the rest of the class.

Plenaries

Txt – Ask students to write a text message to summarise what they have learnt today. You could support students by giving them the text message and ask them to rewrite it in standard English. Students could be extended by being asked to include an example of a chemical reaction that can be affected by a catalyst. *(5 minutes)*

AfL (Assessment for Learning) – To continue 'the poem' activity from the main part of the lesson, ask students to swap their poem with a partner. If students feel that there is some incorrect science, they should amend the work in pencil. Once they have worked on the poem, it should be returned to its owner who should then review any comments that have been made. *(10 minutes)*

Answers to in-text questions

a Increasing temperature, surface area, concentration (if reactants are in solution) or pressure (if reactants are gases).

b It is unaffected.

c It has greater surface area, allowing a higher frequency of collisions between reacting particles on the surface.

Support

- You could support students in drawing the graph by giving them the labelled axis with the scales already written on to it. Encourage students to use a different colour for each catalyst that they consider for the decomposition of hydrogen peroxide.

Extend

- Ask students to discover some other examples of chemical reactions and the catalysts that they can use. You could also ask students to discover about inhibitors (catalysts that are used to slow down the rate of reaction or negative catalysts – e.g. in petrol).

Summary answers

1 increases, used, remains, reaction

2 This increases their surface area.

3 The catalyst is not used up in the reaction.

Practical support

Foam of death

Equipment and materials required

Hydrogen peroxide (100 vol.) (CLEAPSS Hazcard 50 – corrosive), 1000 cm³ measuring cylinder, washing-up bowl, washing-up liquid, manganese(IV) oxide (harmful), spatula, cobalt chloride paper, splints, eye protection and gloves.

Details

Stand a 1000 cm³ measuring cylinder in a washing-up bowl. Add a good dash of washing-up liquid, and about 100 cm³ of 100 vol. H_2O_2. Add a spatula of manganese(IV) oxide and allow the students to observe.

Safety: Wear eye protection and be aware of skin burns. CLEAPSS Hazcard 60 Manganese(IV) oxide – harmful; 25 Cobalt chloride – toxic and harmful.

Investigating catalysis

Equipment and materials required

Stand, boss, gas syringe holder, gas syringe, 10 vol. hydrogen peroxide (irritant), manganese(IV) oxide (harmful), potato, liver, white tile, knife, stopwatch, conical flask, bung, delivery tube, about a 25 cm length of rubber tube, measuring cylinder, spatula, eye protection.

Details

Measure out 25 cm³ of hydrogen peroxide and put it into the conical flask, eye protection should be worn. Finely chop some raw potato and put it into the flask. Quickly connect the bung to the gas syringe and note the volume of gas produced every 10 s for 2 min. Repeat with chopped liver, and repeat with a spatula of manganese(IV) oxide. Other transition metal oxides can also be investigated. You may wish to monitor this reaction by sealing the container and using a carbon dioxide probe or gas pressure probe and data logger.

Safety: Be aware of irritation caused by the hydrogen peroxide. Wash the affected area under cold water and it should dissipate. Make sure the syringe plunger is free moving. Make sure the seal is loose to avoid the glass container exploding. CLEAPSS Hazcard 60 Manganese(IV) oxide – harmful.

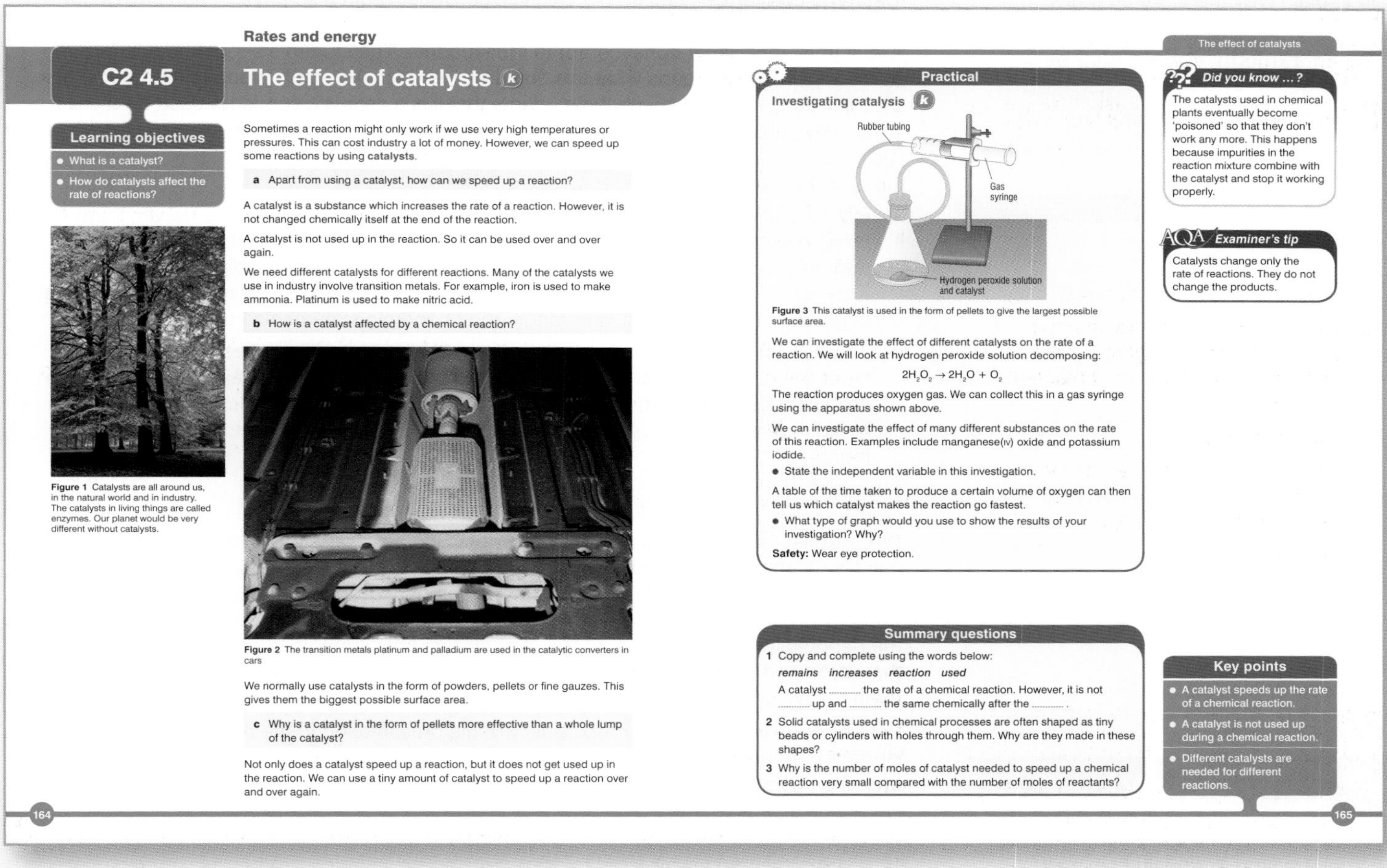

Rates and energy

C2 4.5 The effect of catalysts

Learning objectives

- What is a catalyst?
- How do catalysts affect the rate of reactions?

Sometimes a reaction might only work if we use very high temperatures or pressures. This can cost industry a lot of money. However, we can speed up some reactions by using **catalysts**.

a Apart from using a catalyst, how can we speed up a reaction?

A catalyst is a substance which increases the rate of a reaction. However, it is not changed chemically itself at the end of the reaction.

A catalyst is not used up in the reaction. So it can be used over and over again.

We need different catalysts for different reactions. Many of the catalysts we use in industry involve transition metals. For example, iron is used to make ammonia. Platinum is used to make nitric acid.

b How is a catalyst affected by a chemical reaction?

Figure 1 Catalysts are all around us, in the natural world and in industry. The catalysts in living things are called enzymes. Our planet would be very different without catalysts.

Figure 2 The transition metals platinum and palladium are used in the catalytic converters in cars

We normally use catalysts in the form of powders, pellets or fine gauzes. This gives them the biggest possible surface area.

c Why is a catalyst in the form of pellets more effective than a whole lump of the catalyst?

Not only does a catalyst speed up a reaction, but it does not get used up in the reaction. We can use a tiny amount of catalyst to speed up a reaction over and over again.

164

The effect of catalysts

Practical

Investigating catalysis

Rubber tubing
Gas syringe
Hydrogen peroxide solution and catalyst

Figure 3 This catalyst is used in the form of pellets to give the largest possible surface area.

We can investigate the effect of different catalysts on the rate of a reaction. We will look at hydrogen peroxide solution decomposing:

$$2H_2O_2 \rightarrow 2H_2O + O_2$$

The reaction produces oxygen gas. We can collect this in a gas syringe using the apparatus shown above.

We can investigate the effect of many different substances on the rate of this reaction. Examples include manganese(IV) oxide and potassium iodide.

- State the independent variable in this investigation.

A table of the time taken to produce a certain volume of oxygen can then tell us which catalyst makes the reaction go fastest.

- What type of graph would you use to show the results of your investigation? Why?

Safety: Wear eye protection.

Did you know ...?

The catalysts used in chemical plants eventually become 'poisoned' so that they don't work any more. This happens because impurities in the reaction mixture combine with the catalyst and stop it working properly.

AQA Examiner's tip

Catalysts change only the rate of reactions. They do not change the products.

Summary questions

1 Copy and complete using the words below:
remains increases reaction used
A catalyst .... the rate of a chemical reaction. However, it is not .... up and .... the same chemically after the ....
2 Solid catalysts used in chemical processes are often shaped as tiny beads or cylinders with holes through them. Why are they made in these shapes?
3 Why is the number of moles of catalyst needed to speed up a chemical reaction very small compared with the number of moles of reactants?

Key points

- A catalyst speeds up the rate of a chemical reaction.
- A catalyst is not used up during a chemical reaction.
- Different catalysts are needed for different reactions.

165

Further teaching suggestions

Equations

- Encourage students to represent the reactions studied as balanced symbol equations. However, ensure that the condition of the reaction (including the catalyst) is listed on the arrow.

Hydrogen peroxide storage

- Ask students to consider why hydrogen peroxide is kept in a brown or black bottle away from light. [Light activates decomposition into water and oxygen.]

Questions and answers

- Give students a sheet of A4 paper and ask them to fold it in half (portrait). They should create five questions on the left-hand side about catalysts, using the Student Book for inspiration. Then they should write the answers on the right-hand side.

C2 4.6 Catalysts in action

Learning objectives

Students should learn:

- why catalysts are used in so many industrial processes
- how new catalysts are developed and the reason why there are so many different catalysts
- why there are disadvantages of using catalysts in industry.

Learning outcomes

Most students should be able to:

- state some advantages and disadvantages of using catalysts.
- explain why there are so many catalysts.

Some students should also be able to:

- evaluate the advantages and disadvantages of using catalysts in industry.

Specification link-up: Chemistry C2.4

- Catalysts are important in increasing the rates of chemical reactions used in industrial processes to reduce costs. *[C2.4.1 h)]*
- Explain and evaluate the development, advantages and disadvantages of using catalysts in industrial processes. *[C2.4]*

Lesson structure

Starters

Enzymes – Show students a photograph of a stomach, washing powder, washing-up liquid, and carpet cleaner. Ask students to make the link between the pictures [they all use enzymes]. Support students by giving them the word 'enzyme' as an anagram to help them get the connection. Extend students by asking them to give a definition of an enzyme [biological catalyst]. *(5 minutes)*

Catalytic converter – Get an old catalytic converter (from a scrapyard) and have it cut into slices (maybe in the Technology Department). Clean out the deposits, then show the slices (they look like a honeycomb) to the students. The students should be encouraged to discuss in small groups what this could be used for. Ask each group to feed back their thoughts and then share with the students that it is a catalytic converter used on a car exhaust to remove pollutant gases. *(10 minutes)*

Main

- Most industrial reactions use catalysts to reduce production costs. However, there are some disadvantages of using catalysts. Ask students to draw a table of advantages and disadvantages of using transition metals and their compounds, as well as enzymes. Extend students by asking them to underline the environmental statements in green, social statements in orange and economic statements in black.
- Catalysts are being developed all the time. Students could use secondary research such as books in the school library or the internet to find out current developments in catalysts. Key words would include: zeolites, fullerenes and nanochemistry. Students could then work in groups to make a 3-minute presentation about what they have found out.

Plenaries

Definitions – Ask students to use the Student Book to define the following key terms: catalyst [a chemical that changes the rate of reaction without being used up, e.g. manganese(IV) oxide used in the decomposition of hydrogen peroxide], enzyme [biological catalyst, e.g. lipase], transition metals [metals found in the central block of the periodic table and are often toxic, e.g. chromium]. Support students by giving them the words and the definitions so that they just match them up. Extend students by asking them to give an example for each. *(5 minutes)*

Press conference – Each group could give its presentation to the rest of the class. One student from each group in the audience should then pose a question to the presenters based on what they have seen and heard. The group should then answer the questions. *(10 minutes)*

Support

- Give students a flow chart with missing information to explain how new catalysts are researched, designed and made. You may wish to give the missing information to the students so that they can complete the activity as a 'cut-and-stick' activity.

Extend

- Ask students to research industrial processes and the catalysts that they use, e.g. the contact process to make sulfuric acid, which uses vanadium(V) oxide.

Further teaching suggestions

Debate

- Students could hold a debate about whether the use of catalysts is a good or bad idea for industrial processes.

Enzymes in industry

- Students could research to find out how enzymes are made (often by genetically modifying microorganisms) and consider the environmental, social and economic impact of this industry.

Questions and answers

- Students could copy out the objective questions and use the Student Book to answer them fully.

How Science Works **Rates and energy**

C2 4.6 Catalysts in action

Learning objectives

- Why are catalysts used in so many industrial processes?
- How are new catalysts developed and why are there so many different catalysts?
- What are the disadvantages of using catalysts in industry?

Catalysts are often very expensive precious metals. Gold, platinum and palladium are all costly but are the most effective catalysts for particular reactions. But it is often cheaper to use a catalyst than to pay for the extra energy needed without one. To get the same rate of reaction without a catalyst would require higher temperatures and/or pressures.

So catalysts save money and help the environment. That's because using high temperatures and pressures often involves burning fossil fuels. So operating at lower temperatures and pressures conserves these non-renewable resources. It also stops more carbon dioxide entering the atmosphere.

a Why do catalysts save a chemical company money?

However, many of the catalysts used in industry are transition metals or their compounds. These are often toxic. If they escape into the environment, they build up inside living things. Eventually they poison them. For example, the platinum and palladium used in catalytic converters slowly escape from car exhausts.

So chemists are working to develop new catalysts that are harmless to the environment. The search for the ideal catalyst is often a bit like trial and error. Each reaction is unique. Once a catalyst is found it might be improved by adding small amounts of other chemicals to it. All this takes a lot of time to investigate. However, the research is guided by knowledge of similar catalysed reactions. This knowledge is growing all the time.

Figure 1 Chinese scientists have recently developed a new catalyst for making biodiesel from vegetable oils. It's made from shrimp shells, and is cheaper and more efficient than conventional catalysts. The process that uses the new catalyst also causes less pollution.

Future development

Chemists have developed new techniques to look at reactions. They can now follow the reactions that happen on the surface of the metals in a catalytic converter. These are very fast reactions lasting only a fraction of a second. Knowing how the reactions take place will help them to design new catalysts.

Nanoparticles are also at the cutting edge of work on new catalysts. Scientists can arrange atoms into the best shapes for catalysing a particular reaction they have studied. A small mass of these catalysts has a huge surface area. This has raised hopes that fuel cells will one day take over from petrol and diesel to run cars.

links
For information about nanoparticles, look back to C2 2.6 Nanoscience.

Catalysts in medicine

The catalysts used in making new drugs also contain precious metal compounds. The metal is bonded to an organic molecule. But now chemists can make these catalysts without the metal. The metal was needed to make a stable compound. However, research has resulted in a breakthrough which will mean much cheaper catalysts. There is also no risk of contaminating the drug made with a toxic transition metal.

b Why could it be unsafe to use compounds of transition metals to catalyse reactions to make drugs?

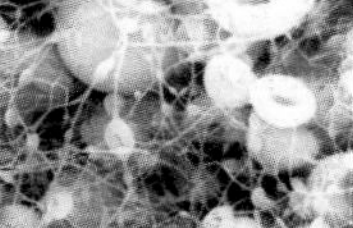

Figure 2 Scientists are developing long nanowires of platinum to use as catalysts in fuel cells. This photo is from an electron microscope. The wires are 1/50 000th of the width of a human hair. The breakthrough has been made in making them over a centimetre in length.

Enzymes

Enzymes are the very efficient catalysts found in living things. For years we've been using enzymes to help clean our clothes. Biological washing powders contain enzymes that help to 'break apart' stain molecules such as proteins at low temperatures. The low temperature washes save energy.

Low-temperature enzyme reactions are the basis of the biotechnology industry. Enzymes are soluble so would have to be separated from the products they make. However, scientists can bind them to a solid. The solution of reactants flows over the solid. No time or money has to be wasted separating out the enzymes to use again. The process can run continuously.

Summary questions

1 Give two ways in which catalysts are beneficial to the chemical industry.
2 What are the disadvantages of using transition metals or their compounds as catalysts?
3 Do some research to find out four industrial processes that make products using catalysts. Write a word equation for each reaction and name the catalyst used.

Key points

- Catalysts are used whenever possible in industry to increase rate of reaction and reduce energy costs.
- Traditional catalysts are often transition metals or their compounds, which can be toxic and harm the environment if they escape.
- Modern catalysts are being developed in industry which result in less waste and are safer for the environment.

Answers to in-text questions

a They save energy costs by allowing reactions to be conducted at lower temperatures or pressures.

b Drugs might be toxic if contaminated with transition metal compounds.

Summary answers

1 For example: they reduce energy costs and speed up the rate of production.

2 They can be toxic, so are harmful to living things if they escape into the environment. Some can be expensive.

3 For example:
Haber process, iron: nitrogen + hydrogen $\rightleftharpoons$ ammonia
Contact process, vanadium(V) oxide: sulfur dioxide + oxygen $\rightleftharpoons$ sulfur trioxide
Hydrogenation of oils, nickel: unsaturated oil + hydrogen $\rightarrow$ more saturated oil/fat
Making nitric acid, platinum/rhodium: ammonia + oxygen $\rightleftharpoons$ nitrogen(II) oxide + water

C2 4.7 Exothermic and endothermic reactions

Learning objectives

Students should learn:

- that energy changes are involved in chemical reactions
- what is meant by exothermic and endothermic reactions
- that energy changes in a chemical reaction can be measured.

Learning outcomes

Most students should be able to:

- state a definition of exothermic and endothermic reactions
- list one example of an exothermic reaction and one of an endothermic reaction
- recognise an endothermic or an exothermic reaction when data are given
- describe how energy change in a reaction can be monitored.

Some students should also be able to:

- explain in detail the difference between exothermic and endothermic reactions.

Answers to in-text questions

a exothermic
b endothermic
c Respiration/oxidation; combustion/burning; neutralisation (or other reaction releasing energy).
d Thermal decomposition, photosynthesis (or other reaction that absorbs energy).

Support

- Give students information to incorporate into their posters about exothermic and endothermic reactions. However, they would need to decide which poster the information is referring to before copying it into their work.

Extend

- Introduce the idea that exothermic and endothermic reactions are usually monitored by temperature changes and these indicate that energy is given out or taken in. Encourage students to consider other ways of monitoring a reaction for energy change (e.g. light or temperature sensor).

Specification link-up: Chemistry C2.5

- When chemical reactions occur, energy is transferred to or from the surroundings. *[C2.5.1 a)]*
- An exothermic reaction is one that transfers energy to the surroundings. Examples of exothermic reactions include combustion, many oxidation reactions and neutralisation. Everyday uses … . *[C2.5.1 b)]*
- An endothermic reaction is one that takes in energy from the surroundings. Endothermic reactions include thermal decompositions. Some sports … . *[C2.5.1 c)]*

Controlled Assessment: C4.5 Analyse and interpret primary and secondary data. *[C4.5.4 d)]*

Lesson structure

Starters

Sherbet – Give students a sherbet sweet before they enter the room. Ask students to detail what their observations are as they eat it. Then, using questions and answers, get feedback from the students and ask them if they think the reaction is chemical or physical and exothermic or endothermic. Support students by having these words already defined on the board for them to refer to. Extend students by telling them the reactants for this reaction (citric acid and sodium hydrogencarbonate), and ask them to write a word equation for this reaction and classify the reaction [citric acid + sodium hydrogencarbonate → sodium citrate + carbon dioxide + water, neutralisation]. *(5 minutes)*

Cut and stick – Give students photographs of different exothermic or endothermic processes, such as a fire burning, a sports cold pack being used and a match burning. They first need to make themselves aware of the definitions of 'exothermic' and 'endothermic'. They could then cut up the pictures and arrange them in a table to detail the energy changes shown in the reactions. *(10 minutes)*

Main

- The energy changes of a reaction can be recorded using a coffee-cup calorimeter. Explain to the students that most reactions show their energy change by getting hotter or colder and that the reaction needs to be insulated to prevent energy loss to the surroundings.
- Then ask students to complete the displacement reaction between zinc powder and copper sulfate solution. Students should design their own results table and record their results in order to draw a graph. They should be reminded that the scales do not have to start at zero (and the *y*-axis will probably start at about 15 °C). Students may struggle in drawing the line of best fit for this reaction; you could show them how to do this on the board. There are many whole investigations that this can be developed into to extend the Controlled Assessment concepts already covered. Consider concentrating on evaluating methodology in terms of the reproducibility and repeatability of the evidence generated.
- Give the students an A5 sheet of blue paper. They should write a definition of 'endothermic' on it and include examples of endothermic reactions. A similar poster could then be created for exothermic reactions on red paper.

Plenaries

Exo-/endothermic – Give the students a blue card with the word 'endothermic' written on, and a red card with the word 'exothermic' printed on. Then read out these reactions and ask the students to decide if they are exo- or endothermic, displaying the card to represent their answer:

- Thermal decomposition of marble. [Endothermic]
- Combustion of methane. [Exothermic]
- Neutralisation of hydrochloric acid and sodium hydroxide. [Exothermic]
- Rusting of an iron nail. [Exothermic]
- Thermal decomposition of copper carbonate. [Endothermic]

Support students by giving them the definitions on the board with an example to refer to. Extend students by asking them to complete balanced symbol equations for the reactions. You could demonstrate these reactions, show photographs or a video via a digital projector. *(5 minutes)*

Demonstration – Use a data logger to plot the temperature changes in a neutralisation reaction. Display the temperature graph using a digital projector. In small groups, students should decide whether the reaction is exothermic or endothermic and say how they could tell. Choose a few students to feed back to the class. *(10 minutes)*

Practical support

Investigating energy changes

Equipment and materials required

Polystyrene coffee cup, polystyrene lid with two holes in, a mercury thermometer (0–50 °C), 1 mol/dm^3 copper sulfate solution (harmful), zinc powder (highly flammable), spatula, balance, measuring cylinder, stopwatch, stirrer, eye protection.

Details

Wear eye protection and measure 25 cm^3 of copper sulfate solution into the coffee cup. Measure the temperature every 30 s for 5 min. Then add 1 g of zinc to the cup and quickly put on the lid and stir constantly. Take the temperature every 10 s for 10 min.

Safety: Make students aware that they are using a mercury thermometer for accuracy but that this involves a risk and they should be careful not to leave it by the edge of the bench. You should be aware of where the mercury spillage kit is and how to use it. CLEAPSS Hazcard 27C Copper sulfate – harmful; 107 Zinc powder – highly flammable.

Demonstration of neutralisation reaction

Equipment and materials required

Burette, measuring cylinder, burette holder, stand, 1 mol/dm^3 sodium hydroxide (corrosive), 1 mol/dm^3 hydrochloric acid, universal indicator (highly flammable), magnetic stirrer, conical flask, magnetic stirrer bar, temperature probe, interface, computer, digital projector, white tile, filter funnel, eye protection (chemical splashproof goggles).

Details

Measure 25 cm^3 of sodium hydroxide into a conical flask and add a few drops of indicator. Place the flask on to the magnetic stirrer and add the bar. Fill the burette with hydrochloric acid using the filter funnel. Position the burette over the conical flask and add the temperature probe to the flask, taking care that it doesn't hit the stirrer. Set the graph to take data for about 2 min and begin stirring. Start the data collection. Turn on the flow of acid to the flask and observe.

This activity can be extended by adding a pH probe and comparing the temperature rise with the pH of the solution.

Safety: CLEAPSS Hazcard 91 Sodium hydroxide – corrosive; 47A Hydrochloric acid – corrosive; 32 Universal indicator – highly flammable/harmful.

Rates and energy

C2 4.7

Exothermic and endothermic reactions

Learning objectives

- How is energy involved in chemical reactions?
- How can we measure the energy transferred in a chemical reaction?

Whenever chemical reactions take place, energy is involved. That's because energy is always transferred as chemical bonds are broken and new ones are made.

Some reactions transfer energy **from** the reacting chemicals **to** their surroundings. We call these **exothermic** reactions. The energy transferred from the reacting chemicals often heats up the surroundings. This means that we can measure a rise in temperature as the reaction happens.

Some reactions transfer energy **from** the surroundings **to** the reacting chemicals. We call these **endothermic** reactions. As they take in energy from their surroundings, these reactions cause a drop in temperature as they happen.

a What do we call a reaction that releases energy to its surroundings?
b What do we call a reaction that absorbs energy from its surroundings?

Figure 1 When a fuel burns in oxygen, energy is transferred to the surroundings. We usually don't need a thermometer to know that there is a temperature change!

Exothermic reactions

Fuels burning are an obvious example of exothermic reactions. For example, when methane (in natural gas) burns it gets oxidised and releases energy.

Respiration is a very special kind of oxidation. It involves reacting sugar with oxygen inside the cells of every living thing. The reaction produces water and carbon dioxide as waste products. Respiration is another exothermic reaction.

Neutralisation reactions between acids and alkalis are also exothermic. We can easily measure the rise in temperature using simple apparatus (see the practical on the next page).

c Give two examples of exothermic reactions.

Figure 2 All warm-blooded animals rely on exothermic reactions to keep their body temperatures steady

links

For more information on energy transfers in chemical reactions, see C3 3.2 Energy transfers in solution.

168

Exothermic and endothermic reactions

Endothermic reactions

Endothermic reactions are much less common than exothermic ones.

Thermal decomposition reactions are endothermic. An example is the decomposition of calcium carbonate. When heated it forms calcium oxide and carbon dioxide. This reaction only takes place if we keep heating the calcium carbonate strongly. It takes in a great deal of energy from the surroundings.

d Give an example of an endothermic reaction.

Figure 3 When we eat sherbet we can feel an endothermic reaction. Sherbet dissolving in the water in your mouth takes in energy. It provides a slight cooling effect.

Practical

Investigating energy changes

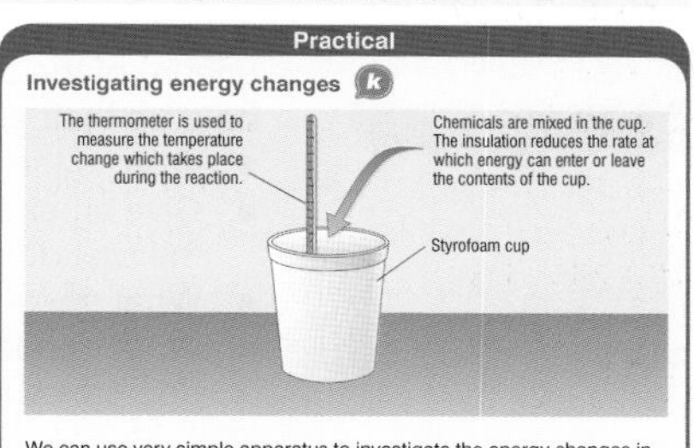
The thermometer is used to measure the temperature change which takes place during the reaction.
Chemicals are mixed in the cup. The insulation reduces the rate at which energy can enter or leave the contents of the cup.
Styrofoam cup

We can use very simple apparatus to investigate the energy changes in reactions. Often we don't need to use anything more complicated than a styrofoam cup and a thermometer.

- State two ways in which you could make the data you collect more accurate.

AQA

Examiner's tip

Remember that exothermic reactions involve energy EXiting (leaving) the reacting chemicals, so the surroundings get hotter.

In endothermic reactions energy moves INTO (sounds like 'endo'!) the reacting chemicals, so the surroundings get colder.

Summary questions

1 Copy and complete using the words below:

endothermic exothermic changes neutralisation oxidation decomposition

Chemical reactions involve energy .... When a reaction releases energy we say that it is an .... reaction. Two important examples of this type of reaction are .... and .... When a reaction takes in energy we say that it is an .... reaction. An important example of this type of reaction is thermal ....

2 Potassium chloride dissolving in water is an endothermic process. What would you expect to observe when potassium chloride dissolves in a test tube of water?

Key points

- Energy may be transferred to or from the reacting substances in a chemical reaction.
- A reaction in which energy is transferred from the reacting substances to their surroundings is called an exothermic reaction.
- A reaction in which energy is transferred to the reacting substances from their surroundings is called an endothermic reaction.

169

Further teaching suggestions

Poster

- Ask the students to imagine that they work for a marketing company. They are to make a poster to encourage students to think about science in their everyday lives (such as the RSC posters 'Scientists don't always wear white coats'). Their poster should include all the key points.

Other reactions

- Students could use temperature probes, data loggers or thermometers to monitor the temperature change of a variety of exothermic or endothermic changes and classify them. Reactions include dissolving ammonium nitrate, citric acid and sodium hydrogencarbonate and adding ammonium nitrate to barium hydroxide.

Summary answers

1 changes, exothermic, neutralisation/oxidation (either order), endothermic, decomposition

2 The test tube would feel colder/temperature of the solution would fall.

C2 4.8 Energy and reversible reactions

Learning objectives

Students should learn:

- what happens in the energy transfers in reversible reactions.

Learning outcomes

Most students should be able to:

- recognise that if the forward reaction is exothermic, the reverse reaction will be endothermic
- recognise that if the forward reaction is endothermic, the reverse reaction will be exothermic
- recognise that the same amount of energy is taken in or released in either direction.

Some students should also be able to:

- explain why the same amount of energy is taken in or released in either direction.

Specification link-up: Chemistry C2.5

- If a reversible reaction is exothermic in one direction, it is endothermic in the opposite direction. The same amount of energy is transferred in each case. For example:

$$\text{hydrated copper sulfate (blue)} \underset{\text{exothermic}}{\overset{\text{endothermic}}{\rightleftharpoons}} \text{anhydrous copper sulfate (white)} + \text{water}$$

[C2.5.1 d)]

Lesson structure

Starters

Reversible reaction – Show students a solution of potassium dichromate(VI), $K_2Cr_2O_7$ (oxidising/very toxic), in a beaker, then add sodium hydroxide solution. Explain that the solution changes colour from orange to yellow as potassium chromate(VI), K_2CrO_4, forms, then restore the orange colour by adding dilute hydrochloric acid. Add the alkali and acid again to show that the reactions are reversible. Ask the students to write equations to represent the reactions. Support students by giving them a list of compounds present so they can write the word equations. Extend students by asking them to write the balanced symbol equations. *(5 minutes)*

Questions – Give each student an A4 whiteboard (or laminated sheet of paper), a washable pen and eraser. Then ask them the following series of questions. The students should note down their answers and show you for immediate assessment. If students are unsure of the answer, they could refer to the Student Book or wait for other students to hold up their answer and then use these responses to inform their answer.

- What is the symbol to show a reversible reaction? [$\rightleftharpoons$]
- Give an example of a reversible reaction. [Hydration of anhydrous copper sulfate, thermal decomposition of ammonium chloride.]
- What does 'exothermic' mean? [Energy is given out in the reaction.]
- What happens to the temperature in an endothermic reaction? [Temperature decreases.] *(10 minutes)*

Main

- The students can experimentally complete the reversible reaction of hydration/dehydration of copper sulfate. Before the experiment is completed, encourage the students to think about how they will record their results (table, diagram, flow chart, paragraphs, bullet points, etc.).
- Once the practical is complete, show the exemplar work to the rest of the class and explain why it is a good way to record the results. You may wish to set up a flexi-cam or video camera. This can be used to show exemplar work quickly and easily to the rest of the class.
- Ask students to imagine that a top publisher has commissioned them to create a GCSE science revision book. Show students a selection of revision materials and ask them to discuss in groups what they like and dislike about the material.
- Explain that they have an A4-page spread in such a book to explain energy and reversible reactions. They must include a worked examination question and an extra question for the reader to attempt, with the answers upside-down on the page. Students could work in small teams to complete this, allowing them to distribute the tasks as they desire.

Plenaries

Objectives – Ask students to try to answer the questions posed by the objectives. *(5 minutes)*

Crossword – Create a crossword with the answers taken from this double-page spread (words could include: 'reversible', 'endothermic', 'exothermic', 'energy', 'hydrated', 'water'). There are many free sites on the internet that can be used to create your own crossword. Then ask students to complete the crossword. Students could be supported by being given both the words and the clues, so they match them up. *(10 minutes)*

Answers to in-text questions

- **a** Amount of energy released in one direction is the same as the amount of energy absorbed in the other direction.
- **b** water
- **c** It absorbs water from the air.
- **d** released

Support

- Give students the word equation for the reversible dehydration of hydrated copper sulfate. Ask them to annotate the equation to explain the observations and what the equation shows. You could support them further by giving them the labels and making this a 'cut-and-stick' exercise.

Extend

- Ask students to discover other examples of energy changes in reversible reactions.

Practical support

Energy changes in a reversible reaction

Equipment and materials required

Hydrated copper sulfate (CLEAPSS Hazcard 27C – harmful), spatula, Bunsen burner and safety equipment, dropping pipette, water, boiling tube, boiling-tube holder, eye protection.

Details

Eye protection should be worn throughout this practical. Put a spatula of copper sulfate crystals into a boiling tube. Using the boiling-tube holder, hold the boiling tube just above the blue flame of the Bunsen burner. The tube should be held at an angle and pointing away from people's faces. Do not overheat. Once the visible change is complete, allow the tube to cool. Add a few drops of water. Be aware that water added directly to the boiling tube may crack it.

Demonstration of making cobalt chloride paper

Equipment and materials required

Filter paper, cobalt chloride (toxic), 50 cm^3 beaker, stirring rod, wash bottle and water and a spatula, Bunsen burner, (desiccator), tweezers.

Details

Add half a spatula of cobalt chloride crystals to the beaker. Add water and stir until the crystals dissolves. Soak some filter paper in the solution. Take care drying the paper, using a yellow Bunsen flame. The paper will become blue (dehydrated), add water and it will become pink (hydrated).

Explain that the paper should be kept in a desiccator because the air contains water and will turn the paper pink. A desiccator could be shown to the students and they could research how it works.

Safety: Wear chemical splashproof eye protection. Keep cobalt chloride off skin (avoid handling papers with fingers). Wash hands after use. (See CLEAPSS Hazcard 25.)

Rates and energy

C2 4.8 Energy and reversible reactions

Learning objectives

- What happens in the energy transfers in reversible reactions?

Energy changes are involved in reversible reactions too. Let's consider an example.

Figure 1 shows a reversible reaction where A and B react to form C and D. The products of this reaction (C and D) can then react to form A and B again.

If the reaction between A and B is exothermic, energy will be released when the reaction forms C and D.

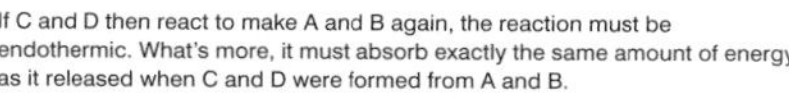
If the reaction *releases* energy when it goes in this direction ...
A+B ⇌ C+D
... it will *absorb* exactly the same amount of energy when it goes in this direction

Figure 1 A reversible reaction

If C and D then react to make A and B again, the reaction must be endothermic. What's more, it must absorb exactly the same amount of energy as it released when C and D were formed from A and B.

Energy cannot be created or destroyed in a chemical reaction. The amount of energy released when we go in one direction in a reversible reaction must be exactly the same as the energy absorbed when we go in the opposite direction.

a How does the energy change for a reversible reaction in one direction compare with the energy change for the reaction in the opposite direction?

We can see how this works if we look at what happens when we heat blue copper sulfate crystals. The crystals contain water as part of the lattice formed when the copper sulfate crystallised. We say that the copper sulfate is **hydrated**. Heating the copper sulfate drives off the water from the crystals, producing white **anhydrous** ('without water') copper sulfate. This is an endothermic reaction.

$$CuSO_4{\cdot}5H_2O \rightleftharpoons CuSO_4 + 5H_2O$$

hydrated copper sulfate (blue) ⇌ anhydrous copper sulfate (white) + water

When we add water to anhydrous copper sulfate we form hydrated copper sulfate. The colour change in the reaction is a useful test for water. The reaction in this direction is exothermic. In fact, so much energy may be produced that we may see steam rising as the water boils.

Figure 2 Hydrated copper sulfate and white anhydrous copper sulfate

Practical

Energy changes in a reversible reaction

Try these reactions yourself. Gently heat a few copper sulfate crystals in a test tube. Observe the changes. When the crystals are completely white allow the tube to cool to room temperature (this takes several minutes). Add two or three drops of water from a dropper and observe the changes. Carefully feel the bottom of the test tube.

- Explain the changes you have observed.

You can repeat this with the same solid, as it is a reversible reaction or try with other hydrated crystals, such as cobalt chloride. Some are not so colourful but the changes are similar.

Safety: Wear eye protection. Avoid skin contact with cobalt chloride.

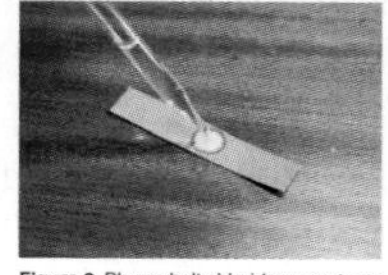

Figure 3 Blue cobalt chloride paper turns pink when water is added

b What can anhydrous copper sulfate be used to test for?

We can soak filter paper in cobalt chloride solution and allow it to dry in an oven. The blue paper that is produced is called cobalt chloride paper. The paper turns pale pink when water is added to the paper.

c Why does blue cobalt chloride turn pink if left out in the open air?

d When water is added to blue cobalt chloride is energy released or absorbed?

Summary questions

1 A reversible reaction gives out 50 kilojoules (kJ) of energy in the forward reaction. In this reaction W and X react to give Y and Z.
   a Write an equation to show the reversible reaction.
   b What can you say about the energy transfer in the reverse reaction?

2 Blue cobalt chloride crystals turn pink when they become damp. The formula for the two forms can be written as $CoCl_2{\cdot}2H_2O$ and $CoCl_2{\cdot}6H_2O$.
   a How many moles of water will combine with 1 mole of $CoCl_2{\cdot}2H_2O$?
   b Write a balanced chemical equation for the reaction, which is reversible. [H]
   c How can pink cobalt chloride crystals be changed back to blue cobalt chloride crystals?

Key points

- In reversible reactions, one reaction is exothermic and the other is endothermic.
- In any reversible reaction, the amount of energy released when the reaction goes in one direction is exactly equal to the energy absorbed when the reaction goes in the opposite direction.

Further teaching suggestions

Making cobalt chloride paper

- Cobalt chloride is usually made into paper for use in practicals to test for water. This can be shown to the students. See 'Practical support'.

Flash cards

- Encourage the students to copy out the key points on to flash cards. They can then create a bank of the key points on separate cards to use for revision.

Summary answers

1 a $W + X \rightleftharpoons Y + Z$

b 50 kJ of energy is absorbed.

2 a 4

b $CoCl_2.2H_2O + 4H_2O \rightleftharpoons CoCl_2.6H_2O$

c Heat them (gently) to drive off the water.

C2 4.9 Using energy transfers from reactions

Learning objectives

Students should learn:

- how the energy from exothermic reactions can be used
- how the cooling effect from endothermic reactions can be used
- that there are advantages and disadvantages of using energy changes in a chemical reaction.

Learning outcomes

Most students should be able to:

- state a use for an exothermic reaction
- state a use for an endothermic reaction
- list advantages and disadvantages of using energy changes from a chemical reaction.

Some students should also be able to:

- explain in detail how an exothermic reaction can be used
- explain in detail how an endothermic reaction can be used
- evaluate the advantages and disadvantages of using energy changes from a chemical reaction.

Specification link-up: Chemistry C2.5

- Evaluate everyday uses of exothermic and endothermic reactions. *[C2.5]*

Lesson structure

Starters

Hand warmers – Give each group a hand warmer and ask them to pop the metal clip inside and observe. Ask students to share their observations with the whole class. *(5 minutes)*

Sports packs – Ask for a volunteer and give him or her a cooling sports pack. Allow the volunteer to break the inner bag and mix the chemicals. Encourage the student to describe what he or she is doing and observing. Allow the cool pack to be passed around the class. Explain to students that the change is between ammonium nitrate and water and that it creates a solution. Extend students by asking them to use their Key Stage 3 knowledge as well as GCSE knowledge to identify the solute, solvent and name of solution and draw a particle model of the solution. [The solute is ammonium nitrate, the solvent is water and the solution is ammonium nitrate mixed with water.] *(10 minutes)*

Main

- Split the class into small groups. Give each group a self-heating can and ask them to think about how it could be self-heating. Ask groups to feed back their ideas to the rest of the class. Show students one example of a used self-heating can cut open – complete this using appropriate protective clothing as the chemicals are hazardous.
- Give each table of students a different commercial product that uses an exothermic or endothermic reaction. Ask the students to classify the reaction type that has been used and suggest some chemicals that would produce the desired effect. Then ask students to brainstorm the advantages and disadvantages of this commercial use.
- Demonstrate the crystallisation of a supersaturated solution (see 'Practical support'). Link this demonstration with the 'Hand warmers' Starter. Explain to students that this is a physical change and ask them to suggest how they could easily reverse it [reheat the crystals].

Plenaries

Other uses of exothermic/endothermic reactions – Ask students to brainstorm other uses of exothermic or endothermic reactions – e.g. combustion is an exothermic reaction used to cook food. Encourage volunteers to share their ideas. *(5 minutes)*

Mark scheme – Give students an examination question and ask them to create the mark scheme rather than answer the question. Students should consider all acceptable answers and also answers that would not be worthy of credit. Use a question from the same tier of entry as the students in the class. You could further support students by allowing them to work in small groups. *(10 minutes)*

Support

- You can support students by giving them the advantages and disadvantages of using chemical reactions to generate an energy change. Students could then use these to generate a table to list the advantages and disadvantages for specific products such as the self-heating can.

Extend

- You can extend students by asking them to write balanced symbol equations for some of the chemical reactions that they have studied in this double-page spread.

Practical support

Demonstration: crystallisation of a supersaturated solution

Equipment and materials required

700 g sodium ethanoate, 50 cm^3 of hot water, 250 cm^3 glass conical flask, a crystal of sodium ethanoate, tweezers, magnetic hot plate, magnetic stirrer bar, eye protection.

Details

Eye protection should be worn throughout this practical. Put 50 cm^3 of hot water into the conical flask. Add 700 g of sodium ethanoate and a magnetic stirrer bar. Stir on a warm hot plate until all the solid has dissolved. Remove the flask from the heat and allow it to cool to room temperature. Avoid 'shocking' – for example, violently moving it – or continued stirring of the liquid or it will start to crystallise at once. Seed the solution with one crystal and see how it solidifies, releasing the heat of crystallisation – an exothermic physical change.

You could alternatively demonstrate crystallisation by pouring the hot solution into a Petri dish on an overhead projector, and seeding the crystals there.

Safety: Refer to CLEAPSS Hazcard 38A.

Rates and energy

Using energy transfers from reactions

C2 4.9 Using energy transfers from reactions

Learning objectives

- How can we use the energy from exothermic reactions?
- How can we use the cooling effect of endothermic reactions?
- What are the advantages and disadvantages of using exothermic and endothermic reactions in the uses described?

Practical

Crystallisation of a supersaturated solution

Dissolve 700 g of sodium ethanoate in 50 cm^3 of hot water in a conical flask. Then let the solution cool to room temperature. Now add a small crystal of sodium ethanoate.

- What do you see happen? What does the outside of the flask feel like?

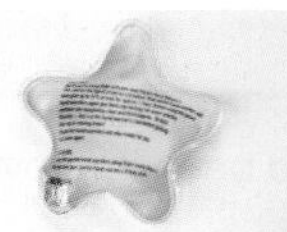

Figure 1 Here is a hand warmer based on the recrystallisation of sodium ethanoate

Warming up

Chemical hand and body warmers can be very useful. These products use exothermic reactions to warm you up. People can take hand warmers to places they know will get very cold. For example, spectators at outdoor sporting events in winter can warm their hands up. People usually use the body warmers to help ease aches and pains.

Some hand warmers can only be used one. An example of this type uses the oxidation of iron to release energy. Iron turns into hydrated iron(III) oxide in an exothermic reaction. The reaction is similar to rusting. Sodium chloride (common salt) is used as a catalyst. This type of hand warmer is disposable. It can be used only once but it lasts for hours.

Other hand warmers can be reused many times. These are based on the formation of crystals from solutions of a salt. The salt used is often sodium ethanoate. A supersaturated solution is prepared. We do this by dissolving as much of the salt as possible in hot water. The solution is then allowed to cool.

A small metal disc in the plastic pack is used to start the exothermic change. When you press this a few times small particles of metal are scraped off. These 'seed' (or start off) the crystallisation. The crystals spread throughout the solution, giving off energy. They work for about 30 minutes.

To reuse the warmer, you simply put the solid pack into boiling water to re-dissolve the crystals. When cool, the pack is ready to activate again.

a Common salt is used as a *catalyst* in some disposable hand warmers. What does this mean?

Exothermic reactions are also used in self-heating cans (see Figure 2). The reaction used to release the energy is usually:

calcium oxide + water → calcium hydroxide

You press a button in the base of the can. This breaks a seal and lets the water and calcium oxide mix. Coffee is available in these self-heating cans.

Development took years and cost millions of pounds. Even then, over a third of the can was taken up with the reactants to release energy. Also, in some early versions, the temperature of the coffee did not rise high enough in cold conditions.

b Which solid is usually used in the base of self-heating coffee cans?

Activity

Hot food

Mountaineers and explorers can take 'self-heating' foods with them on their journeys. One uses the energy released when calcium oxide reacts with water to heat the food.

Design a self-heating, disposable food container for stew.

- Draw a labelled diagram of your container and explain how it works.
- What are the safety issues involved in using your product?

172

Cooling down

Endothermic processes can be used to cool things down. For example, chemical cold packs usually contain ammonium nitrate and water. When ammonium nitrate dissolves it takes in energy from its surroundings, making them colder. These cold packs are used as emergency treatment for sports injuries. The coldness reduces swelling and numbs pain.

The ammonium nitrate and water (sometimes as a gel) are kept separate in the pack. When squeezed or struck the bag inside the water pack breaks releasing ammonium nitrate. The instant cold packs work for about 20 minutes.

They can only be used once but are ideal where there is no ice available to treat a knock or strain.

The same endothermic change can also be used to chill cans of drinks.

Figure 2 Development of this self-heating can in the USA took about 10 years. The pink circle on the can turns white when the coffee is hot enough. This takes 6–8 minutes.

Figure 3 Instant cold packs can be applied as soon as an injury occurs to minimise damage to the sportsperson

Summary questions

1 a Describe how a disposable hand warmer works.
  b Describe how a re-usable hand warmer works.
  c Give an advantage and a disadvantage of each type of hand warmer.
  d Name one use of an exothermic reaction in the food industry.
2 a Give two uses of endothermic changes.
  b Which endothermic change is often used in cold packs?

Key points

- Exothermic changes can be used in hand warmers and self-heating cans. Crystallisation of a supersaturated solution is used in reusable warmers. However, disposable, one-off warmers can give off heat for longer.
- Endothermic changes can be used in instant cold packs for sports injuries.

173

Further teaching suggestions

Design hot and cold packs

- Ask students to design their own sports cold pack and hot packs. They should create a detailed diagram including the chemicals that would be contained within it.

Reuse the hand warmer

- Allow students to boil the hand warmer and then reuse it.

Answers to in-text questions

a The sodium chloride speeds up the reaction but remains unchanged at the end of the reaction.

b calcium oxide

Summary answers

1 **a** Iron is oxidised to give out energy. Iron turns into hydrated iron(III) oxide in an exothermic reaction. Sodium chloride is used as a catalyst.
  b A supersaturated solution is made to crystallise by pressing a small metal disc. The crystals spread throughout the solution, giving off energy. The crystals are re-dissolved in hot water to use the warmer again.
  c The disposable hand warmer lasts longer when activated than the reusable warmer. However, it can only be used once. The opposite applies to the re-usable hand warmers.
  d Self-heating cans.

2 **a** To treat injuries with cold packs; to chill drinks in cans.
  b Ammonium nitrate dissolving in water.

Summary answers

1 **a** A and C
b A
c B
d B

2 **a** Measure volume of gas or mass of reaction mixture over time.

b **i** Three of: increase concentration of acid; increase surface area of magnesium; increase temperature of reaction mixture; add a catalyst.

ii Increasing concentration/surface area increases number of collisions between reactants; increasing temperature increases number of collisions between reactants and the energy possessed by reacting particles; catalyst lowers activation energy.

3 **a** The mass of gas produced at each minute is the difference between the initial mass of the reaction flask and reactants (at the start of the reaction) minus the mass at each minute.

b

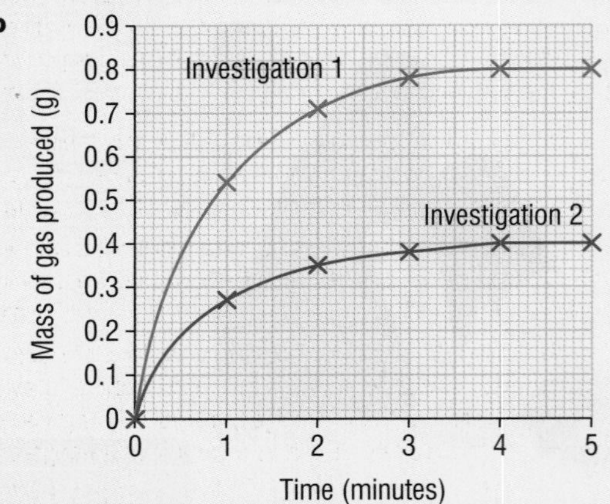

c The rate is less than in investigation 1.

d Half the mass of investigation 1.

e Investigation 2 uses acid with twice the concentration of that in investigation 1 (rate and volume of gas double), with at least enough marble chips to react fully with the acid in each case.

4 [Students should describe a way in which the temperature change can be measured when known amounts of sherbet dissolve in water.]

5 **a** and **b**

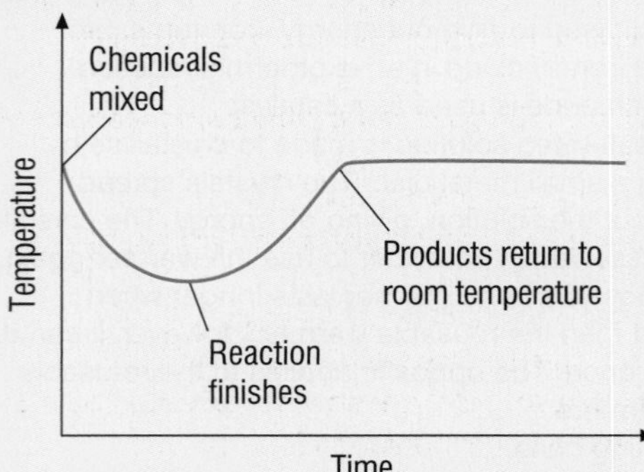

6 **a** The higher the temperature, the more quickly the cross will disappear.

b Wear eye protection or do not heat solution above 50 °C or dispose of solutions in fume cupboard.

Rates and energy: C2 4.1–C2 4.9

Summary questions

1 Select from A, B and C to show how the rate of each reaction, **a** to **d**, could be measured.

a	Gas evolved from reaction mixture	A	Measure mass
b	Mass of reaction mixture changes	B	Measure light transmitted
c	Precipitate produced	C	Measure volume
d	Colour of solution changes		

2 A student carried out a reaction in which she dropped a piece of magnesium ribbon in sulfuric acid with a concentration of 0.5 mol/dm³.

a Suggest **one** way in which the student could measure the rate of this reaction.

b **i** Suggest **three** ways in which the student could increase the rate of this reaction.

ii Explain how each of these methods changes the rate of the reaction.

3 The following results show what happened when two students investigated the reaction of some marble chips with acid.

Time (minutes)	Investigation 1 Mass of gas produced (g)	Investigation 2 Mass of gas produced (g)
0	0.00	0.00
1	0.54	0.27
2	0.71	0.35
3	0.78	0.38
4	0.80	0.40
5	0.80	0.40

a The students were investigating the effect of concentration on rate of reaction. How did the students get the data for their table above?

b Plot a graph of these results with time on the *x*-axis.

c After one minute, how does the rate of the reaction in Investigation 2 compare with the rate of reaction in Investigation 1?

d How does the final mass of gas produced in Investigation 2 compare with that produced in Investigation 1?

e From the results, what can you say about the concentration of the acids in Investigations 1 and 2?

4 'When sherbet sweets dissolve in your mouth this is an endothermic process.' Devise an experiment to test your statement. Use words and diagrams to describe clearly what you would do.

5 Two chemicals are mixed and react endothermically. When the reaction has finished, the reaction mixture is allowed to stand until it has returned to its starting temperature.

a Sketch a graph of temperature (*y*-axis) against time (*x*-axis) to show how the temperature of the reactio mixture changes.

b Label the graph clearly and explain what is happer wherever you have shown the temperature is changing.

6 This student's account of an investigation into the effe of temperature on the rate of a reaction was found on internet:

I investigated the effect of temperature on the rate of reaction. The reaction was between sodium thiosulfat and hydrochloric acid. I set up my apparatus as in th diagram.

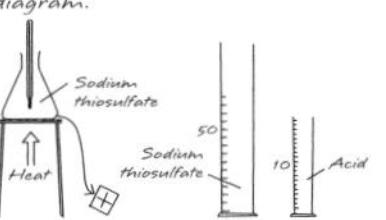

The cross was put under the flask. I heated the sodiu thiosulfate to the temperature I wanted and then ad the hydrochloric acid to the flask. I immediately star the watch and timed how long it took for the cross to disappear.

My results are below.

Temperature of the sodium thiosulfate	*Time taken for the cros disappear*
15	110
30	40
45	21

My conclusion is that the reaction goes faster the hig the temperature.

a Suggest a suitable prediction for this investigation.

b Describe one safety feature that is not mentioned i the method.

c Suggest some ways in which this method could be improved. For each suggestion, say why it is a improvement.

d Suggest how the table of results could be improve

e Despite all of the problems with this investigation, the conclusion appropriate? Explain your answer.

174

c Ways in which the method could have been improved include:

- There should have been more temperatures chosen, so that the pattern could have been seen in the results.
- The range could have been wider, so that the effect of higher and lower temperatures could have been noted.
- The volume and concentration of the two reactants should be known, to make sure that the method is valid.
- The hydrochloric acid should have been heated to the desired temperature as well, to ensure that the reaction took place at the stated temperature.
- Data logging could have been used to detect the end point. It is difficult to tell accurately when the cross disappears.
- The solutions should be continually stirred, to ensure validity.
- A water bath should have been used to control the temperature.

The results would have been more valid if repeated values of time had been taken at each temperature and the mean values calculated.

d Include units in the table.

e It is not possible to tell because the evidence is not repeatable and reproducible. (Also accept an answer that indicates that the conclusion is appropriate because there are large differences between the results at different temperatures – assuming the timings were taken in seconds.)

AQA Examination-style questions

1 A glue is made by mixing together two liquids.

a When the liquids are mixed an exothermic reaction takes place. Complete the sentence below using a word or phrase from the list.

decrease increase stay the same

During the reaction the temperature of the mixture will .... (1)

b The time taken for the glue to set at different temperatures is given in the table below.

Temperature (°C)	Time taken for the glue to set
20	3 days
60	6 hours
90	1 hour

Complete the sentences below using words or phrases from the list.

decreases increases stays the same

i When the temperature is increased the time taken for the glue to set .... (1)

ii When the temperature is increased the rate of the setting reaction .... (1)

c Which **two** of the following are reasons why an increase in temperature affects the rate of reaction?

It gives the particles more energy.

It increases the concentration of the particles.

It increases the surface area of the particles.

It makes the particles move faster. (2)

AQA, 2009

2 Instant cold packs are used to treat sports injuries.

One type of cold pack has a plastic bag containing water. Inside this bag is a smaller bag containing ammonium nitrate.

The outer bag is squeezed so that the inner bag bursts. The pack is shaken and quickly gets very cold as the ammonium nitrate dissolves in the water.

a Explain why the pack gets cold. (2)

b Suggest and explain why the pack is shaken after the inner bag has burst. (2)

AQA, 2008

3 A student reacted small pieces of zinc with dilute acid to make hydrogen gas. The graph shows how the volume of hydrogen gas produced changed with time.

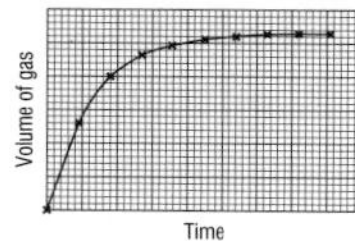

a Describe, as fully as you can, how the rate of this reaction changes with time. (2)

b The student wanted to make the reaction go faster.

Which suggestion would make the reaction go faster?

Use bigger pieces of the same total mass of zinc.

Use more of the dilute acid.

Use zinc powder. (1)

c The student decided to increase the concentration of the acid. Explain, in terms of particles, why increasing the concentration of the acid increases the rate of reaction. (2)

d The student increased the temperature of the reaction by 10 °C. The student found that the reaction went twice as fast. Explain, as fully as you can, why an increase in temperature increases the rate of the reaction. (3)

AQA, 2008

4 Platinum is used as a catalyst in many industrial processes. Platinum is a very expensive metal. The catalysts often contain only about 1% platinum dispersed on an inert support such as aluminium oxide to give a surface area of about 200 m² per gram. Cobalt catalysts with nanosized particles have been developed as an alternative to platinum catalysts for use in some industrial processes.

a Suggest two reasons why platinum is used as a catalyst, even though it is very expensive. (2)

b Explain, in terms of particles, why catalysts like platinum should have a very large surface area. (2)

c Suggest an economic reason and an environmental reason why cobalt catalysts have been developed as alternatives to platinum catalysts. (2)

d Suggest **three** reasons why the use of catalysts is important in industrial processes. (3)

Practical suggestions

Practicals	AQA	k	📖	⚙
Designing and carrying out investigations into factors that affect the rate of reaction …	✓	✓	✓	
Investigating temperature changes of neutralisations and displacement reactions, e.g. zinc and copper sulfate.	✓	✓	✓	
Investigating temperature changes when dissolving ammonium nitrate, or reacting citric acid and sodium hydrogencarbonate.	✓		✓	
Adding ammonium nitrate to barium hydroxide.	✓		✓	
Demonstration of the addition of concentrated sulfuric acid to sugar.	✓		✓	
Demonstration of the reaction between iodine and aluminium after activation by a drop of water.	✓		✓	
Demonstration of the screaming jelly baby.	✓		✓	
Demonstration of the thermite reaction, i.e. aluminium mixed with iron(III) oxide.	✓		✓	
Investigation of hand warmers, self-warming cans, sports injury packs.	✓		✓	

Examination-style answers

1 a increase *(1 mark)*

b i decreases *(1 mark)*

ii increases *(1 mark)*

c It gives the particles more energy. It makes the particles move faster. *(2 marks)*

2 a The bag gets cold because heat is taken in from the surroundings. *(2 marks)*

b Two from: mix/spread (the ammonium nitrate and water), so the whole bag gets cold; dissolve faster; get cold faster; particles collide more or more collisions, *(allow increase rate or quicker reaction)*. *(2 marks)*

3 a Two from: rate is high at the start *(allow fast at the start)*, decreases with time until it becomes zero. *(2 marks)*

b Use zinc powder. *(1 mark)*

c More particles in given volume or particles closer together/more crowded; particles collide more frequently or more often. *(2 marks)*

d The speed of particles increases and so there are more frequent collisions and the collisions are more energetic, therefore more particles have the activation energy or minimum energy to react. *(3 marks)*

4 a Two from: not used up in reaction or does not need replacing very often; only a small amount is needed or very effective; catalyses many reactions. *(2 marks)*

b There is a greater surface area of catalyst/metal and so particles of reactants collide with the surface (of catalyst/metal) more frequently. *(2 marks)*

c Economic: low(er) cost of cobalt (relative to platinum). Environmental: sensible suggestion, e.g. cobalt (may be) less toxic/harmful (to living things); cobalt mining causes less damage (because there is a higher percentage of cobalt in ores). *(2 marks)*

d Three from: increases rate of reaction; more product in less time; reduces costs; less energy needed; less fossil fuel needed (for energy/heating/pressure); smaller workforce. *(3 marks)*

Kerboodle resources

Resources available for this chapter on Kerboodle are:

- Chapter map: Rates and energy
- Practical: The effect of surface area on reaction rate (C2 4.2)
- Practical: The effect of temperature on reaction rate (C2 4.3)
- Simulation: How can I change the rate of a reaction? (C2 4.2–4.4)
- Data handling skills: Effect of concentration on reaction rate (C2 4.4)
- Support: Changing the rate (C2 4.4)
- Practical: The effect of concentration on reaction rates (C2 4.4)
- How Science Works: The state of the rates (C2 4.4)
- Extension: How do catalysts work? (C2 4.5)
- Practical: The effect of catalysts on reaction rates (C2 4.5)
- WebQuest: Catalytic nanoparticles (C2 4.6)
- Practical: Exothermic and endothermic reactions (C2 4.7)
- Interactive activity: Rates and energy
- Revision podcast: Collision theory
- Test yourself: Rates and energy
- On your marks: Rates and energy
- Examination-style questions: Rates and energy
- Answers to examination-style questions: Rates and energy

C2 5.1 Acids and alkalis

Learning objectives

Students should learn:

- why solutions are acidic or alkaline
- what bases and alkalis are
- how acidity or alkalinity can be measured.

Learning outcomes

Most students should be able to:

- list some properties of acids and alkalis
- explain in terms of ions what acids and alkalis are
- give an example of an acid, an alkali and a base
- state the ions formed by acids and alkalis in solution
- recognise whether a solution is acidic or alkaline if the pH is given.

Some students should also be able to:

- explain the differences and similarities between alkalis and bases.

Specification link-up: Chemistry C2.6

- The state symbols in equations are (s), (l), (g) and (aq). *[C2.6.1 a)]*
- Metal oxides and hydroxides are bases. Soluble hydroxides are called alkalis. *[C2.6.2 a)]*
- Hydrogen ions, H^+ (aq), make solutions acidic and hydroxide ions, OH^- (aq), make solutions alkaline. The pH scale is a measure of the acidity or alkalinity of a solution. *[C2.6.2 d)]*

  Controlled Assessment: C4.1 Plan practical ways to develop and test candidates' own scientific ideas. *[C4.1.1 b) c)]*

Lesson structure

Starters

Table – Ask students to draw a three-column table, with titles: 'acid', 'base', 'neutral'. They should then list as many things about each category as they can, including examples. Then draw a similar table on the board, and ask each student in turn to write a piece of information into the table. Support students by giving them statements that they sort. Extend students by asking them to include a fourth column for alkalis. *(5 minutes)*

Think – Ask students to work in pairs to brainstorm everything that they can remember from KS3 about acids, alkalis and the pH scale. Feedback ideas from the class. *(10 minutes)*

Main

- It is important that students are clear about the definitions and some common examples of acids, alkalis and neutral substances. You may wish to supply students with a tray of examples containing acids (oranges, vinegar, a sealed bottle of water labelled as hydrochloric acid), alkalis (soap, washing powder, a sealed bottle of water labelled as sodium hydroxide) and neutral chemicals (two sealed bottles of water, one labelled 'water' and the other labelled 'alcohol', a sealed bottle of brine labelled 'sodium chloride solution'). Ask students to sort them into acids, alkalis and neutral chemicals and try to come up with a bullet-point definition for each key word.
- If you have not already introduced state symbols to students, it is important that you do so now. You could give students a simple activity in which the state symbols are listed in the left-hand column of a table and the name of the states are listed in the right-hand column. Students can then use a rule to match the state symbol with its state.
- Students could test a variety of solutions using universal indicator to find out their pH. Encourage them to design their own results table for the experiment. Point out to students that the independent variable (chemical) should be in the first column and the dependent variable (pH) in the second column (Controlled Assessment AS4.4.2 a). Then ask them to draw conclusions from their results. They should realise that some everyday acids can be eaten, whereas most alkalis are cleaning products.
- Students could design and carry out their own investigation to determine the alkalinity of different household cleaning products. Encourage students to work in pairs to plan their experiment. They could use a variety of methods, including universal indicator, digital pH probes or data loggers. This could lead to a useful discussion of the Controlled Assessment idea that technology such as data logging may provide a better means of obtaining data, and explaining why a particular technology is the most appropriate. The concept of accuracy in measurements (Controlled Assessment AS4.3.2 c) could also be included.

Plenaries

5, 4, 3, 2, 1 – Ask students to list the names of five everyday acids, four lab-based bases, three properties of acids, two properties of alkalis and one example of a neutral chemical. Support students by asking them to work in small teams. They could show you their answers on A4 whiteboards for instant feedback and praise. Extend students by encouraging them to use correct chemical formulae for their answers where possible. *(5–10 minutes)*

Review – Show students the table that the class made. Ask them to look carefully and consider any mistakes. Discuss the mistakes and make amendments. Then ask students to copy out three facts from the table and add a new fact that they have learned from the lesson into their notes. *(10 minutes)*

Answers to in-text questions

- **a** Any soluble hydroxide, e.g. sodium hydroxide.
- **b** A base is a substance that can neutralise acids.
- **c** H^+ ions/hydrogen ions
- **d** OH^- ions/hydroxide ions
- **e** The substance is a solid.

Support

- Students may need support during the practical. The method to test the alkalinity of different household chemicals could be given to these students in labelled diagrams but in the wrong order. They could sort out the steps before completing the practical.

Extend

- Students will be familiar with the use of indicators to classify chemicals as acid, alkali or neutral from KS3. Encourage students to suggest how electricity can be used to measure pH. Encourage them to think about the ions in solution.

Practical support

Testing pH of various chemicals

Equipment and materials required

Dimple tiles, a beaker of water, dropping pipettes, samples of acids (irritants/harmful), alkalis (irritants/harmful), neutral chemicals and buffer solutions (irritants/harmful), universal indicator solution (highly flammable/harmful), universal indicator paper, scissors.

Details

Wear eye protection throughout the practical and wash hands in cold water if any of the solutions touch the skin. Be aware that universal indicator will stain the skin for about three days.

Put a few drops of each type of solution in separate dimples. Add either a few drops of universal indicator solution or a small square of universal indicator paper. Compare the colour of the paper or solution with the given colour chart (each brand of universal indicator will change to a different colour at each pH value, so it is important to compare with the appropriate chart). Put the dropping pipette into a beaker of water when finished.

Safety: Wear eye protection.

Investigating household cleaning products

Equipment and materials required

Different cleaning products (irritant/harmful), beakers, stirring rods, pH probes, pH data logger and equipment, universal indicator (highly flammable/harmful), dropping pipette, dimple tile, eye protection.

Details

Wear eye protection and put a small amount of cleaning product on a dimple tile. Add a few drops of water if the product is a solid. Then add universal indictor and compare with the colour chart. Alternatively, put the cleaning product into a beaker and submerge the calibrated pH probe (either digital or data logger) and note the reading.

Safety: If the cleaning product touches the skin, wash well with cold water. Be aware that some cleaning products may be 'corrosive' and 'toxic' – avoid these.

Salts and electrolysis

C2 5.1 Acids and alkalis

Learning objectives

- Why are solutions acidic or alkaline?
- What are bases and alkalis?
- How do we measure acidity?

Acids and bases are an important part of our understanding of chemistry. They play an important part inside us and all other living things.

What are acids and bases?

When we dissolve a substance in water we make an **aqueous solution**. The solution may be acidic, alkaline or neutral. That depends on which substance we have dissolved.

- Soluble hydroxides are called **alkalis**. Their solutions are alkaline. An example is sodium hydroxide solution.
- **Bases**, which include alkalis, are substances that can neutralise **acids**. Metal oxides and metal hydroxides are bases. Examples include iron oxide and copper hydroxide.
- Acids include citric acid, sulfuric acid and ethanoic acid. All acids taste very sour, although many acids are far too dangerous to put in your mouth. Ethanoic acid (in vinegar) and citric acid (in citrus fruit and fizzy drinks) are acids which we regularly eat.
- Pure water is **neutral**.

a Name an alkali.
b What is a base?

One acid that we use in science labs is hydrochloric acid. This is formed when the gas hydrogen chloride (HCl) dissolves in water:

$$HCl(g) \xrightarrow{\text{water}} H^+(aq) + Cl^-(aq)$$

All acids form H^+ ions when we add them to water. It is these H^+ ions that make a solution acidic. Hydrogen chloride also forms chloride ions (Cl^-). The '(aq)' in the equation above is called a **state symbol**. It shows that the ions are in an 'aqueous solution'. In other words, they are dissolved in water.

c What ions do all acids form when we add them to water?

Because alkalis are bases which dissolve in water, they are the bases we often use in experiments. Sodium hydroxide solution is often found in school labs. We get sodium hydroxide solution when we dissolve solid sodium hydroxide in water:

$$NaOH(s) \xrightarrow{\text{water}} Na^+(aq) + OH^-(aq)$$

All alkalis form hydroxide ions (OH^-) when we add them to water. It is these hydroxide ions that make a solution alkaline.

d What ions do all alkalis form when we add them to water?
e What does the state symbol '(s)' stand for?

Figure 1 Acids and bases are all around us, in many of the things we buy at the shops, in our schools and factories – and in our bodies too

176

Acids and alkalis

Measuring acidity or alkalinity

Indicators are substances which change colour when we add them to acids and alkalis. Litmus paper is a well-known indicator, but there are many more.

We use the **pH scale** to show how acidic or alkaline a solution is. The scale runs from 0 (most acidic) to 14 (most alkaline). We can use **universal indicator (UI)** to find the pH of a solution. It is a very special indicator made from a number of dyes. It turns a range of colours as the pH changes. Anything in the middle of the pH scale (pH 7) is neutral, neither acid nor alkali.

Practical

Which is the most alkaline product?

Compare the alkalinity of various cleaning products.

You can test washing-up liquids, shampoos, soaps, hand-washing liquids, washing powders/liquids and dishwasher powders/tablets.

You could use a pH sensor and data logger to collect your data.

- What are the advantages of using a pH sensor instead of universal indicator solution or paper?

Safety: Wear eye protection.

Maths skills

We can use the mathematical symbols ' $>$ ' (read as 'is greater than') and ' $<$ ' ('is less than') when interpreting pH values.
We can say:
$\text{pH} < 7$ indicates an acidic solution.
i.e. pH values less than 7 are acidic.

$\text{pH} > 7$ indicates an alkaline solution.
i.e. pH values greater than 7 are alkaline.

pH	Universal indicator solution		
0			Very acidic
1		Hydrochloric acid	
2		Lemon juice	
3		Orange juice Vinegar	
4			
5		Black coffee	Slightly acidic
6		Rainwater	
7		Pure water	Neutral
8		Seawater Baking soda	Slightly alkaline
9		Milk of magnesia Soap	
10			
11			
12		Washing soda	
13			
14		Oven cleaner Sodium hydroxide	Very alkaline

Figure 2 The pH scale tells us how acidic or alkaline a solution is

Summary questions

1 Match the halves of the sentences together:

a	A base that is soluble in water	A	.... a pH of exactly 7.
b	Pure water is neutral with	B	.... form OH^- ions when they dissolve in water.
c	Acids are substances that	C	.... is called an alkali.
d	Alkalis are substances that	D	.... is acidic.
e	Indicators are substances that	E	.... form H^+ ions when they dissolve in water.
f	A solution with a pH less than 7	F	.... change colour when we add them to acids and alkalis.

2 How could you use universal indicator paper as a way of distinguishing between pure water, sodium hydroxide solution and citric acid solution?

Key points

- Acids are substances which produce H^+ ions when we add them to water.
- Bases are substances that will neutralise acids.
- An alkali is a soluble hydroxide. Alkalis produce OH^- ions when we add them to water.
- We can use the pH scale to show how acidic or alkaline a solution is.

177

Further teaching suggestions

Homemade indicator

- Ask students to find out how to make an indicator at home and write a bullet-pointed method on how to make it, e.g. using elderberries, cranberries or red cabbage.

Interactive pH scale

- Make a coloured universal indicator pH scale on interactive whiteboard software or PowerPoint. Invite the students to label weak/strong acid/alkali and neutral chemicals. Then show images of different things, e.g. stomach acid, then drag and drop the images on to the pH scale.

Other indicators

- Ask students to research and find out different indicators that can be used to test for acids and bases.

Summary answers

1 **a** C **b** A **c** E **d** B **e** F **f** D

2 The paper would turn green in water (neutral), blue/purple in sodium hydroxide solution (strong alkali) and red/orange in citric acid solution (weak acid).

C2 5.2 Making salts from metals or bases

Learning objectives

Students should learn:

- the products of the reaction between acids and metals
- the products of the reaction between acids and bases
- that salts can be made using neutralisation.

Learning outcomes

Most students should be able to:

- state a definition of neutralisation
- state the general word equation when a metal reacts with an acid
- state the general word equation when an acid reacts with a base
- name the salt formed if the acid and alkali are given
- write the ionic equation for neutralisation.

Some students should also be able to:

- construct balanced symbol equations including state symbols.

Specification link-up: Chemistry C2.6

- The state symbols in equations are (s), (l), (g) and (aq). *[C2.6.1 a)]*
- Soluble salts can be made from acids by reacting them with:
  - metals – not all metals are suitable; some are too reactive and others are not reactive enough
  - insoluble bases – the base is added to the acid until no more will react and the excess solid is filtered off … *[C2.6.1 b)]*
- Salt solutions can be crystallised to produce solid salts. *[C2.6.1 c)]*
- The particular salt produced in any reaction between an acid and a base or alkali depends on:
  - the acid used (hydrochloric acid produces chlorides, nitric acid produces nitrates, sulfuric acid produces sulfates)
  - the metal in the base or alkali. *[C2.6.2 b)]*

Lesson structure

Starters

Table – Ask students to complete the following table:

Name of acid	Name in salt	Example
Hydrochloric	[Chloride]	[e.g. Sodium chloride]
[Nitric]	Nitrate	[e.g. Copper nitrate]
Sulfuric	Sulfate	[e.g. Calcium sulfate]

Show the incomplete table on the board and ask different students to fill in the missing data. Support students by giving them the missing words for them to slot into the table. Extend students by asking them to write the formulae of the ions and compounds. *(5 minutes)*

Neutralisation experiment – Ask students to complete a neutralisation reaction between an acid and alkali in pairs. The first group to get exactly green is the winner. You might wish to use a pH probe and data logger so that students could chart the change in pH during their experiment. *(10 minutes)*

Main

- The word 'salt' is used in everyday life to mean sodium chloride. However, this is actually a chemical classification.
- Ask students to suggest names of acids that they have used in their school science lessons. You should write these formulae on the board. Ask the students to suggest what they all have in common [they all contain hydrogen]. Remind students that the H^+ ion is released when an acid is put into water. Then explain that the hydrogen ion can be swapped for a metal. Ask a student to name any metal and choose an acid and write the metal salt it would produce. Repeat this a few times with the other acids written on the board. Explain to students that when the hydrogen ion has been swapped for the metal ion, a new substance is made and this is known as a salt. Encourage students to write their own definition of a salt into their notes and give the name and formula of two metals salts of their choice.
- At KS3, most students will have had experience of neutralisation with an acid and an alkali. However, they may not have used an insoluble base before. Show the students some copper(II) oxide (harmful) and allow them to try to dissolve it in water. Then allow students to prepare copper sulfate crystals (harmful) using the method in the Student Book. Encourage the students to write up the method in a brief bullet-pointed format and summarise the reaction in a general word equation.
- Give each student three index cards. These can be made out of three different coloured pieces of card about A6 in size. Punch all the cards in the top right-hand corner hole. On the front of each card the student should write a general word equation. Then on the reverse a specific example, including a method and an equation for the reaction. After a few lessons when all of the cards have been completed, they can be joined together with a treasury tag or a piece of string. These can then be tied to their notes and used for revision for examination.

Answers to in-text questions

- **a** A salt and hydrogen.
- **b** The substance is a gas.
- **c** A salt and water.
- **d** zinc sulfate
- **e** The substance is a liquid.

Support

- Some students may find it difficult to represent chemical reactions in equations. Provide these students with a skeleton structure of the equations. Then each additional piece of information could be made into a card. Each card should have a separate chemical and then the students try to create the equations, using sticky-tac to secure the cards.

Extend

- Show students how to generate the ionic equation for neutralisation from first principles. They could then use the method to check other examples and prove that the same ionic equation is always generated.

Plenary

Chemical equations – Ask students to complete the following equations (they get progressively more difficult). Time the students for 5 minutes and assure them that it doesn't matter how far they get.

- acid + alkali → [salt] + [water]
- [acid] + [metal] → metal salt + hydrogen
- acid + base → [salt] + [water]
- [sodium hydroxide] + [nitric acid] → sodium nitrate + water
- sulfuric acid + zinc → [zinc sulfate] + hydrogen
- [2HCl(aq)] + [Ca(s)] → $CaCl_2$ [(aq]) + H_2 [(g)]

When the notes are marked it will give an idea of the level that each student is working at. *(10 minutes)*

Summary answers

1 neutralisation, salt, water, metals, hydrogen

2 a Copper is not reactive enough to react with acid.

b Potassium is too reactive and would explode on contact with acid.

c Heat the solution in an evaporating basin on a water bath until the point of crystallisation. Then leave the liquid for a few days to allow the remainder of the water to evaporate off and the crystals to form.

Practical support

Preparing copper sulfate crystals

Equipment and materials required

Copper oxide (harmful), 1 mol/dm^3 sulfuric acid (irritant), stirring rod, beaker (100 cm^3), Bunsen burner, tripod, gauze, filter funnel, filter paper, conical flask, evaporating basin, beaker (250 cm^3), spatula, measuring cylinder, conical flask, eye protection.

Details

Add a spatula of copper oxide to a beaker, then add 25 cm^3 of sulfuric acid. Stir the reaction mixture well and note any observations. Warm the mixture gently on a tripod and gauze. Do not allow to boil. Let the mixture containing excess black copper oxide cool down. Fold the filter paper and put into the funnel in the neck of a conical flask. Filter the mixture. Collect the filtrate and put the solution into an evaporating basin. Heat it on a water bath until the point of crystallisation, then leave the liquid in a warm place for a few days to allow the crystals to form.

Safety: Eye protection should be worn throughout this practical. This reaction makes copper sulfate, which is harmful. CLEAPSS Hazcard 27C Copper sulfate – harmful; 98A Sulfuric acid–irritant.

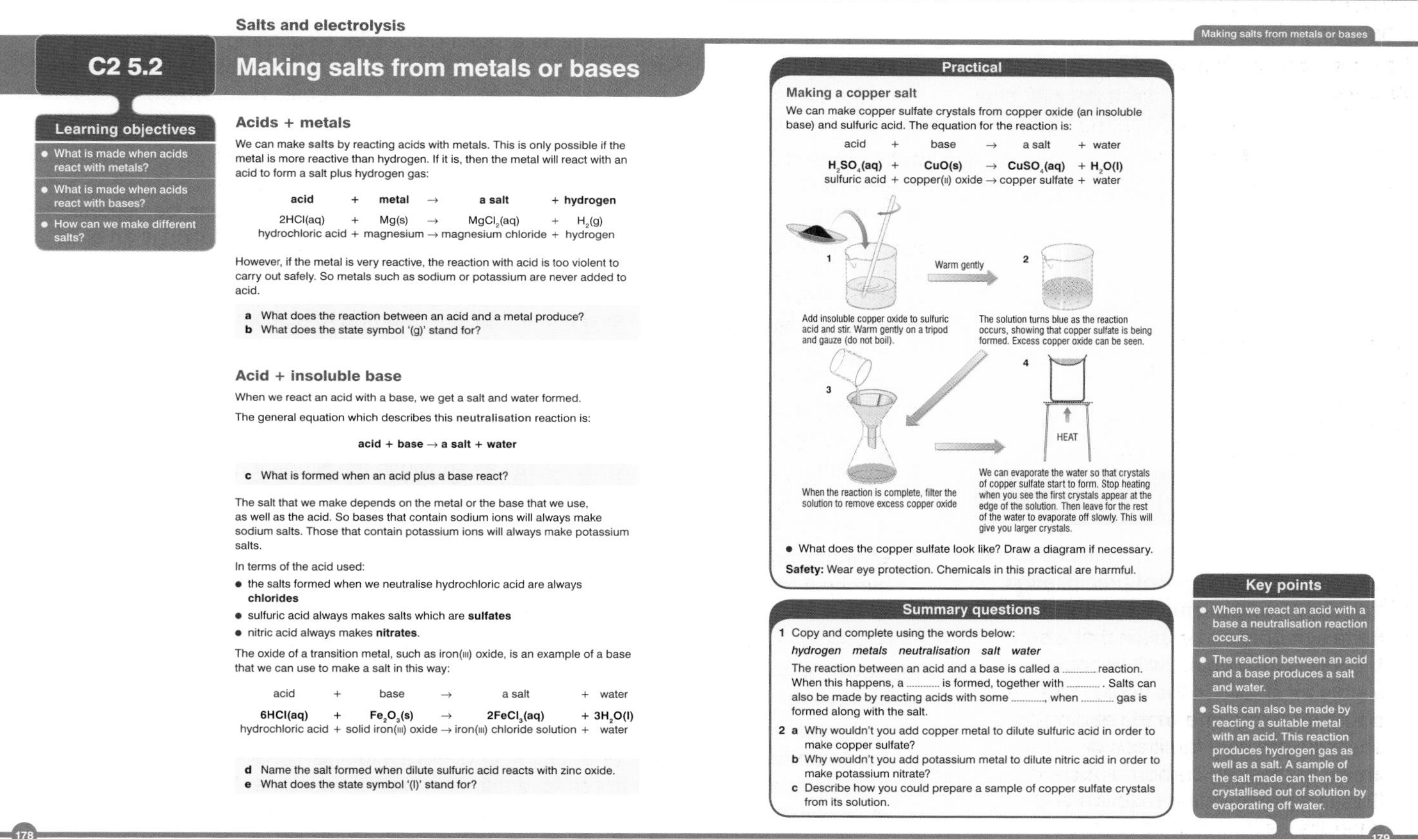

Salts and electrolysis

Making salts from metals or bases

C2 5.2 Making salts from metals or bases

Learning objectives
- What is made when acids react with metals?
- What is made when acids react with bases?
- How can we make different salts?

Acids + metals

We can make salts by reacting acids with metals. This is only possible if the metal is more reactive than hydrogen. If it is, then the metal will react with an acid to form a salt plus hydrogen gas:

acid + metal → a salt + hydrogen

2HCl(aq) + Mg(s) → $MgCl_2$(aq) + H_2(g)
hydrochloric acid + magnesium → magnesium chloride + hydrogen

However, if the metal is very reactive, the reaction with acid is too violent to carry out safely. So metals such as sodium or potassium are never added to acid.

a What does the reaction between an acid and a metal produce?
b What does the state symbol '(g)' stand for?

Acid + insoluble base

When we react an acid with a base, we get a salt and water formed.

The general equation which describes this **neutralisation** reaction is:

acid + base → a salt + water

c What is formed when an acid plus a base react?

The salt that we make depends on the metal or the base that we use, as well as the acid. So bases that contain sodium ions will always make sodium salts. Those that contain potassium ions will always make potassium salts.

In terms of the acid used:
- the salts formed when we neutralise hydrochloric acid are always **chlorides**
- sulfuric acid always makes salts which are **sulfates**
- nitric acid always makes **nitrates**.

The oxide of a transition metal, such as iron(III) oxide, is an example of a base that we can use to make a salt in this way:

acid + base → a salt + water
6HCl(aq) + Fe_2O_3(s) → $2FeCl_3$(aq) + $3H_2O$(l)
hydrochloric acid + solid iron(III) oxide → iron(III) chloride solution + water

d Name the salt formed when dilute sulfuric acid reacts with zinc oxide.
e What does the state symbol '(l)' stand for?

Practical

Making a copper salt

We can make copper sulfate crystals from copper oxide (an insoluble base) and sulfuric acid. The equation for the reaction is:

acid + base → a salt + water
H_2SO_4(aq) + CuO(s) → $CuSO_4$(aq) + H_2O(l)
sulfuric acid + copper(II) oxide → copper sulfate + water

1 Add insoluble copper oxide to sulfuric acid and stir. Warm gently on a tripod and gauze (do not boil).

2 The solution turns blue as the reaction occurs, showing that copper sulfate is being formed. Excess copper oxide can be seen.

3 When the reaction is complete, filter the solution to remove excess copper oxide

4 We can evaporate the water so that crystals of copper sulfate start to form. Stop heating when you see the first crystals appear at the edge of the solution. Then leave for the rest of the water to evaporate off slowly. This will give you larger crystals.

- What does the copper sulfate look like? Draw a diagram if necessary.

Safety: Wear eye protection. Chemicals in this practical are harmful.

Summary questions

1 Copy and complete using the words below:
hydrogen metals neutralisation salt water
The reaction between an acid and a base is called a .... reaction. When this happens, a .... is formed, together with .... Salts can also be made by reacting acids with some ...., when .... gas is formed along with the salt.

2 a Why wouldn't you add copper metal to dilute sulfuric acid in order to make copper sulfate?
b Why wouldn't you add potassium metal to dilute nitric acid in order to make potassium nitrate?
c Describe how you could prepare a sample of copper sulfate crystals from its solution.

Key points
- When we react an acid with a base a neutralisation reaction occurs.
- The reaction between an acid and a base produces a salt and water.
- Salts can also be made by reacting a suitable metal with an acid. This reaction produces hydrogen gas as well as a salt. A sample of the salt made can then be crystallised out of solution by evaporating off water.

178 179

Further teaching suggestions

Calculations

- Higher Tier students could complete mole calculations on the balanced symbol equations.

Uses of neutralisation

- Students could find three uses for a neutralisation reaction (e.g. antacids, reduction of acidity of soils, to remove harmful gases from factory emissions).

Complete the table

- Ask students to draw a table where they list common acids in the first column, then the name of a base or metal in the second column and the third column should be the name of the salt. Support students by giving them the table with some missing information so they just need to complete one piece of information per row.

C2 5.3 Making salts from solutions

Learning objectives

Students should learn:

- how salts can be made from an acid and alkali
- how insoluble salts can be made
- how unwanted ions can be removed from solutions.

Learning outcomes

Most students should be able to:

- record a method to make soluble salts
- record a method to make insoluble salts
- state what a precipitation reaction is and what they can be used for
- suggest a method for making a named salt.

Some students should also be able to:

- explain in detail what precipitation is in terms of the ions involved.

Specification link-up: Chemistry C2.6

- Soluble salts can be made from acids by reacting them with: …
  - alkalis – an indicator can be used to show when the acid and alkali have completely reacted to produce a salt solution. *[C2.6.1 b)]*
- Insoluble salts can be made by mixing appropriate solutions of ions so that a precipitate is formed. Precipitation can be used to remove unwanted ions from solutions, for example in treating water for drinking or in treating effluent. *[C2.6.1 d)]*
- Ammonia dissolves in water to produce an alkaline solution. It is used to produce ammonium salts. Ammonium salts are important as fertilisers. *[C2.6.2 c)]*
- In neutralisation reactions, hydrogen ions react with hydroxide ions to produce water. This reaction can be represented by the equation:

  $H^+ (aq) + OH^- (aq) + H_2O (l)$ *[C2.6.2 e)]*
- Select an appropriate method for making a salt, given appropriate information. *[C2.6]*

Lesson structure

Starters

Definitions – Ask students to define the terms 'soluble' and 'insoluble'. Then ask them to write two sentences, each using one of the key words. *(5 minutes)*

Demonstration – The solubility of ammonia can be shown using the fountain experiment (see RSC Classic Chemistry demonstrations and CLEAPSS Handbook). Show the students and use their observations as a starting point for discussing solubility. Support students by having a list of prompt words to help them focus their observations. Extend students by asking them to classify the reaction and write a balanced symbol equation for this [$NH_3 + H_2O \rightarrow NH_4OH$, hydrolysis]. *(10 minutes)*

Main

- Students could make a flow chart to show how ammonia solution can be used to make ammonium nitrate. Students should consider that the ammonia solution is an alkali and undergoes a neutralisation reaction with nitric acid to make ammonium nitrate. They should include word equations in their flow chart. You could also encourage students to use a red pen every time they are referring to acids, a blue one for bases and a green one for neutral. This colour code will reinforce previous work on the pH scale and universal indicator.
- Extend students by asking them to find out the industrial conditions for this process and write a balanced symbol equation, including state symbols.
- Students could make an insoluble salt as shown in the Student Book. Encourage them to record this in a step-by-step method including an equipment list. The students could also generate the word equation for the reaction.
- Extend students by encouraging them to write a balanced symbol equation with state symbols for the reaction.

Plenaries

Method – Ask students to explain briefly how they would make one of the following salts:

- Sodium chloride. [Neutralisation between sodium hydroxide and hydrochloric acid. Use indicator to find the quantities needed. Repeat without indicator, then evaporate the water to the point of crystallisation and leave it to finish crystallising.]
- Lead iodide. [Reaction between lead nitrate and sodium iodide. Filter and collect the solid, wash it with distilled water and dry it.] *(5 minutes)*

Word search – Create a word search, but students need to answer questions to determine the words that they need to find:

- A method for removing pollutants from water. [Precipitation]
- When a solute will dissolve into a solvent, it is described as … [soluble]
- Chalk is described as this because it will not dissolve in water. [Insoluble]
- A chemical with a pH < 7 [Acid]
- A soluble base. [Alkali]
- The name of the chemical reaction between a hydrogen ion and hydroxide ion. [Neutralisation]

Support students by giving them both the word and clue. Encourage them to match the clue with the key word and then find the word in the word search. *(10 minutes)*

Support

- Some students may find it difficult to understand the cause of precipitation. You may find it useful to focus on state symbols and explain that when the salt is insoluble it will become visible as a solid in the liquid. You could refer to a chemical reaction that students are very familiar with – the limewater test for carbon dioxide. This is a precipitation reaction and a neutralisation reaction (between acidic carbon dioxide and the calcium hydroxide alkali). The cloudy parts are solid calcium carbonate.

Extend

- Ask students to find some other examples of precipitation reactions and to write balanced symbol equations for these reactions, including the state symbols.

Practical support

Making an insoluble salt

Equipment and materials required

Lead nitrate solution (toxic) (0.01 mol/dm^3), potassium iodide solution, beaker, measuring cylinder, conical flask, filter funnel, filter paper, distilled water, eye protection.

Details

Measure 10 cm^3 of potassium iodide and 5 cm^3 of lead nitrate into a small beaker. Gently shake, and then filter the mixture and wash through with distilled water. Scrape the solid on to some fresh filter paper and allow it to dry. This is the insoluble salt.

Safety: Chemical splashproof eye protection should be worn throughout, and hands should be thoroughly washed after the experiment has been completed. Pregnant women should be aware that lead salts can affect unborn children. CLEAPSS Hazcard 57A – toxic. Dispose of waste following CLEAPSS guidelines.

Salts and electrolysis

C2 5.3 Making salts from solutions

Learning objectives

- How can we make salts from an acid and an alkali?
- How can we make insoluble salts?
- How can we remove unwanted ions from solutions?

There are two other important ways of making salts from solutions.

- We can react an acid and an alkali together to form a soluble salt.
- We can make an *insoluble* salt by reacting solutions of two soluble salts together.

Acid + alkali

When an acid reacts with an alkali, a neutralisation reaction takes place.

Hydrochloric acid reacting with sodium hydroxide solution is an example:

acid + alkali → a salt + water

$$\mathbf{HCl(aq) + NaOH(aq) \rightarrow NaCl(aq) + H_2O(l)}$$

hydrochloric acid + sodium hydroxide solution → sodium chloride + water

We can think about neutralisation in terms of $H^+(aq)$ ions reacting with OH^- (aq) ions. They react to form water:

$$H^+(aq) + OH^-(aq) \rightarrow H_2O(l)$$

When we react an acid with an alkali we need to know when the acid and alkali have completely reacted. We can use an indicator for this.

We can make ammonium salts, as well as metal salts, by reacting an acid with an alkali. Ammonia reacts with water to form a weakly alkaline solution:

$$NH_3(aq) + H_2O(l) \rightleftharpoons NH_4^+(aq) + OH^-(aq)$$

Ammonia solution reacts with an acid (for example, nitric acid):

acid + ammonia solution → an ammonium salt + water

$$\mathbf{HNO_3(aq) + NH_4^+(aq) + OH^-(aq) \rightarrow NH_4NO_3(aq) + H_2O(l)}$$

nitric acid + ammonia solution → ammonium nitrate + water

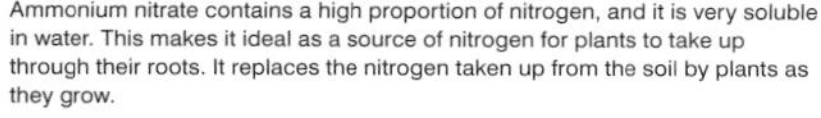

Figure 1 Ammonium nitrate is used as a fertiliser

Ammonium nitrate contains a high proportion of nitrogen, and it is very soluble in water. This makes it ideal as a source of nitrogen for plants to take up through their roots. It replaces the nitrogen taken up from the soil by plants as they grow.

Ammonium salts are made by adding ammonia solution to an acid until there is a small excess of ammonia. We can detect the excess ammonia by using universal indicator. We then crystallise the ammonium salt from its solution. The excess ammonia evaporates off.

a Write down a general equation for the reaction between an acid and an alkali.
b Name a salt which is used as a fertiliser to provide crops with nitrogen.

Making insoluble salts

We can sometimes make salts by combining two solutions that contain different soluble salts. When the soluble salts react to make an insoluble salt, we call the reaction a precipitation reaction. That's because the insoluble solid formed is called a **precipitate**.

180

Making salts from solutions

$$\mathbf{Pb(NO_3)_2(aq) + 2KI(aq) \rightarrow PbI_2(s) + 2KNO_3(aq)}$$

lead nitrate solution + potassium iodide solution → lead iodide precipitate + potassium nitrate solution

Each of the reactant solutions contains one of the ions of the insoluble salt. In this case, they are lead ions in lead nitrate and iodide ions in potassium iodide. Lead iodide forms a yellow precipitate that we can filter off from the solution.

Practical

Making an insoluble salt

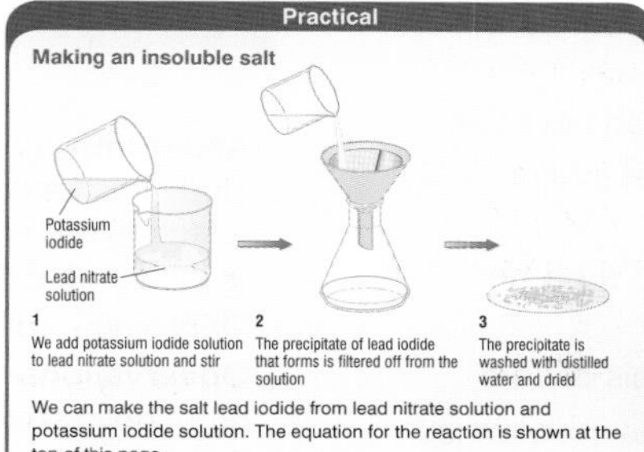

Potassium iodide
Lead nitrate solution
1
We add potassium iodide solution to lead nitrate solution and stir
2
The precipitate of lead iodide that forms is filtered off from the solution
3
The precipitate is washed with distilled water and dried

We can make the salt lead iodide from lead nitrate solution and potassium iodide solution. The equation for the reaction is shown at the top of this page.

- Why is the precipitate of lead iodide washed with distilled water?

Using precipitation

We use precipitation reactions to remove pollutants from the wastewater from factories. The effluent must be treated before it is discharged into rivers and the sea.

Precipitation is used in the removal of metal ions from industrial wastewater. By raising the pH of the water, we can make insoluble metal hydroxides precipitate out. This produces a sludge which we can easily remove from the solution.

The cleaned-up water can then be discharged safely into a river or the sea.

Precipitation can be also used to remove unwanted ions from drinking water.

Figure 2 Water treatment plants use chemical treatments to precipitate out metal compounds which can then be removed by filtering the solution

Summary questions

1 Copy and complete using the words below:

acid alkali insoluble metal polluted precipitation solid neutralisation soluble water indicator

We can make salts by reacting an .... with an .... This makes the salt and .... and is called a .... reaction. We need an .... to tell us when the reaction is complete. We can also make .... salts by reacting two .... salts together. We call this a .... reaction because the salt is formed as a .... This type of reaction is also important when we want to remove .... ions from .... water.

2 Write word equations and a brief method to show how to make the following salts:
a potassium nitrate (a soluble salt)
b silver chloride (an insoluble salt). Hint: all nitrates are soluble in water.

Key points

- An indicator is needed when a soluble salt is prepared by reacting an alkali with an acid.
- Insoluble salts can be made by reacting two solutions to produce a precipitate.
- Precipitation is an important way of removing some metal ions from industrial wastewater.

181

Further teaching suggestions

Role play

- Students could act out a precipitation reaction, where each student represents an ion.

Calculations

- Mole calculations could be completed from balanced symbol equations.

Solubility curves

- Students could test the solubility of salts at different temperatures and plot solubility curves (see 'Practical support'). Secondary data could be obtained to compare the solubility of different chemicals, e.g. from RSC Data book or search the internet for 'solubility data'.

Other reactions

- You may wish to allow students to make a different insoluble salt. One example is barium sulfate, prepared by mixing solutions of barium chloride and sodium sulfate (see www.practicalchemistry.org and search for 'insoluble salt').

Answers to in-text questions

a acid + alkali → salt + water
b ammonium nitrate/ammonium sulfate

Summary answers

1 acid (alkali), alkali (acid), water, neutralisation, indicator, insoluble, soluble, precipitation, solid, metal, polluted

2 **a** nitric acid + potassium hydroxide → potassium nitrate + water

Use an indictor to find how much of each solution is required. Repeat with the correct quantities. Evaporate the solution to the point of crystallisation and then leave it for the rest of the water to evaporate off to get crystals of potassium nitrate.

b silver nitrate + a soluble chloride, e.g. sodium chloride → silver chloride + e.g. sodium nitrate

Filter off the silver chloride precipitate, wash with distilled water and leave (or warm in an oven) to dry.

C2 5.4 Electrolysis

Learning objectives

Students should learn:

- what electrolysis is
- which types of substance can be electrolysed
- the products of electrolysis.

Learning outcomes

Most students should be able to:

- state a definition for electrolysis
- recognise which compounds will undergo electrolysis
- add state symbols to an equation
- predict the products of molten electrolysis.

Some students should also be able to:

- explain how electrolysis occurs
- summarise electrolysis in balanced symbol equations.

Answers to in-text questions

a Using an electric current to break down a substance.
b electrolyte
c negative electrode
d positive electrode

Support

- Give students the parts of the flow chart for the decomposition of lead bromide but in the wrong order. The students can then cut them up and stick them into their own diagram in their notes.

Extend

- Ask students to find out what materials could be used to make electrodes, e.g. carbon, and to list the benefits and drawbacks of using them.

Specification link-up: Chemistry C2.7

- When an ionic substance is melted or dissolved in water, the ions are free to move about within the liquid or solution. *[C2.7.1 a)]*
- Passing an electric current through ionic substances that are molten, for example lead bromide, or in solution breaks them down into elements. This process is called electrolysis and the substance that is broken down is called the electrolyte. *[C2.7.1 b)]*
- During electrolysis, positively charged ions move to the negative electrode, and negatively charged ions move to the positive electrode. *[C2.7.1 c)]*

Lesson structure

Starters

Anagram – As a title write: 'cysistrollee'. Explain to the students that they are going to study this topic, but the letters are jumbled up. Encourage them to find out the word [electrolysis] and write this as the title. Support students by giving them the first letter of the word. Extend students by asking them to write a brief definition [splitting up a compound using electricity]. *(5 minutes)*

Observations – Show the students an ampoule of bromine, and samples of lead and lead bromide in sealed containers. Ask students to work in pairs to make a list of everything they know about these chemicals and encourage them to use the periodic table and list the masses etc. Draw a three-column table on the board, each headed with a different one of the chemicals and ask each pair for pieces of information to fill in the table. Then ask the students how lead and bromine could be made from lead bromide. *(10 minutes)*

Main

- A classic demonstration is electrolysis of molten lead bromide. This is chosen as it is an ionic solid with a relatively low melting point. The reaction produces lead and bromine and therefore should be completed in a fume cupboard. Demonstrate the experiment and use questions and answers to extract observations from the students. Then ask them to draw a diagrammatic flow chart, including a symbol equation to represent the demonstration.
- Often it is difficult for students to see a demonstration in a fume cupboard. If the reaction can be filmed beforehand, it could be shown to students. Alternatively set up a flexicam or a camcorder connected to a TV or digital projector to show the demonstration magnified in real time.
- Students could complete their own electrolysis experiment. However, as a molten liquid is hazardous to use, they must use a solution instead. To prevent any confusion due to water producing oxygen or hydrogen, copper chloride solution should be used. Students could be given a set of questions to consider as they complete the reaction to channel their thoughts.

Plenaries

Definitions – Ask students to define the key words: 'electrolysis', 'electrolyte', 'decompose', 'electrode' in their books. Support students by giving them the definitions and the key words and they have to match them up. Extend students by asking them to define the term, then use it correctly in a sentence. *(5 minutes)*

Taboo – Create a set of cards with the key words: 'electricity', 'electrolysis', 'electrolyte', 'electrode', 'decompose'. Below each key word list three further words that would aid in explaining the main word. These will be the 'taboo' words. Give the pack of cards to groups of three. Each student should take it in turns to pick a card and try to explain the main word, without using the taboo words. The person who managed to explain the most words (without using any taboo words) is the winner. *(10 minutes)*

Practical support

Demonstration of the electrolysis of lead bromide

Equipment and materials required

Ceramic evaporating basin, lead bromide (CLEAPSS Hazcard 57 – toxic), spatula, tongs, Bunsen burner and safety equipment, tongs, two carbon electrodes, lamp, three wires, two crocodile clips, low voltage power supply (0–12 V), fume cupboard, tripod, pipe-clay triangle, eye protection, protective gloves.

Details

Half-fill the evaporating basin with lead bromide and submerge the ends of the electrodes. Connect the electrodes into the circuit involving the lamp and the power supply. Put the evaporating basin on the pipe-clay triangle above the Bunsen burner. Ignite the Bunsen burner and heat the lead bromide strongly, turn on the power supply and observe. Once the lamp is on, the electricity is flowing. This will only occur when the ions are free to move, i.e. the lead bromide is molten. Point out the vapour (CLEAPSS Hazcard 15 Bromine – toxic/corrosive). The molten salt is at a higher temperature than the boiling point of bromine, so the bromine is released as a gas not as a liquid, which is its state at room temperature. The molten lead collects at the bottom of the basin. Switch off the Bunsen and use the tongs to tip the molten lead on to the flameproof mat to show the students.

Safety: Chemical splashproof eye protection should be worn during this demonstration and it should be completed in a fume cupboard. Pregnant women should not use lead bromide. Anhydrous zinc chloride (CLEAPSS Hazcard 108A – corrosive) melts at a lower temperature than lead bromide, so can be used as an alternative.

Electrolysis of copper chloride solution

Equipment and materials required

A beaker, two carbon electrodes, lamp, three wires, two crocodile clips, low-voltage power supply (0–12 V), 1 mol/dm^3 copper chloride solution (CLEAPSS Hazcard 27A – harmful), eye protection.

Details

Half-fill the beaker with copper chloride solution and immerse the tips of the electrodes. Connect the electrodes in a simple circuit with the low-voltage power supply and lamp. Start the current and observe. Chlorine (toxic) should be smelt at the positive electrode, do not smell directly at source, and copper should be deposited at the negative electrode. As soon as the observations are complete, the low-voltage power supply should be switched off.

Safety: Chemical splashproof eye protection should be worn throughout the practical. This experiment should be completed in a well ventilated room as the chlorine could irritate asthmatics. CLEAPSS Hazcard 22A Chlorine – toxic.

C2 5.4 Electrolysis

Learning objectives

- What is electrolysis?
- What types of substance can we electrolyse?
- What is made when we electrolyse substances?

Figure 1 The first person to explain electrolysis was Michael Faraday. He worked on this and many other problems in science nearly 200 years ago.

The word electrolysis means 'splitting up using electricity'. In electrolysis we use an electric current to break down an ionic substance. We call the substance that is broken down by electrolysis the **electrolyte**.

a What is electrolysis?
b What do we call the substance broken down by electrolysis?

To set up an electrical circuit for electrolysis, we have two electrodes which dip into the electrolyte. The electrodes are conducting rods. One of these is connected to the positive terminal of a power supply. The other electrode is connected to the negative terminal.

The electrodes are often made of an unreactive (or **inert**) substance. This is often graphite or sometimes platinum. This is so the electrodes do not react with the electrolyte or the products made in electrolysis.

During electrolysis, positively charged ions move to the negative electrode. At the same time, the negative ions move to the positive electrode.

When the ions reach the electrodes they lose their charge and become elements. Gases may be given off or metals deposited at the electrodes. This depends on the compound used and whether it is molten or dissolved in water.

Demonstration

The electrolysis of molten lead bromide

- This demonstration needs a fume cupboard because bromine is toxic and corrosive.
- When does the bulb light up?

Bromine gas
Molten lead
Molten lead bromide
Heat

Figure 2 Passing electricity through molten lead bromide. It forms molten lead and brown bromine gas as the electrolyte is broken down by the electricity.

Figure 2 above shows how electricity breaks down lead bromide into lead and bromine:

$$\text{lead bromide} \rightarrow \text{lead} + \text{bromine}$$
$$PbBr_2(l) \rightarrow Pb(l) + Br_2(g)$$

Lead bromide is an ionic substance. Ionic substances do not conduct electricity when they are solid. But once we melt them, the ions are free to move and carry their charge towards the electrodes.

The positive lead ions (Pb^{2+}) move towards the negative electrode. At the same time, the negatively charged bromide ions (Br^-) move towards the positive electrode.

Did you know ...?

Electrolysis is also a way of getting rid of unwanted body hair. A small electric current is passed through the base of each individual hair to be removed. The hair is destroyed through chemical changes caused by the electric current, which destroy the cells that make the hair grow.

Notice the state symbols in the equation. They tell us that the lead bromide and the lead are molten at the temperature in the dish. The '(l)' stands for 'liquid'. The bromine is given off as a gas, shown as '(g)'.

c Which electrode do positive ions move towards during electrolysis?
d Which electrode do negative ions move towards during electrolysis?

Electrolysis of solutions

Many ionic substances have very high melting points. This can make electrolysis very difficult. But some ionic substances dissolve in water. When this happens, the ions also become free to move around.

However, when electrolysing solutions it is more difficult to predict what will be formed. This is because water also forms ions. So the products at each electrode are not always exactly what we expect.

When we electrolyse a solution of copper bromide, copper ions (Cu^{2+}) move to the negative electrode. The bromide ions (Br^-) move to the positive electrode. Copper bromide is split into its elements at the electrodes (see Figure 3):

$$\text{copper bromide} \rightarrow \text{copper} + \text{bromine}$$
$$CuBr_2(aq) \rightarrow Cu(s) + Br_2(aq)$$

In this case the state symbols in the equation tell us that the copper bromide is dissolved in water. This is shown as '(aq)'. The copper is formed as a solid, shown as '(s)'. The bromine formed remains dissolved in the water – '(aq)'.

Covalent compounds cannot usually be electrolysed unless they react in water to form ions, e.g. acids in water.

links

For more information about the effect of water in electrolysis, see C2 5.5 Changes at the electrodes and C2 5.7 Electrolysis of brine.

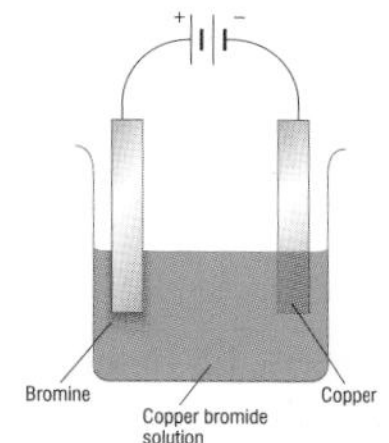

Figure 3 If we dissolve copper bromide in water, we can decompose it by electrolysis. Copper metal is formed at the negative electrode. Brown bromine appears in solution around the positive electrode.

Summary questions

1 Copy and complete using the words below:
ions molten move solution
For the current to flow in electrolysis, the .... must be able to .... between the electrodes. This can only happen if the substance is in .... or if it is ....

2 Predict the products formed at each electrode when the following compounds are melted and then electrolysed:
a zinc iodide
b lithium bromide
c iron(III) fluoride.

3 Solid ionic substances do not conduct electricity. Using words and diagrams explain why they conduct electricity when molten or in solution.

Key points

- Electrolysis breaks down a substance using electricity.
- Ionic compounds can only be electrolysed when they are molten or in solution. That's because their ions are then free to move to the electrodes.
- In electrolysis, positive ions move to the negative electrode while negative ions move to the positive electrode.

Further teaching suggestions

Simulation

- Show students a simulation of the reaction detailing the particles.

Hoffman voltameter

- You may wish to demonstrate the electrolysis of water using a Hoffman voltameter – see CLEAPSS 11.4.2. A microscale version of this experiment that students can do is found at www.chemistryteachers.org – search for 'electrolysis of water'.

Summary answers

1 ions, move, solution, molten

2 **a** Zinc at −; iodine at +.
b Lithium at −; bromine at +.
c Iron at −; fluorine at +.

3 [Words/diagrams explain how ions carry charge.] If ions are not free to move (as they are not in a solid because they are held in position by strong electrostatic forces), no current can flow in the circuit.

C2 5.5 Changes at the electrodes

Learning objectives

Students should learn:

- what happens to the ions during electrolysis
- that electrolysis can be represented in half-equations **[HT only]**
- that water affects electrolysis
- how to predict the products of electrolysis.

Learning outcomes

Most students should be able to:

- describe the transfer of electrons at the electrodes
- recognise oxidation and reduction at electrodes
- predict the products of electrolysis.

Some students should also be able to:

- explain in detail the transfer of electrons in electrolysis
- construct half equations **[HT only]**
- explain how water affects the products of electrolysis.

Answers to in-text questions

a Electron(s) are transferred from the ion to the electrode (electron loss).

b Electron(s) are transferred to the ion from the electrode (electron gain).

Support

- Some students may find it difficult to understand why oxygen or hydrogen can sometimes be produced. Use questions and answers to lead students through all of the particles that are in the electrolyte (water, metal ions, etc.). Then explain that water can split into H^+ and OH^-, which are present in low concentration. Explain that it is these ions that give rise to the oxygen or hydrogen.

Extend

- Give students the reactivity series and refer to the order of discharge of negatively charged ions. Students could then use these to explain how you can predict the chemicals that are discharged at each electrode.

Specification link-up: Chemistry C2.7

- At the negative electrode, positively charged ions gain electrons (reduction) and at the positive electrode, negatively charged ions lose electrons (oxidation). *[C2.7.1 e)]*
- If there is a mixture of ions, the products formed depend on the reactivity of the elements involved. *[C2.7.1 f)]*
- Reactions at electrodes can be represented by half equations, for example:

  $2Cl^- \rightarrow Cl_2 + 2e^-$

  or

  $2Cl^- - 2e^- \rightarrow Cl_2$. *[C2.7.1 g)]* **[HT only]**
- Predict the products of electrolysing solutions of ions. *[C2.7]*

Lesson structure

Starters

Card sort – Give the key words (oxidation, reduction and redox) and their definitions on separate cards. Students should sort the cards to match the key words with their definitions. Support students by encouraging them to complete the exercise in small groups. Extend students by asking them to represent each of these key words as symbol or ionic equations. *(5 minutes)*

Poem – Encourage the students to create a little poem or saying to help them remember that oxidation is the loss of electrons, which happens at the positive electrode, and that reduction is the gain of electrons, which occurs at the negative electrode. The best one could then be copied by all the students into their notes. *(10 minutes)*

Main

- Show the students a sample of potassium chloride and ask them to predict the products of electrolysing it when molten. Encourage a student to write the balanced symbol equation on the board. Now ask pairs of students to predict the products of electrolysing a solution of potassium chloride. Ask each pair their thoughts and why they came to this idea. Then allow the students to complete the experiment to find out it they were correct. Encourage students to note their work in the form of a fully labelled diagram, including half equations and brief notes to explain where the hydrogen comes from. (Note that half equations are Higher Tier only.)
- Students could act out an electrolysis experiment. They could wear black bibs and make a line to represent the electrodes and wires in the circuit. Polystyrene balls could be used as electrons, two students could stand by a bucket of balls – one student giving them out and one putting them into the bucket (this represents the power source). Different-coloured bibs (red for positively charged ions, blue for negatively charged ions) could be used to represent the solution. The circuit could then 'run' under your instructions. Students could use the play to describe what happens in terms of particles at the electrodes.
- Take digital photographs of the electrolysis play. These could then be used in the classroom to remind students of the play. If a photograph is displayed on an interactive whiteboard, then annotations could be added in front of the class.

Plenaries

Half equations – Ask students to complete the half equations as detailed in Question 2 of the summary questions. Then encourage the students to create a further example of a half equation of their choice. **[HT only]** *(5 minutes)*

What am I? – On sticky labels write the following words: 'redox', 'reduction', 'oxidation', 'reduced', 'oxidised', 'half equation'. Split the students into teams of six and give each a word and ask them to stick it on their forehead (but they should not know what their word is). Each student then takes it in turns to ask his or her group questions, to which the team can only respond with 'yes' or 'no'. The aim is for each student to guess their word. Support students by giving them a set of statements to prompt their questions to the other team members. Extend students by asking them to write a balanced symbol equation or ionic equation to illustrate each word. *(10 minutes)*

Practical support

Electrolysis of potassium chloride solution

Equipment and materials required

Beaker, two carbon electrodes, lamp, three wires, two crocodile clips, low-voltage power supply (0–12 V), saturated solution of potassium chloride, test tube, splint, eye protection.

Details

Half-fill the beaker with potassium chloride solution and immerse the tips of the electrodes. Connect the electrodes in a series circuit with the power supply and lamp. Start the current and observe: (CLEAPSS Hazcard 22A Chlorine – toxic) should be smelt at the anode, and bubbles (CLEAPSS Hazcard 48 Hydrogen – extremely flammable) should be observed at the cathode. As soon as the observations are complete, the power supply should be switched off. The hydrogen could be collected in a test tube under displacement, and tested with a lighted splint.

Safety: Chemical splashproof eye protection should be worn throughout the practical. This experiment should be completed in a well-ventilated room as the chlorine could irritate asthmatics. Do not smell the chlorine directly at its source.

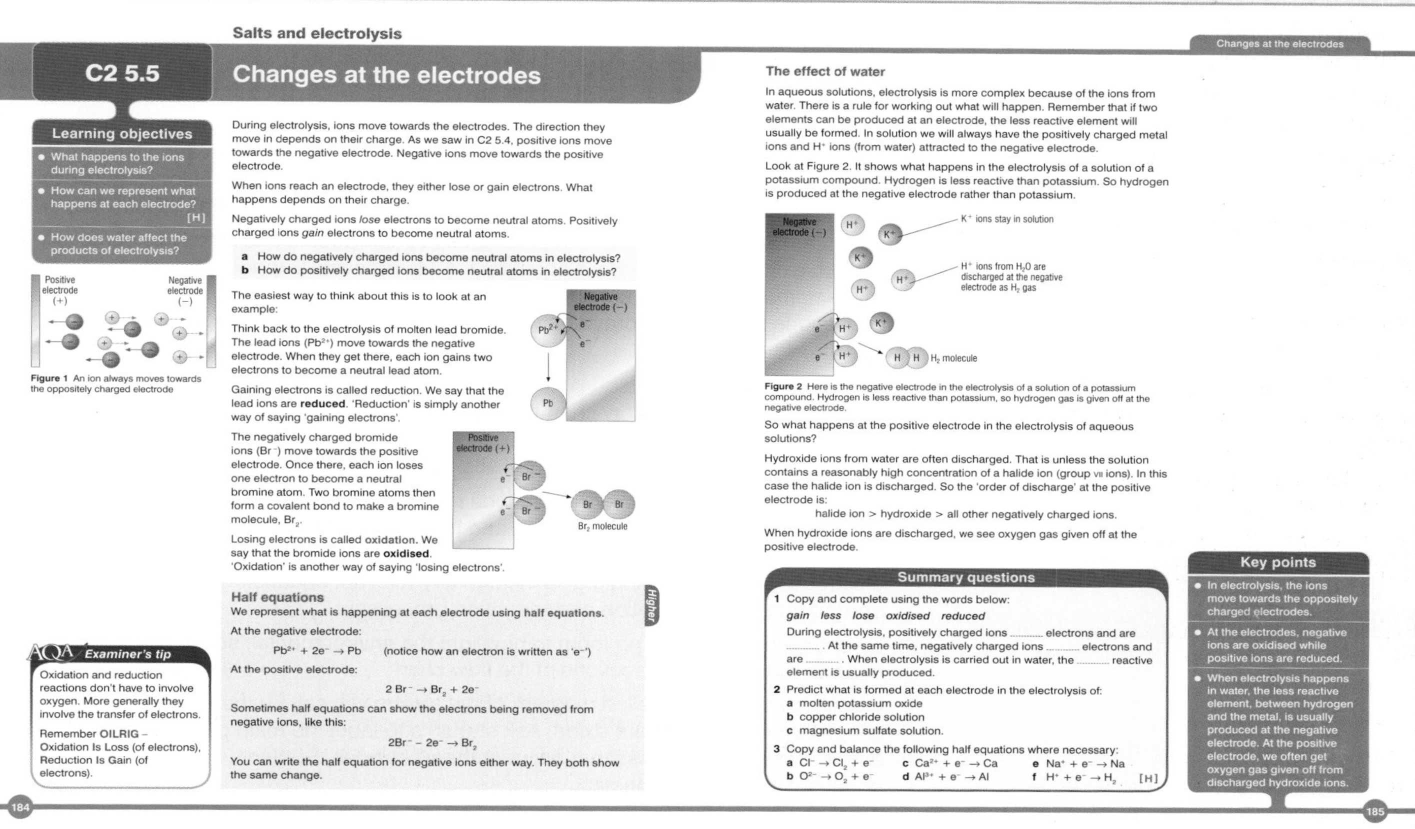

Salts and electrolysis

C2 5.5 Changes at the electrodes

Learning objectives

- What happens to the ions during electrolysis?
- How can we represent what happens at each electrode? [H]
- How does water affect the products of electrolysis?

Figure 1 An ion always moves towards the oppositely charged electrode

During electrolysis, ions move towards the electrodes. The direction they move in depends on their charge. As we saw in C2 5.4, positive ions move towards the negative electrode. Negative ions move towards the positive electrode.

When ions reach an electrode, they either lose or gain electrons. What happens depends on their charge.

Negatively charged ions *lose* electrons to become neutral atoms. Positively charged ions *gain* electrons to become neutral atoms.

a How do negatively charged ions become neutral atoms in electrolysis?
b How do positively charged ions become neutral atoms in electrolysis?

The easiest way to think about this is to look at an example:

Think back to the electrolysis of molten lead bromide. The lead ions (Pb^{2+}) move towards the negative electrode. When they get there, each ion gains two electrons to become a neutral lead atom.

Gaining electrons is called reduction. We say that the lead ions are **reduced**. 'Reduction' is simply another way of saying 'gaining electrons'.

The negatively charged bromide ions (Br^-) move towards the positive electrode. Once there, each ion loses one electron to become a neutral bromine atom. Two bromine atoms then form a covalent bond to make a bromine molecule, Br_2.

Losing electrons is called **oxidation**. We say that the bromide ions are **oxidised**. 'Oxidation' is another way of saying 'losing electrons'.

Half equations (Higher)

We represent what is happening at each electrode using **half equations**.

At the negative electrode:

$$Pb^{2+} + 2e^- \rightarrow Pb$$ (notice how an electron is written as 'e^-')

At the positive electrode:

$$2\,Br^- \rightarrow Br_2 + 2e^-$$

Sometimes half equations can show the electrons being removed from negative ions, like this:

$$2Br^- - 2e^- \rightarrow Br_2$$

You can write the half equation for negative ions either way. They both show the same change.

AQA Examiner's tip

Oxidation and reduction reactions don't have to involve oxygen. More generally they involve the transfer of electrons.

Remember OILRIG – Oxidation Is Loss (of electrons), Reduction Is Gain (of electrons).

184

Changes at the electrodes

The effect of water

In aqueous solutions, electrolysis is more complex because of the ions from water. There is a rule for working out what will happen. Remember that if two elements can be produced at an electrode, the less reactive element will usually be formed. In solution we will always have the positively charged metal ions and H^+ ions (from water) attracted to the negative electrode.

Look at Figure 2. It shows what happens in the electrolysis of a solution of a potassium compound. Hydrogen is less reactive than potassium. So hydrogen is produced at the negative electrode rather than potassium.

Figure 2 Here is the negative electrode in the electrolysis of a solution of a potassium compound. Hydrogen is less reactive than potassium, so hydrogen gas is given off at the negative electrode.

So what happens at the positive electrode in the electrolysis of aqueous solutions?

Hydroxide ions from water are often discharged. That is unless the solution contains a reasonably high concentration of a halide ion (group VII ions). In this case the halide ion is discharged. So the 'order of discharge' at the positive electrode is:

halide ion > hydroxide > all other negatively charged ions.

When hydroxide ions are discharged, we see oxygen gas given off at the positive electrode.

Summary questions

1 Copy and complete using the words below:
gain less lose oxidised reduced
During electrolysis, positively charged ions ………… electrons and are ………… . At the same time, negatively charged ions ………… electrons and are ………… . When electrolysis is carried out in water, the ………… reactive element is usually produced.

2 Predict what is formed at each electrode in the electrolysis of:
a molten potassium oxide
b copper chloride solution
c magnesium sulfate solution.

3 Copy and balance the following half equations where necessary:
a $Cl^- \rightarrow Cl_2 + e^-$
b $O^{2-} \rightarrow O_2 + e^-$
c $Ca^{2+} + e^- \rightarrow Ca$
d $Al^{3+} + e^- \rightarrow Al$
e $Na^+ + e^- \rightarrow Na$
f $H^+ + e^- \rightarrow H_2$ [H]

Key points

- In electrolysis, the ions move towards the oppositely charged electrodes.
- At the electrodes, negative ions are oxidised while positive ions are reduced.
- When electrolysis happens in water, the less reactive element, between hydrogen and the metal, is usually produced at the negative electrode. At the positive electrode, we often get oxygen gas given off from discharged hydroxide ions.

185

Further teaching suggestions

Writing half-equations

- Higher Tier students could be asked to write the following symbol equation and half equations for the electrolysis of a concentrated solution of iron(III) chloride:
  $2FeCl_3(aq) \rightarrow [2Fe(s)] + [3Cl_2(g)]$
  Positive electrode: $[2Cl^-(aq) \rightarrow Cl_2(g) + 2e^-]$
  Negative electrode: $[Fe^{3+}(aq) + 3e^- \rightarrow Fe(s)]$

Electrolysis circus

- Students could be encouraged to electrolyse a circus of different solutions and record their observations in a table. They could electrolyse solutions of other halides, e.g. sodium chloride, zinc bromide and zinc iodide. The experiments are all similar to the one with copper chloride solution, as detailed in 'Practical support' in C2 5.4 'Electrolysis'.
  They could then write word equations for the reactions. Higher Tier students could write half equations and balanced symbol equations for the reactions. You may wish to use conductivity sensors to monitor changes in conductivity.

Positive and negative

- Ask students to copy the key points and to write any word that has a positive charge in red (e.g. oxidation, positive,) and any with a negative charge in blue.

Moving ions

- You may wish to demonstrate the movement of ions to each electrode by demonstrating by the electrolysis of a crystal of $KMnO_4$ on filter paper dampened with sodium chloride solution, or the electrolysis of $CuCrO_4$ in a saturated urea solution using a U-tube. Another example of the electrolysis of copper chromate, can be found in CLEAPSS PS67-13. It is possible for students to complete this experiment.

Summary answers

1 gain, reduced, lose, oxidised, less

2 **a** potassium at −, oxygen at +
b copper at −, chlorine at +
c hydrogen at −, oxygen at +

3 **a** $2Cl^- \rightarrow Cl_2 + 2e^-$
b $2O^{2-} \rightarrow O_2 + 4e^-$
c $Ca^{2+} + 2e^- \rightarrow Ca$
d $Al^{3+} + 3e^- \rightarrow Al$
e $Na^+ + e^- \rightarrow Na$
f $2H^+ + 2e^- \rightarrow H_2$

C2 5.6 The extraction of aluminium

Learning objectives

Students should learn:

- how aluminium is extracted from aluminium oxide
- why cryolite is added to the melt
- what happens at each electrode in the process.

Learning outcomes

Most students should be able to:

- recall the products of the electrolysis of aluminium oxide
- explain why cryolite is added to the melt
- label a simple diagram of the electrolytic cell used for the extraction of aluminium
- explain how the products form at each electrode.

Some students should also be able to:

- generate half equations for the electrolysis of aluminium oxide. [HT only]

Support

- Encourage students to ensure that they know the main processes involved in extracting aluminium [mine aluminium ore, purify into aluminium oxide, electrolyse to make aluminium at the negative electrode]. Give the students these statements and allow them to flesh them out as a spider diagram.

Extend

- Ask students to explain why the carbon positive electrodes need to be replaced frequently [oxygen gas produced at the electrode immediately reacts with the carbon]. Students could represent this process as a balanced symbol equation, including state symbols [$C(s) + O_2(g) \rightarrow CO_2(g)$]. Ask students to suggest why inert electrodes such as platinum are not used [expense].

Specification link-up: Chemistry C2.7

- Aluminium is manufactured by the electrolysis of a molten mixture of aluminium oxide and cryolite. Aluminium forms at the negative electrode and oxygen at the positive electrode. The positive electrode is made of carbon, which reacts with the oxygen to produce carbon dioxide. *[C2.7.1 h)]*

Lesson structure

Starters

Uses – Ask students to reflect back on their work in C1 3.4 ‘Aluminium and titanium’, and to recall a use of aluminium and the property that makes this material fit for its purpose [e.g. aeroplane fuselage as its alloys have a low density and are strong]. *(5 minutes)*

Predict – Ask students to work in pairs to predict the products of electrolysis of molten aluminium oxide and write a word equation for this process. Support students by giving them the names of the reactant and products. Higher Tier students could also complete half-equations to show the formation of the products. *(10 minutes)*

Main

- Students could make a flow chart to explain how aluminium ore is mined and then the metal extracted and finally used. Support students by giving them the stages in the wrong order. They could then cut and stick to make their flow chart.
- Students should ensure that they include the name of the raw materials [bauxite, cryolite], conditions [electrolytic cell] and how energy consumption is reduced [addition of cryolite to reduce the melting point and recycling of the carbon electrodes]. Students should also include any relevant equations and where appropriate, detail oxidation and reduction processes.
- Extend students by asking them to highlight the environmental, social and economic impacts of the different sections of the flow chart.
- Give students an unlabelled diagram of the electrolytic cell for the extraction of aluminium from aluminium oxide. Ask students to label the main parts of the diagram and to explain the process that occur at the electrodes. Students should include equations where appropriate.

Plenaries

Key points – Ask students to write questions that could be answered by each of the key points. Support students by supplying them with a selection of questions. They have to choose the one that matches each key point. Extend students by asking them to form the question and then rewrite the key point to include the formula, half equations or symbol equations as appropriate. *(5 minutes)*

Flow chart summary – Ask students to summarise the extraction of aluminium from bauxite, including any equations in a flow chart. Support students by making this a ‘cut-and-stick’ exercise. Extend students by encouraging them to include symbol equations in their flow chart. They could also write all reduction processes in blue and oxidation processes in red. *(10 minutes)*

Further teaching suggestions

Handling samples

- You may have samples of cryolite, aluminium oxide, bauxite and aluminium that students could handle and observe. Students could create a table to detail their observations.

Video

- The RSC Industrial reactions have a video on aluminium extraction that you could play to students.

Cartoon strip

- Ask students to draw a cartoon strip using particles to demonstrate what happens at the electrodes. You could support the students by supplying the images or the words for them to complete the cartoon strip.

Animation

- Show an animation of what happens at each electrode in terms of the ions, electrons and atoms. Ask students to work in small groups and to create a 'voice-over' for the animation. Animations are freely available on the internet. Use a search engine.

C2 5.6 The extraction of aluminium

Learning objectives

- How is aluminium obtained from aluminium oxide?
- Why is cryolite used in the process?
- What happens at each electrode in the process?

You already know that aluminium is a very important metal. The uses of the metal or its alloys include:

- pans
- overhead power cables
- aeroplanes
- cooking foil
- drink cans
- window and patio door frames
- bicycle frames and car bodies.

a Why is aluminium used to make overhead power cables?

Aluminium is quite a reactive metal. It is less reactive than magnesium but more reactive than zinc or iron. Carbon is not reactive enough to use in its extraction so we must use electrolysis. The compound electrolysed is aluminium oxide, Al_2O_3.

Figure 1 Aluminum alloys have a low density but are very strong

We get aluminium oxide from bauxite ore. The ore is mined by open cast mining. Bauxite contains mainly aluminium oxide. However, it is mixed with other rocky impurities. So the first step is to separate aluminium oxide from the ore. The impurities contain a lot of iron(III) oxide. This colours the waste solution from the separation process rusty brown. The solution has to be stored in large lagoons.

Figure 2 Extracting aluminium from its ore. This process requires a lot of energy. The purification stage makes aluminium hydroxide. This is separated from the impurities but then must be heated to turn it back to pure aluminium oxide. Then even more energy is needed melting and electrolysing the oxide.

Electrolysis of aluminium oxide

To electrolyse the aluminium oxide we must first melt it. This enables the ions to move to the electrodes.

Unfortunately aluminium oxide has a very high melting point. It melts at 2050 °C. However, chemists have found a way of saving at least some energy. This is done by mixing the aluminium oxide with molten cryolite. Cryolite is another ionic compound. The molten mixture can be electrolysed at about 850 °C. The electrical energy transferred to the electrolysis cells keeps the mixture molten.

b Why must aluminium oxide be molten for electrolysis to take place?

The overall reaction in the electrolysis cell is:

$$\text{aluminium oxide} \xrightarrow{\text{electrolysis}} \text{aluminium} + \text{oxygen}$$
$$2Al_2O_3(l) \longrightarrow 4Al(l) + 3O_2(g)$$

Figure 3 The extraction of aluminium by electrolysis

At the negative (–) electrode:

Each aluminium ion (Al^{3+}) gains 3 electrons. The ions turn into aluminium atoms. We say that the Al^{3+} ions are reduced to form Al atoms.

The aluminium metal formed is molten at the temperature of the cell and collects at the bottom. It is siphoned or tapped off.

At the positive (+) electrode:

Each oxide ion (O^{2-}) loses 2 electrons. The ions turn into oxygen atoms. We say that the O^{2-} ions are oxidised to form oxygen atoms. These bond in pairs to form molecules of oxygen gas (O_2).

The oxygen reacts with the hot, positive carbon electrodes, making carbon dioxide gas. So the positive electrodes gradually burn away. They need to be replaced in the cells regularly.

c Are the oxide ions reduced or oxidised in the electrolysis of molten aluminium oxide?

Summary questions

1 Copy and complete using the words below:

positive oxygen extraction carbon cryolite negative energy

In the .... of aluminium, aluminium oxide is dissolved in molten .... in order to use less .... to melt it. The aluminium metal is collected at the .... electrode in the cells, while oxygen is formed at the .... electrode. The electrodes used are made of .... The positive electrodes burn away as they react with .... and form carbon dioxide gas.

2 **a** Explain which ions are oxidised and which ions are reduced in the electrolysis of molten aluminium oxide.

b Why are the positive electrodes replaced regularly in the industrial electrolysis of aluminium oxide? Include a word equation.

3 Write half equations for the changes at each electrode in the electrolysis of molten aluminium oxide. [H]

Key points

- Aluminium oxide is electrolysed in the manufacture of aluminium metal.
- The aluminium oxide is mixed with molten cryolite to lower its melting point.
- Aluminium forms at the negative electrode and oxygen at the positive electrode.
- The positive carbon electrodes are replaced regularly as they gradually burn away.

Answers to in-text questions

a Aluminium is a good conductor of electricity and has a low density.

b Its ions must be free to move to the electrodes.

c Oxidised (ions lose electrons to become neutral atoms).

Summary answers

1 extraction, cryolite, energy, negative, positive, carbon, oxygen

2 **a** Aluminium ions (Al^{3+}) are reduced (electrons gained); oxide ions (O^{2-}) are oxidised (electrons lost).

b The carbon in the positive electrodes reacts with oxygen produced to give off carbon dioxide gas, burning the electrodes away:

carbon + oxygen → carbon dioxide

3 At negative electrode: $Al^{3+} + 3e^- \rightarrow Al$;
At positive electrode: $2O^{2-} \rightarrow O_2 + 4e^-$

C2 5.7 Electrolysis of brine

Learning objectives

Students should learn:

- the products of the electrolysis of brine
- how the products are used.

Learning outcomes

Most students should be able to:

- state the products of the electrolysis of brine
- describe how brine can be electrolysed
- list some uses of the products of the electrolysis of brine.

Some students should also be able to:

- generate half equations for the electrolysis of brine. **[HT only]**

Specification link-up: Chemistry C2.7

- The electrolysis of sodium chloride solution produces hydrogen and chlorine. Sodium hydroxide solution is also produced. These are important reagents for the chemical industry, e.g. sodium hydroxide for the production of soap and chlorine for the production of bleach and plastics. *[C2.7.1 i)]*

Lesson structure

Starters

Spot the odd one out – Show students a picture of a bar of soap, rayon, paper, a bottle marked brine and a bottle of detergent. Ask students to suggest the odd one out [brine as all the others are products made using sodium hydroxide]. *(5 minutes)*

Predict – Ask students to work in pairs to predict the products of electrolysis of molten sodium chloride and a solution of sodium chloride. Support students by giving them a list of the particles present in each of the electrolytes. Higher Tier students could also complete half equations to show the formation of the products. *(10 minutes)*

Main

- Students can complete their own electrolysis of sodium chloride. They could record their observations on a diagram of the apparatus. They could also annotate the formation of the products and detail their uses.
- Show students different popular science publications, from *New Scientist* to *Horrible Science*. Ask the students to write a magazine article for a popular science magazine to explain the importance of the chloro-alkali industry (search for 'electrolysis of salt' or 'chlor-alkali').
- Students could use desktop publishing packages to produce their article. They could be encouraged to use photographs of the industrial processes using the internet, or they could use a digital camera to take images in the lab to be used in their article.

Plenaries

Uses – Give the students separate cards with the words 'hydrogen', 'chlorine' and 'sodium hydroxide' printed on them. Read out the following uses of the products of electrolysis of sodium chloride solution, then students hold up the card that shows which product is used for that specific use:

- Margarine [hydrogen]
- PVC [chlorine]
- Bleach [sodium hydroxide, chlorine]
- Soap [sodium hydroxide]
- Paper [chlorine, sodium hydroxide]
- Rayon fibres [sodium hydroxide]
- Detergents [sodium hydroxide, chlorine]
- Purification of aluminium ore [sodium hydroxide]
- Hydrochloric acid manufacture [chlorine, hydrogen] *(5 minutes)*

Demonstration – Show the electrolysis of sodium chloride solution in a Petri dish. This could be completed on an overhead projector and projected on to a whiteboard. The colours of the universal indicator solution can be clearly seen. Take care not to spill solution into the projector. Ask the students to explain their observations. Support students by giving them statements that they can use to label directly the image that is being projected. Extend students by encouraging them to write half equations next to each electrode. *(10 minutes)*

Support

- Students may need support to understand why there are three different products from the electrolysis of brine. You may wish to ask them again to list all of the particles present in the electrolyte [H^+, OH^-, Na^+, Cl^- and H_2O]. State that hydrogen is given off at the negative electrode and ask a volunteer to come to the board and rub out the ion that would make hydrogen. Then repeat this for chlorine. Ask students what are left [OH^-, Na^+ and H_2O]. Ask students what would be left if the water was removed [NaOH or sodium hydroxide]. You could ask students to suggest how the water could be removed [evaporation].

Extend

- Extend students by asking them to research the Solvay process, which uses sodium chloride. Students could research the industrial production of sodium carbonate and sodium hydrogencarbonate using this method.

Answers to in-text questions

- **a** Chlorine, hydrogen, sodium hydroxide solution.
- **b** Water treatment, making bleach, making plastics (PVC).
- **c** For example, making margarine.
- **d** Making soap and paper, making bleach (with chlorine).

Practical support

Electrolysing brine in the lab

Equipment and materials required

Beaker or electrolysis cell, saturated sodium chloride solution, two carbon electrodes, two crocodile clips, two wires, low-voltage power supply (0–12 V), litmus paper, water, two test tubes, splint, matches, gloves, eye protection (chemical splashproof).

Details

Half-fill the beaker with the sodium chloride solution and submerge one end of the carbon electrodes. Using the wires and crocodile clips, connect to the power supply. Wearing gloves, fill the test tube with the solution and hold, inverted with the neck in the solution over the negative electrode (the carbon electrode attached to the black terminal of the power supply). Put on eye protection and start the electrolysis. Once the test tube is full of gas, put a gloved finger over the end of the tube to seal it, and remove it from the water. Test the gas with a lighted splint (a pop should be heard). While the gas is being collected, test the gas at the positive electrode by holding a damp piece of litmus paper over the electrode and observe.

Safety: This practical produces chlorine gas (CLEAPSS Hazcard 22A – toxic, could irritate asthmatics) and should only be completed in a well-ventilated area. The equipment should be switched off when the products have been tested. The solution produces sodium hydroxide and this is why gloves should be worn to collect the gas by displacement.

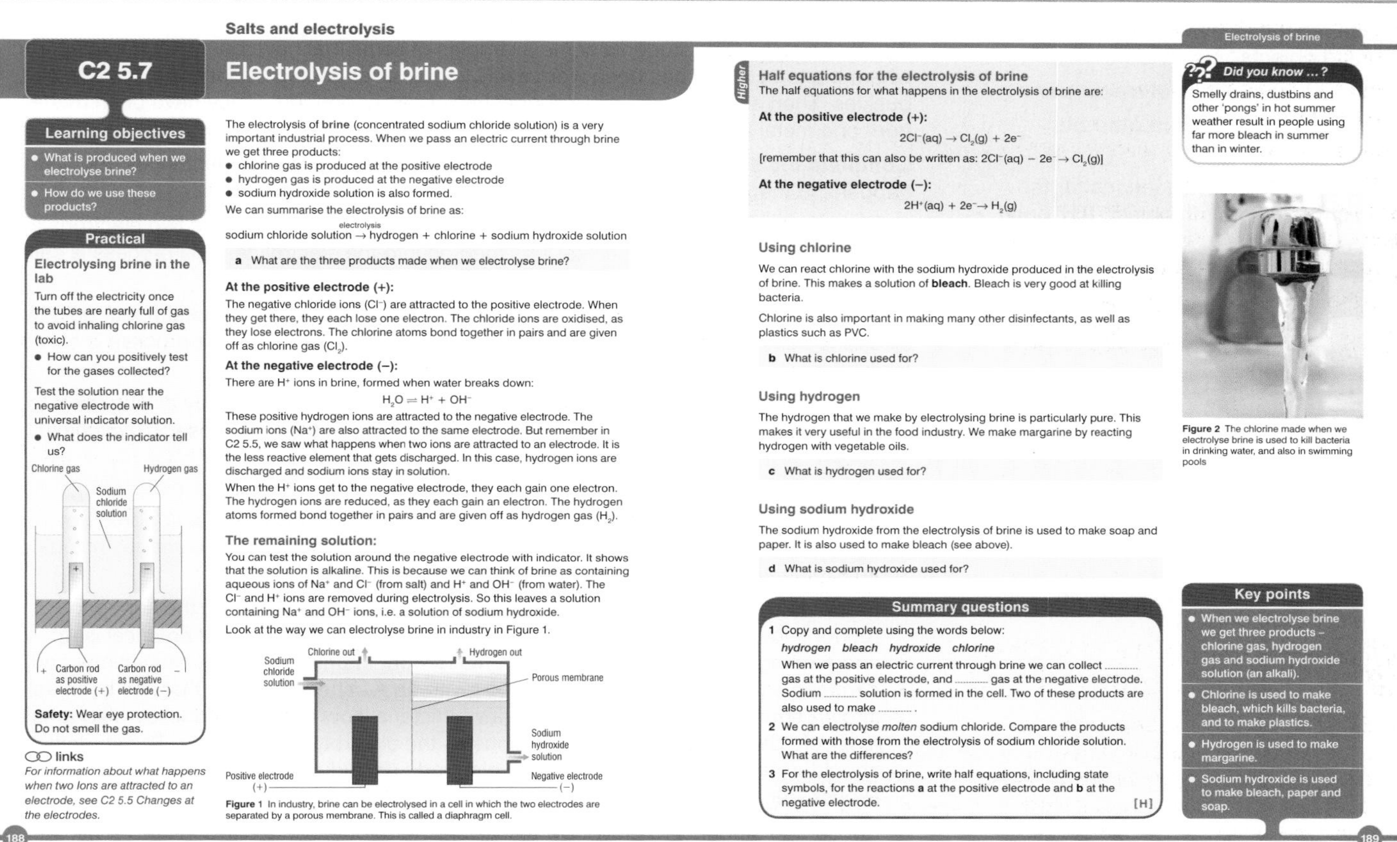

Salts and electrolysis

C2 5.7 Electrolysis of brine

Learning objectives

- What is produced when we electrolyse brine?
- How do we use these products?

The electrolysis of **brine** (concentrated sodium chloride solution) is a very important industrial process. When we pass an electric current through brine we get three products:

- chlorine gas is produced at the positive electrode
- hydrogen gas is produced at the negative electrode
- sodium hydroxide solution is also formed.

We can summarise the electrolysis of brine as:

$$\text{sodium chloride solution} \xrightarrow{\text{electrolysis}} \text{hydrogen} + \text{chlorine} + \text{sodium hydroxide solution}$$

a What are the three products made when we electrolyse brine?

At the positive electrode (+):

The negative chloride ions (Cl^-) are attracted to the positive electrode. When they get there, they each lose one electron. The chloride ions are oxidised, as they lose electrons. The chlorine atoms bond together in pairs and are given off as chlorine gas (Cl_2).

At the negative electrode (–):

There are H^+ ions in brine, formed when water breaks down:

$$H_2O \rightleftharpoons H^+ + OH^-$$

These positive hydrogen ions are attracted to the negative electrode. The sodium ions (Na^+) are also attracted to the same electrode. But remember in C2 5.5, we saw what happens when two ions are attracted to an electrode. It is the less reactive element that gets discharged. In this case, hydrogen ions are discharged and sodium ions stay in solution.

When the H^+ ions get to the negative electrode, they each gain one electron. The hydrogen ions are reduced, as they each gain an electron. The hydrogen atoms formed bond together in pairs and are given off as hydrogen gas (H_2).

The remaining solution:

You can test the solution around the negative electrode with indicator. It shows that the solution is alkaline. This is because we can think of brine as containing aqueous ions of Na^+ and Cl^- (from salt) and H^+ and OH^- (from water). The Cl^- and H^+ ions are removed during electrolysis. So this leaves a solution containing Na^+ and OH^- ions, i.e. a solution of sodium hydroxide.

Look at the way we can electrolyse brine in industry in Figure 1.

Figure 1 In industry, brine can be electrolysed in a cell in which the two electrodes are separated by a porous membrane. This is called a diaphragm cell.

Practical

Electrolysing brine in the lab

Turn off the electricity once the tubes are nearly full of gas to avoid inhaling chlorine gas (toxic).

- How can you positively test for the gases collected?

Test the solution near the negative electrode with universal indicator solution.

- What does the indicator tell us?

Safety: Wear eye protection. Do not smell the gas.

links

For information about what happens when two ions are attracted to an electrode, see C2 5.5 Changes at the electrodes.

188

Electrolysis of brine

Higher

Half equations for the electrolysis of brine

The half equations for what happens in the electrolysis of brine are:

At the positive electrode (+):

$$2Cl^-(aq) \rightarrow Cl_2(g) + 2e^-$$

[remember that this can also be written as: $2Cl^-(aq) - 2e^- \rightarrow Cl_2(g)$]

At the negative electrode (–):

$$2H^+(aq) + 2e^- \rightarrow H_2(g)$$

Using chlorine

We can react chlorine with the sodium hydroxide produced in the electrolysis of brine. This makes a solution of **bleach**. Bleach is very good at killing bacteria.

Chlorine is also important in making many other disinfectants, as well as plastics such as PVC.

b What is chlorine used for?

Using hydrogen

The hydrogen that we make by electrolysing brine is particularly pure. This makes it very useful in the food industry. We make margarine by reacting hydrogen with vegetable oils.

c What is hydrogen used for?

Using sodium hydroxide

The sodium hydroxide from the electrolysis of brine is used to make soap and paper. It is also used to make bleach (see above).

d What is sodium hydroxide used for?

Did you know ...?

Smelly drains, dustbins and other 'pongs' in hot summer weather result in people using far more bleach in summer than in winter.

Figure 2 The chlorine made when we electrolyse brine is used to kill bacteria in drinking water, and also in swimming pools

Summary questions

1 Copy and complete using the words below:

hydrogen bleach hydroxide chlorine

When we pass an electric current through brine we can collect .... gas at the positive electrode, and .... gas at the negative electrode. Sodium .... solution is formed in the cell. Two of these products are also used to make ....

2 We can electrolyse *molten* sodium chloride. Compare the products formed with those from the electrolysis of sodium chloride solution. What are the differences?

3 For the electrolysis of brine, write half equations, including state symbols, for the reactions **a** at the positive electrode and **b** at the negative electrode. [H]

Key points

- When we electrolyse brine we get three products – chlorine gas, hydrogen gas and sodium hydroxide solution (an alkali).
- Chlorine is used to make bleach, which kills bacteria, and to make plastics.
- Hydrogen is used to make margarine.
- Sodium hydroxide is used to make bleach, paper and soap.

189

Further teaching suggestions

Video

- Show students the industrial electrolysis of brine using a video, e.g. RSC Industrial Chemistry.

Internet search

- Electrolysis of brine occurs in two main ways: membrane cell and mercury cathode cell. Students could research these methods using the internet.

Displaying data

- Ask the students to find out the different uses of chlorine in the UK, represented as percentages. Then ask the students to display the information as a pie chart. Support some students by giving them the data, then they could plot a bar chart.

Diagram annotation

- Give students a diagram of the electrolysis of brine. They should then annotate their diagram so that it contains all the information from the key points.

Summary answers

1 chlorine, hydrogen, hydroxide, bleach

2 With molten sodium chloride we would get sodium metal produced at the negative electrode and chlorine gas produced at the positive electrode. With sodium chloride solution we get hydrogen gas given off. With sodium chloride solution we also get a solution of sodium hydroxide formed in the electrolysis cell but not with molten sodium chloride.

3 **a** $2Cl^-(aq) \rightarrow Cl_2(g) + 2e^-$ or $2Cl^-(aq) - 2e^- \rightarrow Cl_2(g)$

b $2H^+(aq) + 2e^- \rightarrow H_2(g)$

C2 5.8 Electroplating

Learning objectives

Students should learn:

- the reasons for electroplating objects
- how a metal object is electroplated.

Learning outcomes

Most students should be able to:

- state a definition of electroplating
- give reasons why we electroplate some objects
- explain the process of electroplating.

Some students should also be able to:

- construct the half equations for the electroplating of an object. **[HT only]**

Specification link-up: Chemistry C2.7

- Electrolysis is used to electroplate objects. This may be for a variety of reasons and includes copper plating and silver plating. *[C2.7.1 d)]*
- Explain and evaluate processes that use the principles described in this unit, including the use of electroplating. *[C2.7]*

Lesson structure

Starters

Show and tell – Give students some examples of items that have been electroplated and where the top surface has been worn away, e.g. a silver-coated copper ring, cutlery, belt buckles. Then ask the students to suggest what has happened. [They have been made from one metal and coated in another.] *(5 minutes)*

Complete the sentences – Ask students to complete the following sentences about what happens to metal ions in electroplating:

If a [positive] electrode is made from nickel, the nickel [atoms] can lose two electrons to become nickel [ions]. These nickel ions then go into the electrolyte. This process is called [oxidation].

If a [negative] electrode attracts nickel ions from the electrolyte, the nickel [ions] gain two electrons from the negative electrode and become nickel [atoms]. This process is called [reduction].

Support students by giving them the missing words. Extend students by asking them to write two half equations, including state symbols to summarise the prose. *(10 minutes)*

Main

- Supply students with the parts of a diagram of the electroplating of a spoon. They should assemble the full diagram from the pieces and then label the image, including an explanation of why electroplating is done.
- Students can complete their own electroplating of copper. Allowing the set-up to run for a long time at low current will give the best results. Once the practical is up and running, it is often useful for students to work in pairs to make observations and comments together. Students could be given an A4 sheet of paper that has been split into six equal sections. They could draw a pictorial flow chart with labelled diagrams to record their method and observations as the electrolysis continues. The students should explain what they see at each electrode (including half equations if they are Higher Tier students).
- Encourage students to draw how electroplating happens in a cartoon-strip style. To guide students, ask them to use the key words, e.g. electrolysis, electrode, electron, at least once. More artistic students may wish to personify the ions and electrons to make the cartoon more amusing.
- The electroplating practical provides a good opportunity to develop investigative aspects of Controlled Assessment (AS4.1.1). For example, students could investigate the factors that might affect the rate of electroplating.

Plenaries

I went to the shops … – Sit the students in a circle, the first student says 'I went to the shops to buy ...' (insert an electroplated object here). The next student repeats the first and adds another electroplated item to the list. This continues around the circle. *(5 minutes)*

AfL (Assessment for Learning) – Lay the cartoons or observation flow charts on the side bench of the room, with an A4 sheet of paper at one side. Ask each student to study some of the work and comment on the science content, by noting his or her thoughts on the paper. Support students by giving them a marking scheme with statements that they need to look for in the work. Extend students by asking them to suggest how the work could be improved. The owners of the work could then make the suggested changes for homework before handing their materials in for assessment by the teacher. *(10 minutes)*

Answers to in-text questions

a Chromium, tin, gold, silver (copper, nickel).

b At the positive electrode (atoms lose electrons).

c Nickel ions (Ni^{2+}).

Support

- Some students may have difficulty selecting the important information to include on a diagram. You could give these students the labels to cut and stick on to their diagram of the electrolysis of copper.

Extend

- Students could suggest an experimental set-up for electroplating with precious metals as used by the jewellery industry. Students could write a half equation for each electrode. Encourage students to use secondary sources such as search engines on the internet to find out how electroplating is actually done in the industrial manufacture of jewellery.

Practical support

Nickel plating copper metal

Equipment and materials required

A 250 cm^3 beaker, nickel sulfate solution (dissolve 5 g of nickel ammonium sulfate in 100 cm^3 of water), wax in a 100 cm^3 beaker held in a water bath set at 80 °C, low voltage power supply (0–12 V), two wires, two crocodile clips, copper foil strips cut small enough to fit into the beaker, nickel electrode, mounted needle, sandpaper, eye protection.

Details

Wear eye protection throughout this practical. Using sandpaper, clean the surface of the copper foil. Dip the copper foil into the molten wax and allow it to dry. Scratch a design into the wax using the mounted needle. Half-fill the beaker with nickel sulfate solution. Add the nickel electrode and connect to the positive terminal of the power supply using the wires and crocodile clips. Add the wax-coated copper foil and connect to the negative terminal. Ensure that one end of each electrode is submerged in the solution, while the other is out of it. The slower the rate of deposition, the better the nickel will adhere to the copper. Turn off the power supply and remove the copper foil. Rinse off the nickel sulfate solution and remove the excess wax.

Safety: CLEAPSS Hazcard 65B Nickel sulfate – harmful. Hot wax can cause burns.

Salts and electrolysis

C2 5.8 Electroplating

Learning objectives

- Why do we electroplate objects?
- How can we electroplate a metal object?

Most of us will use an electroplated objected at some time each day. You might use a chromium-plated kettle to boil water or ride a bicycle with chromium-plated handlebars. You could open a tin-plated steel can for a meal or put on some gold- or silver-plated jewellery.

An electroplated object is coated with a thin layer of metal by electrolysis.

a Name four metals that we can use to electroplate another metal.

Why do we electroplate objects?

There can be different reasons why we electroplate objects. These include:

- to protect the metal beneath from corroding
- to make the object look more attractive
- to increase the hardness of a surface and its resistance to scratching
- to save money by using a thin layer of a precious metal instead of the pure expensive metal. This also helps people who are allergic to nickel – a metal often used to make cheap jewellery.

Electroplating saves money in making cheaper jewellery. However, using electroplating to protect large metal surfaces against rusting and damage makes things more expensive. In the long term, though, this can still make economic sense because we don't have to replace objects so often.

Figure 1 Chromium-plated objects look very shiny and attractive. The chromium layer does not corrode away so it protects the steel beneath from rusting.

Electroplating a metal object

You can try to nickel plate some copper foil in the experiment below.

Practical

Nickel plating copper metal

Your teacher will melt some wax in a metal tray. Using tongs you can dip in a piece of copper foil.

Let the wax set. Then scratch a simple design in the wax. You want the design to be plated with nickel so get this area as free from wax as possible.

Set up the apparatus as shown in the diagram. Using a small current for a long time will give best results.

When you have finished, rinse the copper foil in water, dry, then scrape off the rest of the wax.

- What happens at the negative electrode?

Wax
Nickel anode
Copper to be plated
Nickel sulfate solution

Figure 2 So-called 'tin' cans actually contain very little tin. The layer on the steel can be only a few thousandths of a millimetre thick! The tin keeps air and water away from the iron in steel and stops it rusting – at least until the tin gets scratched! Tin is quite a soft metal, unlike chromium.

190

Electroplating

Explaining electroplating

The metal object to be plated (the copper foil in this case) is used as the negative electrode. The positive electrode is made from the plating metal (nickel). The electrolysis takes place in a solution containing nickel ions. In the previous experiment we use nickel sulfate solution.

At the positive electrode made of the plating metal:

Nickel atoms in the electrode are oxidised. They lose 2 electrons each and form nickel ions (Ni^{2+}) which go into the solution.

At the negative electrode to be plated:

Nickel ions (Ni^{2+}) from the solution are reduced. They gain 2 electrons and form nickel atoms which are deposited on the copper electrode.

b Where are the nickel atoms oxidised?
c What is formed when nickel atoms are oxidised?

Higher

Electroplating half equations

Here are the half equations at each electrode in electroplating by nickel:

At the positive nickel electrode:

$$Ni(s) \rightarrow Ni^{2+}(aq) + 2e^-$$

At the negative electrode to be plated:

$$Ni^{2+}(aq) + 2e^- \rightarrow Ni(s)$$

Did you know ...?

It is not only metal objects that can be electroplated. We can now electroplate plastic objects as well. The object to be plated is first coated in a 'paint' containing tiny particles of graphite. Once dry, the object has a layer of graphite that will conduct electricity. It can then be electroplated.

Summary questions

1 Copy and complete using the words below:

atoms negative nickel plating deposited electrons oxidised reduced

In electroplating, a solution of the .... metal is electrolysed. In the case of nickel plating, the positive electrode is made of .... The nickel atoms are .... and go into the solution. At the .... electrode, the nickel ions gain .... and are .... They form nickel .... and are .... on the object to be plated.

2 What are the economic advantages and disadvantages of electroplating a metal object?

3 In making 'chrome' objects, chromium metal is used to electroplate a steel object. The steel is first electroplated with nickel because chromium does not stick well on steel. Give the half equation at the negative electrode for the nickel, then the chromium, plating processes. Include state symbols in your answer, (Chromium ions are Cr^{3+}.) [H]

Key points

- We can electroplate objects to improve their appearance, protect their surface and to use smaller amounts of precious metals.
- The object to be electroplated is made the negative electrode in an electrolysis cell. The plating metal is made the positive electrode. The electrolyte contains ions of the plating metal.

191

Further teaching suggestions

Electroplating research

- Students could research into the use of electrolysis to plate other metals (electroplating) or for colouring aluminium by anodising.

Advert

- Students could be asked to imagine that they work in the marketing department of an industrial electroplating firm. They have been asked to make radio adverts to see their services. Students could record their adverts as MP3 files and the best ones could be uploaded on to the school website to showcase exemplar work.

Summary answers

1 plating, nickel, oxidised, negative, electrons, reduced, atoms, deposited

2 The plating process itself will cost money in terms of energy and chemicals. However, in jewellery this will still make it cheaper than an object made of a pure precious metal. Plating metals to protect them from corrosion or damage again costs money but the object will need replacing less frequently.

3 $Ni^{2+}(aq) + 2e^- \rightarrow Ni(s)$ then $Cr^{3+}(aq) + 3e^- \rightarrow Cr(s)$

Summary answers

1 a **i** zinc oxide/zinc hydroxide
ii zinc oxide + sulfuric acid → zinc sulfate + water
or
zinc hydroxide + sulfuric acid → zinc sulfate + water

b Add zinc oxide/zinc hydroxide to the acid. Warm the mixture and stir. Filter to remove the excess zinc oxide/zinc hydroxide from the zinc sulfate solution. Heat the solution in an evaporating basin on a water bath until the point of crystallisation. Then leave the liquid for a few days to allow the rest of the water to evaporate off and the crystals to form.

2 a $2KOH(aq) + H_2SO_4(aq) \rightarrow K_2SO_4(aq) + 2H_2O(l)$
b $ZnO(s) + 2HNO_3(aq) \rightarrow Zn(NO_3)_2(aq) + H_2O(l)$
c $Ca(s) + 2HCl(aq) \rightarrow CaCl_2(aq) + H_2(g)$
d $Ba(NO_3)_2(aq) + Na_2SO_4(aq) \rightarrow BaSO_4(s) + 2NaNO_3(aq)$

3 a B
b A
c B
d A
e B
f A

4 Negative electrode: sodium, calcium, zinc, aluminium.
Positive electrode: iodide, fluoride, oxide, bromide.

5 **A** chlorine gas, **B** hydrogen gas, **C** sodium hydroxide solution

6 a $2H_2O \rightarrow 2H_2 + O_2$
b Negative electrode (−): $2H^+ + 2e^- \rightarrow H_2$
Positive electrode (+): $4OH^- \rightarrow O_2 + 2H_2O + 4e^-$
c 1 mole.
d The power supply.

7 a $K^+ + e^- \rightarrow K$
b $Ba^{2+} + 2e^- \rightarrow Ba$
c $2I^- \rightarrow I_2 + 2e^-$
d $2O_2- \rightarrow O_2 + 4e$

8 Description should include the object to be plated as the negative electrode in a solution containing metal ions (e.g. $CuSO_4$ solution for copper plating). To keep the concentration of metal ions in solution constant, a positive electrode made of the metal to be plated should be used. Half equation: $Cu^{2+} + 2e^- \rightarrow Cu$

Kerboodle resources

Resources available for this chapter on Kerboodle are:

- Chapter map: Salts and electrolysis
- Practical: Making salts from solutions (C2 5.3)
- Support: Making salts (C2 5.3)
- Animation: Electrolysis (C2 5.4)
- Extension: Explaining electrolysis (C2 5.4)
- WebQuest: Extracting aluminium (C2 5.6)
- How Science Works: Wouldn't give a nickel for the whole process (C2 5.8)
- Interactive activity: Salts and electrolysis
- Revision podcast: Electrolysis of brine
- Test yourself: Salts and electrolysis
- On your marks: Salts and electrolysis
- Examination-style questions: Salts and electrolysis
- Answers to examination-style questions: Salts and electrolysis

Salts and electrolysis: C2 5.1–C2 5.8

Summary questions

1 Zinc sulfate crystals can be made from an insoluble base and sulfuric acid.

a i Name the insoluble base that can be used to make zinc sulfate.
ii Write a word equation to show the reaction.

b Describe how you could make crystals of zinc sulfate from the reaction in **a ii**.

2 Write balanced symbol equations, including state symbols, to describe the reactions below. (Each reaction forms a salt.)

a Potassium hydroxide (an alkali) and sulfuric acid.
b Zinc oxide (an insoluble base) and nitric acid.
c Calcium metal and hydrochloric acid.
d Barium nitrate and sodium sulfate (this reaction produces an insoluble salt – Hint: all sodium salts are soluble). [H]

3 Select A or B to describe correctly what happens at the positive electrode (+) and negative electrode (–) in electrolysis for **a** to **f**.

A Positive electrode
B Negative electrode

a Positive ions move towards this.
b Negative ions move towards this.
c Reduction happens here.
d Oxidation happens here.
e Connected to the negative terminal of the power supply.
f Connected to the positive terminal of the power supply.

4 Make a table to show which of the following ions would move towards the positive electrode and which towards the negative electrode during electrolysis. (You may need to use a copy of the periodic table to help you.)

sodium iodide calcium fluoride
oxide zinc aluminium bromide

5 The diagram shows an industrial process used for the electrolysis of sodium chloride solution.

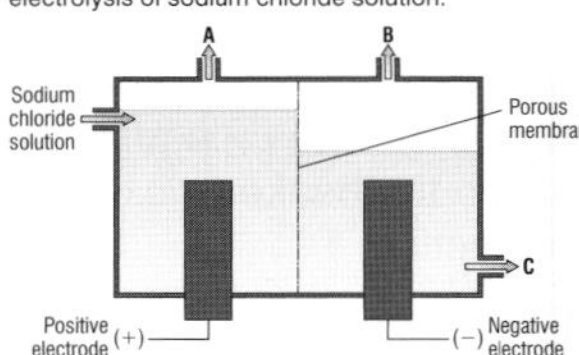

Identify the products **A**, **B** and **C** on the diagram using substances from the list.

chlorine gas oxygen gas hydrogen gas
sodium hydroxide solution sodium metal

6 Water can be split into hydrogen and oxygen using electrolysis. The word equation for this reaction is:

water → hydrogen + oxygen

a Write a balanced symbol equation for this reaction using the correct chemical symbols.
b Write half equations to show what happens at the positive and negative electrodes.
c When some water is electrolysed it produces 2 mole of hydrogen. How much oxygen is produced?
d Where does the energy needed to split water into hydrogen and oxygen come from during electrolysis?

7 Copy and complete the following half equations:

a $K^+ \rightarrow K$
b $Ba^{2+} \rightarrow Ba$
c $I^- \rightarrow I_2$
d $O^{2-} \rightarrow O_2$

8 Electrolysis can be used to produce a thin layer of metal on the surface of a metal object. Using words and diagrams, describe how you would cover a small piece of steel with copper. Make sure that you write down the half equation that describes what happens at the surface of the steel.

192

Practical suggestions

Practicals	AQA	k	📖	⚙
The preparation of soluble salts.	✓		✓	
The preparation of insoluble salts.	✓		✓	
The electrolysis of molten lead bromide or zinc chloride.	✓		✓	
Investigation of the electrolysis of any solutions of a soluble ionic compound, e.g. copper chloride, sodium chloride, zinc bromide, zinc iodide.	✓		✓	
A demonstration of the Hoffman voltameter.	✓		✓	
The electroplating of copper foil with nickel in a nickel sulfate solution.	✓		✓	✓
The movement of ions, e.g. by the electrolysis of a crystal of $KMnO_4$ on filter paper dampened with sodium chloride solution, or the electrolysis of $CuCrO_4$ in a saturated urea solution using a U-tube.	✓		✓	
Using conductivity sensors to monitor conductivity and changes in conductivity.	✓		✓	

Examination-style questions

ydrogen chloride gas reacts with water to make ydrochloric acid. The equation for the reaction is:

$Cl(g) \rightarrow H^+(aq) + Cl^-(aq)$

Which of the following shows that an acid has been made?

A An aqueous solution has been made.

B Hydrogen ions have been made.

C Chloride ions have been made. (1)

Choose a number from the list for the pH of hydrochloric acid.

1 7 12 (1)

Hydrochloric acid reacts with sodium hydroxide solution to produce a salt and water.

i Choose a word from the list that describes sodium hydroxide. (1)

alcohol alkali insoluble

ii Choose a word from the list to complete the sentence.
The reaction between hydrochloric acid and sodium hydroxide is an example of .... (1)

combustion neutralisation oxidation

iii Name the salt made when hydrochloric acid reacts with sodium hydroxide. (1)

ead chloride is a white insoluble salt. It can be made mixing lead nitrate solution with sodium chloride olution. Both of these solutions are colourless.

What would you **see** when lead nitrate solution is mixed with sodium chloride solution? (1)

Write a word equation for the reaction. (2)

A mining company produces wastewater that contains dissolved lead ions. Suggest how the company could treat the wastewater to reduce the concentration of lead ions. (2)

this question you will be assessed on using good nglish, organising information clearly and using ecialist terms where appropriate.

opper(II) oxide is an insoluble base.

escribe how you could make crystals of copper(II) lfate from copper(II) oxide. (6)

4 The diagram shows a nickel spoon being coated with silver.

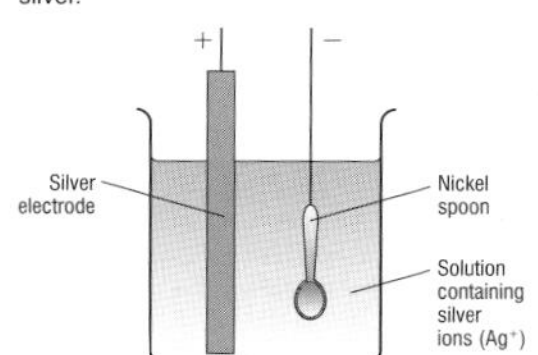

a Explain why silver ions in the solution move towards the spoon. (2)

b Use words from the list to complete the sentence.

gaining losing sharing electron neutron proton

When silver ions reach the spoon they change into silver atoms by .... an .... (2)

c Suggest one reason why spoons made from nickel are coated with silver. (1)

AQA, 2002

5 Magnesium is manufactured by the electrolysis of molten magnesium chloride. The container is made of steel, which is the negative electrode. Carbon (graphite) is used for the positive electrode.

a Steel and carbon (graphite) both conduct electricity.

i Suggest one other reason why the negative electrode is made of steel. (1)

ii Suggest one other reason why the positive electrode is made of carbon (graphite). (1)

b Magnesium chloride melts at 950 °C. It is mixed with sodium and calcium chloride so that it can be electrolysed at 750 °C.

i Suggest one way this benefits the manufacturer. (1)

ii Suggest one way this benefits the environment. (1)

c Complete and balance the equations for the reactions at the electrodes.

i At the negative electrode: $Mg^{2+} + e^- \rightarrow Mg$ (1)

ii At the positive electrode: $Cl^- \rightarrow Cl_2 + e^-$ [H] (1)

Examination-style answers

1 a B *(1 mark)*

b 1 *(1 mark)*

c i alkali *(1 mark)*

ii neutralisation *(1 mark)*

iii sodium chloride *(1 mark)*

2 a A white precipitate or a white solid. *(1 mark)*

b lead nitrate + sodium chloride → lead chloride + sodium nitrate *(reactants for one mark, products for one mark)* *(2 marks)*

c Add sodium chloride or any other soluble salt that will give a precipitate with lead ions or a named alkali; allow to settle or filter. *(2 marks)*

3 There is a clear and detailed scientific description of how to prepare copper(II) sulfate solution from copper(II) oxide and dilute sulfuric acid and how to obtain crystals from the solution.The answer shows almost faultless spelling, punctuation and grammar. It is coherent and in an organised, logical sequence. It contains a range of appropriate and relevant specialist terms used accurately. *(5–6 marks)*

There is a scientific description of the preparation of copper(II) sulfate. There are some errors in spelling, punctuation and grammar. The answer has some structure and organisation. The use of specialist terms has been attempted but not always accurately. *(3–4 marks)*

There is a brief description of the reaction of copper(II) oxide with an acid or of crystallisation of a solution of copper(II) sulfate. The spelling, punctuation and grammar are very weak. The answer is poorly organised with almost no specialist terms and/or their use demonstrating a general lack of understanding of their meaning. *(1–2 marks)*

No relevant content. *(0 marks)*

Examples of chemistry points made in the response:

- Use dilute sulfuric acid.
- Place acid in a beaker.
- Warm the acid.
- Add copper(II) oxide.
- In small amounts.
- Until in excess or there is no further reaction.
- Filter (to remove excess copper(II) oxide).
- Heat filtrate/solution to evaporate some water.
- Allow solution to cool and crystallise or allow to evaporate slowly at room temperature.
- Remove/filter crystals from remaining solution.

4 a Ions can move in solution, silver ions are positive and are attracted to the negatively charged spoon. *(2 marks)*

b gaining, electron *(2 marks)*

c Any sensible suggestion – e.g. better appearance; less toxic; less likely to corrode; less expensive than solid silver. *(1 mark)*

5 a i Sensible suggestion – e.g. high melting point; strong; can be shaped; holds (hot) liquid, does not react with liquid (magnesium chloride/magnesium); low cost, at high temperature carbon electrode would burn with oxygen from the air to produce carbon dioxide. *(1 mark)*

ii Sensible suggestion – e.g. does not react with chlorine; high melting point; low cost (but do not accept if low cost already allowed in **i**). *(1 mark)*

b i Less heat/energy needed or lower cost of energy *(accept less heat lost)*. *(1 mark)*

ii Less fossil fuel burned (for heat/energy) so less pollution; less global warming; resources conserved; less mining (must be linked) or less thermal; heat pollution. *(1 mark)*

c i $Mg^{2+} + 2e^- \rightarrow Mg$ *(1 mark)*

ii $Cl^- \rightarrow Cl_2 + 2e^-$ *(1 mark)*

Examination-style questions

1 Calcium chloride is an ionic compound.
Use a table showing the charges on ions to help you answer this question.

a Which of these is the formula of calcium chloride?

CaCl **Ca_2Cl** **$CaCl_2$** **Ca_2Cl_2** (1)

b A sodium ion can be represented in the following way:

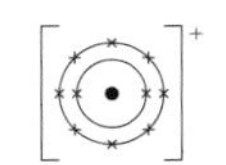

Draw diagrams like this to show the ions in calcium chloride. (4)

c Calcium chloride is a crystalline solid with a high melting point.
Explain why calcium chloride has these properties. (3)

2 A company extracts copper from an ore that contains the mineral chalcopyrite, $CuFeS_2$.

a Calculate the relative formula mass (M_r) of $CuFeS_2$. *(Relative atomic masses (A_r): Cu = 63.5; Fe = 56; S = 32)* (2)

b What is the percentage by mass of copper in chalcopyrite? (2)

c Suggest one reason why the company might prefer to use an ore containing the mineral chalcocite, Cu_2S. (1)

d The company uses the copper it produces to restore items made of copper. It electroplates the copper items with a new coating of copper. Pure copper is used as the positive electrode and the copper item is the negative electrode. The electrolyte is a solution containing copper(II) sulfate.

i Draw a diagram showing how you could electroplate a copper spoon in the laboratory. (3)

ii Explain how copper is deposited onto the spoon by electrolysis. (3)

3 A blackcurrant-flavoured drink was analysed for artificial colours and flavours.
A scientist used paper chromatography to identify the artificial colours in the drink.
The result of the chromatography is shown in the diagram.

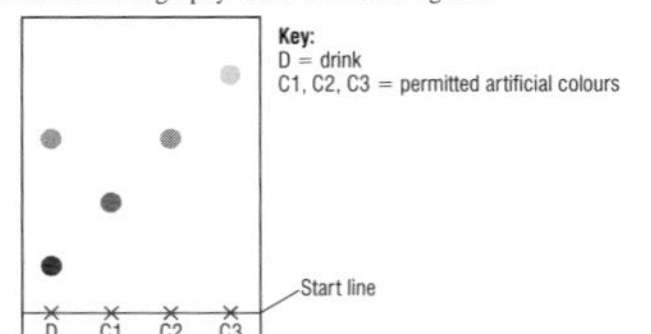

a *In this question you will be assessed on using good English, organising information clearly and using specialist terms where appropriate.*
Describe how the chromatography was done to produce this result. (6)

b What conclusions can you make about the colours in the drink? (3)

c To identify the flavour compounds in the drink the scientist put a sample of the drink into a gas chromatography column linked to a mass spectrometer. The output of the column showed five main peaks.

i What was the purpose of the gas chromatography column? (1)

ii What was the purpose of the mass spectrometer? (1)

AQA Examiner's tip

The AQA data sheet that you will have in the exam has a table showing the charges on some common ions.

AQA Examiner's tip

Do not be put off by unusual formulae that you may not have seen before (**Q2** and **Q4**). Some questions are testing your ability to apply what you know in new situations.

AQA Examiner's tip

Q3 a i requires a description in a logical order so that someone else could do the experiment. Think about your answer before writing. Make a brief list of the key steps and number them in the correct sequence. Read them through in your head to check the order. Then write your answer. Cross out any notes that you do not want to be marked.

194

Examination-style answers

1 a $CaCl_2$ *(1 mark)*

b calcium ion with 2,8,8 electrons on three circles square bracket with 2+ charge outside top right chloride ion with 2,8,8 electrons on three circles square bracket with − charge outside top right *(4 marks)*

c The (electrostatic) forces/bonds between oppositely charged ions are strong, and there are many bonds/forces or it has a giant structure/lattice, and so a large amount of energy is needed to break the bonds so that ions can move freely. *(3 marks)*

2 a 183.5 or 184 *(2 marks)*
(if answer incorrect, correct working 63.5 or 64 + 56 + 32 × 2 gains one mark).

b 34.6 or 34.8 or 34.78 *(2 marks)*
(if answer incorrect (63.5 or 64 × 100) ÷ (183.5 or 184) gains one mark).

c Any **one** from: contains a higher percentage *(accept greater amount or more)* of copper; contains no iron or fewer impurities. *(1 mark)*

d **i** Diagram showing any **three** from: spoon as negative electrode; pure copper as positive electrode; in (beaker/container) copper(II) sulfate solution; power supply or battery or cell symbol or electrodes labelled + and −. *(3 marks)*

ii Copper ions are positively charged and so are attracted to or move to the negative electrode, where they gain electrons or are reduced, forming copper atoms which are deposited on the spoon. *(3 marks)*

3 a There is a clear, logical and detailed scientific description of how to do paper chromatography. The answer shows almost faultless spelling, punctuation and grammar. It is coherent and in an organised, logical sequence. It contains a range of appropriate and relevant specialist terms used accurately. *(5–6 marks)*

There is a scientific description of how to do paper chromatography. There are some errors in spelling, punctuation and grammar. The answer has some structure and organisation. The use of specialist terms has been attempted, but not always accurately. *(3–4 marks)*

There is a brief description of how to do paper chromatography with. The spelling, punctuation and grammar are very weak. The answer is poorly organised with almost no specialist terms and/or their use demonstrating a general lack of understanding of their meaning. *(1–2 marks)*

No relevant content. *(0 marks)*

Examples of chemistry points made in the response:

- Draw a pencil line on paper (close to bottom).
- Mark/label starting points (in pencil).
- Put spots of colours/samples onto starting points.
- Put paper into solvent/water.
- Solvent/water below (pencil) line/spots.
- Use cover/lid on beaker/container.
- Allow solvent to run to (near to) top of paper.
- Remove from solvent/allow paper to dry.
- Any other technical detail, e.g. use capillary tube for sample spots.

b Any **three** from: drink contains two colours; one colour in drink is C2; one colour in drink is unknown *(allow not permitted)*; drink does not contain C1 and/or C3. *(3 marks)*

c **i** To separate the (flavour) compounds *(allow to show how many flavours/compounds there are)*. *(1 mark)*

ii To identify the compounds *(allow to find the M_r or relative masses of the compounds)*. *(1 mark)*

4 a 2.61 (g) *(2 marks)*
If the answer is incorrect then correct working, e.g. 2/138 or (180/138) × 2 or 1 g → 180/138 gains one mark.

b 42.1 (per cent) *(2 marks)*
If answer incorrect then or (1.1/ecf from a) × 100 correctly evaluated gains two marks, or correct working, e.g. (1.1/2.61) × 100 gains one mark.

c Any **one** from: errors in weighing; some (of the aspirin) lost; not all of the reactant may have been converted to product or reaction didn't go to completion; the reaction is reversible; side reactions; reactants impure; not heated for long enough; not hot enough for reaction to take place. *(1 mark)*

d

C	H	O	
$\frac{75.7}{12}$	$\frac{8.80}{1}$	$\frac{15.5}{16}$	divide per cent by M_r
= 6.308 or 6.31	= 8.80	= 0.969 or 0.97	proportions
6.501	9.08		divide by 0.969
or 6.51	or 9.1		or 0.97
or 6.5	or 9	1	
or 13	18	2	

empirical formula = $C_{13}H_{18}O_2$ *(4 marks)*
(correct answer with no working gains only 1 mark)

5 a Diamond is very hard because it has a giant (covalent) structure/lattice, in which each carbon atom forms four (covalent) bonds, and so there are many strong

4 A student carried out an experiment to make aspirin. The method is given below.

1. Weigh 2.00 g of salicylic acid.
2. Add 4 cm^3 of ethanoic anhydride (an excess).
3. Add 5 drops of concentrated sulfuric acid.
4. Warm the mixture for 15 minutes.
5. Add ice cold water to remove the excess ethanoic anhydride.
6. Cool the mixture until a precipitate of aspirin is formed.
7. Collect the precipitate and wash it with cold water.
8. The precipitate of aspirin is dried and weighed.

a The equation for this reaction is:

$C_7H_6O_3 + C_4H_6O_3 \rightarrow C_9H_8O_4 + CH_3COOH$
salicylic acid aspirin

The relative formula mass (M_r) of salicylic acid, $C_7H_6O_3$, is 138.
The relative formula mass (M_r) of aspirin, $C_9H_8O_4$, is 180.
Calculate the maximum mass of aspirin that could be made from 2.00 g of salicylic acid. (2)

b The student made 1.10 g of aspirin from 2.00 g of salicylic acid.
Calculate the percentage yield of aspirin for this experiment. (2)

c Suggest **one** possible reason why this method does not give the maximum amount of aspirin. (1)

d The student made another compound with properties similar to aspirin. The student sent this compound to a laboratory for analysis. The analysis showed that the compound contained 75.7% C, 8.80% H and 15.5% O.
Calculate the empirical formula of this compound. You must show all of your working to gain full marks. (4)

[H]

AQA, 2009

AQA Examiner's tip

Always show your working when you do a calculation. If you make a mistake calculating the final answer you may still gain some marks if you show that you know how to do the calculation.

5 The element carbon has several forms.

a Diamond is one form of carbon. Explain, in terms of structure and bonding, why diamond is very hard. (3)

b *In this question you will be assessed on using good English, organising information clearly and using specialist terms where appropriate.*
Another form of carbon is graphite. Graphite is used for the contacts in electric motors because it conducts electricity and is soft and slippery. Explain, in terms of structure and bonding, why graphite has these properties. (6)

c Carbon can also form fullerenes. The first fullerene to be discovered has a structure that contains 60 carbon atoms. Other fullerenes contain a few hundred atoms.

i What is the basic unit of the structure of fullerenes? (1)

ii Give two reasons why there has been much research interest in the fullerenes since their discovery in 1985. (2)

[H]

AQA Examiner's tip

Q5 b requires you to describe the structure and bonding in graphite and use this to explain the properties given in the question. You need to link the properties clearly to particular points in your description. Before writing your answer, briefly list the key points that you know about the structure and bonding in graphite and then link these to the properties. Then write your answer in a logical order, giving as much detail as you can.

(covalent) bonds to break. (Covalent must be mentioned at least once for full marks.) *(3 marks)*

b There is a clear, logical and detailed scientific description of the structure and bonding in graphite correctly explaining both of the properties. The answer shows almost faultless spelling, punctuation and grammar. It is coherent and in an organised, logical sequence. It contains a range of appropriate and relevant specialist terms used accurately. *(5–6 marks)*

There is a scientific description of the structure and bonding in graphite and some explanation of at least one the properties. There are some errors in spelling, punctuation and grammar. The answer has some structure and organisation. The use of specialist terms has been attempted, but not always accurately. *(3–4 marks)*

There is a brief description of the structure and bonding in graphite and an attempt at an explanation of one the properties. The spelling, punctuation and grammar are very weak. The answer is poorly organised with almost no specialist terms and/or their use demonstrating a general lack of understanding of their meaning. *(1–2 marks)*

No relevant content. *(0 marks)*

Examples of chemistry points made in the response:

- Each carbon/atom joined/bonded to three other carbon/atoms.
  or Each carbon forms three bonds.
- In layers.
- Only weak forces (of attraction)/bonds between layers.
- Layers/atoms can slide over each other.
- One electron on each carbon is not used for bonding.
- Electrons delocalised **or** electrons free.
- Electrons carry the charge/current.
- Giant structure/lattice.
- Covalent (bonds).
- Strong bonds **or** a lot of energy needed to break bonds.
- Diagrams could be used: to show layered structure, to show that each carbon is bonded to three other carbon atoms, to show giant structure (at least three rings required).

c **i** Hexagons of carbon atoms or rings of six carbon atoms. *(1 mark)*

ii Any **two** correct reasons, e.g. they have special properties as they are nanoparticles; can form nanotubes; can be used for drug delivery; as catalysts; as lubricants; for reinforcing materials; or any other specific correct use. *(2 marks)*

Bump up your grades

Examiners are expected to challenge candidates at all levels with open questions requiring prose answers. Responses to these questions should be clearly and logically written using correct scientific terms. You should encourage your students to plan their answers before writing, using space on the question paper for notes.

Bump up your grades

Encourage your students to show their working when attempting calculations. Some students seem reluctant to do this, but it may gain them marks if they make an arithmetical error.

C3 1.1 The early periodic table

Learning objectives

Students should learn:

- how chemists such as Newlands and Mendeleev developed the periodic table
- why the periodic table is useful to chemists.

Learning outcomes

Most students should be able to:

- describe how the periodic table was first constructed
- explain why the periodic table is important
- explain how the periodic table was discovered.

Some students should also be able to:

- explain in detail why Mendeleev's ordering of the elements was so innovative.

Specification link-up: Chemistry C3.1

- Newlands, and then Mendeleev, attempted to classify the elements by arranging them in order of their atomic weights. The list can be arranged in a table so that elements with similar properties are in columns, known as groups. The table is called a periodic table because similar properties occur at regular intervals. *[C3.1.1 a)]*
- The early periodic tables were incomplete and some elements were placed in inappropriate groups if the strict order of atomic weights was followed. Mendeleev overcame some of the problems by leaving gaps for elements that he thought had not been discovered. *[C3.1.1 b)]*
- Evaluate the work of Newlands and Mendeleev in terms of their contributions to the development of the modern periodic table. *[C3.1]*
- Explain why scientists regarded a periodic table of the elements first as a curiosity, then as a useful tool and finally as an important summary of the structure of atoms. *[C3.1]*

Lesson structure

Starters

True or false? – Ask students to show if they think a statement is true (put their thumbs up), or false (thumbs down), or if they are unsure (thumbs horizontal). You could read out the following statements:

- The periodic table is the only way to classify the elements. [false]
- Most elements are metals. [true]
- Most elements are gases at room temperature. [false]
- The periodic table can be used to make predictions about the products formed in chemical reactions. [true]
- Most non-metallic elements are gases at 20 °C. [true]
- The metals are on the right of the periodic table. [false] *(5 minutes)*

List – Show the students a selection of different elements in sealed Petri dishes. Perhaps put a different selection of about 10 on each table. Ask the students to sort the elements into different groups. They might select colour, metal/non-metals, state of matter at 20 °C, hazard, increasing atomic number. Ask each group to feed back on how it ordered its elements. Support students by giving them the categories to sort the elements into. Extend students by giving them samples of an alloy. You could ask students why it is an 'odd one out' and how it could still be categorised with the other substances. *(10 minutes)*

Main

- There have been many attempts to classify the elements, such as Dalton's table, Newlands' law of octaves and Mendeleev's table. Ask the students to imagine that they work for an advertising agency. They are to make an advert for each method of listing elements. Their advert needs to be persuasive as they want scientists to subscribe to their method of listing the elements.
- In pairs, give the students 20 cards of A6 size. Ask the students to make a separate card for each of the first 20 elements. The card should include the name, symbol, relative atomic mass of the element, its state at 20 °C and a little information about its chemical reactions. Then ask the students to sort the cards.
- Circulate around the different pairs to find out how the students have ordered the cards. Students may choose alphabetical order, increasing mass, groupings in terms of chemical properties. Allow the students to visit other pairs to find out how they ordered their cards.
- Ask the students to make a decision on how they would order the first 20 elements and ask them to jot their answer in their notes with an explanation of their choice. Then give the students a mystery element card, only including chemical and physical data. Ask the students to place this into their table. Hold a discussion about how the elements were ordered by each pair and draw into the discussion how Mendeleev discovered the periodic table. Allow students to contrast their ordering with Mendeleev's table and discuss the leaving of gaps in his table.

Support

- Students could make a timeline to explain how the periodic table was generated. You could give them information to 'cut and stick' and add to the timeline. The information could include pictures of the scientists, a sentence to explain their input to the periodic table and an image of their ordering of the elements.

Extend

- Ask students to act out Mendeleev trying to communicate his ideas. The scientific community laughed at Mendeleev when he first went to the them with gaps in his table. Ask the students to act out a scene where great scientists were ridiculing Mendeleev at a meeting.

Plenaries

Spot the difference – Show the students Mendeleev's original periodic table. Ask students to contrast this with the current periodic table and make a list of differences. Support students by encouraging them to work in small groups. Extend students by giving them other representations of the periodic table and ask them to contrast these. *(5 minutes)*

Think, pair, square – Ask students to think about why the periodic table is important. Then ask them to discuss their answers in pairs and finally in groups of four. Then ask each group of four for a reason why the periodic table is important and list the reasons on the board. *(10 minutes)*

C3 1.1 The early periodic table

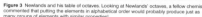

Learning objectives

- How did chemists, such as Newlands and Mendeleev, help develop the periodic table?
- Why is the periodic table so useful to chemists?

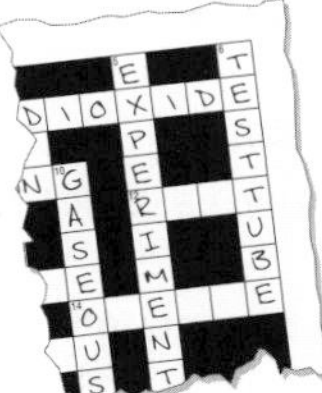

Figure 1 Looking for patterns in the chemical elements in the early part of the 19th century was a bit like solving a crossword puzzle. Some answers were clear (as they did have some correctly identified elements); they only had a vague idea about some (as some compounds were wrongly thought to be elements). However, they didn't even know the clues for other answers (as there were still undiscovered elements).

Imagine trying to understand chemistry:

- without knowing much about atoms
- with each chemical compound having lots of different names
- without knowing a complete list of the elements.

Not an easy job! But that's the task that faced scientists at the start of the 1800s.

During the 19th century, chemists were finding new elements almost every year. They were also trying very hard to find patterns in the behaviour of the elements. This would allow them to organise the elements and understand more about chemistry.

One of the first suggestions came from John Dalton. He arranged the elements in order of their masses, which had been measured in various chemical reactions. In 1808 he published a table of elements in his book *A New System of Chemical Philosophy*. Look at his list in Figure 2.

ELEMENTS

Hydrogen	1	Strontian	46
Azote	5	Barytes	68
Carbon	5.4	Iron	50
Oxygen	7	Zinc	56
Phosphorus	9	Copper	56
Sulphur	13	Lead	90
Magnesia	20	Silver	190
Lime	24	Gold	190
Soda	28	Platina	190
Potash	42	Mercury	167

Figure 2 Dalton and his table of elements

a How did Dalton put the elements in order?

In 1864, John Newlands built on Dalton's ideas with his 'law of octaves' (an octave in music is eight notes). Newlands arranged the known elements in order of mass. He noticed that the properties of every *eighth* element seemed similar.

He produced a table of his octaves (see Figure 3). However, he assumed that all the elements had been found. He did this even though chemists were still discovering new ones. So he filled in his octaves even though some of his elements were not similar at all. His table only really worked for the known elements up to calcium before the pattern broke down.

Other scientists ridiculed his ideas, and refused to accept them.

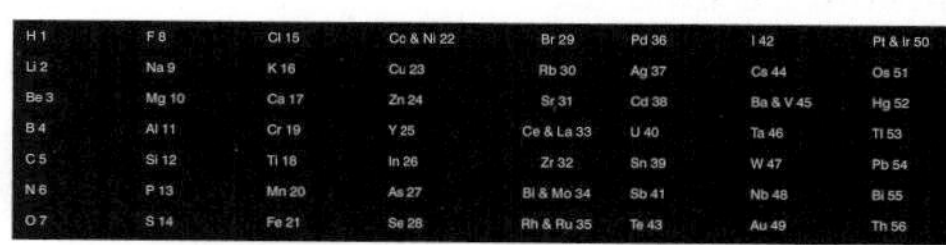

H 1	F 8	Cl 15	Co & Ni 22	Br 29	Pd 36	I 42	Pt & Ir 50
Li 2	Na 9	K 16	Cu 23	Rb 30	Ag 37	Cs 44	Os 51
Be 3	Mg 10	Ca 17	Zn 24	Sr 31	Cd 38	Ba & V 45	Hg 52
B 4	Al 11	Cr 19	Y 25	Ce & La 33	U 40	Ta 46	Tl 53
C 5	Si 12	Ti 18	In 26	Zr 32	Sn 39	W 47	Pb 54
N 6	P 13	Mn 20	As 27	Bi & Mo 34	Sb 41	Nb 48	Bi 55
O 7	S 14	Fe 21	Se 28	Rh & Ru 35	Te 43	Au 49	Th 56

Figure 3 Newlands and his table of octaves. Looking at Newlands' octaves, a fellow chemist commented that putting the elements in alphabetical order would probably produce just as many groups of elements with similar properties!

b What were Newlands' octaves?

Mendeleev's breakthrough

Then, in 1869, the Russian chemist Dmitri Mendeleev cracked the problem. At this time around 50 elements had been identified. Mendeleev arranged all of these in a table. He placed them in the order of their atomic weights. Then he arranged them so that a periodic pattern in their properties could be seen. ('Periodic' means 'regularly reoccurring'.)

His brilliant idea was to leave gaps for elements that had not yet been discovered. Then he used his table to predict what their properties should be. A few years later, new elements were discovered with properties that closely matched Mendeleev's predictions. Then there were not many doubts left that his table was a breakthrough in scientific understanding.

Dmitri Mendeleev is remembered as the father of the modern periodic table. With this tool, chemists could now make sense of the chemical elements.

Figure 4 Dmitri Mendeleev together with a Russian stamp issued in his honour in 1969

Summary questions

1 Copy and complete using the words below:

discovered gaps Mendeleev mass periodic properties Newlands

The chemical elements can be arranged in a .... table. Within the table, elements with similar .... are placed together in groups. Like other chemists, such as ...., ... listed elements in order of ...., but he realised that he needed to leave .... for elements that had not yet been ....

2 **a** Why did Newlands' fellow scientists refuse to accept his 'law of octaves'?

b How did Mendeleev persuade any doubters that his periodic table really was a useful tool in understanding the chemical elements?

Key points

- The periodic table of elements developed as chemists tried to classify the elements. It arranges them in a pattern in which similar elements are grouped together.
- Newlands' table put the elements in order of atomic mass but failed to take account of elements that were unknown at that time.
- Mendeleev's periodic table left gaps for the unknown elements, and so provided the basis for the modern periodic table.

Further teaching suggestions

Periodic gaps

- Encourage students to consider why there were gaps in the periodic table proposed by Mendeleev and how he managed to predict the properties of the missing chemicals. Ask students to write a letter from Mendeleev to explain his great discovery to a friend. Students may even wish to 'age' the paper using tea bags and singe the edges with a candle. They could even melt some candle wax and make a seal.

Other lists of elements

- Ask students to find out another way that elements were ordered in the past, e.g. the law of triads.

Summary answers

1 periodic, properties, Newlands, Mendeleev, mass, gaps, discovered

2 **a** His pattern soon broke down because of undiscovered elements, so only some of the elements with similar properties lined up; others just did not make sense.

b He left gaps for elements that were as yet undiscovered, and predicted their properties. When the elements were discovered (e.g. germanium) his predictions were remarkably accurate and this persuaded his doubters of the usefulness of his periodic table.

Answers to in-text questions

a In order of their mass.

b Groups of elements formed by arranging the elements in order of mass/atomic weights and selecting every eighth element.

C3 1.2 The modern periodic table

Learning objectives

Students should learn:

- how atomic structure is linked to the periodic table
- how electronic structure can explain trends in reactivity in groups. **[HT only]**

Learning outcomes

Most students should be able to:

- label the main groups of the periodic table
- determine which group a given element belongs to
- determine the number of outer shell electrons an element has when its group number is given
- explain why the chemical properties are similar in a group.

Some students should also be able to:

- explain trends in reactivity in groups. **[HT only]**

Support

- Some students will need guidance and further support to ensure that all the information they need is on their periodic table. You could give these students a prompt card with questions/statements to help them complete their periodic table. Their outline could also include a key, to which they just need to add colour.

Extend

- Ask students to predict and explain the trends in reactivity for all the groups in the periodic table. Students should use the information in the Higher Tier box in the Student Book and think back to their work on bonding in C2 to help them.

Specification link-up: Chemistry C3.1

- When electrons, protons and neutrons were discovered early in the 20th century, the periodic table was arranged in order of atomic (proton) numbers. When this was done, all elements were placed in appropriate groups. *[C3.1.2 a)]*
- The modern periodic table can be seen as an arrangement of the elements in terms of their electronic structures. Elements in the same group have the same number of electrons in their highest occupied energy level (outer shell). *[C3.1.2 b)]*
- The trends in reactivity within groups in the periodic table can be explained because the higher the energy level of the outer electrons:
  - the more easily electrons are lost
  - the less easily electrons are gained. *[C3.1.3 h)]* **[HT only]**

Lesson structure

Starters

Definitions – Ask students to define the following terms:

- Atomic number [same as proton number, the number of protons in an atom].
- Mass number [number of protons plus the number of neutrons in an atom].
- Isotope [atoms with the same proton number but different mass number, i.e. a different number of neutrons]. *(5 minutes)*

Labelling an atom – Give the students a diagram of a helium atom, just big enough to fit on an A6 sheet of paper. All the students then add as many labels and as much information as they can. Then ask them to swap with a partner, who should mark the information and make any amendments. Students can then stick this into their notes. Support students by giving them a list of information to label their diagram correctly. Extend students by asking them to draw a fully labelled diagram of a lithium atom, using the helium atom diagram to help them. *(10 minutes)*

Main

- Give students a blank A3 periodic table and a selection of colours. Ask them to label each group (1 to 0) and add extra names if they can (e.g. alkali metals).
- Then ask the students to detail which elements are metals, non-metals, liquids and gases by developing their own key.
- Also ask students to work out the electronic structure of the first 20 elements and note them in the appropriate box, e.g. Ca 2.8.8.2.
- Finally, ask students to determine any relationship between electronic structure and position in the periodic table. They should realise that the number of outer shell electrons is the same as the group number.
- Alternatively, give the students a pack of cards with the electronic structure of the first 20 elements. Ask them to order the cards in the form of the periodic table. Then challenge the students to work out the connection between the group number and the electronic structure.
- State the trends in reactivity in Groups 1 and 7. Then ask the students to explain the trends in prose, using electronic structures. The explanation for the trends in reactivity is Higher Tier content. **[HT only]**

Plenaries

Copy and complete – Ask students to copy and complete the following four sentences with as much scientific detail and language as possible:

- Elements in the periodic table …
- The group number is …
- The reactivity in Group 1 ...
- The reactivity in Group 7 ...

Then ask a few students to read out and share their sentences. The first two statements are suitable for Foundation Tier; the final two sentences are suitable for Higher Tier only. Support students by giving them the start and end of the sentences. The students match them up. Extend students by asking them to illustrate each sentence with a labelled diagram. *(5 minutes)*

Answering objectives – Ask the students, in pairs, to read the learning objectives from the Student Book. Then ask the pairs to generate an answer to each of the questions, asking each pair to note their answer on the board. Then allow the students time to read all the answers and take a vote as to which is the best response. *(10 minutes)*

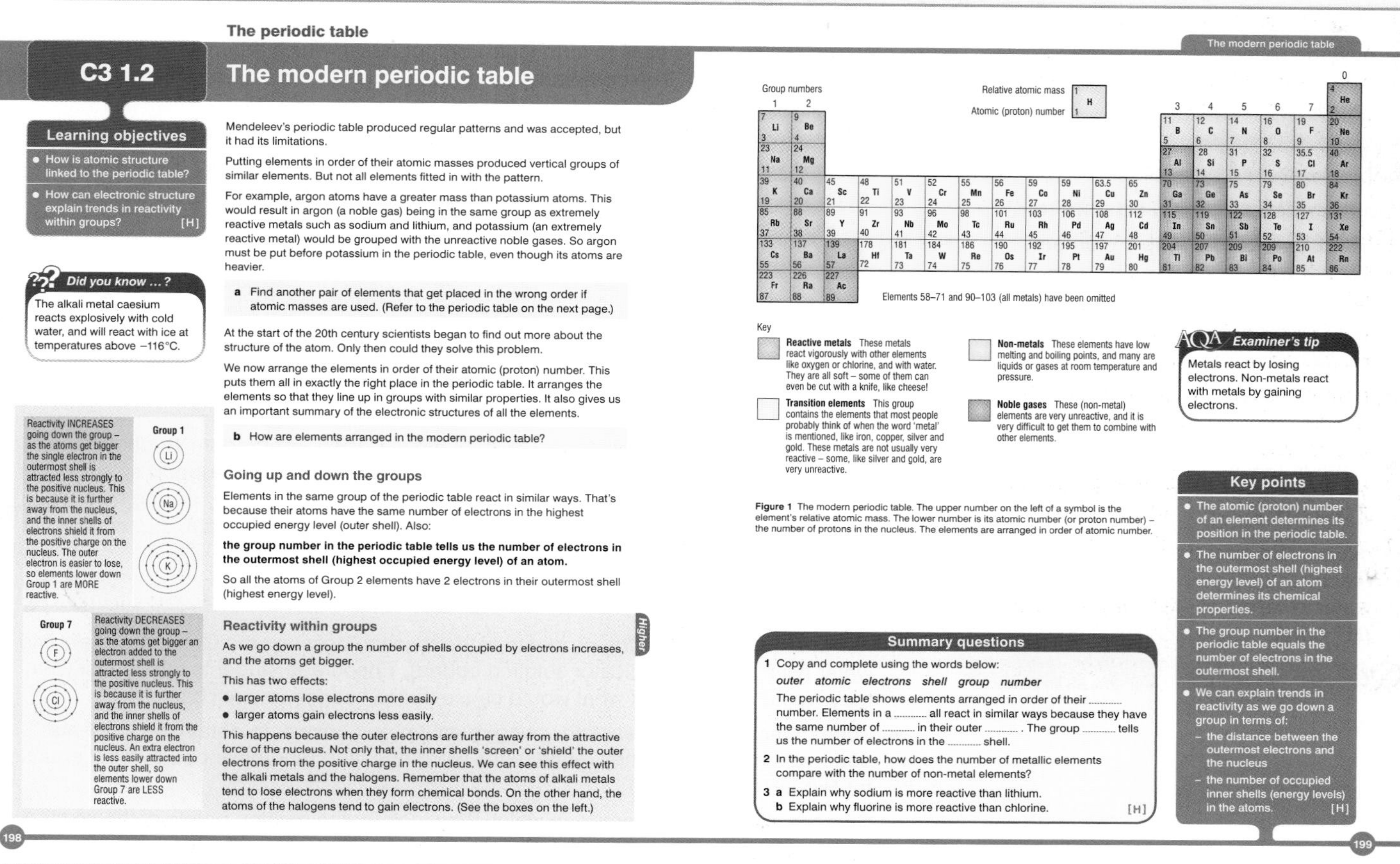

The periodic table

C3 1.2 The modern periodic table

Learning objectives

- How is atomic structure linked to the periodic table?
- How can electronic structure explain trends in reactivity within groups? [H]

Did you know ...?

The alkali metal caesium reacts explosively with cold water, and will react with ice at temperatures above −116°C.

Mendeleev's periodic table produced regular patterns and was accepted, but it had its limitations.

Putting elements in order of their atomic masses produced vertical groups of similar elements. But not all elements fitted in with the pattern.

For example, argon atoms have a greater mass than potassium atoms. This would result in argon (a noble gas) being in the same group as extremely reactive metals such as sodium and lithium, and potassium (an extremely reactive metal) would be grouped with the unreactive noble gases. So argon must be put before potassium in the periodic table, even though its atoms are heavier.

a Find another pair of elements that get placed in the wrong order if atomic masses are used. (Refer to the periodic table on the next page.)

At the start of the 20th century scientists began to find out more about the structure of the atom. Only then could they solve this problem.

We now arrange the elements in order of their atomic (proton) number. This puts them all in exactly the right place in the periodic table. It arranges the elements so that they line up in groups with similar properties. It also gives us an important summary of the electronic structures of all the elements.

b How are elements arranged in the modern periodic table?

Going up and down the groups

Elements in the same group of the periodic table react in similar ways. That's because their atoms have the same number of electrons in the highest occupied energy level (outer shell). Also:

the group number in the periodic table tells us the number of electrons in the outermost shell (highest occupied energy level) of an atom.

So all the atoms of Group 2 elements have 2 electrons in their outermost shell (highest energy level).

Reactivity INCREASES going down the group – as the atoms get bigger the single electron in the outermost shell is attracted less strongly to the positive nucleus. This is because it is further away from the nucleus, and the inner shells of electrons shield it from the positive charge on the nucleus. The outer electron is easier to lose, so elements lower down Group 1 are MORE reactive.

Reactivity DECREASES going down the group – as the atoms get bigger an electron added to the outermost shell is attracted less strongly to the positive nucleus. This is because it is further away from the nucleus, and the inner shells of electrons shield it from the positive charge on the nucleus. An extra electron is less easily attracted into the outer shell, so elements lower down Group 7 are LESS reactive.

Reactivity within groups

Higher

As we go down a group the number of shells occupied by electrons increases, and the atoms get bigger.

This has two effects:

- larger atoms lose electrons more easily
- larger atoms gain electrons less easily.

This happens because the outer electrons are further away from the attractive force of the nucleus. Not only that, the inner shells 'screen' or 'shield' the outer electrons from the positive charge in the nucleus. We can see this effect with the alkali metals and the halogens. Remember that the atoms of alkali metals tend to lose electrons when they form chemical bonds. On the other hand, the atoms of the halogens tend to gain electrons. (See the boxes on the left.)

198

The modern periodic table

Group numbers

Relative atomic mass / Atomic (proton) number: 1 H 1

1	2											3	4	5	6	7	0
																	4 He 2
7 Li 3	9 Be 4											11 B 5	12 C 6	14 N 7	16 O 8	19 F 9	20 Ne 10
23 Na 11	24 Mg 12											27 Al 13	28 Si 14	31 P 15	32 S 16	35.5 Cl 17	40 Ar 18
39 K 19	40 Ca 20	45 Sc 21	48 Ti 22	51 V 23	52 Cr 24	55 Mn 25	56 Fe 26	59 Co 27	59 Ni 28	63.5 Cu 29	65 Zn 30	70 Ga 31	73 Ge 32	75 As 33	79 Se 34	80 Br 35	84 Kr 36
85 Rb 37	88 Sr 38	89 Y 39	91 Zr 40	93 Nb 41	96 Mo 42	98 Tc 43	101 Ru 44	103 Rh 45	106 Pd 46	108 Ag 47	112 Cd 48	115 In 49	119 Sn 50	122 Sb 51	128 Te 52	127 I 53	131 Xe 54
133 Cs 55	137 Ba 56	139 La 57	178 Hf 72	181 Ta 73	184 W 74	186 Re 75	190 Os 76	192 Ir 77	195 Pt 78	197 Au 79	201 Hg 80	204 Tl 81	207 Pb 82	209 Bi 83	209 Po 84	210 At 85	222 Rn 86
223 Fr 87	226 Ra 88	227 Ac 89															

Elements 58–71 and 90–103 (all metals) have been omitted

Key

Reactive metals These metals react vigorously with other elements like oxygen or chlorine, and with water. They are all soft – some of them can even be cut with a knife, like cheese!

Transition elements This group contains the elements that most people probably think of when the word 'metal' is mentioned, like iron, copper, silver and gold. These metals are not usually very reactive – some, like silver and gold, are very unreactive.

Non-metals These elements have low melting and boiling points, and many are liquids or gases at room temperature and pressure.

Noble gases These (non-metal) elements are very unreactive, and it is very difficult to get them to combine with other elements.

Figure 1 The modern periodic table. The upper number on the left of a symbol is the element's relative atomic mass. The lower number is its atomic number (or proton number) – the number of protons in the nucleus. The elements are arranged in order of atomic number.

AQA Examiner's tip

Metals react by losing electrons. Non-metals react with metals by gaining electrons.

Key points

- The atomic (proton) number of an element determines its position in the periodic table.
- The number of electrons in the outermost shell (highest energy level) of an atom determines its chemical properties.
- The group number in the periodic table equals the number of electrons in the outermost shell.
- We can explain trends in reactivity as we go down a group in terms of:
  - the distance between the outermost electrons and the nucleus
  - the number of occupied inner shells (energy levels) in the atoms. [H]

Summary questions

1 Copy and complete using the words below:

outer atomic electrons shell group number

The periodic table shows elements arranged in order of their .... number. Elements in a .... all react in similar ways because they have the same number of .... in their outer .... The group .... tells us the number of electrons in the .... shell.

2 In the periodic table, how does the number of metallic elements compare with the number of non-metal elements?

3 **a** Explain why sodium is more reactive than lithium.
b Explain why fluorine is more reactive than chlorine. [H]

199

Further teaching suggestions

Trends

- Ask students to find the trends in reactivity of some other groups of the periodic table.

Remembering reactivity

- Ask students to create a rhyme or mnemonic to help them remember the patterns of reactivity in Groups 1 and 7.

Word search

- Give the students a word search, but not a list of words to find. All the missing words should be found in the key points or group names from the periodic table. Students could then find the words and write definitions for each of them.

Drama

- Students could wear sports bibs to represent the different subatomic particles in an atom. This drama then can be used to explain physically the effect of shielding on the outer electrons and its effect on reactivity. **[HT only]**

Summary answers

1 atomic, group, electrons, shell, number, outer

2 There are many more metallic elements than non-metallic elements.

3 **a** Sodium's outer electron is further from the attractive force of the nucleus and is therefore more easily lost than lithium's. (Sodium's outer electron is also shielded by more inner electrons than lithium's.)

b An extra electron is attracted into the outer shell more strongly in fluorine's small atoms where the outer shell is nearer the attractive force of the nucleus than in chlorine.

Answers to in-text questions

a Iodine and tellurium.

b In order of their atomic (proton) number.

C3 1.3 Group 1 – the alkali metals

Learning objectives

Students should learn:

- how the elements of Group 1 behave
- how the reactivity of the Group 1 elements changes going down the group.

Learning outcomes

Most students should be able to:

- list examples of Group 1 elements
- list the properties of Group 1 elements
- state the trends in physical and chemical properties going down the group.

Some students should also be able to:

- write balanced symbol equations for the reaction of Group 1 elements with water **[HT only]**
- explain the observations for the reactions of Group 1 elements with water in terms of electronic structures. **[HT only]**

Specification link-up: Chemistry C3.1

- The elements in Group 1 of the periodic table (known as the alkali metals):
  - are metals with low density (the first three elements in the group are less dense than water)
  - react with non-metals to form ionic compounds in which the metal ion carries a charge of +1. The compounds are white solids that dissolve in water to form colourless solutions
  - react with water, releasing hydrogen
  - form hydroxides that dissolve in water to give alkaline solutions. *[C3.1.3 a)]*
- In Group 1, the further down the group an element is:
  - the more reactive the element
  - the lower its melting point and the boiling point. *[C3.1.3 b)]*

Lesson structure

Starter

List – Ask students to list all the properties of metals. Then ask random students to name one of the properties, and draw up an exhaustive list on the board. Support students by giving them a list of properties to choose from. Extend students by asking them to suggest a metal and a use for which the particular property is important, e.g. conducts electricity, copper used in electronics. *(10 minutes)*

Main

- Ask students to predict the products of the reaction between the Group 1 metals and water (with universal indicator solution added). Encourage students to suggest what they would observe and what would cause this to happen in each case.
- Then demonstrate the reaction, using question-and-answer to obtain the students' observations and how they relate to the word equation (and balanced-symbol equation for Higher Tier students).
- Give the students an exercise in which they look up the melting points of the alkali metals and then present the data in a table. Then ask them to display the data graphically, justifying their choice [bar chart]. Extend students by asking them to plot melting points against atomic numbers. Use this to revise 'How Science Works' concepts of 'types of variable' and 'how to present data'.
- Show students some revision guides and examination papers. Ask students to work in pairs to create an A3 spread for a revision book to review the objectives detailed. Students should include a set of examination questions and their answers in their work. Once the students have planned their work, they could use a desktop publishing package to present it.
- Give pairs of students a review sheet containing a table. Each team should rate another pair's revision page for its layout, attractiveness, ease of reading, scientific correctness, whether it meets the objectives and examination practice. Each area should be awarded a mark out of 10 and the total mark recorded. Once the assessment has been completed, the review sheet and work should be passed back to the 'authors' for them to consider their work.

Plenaries

Reflection – Ask students to review their table from the Starter. Using a different colour pen, they should make any amendments in the light of the lesson and write the answers to the questions posed in the second column. Then ask the students to fill in the final column with any other facts that they have learned. *(5 minutes)*

Questions – Give students 10 1-mark questions, becoming progressively harder. Using a stopwatch, give the students 5 minutes to answer as many questions as they can. Explain that the questions grow progressively harder and they should just do as many as they can. After 5 minutes the students can mark their own work, with you reading out the answers. Then ask the students to put up their hands if they attempted 10 questions, attempted nine, etc., and finally if they answered 10 questions correctly, or nine, etc. This gives instant feedback about understanding. Questions can be selected from Foundation or Higher Tier papers. *(10 minutes)*

Support

- Give students a worksheet with half-completed tables, and diagrams. Students should then observe the demonstrations of the reactions of lithium, sodium and potassium with water and use the Student Book to complete the poster, so that it contains all the information needed.

Extend

- Ask students to find out three uses of Group 1 metals. They should detail which metal is used for what purpose and if it is in a compound; the formula should be included.

Practical support

Reactions of alkali metals with water

Equipment and materials required

Group 1 metals (lithium, sodium, potassium – highly flammable, corrosive), white tile, filter paper, spatula, tweezers, knife, water trough, water, universal indicator solution (flammable), safety screen.

Details

Half-fill the trough of water and add enough universal indicator solution to make the colour clearly visible. Cut a small piece of lithium (no more than 3 mm). Wipe off the excess oil and put into the solution of universal indicator. Repeat with the other metals.

Safety: You should wear eye protection and there should be safety screens between the water trough and the students. Students should also wear eye protection. Once the white tile, spatula, tweezers and knife have been finished with, place in water trough to react any residue metal with water. Be sure not to get the metal on the skin – if this does happen, remove with paper or tweezers and wash the affected area well with cold water. CLEAPSS Hazcards: Lithium 58, Sodium 88, Potassium 76.

The periodic table

C3 1.3 Group 1 – the alkali metals

Learning objectives

- How do the Group 1 elements behave?
- How do the properties of the Group 1 elements change going down the group?

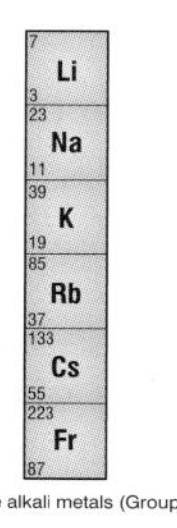

Figure 1 The alkali metals (Group 1)

We call the first group (Group 1) of the periodic table the alkali metals. This group consists of the metals lithium (Li), sodium (Na), potassium (K), rubidium (Rb), caesium (Cs) and francium (Fr). You will probably only see the first three of these as the others are extremely reactive. Francium is radioactive as well!

Properties of the alkali metals

All the alkali metals are very reactive. They have to be stored in oil (see Figure 2). This stops them reacting with oxygen in the air. Their reactivity increases as we go down the group. So lithium is the least reactive alkali metal and francium is the most reactive.

All the alkali metals have a very low density compared with other metals. In fact lithium, sodium and potassium all float on water. The alkali metals are also all very soft. They can be cut with a knife. They have a silvery, shiny surface when we first cut them. However, this quickly goes dull as the metals react with oxygen in the air. This forms a layer of oxide on the shiny surface.

The properties of this unusual group of metals result from their electronic structure. The atoms of alkali metals all have one electron in their outermost shell (highest energy level). This gives them similar properties. It also makes them very reactive. That's because they need to lose just one electron to get a stable (noble gas) electronic structure.

They react with non-metals, losing their single outer electron. They form a metal ion carrying a 1+ charge, e.g. Na^+, K^+. They always form ionic compounds.

a What is the formula of a lithium ion?

Melting points and boiling points

The Group 1 metals melt and boil at relatively low temperatures for metals. Going down the group, the melting points and boiling points get lower and lower. In fact, caesium turns into a liquid at just 29 °C.

Figure 2 The alkali metals have to be stored in oil

Reaction with water

When we add lithium, sodium or potassium to water the metal floats on the water, moving around and fizzing. The fizzing happens because the metal reacts with the water to form hydrogen gas. Potassium reacts so vigorously with the water that the hydrogen produced catches fire. It burns with a lilac flame.

The reaction between an alkali metal and water also produces a metal hydroxide, hence the name 'alkali metals'.

200

Group 1 – the alkali metals

The hydroxides of the alkali metals are all soluble in water. The solution is colourless with a high pH. (Universal indicator turns purple.)

sodium + water → sodium hydroxide + hydrogen

$$2Na(s) + 2H_2O(l) \rightarrow 2NaOH(aq) + H_2(g)$$

potassium + water → potassium hydroxide + hydrogen

$$2K(s) + 2H_2O(l) \rightarrow 2KOH(aq) + H_2(g)$$

b Write the word equation for the reaction of lithium with water.

Demonstration

Reactions of alkali metals with water

The reaction of the alkali metals with water can be demonstrated by dropping a small piece of the metal into a trough of water. This must be done with great care. The reactions are vigorous, releasing a lot of energy. Hydrogen gas is also given off.

- Describe your observations.

Very small piece of alkali metal
Forceps
Trough of water

Figure 3 Lithium, sodium and potassium reacting with water (the lithium is on the left of the trough, the potassium has burning hydrogen above it, and the sodium is the molten silvery ball on the right)

links

For further information on reactivity within groups, look back at C3 1.2 The modern periodic table.

Other reactions

The alkali metals also react vigorously with non-metals such as chlorine. They produce metal chlorides, which are white solids. Their chlorides all dissolve readily in water to form colourless solutions.

The reactions get more and more vigorous as we go down the group. That's because it becomes easier to lose the single electron in the outer shell to form ions with a 1+ charge.

sodium + chlorine → sodium chloride

$$2Na(s) + Cl_2(g) \rightarrow 2NaCl(s)$$

They react in a similar way with fluorine, bromine and iodine. All of these ionic compounds of the alkali metals are also white and dissolve easily in water. The solutions are all colourless.

Summary questions

1. Copy and complete using the words below:
   alkali alkaline bottom hydrogen ions less reactive top
   The Group 1 elements are also known as the .... metals. They are very .... with water, producing .... solutions and .... gas. Lithium, at the .... of the group, is .... reactive than potassium, nearer the .... of the group. These elements always react to make compounds in which they form singly charged positive ....
2. Caesium is almost at the bottom of Group 1. What do you think would happen if it was dropped into water?
3. Write a balanced symbol equation for the reaction of caesium with: **a** iodine **b** water [H]

Key points

- The elements in Group 1 of the periodic table are called the alkali metals.
- Their melting points and boiling points decrease going down the group.
- The metals all react with water to produce hydrogen and an alkaline solution containing the metal hydroxide.
- They form 1+ ions in reactions to make ionic compounds. These are generally white and dissolve in water, giving colourless solutions.
- The reactivity of the alkali metals increases going down the group.

201

Further teaching suggestions

Flame test

- Students can complete flame tests with solutions of Group 1 metal salts (e.g. sodium chloride), on an inoculating loop to determine the metal ion by the flame colour. See 'Practical support' in C3 4.1 Tests for positive ions.

Samples

- Put small pieces of Group 1 metals in oil-filled vials. These should then be placed on a cotton-wool bed in a sealed and labelled glass dish. Ask the students to look at the samples. Then ask them how these are hazardous and how they are stored for safety. Use a question-and-answer session to get feedback from the students about why the metals are kept under oil to stop them reacting with the oxygen and water in the air. You can also explain that they are stored in a flame-resistant lockable cupboard in school. **Safety:** make sure these samples are returned to the technician at the end of the lesson.

Videos

- You could show videos of the reactions of rubidium and caesium with water.

Other reactions

- You may wish to demonstrate the combustion of sodium or potassium with air and sodium with oxygen see CLEAPSS 13.2.3 table 13.4 and Hazcard 76. The reaction between sodium or potassium with chlorine can also be demonstrated. See CLEAPSS 13.2.5 and L195.

Did you know ... ?

Francium is also a Group 1 metal but very little is known about it. This is because its nuclei are unstable and quickly change by radioactive decay. It should react with water even more violently than caesium does but so far no one has been able to get a large enough sample to try this!

Summary answers

1. alkali, reactive, alkaline, hydrogen, top, less, bottom, ions
2. It would explode/react violently.
3. **a** $2Cs + I_2 \rightarrow 2CsI$
   b $2Cs + 2H_2O \rightarrow 2CsOH + H_2$

Answers to in-text questions

a Li^+. **b** Lithium + water → lithium hydroxide + hydrogen.

C3 1.4 The transition elements

Learning objectives

Students should learn:

- the properties of transition elements
- how the transition elements compare with the alkali metals.

Learning outcomes

Most students should be able to:

- list examples of transition elements
- list the properties of transition elements
- compare the general properties of the transition elements with those of the Group 1 elements
- state that transition elements can form more than one type of ion and work out the formula of their compounds from information provided.

Support

- Give students a cube template that could contain some information such as fill-in-the-gaps prose, unfinished bullet points and unlabelled diagrams. Students can then complete the cube rather than generate all the material themselves.

Extend

- Encourage students to use a graphing package, e.g. Excel, to plot data in a variety of forms. They should then draw conclusions. For example, students could generate a scatter graph of conductivity of transition elements (atomic/proton number against conductivity). Data can be sourced from the internet or the *RSC Electronic Data Book*.

Specification link-up: Chemistry C3.1

- Compared with the elements in Group 1, transition elements:
  - have higher melting points (except for mercury) and higher densities
  - are stronger and harder
  - are much less reactive and so do not react as vigorously with water or oxygen. *[C3.1.3 c)]*
- Many transition elements have ions with different charges, form coloured compounds and are useful as catalysts. *[C3.1.3 d)]*

Lesson structure

Starters

Images – Show the students different and interesting images of transition elements being used, e.g. gold covering the sun visor of an astronaut, titanium hip joints, coloured gemstones. Ask the students to look at the pictures and find the connection. *(5 minutes)*

Crossword – Give the students a crossword with all the words being the physical properties of metals. Support students by giving them a list of words to choose from. Extend students by giving clues for both the physical and chemical properties of transition elements. These students could then colour-code the clues to show whether they are chemical or physical properties. *(10 minutes)*

Main

- Give the students a template of a cube on an A4 piece of card. Encourage them to put a different piece of information on each of the six faces (physical properties, chemical properties, specific uses, information about transition compounds, position in the periodic table, names). Students should be encouraged to use colour and summarise the information in an easy-to-digest format, then to cut out the cube template, scoring in the appropriate positions and to stick the cube together using glue.
- Students need to be aware of the properties of transition elements and their compounds. On a large poster of the periodic table, put samples of different transition elements and their compounds. Allow the students to look at the exhibition and to draw visible conclusions.
- Let students use different metals to test their properties and contrast them with Group 1 metals. Set up a circus of five stations with different transition elements, allowing the students to rotate around the circus using an activity card placed at each station with the method to complete the practical.
- Once all the activities have been tried, encourage the students to write a bullet-point list of the properties of transition elements. Then ask the students to contrast these properties with those of Group 1 metals and write a few sentences to explain the similarities and differences.
- You may wish to demonstrate the different colours of vanadium ions (see 'Practical support').
- To illustrate the catalytic properties, you could investigate the effectiveness of different transition metal oxides on the decomposition of hydrogen peroxide solution (see C2 4.5 The effect of catalysts for more details if this investigation was not attempted in C2).

Plenaries

5, 4, 3, 2, 1 – Ask students to create the following list: five properties of transition elements, four symbols of transition elements, three magnetic transition elements, two different copper ions, one example of a transition element ion with a 3+ charge. Support students by giving them a copy of the periodic table with pictures of the metals on to help them. Extend students by being asking them to use symbols, where possible, for the elements and ions. *(5 minutes)*

Facts – Ask students to think about a fact from the lesson. Choose one student to say their fact, then the next student should repeat the first fact and add their own fact. This activity can continue around the classroom. *(10 minutes)*

Practical support

Demonstrating the colours of vanadium

Equipment and materials required

Ammonium metavanadate 0.1 mol/dm³ solution (toxic), zinc chips, concentrated hydrochloric acid (corrosive), test tube, test-tube rack, spatula, 2 × dropping pipettes, eye protection, gloves.

Details

Quarter-fill a test tube with ammonium metavanadate and encourage students to note the colour (VO_3^- orange). Half-fill the test tube with concentrated hydrochloric acid and observe the colour (VO_2^+, yellow). Add one piece of zinc and observe the colour change to blue (VO^{2+}), green (V^{3+}) and finally to violet (V^{2+}).

Safety: CLEAPSS Hazcard 47A Hydrochloric acid – corrosive. CLEAPSS Hazcard 9B Ammonia vanadate (V) – toxic.

Investigating transition elements

Equipment and materials required

Samples of different transition elements, two crocodile clips, three wires, lamp, low-voltage power supply (0–12 V), boiling tubes, boiling-tube holders, Bunsen burner and safety equipment, beaker, water, 1 mol/dm³ hydrochloric acid, white tile, dropping pipette, spatula, tongs, magnets, tin lid, Bunsen burner and safety equipment, eye protection.

Details

Allow students to handle the metals to find out that they can be bent and are shiny. They can check if a metal is magnetic using the magnet. Allow students to place the metals in water and observe whether there is any reaction and comment on the density. Wearing eye protection, put the metal on a white tile, then add a drop of acid and observe. Still wearing eye protection, connect the metal with the crocodile clip into a simple circuit. If the lamp shines, this is a positive result for conduction. Be aware that the metal might grow hot in the circuit. Then take a sample of each metal and heat on a tin lid using a Bunsen burner. Students should reflect on the melting point of these metals compared to that of Group 1 metals.

Safety: CLEAPSS Hazcard 47A Hydrochloric acid – corrosive.

C3 1.4 The transition elements

Learning objectives

- What are the properties of the transition elements?
- How do the transition elements compare with the alkali metals?

In the centre of the periodic table there is a large block of metallic elements. We call these the **transition elements or transition metals**.

45 Sc 21	48 **Ti** 22	51 V 23	52 **Cr** 24	55 **Mn** 25	56 **Fe** 26	59 **Co** 27	59 **Ni** 28	63 **Cu** 29	64 **Zn** 30
89 Y 39	91 Zr 40	93 Nb 41	96 Mo 42	99 Tc 43	101 Ru 44	103 Rh 45	106 Pd 46	108 **Ag** 47	112 Cd 48
	178 Hf 72	181 Ta 73	184 W 74	186 Re 75	190 Os 76	192 Ir 77	195 **Pt** 78	197 **Au** 79	201 **Hg** 80

Figure 1 The transition elements. The more common metals are shown in **bold type**. (Strictly speaking, the metals in the first and last columns above should not be called transition elements. This is because of the electronic structure of their ions).

Physical properties

The transition elements have the properties of 'typical' metals. Their metallic bonding and giant structures explain most of their properties.

Transition elements:

- are good conductors of electricity and energy
- are hard and strong
- have high densities
- have high melting points (with the exception of mercury, which is a liquid at room temperature).

a Which metallic element has the lowest melting point?

The transition elements have very high melting points compared with those of the alkali metals in Group 1. They are also harder, stronger and much denser.

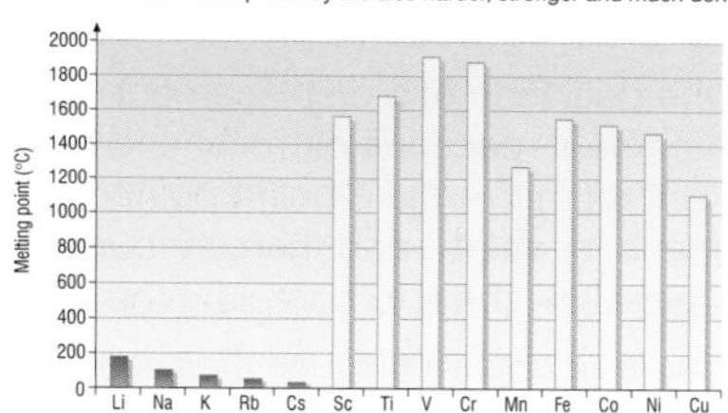

Figure 2 The melting points of the transition elements are much higher than those of the Group 1 elements

Chemical properties

The transition elements are much less reactive than the metals in Group 1. This means they do not react as easily with oxygen or water as the alkali metals do. So if they corrode, they do so very slowly. Together with their physical properties, this makes the transition elements very useful as structural materials.

Demonstration

The colours of vanadium ions

Your teacher can show you the range of colours that different ions of vanadium can have.

Compounds of transition elements

Many of the transition elements form coloured compounds. These include some very common compounds that we use in the laboratory. For example, copper(II) sulfate is blue (from the copper ions, Cu^{2+}).

Figure 3 Compounds of transition elements are coloured (as opposed to the mainly white compounds of the alkali metals). The colours of many minerals, rocks and gemstones are due to transition element ions. The reddish-brown colour in a rock is often due to iron ions, Fe^{3+}. The blue colour of sapphires and the green of emeralds are both due to transition element ions in the structures of their crystals.

Notice that the name of a compound containing a transition element usually includes a Roman number. For example, you will have seen copper(II) sulfate or iron(III) oxide.

This is because transition elements can form more than one ion. For example, iron may exist as Fe^{2+} or Fe^{3+}. Copper can form Cu^+ and Cu^{2+}, and chromium Cr^{2+} and Cr^{3+}. Compounds of these ions are different colours. For example, iron(II) ions (Fe^{2+}) give compounds a green colour, but iron(III) (Fe^{3+}) ions give a reddish-brown colour.

AQA Examiner's tip

The charge on the ion is given in the name of the transition element compound, e.g. copper(II) sulfate contains Cu^{2+} ions.

b Write down the full name of this transition element compound: $NiCl_2$.

Transition elements and their compounds are also very important in the chemical industry as catalysts. For example, you have met the use of nickel as a catalyst in the manufacture of margarine.

Summary questions

1 Copy and complete using the words below:

catalysts coloured conductors densities less melting

The transition elements nearly all have high ...., high .... points and are good .... They are .... reactive than the alkali metals, and often form .... compounds. The transition elements and their compounds are useful .... in the chemical industry.

2 Iron (Fe) can form ions that carry a 2+ or a 3+ charge. Write down the name and formula of each compound iron can form with chlorine.

3 Vanadium (symbol V) reacts with oxygen gas to form the compound vanadium(V) oxide. Write a balanced symbol equation for the reaction. [H]

Key points

- Compared with the alkali metals, transition elements have much higher melting points and densities. They are also stronger and harder, but are much less reactive.
- The transition elements do not react vigorously with oxygen or water.
- A transition element can form ions with different charges, in compounds that are often coloured.
- Transition elements and their compounds are important industrial catalysts.

Further teaching suggestions

Uses of transition elements

- Ask students to find out which transition elements or their compounds are used in the following: Haber process [Fe], Contact process [vanadium oxide], making margarine [Ni] and nitric acid production [Pt/Rh].

Gemstones

- Students could find out the transition elements that cause the colours of gemstones.

Other reactions

- You may wish to demonstrate the reaction between iron and steam, see CLEAPSS 13.2.5. You may wish to demonstrate transition metals or their compounds working as catalysts such as the decomposition of hydrogen peroxide using manganese dioxide as detailed in C2 4.5 The effect of catalysts.

Answers to in-text questions

a Mercury.

b Nickel(II) chloride.

Summary answers

1 densities, melting, conductors, less, coloured, catalysts

2 **a** $FeCl_2$, $FeCl_3$

b iron(II) chloride, iron(III) chloride

3 $4V + 5O_2 \rightarrow 2V_2O_5$

C3 1.5 Group 7 – the halogens

Learning objectives

Students should learn:

- how the Group 7 elements behave
- how the properties of the Group 7 elements change going down the group.

Learning outcomes

Most students should be able to:

- list examples of Group 7 elements
- list the properties of Group 7 elements
- state the trends in physical and chemical properties in the group.

Some students should also be able to:

- generate balanced symbol equations for the reaction of Group 7 elements with another non-metal, e.g. hydrogen and in displacement reactions **[HT only]**
- explain the relative reactivity of Group 7 elements in terms of their electronic structures. **[HT only]**

Specification link-up: Chemistry C3.1

- The elements in Group 7 of the periodic table (known as the halogens) react with metals to form ionic compounds in which the halide ion carries a charge of −1. *[C3.1.3 e)]*
- In Group 7, the further down the group an element is:
  - the less reactive the element
  - the higher its melting point and boiling point. *[C3.1.3 f)]*
- A more reactive halogen can displace a less reactive halogen from an aqueous solution of its salt. *[C3.1.3 g)]*

Lesson structure

Starters

Definitions – Write these definitions on the board and ask the students to determine which word the statements are describing:

- The other name for Group 7 elements. [Halogen]
- A 1– ion made from halogens. [Halide ion]
- A reaction in which a more reactive element takes the place of a less reactive element in its compound. [Displacement]
- Halogens exist as 'two-atom' molecules – another name for this. [Diatomic molecules] *(5 minutes)*

Explain – Ask the students to explain the difference between halogens and halides. Ask some students to write their explanation on the board. Then, with the help of the students, select the most appropriate explanation for all the students to note in their books. Support students by encouraging them to work in small groups. Extend students by asking them to give some examples of halogens and halides. *(10 minutes)*

Main

- Students can compare the reactivity of halogens using the displacement reaction between halogen and solutions of iodide, bromide and chloride.
- Ask students to think about the definition of a displacement reaction. (You could remind students of the displacement reactions of metal and metal ion solutions they probably met in KS3.) Then ask them to write a bullet-point method and equipment list for this reaction. Check the plans for safety, and then encourage the students to design an appropriate results table and allow them to complete the experiment.
- In small groups, encourage the students to discuss their results. Use a question-and-answer session to determine a class-wide conclusion. [Reactivity decreases as you go down the group and a more reactive halogen will displace a less reactive halogen from a halide.]
- Ask students to imagine that they are part of the advertising branch of a large chemical company. They must design a magazine advert for a science publication explaining what halogens are, their properties and their uses, in order to increase sales.

Plenaries

Reflection – Ask the students to consider one fact that they revised in the lesson and one new fact that they learned in the lesson. Select a few students to share their facts with the rest of the class. *(5 minutes)*

Chemical equations – Ask students to copy and complete the following equations (the equations get progressively more difficult):

bromine + hydrogen → [hydrogen bromide]

F_2 + [H_2] → 2HF

Cl_2 + [2KI] → 2KCl + I_2

Br_2 + 2KCl → [no reaction]

Support students by giving them the missing information to put into the equations. Extend students by encouraging them to include state symbols. *(10 minutes)*

Support

- You could give students the steps for the method in the practical as a pictorial flowchart. The students should cut out and stick the steps into the correct order to generate their own method.

Extend

- Ask students to explain the trend in reactivity in Group 7. They could be encouraged to illustrate their answer with balanced symbol equations and information about the electronic structure of the atoms/ions.

Practical support

Displacement reactions between halogens

Equipment and materials required

Dimple tiles, 6 × dropping pipettes, 0.1 per cent chlorine water (toxic gas may be released), 0.1 per cent bromine water (harmful), solution of iodine in potassium iodide, 0.1 mol/dm^3 solutions of potassium chloride, potassium bromide, potassium iodide.

Details

Put three drops of potassium chloride solution into three separate dimples, then put three drops of each halogen solution into each of the dimples and observe. Note that the same halogen and halide will not react but some students may try to complete this experiment. Repeat the above with potassium bromide and potassium iodide.

Safety: Wear eye protection and keep the halogen solutions in a fume cupboard. These reactions may produce toxic fumes and should be completed in a well ventilated area. CLEAPSS Recipe Cards: potassium iodide 55; potassium chloride 51; gas solutions 28. CLEAPSS Hazcard 15A Bromine – very toxic/corrosive. CLEAPSS Hazcard 22A Chlorine – toxic. CLEAPSS Hazcard 15B Bromine water – toxic.

The periodic table

C3 1.5 Group 7 – the halogens

Learning objectives

- How do the Group 7 elements behave?
- How do the properties of the Group 7 elements change going down the group?

19 F 9
35 Cl 17
80 Br 35
127 I 53
210 At 85

Figure 1 The Group 7 elements

Properties of the halogens

The Group 7 elements are called the halogens. They are a group of poisonous non-metals that have coloured vapours. They have fairly typical properties of non-metals.

- They have low melting points and boiling points.
- They are also poor conductors of energy and electricity.

As elements, the halogens all exist as molecules made up of pairs of atoms. The atoms are joined to each other by covalent bonds.

	F—F (F_2)	Cl—Cl (Cl_2)	Br—Br (Br_2)	I—I (I_2)
Melting Point (°C)	−220	−101	−7	114
Boiling Point (°C)	−188	−35	59	184

Figure 2 The halogens all form molecules made up of a pair of atoms, joined by a covalent bond. We call this type of molecule a diatomic molecule.

a What patterns can you spot in the properties of the halogens going down Group 7?

Reactions of the halogens

The electronic structure of the halogens determines the way they react with other elements. They all have seven electrons in their outermost shell (highest energy level). So they need to gain just one more electron to achieve the stable electronic structure of a noble gas. This means that the halogens take part in both ionic and covalent bonding.

	How the halogens react with hydrogen
$F_2(g) + H_2(g) \rightarrow 2HF(g)$	Explosive even at −200 °C and in the dark
$Cl_2(g) + H_2(g) \rightarrow 2HCl(g)$	Explosive in sunlight / slow in the dark
$Br_2(g) + H_2(g) \rightarrow 2HBr(g)$	300 °C + platinum catalyst
$I_2(g) + H_2(g) \rightleftharpoons 2HI(g)$	300 °C + platinum catalyst (very slow, reversible)

b Look at the reactions of the halogens in the table above. What is the pattern in reactivity of the halogens going down Group 7?

The halogens all react with metals. They gain a single electron to give them a stable arrangement of electrons. They form ions with a 1− charge, e.g. F^-, Cl^-, Br^-. Examples of their ionic compounds include sodium chloride, NaCl, and iron(III) bromide, $FeBr_3$.

AQA Examiner's tip

In Group 7, reactivity *decreases* as you go down the group. However, in Group 1 reactivity *increases* going down the group.

204

Group 7 – the halogens

Look at the dot and cross diagram of calcium chloride, $CaCl_2$, below:

$[Cl]^-$ $[Ca]^{2+}$ $[Cl]^-$

c Write down the formula of an iodide ion.

When a halogen reacts with another non-metal, the atoms of the halogen share electrons with the atoms of the other element. This gives the atoms of both elements a stable electronic structure. Therefore the compounds of halogens with non-metals contain covalent bonds.

Look at the dot and cross diagram of hydrogen chloride, HCl, below:

H Cl

(Note that the circles need not be shown in dot and cross diagrams.)

Displacement reactions between halogens

We can use a more reactive halogen to displace a less reactive halogen from solutions of its salts.

Bromine displaces iodine from solution because it is more reactive than iodine. Chlorine will displace both iodine and bromine.

For example, chlorine will displace bromine:

chlorine + potassium bromide → potassium chloride + bromine

$$Cl_2(aq) + 2KBr(aq) \rightarrow 2KCl(aq) + Br_2(aq)$$

Obviously fluorine, the most reactive of the halogens, would displace all of the others. However, it reacts so strongly with water that we cannot carry out any reactions in aqueous solutions.

Summary questions

1 Copy and complete using the words below:

covalent halogens ionic less most top

Group 7 elements are also called the ………… . Fluorine, at the ………… of the group, is the ………… reactive, while iodine is much ………… reactive. They react with other non-metals to form compounds which have ………… bonds. With metals they react to form ………… compounds.

2 **a** Write a word equation for the reaction of sodium with bromine.
b Write a word equation for the reaction of bromine water with potassium iodide solution.

3 Write a balanced symbol equation, including state symbols, for the reaction of:
a potassium metal with iodine vapour [H]
b chlorine water with sodium iodide solution. [H]

Did you know …?

Many early chemists were badly hurt or even killed as they tried to make pure fluorine. They called it 'the gas of Lucifer'. It was finally produced by the French chemist Henri Moissan, who died aged just 55. His life was almost certainly shortened by his work with fluorine.

Practical

Displacement reactions

Add bromine water to potassium iodide solution in a test tube. Then try some other combinations of solutions of halogens and potassium halides.

- Record your results in a table.
- Explain your observations.

Key points

- The halogens all form ions with a single negative charge in their ionic compounds with metals.
- The halogens form covalent compounds by sharing electrons with other non-metals.
- A more reactive halogen can displace a less reactive halogen from a solution of one of its salts.
- The reactivity of the halogens decreases going down the group.

205

Further teaching suggestions

Bleaching

- Students could complete an experiment to research into the acidity and bleaching affect of halogens. Put three pieces of universal indicator paper on to a white tile, and add a few drops of a weak solution of chlorine, bromine and iodine solution and observe.

Displacement

- Using halogens is quite hazardous and you may not wish to allow the students to complete the displacement reactions themselves. You could demonstrate them in test tubes and project the image using a flexicam connected to a digital projector or TV. This allows the image to be enlarged for the whole class to see easily.

Other reactions

- You may wish to demonstrate the reactions of chlorine and bromine with iron wool, see CLEAPSS 13.2.5.

Answers to in-text questions

a Their melting points and boiling points increase.

b The reactivity decreases going down the group.

c I^-.

Summary answers

1 halogens, top, most, less, covalent, ionic

2 a Sodium + bromine → sodium bromide
b Bromine + potassium iodide → iodine + potassium bromide

3 a $2K(s) + I_2(g) \rightarrow 2KI(s)$
b $Cl_2(aq) + 2NaI(aq) \rightarrow I_2(aq) + 2NaCl(aq)$

Summary answers

1 a Mendeleev
  b atomic weight
  c groups
  d undiscovered
  e compound

2 a i The alkali metals.
  ii The halogens.
  iii The noble gases.
  iv The transition elements/metals.
  b i The transition elements/metals.
  ii The alkali metals Group 1.
  iii The halogens/Group 7.
  c It has seven electrons in its outermost shell because the group number equals the number of electrons in the outermost shell and it is in Group 7.

3 He arranged the elements in order of relative atomic masses, lining up elements with similar properties in columns. Where the pattern of similar elements broke down, he left gaps for elements that he argued had not yet been discovered. He then made predictions about the properties of these elements which turned out to be accurate, convincing any doubters that his table was correct.

4 a lithium + water → lithium hydroxide + hydrogen
  b rubidium + chlorine → rubidium chloride
  c aluminium + fluorine → aluminium fluoride
  d bromine + potassium iodide → potassium bromide + iodine

5 a $2Li(s) + 2H_2O(l) \rightarrow 2LiOH(aq) + H_2(g)$
  b $2Rb(s) + Cl_2(g) \rightarrow 2RbCl(s)$
  c $2Al(s) + 3F_2(g) \rightarrow 2AlF_3(s)$
  d $Br_2(aq) + 2KI(aq) \rightarrow 2KBr(aq) + I_2(aq)$

6 a 

  b 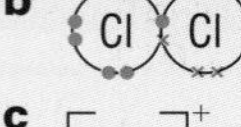

  c 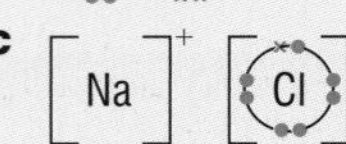

(N.B. Circles are not essential in dot and cross diagrams.)

7 a Alkali metals are not very dense, they are soft, reactive metals, compared to the transition metals, which are denser, harder and less reactive.
  b The transition metals are stronger and less reactive.

8 a Potassium's outer electron is further from the attractive force of the nucleus and is therefore more easily lost than sodium's. Potassium's outer electron is also shielded by more inner shells full of electrons than sodium's.
  b An extra electron is attracted into the outer shell more strongly in chlorine's smaller atoms where the outer shell is nearer the attractive force of the nucleus, than in bromine. Chlorine's outer shell electrons also experience less shielding from inner shells of electron than bromine's do.

The periodic table: C3 1.1–C3 1.5

Summary questions 

1 Choose the correct word to complete each sentence.
  a The Russian chemist who introduced the periodic table in 1869 was ....
  Dalton Mendeleev Newlands
  b He put the elements in order of their ....
  atomic weight boiling point density electrical conductivity
  c He put elements with similar chemical reactions in columns, known as ....
  groups periods rows sets
  d He left gaps for elements that were ....
  insoluble unreactive undiscovered
  e He did **not** put water, H_2O, into the periodic table because water is a ....
  compound liquid mixture

AQA, 2009

2 a What is the general name for the elements in:
  i Group 1
  ii Group 7
  iii Group 0
  iv the block between Group 2 and Group 3?
  b Which group does the element described belong to?
  i It is a hard, dense metal which reacts very slowly with water but will react well with steam.
  ii It is a soft metal that reacts violently with water and forms ions with a 1+ charge.
  iii It is the most reactive of the non-metallic elements. It forms ions with a 1− charge and will also form covalent compounds.
  c Iodine is a halogen whose atomic number is 53. How many electrons occupy its outermost shell (highest energy level)? How did you figure out the answer?

3 Explain how Mendeleev used atomic masses to construct the periodic table. What made other chemists believe in his ideas?

4 Write down word equations for the following reactions:
  a lithium and water
  b rubidium (Rb) and chlorine
  c fluorine and aluminium
  d bromine and potassium iodide solution.

5 Write down balanced symbol equations for the reaction in Question 3.

6 Draw dot and cross diagrams to show the bonding in the following substances. (Draw only the outer shell electrons.)
  a hydrogen chloride
  b chlorine
  c sodium chloride.

7 a How do the properties of the transition elements differ from the properties of the alkali metals?
  b Why are the transition elements more useful than the alkali metals as structural materials (materials used to make things)?

8 Explain why:
  a potassium is more reactive than sodium
  b chlorine is more reactive than bromine.

206

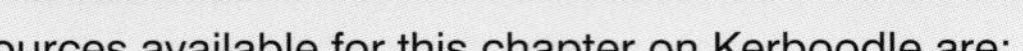

Kerboodle resources

Resources available for this chapter on Kerboodle are:

- Chapter map: The periodic table
- WebQuest: The early periodic table (C3 1.1)
- Audio slideshow: The early periodic table (C3 1.1)
- Extension: Predicting the future (C3 1.1)
- Support: Patterns and trends (C3 1.3)
- Animation: Reactivity trends in the periodic table (C3 1.5)
- Practical: Investigating the reactivity of halogens (C3 1.5)
- Interactive activity: The periodic table and its development
- Revision podcast: The development of the periodic table
- Test yourself: The periodic table
- On your marks: The halogens
- Examination-style questions: The periodic table
- Answers to examination-style questions: The periodic table

End of chapter questions

AQA Examination-style questions

a periodic table to help you to answer these questions.

odium is in Group 1 of the periodic table.

he equation for the reaction of sodium with water is:

$$2Na(s) + 2H_2O(l) \rightarrow 2NaOH(aq) + H_2(g)$$

Name the gas that is produced when sodium reacts with water. (1)

Some water and a few drops of universal indicator were put into a glass trough. A small piece of sodium was added. Describe what you would observe. (4)

Give **two** reasons why the elements in Group 1 are called the *alkali metals*. (2)

he transition elements are between Group 2 and roup 3 in the periodic table.

Which of the following statements are correct comparisons of the transition elements with the metals in Group 1?

Compared with the metals in Group 1, transition elements:

- have lower melting points (except for mercury)
- are stronger and harder
- react more vigorously with oxygen
- react less vigorously with water
- have higher densities (3)

Iron reacts with dilute sulfuric acid to produce a solution of iron(II) sulfate. The solution is pale green. When sodium hydroxide solution is added, a dark green precipitate of iron(II) hydroxide is formed. When left in the air, the precipitate oxidises to iron(III) hydroxide which is red-brown in colour.

What is the evidence from this information that iron is a transition element? (2)

he table gives information about some of the Group 7 ements.

Name of element	Melting point in °C	Boiling point in °C	Electronic structure
Fluorine	−220	−188	2,7
Chlorine	−101	−35	2,8,7
Bromine	−7	58	2,8,18,7
odine	114	183	2,8,18,18,7

What name is given to the Group 7 elements?

Halogens Noble gases Transition elements (1)

Which Group 7 element is liquid at 20 °C? (1)

c Which two statements correctly describe Group 7 elements?
They are metals.
They consist of molecules.
They have coloured vapours.
They have high melting points. (2)

d In terms of their electronic structure, explain why these elements are in Group 7. (1)

e The reactivity of the elements in Group 7 decreases down the group.

This can be shown by reacting chlorine with an aqueous solution of potassium bromide.

i Complete and balance the equation for this reaction:
$Cl_2 + KBr \rightarrow$ (2)

ii Explain the trend in reactivity of the elements in Group 7 in terms of their electronic structure.
[H] (3)
AQA, 2009

4 In 1864 John Newlands suggested the law of octaves. He put the elements known in 1864 in order of increasing atomic mass. He said that every eighth element showed similar chemical properties. He put the elements into a table to show the groups of elements.

The first three rows of his table are shown below.

H	Li	Be	B	C	N	O
F	Na	Mg	Al	Si	P	S
Cl	K	Ca	Cr	Ti	Mn	Fe

Use a modern periodic table to help you to answer these questions.

a i Which group of elements in the modern periodic table is missing from Newlands' table? (1)

ii Why is this group of elements missing from his table? (1)

b In Newlands' table the first two elements in several groups are the same as the first two elements in groups in the modern periodic table. How many of these groups are there in Newlands' table? (1)

c Newlands put fluorine and chlorine in the same group as hydrogen.

i Give one way in which the chemical properties of hydrogen are similar to those of fluorine and chlorine. (1)

ii Give one way in which the chemical properties of hydrogen are different from those of fluorine and chlorine. (1)

207

Examination-style answers

1 a hydrogen *(1 mark)*

b Any **four** from: sodium floats, moves around, fizzes/bubbles; melts/forms a ball; disappears/dissolves; UI/solution/water turns purple/blue; any other correct observation *(4 marks)*

c Any **two** from: all metals; all react to form positive ions; all have any stated metallic property; all form hydroxides; their hydroxides are alkali(s)/alkaline/soluble bases/form solutions with high pH *(2 marks)*

2 a are stronger and harder; react less vigorously with water; have higher densities *(3 marks)*

b Coloured compounds; ions with different charges or Fe(II) *and* Fe(III) compounds. *(2 marks)*

3 a halogens *(1 mark)*

b bromine *(1 mark)*

c They consist of molecules; they have coloured vapours. *(2 marks)*

d All have seven electrons in their outer shell/highest energy level. *(1 mark)*

e i $Cl_2 + 2KBr \rightarrow 2KCl + Br_2$ (one for correct formulae, one for balancing). *(2 marks)*

ii The idea that reaction/reactivity of Group 7 elements involves gaining electrons *gains one mark*; going down the group atoms have less attraction for electrons (or the opposite going up the group) *gains one mark*; because going down the group atoms have more electrons or more occupied energy levels/shells or the atoms get larger *gains one mark*. *(3 marks)*

4 a i Group 0 or the noble/inert gases. *(1 mark)*

ii They had not been discovered. *(1 mark)*

b 6 *(1 mark)*

c i Reacts with other non-metals to form covalent compounds or compounds with (one) shared (pair of) electrons or small molecules or has diatomic molecules or specific example of molecule formed, e.g. H_2, HCl, or reacts with metals to form ionic compounds/compounds with negative hydrogen ions/hydrides/H^- ions or reacts by gaining electrons. *(1 mark)*

ii Forms positive ions/hydrogen ions/H^+ ions or reacts by losing electrons (Halogens gain electrons). *(1 mark)*

Practical suggestions

Practicals	AQA	k	Book	Gears
Demonstration of the combustion of reactions of sodium and potassium.	✓		✓	
Demonstration of the reactions of sodium and potassium with chlorine.	✓		✓	
Demonstration of the reactions of lithium, sodium and potassium with water.	✓		✓	
Demonstration of the reactions of the halogens with iron wool.	✓		✓	
Investigation of the displacement of halogens from solutions of their salts by more reactive halogens.	✓	✓	✓	
Heating transition metals in air (any of Ti, Cr, Co, Ni, Fe, Cu) to compare reactivity and melting points with Group 1.	✓		✓	
Demonstration of the reaction of iron wool with steam.	✓		✓	

Practicals	AQA	k	Book	Gears
Demonstrations of transition metals and their salts as catalysts.	✓		✓	
Investigation of the catalysis of hydrogen peroxide decomposition by different transition metals and their compounds.	✓		✓	

C3 2.1 Hard water

Learning objectives

Students should learn:

- what hard water is
- how hard water is formed
- what the advantages and disadvantages of hard water are.

Learning outcomes

Most students should be able to:

- state a definition of hard water
- state the ions that make water hard
- list advantages of hard water
- list disadvantages of hard water.

Some students should also be able to:

- explain what happens to soap in hard water
- explain using a balanced equation how scale can form in hard water.

Support

- Students often find it difficult to understand that dissolved chemicals are present in water that is fit for drinking. They also find it difficult to understand that the amounts of these substances correlate with the hardness. You could support students by boiling samples of water and measuring the mass of the solute. See 'Practical support' for more details.

Extend

- Give students a selection of different samples of water. They could plan an investigation to discover which water is the hardest.

Specification link-up: Chemistry C3.2

- Soft water readily forms lather with soap. Hard water reacts with soap to form scum and so more soap is needed to form lather. Soapless detergents do not form scum. *[C3.2.1 a)]*
- Hard water contains dissolved compounds, usually of calcium or magnesium. The compounds are dissolved when water comes into contact with rocks. *[C3.2.1 b)]*
- Using hard water can increase costs because more soap is needed. When temporary hard water is heated it can produce scale that reduces the efficiency of heating systems and kettles. *[C3.2.1 e)]*
- Hard water has some benefits because calcium compounds are good for the development and maintenance of bones and teeth and also help to reduce heart disease. *[C3.2.1 f)]*
- Consider and evaluate the environmental, social and economic aspects of water quality and hardness. *[C3.2]*

Controlled Assessment: C4.4 Select and process primary and secondary data. *[C4.4.2 b) c)]*; C4.5 Analyse and interpret primary and secondary data. *[C4.5.4 d)]*

Lesson structure

Starters

Question – Ask students to write the symbols of calcium and magnesium ions [Ca^{2+} and Mg^{2+}]. Support students by giving them the data sheet that they will have access to in the examination. Students could then find the symbols on the list of common ions. Extend students by asking them to draw the dot and cross diagrams from these two ions. *(5 minutes)*

Soap up – Give students samples of hard and soft water. Ask them to wash their hands with soap and each type of water in turn and to note their observations. Encourage students to share their ideas in pairs and then in groups of four. Ask each group to feed back its observations to the class. *(10 minutes)*

Main

- The hardness of water could be investigated by the amount of standard soap solution that is needed to produce a permanent soap lather. A set volume of the water to be tested should be measured into a conical flask (or a boiling tube if a graduated dropping pipette is used instead of a burette). Then add 1 cm^3 of soap and shake the solution. Keep adding 1 cm^3 of soap and shaking until a lather of bubbles persists for 10 s. This experiment can be carried out using microscale equipment. See 'Practical support' for more details. Extend students by encouraging them to design their own experiment. You could supply others with the method. Students can decide how best to display their results and evaluate their method used ('Controlled Assessment').
- Students sometimes find it difficult to visualise how substances dissolve in water. Encourage students to use the particle model that they met in KS3 to help with their understanding. Take a very large glass beaker, table-tennis balls, marbles and a large glass rod. Put the table-tennis balls into the glass beaker and carefully add the marbles and stir. Ask the students to think about what they have just been shown and ask them to work in pairs to explain what the model shows. Ask a few groups to feed back to the class about what they think the model shows. [The table-tennis balls represent the solvent and the marbles represent the solute that dissolves to form a mixture of the two spheres, the solution.] Ask about the limitations of the model and how it could be improved.
- Ask the students to imagine that they have been commissioned by a water company to create a fact-file card to be put into an information pack given to schools. Many companies provide educational resources. Show students some information from companies – maybe hard copies or websites. They can use the information on the pages, and their card should be an introduction to what causes hard water, its benefits and the problems that it can cause. They should also define 'scale' and 'scum'.

Plenaries

AfL (Assessment for Learning) – Put out the fact-file cards on to the side benches. Next to each piece of work leave a table made of two columns headed 'mark' and 'comment'. Ask the students to choose one piece of work, give it a mark in line with the school's marking policy and give a comment. The comment should have at least one piece of praise and one comment for improvement of the work. Then give individual students their comment sheets to review their own work. *(5 minutes)*

Think – Explain to students that the dissolved chemicals in the water produce 'scale' and 'scum'. Ask students to generate a list of problems that this may cause. Ask each student to write one problem onto the board (it must be new, i.e. not already listed on the board). You may wish to support students by showing them a collage of images, e.g. limescale on household applicances or a scum mark in a bathtub, to help students make suggestions. Ask the students to pick three of the problems to put into their notes. Extend students by asking them to illustrate their answer with balanced symbol equations. *(10 minutes)*

Practical support

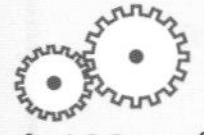

Investigating the hardness of water samples

Equipment and materials required

A 100 cm^3 conical flask with a stopper (or a boiling tube plus stopper), 10 cm^3 measuring cylinder, 50 cm^3 burette (or graduated dropping pipette), soap solution, samples of hard and soft water, eye protection.

Details

Add 10 cm^3 of a water sample to the conical flask (or boiling tube). Add 1 cm^3 of soap solution from the burette (or graduated dropping pipette) into the water sample. Stopper, shake and observe. Add 1 cm^3 of soap solution at a time and shake until a permanent lather is observed. Record the volume of soap solution that is needed for a foam to form. Repeat the experiment with other water samples.

Water

C3 2.1 Hard water

Learning objectives

- What is hard water?
- How is hard water formed?
- What are the advantages and disadvantages of hard water?

Figure 1 Clean water may look the same wherever you are – but appearances can be deceptive

Most people in the developed world have fresh, clean water piped to our homes. The water from our taps looks similar no matter where you live in this country. However, differences become obvious when we use soap to wash.

The water in some areas of the country easily forms a rich, thick lather with soap. This water is called **soft water**. But in some other places it is more difficult to form a lather with soap and water. This is because these areas have **hard water**.

Hard water makes it difficult to wash ourselves. As well as this, it also makes it difficult to clean the bath or sink when we have finished. This is because hard water contains dissolved compounds that react with the soap to form **scum**. The scum floats on the water and sticks to the bath.

The scum formed by hard water is not a problem when washing clothes. That's because modern detergents do not produce scum in hard water. They are called **soapless detergents**.

a Why is it difficult to wash with hard water?

Practical

Investigating the hardness of water samples

You can use soap solution to test how hard a sample of water is. You add soap solution 1 cm^3 at a time. Stopper and shake the tube after each addition. Continue until you get a permanent lather formed.

The more soap needed, the harder the water.

- What type of graph would you use to display the results from the different water samples?
- Evaluate your investigation.

How hard water forms

Most hard water contains dissolved calcium and magnesium ions. These dissolve when streams and rivers flow over rocks containing calcium and/or magnesium compounds. Gypsum, which is calcium sulfate, is an example of such a rock.

Limestone is another rock which causes hardness in water. It contains mainly calcium carbonate. This is practically insoluble in water. However, as rain falls through the air, carbon dioxide dissolves in it. This dissolved carbon dioxide makes rainwater slightly acidic. So the water in streams and rivers becomes slightly acidic too.

Calcium carbonate reacts with the weakly acidic solution. The calcium hydrogencarbonate formed is soluble in water. Therefore calcium ions, $Ca^{2+}(aq)$, get into the water, making it hard. This equation describes what happens:

$$CaCO_3(s) + H_2O(l) + CO_2(aq) \rightarrow \underset{\text{('hardness')}}{Ca^{2+}(aq)} + 2HCO_3^-(aq)$$

Figure 2 It isn't always easy to get bubbles like this

208

Hard water

The dissolved ions are carried into the reservoirs and on into our domestic water supply. It is the dissolved calcium and magnesium ions that react with soap to form scum.

b What does hard water form when we put soap in it?

How hard water wastes soap

Using hard water is expensive because we need to use much more soap. Before soap can do its job of removing dirt, some of it is wasted reacting with the calcium and magnesium ions in the water. It forms insoluble salts that appear as scum. It is only once all the calcium and magnesium ions have reacted that the soap can form a lather. Here is how the calcium ions react with soap:

sodium stearate (soap) + Ca^{2+} ions (calcium ions) → calcium stearate precipitate (scum) + Na^+ ions (soluble in water)

How scale (limescale) is formed

As well as scum, hard water often leads to **scale** (limescale) forming. This insoluble solid can form when we heat one type of hard water. For example, it forms in washing machines, pipes, immersion heaters and other parts of our hot water systems. Pipes can eventually block up.

The same scale forms in our kettles. This 'furring up' of the heating element makes them much less efficient. The scale is a very poor conductor of energy. So it takes longer to boil the water and uses more energy. This costs more money, every time you use a kettle.

c What does hard water form when we heat it?

Figure 3 As scale builds up in heating systems and kettles it not only makes them less efficient – it can stop them working completely

We have seen the disadvantages of hard water but there are also some advantages.

- Calcium ions in drinking water help in the development of strong bones and teeth.
- There is also evidence which suggests that hard water helps to reduce heart disease.

Summary questions

1 Copy and complete using the words below:

calcium heart conductor efficient bones scale scum

Hard water contains .... and/or magnesium ions. These react with soap to form .... Hard water can also produce .... when it is heated. This is a poor .... of energy, and makes kettles and water heaters less .... Hard water is better for teeth and .... than soft water. It may also help reduce .... disease.

2 How can the calcium ions in limestone become dissolved in water?

3 Explain the difference between *scale* and *scum*.

Key points

- Hard water contains dissolved compounds such as calcium and magnesium salts.
- The calcium and/or magnesium ions in hard water react with soap producing a precipitate called scum.
- One type of hard water can produce a solid scale when it is heated, reducing the efficiency of heating systems and kettles.
- Hard water is better than soft water for developing and maintaining teeth and bones. It may also help to prevent heart disease.

209

Further teaching suggestions

Label design

- Students could design a label for a bottle of mineral water explaining the health benefits of dissolved minerals and how they get into the water.

Mineral water composition

- Students could study the packaging of different mineral waters. They could then rate the science content of the labelling. They could also draw pie charts to show the mineral composition of the different waters.

Ions and water hardness

- You may wish to use the method outlined in 'Practical support' for students to investigate which metal ions cause hardness. Supply students with saturated solutions of sodium chloride, calcium chloride, potassium chloride and magnesium chloride for testing.

Answers to in-text questions

a Because the soap reacts with the dissolved ions in the water, which reduces the amount of soap available to wash with.

b scum

c scale

Summary answers

1 calcium, scum, scale, conductor, efficient, bones, heart

2 Because dissolved carbon dioxide makes the water slightly acidic – calcium carbonate in rocks reacts with it to form soluble calcium hydrogencarbonate.

3 Scum – produced when soap reacts with the calcium and/or magnesium salts in hard water.
Scale – produced when one type of hard water is heated.

C3 2.2 Removing hardness

Learning objectives

Students should learn:

- the difference between permanent and temporary hard water
- how hard water can be softened
- what happens to the hydrogencarbonate ions and calcium/magnesium ions in temporary hard water when it is boiled. **[HT only]**

Learning outcomes

Most students should be able to:

- list methods for removing hardness from water
- describe the difference between permanent and temporary hardness.

Some students should also be able to:

- explain how temporary hardness can be removed by boiling the water. **[HT only]**

Did you know ... ?

Romans used lead water pipes; some historians believe that the fall of the Roman Empire was due to lead toxicity.

Answers to in-text questions

a It does not contain the calcium and/or magnesium ions, which make the water hard.
b It would be too expensive because of the large energy requirements to heat huge volumes of water.
c To replenish supply of sodium ions in the ion exchange resin.

Support

- You could simplify the circus of experiments by demonstrating the ion-exchange column and by allowing the students to complete the sodium carbonate practical themselves. The methods, advantages and uses could be supplied to the students as a 'cut and stick' activity.

Extend

- Encourage students to design their own experiment to test the effectiveness of the different methods of softening water.

Specification link-up: Chemistry C3.2

- There are two types of hard water. Permanent hard water remains hard when it is boiled. Temporary hard water is softened by boiling. *[C3.2.1 c)]*
- Temporary hard water contains hydrogencarbonate ions (HCO_3^-) that decompose on heating to produce carbonate ions which react with calcium and magnesium ions to form precipitates. **[HT only]** *[C3.2.1 d)]*
- Hard water can be made soft by removing the dissolved calcium and magnesium ions. This can be done by:
  - adding sodium carbonate, which reacts with the calcium and magnesium ions, forming a precipitate of calcium carbonate and magnesium carbonate
  - using commercial water softeners such as ion-exchange columns containing hydrogen ions or sodium ions which replace the calcium and magnesium ions when hard water passes through the column. *[C3.2.1 g)]*
- Evaluate the use of commercial water softeners. *[C3.2]*
- Consider and evaluate the environmental, social and economic aspects of water quality and hardness. *[C3.2]*

  Controlled Assessment: C4.1 Plan practical ways to develop and test candidates' own scientific ideas. *[C4.1.1 a) b)]*

Lesson structure

Starter

Think 'fur' – Show the students a furred-up element, such as one in the base of an old kettle. Ask students to think which ions might have caused this. Ask groups to feed back to the class. *(10 minutes)*

Main

- Hardness in water can be permanent or temporary. Students can make a sample of temporary hard water by reacting excess carbon dioxide with limewater. See 'Practical support' for more details. Support students by giving them a pictorial method for making and testing their hard water. Extend students by asking them to suggest a method of testing for temporary hardness. Ask students to predict what will happen in their tests.
- There is one method for removing temporary hardness and there are two methods for removing permanent or temporary hardness. Create a circus of three different methods to remove hardness from water. See 'Practical support' for more details.
- Allow students to circulate around the different stations, removing the hardness using a method card at each station. Each student should record in his or her notes a brief explanation of the method, how successful it is, its advantages and a use for that method.
- Students could be encouraged to write a report on the different methods, including a results table to compare the different methods ('Controlled Assessment' opportunity).
- Show students a selection of science magazines. Ask them to imagine that they have been commissioned to write a feature about water softening. Their article should be aimed at a scientific audience and should include diagrams.

Plenaries

Diagram – Give the students an unlabelled diagram of an ion-exchange column and a sheet of information to explain how it works. Ask them to annotate the diagram using the labels provided to explain how this water-softening method works. *(5 minutes)*

Earn the right to leave – Ask all the students to sit down. Then ask each student in turn a question. If a student answers it correctly, he or she can leave the classroom on the bell. If a student answers it incorrectly, pass the question to hands up. Give the students as many questions as they need to get one correct. Support students by asking them questions with simple one-word answers. These key words could be written on the board for students to refer to. Extend students by giving them mini-whiteboards and encouraging them to answer questions that require symbols, formulae or balanced symbol equations. *(10 minutes)*

Practical support

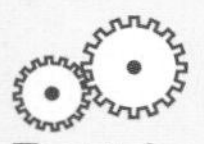

Making and testing temporary hard water

Equipment and materials required

Test tube, straw, limewater, eye protection.

Details

Half-fill a test tube with limewater. Blow exhaled air through the straw into the limewater and observe what happens. First, the limewater should turn from colourless to cloudy and then colourless again. See Station 1 of 'Methods of removing hardness' opposite for details of how to show this is temporary hardness.

Safety: CLEAPSS Hazcard 18 Limewater – irritant.

Methods of removing hardness

Equipment and materials required

Ion-exchange column, hard water, conical flask, soap solution (irritant, may be flammable), sodium carbonate (irritant), filter funnel, filter paper, conical flasks, beaker, tripod, gauze, Bunsen burner and safety equipment, measuring cylinder, spatula. CLEAPSS Recipe Cards: hard water 72; soap solution 60.

Details

To test the hardness of water, add a measured amount of water into a conical flask. Add 1 cm^3 of soap solution and shake. Repeat, until the foam lasts for 10 s or more.

Station 1 – Measure 25 cm^3 of hard water into a beaker. Wearing eye protection, boil the water for about 2 min and observe it. Allow the water to cool and test to compare hardness.

Station 2 – Measure 25 cm^3 of hard water into a beaker. Add excess sodium carbonate and filter the mixture. Test to compare hardness.

Station 3 – Measure 25 cm^3 of hard water and run through an ion-exchange column. Test to compare hardness.

Safety: CLEAPSS Hazcard 18 Limewater – irritant; CLEAPSS Hazcard 95A Sodium carbonate – irritant.

Water

C3 2.2 Removing hardness

Learning objectives

- What is the difference between permanent and temporary hard water?
- How can we make hard water better to use?
- How can we explain what happens when we heat temporary hard water? [H]

Soft water does not contain the dissolved salts that produce scum and scale.

We can soften hard water by removing the calcium and magnesium ions which give it its 'hardness'. Softening water has big benefits. We don't waste soap or get scum formed when washing, and heating our water won't make scale. But people are advised to continue drinking hard water if they can. Scientists think that it is better for your health.

Soft water is also important in many industrial processes. Here hardness can produce scale in boilers, making them more expensive to run. Hardness may also interfere with chemical processes such as dyeing.

a How is soft water different from hard water?

Temporary and permanent hardness

We have seen how scale can form inside a kettle. This has the effect of softening the hard water as the calcium and/or magnesium ions are removed. Water that can have its hardness removed by boiling is called **temporary hard water**.

Calcium and magnesium ions from some salts, such as sulfates, are not removed by heating. Their solutions cause **permanent hard water**. The water remains hard even after boiling. But despite its name, we can also soften this type of hard water.

Practical

Making and testing temporary hard water

You can try to make a sample of temporary hard water. You pass excess carbon dioxide into limewater until the cloudiness disappears again.

- Devise a test to see whether this is temporary hard water or not.
- Predict what will happen in your experiment.

Explaining the effect of heating temporary hard water (Higher)

This is a summary of what happens when we heat temporary hard water:

$$\underset{\text{('hardness')}}{Ca^{2+}(aq)} + 2HCO_3^-(aq) \xrightarrow{\text{heat}} \underset{\text{(scale)}}{CaCO_3(s)} + H_2O(l) + CO_2(g)$$

We can think of the reaction that makes scale in two steps.

1 When we boil hard water containing hydrogencarbonate ions, they decompose:

$$2HCO_3^-(aq) \xrightarrow{\text{heat}} CO_3^{2-}(aq) + H_2O(l) + CO_2(g)$$

2 The carbonate ions, $CO_3^{2-}(aq)$, react with calcium and/or magnesium ions in the hard water:

$$Ca^{2+}(aq) + CO_3^{2-}(aq) \rightarrow \underset{\text{scale}}{CaCO_3(s)}$$

This process removes $Ca^{2+}(aq)$ or $Mg^{2+}(aq)$ ions from hard water and softens it. However, removing hardness by heating would waste energy and be very expensive.

Did you know ...?

Brewers of beer favour areas with permanent hard water (having a high sulfate content) for making bitter and pale ale. Lager and stout are best brewed in areas with soft or temporary hard water (which can be boiled to soften it). The hardness of the water affects the taste of the beer.

b Why is heating temporary hard water not used as a method to soften water on a large scale?

There are two important ways to soften either type of hard water.

210

Removing hardness

Method 1 – Using washing soda

One way to soften water is to add sodium carbonate to it. Sodium carbonate is also called 'washing soda'. It has been used for many years when washing clothes. When we add washing soda to hard water, a reaction takes place. Its soluble carbonate ions precipitate out calcium and magnesium ions. The dissolved metal ions form insoluble carbonates:

$$\underset{\text{('hardness')}}{Ca^{2+}(aq)} + \underset{\text{(from sodium carbonate)}}{CO_3^{2-}(aq)} \rightarrow CaCO_3(s)$$

This reaction is similar to the formation of scale when temporary hard water is heated. However, here it happens quickly, where and when we want it to happen, without having to waste energy.

Method 2 – Using an ion-exchange column

Water can also be softened by removing the $Ca^{2+}(aq)$ or $Mg^{2+}(aq)$ ions using an **ion-exchange column**. A column usually contains a resin packed with sodium ions (Na^+). The hard water passes through the column. The sodium ions from the resin are *exchanged* for the $Ca^{2+}(aq)$ or $Mg^{2+}(aq)$ ions in hard water. Some columns work by swapping hydrogen ions for the aqueous calcium or magnesium ions. This is how domestic water-softening units work.

The resin can be recharged with sodium ions after they have been exchanged for calcium and magnesium ions. The resin is washed with salt (sodium chloride) solution. This puts the sodium ions back in. This is why water softeners must be kept topped up with salt.

c Why do water softeners need to have salt added to them?

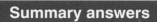

SODA CRYSTALS

Figure 1 Washing soda is a simple way to soften water without the need for any complicated equipment

Summary questions

1 Copy and complete using the words below:

calcium ion-exchange permanent scum softener sodium temporary

Soft water does not contain .... or magnesium ions. This means that it does not produce .... or scale.
.... hard water can be softened by heating it whereas .... hard water cannot. Either type of hard water can be softened by adding .... carbonate or by using a water .... which contains an .... column.

2 Explain how sodium carbonate (washing soda) softens hard water. You should include an equation in your answer.

3 Explain using balanced symbol equations how temporary hard water is softened by heating. [H]

Key points

- Soft water does not contain ions that produce scum or scale.
- Hard water can be softened by removing the ions that produce scum and scale.
- Temporary hardness is removed from water by heating it. Permanent hardness is not changed by heating.
- The hydrogencarbonate ions in temporary hard water decompose on heating. The carbonate ions formed react with $Ca^{2+}(aq)$ and $Mg^{2+}(aq)$ ions, making precipitates. [H]
- Either type of hard water can be softened by adding washing soda or by using an ion-exchange resin to remove calcium and magnesium ions.

211

Summary answers

1 calcium, scum, temporary, permanent, sodium, softener, exchange

2 The aqueous carbonate ions from washing soda (sodium carbonate) precipitate calcium and/or magnesium ions out of the hard water. For example:

$$\underset{\text{('hardness')}}{Ca^{2+}(aq)} + \underset{\text{(from sodium carbonate)}}{CO_3^{2-}(aq)} \rightarrow \underset{\text{(calcium carbonate)}}{CaCO_3(s)}$$

3 When hard water containing hydrogencarbonate ions is boiled, the HCO_3^- ions decompose:

$$2HCO_3^-(aq) \xrightarrow{\text{heat}} CO_3^{2-}(aq) + H_2O(l) + CO_2(g)$$

Then the carbonate ions, $CO_3^{2-}(aq)$, precipitate calcium and/or magnesium ions out of the hard water:

$$Ca^{2+}(aq) + CO_3^{2-}(aq) \rightarrow \underset{\text{scale}}{CaCO_3(s)}$$

C3 2.3 Water treatment

Learning objectives

Students should learn:

- how water is made safe to drink
- what the difference is between pure water and water that is fit to drink.

Learning outcomes

Most students should be able to:

- list places from where drinking water is obtained
- state the stages that water goes through to make it fit to drink
- describe how a domestic water filter works.

Some students should also be able to:

- explain how water is treated to make it fit to drink
- explain in detail the difference between pure water and drinking water.

Specification link-up: Chemistry C3.2

- Water of the correct quality is essential for life. For humans, drinking water should have sufficiently low levels of dissolved salts and microbes. *[C3.2.2 a)]*
- Water filters containing carbon, silver and ion exchange resins can remove some dissolved substances from tap water to improve the taste and quality. *[C3.2.2 b)]*
- Chlorine may be added to drinking water to reduce microbes and fluoride may be added to improve dental health. *[C3.2.2 c)]*
- Pure water can be produced by distillation. *[C3.2.2 d)]*

Lesson structure

Starters

Think 'pure' – Show students a bottle of mineral water and a bottle of distilled water. Explain to the students that both of these waters are fit to drink but only one is pure. Ask the students to work in pairs to decide which water is pure and why. Ask a few groups to feed back their thoughts to the class. *(5 minutes)*

Drinking water – Show students a selection of images that show different places where drinking water comes from, e.g. spring, borehole, reservoir, stream. Ask students to suggest what all the images have in common [they are different sources of clean drinking water]. Support students by using images with which they would be more familiar, e.g. a running tap. Extend students by suggesting things that would make water unfit to drink [microbes]. *(10 minutes)*

Main

- Show students a water-filter cartridge that is used in the domestic environment. Ask the students to work in small groups to generate a TV advert to promote the product, explain its benefits and how it works. Impress on the students that advertising is expensive, so the advert cannot be longer than a minute. The adverts could be recorded on a camcorder.
- This material could then be transferred to Windows Movie Maker. Groups could then add text and manipulate the images to make a discrete advert. The final product could be shown to the students or presented on the school website. Students could then show their advert and the class could vote for which is their favourite.
- Water treatment has a number of stages. Ask the students to draw a flow chart to show how groundwater is turned into drinking water. You could encourage students to add a colour code, e.g. colouring the chemical stages in red and the physical stages in yellow.

Plenaries

Sort – Give the students a list of sentences on separate cards to explain the process of water treatment. Time the students as they put the sentences in order. The quickest, correct student could be given a prize. *(5 minutes)*

Find – Explain to students that to allow water to be fit for drinking, there are three main stages: sedimentation, filtration, disinfection. Ask the students to find the chemicals associated with each stage [sedimentation – aluminium sulfate, lime; filter – sand; disinfection – chlorine]. Support students by giving them the stages and chemicals as a card sort. Extend students by asking them to make a mnemonic to remember the key stages and chemicals involved in water treatment. *(10 minutes)*

Support

- Students often find it difficult to remember the stages of water treatment. If you ask the students to make a flow chart for this process, you could give the sections of the flow chart to the students in the wrong order. They could then cut and stick the information into the correct order and add an image for each step to show understanding.

Extend

- Ask students to contrast the advantages and disadvantages of using free chlorine, ozone or chlorine compounds to kill microorganisms in the water treatment process.

Further teaching suggestions

Water treatment research

- In some countries the water in the taps is not fit to drink, or water may need to be collected by each family from boreholes or rivers. Ask students to find out how this water is made fit to drink. Students may also want to consider desalination as a method of getting drinking water. This uses a lot of energy and is an expensive method, but it is useful for some situations such as cruise ships. You may wish to demonstrate the distillation of seawater and then test the distillate via boiling point to show its purity.

Treatment plant visit

- Many water boards offer free visits to water treatment plants.

Testing for microorganisms

- Students could use agar plates to test for microorganisms in water. Water collected from puddles or another 'dirty' source could be cleaned using different methods for purification of water, e.g. sterilisation tablets, boiling and filtering. The water should be swabbed and spread over an agar plate, taped and left in a warm place for a week to see if microbes grow. Follow aseptic techniques. See CLEAPSS Handbook CD-ROM section 15.2.

Find local water source

- Ask students to discover where the water comes from in their area, e.g. in the Midlands it comes from the Welsh mountains and is stored in reservoirs.

C3 2.3 Water treatment

Learning objectives

- How do we make water safe to drink?
- What is the difference between pure water and water that is fit to drink?

Water is a vital and useful resource. We use it in industry as an important raw material, solvent and coolant.

Other uses of water are for washing and cleaning – and of course, for drinking. Providing people with clean drinking water is a major issue all over the world.

Water that comes from boreholes is usually fairly clean. It has been filtered as it passes through the rocks around the borehole. Normally we just need to sterilise this water with chlorine to make it safe to drink. Chlorine kills microbes in the water.

When we take water from rivers and reservoirs we need to give it more treatment than this. This treatment involves several physical and chemical processes. The water source is chosen so that it contains as few dissolved chemicals as possible. The water then passes through five stages. (See Figure 4.)

a How do we normally need to treat water taken from a borehole?
b Why is chlorine added to water at the end of the treatment process?

Figure 1 Good, clean water is a precious resource. Those of us lucky enough to have it can too easily take it for granted.

Some people use filter jugs in their homes. These usually have a top part into which you put tap water.

As the water goes from the top part of the jug to the lower part it passes through a filter cartridge. This usually contains activated carbon, an ion-exchange resin and silver.

- The carbon in the filter reduces the levels of chlorine, as well as pesticides and other organic impurities in the water.
- The ion-exchange resin removes calcium, magnesium, lead, copper and aluminium ions.
- Some filters may contain silver. Silver particles discourage the growth of bacteria within the filter.

In most jugs the filter cartridge needs to be changed every few weeks.

Pure – or just fit to drink?

Even water that has been treated and then passed through a jug filter is not pure. It will still contain some substances dissolved in it. But despite this, it is definitely fit to drink.

We can get pure water by distilling it. This involves boiling the impure water. The liquid water turns into gaseous steam. Then we cool the steam to condense and collect the pure water.

We use distilled water in chemistry practical work because it is pure. It contains nothing but water. However, distillation is not a method usually used to make drinking water. Distilling large volumes of impure water would need vast amounts of energy. This would make the process very expensive.

Figure 2 Many people claim that water filters like these improve the taste of water and ensure it is safe to drink

Figure 3 Mineral waters contain dissolved minerals, which is what gives them their taste and 'character'

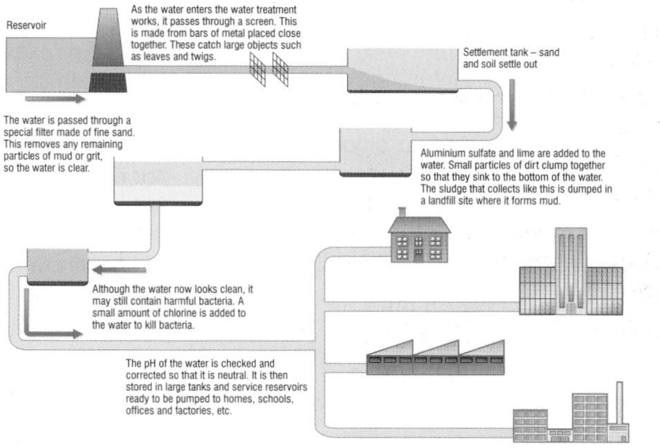

Reservoir

As the water enters the water treatment works, it passes through a screen. This is made from bars of metal placed close together. These catch large objects such as leaves and twigs.

Settlement tank – sand and soil settle out

Aluminium sulfate and lime are added to the water. Small particles of dirt clump together so that they sink to the bottom of the water. The sludge that collects like this is dumped in a landfill site where it forms mud.

The water is passed through a special filter made of fine sand. This removes any remaining particles of mud or grit, so the water is clear.

Although the water now looks clean, it may still contain harmful bacteria. A small amount of chlorine is added to the water to kill bacteria.

The pH of the water is checked and corrected so that it is neutral. It is then stored in large tanks and service reservoirs ready to be pumped to homes, schools, offices and factories, etc.

Figure 4 The treatment of water to make it fit to drink. Fluoride is also added to drinking water in some areas. This helps to prevent tooth decay.

links
See C3 2.4 Water issues for more information on fluoridation.

Summary questions

1 Copy and complete using the words below:
microbes chlorine silver metal filter
Making water fit to drink involves passing it through a .... to remove solids. We then add .... to it to kill .... Household water filters contain ion-exchange resin to remove .... ions. They also include carbon, and often ...., which discourages growth of bacteria in the filter.

2 Water that looks colourless and clean may not be fit to drink. Why not?

3 Water that has passed through a filter jug should not form scale (limescale) in your kettle. Why not?

4 Explain why bottles of water sold in the supermarket should not be described as 'pure' water?

Key points

- Water for drinking should contain only low levels of dissolved substances and microbes.
- Water is made fit to drink by filtering it to remove solids and adding chlorine to reduce the number of microbes.
- We can make pure water by distillation but this requires large amounts of energy which makes it expensive.

Answers to in-text questions

a Sterilise it with chlorine.

b To kill microorganisms that can cause disease.

Summary answers

1. filter, chlorine, microbes, metal, silver
2. It may contain microorganisms, e.g. bacteria, which can cause disease (or dissolved colourless, but toxic, compounds).
3. Calcium and magnesium ions have been removed from the water.
4. They are not pure in the chemical sense (made up of a single substance) because they contain dissolved solids, as shown on their labels.

C3 2.4 Water issues

Learning objectives

Students should learn:

- that there are advantages and disadvantages to softening water
- that there are advantages and disadvantages to adding chlorine to drinking water
- that there are advantages and disadvantages to adding fluoride to drinking water.

Learning outcomes

Most students should be able to:

- state some advantages and disadvantages of softening water
- state some advantages and disadvantages of adding chlorine to drinking water
- state some advantages and disadvantages of adding fluoride to drinking water.

Some students should also be able to:

- evaluate the advantages and disadvantages of softening water
- evaluate the advantages and disadvantages of adding chlorine and fluoride to water.

Support

- Give students a writing frame to list the advantages and disadvantages of a water additive. You could give them a list of statements that they could sort into the correct part of the writing frame to provide further support.

Extend

- Give students a map of the UK, which they can then colour code to show the areas where fluoride is naturally found in the water, where fluoride is added and where there is no fluoride in the water.

Specification link-up: Chemistry C3.2

- Evaluate the use of commercial water softeners. *[C3.2]*
- Consider and evaluate the environmental, social and economic aspects of water quality and hardness. *[C3.2]*
- Consider the advantages and disadvantages of adding chlorine and fluoride to drinking water. *[C3.2]*

  Controlled Assessment: C4.5 Analyse and interpret primary and secondary data. *[C4.5.4 d)]*

Lesson structure

Starters

Uses – Give students a list of chemicals and ask them to suggest why they are added to water. For example, chlorine [to kill microbes], fluoride [to prevent tooth decay], sodium carbonate [to remove hardness]. *(5 minutes)*

Definitions – Ask students to define each of the following key words and give an example of each: fluorine [the first element in Group 7], fluoride [the negative ion made from a fluorine atom], chlorine [the second element in Group 7], chloride [the negative ion made from a chlorine atom]. Support students by giving them the words and the definitions, then they have to match them up. Extend students by asking them to write a half-equation to show how fluoride and chloride ions are made from their elements. *(10 minutes)*

Main

- In the activity on studying the effects of fluoridation, encourage students to think about how they could discover reliable data about the positive and negative effects of the fluoridation of drinking water. You could support students by discussing the 'How Science Works' concepts of repeatability and reproducibility.
- You may wish to give students cards that contain different methods for this type of study. They could work in groups to suggest which method they would choose and why. Encourage each team of students to feed back its ideas to the whole class. Use this feedback to contrast and compare the thoughts of the different groups and generate a 'class' approach to the question by modifying one of the methods to take all the positive features from each group.
- Split the class into three groups. Each group should be given one additive from the following list:
  - water softening agents, e.g. sodium carbonate
  - chlorine
  - fluoride.
- Each group should use secondary sources of information, including the Student Book, to find why this chemical may be added to water. They should also discover the advantages and disadvantages of each additive. They should consider where they are getting their information from and whether the source is reliable and credible. Each group should then decide if it is for or against this additive to water.
- Encourage each group to report its thoughts back to the whole class, including a conclusion where it says if it is for or against the use of this additive in water.

Plenaries

Advantage or disadvantage – Give students a yellow (representing fluoride), green (representing chlorine) and white (representing water softening) card. On one side of each card write 'advantage' and on the reverse write 'disadvantage'. Read out statements and students have to hold up the appropriate card. For example, for the statement 'chlorine kills microorganisms that live in water' students would hold up the green card with the advantage sign facing the teacher. *(5 minutes)*

AfL (Assessment for Learning) – Give the students an examination question with three fictitious student answers. Ask the students to work in small groups and put the answers in order from weakest to most comprehensive. Students should think about why they have put the answers in that order. Each group feeds back about which they feel is the order of the answers and why. The appropriate tier of entry could be chosen based on the ability of the students in the class. *(10 minutes)*

Further teaching suggestions

Other sources of fluoride

- Students could find out other methods of taking fluoride other than via drinking water.

Debate

- You could have a debate on the motion 'Fluoride should be added to drinking water'. Split the class into two groups, one arguing for and the other against. Allow students to prepare their arguments and then hold the debate.

Card sort

- You may wish to give students a list of advantages and disadvantages of each of the types of additives to water as a card sort. Students could then group them for each additive. They could then write a summary paragraph for each additive into their notes. The summary should include whether they think the additive should be used with water.

Soap titration

- Using a simple titration set-up, a monitored amount of soap solution can be added to a sample of water. By shaking the sample and looking for a permanent foam, this method can be used to test for the effectiveness of commercial water-softeners. Take a measured sample of hard water and then add the manufacturer's recommended amount to the water and measure the volume of soap solution needed for a lather.

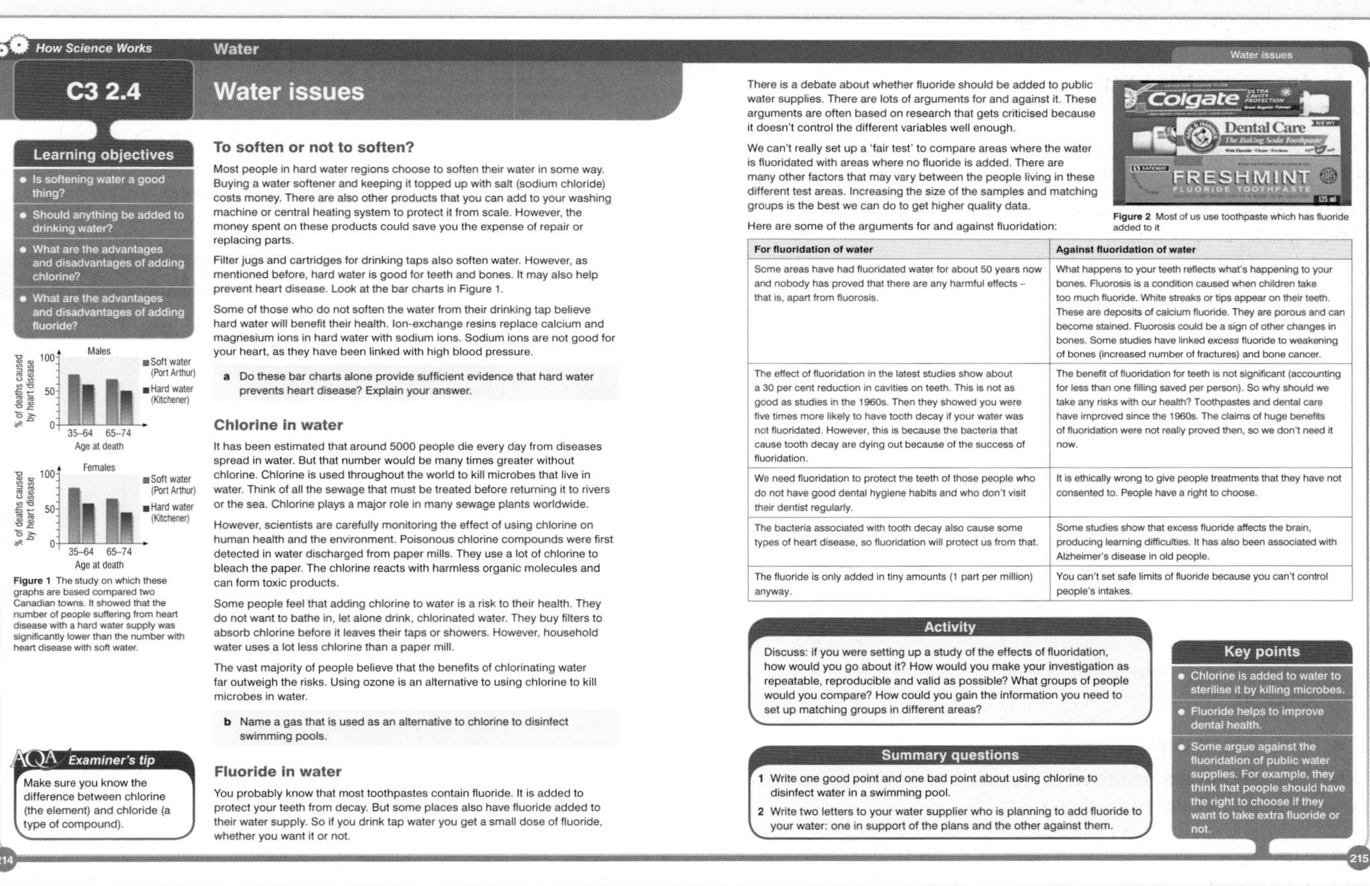

How Science Works Water

C3 2.4 Water issues

Learning objectives

- Is softening water a good thing?
- Should anything be added to drinking water?
- What are the advantages and disadvantages of adding chlorine?
- What are the advantages and disadvantages of adding fluoride?

To soften or not to soften?

Most people in hard water regions choose to soften their water in some way. Buying a water softener and keeping it topped up with salt (sodium chloride) costs money. There are also other products that you can add to your washing machine or central heating system to protect it from scale. However, the money spent on these products could save you the expense of repair or replacing parts.

Filter jugs and cartridges for drinking taps also soften water. However, as mentioned before, hard water is good for teeth and bones. It may also help prevent heart disease. Look at the bar charts in Figure 1.

Some of those who do not soften the water from their drinking tap believe hard water will benefit their health. Ion-exchange resins replace calcium and magnesium ions in hard water with sodium ions. Sodium ions are not good for your heart, as they have been linked with high blood pressure.

a Do these bar charts alone provide sufficient evidence that hard water prevents heart disease? Explain your answer.

Figure 1 The study on which these graphs are based compared two Canadian towns. It showed that the number of people suffering from heart disease with a hard water supply was significantly lower than the number with heart disease with soft water.

Chlorine in water

It has been estimated that around 5000 people die every day from diseases spread in water. But that number would be many times greater without chlorine. Chlorine is used throughout the world to kill microbes that live in water. Think of all the sewage that must be treated before returning it to rivers or the sea. Chlorine plays a major role in many sewage plants worldwide.

However, scientists are carefully monitoring the effect of using chlorine on human health and the environment. Poisonous chlorine compounds were first detected in water discharged from paper mills. They use a lot of chlorine to bleach the paper. The chlorine reacts with harmless organic molecules and can form toxic products.

Some people feel that adding chlorine to water is a risk to their health. They do not want to bathe in, let alone drink, chlorinated water. They buy filters to absorb chlorine before it leaves their taps or showers. However, household water uses a lot less chlorine than a paper mill.

The vast majority of people believe that the benefits of chlorinating water far outweigh the risks. Using ozone is an alternative to using chlorine to kill microbes in water.

b Name a gas that is used as an alternative to chlorine to disinfect swimming pools.

AQA Examiner's tip

Make sure you know the difference between chlorine (the element) and chloride (a type of compound).

Fluoride in water

You probably know that most toothpastes contain fluoride. It is added to protect your teeth from decay. But some places also have fluoride added to their water supply. So if you drink tap water you get a small dose of fluoride, whether you want it or not.

214

Water issues

There is a debate about whether fluoride should be added to public water supplies. There are lots of arguments for and against it. These arguments are often based on research that gets criticised because it doesn't control the different variables well enough.

We can't really set up a 'fair test' to compare areas where the water is fluoridated with areas where no fluoride is added. There are many other factors that may vary between the people living in these different test areas. Increasing the size of the samples and matching groups is the best we can do to get higher quality data.

Figure 2 Most of us use toothpaste which has fluoride added to it

Here are some of the arguments for and against fluoridation:

For fluoridation of water	Against fluoridation of water
Some areas have had fluoridated water for about 50 years now and nobody has proved that there are any harmful effects – that is, apart from fluorosis.	What happens to your teeth reflects what's happening to your bones. Fluorosis is a condition caused when children take too much fluoride. White streaks or tips appear on their teeth. These are deposits of calcium fluoride. They are porous and can become stained. Fluorosis could be a sign of other changes in bones. Some studies have linked *excess* fluoride to weakening of bones (increased number of fractures) and bone cancer.
The effect of fluoridation in the latest studies show about a 30 per cent reduction in cavities on teeth. This is not as good as studies in the 1960s. Then they showed you were five times more likely to have tooth decay if your water was not fluoridated. However, this is because the bacteria that cause tooth decay are dying out because of the success of fluoridation.	The benefit of fluoridation for teeth is not significant (accounting for less than one filling saved per person). So why should we take any risks with our health? Toothpastes and dental care have improved since the 1960s. The claims of huge benefits of fluoridation were not really proved then, so we don't need it now.
We need fluoridation to protect the teeth of those people who do not have good dental hygiene habits and who don't visit their dentist regularly.	It is ethically wrong to give people treatments that they have not consented to. People have a right to choose.
The bacteria associated with tooth decay also cause some types of heart disease, so fluoridation will protect us from that.	Some studies show that excess fluoride affects the brain, producing learning difficulties. It has also been associated with Alzheimer's disease in old people.
The fluoride is only added in tiny amounts (1 part per million) anyway.	You can't set safe limits of fluoride because you can't control people's intakes.

Activity

Discuss: if you were setting up a study of the effects of fluoridation, how would you go about it? How would you make your investigation as repeatable, reproducible and valid as possible? What groups of people would you compare? How could you gain the information you need to set up matching groups in different areas?

Summary questions

1. Write one good point and one bad point about using chlorine to disinfect water in a swimming pool.
2. Write two letters to your water supplier who is planning to add fluoride to your water: one in support of the plans and the other against them.

Key points

- Chlorine is added to water to sterilise it by killing microbes.
- Fluoride helps to improve dental health.
- Some argue against the fluoridation of public water supplies. For example, they think that people should have the right to choose if they want to take extra fluoride or not.

215

Answers to in-text questions

a No, we would need to know that other factors that might cause heart disease were similar in the two towns.

b Ozone.

Summary answers

1. For example: advantage – kills microorganism/bacteria that cause disease.
   Disadvantage – can form toxic products when discharged into environment.
2. Letters based on the points raised in the table on the second side of the spread.

Summary answers

1 a Water that is difficult to form a lather with soap. A scum is formed with soap and scale forms when hard water is boiled.

b For example, wastes soap, scale forms in kettles/hot water systems.

c For example, good for teeth and bones, prevents heart disease.

d A cleaning agent that does not form a scum with hard water.

2 a Put soap solution in the burette; titrate against equal volumes of hard water that has been passed through each of the three resins; record when enough soap has been added to form a permanent lather; repeat as necessary to obtain a constant reading for each; see which requires the least volume of soap to make a permanent lather.

b **i** type of resin
ii volume of soap solution needed

c It is difficult to decide exactly when a permanent lather has formed.

d Repeat the tests until you get two results that agree/ Repeat using an alternative method/Check your results against other groups doing the same investigation.

e Look at the spread of the repeat results for each resin tested.

f According to the class results, C was the least effective softener but we could not decide whether A or B was the most effective.

3 a Na_2CO_3

b **i** One of: calcium or magnesium.

ii

$Ca^{2+}(aq)$ or $Mg^{2+}(aq)$ ('hardness') + $CO_3^{2-}(aq)$ (from sodium carbonate) → $CaCO_3(s)$ (calcium carbonate) or $MgCO_3(s)$ (magnesium carbonate)

Hardness is removed as calcium/magnesium ions are precipitated out from the solution by the carbonate ions from the sodium carbonate.

4 a Temporary hardness can be removed by heating, but permanent hardness cannot.

b The calcium or magnesium ions in the hard water are exchanged for sodium (or hydrogen) ions in the ion exchange resin, as the water passes through the column.

5 a $Ca^{2+}(aq)$ and $HCO_3^-(aq)$.

b The aqueous calcium ions are removed from the water as solid calcium carbonate (scale).

6 a The water boils at a lower temperature, which reduces the energy costs of heating.

b Salty water causes iron to rust more quickly than fresh water.

c No energy is needed to heat the water.

7 a Balanced argument based on the points raised in the table on the second page of C3 2.4 Water issues.

b For example, advantage – kills microorganism/bacteria that cause disease; disadvantage – can form toxic products when discharged into environment/costs more money.

Water: C3 2.1–C3 2.4

Summary questions 

1 a What do we mean by 'hard water'?
b Give two disadvantages of hard water.
c Give two advantages of hard water.
d What is a 'soapless detergent'?

2 a Commercial water softeners contain an ion-exchange resin. Describe how you could use the following equipment to plan an investigation to see which of three different resins is most effective at softening hard water.
Ensure you describe how to make it a fair test.

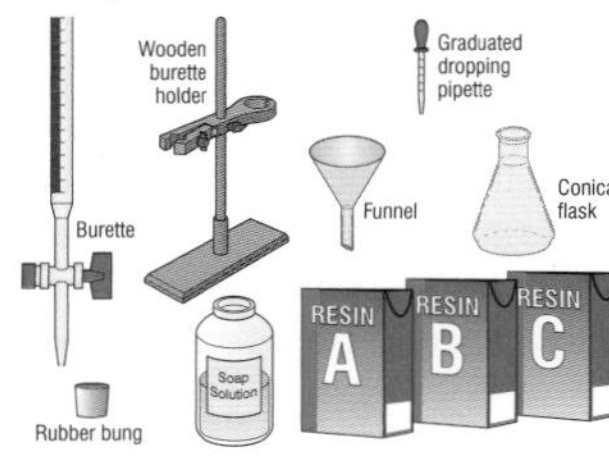

b i Name the independent variable in your investigation.
ii Name the dependent variable.
c Why is it difficult to judge which resin is most effective?
d How can you plan to get results that are as reproducible and repeatable as possible?
e How can you judge the precision of the data you collect?
f A class of students, working in 10 groups, did this investigation. Five groups found the order of effectiveness was A, B then C. The other five groups got B, A then C. Write a conclusion using the class results.

3 Washing soda is used to soften water, making it easier to get clothes clean. It consists of crystals of sodium carbonate.
a What is the formula of sodium carbonate?
b i Name one ion that makes water hard.
ii Write a balanced symbol equation for the reaction of washing soda with this ion, showing how the washing soda makes the water soft. [H]

4 a What is the difference between permanent and temporary hard water?
b Explain how an ion-exchange column can be used t soften hard water.

5 a Which ions are present in the temporary hard water formed when water passes over limestone rock?
b Explain how temporary hard water is softened in a kettle. Include balanced symbol equations.

6 In some hot countries getting sufficient fresh water is difficult. Some have large coastlines and so have plenty of seawater. Here they construct desalination plants. These use a process called 'flash distillation' to turn the salty water into drinking water. Inside the desalination plant water is boiled under reduced pressure, then cooled and condensed. There are even plans to build one in London to help cope in times of water shortage.

a Why is the pressure reduced before boiling the wate How does this keep costs down?
b Rusting of the steel vessels and pipework in the desalination plant is a big problem. Why?
An alternative process uses 'reverse osmosis' to remov the salts from seawater. This passes seawater through a membrane. The latest membranes can remove 98 pe cent of the salts from seawater.
c Why is reverse osmosis a better option than flash distillation for obtaining drinking water?

7 a What are your views on the fluoridation of public wat supplies? Write a balanced argument and state your conclusions.
b State an advantage and a disadvantage of chlorinati our water supplies.

216

Practical suggestions

Practicals	AQA	k	Book	Gears
Investigation of which ions cause hard water, e.g. adding soap solutions to solutions of NaCl, $CaCl_2$, KCl, and $MgCl_2$ with soap solution.	✓		✓	
Making temporary hard water by adding excess carbon dioxide to limewater.	✓		✓	
Determining hardness of samples of water.	✓	✓	✓	
The removal of hardness.	✓	✓	✓	
Testing hard water before and after passing through an ion-exchange column.	✓		✓	
Using conductivity sensors to analyse different samples of hard and soft water.	✓		✓	
Design and carry out an investigation to compare the effectiveness of commercial water softeners using soap titration.	✓		✓	
Investigating the various types of water 'filters' that are commercially available.	✓		✓	
Distillation of seawater …	✓		✓	

Examination-style questions

ost water from natural sources contains dissolved ubstances.

hoose a word from each list to complete each entence.

Water for drinking is filtered to remove ………… . (1)

chlorides ions solids

Water for drinking should contain dissolved salts at levels that are ………… . (1)

high low variable

Water for drinking should contain no ………… . (1)

microbes molecules oxygen

Pure water can be produced by ………… . (1)

compression distillation electrolysis

ard water has some benefits but also causes problems.

Which of the following is a benefit of hard water?

A Bathing in hard water helps develop strong bones.

B Drinking hard water may help to reduce heart disease.

C Washing clothes in hard water uses less washing powder. (1)

Two problems of hard water are *scale* and *scum*, as shown in the pictures of a heating element and a wash basin.

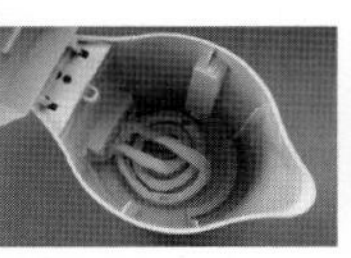

Describe as simply as possible the difference between *scale* and *scum*. (2)

Describe how hard water is made soft by passing it through an ion-exchange column. (2)

AQA, 2008

this question you will be assessed on using good nglish, organising information clearly and using pecialist terms where appropriate.

escribe how drinking water supplies should be selected nd treated and explain how these processes make the vater safe to drink. (6)

4 An advertising leaflet from a company that sells water softeners included the following information.
Everyone needs a water softener because hard water causes:
- damaged boilers
- burnt-out kettles and immersion heaters
- blocked showers
- greater costs for heating water
- extra soap for washing and bathing

Suggest **three** ways in which this information is biased. (3)

5 The table shows the concentrations of some ions in two samples of hard water.

Name and formula of ion	Concentration of ion in mg per litre	
	Sample A	Sample B
Sodium Na^+	12	7
Potassium K^+	2	3
Calcium Ca^{2+}	135	95
Magnesium Mg^{2+}	15	63
Chloride Cl^-	7	18
Hydrogencarbonate HCO_3^-	278	4
Nitrate NO_3^-	2	1
Sulfate SO_4^{2-}	6	165

Sample A is temporary hard water and sample B is permanent hard water.

a Which ions make these two samples of water hard? (2)

b Describe how you could test the samples to show that A is temporary hard water and B is permanent hard water. (4)

c Give **two** methods that would remove the hardness from both samples of water. (2)

d Use the data to explain why sample A is temporary hard water and sample B is permanent hard water. [H] (6)

Kerboodle resources

Resources available for this chapter on Kerboodle are:

- Chapter map: Water
- How Science Works: Investigating the hardness of water samples (C3 2.1)
- Extension: Removing hardness (C3 2.2)
- Practical: Hard and soft water (C3 2.2)
- Support: Safe to drink (C3 2.3)
- Viewpoints: Fluoridation of water (C3 2.4)
- Interactive activity: The water we drink
- Revision podcast: Hard and soft water
- Test yourself: Water
- On your marks: Water treatment
- Examination-style questions: Water
- Answers to examination-style questions: Water

Examination-style answers

1 a solids *(1 mark)*

b low *(1 mark)*

c microbes *(1 mark)*

d distillation *(1 mark)*

2 a B – drinking hard water helps to reduce heart disease. *(1 mark)*

b Scale is produced by heating (temporary hard) water, scum is produced by soap and hard water. *(2 marks)*

c Sodium ions from the resin go into the water/solution, (replacing) calcium and/or magnesium ions from the water/solution (that go into the resin). *(2 marks)*

3 Marks awarded for this answer will be determined by the quality of written communication (QWC) as well as the standard of the scientific response.

There is a clear and detailed scientific description of how water supplies should be selected and treated and an explanation of how these processes make the water safe to drink. The answer shows almost faultless spelling, punctuation and grammar. It is coherent and in an organised, logical sequence. It contains a range of appropriate and relevant specialist terms used accurately. *(5–6 marks)*

There is a scientific description of the selection and treatment of water supplies and some explanation of how these processes make the water safe to drink. There are some errors in spelling, punctuation and grammar. The answer has some structure and organisation. The use of specialist terms has been attempted, but not always accurately. *(3–4 marks)*

There is a brief description of the selection and/or treatment of water supplies and/or an explanation of how these processes make the water safe to drink. The spelling, punctuation and grammar are very weak. The answer is poorly organised with almost no specialist terms and/or their use demonstrating a general lack of understanding of their meaning. *(1–2 marks)*

No relevant content. *(0 marks)*

Examples of chemistry points made in the response:

- Sources include rivers, lakes, wells, boreholes, springs.
- Source chosen for low levels of dissolved salts, microbes, solids, pollution, toxic substances.
- Colour and/or turbidity and/or odour indicate contamination/impurities.
- May be left to settle/clear.
- May be treated with flocculating agents.
- Solids/plants/algae/animals/water life removed by filtering, passing through filter beds, passing through rocks.
- Filtering may not remove microorganisms.
- Treated with chlorine, UV light, ozone to kill microbes.
- pH should be about 7–8, pH checked/adjusted.
- High levels of salt(s) are harmful to health.
- Microbes may be harmful/pathogens.
- Drinking water should be colourless, odourless and have low levels of salts and microbes.

4 Any three from: not everyone has hard water; not everyone has temporary hard water; only temporary hard water produces scale; only THW damages boilers, kettles, immersion heaters, showers; only electric showers affected; only THW increases heating costs; many people use soapless detergents for washing/bathing; any other scientifically valid point. *(3 marks)*

5 a Ca^{2+}, Mg^{2+} *(2 marks)*

b Find volume/number of drops of soap solution to produce lather with both solutions; boil both solutions; find volume of soap to produce lather after boiling; volume/amount of soap solution for A is less after boiling and B remains the same. *(4 marks)*

c Add sodium carbonate/washing soda; use ion exchange (resin/column); (allow distillation). *(2 marks)*

d A has high concentration/amount of Ca^{2+} (and Mg^{2+}), high concentration of HCO_3^-; HCO_3^- decomposes on heating; $CaCO_3$ precipitates; Ca^{2+} (and Mg^{2+}) removed from solution/water becomes soft (max. 3), B has low concentration/amount HCO_3^-; high concentration/amount Ca^{2+} (and Mg^{2+}); high conc/amount SO_4^- (and Cl^-); these ions do not decompose on heating (max. 3). *(6 marks)*

C3 3.1 Comparing the energy released by fuels

Learning objectives

Students should learn:

- how the energy released by different fuels can be measured
- the unit of energy.

Learning outcomes

Most students should be able to:

- describe a method for measuring the energy in fuels
- state that food contains energy
- recall the unit of energy
- calculate the energy content of a fuel with some support.

Some students should also be able to:

- calculate the energy content of a fuel independently.

Answers to in-text questions

a As a source of energy.
b Same quantity of water to be heated, starting at the same temperature for the same time, known quantity of fuel.

Support

- Some students may find it difficult to read scales on a thermometer. These students could be supported in the calorimetry practical by using a temperature probe and data logger to monitor the temperature of the reaction.

Extend

- In the calorimetry practical, encourage students to take the mass of the spirit burner before and after it has been lit in order to calculate the mass of fuel that has been used. Then see if they can work out the energy released by the fuel in joules per gram.

  The formula:
  energy (J/g) = (50 × 4.2 × change in temperature (°C))/mass of fuel (g), will give the amount of energy given per gram of fuel. This may give a different order of fuels to the amount of energy given out in a set amount of time.
- Students could be given other calculations to complete in order to compare fuels and foods.

Specification link-up: Chemistry C3.3

- The relative amounts of energy released when substances burn can be measured by simple calorimetry, e.g. by heating water in a glass or metal container. This method can be used to compare the amount of energy released by fuels and foods. *[C3.3.1 a)]*
- Energy is normally measured in joules (J). *[C3.3.1 b)]*

  Controlled Assessment: C4.3 Collect primary and secondary data. *[C4.3.2 a) c) e)]*

Lesson structure

Starters

5, 4, 3, 2, 1 – Ask students to list five fuels, four high-energy foods, three units to measure temperature, two products of the combustion of a hydrocarbon and one gas needed for combustion. *(5 minutes)*

Definitions – Read out the following definitions. The students' task is to give the correct key word that they relate to.

- The scientific word for burning. [combustion]
- The unit for measuring energy. [joule]
- A chemical change that involves energy being released. [exothermic]
- A chemical change that involves the gain of oxygen, loss of hydrogen or loss of electrons. [oxidation]
- A chemical that is burnt to release usable energy. [fuel]

You can support students by giving a list of the words to match to the definition. Extend students by asking them to write a paragraph that uses all the key words in the correct context at least once. *(10 minutes)*

Main

- Show students the equipment for calorimetry and ask them to consider what is the independent variable [fuel] and dependent variable [temperature]. As a class, using question-and-answer, generate the values for the control variables, e.g. amount of water (50 cm^3), distance from flame to calorimeter (5 cm), time (2 min). (This relates to: 'How Science Works': types of variable and fair testing.)
- Allow the students to compare the energy content of two different fuels experimentally. By measuring the mass of the spirit burner before and after the experiment, they can calculate the energy released by that mass of fuel and use the data to determine the amount of energy released by 1 g of that fuel. Students could check the accuracy of their results by comparing their value with reported values.
- Give students a diagram of a bomb calorimeter. They should then add labels to explain how it can be used to monitor energy changes in a chemical reaction, e.g. combustion. It is not necessary for students to remember this for their examination.
- Give students the equation for calculating the heat energy given out by an experiment. Encourage students to annotate the equation with the units of each term in red and the name of each term in blue. Go through the worked example in the Student Book and ask students to write a step-by-step flow chart showing how to use data from an experiment to work out the energy change in the experiment.
- You may wish to give students some past paper questions and allow them to work in groups to answer them.

Plenaries

Card sort – Provide the students with a set of images that, when they are sorted into the correct order, produce a pictorial description of the method of calorimetry. Ask the students to work in pairs to sort the images into the correct order. *(5 minutes)*

Evaluation – Ask students to reflect on the calorimetry experiment that they have completed. Tell them that the experiment is not very accurate, then ask why this is so. [There are a number of reasons, which could include 'a lot of energy is lost to the surroundings' or 'the can is heated but this is not taken into account in the calculation'.] Ask students to work in pairs to generate amendments to the equipment to minimise this problem [e.g. insulation – around the calorimeter, lid on the calorimeter, use shields to reduce draughts]. Then ask each group to feed back its ideas to the class. Support students by giving them a selection of statements to choose from. Extend students by asking them to comment on the accuracy and precision of their measurements and the repeatability and reproducibility of their results. *(10 minutes)*

Practical support

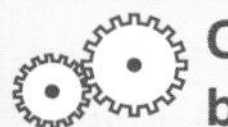

Comparing the energy released when fuels burn

Equipment and materials required

Glass beaker/metal can/calorimeter, tripod, gauze, flameproof mat, spirit burner, paraffin (harmful), ethanol (harmful/highly flammable), thermometer, measuring cylinder, water, stopwatch, top-pan balance, eye protection.

Details

Using the top-pan balance, measure the mass of the spirit burner. Measure 50 cm^3 of water into a calorimeter and place on the tripod. Take the temperature of the water. Light the spirit burner and put it on the flameproof mat, positioned under the calorimeter. Time for 2 min, then remove the burner and take the temperature of the water and the mass of the spirit burner again.

Safety: Know how to extinguish a bench fire. See CLEAPSS Handbook CD-ROM section 9.4.3 (spirit burners).
CLEAPSS Hazcard 40A Ethanol – harmful/highly flammable;
CLEAPSS Hazcard 45B Paraffin – harmful.

Energy calculations

C3 3.1 Comparing the energy released by fuels

Learning objectives

- How can we measure the energy released by different fuels?
- What is the unit of energy?

links

For more information on exothermic reactions, look back at C2 4.7 Exothermic and endothermic reactions.

Figure 1 Keeping warm or moving about – we need exothermic reactions

We have already seen that exothermic reactions release energy. When we burn a fuel, we use this energy to keep ourselves warm or we use it for transport.

a Why do we use fuels?

Not all fuels release the same amount of energy when they burn. Some combustion reactions are more exothermic than others. It is often very important to know how much energy a fuel releases when it burns.

In a school chemistry lab it is very difficult to measure accurately the energy released by fuels when they burn.

The actual amount of energy released by a burning fuel in an experiment is related to the rise in temperature of the water in a calorimeter. The larger the rise in temperature, the more energy has been released (see Figure 2).

Maths skills

Calculating the energy released

There is an equation we can use to work out the energy released, measured in joules (J):

energy released = mass of water heated × specific heat capacity of water × rise in temperature

This is sometimes written as:

$$Q = mc\Delta T$$

Where:
Q is the energy released by the fuel
m is the mass of water heated in the calorimeter (1 cm^3 of water has a mass of 1 g)
c is the specific heat capacity of water (this is the amount of energy needed to raise 1 g of water by 1 °C)
ΔT is the rise in temperature (i.e. the final temperature of the water minus the initial temperature).

Here is an example of how to do the calculation:

Worked example

In an experiment a burning fuel raised the temperature of 50 cm^3 of water from 16 °C to 41 °C. How much energy was released by the fuel? (4.2 J of energy raise the temperature of 1 g of water by 1 °C).

Write down the data given:
Mass of water heated in the calorimeter = 50 g
Specific heat capacity of water = 4.2 J/g °C
Rise in temperature of the water = 41 °C − 16 °C = 25 °C

Now use the data in the $Q = mc\Delta T$ equation (which will always be provided for you in the exam):

$$Q = mc\Delta T$$

$Q = 50 \times 4.2 \times 25$ J
$Q =$ **5250 J (or 5.25 kJ)**

It is useful to be able to compare the energy content of different fuels. Their combustion data is given as the number of joules (or kilojoules) of energy released per gram or per mole of fuel burned.

In the previous experiment, the fuel was weighed before and after burning. It was found that the mass had decreased by 0.2 g. So the energy given out in the experiment will be multiplied by the number of 0.2 g there are in 1 g.

$$\text{energy released per gram} = 5.25 \times \frac{1}{0.2}\text{ kJ/g} = \mathbf{26.25\ kJ/g}$$

Suppose we know that the relative formula mass of the fuel is 46. We can now also work out the energy released per mole. One mole of the fuel has a mass of 46 g. So the energy given out in the experiment will be multiplied by the number of 0.2 g there are in 46 g:

$$\text{energy released per mole} = 5.25 \times \frac{46}{0.2}\text{ kJ/mol} = \mathbf{1207.5\ kJ/mol}$$

Figure 2 The energy released by fuels and foods when they burn can be compared using a simple calorimeter. The energy content of food is sometimes given in units called calories. However, the joule is the usual scientific unit of energy.

Practical

Comparing the energy released when fuels burn

We can compare the energy released by different fuels when they burn. The fuels are burned to heat water in a copper can or a glass beaker. We measure the temperature changes produced by different fuels, so then we can compare the energy they release when they burn. Use the worked example to help with your calculations.

b What variables must we control in order to compare the energy released by different fuels?

AQA Examiner's tip

Measurements using simple calorimeters are not accurate because of energy losses, but they can be used to compare the amounts of energy released.

Summary questions

1 Copy and complete using the words below:

calorimeter energy exothermic oxygen

When fuels burn they react with .... This reaction releases .... The reaction is .... The energy content of foods and fuels can be measured using a ....

2 A simple calorimeter was used to compare the energy released by three different fuels. The results were as follows:

	Mass of fuel burned (g)	Volume of water (cm^3)	Temperature change (°C)
Fuel A	0.24	160	10.0
Fuel B	0.18	100	8.0
Fuel C	0.27	150	9.0

a Using the equation $Q = m\,c\,\Delta T$, calculate the energy released by each fuel in the three tests. (4.2 J of energy raise the temperature of 1 g of water by 1 °C.)
b Calculate the energy released per gram for each fuel.
c Arrange the fuels in order of the amount of energy they release per gram.
d The relative formula mass of Fuel A is 48. What in the energy released by A in kilojoules per mole (kJ/mol)?

Key points

- When fuels and food react with oxygen, energy is released in an exothermic reaction.
- A simple calorimeter can be used to compare the energy released by different fuels or different foods in a school chemistry lab.

Further teaching suggestions

Graph of combustion data

- Give the students data for the amount of energy released by burning different alkanes. Encourage them to show this in a graph (number of carbon atoms, energy released) and draw conclusions from this.

Comparing units

- Students could discover all the different units for measuring energy and how they relate to each other. They could then change the energy content data of fuels into different units.

Summary answers

1 oxygen, energy, exothermic, calorimeter

2 **a** A: 6720 J (6.72 kJ)
B: 3360 J (3.36 kJ)
C: 5670 J (5.67 kJ)

b A: 28 000 J/g (28 kJ/g)
B: 18 670 J/g (18.67 kJ/g)
C: 21 000 J/g (21 kJ/g)

c A > C > B

d 1344 kJ/mol

C3 3.2 Energy transfers in solutions

Learning objectives

Students should learn:

- how to measure energy changes in solution.

Learning outcomes

Most students should be able to:

- calculate simple energy changes in solution.

Some students should also be able to:

- use data from experiments to calculate energy changes in solution in kJ/mol.

Specification link-up: Chemistry C3.3

- The amount of energy produced by a chemical reaction in solution can be calculated from the measured temperature change of the solution when the reagents are mixed in an insulated container. This method can be used for reactions of solids with water or for neutralisation reactions. *[C3.3.1 c)]*

Lesson structure

Starters

Measuring temperature – Ask students to suggest how you could measure temperature in a science experiment [e.g. spirit thermometer, mercury thermometer, temperature probe, data logger, thermochromic strip, thermocouple]. You may wish to support students by having a tray of different pieces of equipment that they can look at. You could extend students by asking them to suggest the most appropriate piece of temperature-measuring equipment for different experiments. *(5 minutes)*

Calculations – Give students the formula:

$$Q = m\,c\,\Delta T$$

Ask students to define each term and its common units.

[energy (J) = mass (g) × specific heat capacity (J/g °C) × change in temperature (°C)]

Then ask students to calculate the energy released when a liquid hydrocarbon fuel is burnt for 5 min and heats 50 cm^3 of water by 12 °C [2520 J]. *(10 minutes)*

Main

- Create a circus of different experiments where students will use a coffee cup calorimeter to determine the temperature change of the reaction. See 'Practical support' for more details.
- Set up each station with an information card including a method for the experiment. These could include a writing frame so that students can calculate the energy released in the reaction. The information cards could be laminated and then students could use a whiteboard pen to complete their working out on the card before recording it in their notes. Students can then erase their working for the next team.
- Show students a revision guide. They then work in small groups to make a page for a revision guide about this topic. Their work should include the key points, a worked example and an exam-style question to check that students have read and understood their work. On the reverse of their page, they could include the answer to their exam question. The best pages could be photocopied and used for homework.

Plenaries

Evaluation – Encourage students to reflect on their practical and to comment on the reliability and accuracy of their results. *(5 minutes)*

AfL (Assessment for Learning) – Give the students an examination question on this topic. Using a stopwatch, time each section (a minute a mark) and give the students time checks for each section of the question. Once all the time has been used, give the students the mark scheme and ask them to mark their own work in a different colour and jot down any amendments. Either Foundation or Higher Tier questions can be used depending on the tier of entry of the students. *(10 minutes)*

Support

- You could consider only neutralisation as this is a reaction that students are familiar with from KS3 and C2. Students could also use a data logger to monitor the temperature changes as these are often quite small and difficult to determine.

Extend

- Encourage students to draw an energy-level diagram for each of the reactions that they are studying. They could also put the energy change that they have calculated on to their diagram.

Practical support

Measuring energy changes in reactions

Equipment and materials required

Polystyrene cup with lid and two holes in the top, −10 to 110 °C spirit thermometer, stirring rod, 20 cm^3 measuring cylinder, spatula, iron filings, magnesium ribbon, potassium nitrate powder, anhydrous copper sulfate powder, dilute (<2 mol/dm^{-3}) hydrochloric acid, dilute (<0.5 mol/dm^3) sodium hydroxide, eye protection.

Details

Iron filings and copper sulfate – Measure 20 cm^3 of copper sulfate solution into the coffee cup. Take and record the initial temperature. Put a spatula of iron filings into the coffee cup. Quickly replace the lid and push through the thermometer and stirring rod into different holes. Stir and monitor the temperature, recording the highest temperature.

Magnesium ribbon and hydrochloric acid – Measure 20 cm^3 of hydrochloric acid into the coffee cup. Take and record the initial temperature. Put a small coiled piece of magnesium ribbon into the coffee cup. Quickly replace the lid and push through the thermometer and stirring rod into different holes. Stir and monitor the temperature, recording the highest temperature.

Sodium hydroxide and hydrochloric acid – Measure 20 cm^3 of sodium hydroxide solution into the coffee cup. Take and record the initial temperature. Add 20 cm^3 of hydrochloric acid into the coffee cup. Quickly replace the lid and push through the thermometer and stirring rod into different holes. Stir and monitor the temperature, recording the highest temperature.

Potassium nitrate and water – Measure 20 cm^3 of water into the coffee cup. Take and record the initial temperature. Put a spatula of potassium nitrate into the coffee cup. Quickly replace the lid and push through the thermometer and stirring rod into different holes. Stir and monitor the temperature, recording the highest temperature.

Safety: CLEAPSS Hazcard 27C Copper sulfate – harmful; CLEAPSS Hazcard 47A Hydrochloric acid – corrosive; CLEAPSS Hazcard 91 Sodium hydroxide – corrosive; CLEAPSS Hazcard 82 Potassium nitrate – oxidising.

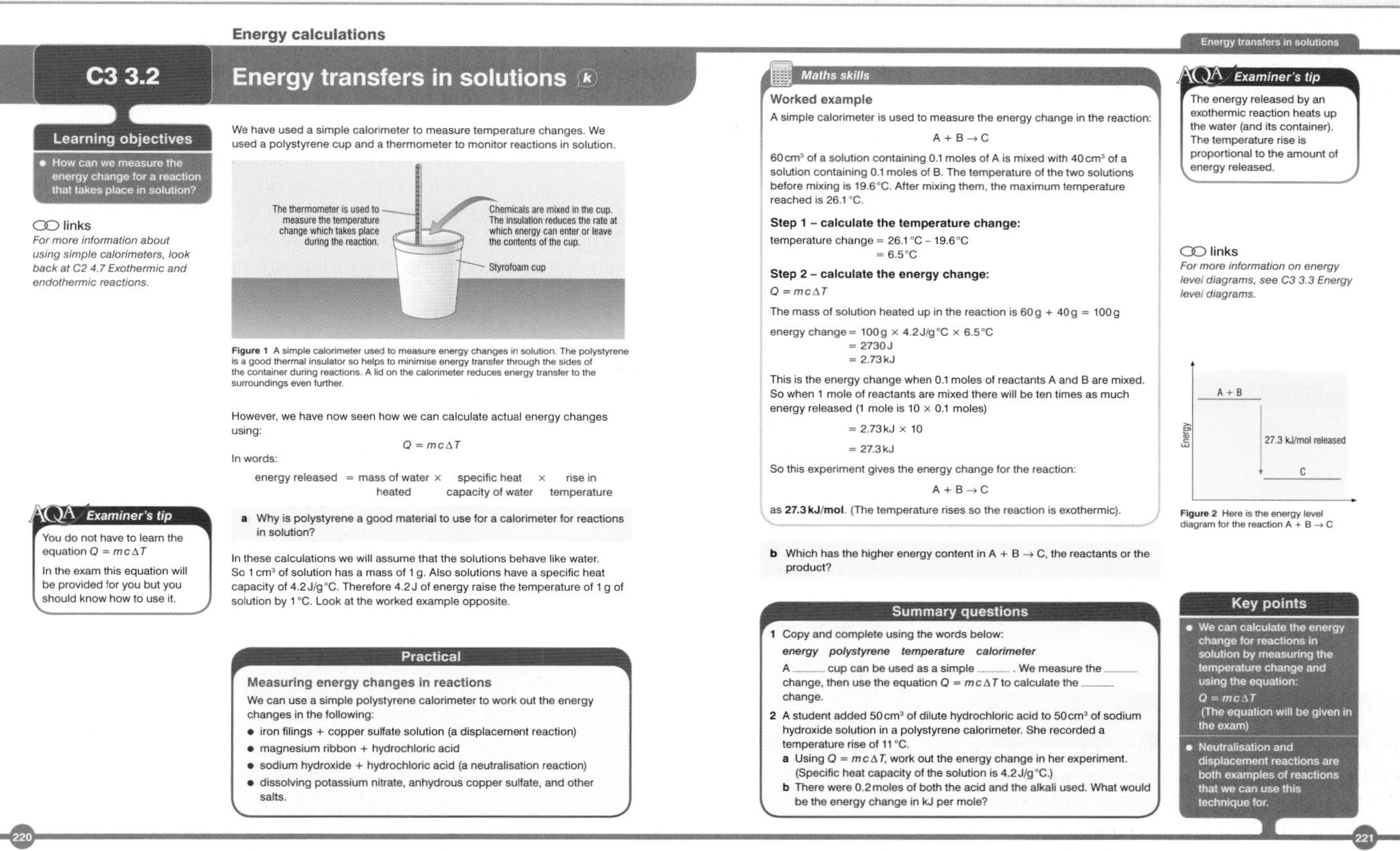

Energy calculations

C3 3.2 Energy transfers in solutions

Learning objectives

- How can we measure the energy change for a reaction that takes place in solution?

links
For more information about using simple calorimeters, look back at C2 4.7 Exothermic and endothermic reactions.

We have used a simple calorimeter to measure temperature changes. We used a polystyrene cup and a thermometer to monitor reactions in solution.

The thermometer is used to measure the temperature change which takes place during the reaction.
Chemicals are mixed in the cup. The insulation reduces the rate at which energy can enter or leave the contents of the cup.
Styrofoam cup

Figure 1 A simple calorimeter used to measure energy changes in solution. The polystyrene is a good thermal insulator so helps to minimise energy transfer through the sides of the container during reactions. A lid on the calorimeter reduces energy transfer to the surroundings even further.

However, we have now seen how we can calculate actual energy changes using:

$$Q = mc\Delta T$$

In words:

energy released = mass of water heated × specific heat capacity of water × rise in temperature

a Why is polystyrene a good material to use for a calorimeter for reactions in solution?

AQA Examiner's tip
You do not have to learn the equation $Q = mc\Delta T$
In the exam this equation will be provided for you but you should know how to use it.

In these calculations we will assume that the solutions behave like water. So 1 cm^3 of solution has a mass of 1 g. Also solutions have a specific heat capacity of 4.2 J/g °C. Therefore 4.2 J of energy raise the temperature of 1 g of solution by 1 °C. Look at the worked example opposite.

Practical

Measuring energy changes in reactions
We can use a simple polystyrene calorimeter to work out the energy changes in the following:
- iron filings + copper sulfate solution (a displacement reaction)
- magnesium ribbon + hydrochloric acid
- sodium hydroxide + hydrochloric acid (a neutralisation reaction)
- dissolving potassium nitrate, anhydrous copper sulfate, and other salts.

220

Energy transfers in solutions

Maths skills

Worked example
A simple calorimeter is used to measure the energy change in the reaction:

$$A + B \rightarrow C$$

60 cm^3 of a solution containing 0.1 moles of A is mixed with 40 cm^3 of a solution containing 0.1 moles of B. The temperature of the two solutions before mixing is 19.6 °C. After mixing them, the maximum temperature reached is 26.1 °C.

Step 1 – calculate the temperature change:
temperature change = 26.1 °C – 19.6 °C
= 6.5 °C

Step 2 – calculate the energy change:
$Q = mc\Delta T$
The mass of solution heated up in the reaction is 60 g + 40 g = 100 g
energy change = 100 g × 4.2 J/g °C × 6.5 °C
= 2730 J
= 2.73 kJ

This is the energy change when 0.1 moles of reactants A and B are mixed. So when 1 mole of reactants are mixed there will be ten times as much energy released (1 mole is 10 × 0.1 moles)
= 2.73 kJ × 10
= 27.3 kJ

So this experiment gives the energy change for the reaction:

$$A + B \rightarrow C$$

as **27.3 kJ/mol**. (The temperature rises so the reaction is exothermic).

b Which has the higher energy content in A + B → C, the reactants or the product?

AQA Examiner's tip
The energy released by an exothermic reaction heats up the water (and its container). The temperature rise is proportional to the amount of energy released.

links
For more information on energy level diagrams, see C3 3.3 Energy level diagrams.

Energy
A + B
27.3 kJ/mol released
C

Figure 2 Here is the energy level diagram for the reaction A + B → C

Summary questions

1 Copy and complete using the words below:
energy polystyrene temperature calorimeter
A .... cup can be used as a simple .... We measure the .... change, then use the equation $Q = mc\Delta T$ to calculate the .... change.

2 A student added 50 cm^3 of dilute hydrochloric acid to 50 cm^3 of sodium hydroxide solution in a polystyrene calorimeter. She recorded a temperature rise of 11 °C.
a Using $Q = mc\Delta T$, work out the energy change in her experiment. (Specific heat capacity of the solution is 4.2 J/g °C.)
b There were 0.2 moles of both the acid and the alkali used. What would be the energy change in kJ per mole?

Key points
- We can calculate the energy change for reactions in solution by measuring the temperature change and using the equation:
$Q = mc\Delta T$
(The equation will be given in the exam)
- Neutralisation and displacement reactions are both examples of reactions that we can use this technique for.

221

Further teaching suggestions

Error explanation

- Students could be encouraged to evaluate the method and their results in order to explain any random and systematic errors that it contains.

Assessment for Learning

- You could give students examination questions and ask them to work in pairs to write a mark scheme rather than answer the question. This will encourage the students to think critically about the most acceptable answer, other answers that would be worthy of credit and answers that are not acceptable at all.

Answers to in-text questions

a It is a good thermal insulator (and it is waterproof).

b The reactants (A + B) since the reaction is exothermic.

Summary answers

1 polystyrene, calorimeter, temperature, energy

2 **a** $Q = 100 \times 4.2 \times 11 = 4620$ J (4.62 kJ)

b $4620 \times \frac{1}{0.2} = 23\,100$ J/mol (23.1 kJ/mol)

C3 3.3 Energy level diagrams

Learning objectives

Students should learn:

- what the energy changes are when bonds are made or broken
- what energy level diagrams show
- the effect of catalysts on the activation energy of a reaction.

Learning outcomes

Most students should be able to:

- complete a simple energy level diagram for a chemical reaction
- draw a simple energy level diagram for a chemical reaction
- show the effect of a catalyst on activation energy on an energy level diagram
- state and explain how we can think of reactions as having exothermic and endothermic stages
- recognise an exothermic and an endothermic change from an energy level diagram.

Some students should also be able to:

- explain in detail the effect of a catalyst on activation energy.

Answers to in-text questions

a Endothermic.

b Exothermic.

Support

- Provide students with the energy level diagrams and explanations about their shape on separate cards. Students can then match the explanation with the image and copy them into their notes.

Extend

- Allow students to use the internet to discover the energy change of a reaction and add this information on to their diagrams, in order to scale the differences in energies between the reactants and products. They could then use their diagrams to determine which reaction would release the most energy or take in most energy.

Specification link-up: Chemistry C3.3

- Simple energy level diagrams can be used to show the relative energies of reactants and products, the activation energy and the overall energy change of a reaction. *[C3.3.1 d)]*
- During a chemical reaction:
  - energy must be supplied to break bonds
  - energy is released when bonds are formed. *[C3.3.1 e)]*
- Catalysts provide a different pathway for a chemical reaction that has a lower activation energy. *[C3.3.1 h)]*
- Interpret simple energy level diagrams in terms of bond breaking and bond formation (including the idea of activation energy and the effect on this of catalysts). *[C3.3]*

Lesson structure

Starters

Odd one out – Show students a variety of pictures of exothermic reactions (combustion, sodium reacting with water, a carbonate reacting with acid, etc.) and one of a road gritter. Ask students to suggest which is the odd one out and why, [Road gritter, as this is a physical change and endothermic; all the other pictures are of chemical reactions and exothermic.] *(5 minutes)*

List – Ask students to brainstorm everything they know about exothermic and endothermic reactions. Then ask each student to come in turn and write one fact on to the board, but without repeating anything that is already there. Support students by encouraging them to work in small groups. Extend students by asking them to look at the finished list and comment on any amendments that they think are needed. *(10 minutes)*

Main

- Create a set of statements about a number of energy transfer diagrams, e.g. these statements relate to the energy level diagram for the formation of water from its elements:
  - Reactants have higher energy than the product.
  - The reactants are hydrogen and oxygen.
  - There is a relatively small activation energy.
  - The *y*-axis is energy.
  - Energy is measured in kJ/mol.
  - The product is water.

  Cut out each statement so that they are on separate cards. Give a pack of statements to pairs of students. They should use them to create and label the energy level diagram that it describes.
- Give the students a selection of different reactions and state whether they are exothermic or endothermic. They could then draw their own energy level diagrams to represent these reactions.
- Ask students to write a mnemonic or rhyme to help them remember that energy is used to break bonds and energy is released on making bonds.
- Revise the effect of catalysis on a chemical reaction and introduce the concept of providing an alternative reaction pathway that speeds up the reaction. You may wish to add these ideas to an energy level diagram, showing the reaction pathway with and without the use of a catalyst.

Plenaries

Diagram – Give the students an unlabelled energy level diagram for the combustion of carbon. They should then add the axis labels, detail which are the reactants and which is the product; annotate the activation energy and finally determine whether the reaction is exothermic or endothermic. *(5 minutes)*

Fill in the missing word – Give the students this passage, with some missing words. Encourage students to complete the prose, using the most scientific word that they can:

'An [energy level diagram] can be used to show how the energy stored in chemicals changes during a chemical reaction. In [endothermic] reactions the products have more energy than the [reactants]. In exothermic reactions the [products] have less energy than the [reactants].'

Support students by giving them the missing words to put into the prose. Extend students by encouraging them to illustrate each sentence with a labelled diagram showing the atoms in a chemical reaction. *(10 minutes)*

C3 3.3 Energy level diagrams

Learning objectives

- What do energy level diagrams show us?
- What effect do catalysts have on the activation energy of a reaction?
- What energy changes do we get when bonds are broken?
- What energy changes do we get when bonds are made?

We can find out more about what is happening in a particular reaction by looking at its energy level diagram. These diagrams show us the relative amounts of energy contained in the reactants and the products. This energy is measured in kilojoules per mole (kJ/mol).

Exothermic reactions

Figure 1 shows the energy level diagram for an exothermic reaction. The products are at a lower energy level than the reactants. Therefore when the reactants form the products, energy is released.

Reactants
Energy
Energy is released
Products

Figure 1 The energy level diagram for an **exothermic** reaction

The difference between the energy levels of the reactants and the products is the energy change during the reaction, measured in kJ/mol.

The difference in energy between the products and the reactants is released to the surroundings. Therefore in exothermic reactions the temperature of the surroundings increases. The surroundings get hotter.

Endothermic reactions

Figure 2 shows the energy level diagram for an endothermic reaction.

Products
Energy
Energy is absorbed
Reactants

Figure 2 The energy level diagram for an **endothermic** reaction.

Here the products are at a higher energy level than the reactants. As the reactants react to form products, energy is absorbed from the surroundings. The temperature of the surroundings decreases because energy is taken in during the reaction. The surroundings get colder. The products are at a higher energy level than the reactants.

Activation energy and catalysis

Think back to your work on rates of reaction; you learned about the collision theory of reactions. This stated that there is a minimum amount of energy needed before colliding particles can react. This energy needed to start a reaction is called the activation energy. We can show this on our energy level diagrams. Look at Figure 3.

Catalysts can increase the rate of a reaction. The way they do this is to provide an alternative pathway to the products, which has a lower activation energy. This means that a higher proportion of reactant particles now have enough energy to react. This is shown on the energy level diagram in Figure 4.

Energy
Reactants
Activation energy (the energy needed to start a reaction)
Products

Figure 3 The minimum amount of energy needed to start a reaction is called its activation energy

Energy
Reactants
Activation energy (the energy needed to start a reaction)
Lower activation energy using a catalyst
Products

Figure 4 A catalyst provides a different reaction pathway with a lower activation energy so that a higher proportion of reactant particles have enough energy to react

Bond breaking and bond making

Think about what happens as a chemical reaction takes place. We can think of the chemical bonds between the atoms or ions in the reactants being broken. Then new chemical bonds can be formed to make the products.

- Energy has to be supplied to break chemical bonds. This means that breaking bonds is an **endothermic** process. Energy is taken in from the surroundings.
- But when new bonds are formed, energy is released. So making bonds is an **exothermic** process.

a What kind of process is breaking bonds?
b What kind of process is making bonds?

Figure 5 Hydrogen and oxygen react to make water. The bonds between hydrogen atoms and between oxygen atoms have to be broken before bonds between oxygen atoms and hydrogen atoms in water can be formed.

AQA Examiner's tip

Remember that **B**reaking bonds a**B**sorbs energy, fo**R**ming bonds **R**eleases energy.

Summary questions

1 Copy and complete using the words below:

activation exothermic lower increases difference level endothermic

Bond breaking is an .... process whereas bond making is exothermic. An energy .... diagram shows the .... in energy between reactants and products. In an .... reaction the products have less energy than the reactants and the temperature of the surroundings ....

The minimum amount of energy needed to start a reaction is its .... energy. A catalyst offers an alternative reaction pathway with a .... activation energy and speeds up the reaction.

2 Draw energy level diagrams for the following reactions:
  a $H_2(g) + Cl_2(g) \rightarrow 2HCl(g)$; The reaction releases 184 kJ/mol
  b $H_2(g) + I_2(g) \rightarrow 2HI(g)$; The reaction absorbs 26.5 kJ/mol

3 **a** Draw an energy level diagram for an endothermic reaction, $X + Y \rightarrow Z$, including its activation energy.
  b Now show the effect of a catalyst on your energy level diagram drawn in part **a**.

Key points

- We can show the relative difference in the energy of reactants and products on energy level diagrams.
- Catalysts lower the activation energy so a greater proportion of reactant particles have enough energy to react.
- Bond breaking is endothermic and bond making is exothermic.

222 223

Summary answers

1 endothermic, level, difference, exothermic, increases, activation, lower

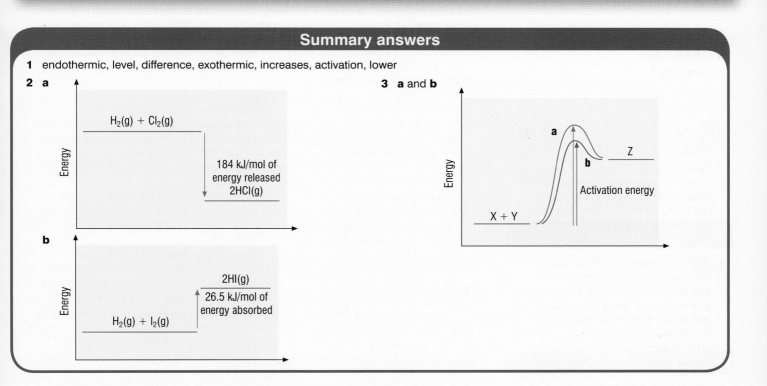

2 a

Energy
$H_2(g) + Cl_2(g)$
184 kJ/mol of energy released
2HCl(g)

b

Energy
2HI(g)
26.5 kJ/mol of energy absorbed
$H_2(g) + I_2(g)$

3 a and **b**

Energy
a
b
Z
Activation energy
X + Y

C3 3.4 Calculations using bond energies

Learning objectives

Students should learn:

- how the balance of energy changes when bonds are broken or made, and how this affects the overall energy change of a reaction **[HT only]**
- how bond energies can be used to calculate energy changes in reactions. **[HT only]**

Learning outcomes

Most students should be able to:

- calculate the energy transferred in reactions using supplied bond energies **[HT only]**
- explain how overall energy change in a reaction is affected by the balance of energy changes when bonds are made or broken. **[HT only]**

Specification link-up: Chemistry C3.3.1

- In an exothermic reaction, the energy released from forming new bonds is greater than the energy needed to break existing bonds. *[C3.3.1 f)]* **[HT only]**
- In an endothermic reaction, the energy needed to break existing bonds is greater than the energy released from forming new bonds. *[C3.3.1 g)]* **[HT only]**
- Interpret simple energy level diagrams in terms of bond breaking and bond formation … *[C3.3]*

Lesson structure

Starters

True or false – Give each student a red card and a green card. Read out the following statements. If the students think a statement is true, they should show a green card. If they think it is false, they should show the red card.

- Bond making is exothermic. [True]
- If the reactants have more energy than the products, the reaction is endothermic. [False]
- Ammonia contains three identical bonds. [True]
- Energy is needed to break bonds. [True]
- Bond energy is measured in joules per gram (J/g). [False]
- Bond energy is the energy needed to break the bond between two moles of atoms. [True] *(5 minutes)*

Sort – Write the following sentences on separate cards:

- Reactant bonds break (endothermic).
- Atoms rearrange.
- Product bonds form (exothermic).

Ask students to sort the sentences into the correct order. Support students by encouraging them to work in small groups. Extend students by giving each group a molecular-model kit. Ask them to choose a reaction (the Student Book could help them, e.g. formation of ammonia from nitrogen and hydrogen). Then use the molecular-model kit to make models of each stage of a chemical reaction (nitrogen molecules and hydrogen molecules for the first stage, free atoms for the second stage and ammonia molecules for the third stage). Then each group could feed back their thoughts to the whole of the class. *(10 minutes)*

Main

- On flip-chart paper, write different examination questions that relate to bond energies. The worked example is one that could be used. Using sticky-tac, attach the sheets to the wall. Split the students into pairs and give each team a marker pen and a calculator. Ask each pair to start answering one question, and time the students for 3 minutes.
- Then ask the students to move on. They should read the previous work, correct if necessary and add to the answer. Once all the students have attempted each question, give each group a sheet of paper and ask them to copy up the question and what they consider to be the correct working into their notes.
- Give the students different energy level diagrams (the diagrams from lesson C3 3.2 Energy transfers in solutions could be used). Encourage them to add on the mathematical information relating to the bond energies for each chemical. Students should also calculate the energy change in kJ/mol and add this to the diagram.
- There is a set number and order of stages to calculate the energy change of a reaction, using bond energies. Encourage students to create a flow chart to explain the process. Then ask students to use their flow chart to answer some bond energy questions.

Plenaries

Relationship – Bond energies and the energy change of a reaction are related. Ask students to work in small groups to explain how these terms can be linked. Then ask each group to share their thoughts. *(5 minutes)*

Support

- Supply students with the answers to bond energy calculation questions in discrete lines of working out, in an incorrect order. Students should cut and stick these sentences into the correct order.

Extend

- You could extend students by using some AS-level bond energy questions.

Calculate – Ask students to work out how much energy would be needed to break a mole of molecules of the following into separate atoms:

- methane [1652 kJ/mol]
- ammonia [1173 kJ/mol]
- hydrogen chloride [432 kJ/mol]
- water [928 kJ/mol].

Support students by giving them a writing frame to help then complete the calculation. Extend students by asking them to draw an energy level diagram with energy data on it for one of the reactions. *(10 minutes)*

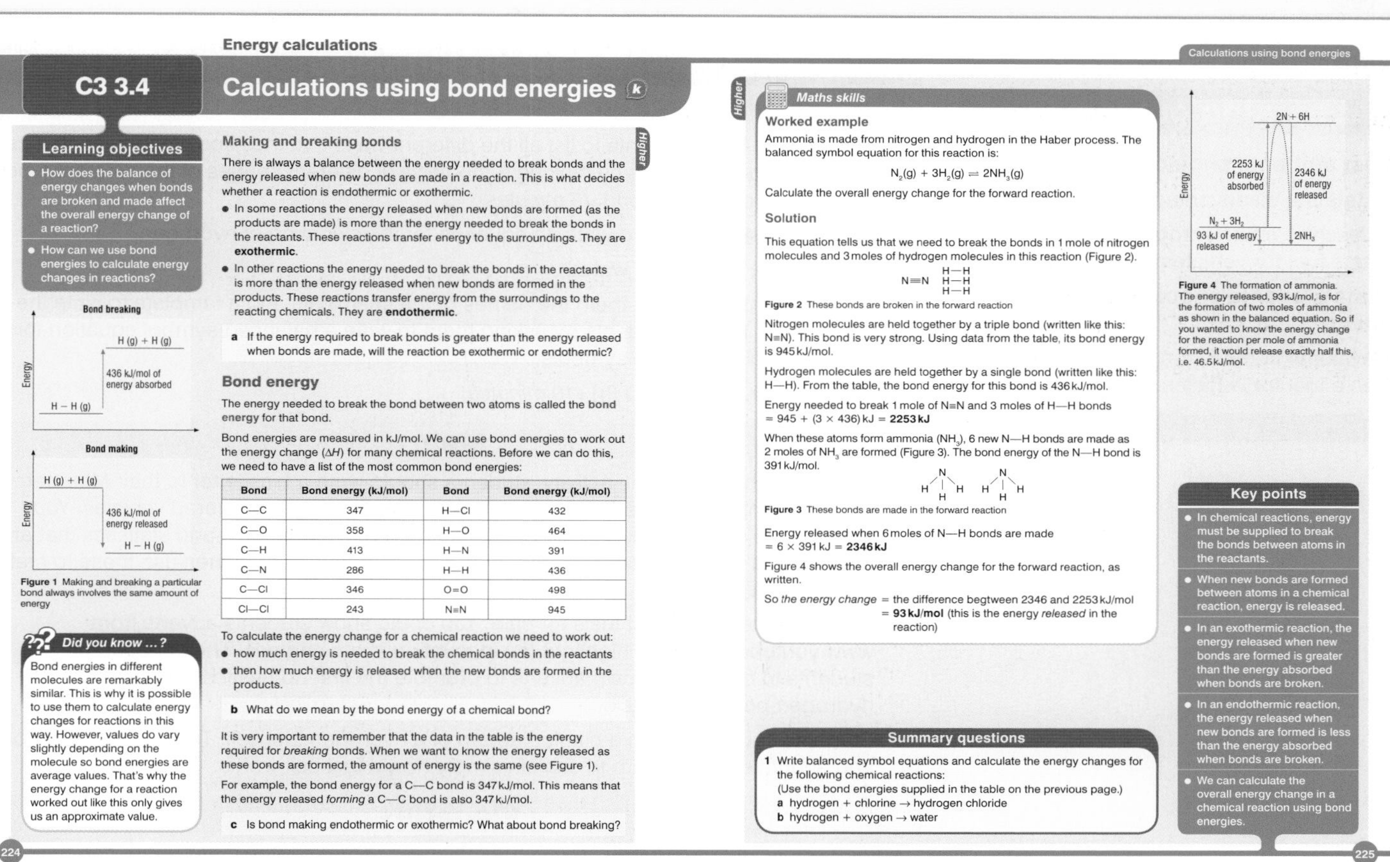

Energy calculations

C3 3.4 Calculations using bond energies

Learning objectives

- How does the balance of energy changes when bonds are broken and made affect the overall energy change of a reaction?
- How can we use bond energies to calculate energy changes in reactions?

Figure 1 Making and breaking a particular bond always involves the same amount of energy

Did you know ...?

Bond energies in different molecules are remarkably similar. This is why it is possible to use them to calculate energy changes for reactions in this way. However, values do vary slightly depending on the molecule so bond energies are average values. That's why the energy change for a reaction worked out like this only gives us an approximate value.

Making and breaking bonds

There is always a balance between the energy needed to break bonds and the energy released when new bonds are made in a reaction. This is what decides whether a reaction is endothermic or exothermic.

- In some reactions the energy released when new bonds are formed (as the products are made) is more than the energy needed to break the bonds in the reactants. These reactions transfer energy to the surroundings. They are **exothermic**.
- In other reactions the energy needed to break the bonds in the reactants is more than the energy released when new bonds are formed in the products. These reactions transfer energy from the surroundings to the reacting chemicals. They are **endothermic**.

a If the energy required to break bonds is greater than the energy released when bonds are made, will the reaction be exothermic or endothermic?

Bond energy

The energy needed to break the bond between two atoms is called the bond energy for that bond.

Bond energies are measured in kJ/mol. We can use bond energies to work out the energy change (ΔH) for many chemical reactions. Before we can do this, we need to have a list of the most common bond energies:

Bond	Bond energy (kJ/mol)	Bond	Bond energy (kJ/mol)
C—C	347	H—Cl	432
C—O	358	H—O	464
C—H	413	H—N	391
C—N	286	H—H	436
C—Cl	346	O=O	498
Cl—Cl	243	N≡N	945

To calculate the energy change for a chemical reaction we need to work out:

- how much energy is needed to break the chemical bonds in the reactants
- then how much energy is released when the new bonds are formed in the products.

b What do we mean by the bond energy of a chemical bond?

It is very important to remember that the data in the table is the energy required for *breaking* bonds. When we want to know the energy released as these bonds are formed, the amount of energy is the same (see Figure 1).

For example, the bond energy for a C—C bond is 347 kJ/mol. This means that the energy released *forming* a C—C bond is also 347 kJ/mol.

c Is bond making endothermic or exothermic? What about bond breaking?

224

Higher

Maths skills

Worked example

Ammonia is made from nitrogen and hydrogen in the Haber process. The balanced symbol equation for this reaction is:

$$N_2(g) + 3H_2(g) \rightleftharpoons 2NH_3(g)$$

Calculate the overall energy change for the forward reaction.

Solution

This equation tells us that we need to break the bonds in 1 mole of nitrogen molecules and 3 moles of hydrogen molecules in this reaction (Figure 2).

Figure 2 These bonds are broken in the forward reaction

Nitrogen molecules are held together by a triple bond (written like this: N≡N). This bond is very strong. Using data from the table, its bond energy is 945 kJ/mol.

Hydrogen molecules are held together by a single bond (written like this: H—H). From the table, the bond energy for this bond is 436 kJ/mol.

Energy needed to break 1 mole of N≡N and 3 moles of H—H bonds
= 945 + (3 × 436) kJ = **2253 kJ**

When these atoms form ammonia (NH_3), 6 new N—H bonds are made as 2 moles of NH_3 are formed (Figure 3). The bond energy of the N—H bond is 391 kJ/mol.

Figure 3 These bonds are made in the forward reaction

Energy released when 6 moles of N—H bonds are made
= 6 × 391 kJ = **2346 kJ**

Figure 4 shows the overall energy change for the forward reaction, as written.

So *the energy change* = the difference begtween 2346 and 2253 kJ/mol
= **93 kJ/mol** (this is the energy *released* in the reaction)

Summary questions

1 Write balanced symbol equations and calculate the energy changes for the following chemical reactions:
(Use the bond energies supplied in the table on the previous page.)
a hydrogen + chlorine → hydrogen chloride
b hydrogen + oxygen → water

Calculations using bond energies

Figure 4 The formation of ammonia. The energy released, 93 kJ/mol, is for the formation of two moles of ammonia as shown in the balanced equation. So if you wanted to know the energy change for the reaction per mole of ammonia formed, it would release exactly half this, i.e. 46.5 kJ/mol.

Key points

- In chemical reactions, energy must be supplied to break the bonds between atoms in the reactants.
- When new bonds are formed between atoms in a chemical reaction, energy is released.
- In an exothermic reaction, the energy released when new bonds are formed is greater than the energy absorbed when bonds are broken.
- In an endothermic reaction, the energy released when new bonds are formed is less than the energy absorbed when bonds are broken.
- We can calculate the overall energy change in a chemical reaction using bond energies.

225

Further teaching suggestions

Graph or chart

- Students could draw a graph or chart to compare the bond energies of different bonds.

Carbon–carbon bonds

- Students could find out how the C—C, C=C and C≡C bonds compare and why they are different.

Measured versus calculated energy changes

- Ask students to explain why the calculated energy change of reaction is never the same as the actual energy change measured in the reaction.

Worked example

- Copy out the key points. Then encourage each student to illustrate them by using a worked example of a calculation.

Answers to in-text questions

a Endothermic

b The energy required to make or break a bond (or a mole of bonds).

c Bond breaking is endothermic; bond making is exothermic.

Summary answers

1 a $H_2(g) + Cl_2(g) \rightarrow 2HCl(g)$ 185 kJ/mol of energy released

b $2H_2(g) + O_2(g) \rightarrow 2H_2O(l)$ 486 kJ/mol of energy released

C3 3.5 Fuel issues

Learning objectives

Students should learn:

- what the consequences of using fuels are
- that hydrogen can be used as a fuel for vehicles
- how hydrogen can be used in fuel cells
- whether hydrogen is a realistic alternative fuel for cars.

Learning outcomes

Most students should be able to:

- state the disadvantages of using fuels
- describe how hydrogen can be used as a fuel for vehicles.

Some students should also be able to:

- evaluate in detail the use of hydrogen as a fuel for vehicles.

Specification link-up: Chemistry C3.3.1

- Hydrogen can be burned as a fuel in combustion engines.

  hydrogen + oxygen → water

  It can also be used in fuel cells that produce electricity to power vehicles. *[C3.3.1 i)]*
- Consider the social, economic and environmental consequences of using fuels. *[C3.3]*
- Evaluate the use of hydrogen to power cars compared to other fuels. *[C3.3]*

Lesson structure

Starters

Fuelling a car – Ask students to list all the different ways that they know of powering a car. You could support students by making a montage of different images of the different types of fuels from around the world. *(5 minutes)*

Reactions – Ask students to write an equation for the oxidation of hydrogen.

[hydrogen + oxygen → water]

Support students by giving them the names of the chemicals and a template to write the word equation. Extend students by asking them to write a balanced-symbol equation for the reaction including state symbols:

[$2H_2(g) + O_2(g) \rightarrow 2H_2O\ (l)$] *(10 minutes)*

Main

- In the 'Hydrogen please!' activity, students should write a bias letter to their Member of Parliament to increase funding into researching hydrogen-powered vehicles. You may wish to support students by providing them with statements and statistics that are positive and negative about hydrogen as a fuel. Students could then use these to craft their argument.
- There are many different fuels for cars. You could show students adverts from www.youtube.com, that highlight the innovation in car engines. Then encourage students to create their own adverts to promote the first domestically available hydrogen-powered vehicle.
- Students could be given a blank diagram of a hydrogen fuel cell. They could then label and annotate the diagram to explain simply how a fuel cell works.

Plenaries

Advantage or disadvantage – Give students a red (disadvantage) and green (advantage) card. Read out statements and students have to hold up the card to classify it as an advantage or disadvantage. Support students by encouraging them to work in small groups before holding up the card. Extend students by encouraging them to classify the statement as an economic, social or environmental statement. *(5 minutes)*

AfL (Assessment for Learning) – Give the students an examination question. Ask them to work in small groups to write a mark scheme. This allows them to consider all the acceptable phrases and where to award marks. It also encourages students to consider what is not worthy of a mark. You could choose a tier of entry suitable for the students in the class. *(10 minutes)*

Support

- Give students a writing frame to list the advantages and disadvantage of using hydrogen as a fuel for a car. You could give them a list of statements that they could sort into the correct part of the writing frame to support them further.

Extend

- Ask students to find out about other fuel cells that do not run on hydrogen, e.g. alcohol fuel cells, which run on methanol. They could contrast the different fuel cells, detailing their similarities and differences.

Further teaching suggestions

Hydrogen fuel research

- Students could find out about the study that the University of Birmingham is running into hydrogen-powered vehicles. For more information go to www.newscentre.bham.ac.uk and search for 'hydrogen fuel'.

Hoffman voltameter

- You may wish to demonstrate the generation of hydrogen from the electrolysis of water using a Hoffman voltameter. This links to C2 Chapter 5 Salts and electrolysis.

Combusting hydrogen

- You may wish to demonstrate the exothermic reaction from combusting hydrogen. See 'Practical support' in C1 4.5 Alternative fuels.

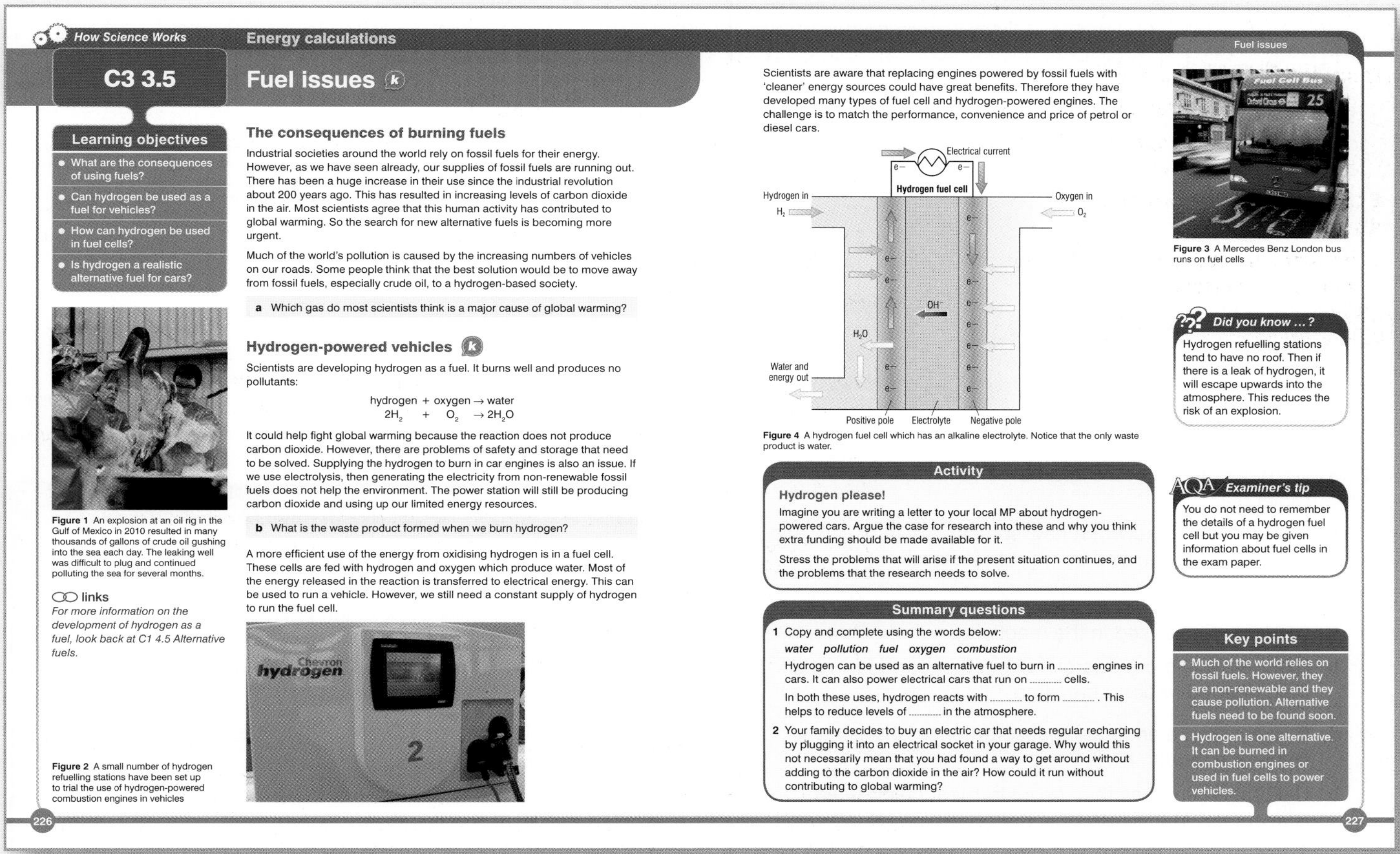

C3 3.5 Fuel issues

Learning objectives

- What are the consequences of using fuels?
- Can hydrogen be used as a fuel for vehicles?
- How can hydrogen be used in fuel cells?
- Is hydrogen a realistic alternative fuel for cars?

Figure 1 An explosion at an oil rig in the Gulf of Mexico in 2010 resulted in many thousands of gallons of crude oil gushing into the sea each day. The leaking well was difficult to plug and continued polluting the sea for several months.

links

For more information on the development of hydrogen as a fuel, look back at C1 4.5 Alternative fuels.

The consequences of burning fuels

Industrial societies around the world rely on fossil fuels for their energy. However, as we have seen already, our supplies of fossil fuels are running out. There has been a huge increase in their use since the industrial revolution about 200 years ago. This has resulted in increasing levels of carbon dioxide in the air. Most scientists agree that this human activity has contributed to global warming. So the search for new alternative fuels is becoming more urgent.

Much of the world's pollution is caused by the increasing numbers of vehicles on our roads. Some people think that the best solution would be to move away from fossil fuels, especially crude oil, to a hydrogen-based society.

a Which gas do most scientists think is a major cause of global warming?

Hydrogen-powered vehicles

Scientists are developing hydrogen as a fuel. It burns well and produces no pollutants:

$$\text{hydrogen} + \text{oxygen} \rightarrow \text{water}$$
$$2H_2 + O_2 \rightarrow 2H_2O$$

It could help fight global warming because the reaction does not produce carbon dioxide. However, there are problems of safety and storage that need to be solved. Supplying the hydrogen to burn in car engines is also an issue. If we use electrolysis, then generating the electricity from non-renewable fossil fuels does not help the environment. The power station will still be producing carbon dioxide and using up our limited energy resources.

b What is the waste product formed when we burn hydrogen?

A more efficient use of the energy from oxidising hydrogen is in a fuel cell. These cells are fed with hydrogen and oxygen which produce water. Most of the energy released in the reaction is transferred to electrical energy. This can be used to run a vehicle. However, we still need a constant supply of hydrogen to run the fuel cell.

Figure 2 A small number of hydrogen refuelling stations have been set up to trial the use of hydrogen-powered combustion engines in vehicles

Scientists are aware that replacing engines powered by fossil fuels with 'cleaner' energy sources could have great benefits. Therefore they have developed many types of fuel cell and hydrogen-powered engines. The challenge is to match the performance, convenience and price of petrol or diesel cars.

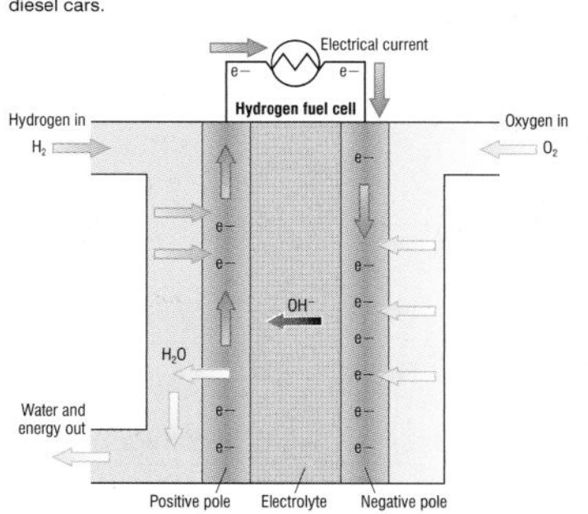

Figure 4 A hydrogen fuel cell which has an alkaline electrolyte. Notice that the only waste product is water.

Figure 3 A Mercedes Benz London bus runs on fuel cells

Did you know ...?

Hydrogen refuelling stations tend to have no roof. Then if there is a leak of hydrogen, it will escape upwards into the atmosphere. This reduces the risk of an explosion.

Activity

Hydrogen please!

Imagine you are writing a letter to your local MP about hydrogen-powered cars. Argue the case for research into these and why you think extra funding should be made available for it.

Stress the problems that will arise if the present situation continues, and the problems that the research needs to solve.

AQA Examiner's tip

You do not need to remember the details of a hydrogen fuel cell but you may be given information about fuel cells in the exam paper.

Summary questions

1 Copy and complete using the words below:

water pollution fuel oxygen combustion

Hydrogen can be used as an alternative fuel to burn in .... engines in cars. It can also power electrical cars that run on .... cells.

In both these uses, hydrogen reacts with .... to form .... This helps to reduce levels of .... in the atmosphere.

2 Your family decides to buy an electric car that needs regular recharging by plugging it into an electrical socket in your garage. Why would this not necessarily mean that you had found a way to get around without adding to the carbon dioxide in the air? How could it run without contributing to global warming?

Key points

- Much of the world relies on fossil fuels. However, they are non-renewable and they cause pollution. Alternative fuels need to be found soon.
- Hydrogen is one alternative. It can be burned in combustion engines or used in fuel cells to power vehicles.

Answers to in-text questions

a Carbon dioxide.

b Water.

Summary answers

1 combustion, fuel, oxygen, water, pollution

2 The electricity generated for your home is likely to be from a power station fuelled by burning a fossil fuel. The power station releases carbon dioxide into the atmosphere. If your electricity was generated by an alternative source, such as wind power or solar power, then you would not be releasing carbon dioxide into the air when running the electric car.

Summary answers

1 a

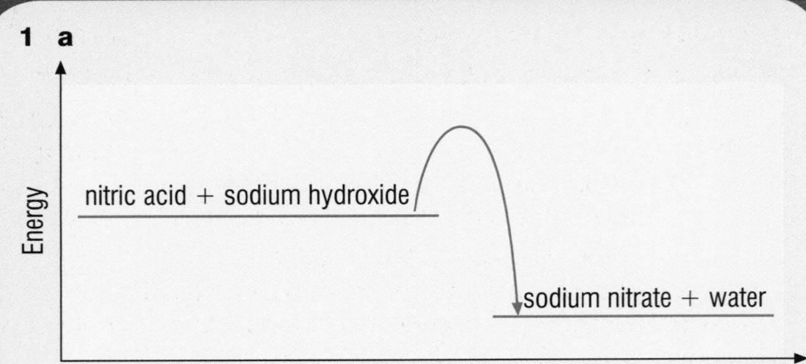

b

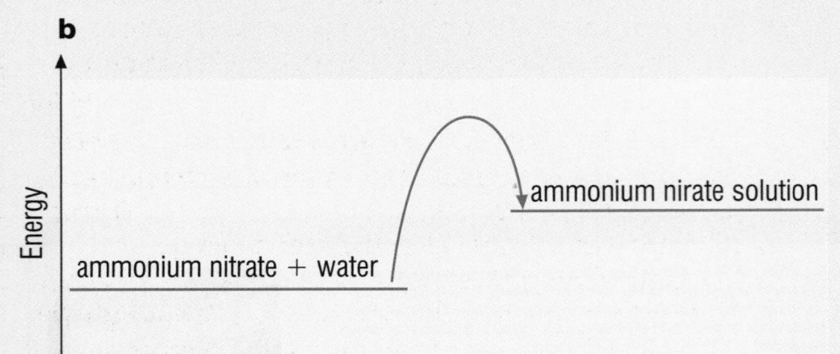

c

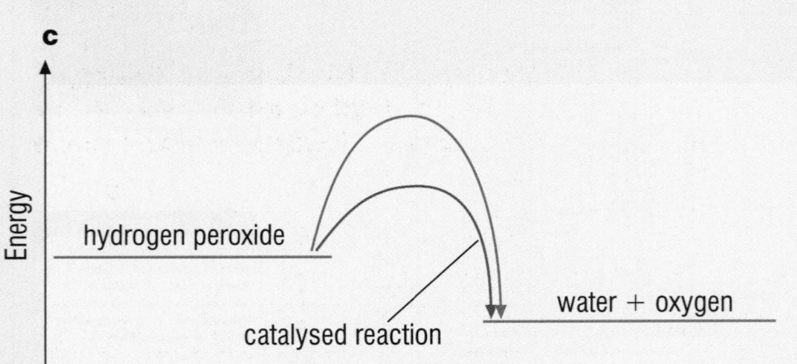

2 a To break the molecule apart/break the bonds in the molecule.

b Bonds are formed when new substances (e.g. CO_2 and water) are formed and energy is released.

c 85 kJ

3 a 210 kJ/mol released in the reaction.

b Exothermic – more energy released in bond formation than needed in bond breaking.

4 a It gives a more accurate measurement of the energy value of a food.

b Calibration ensures that the instrument being used is accurate. In this instance, the bomb calorimeter is tested against a known chemical to check on the accuracy of its measurements.

c $Q = 250 \times 4.2 \times 26 = 27\,300$ J (27.3 kJ)

d $27.3 \times \frac{1}{0.01} = 2730$ kJ/mol

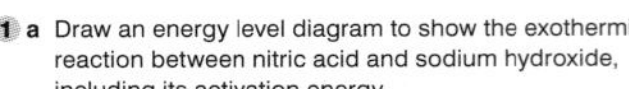

Summary questions

1 a Draw an energy level diagram to show the exothermic reaction between nitric acid and sodium hydroxide, including its activation energy.

b Draw an energy level diagram to show the endothermic change when ammonium nitrate dissolves in water, including its activation energy.

c Show an energy level diagram, including its activation energy, for the exothermic breakdown of hydrogen peroxide into water and oxygen. Then add to your diagram to show what happens when the same reaction is catalysed by manganese(IV) oxide.

2 When we eat sugar we break it down eventually to produce water and carbon dioxide.

$C_{12}H_{22}O_{11} + 12O_2 \rightarrow 12CO_2 + 11H_2O$

a Why must your body *supply* energy in order to break down a sugar molecule?

b When we break down sugar in our bodies, energy is released. Explain where this energy comes from in terms of the bonds in molecules.

c We can get about 1.7×10^3 kJ of energy by breaking down 100 g of sugar.

If a heaped teaspoon contains 5 g of sugar, how much energy does this release when broken down by the body?

3 Hydrogen peroxide has the structure H—O—O—H. It decomposes slowly to form water and oxygen.

$2H_2O_2 \rightarrow 2H_2O + O_2$

The table shows the bond energies for different types of bond.

Bond	Bond energy (kJ/mol)
H—O	464
H—H	436
O—O	144
O=O	498

a Use the bond energies to calculate the energy change for the decomposition of hydrogen peroxide in kJ/mol.

b Is this reaction exothermic or endothermic? Explain your answer. [H]

4 Bomb calorimeters are used in research laboratories to calculate the energy in food. You will have tried to do th at some time in school when you burn a crisp or some other food under a beaker of water and measure the ris in temperature. A bomb calorimeter does the same thin but better!

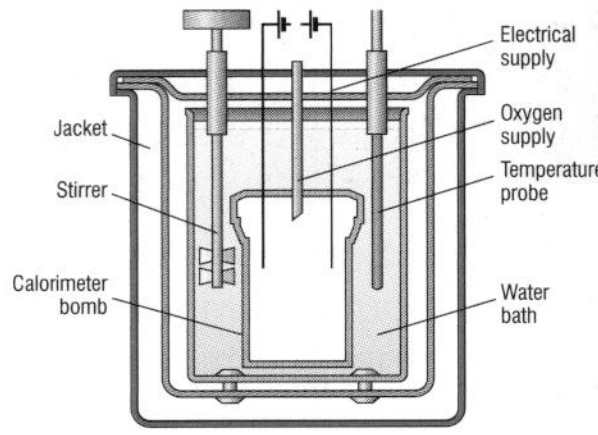

As you can see from this diagram, it is sealed and has i own supply of oxygen. There is a supply of electricity to produce a spark that will ignite the food. The increase i temperature of the water around the bomb is measurec

a Why is the bomb 'better' than the school method?

b The bomb is calibrated before it is used by burning a substance with a known energy value. Use this example to explain what is meant by 'calibration'.

c Using a bomb calorimeter, a food scientist burned a sample of a sugar. There were 250 cm³ of water in the calorimeter and its temperature went from 18 °C to 44 °C.

Use $Q = m\,c\,\Delta T$, where c = 4.2 J /g°C (i.e. 4.2 J of energy raise the temperature of 1 g of water by 1 °C), to calculate the energy released by the burning sugar.

d The scientists burned 0.01 moles of the sugar in the bomb calorimeter. How much energy does this suga release in kJ per mole?

228

Kerboodle resources

Resources available for this chapter on Kerboodle are:

- Chapter map: Energy calculations
- Data handling skills: How much energy is released when a fuel burns? (C3 3.1)
- How Science Works: A burning question (C3 3.1)
- Support: Comparing fuels (C3 3.1)
- Practical: The energy released when fuels burn (C3 3.1)
- Practical: Measuring energy changes in reactions (C3 3.2)
- Animation: Catalysts (C3 3.3)
- Maths skills: Bond energies (C3 3.4)
- Extension: Energy changes (C3 3.4)
- WebQuest: Is hydrogen the perfect fuel? (C3 3.5)
- Interactive activity: Energy calculations
- Revision podcast: Using calorimetry to measure energy in fuels, foods and reactions in solution
- Test yourself: Energy calculations
- On your marks: Energy calculations
- Examination-style questions: Energy calculations
- Answers to examination-style questions: Energy calculations

AQA Examination-style questions

1 The airship caught fire when it was coming in to land in 1937. The airship was filled with hydrogen. A spark or flame ignited the hydrogen. The equation for the reaction for hydrogen burning is:

$2H_2 + O_2 \rightarrow 2H_2O$

a Write a word equation for the reaction of hydrogen with oxygen. (1)

b Complete the sentence by writing the word **released** or **supplied** in each space.

When reactions take place, energy is .... to break the existing bonds and energy is .... when new bonds form. (1)

c An energy level diagram for the reaction of hydrogen and oxygen is shown below.

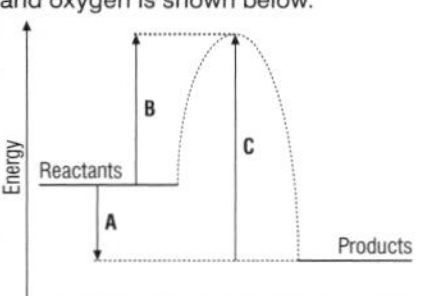

Use the energy level diagram above to help you to answer these questions.

i Which energy change, A, B or C, represents the activation energy? (1)

ii Which energy change, A, B or C, shows that the reaction is exothermic? (1)

iii Why does a mixture of hydrogen and oxygen need a spark or flame to start the reaction? (1)

d Hydrogen can be burned as a fuel. It has been used for aeroplanes and road vehicles. Why would burning hydrogen cause less pollution than burning fossil fuels? (1)

e Hydrogen for use as a fuel can be made by electrolysis of water. Suggest one disadvantage of making hydrogen by electrolysis of water. (1)

f Hydrogen can be used in a fuel cell. Fuel cells can be used in vehicles to produce electricity that is used to drive an electric motor instead of a combustion engine.

i Suggest one advantage of using electric vehicles powered by fuel cells. (1)

ii Suggest one disadvantage of this type of vehicle. (1)

AQA, 2009

2 Propane burns in oxygen. The equation for this reaction is:

$C_3H_8 + 5O_2 \rightarrow 3CO_2 + 4H_2O$

The structural formulae for the substances in the equation are:

H–C–C–C–H (with H above and below each C)   O=O   O=C=O   H–O–H

a Copy and complete the table. The first row has been done for you.

Bond	Bond energy in kJ per mole	Number of bonds broken	Energy required in kJ per mole	Number of bonds formed	Energy required in kJ per mole
C–H	413	8	3304	0	0
C–C	347				
O=O	498				
C=O	805				
H–O	464				

(4)

b Calculate the total energy required to break the bonds in the reactants. (1)

c Calculate the total energy released in forming bonds in the products. (1)

d Calculate the energy change for the reaction. (1)

e In terms of bond energies, explain why this reaction is exothermic. (1)

f i Sketch an energy level diagram for this reaction showing the reaction pathway and activation energy. (3)

ii On the same diagram, draw and label the reaction pathway if a catalyst was used. [H] (1)

Practical suggestions

Practicals	AQA	k	(book icon)	(gears icon)
Design an investigation to compare the energy produced by different liquid fuels and different foods using a simple calorimeter.	✓		✓	
Measuring and calculating the energy change for exothermic reactions (e.g. react acid with Mg ribbon) and endothermic reactions (e.g. dissolving potassium nitrate).	✓		✓	
Carrying out some reactions and measuring the energy produced, assuming that it is only the water in the solution that is being heated and that 4.2 joules will raise the temperature of 1 cm³ of water by 1 °C.	✓	✓	✓	

Examination-style answers

1 a hydrogen + oxygen → water *(1 mark)*

b supplied, released (in this order) *(1 mark)*

c i B *(1 mark)*

ii A *(1 mark)*

iii (Reactants) need to gain the activation energy or need energy B or need to break bonds. *(1 mark)*

d Does not produce carbon dioxide or other correctly-named pollutant gas from fossil fuels; or only produces water; or water is not a pollutant. *(1 mark)*

e Needs a lot of energy; (may) need to burn fossil fuels to produce electricity; efficiency/energy losses in production of electricity/electrolysis; any other valid disadvantage. *(1 mark)*

f i Less pollution (local/at point of use/in towns); quieter; (can be) sustainable/use renewable energy sources; any other valid advantage. *(1 mark)*

ii (Greater) cost of fuel cell/vehicle/technology; hydrogen (fuel) not readily available; performance not as good (as combustion engines); (greater) weight (of fuel cell/fuel tanks); any other valid disadvantage. *(1 mark)*

2 a

Number of bonds broken	Energy required in kJ per mole	Number of bonds formed	Energy required in kJ per mole
8	3304	0	0
2	694	0	0
5	2490	0	0
0	0	6	4830
0	0	8	3712

1 mark for each row correctly completed. *(4 marks)*

b 6488 or error carried forward from table *(1 mark)*

c 8542 or error carried forward from table *(1 mark)*

d 2054 or error carried forward from **b** and **c** *(1 mark)*

e Energy released in forming bonds is greater than energy needed to break bonds. *(1 mark)*

f i Energy level diagram with reactants above products; reaction pathway shown as curve; activation energy labelled. *(3 marks)*

ii Reaction pathway with lower peak/activation energy labelled 'catalyst'. *(1 mark)*

C3 4.1 Tests for positive ions

Learning objectives

Students should learn:

- the tests for common metal ions.

Learning outcomes

Most students should be able to:

- state the positive result for lithium, sodium, potassium, calcium and barium in flame tests
- explain how to carry out a flame test
- state the positive result for aluminium, calcium, copper(II), iron(II)/(III), and magnesium ions with sodium hydroxide.

Some students should also be able to:

- explain how to use sodium hydroxide to test for different positive ions
- interpret results of chemical tests independently.

Summary answers

1 colours, flame, hydroxide, precipitation

2 Dip a wire loop in concentrated hydrochloric acid and heat to clean it → dip the loop in the acid again before placing it in the metal compound to collect a small amount to be tested → hold the loop in the roaring blue flame of a Bunsen burner → match the colour of the flame to the known flame test colours of metal ions

3 **a** brown precipitate
b nothing observed
c nothing observed
d yellow
e potassium
f calcium
g Al^{3+}
h Fe^{2+}

Support

- Have the test tubes with the transition metal solutions set up in advance with labels stuck to them. Students can simply add sodium hydroxide from a pipette to observe the colour change.

Extend

- Allow students to use a spectroscope to see the characteristic patterns produced by each flame colour. See www.sep.org.uk and search for 'forensic chemistry'.

Specification link-up: Chemistry C3.4

- Flame tests can be used to identify metal ions. Lithium, sodium, potassium, calcium and barium compounds produce distinctive colours in flame tests:
  - lithium compounds result in a crimson flame
  - sodium compounds result in a yellow flame
  - potassium compounds result in a lilac flame
  - calcium compounds result in a red flame
  - barium compounds result in a green flame. *[C3.4.1 a)]*
- Aluminium, calcium and magnesium ions form white precipitates with sodium hydroxide solution but only the aluminium hydroxide precipitate dissolves in excess sodium hydroxide solution. *[C3.4.1 b)]*
- Copper(II), iron(II) and iron(III) ions form coloured precipitates with sodium hydroxide solution. Copper forms a blue precipitate, iron(II) a green precipitate and iron(III) a brown precipitate. *[C3.4.1 c)]*
- Interpret results of the chemical tests in this specification. *[C3.4]*

Lesson structure

Starters

Match – Ask students to match the names of the metal ions with their symbols. e.g. [iron(II) = Fe^{2+}, iron(III) = Fe^{3+}, copper(II) = Cu^{2+}] *(5 minutes)*

Images – Show the students pictures of salty water boiling over on a gas hob, a street light, and sodium burning in oxygen or a sodium flame as it reacts with water (with the hydrogen having been ignited). Ask the students to look at the images and decide what they have in common. Support students by encouraging them to discuss the images with other students, then feed the ideas back to the class. Reveal to the students that all of the images contain sodium ions that are heated, causing the yellow colour. Extend students by asking them to draw a dot and cross diagram of a sodium ion. *(10 minutes)*

Main

- Firstly, students could test different known metal solutions to determine the flame colour of lithium, sodium, potassium, calcium and barium.
- Once students can distinguish the differences in the colours, a group of mystery solutions could be given. The students should carry out flame tests to determine the positive ion present.
- Students could use sodium hydroxide to test for copper(II), iron(II) and iron(III) ions in unknown solutions.
- A set of three A6 flash cards could be given to the students. On one side, they could write the chemical test [flame, using sodium hydroxide]. On the back, they could write a step-by-step set of instructions to explain how to conduct each test and what the result would indicate.
- The cards could be laminated and used in a laboratory to help the students to identify 'mystery' solutions.

Plenaries

Conclusion – Ask students to conclude what the following compound contained:

'When sodium hydroxide solution was added to a solution of the compound, a white precipitate was formed. On adding excess sodium hydroxide solution, the precipitate was insoluble. When the solution was tested in a flame, the colour was brick-red.' [calcium] *(5 minutes)*

Crossword – Ask students to create their own crossword, which relates to the chemical tests that have been studied in the lesson. Support students by giving them the outline of the frame with the key words in position, so they just need to generate the clues. Extend students by allowing them to complete each other's crosswords. *(10 minutes)*

Answers to in-text questions

a To identify unknown metal ions. **b** White.

Practical support

Testing with sodium hydroxide

Equipment and materials required

Three test tubes, 0.1 mol/dm^3 sodium hydroxide solution (irritant), three dropping pipettes, test-tube rack, $<$1.4 mol/dm^3 copper(II) chloride (irritant), 1 mol/dm^3 iron(II) chloride (irritant), 1 mol/dm^3 iron(III) chloride (irritant).

Details

Wearing chemical-splashproof eye protection, add about 1 cm^3 of each solution to separate test tubes. Add about 2 cm^3 of sodium hydroxide solution to each test tube, agitate the tube and observe. It is worth noting that the iron(II) precipitate colour changes quickly as it is oxidised in air to iron(III) hydroxide.

Safety: CLEAPSS Hazcard 91 Sodium hydroxide – corrosive; CLEAPSS Hazcard 27A Copper(II) chloride – harmful; CLEAPSS Hazcard 55B Iron(II) chloride – harmful; CLEAPSS Hazcard 55C Iron(III) chloride – harmful.

Flame tests

Equipment and materials required

Inoculating loop (nichrome flame wires, fused on to glass rod handles), Bunsen burner and equipment, nine watch glasses, solutions of lithium chloride (irritant), sodium chloride, potassium chloride, calcium hydroxide (irritant), concentrated hydrochloric acid (corrosive). Be aware of what to do if alkali gets into the eye.

Details

Wearing chemical-splashproof eye protection, put each of the known solutions into six labelled watch glasses. Then choose three of the solutions to be the mystery solutions and put them into the other three watch glasses labelled A–C. Dip the inoculating loop into the hydrochloric acid and put into the top of the blue gas cone in the roaring Bunsen flame (hottest part). This cleans the loop, but it does produce hydrogen chloride fumes and must be completed in a well-ventilated area. Repeated cleaning does corrode the wire and platinum is the best to use. Alternatively, have specific inoculating loops for each solution. This would remove the need to clean them between solutions. It is possible to forgo cleaning and just heat the wire until it is red-hot to burn off any of the previous salt. Put the inoculating loop into one solution and shake gently. Then put it into the hottest part of the blue flame and observe the initial colour. Remove the loop and clean. Repeat with the other solutions.

Safety: CLEAPSS Hazcard 47B Lithium chloride – harmful; CLEAPSS Hazcard 95 Sodium and Potassium salts – harmful; CLEAPSS Hazcard 47A Hydrochloric acid – corrosive; CLEAPSS Hazcard 18 Calcium hydroxide – irritant; CLEAPSS Hazcard 91 Sodium hydroxide – corrosive; CLEAPSS Hazcard 27 Copper(II) chloride – harmful; CLEAPSS Hazcard 55 Iron compounds – harmful.

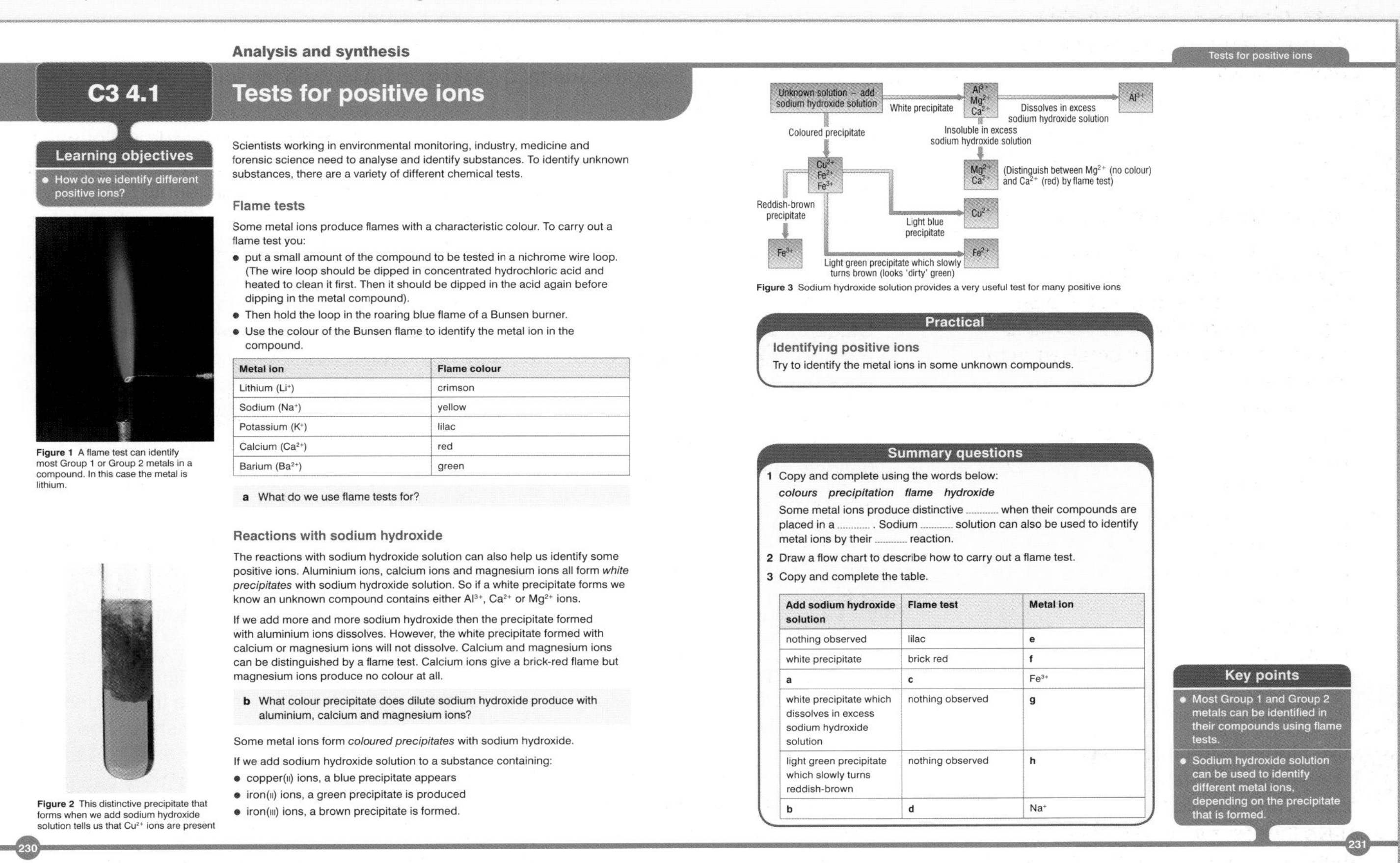

Analysis and synthesis

Tests for positive ions

C3 4.1 Tests for positive ions

Learning objectives

- How do we identify different positive ions?

Scientists working in environmental monitoring, industry, medicine and forensic science need to analyse and identify substances. To identify unknown substances, there are a variety of different chemical tests.

Flame tests

Some metal ions produce flames with a characteristic colour. To carry out a flame test you:

- put a small amount of the compound to be tested in a nichrome wire loop. (The wire loop should be dipped in concentrated hydrochloric acid and heated to clean it first. Then it should be dipped in the acid again before dipping in the metal compound).
- Then hold the loop in the roaring blue flame of a Bunsen burner.
- Use the colour of the Bunsen flame to identify the metal ion in the compound.

Metal ion	Flame colour
Lithium (Li^+)	crimson
Sodium (Na^+)	yellow
Potassium (K^+)	lilac
Calcium (Ca^{2+})	red
Barium (Ba^{2+})	green

a What do we use flame tests for?

Figure 1 A flame test can identify most Group 1 or Group 2 metals in a compound. In this case the metal is lithium.

Reactions with sodium hydroxide

The reactions with sodium hydroxide solution can also help us identify some positive ions. Aluminium ions, calcium ions and magnesium ions all form *white precipitates* with sodium hydroxide solution. So if a white precipitate forms we know an unknown compound contains either Al^{3+}, Ca^{2+} or Mg^{2+} ions.

If we add more and more sodium hydroxide then the precipitate formed with aluminium ions dissolves. However, the white precipitate formed with calcium or magnesium ions will not dissolve. Calcium and magnesium ions can be distinguished by a flame test. Calcium ions give a brick-red flame but magnesium ions produce no colour at all.

b What colour precipitate does dilute sodium hydroxide produce with aluminium, calcium and magnesium ions?

Some metal ions form *coloured precipitates* with sodium hydroxide.

If we add sodium hydroxide solution to a substance containing:

- copper(II) ions, a blue precipitate appears
- iron(II) ions, a green precipitate is produced
- iron(III) ions, a brown precipitate is formed.

Figure 2 This distinctive precipitate that forms when we add sodium hydroxide solution tells us that Cu^{2+} ions are present

Figure 3 Sodium hydroxide solution provides a very useful test for many positive ions

Practical

Identifying positive ions

Try to identify the metal ions in some unknown compounds.

Summary questions

1 Copy and complete using the words below:
colours precipitation flame hydroxide
Some metal ions produce distinctive .... when their compounds are placed in a .... Sodium .... solution can also be used to identify metal ions by their .... reaction.

2 Draw a flow chart to describe how to carry out a flame test.

3 Copy and complete the table.

Add sodium hydroxide solution	Flame test	Metal ion
nothing observed	lilac	e
white precipitate	brick red	f
a	c	Fe^{3+}
white precipitate which dissolves in excess sodium hydroxide solution	nothing observed	g
light green precipitate which slowly turns reddish-brown	nothing observed	h
b	d	Na^+

Key points

- Most Group 1 and Group 2 metals can be identified in their compounds using flame tests.
- Sodium hydroxide solution can be used to identify different metal ions, depending on the precipitate that is formed.

230 231

Further teaching suggestions

Further metals

- Other metal ions could be added to the flame tests, e.g. copper ions. You might even wish to test mixtures of two salts such as a solution of sodium chloride and potassium chloride and look at the resulting flame colour.

Excess sodium hydroxide

- Sodium hydroxide could be used to test for magnesium, aluminium and calcium ions (0.1 mol/dm^3). The sodium hydroxide should be added at 1 cm^3 at a time, up to 5 cm^3 to see if the precipitate redissolves (see 'Practical support').

Photographs

- The Bunsen-flame tests could be photographed. These images could then be printed out and students could add them to their notes.

C3 4.2 Tests for negative ions

Learning objectives

Students should learn:

- the tests for common negative ions.

Learning outcomes

Most students should be able to:

- state the positive result for carbonates
- state the positive result for halides
- state the positive result for sulfates
- explain how to carry out tests for carbonates, halides and sulfates.

Some students should also be able to:

- interpret results of chemical tests independently.

Specification link-up: Chemistry C3.4

- Carbonates react with dilute acids to form carbon dioxide. Carbon dioxide produces a white precipitate with limewater. This turns limewater cloudy. *[C3.4.1 d)]*
- Halide ions in solution produce precipitates with silver nitrate solution in the presence of dilute nitric acid. Silver chloride is white, silver bromide is cream and silver iodide is yellow. *[C3.4.1 e)]*
- Sulfate ions in solution produce a white precipitate with barium chloride solution in the presence of dilute hydrochloric acid. *[C3.4.1 f)]*
- Interpret results of the chemical tests in this specification. *[C3.4]*

Lesson structure

Starters

Find – Ask students to give:

- three examples of an ionic compound. [e.g. barium sulfate, copper carbonate, zinc carbonate, silver halide, copper oxide, zinc oxide, barium chloride]
- two examples of a covalent compound. [e.g. carbon dioxide, ammonia]
- one example of a monoatomic ion. [e.g. any halide]

Encourage students to use the Student Book to help them. Support students by allowing them to work in pairs. Extend students by asking them to write the formulae for each of their examples. *(5 minutes)*

Recall – Ask students to recall the test for carbon dioxide gas [limewater turns cloudy] and explain why it works, [carbon dioxide gas reacts with the dissolved calcium hydroxide to form a precipitate of calcium carbonate]. You may wish to extend students further by asking them to write word equations and symbol equations for this reaction. *(10 minutes)*

Main

- Give the students an unlabelled diagram of the equipment used to test for carbonates. Students should then annotate the diagram. Students could try to write a step-by-step method for testing carbonates. Then they could attempt the practical, after it has been checked for safety and make any amendments to their method as they try to complete the practical. The students could test calcium, copper and zinc carbonates.
- A circus of all the experiments could be set up so that students test for carbonates, halides, sulfates and nitrates. Each station should contain a help card with an outline of the method and a diagram. (See 'Practical support'.) Students should design their own results table and note observations at each station.

Plenaries

Key – Give students a card sort, which can be used to make a key for identifying halides. Encourage students to work in pairs to make the correct key. *(5 minutes)*

AfL (Assessment for Learning) – Students could be given a set of examination questions on this topic. To help students with their exam preparation, time the students (approximately 1 min per mark), and give the students an idea when they should be tackling each section of a question. Then show the students the mark scheme and allow them to mark their own work and make any notes that are needed, so that they can use the material for revision at a later date. The appropriate tier of entry can be used for the students in the class. *(10 minutes)*

Answers to in-text questions

- **a** Carbon dioxide.
- **b** Silver nitrate solution with dilute nitric acid.
- **c** Barium chloride solution with hydrochloric acid.

Support

- Give students the method statements for testing carbonates in the wrong order. They should then sort the sentences to make the method.

Extend

- Ask students to suggest how the positive ion tests and negative ion tests could be used together to prove that a colourless solution was sodium chloride. [A flame test would give a yellow colour, suggesting sodium, and using nitric acid then silver nitrate would give a white precipitate, suggesting chloride ions.]

Practical support

Testing halides

Equipment and materials required

Five different salts containing halides labelled 1–5 (e.g. sodium chloride, sodium iodide, sodium bromide, potassium chloride, potassium bromide), spatula, five test tubes, test-tube rack, 2 × dropping pipettes, 0.5 mol/dm^3 nitric acid (corrosive), 0.05–0.1 mol/dm^3 silver nitrate solution, eye protection.

Details

Wearing chemical-splashproof eye protection, put about a quarter of a spatula of a halide into a test tube. Add about 1 cm^3 of nitric acid, agitate and add more nitric acid until all of the halide has dissolved. Then add about 1 cm^3 of silver nitrate solution and observe the colour of the precipitate to determine the halide in the solution.

Safety: CLEAPSS Hazcard 67 Nitric acid – oxidising/corrosive; CLEAPSS Hazcard 87 Silver nitrate – corrosive.

Testing sulfates

Equipment and materials required

Test tube, test-tube rack, a sulfate solution (e.g. magnesium sulfate), 1 mol/dm^3 hydrochloric acid (irritant), 2 × pipettes, 0.2 mol/dm^3 barium chloride (harmful), eye protection.

Details

Wearing eye protection, put about 1 cm^3 of sulfate solution into a test tube. Add about 1 cm^3 of acid and then 1 cm^3 of barium chloride solution and observe. The mixture may need agitating for the reaction to complete.

Safety: CLEAPSS Hazcard 27A Hydrochloric acid – corrosive; CLEAPSS Hazcard 10 Barium chloride – harmful.

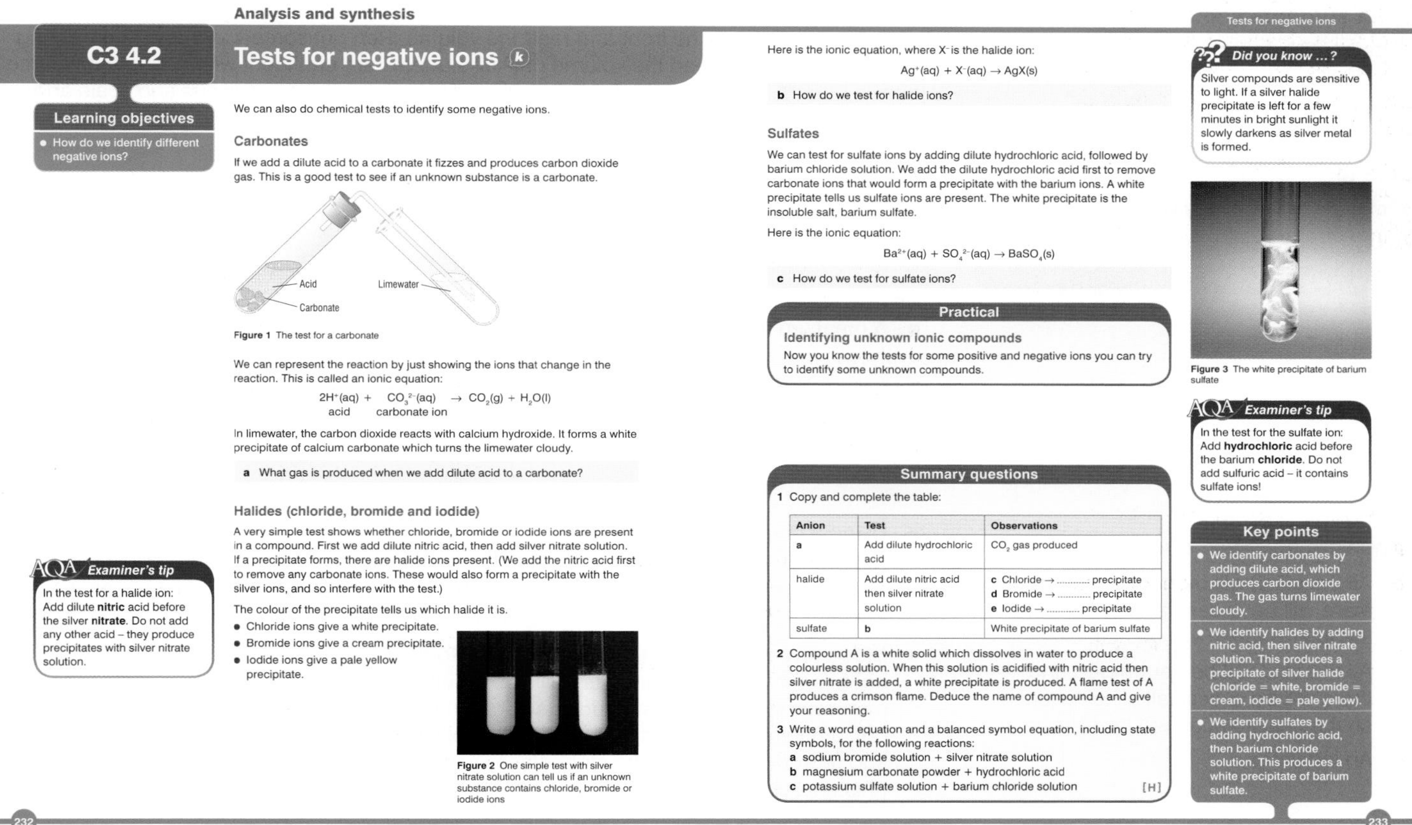

Analysis and synthesis

C3 4.2 **Tests for negative ions**

Learning objectives

- How do we identify different negative ions?

We can also do chemical tests to identify some negative ions.

Carbonates

If we add a dilute acid to a carbonate it fizzes and produces carbon dioxide gas. This is a good test to see if an unknown substance is a carbonate.

Acid
Carbonate
Limewater

Figure 1 The test for a carbonate

We can represent the reaction by just showing the ions that change in the reaction. This is called an ionic equation:

$$2H^+(aq) + CO_3^{2-}(aq) \rightarrow CO_2(g) + H_2O(l)$$

acid carbonate ion

In limewater, the carbon dioxide reacts with calcium hydroxide. It forms a white precipitate of calcium carbonate which turns the limewater cloudy.

a What gas is produced when we add dilute acid to a carbonate?

Halides (chloride, bromide and iodide)

A very simple test shows whether chloride, bromide or iodide ions are present in a compound. First we add dilute nitric acid, then add silver nitrate solution. If a precipitate forms, there are halide ions present. (We add the nitric acid first to remove any carbonate ions. These would also form a precipitate with the silver ions, and so interfere with the test.)

The colour of the precipitate tells us which halide it is.

- Chloride ions give a white precipitate.
- Bromide ions give a cream precipitate.
- Iodide ions give a pale yellow precipitate.

Figure 2 One simple test with silver nitrate solution can tell us if an unknown substance contains chloride, bromide or iodide ions

AQA Examiner's tip

In the test for a halide ion: Add dilute **nitric** acid before the silver **nitrate**. Do not add any other acid – they produce precipitates with silver nitrate solution.

Here is the ionic equation, where X^- is the halide ion:

$$Ag^+(aq) + X^-(aq) \rightarrow AgX(s)$$

b How do we test for halide ions?

Sulfates

We can test for sulfate ions by adding dilute hydrochloric acid, followed by barium chloride solution. We add the dilute hydrochloric acid first to remove carbonate ions that would form a precipitate with the barium ions. A white precipitate tells us sulfate ions are present. The white precipitate is the insoluble salt, barium sulfate.

Here is the ionic equation:

$$Ba^{2+}(aq) + SO_4^{2-}(aq) \rightarrow BaSO_4(s)$$

c How do we test for sulfate ions?

Practical

Identifying unknown ionic compounds

Now you know the tests for some positive and negative ions you can try to identify some unknown compounds.

Summary questions

1 Copy and complete the table:

Anion	Test	Observations
a	Add dilute hydrochloric acid	CO_2 gas produced
halide	Add dilute nitric acid then silver nitrate solution	c Chloride → .... precipitate d Bromide → .... precipitate e Iodide → .... precipitate
sulfate	b	White precipitate of barium sulfate

2 Compound A is a white solid which dissolves in water to produce a colourless solution. When this solution is acidified with nitric acid then silver nitrate is added, a white precipitate is produced. A flame test of A produces a crimson flame. Deduce the name of compound A and give your reasoning.

3 Write a word equation and a balanced symbol equation, including state symbols, for the following reactions:
a sodium bromide solution + silver nitrate solution
b magnesium carbonate powder + hydrochloric acid
c potassium sulfate solution + barium chloride solution [H]

Tests for negative ions

Did you know ... ?

Silver compounds are sensitive to light. If a silver halide precipitate is left for a few minutes in bright sunlight it slowly darkens as silver metal is formed.

Figure 3 The white precipitate of barium sulfate

AQA Examiner's tip

In the test for the sulfate ion: Add **hydrochloric** acid before the barium **chloride**. Do not add sulfuric acid – it contains sulfate ions!

Key points

- We identify carbonates by adding dilute acid, which produces carbon dioxide gas. The gas turns limewater cloudy.
- We identify halides by adding nitric acid, then silver nitrate solution. This produces a precipitate of silver halide (chloride = white, bromide = cream, iodide = pale yellow).
- We identify sulfates by adding hydrochloric acid, then barium chloride solution. This produces a white precipitate of barium sulfate.

232 233

Further teaching suggestions

Spider diagram

- Ask students to summarise the chemical tests in the form of a spider diagram.

Other negative ions

- You may wish to encourage students to test for carbonate ions using microscale techniques. By reacting the metal carbonate with acid, carbon dioxide gas is formed and this can be tested using a drop of limewater on a glass rod.

Synoptic

- You may wish to give students a solution of a metal salt such as sodium chloride and ask them to use the chemical techniques in C3 4.1 and C3 4.2 to identify the positive and negative ions and to use this information to identify the compound present.

Video clips

- On the RSC website (www.chemit.co.uk) there are video clips of some of these experiments.

Summary answers

1 **a** Carbonate
b Add dilute hydrochloric acid, then barium chloride solution.
c Chloride – white precipitate
d Bromide – cream precipitate
e Iodide – pale yellow precipitate

2 Lithium chloride because lithium turns flames crimson and the chloride test is positive if a white precipate is produced with silver nitrate solution.

3 **a** sodium bromide + silver nitrate → sodium nitrate + silver bromide
$NaBr(aq) + AgNO_3(aq) \rightarrow NaNO_3(aq) + AgBr(s)$
b magnesium carbonate + hydrochloric acid → magnesium chloride + water + carbon dioxide
$MgCO_3(s) + 2HCl(aq) \rightarrow MgCl_2(aq) + H_2O(l) + CO_2(g)$
c potassium sulfate + barium chloride → potassium chloride + barium sulfate
$K_2SO_4(aq) + BaCl_2(aq) \rightarrow 2KCl(aq) + BaSO_4(s)$

C3 4.3 Titrations

Learning objectives

Students should learn:

- how the amount of acid and alkali that react together completely can be measured accurately
- how to determine that the reaction is complete.

Learning outcomes

Most students should be able to:

- define titration, end point and indicator
- safely carry out a simple titration.

Some students should also be able to:

- explain qualitatively what is happening in a titration.

Specification link-up: Chemistry C3.4.1

- The volumes of acid and alkali solutions that react with each other can be measured by titration using a suitable indicator. *[C3.4.1 g)]*

  Controlled Assessment: C4.3 Collect primary and secondary data. *[C4.3.2 b) c) e)]*

Lesson structure

Starters

Volume – Set up a burette full of water and a flexicam projecting the water level. Ask for a volunteer to come to the front and read the volume. Run out some water and keep asking random students to read the volume until all students are confident with noting the volume from a burette. By having the flexicam, all students can see the meniscus and scale and can participate in reading the volume. *(5 minutes)*

Crossword – Give students a crossword made from the following clues:

- A substance that can dissolve in water and is a base. [alkali]
- When neutralisation is complete. [end point]
- Acid with formula HNO_3. [nitric acid]
- End points are shown using a solution of one of these. [indicator]
- A measure of acidity. [pH]

Support students by encouraging them to think back to their work in C2 Chapter 5 'Salts and electrolysis'. Students could work in small groups and be given a list of key words to choose from. Extend students by asking them to give a specific example for each definition. *(10 minutes)*

Main

- Using a data logger, you can demonstrate an acid–base titration to the class. Ask students to predict the shape of the pH curve and temperature graph.
- Complete the titration and discuss the shape of the curve and contrast this with the prediction.
- Students can complete their own titration. To help them get used to using the equipment, allow the students to fill the pipettes with water and practise getting the meniscus to the line. Also students can fill the burette with water and practise running it out at different rates and noting the volume. Discuss accuracy, precision, repeatability and reproducibility with the students, as well as designing an appropriate results table (this relates to 'How Science Works').

Plenaries

Results table – Give students a fictitious results table for a titration, with mistakes in it such as units not being in the column heading but after each reading. Ask students to redraw the table correctly into their notes. *(5 minutes)*

Safety – Give students a series of images of titrations being completed unsafely, e.g. no eye protection. Ask the students to discuss the safety issues in the images and explain why they are problems. Pick a few groups to feed back about each image. *(10 minutes)*

Answers to in-text questions

- **a** All the alkali will be neutralised and there will be excess acid left over, so the final solution will be acidic.
- **b** All the acid will be neutralised and there will be excess alkali left over, so the final solution will be alkaline.
- **c** An indicator.

Support

- Some students find it difficult to read volume from a burette. You can help these students by giving them a sheet of white paper. Make two cuts (about 3 cm long and parallel to each other about 2 cm away). Slot this over the top of the burette and this makes the scale easier to read.

Extend

- Ask students to find out some real-life examples where titrations are used.

Further teaching suggestions

Flow chart

- Students could draw a pictorial flow chart to explain the step-by-step method of a titration.

Titration video

- Show the students footage of the use of a pipette, burette and titrations. This is available on the Practical A-level Chemistry CD-ROM from the RSC (or search the internet for 'titration video').

Practical support

Carrying out a titration

Equipment and materials required

Stand, boss, clamp, 1 mol/dm^3 hydrochloric acid (irritant), 1 mol/dm^3 sodium hydroxide (corrosive), pipette, pipette filler, burette, funnel, conical flask, dropping pipette, phenolphthalein, wash bottle, white tile. Titrations: see CLEAPSS Handbook/CD-ROM section 13.8.

Details

Wearing eye protection, rinse all equipment with reagents. Using a pipette and filler, measure 25 cm^3 of sodium hydroxide into a conical flask. Add a few drops of indicator to the conical flask and put it on to a white tile. Fill the burette with acid, remove the funnel and position the tap over the conical flask. Run in the acid, swirling the flask all the time. When the indicator has changed colour, stop the tap and note the volume. This is the rough titration to give an idea of the volume needed. Wash the conical flask thoroughly and add a fresh sample of 25 cm^3 of sodium hydroxide and indicator. Quickly run in the acid (5 cm^3 less than the titre of the rough titration). Add the acid dropwise, swirling between drops. When the colour change begins to persist, using the wash bottle, rinse around the conical flask and the end of the burette. Now add half drops at a time, washing the acid from the bottom of the burette, until the colour change lasts for 10 s. Note the volume. Repeat the titration three times or until there are two concordant results (within 0.1 cm^3 of each other).

Using a data logger to investigate titrations

Equipment and materials required

Temperature probe, pH probe, interface, data logger, stand, boss, clamp, computer, data projector, 1 mol/dm^3 hydrochloric acid (irritant), 1 mol/dm^3 sodium hydroxide (corrosive), pipette, pipette filler, burette, funnel, conical flask, dropping pipette, phenolphthalein, magnetic stirrer, magnetic stirrer bar.

Details

Wearing eye protection, measure 25 cm^3 of sodium hydroxide using the pipette. Add a few drops of phenolphthalein and a magnetic stirrer bar. Put the conical flask on to the magnetic stirrer, and submerge the two data probes. Connect the ICT equipment and set the run to last 2 minutes. Fill the burette with acid and put the tap over the conical flask. Start the stirrer and recording data. Add the acid from the burette at a steady flow rate. (Be aware of what to do if alkali gets into the eye.)

Safety (for both titrations): CLEAPSS Hazcard 47A Hydrochloric acid – corrosive; CLEAPSS Hazcard 91 Sodium hydroxide – corrosive; CLEAPSS Hazcards 32 Phenolphthalein – harmful.

Analysis and synthesis

C3 4.3 Titrations

Learning objectives

- How can we measure accurately the amount of acid and alkali that react together completely?
- How do we know when the reaction is complete?

An acid and an alkali (a soluble base) react together and neutralise each other. They form a salt (plus water) in the process.

Suppose we mix a strong acid and a strong alkali. The solution made will be neutral only if we add exactly the right quantities of acid and alkali.

If we start off with more acid than alkali, then the alkali will be neutralised. However, the solution left after the reaction will be acidic, not neutral. That's because some acid will be left over (it is in excess).

a If there is more acid than alkali to start with in a neutralisation reaction, what will happen?

If we have more alkali than acid to begin with, then all the acid will be neutralised and the solution left will be alkaline.

b If there is more alkali than acid to start with in a neutralisation reaction, what will happen?

We can measure the exact volumes of acid and alkali needed to react with each other by a technique called **titration**. The point at which the acid and alkali have reacted completely is called the **end point** of the reaction. We show this by using an indicator.

c What do we call a chemical used to show the end point of a neutralisation reaction?

links

For more information on neutralisation reactions, look back at C2 5.2 Making salts from metals or bases and C2 5.3 Making salts from solutions.

AQA Examiner's tip

We use a **pipette** to measure out a fixed volume of solution. We use a **burette** to measure the volume of the solution added.

Practical

Carrying out a titration

In this experiment you can carry out a titration. You will find out how much acid is needed to completely react with an alkali:

1 Measure a known volume of alkali into a conical flask using a **pipette**. Before doing this, you should first wash the pipette with distilled water, and then with some of the alkali.

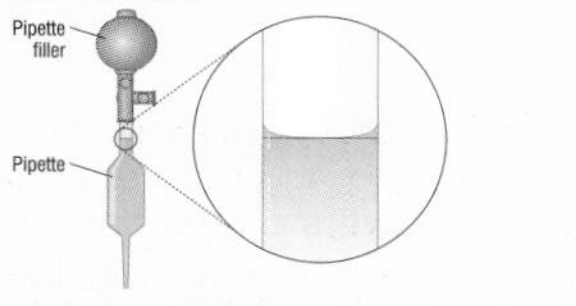

Pipette filler
Pipette

Figure 1 A pipette and pipette filler. Fill the pipette until the bottom of the meniscus (curved surface of the solution) coincides with the mark. Allow the liquid to run out of the pipette and touch the tip on the side of the flask to drain out the solution. It is normal for a tiny amount of solution to remain in the pipette.

2 Now add an indicator solution to the solution in the flask.

3 Pour the acid you are going to use into a **burette**. This is a long tube with a tap on one end. The tube has markings on it to enable you to measure volumes accurately (often to the nearest 0.05 cm^3). Before doing this, you should first wash the burette with distilled water, and then with some of the acid.

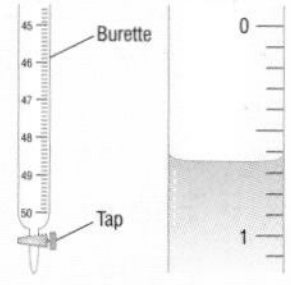

Burette
Tap
45 46 47 48 49 50
0
1

Figure 2 A burette – use the bottom of the meniscus to read the scale. The reading here is 0.65 cm^3.

4 Record the reading on the burette. Then open the tap to release a small amount of acid into the flask. Swirl the flask to make sure that the two solutions are mixed.

5 Keep on repeating step 4 until the indicator in the flask changes colour. This shows when the alkali in the flask has completely reacted with the acid added from the burette. Record the reading on the burette and calculate the volume of acid run into the flask. (On your first go at doing this you will probably run too much acid into the flask, so treat this as a rough estimate of how much acid is needed.)

6 Repeat the whole process at least three times. Discard any anomalous results. Then calculate an average value to give the most accurate results possible. Alternatively, repeat the titration until you get two identical results.

7 Now you can use your results to calculate the concentration of the alkali.

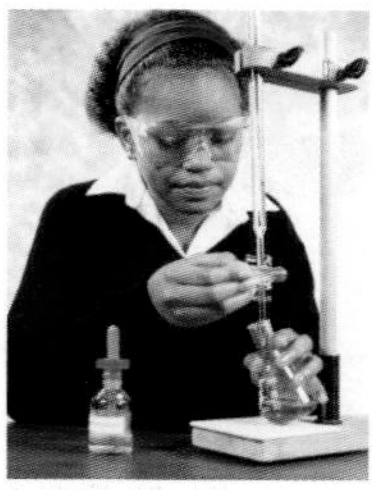

Figure 3 Titrations can be used to measure the exact volumes of acid and alkaline needed to react completely

links

For more information on calculations, see C3 4.4 Titration calculations.

Summary questions

1 Copy and complete using the words below:

acid end indicator neutralisation

Adding an acid to an alkali results in a .... reaction. The point at which enough .... has been added to completely react with the alkali is called the .... point. This can be shown using a chemical called an ....

2 Draw a flowchart to show how to carry out a titration between an acid of unknown concentration and an alkali of known concentration.

3 Indicator A changes from deep blue in acid solution to colourless in alkaline solution. Indicator B changes from pale green in acid solution to pale blue in alkaline solution. Which is the better indicator to use in an acid–base titration and why?

Key points

- Titration is used to measure accurately how much acid and alkali react together completely.
- The point at which an acid–alkali reaction is complete is called the end point of the reaction.
- We use an indicator to show the end point of the reaction between an acid and an alkali.

234 235

Summary answers

1 neutralisation, acid, end, indicator

2 The student flowchart should be based on the steps described in this section, i.e.

Step 1: Measure alkaline solution into conical flask, using a pipette.

Step 2: Add indicator solution.

Step 3: Put acid solution into burette.

Step 4: Record reading on burette.

Step 5: Add small amount of acid to flask from burette.

Step 6: Check colour of indicator – if it has changed, go to **Step 7**, if not, go back to **Step 5**.

Step 7: Read burette.

Step 8: Repeat **steps 1 to 7** until consistent readings are obtained.

3 Indicator A is better because the colour change is much easier to see.

C3 4.4 Titration calculations

Learning objectives

Students should learn:

- how to calculate concentrations from reacting volumes of solutions **[HT only]**
- how to calculate the amount of acid or alkali needed in a neutralisation reaction. **[HT only]**

Learning outcomes

Most students should be able to:

- complete straightforward calculations involving concentration, expressed in mol/dm^3 and in g/dm^3 **[HT only]**
- complete calculations with results from a titration involving simple ratios. **[HT only]**

Some students should also be able to:

- carry out more complex titration calculations. **[HT only]**

Specification link-up: Chemistry C3.4

- If the concentration of one of the reactants is known, the results of a titration can be used to find the concentration of the other reactant. *[C3.4.1 h)]* **[HT only]**

Lesson structure

Starters

Anagrams – Write the following anagrams on to the board for the students to unscramble:

- tittirason [titrations]
- sleom [moles]
- eumolv [volume]
- rationcentnoc [concentration]
- ionculatcal [calculation]

Support students by giving them the words and they have to match them to the anagram. This activity can be extended by encouraging students to define each of the key words. *(5 minutes)*

Find – Ask students to use the Student Book to find out all the formulae that will be used in this topic. Support students by asking them to reflect on the work that they completed in C2 Chapter 3, How much? Then ask a few students to write on the board the formula and the unit that each variable is measured in, e.g. number of moles (mol) = mass (g)/M_r. *(10 minutes)*

Main

- Create a card sort made up of 30 questions and their answers on separate cards. The questions should range from easy questions, such as 'what unit is concentration measured in?', to calculation questions from the worked examples and summary questions in the Student Book. This activity can be differentiated by giving fewer questions to some students, or students could work in teams. Students should match the questions with their answers.
- Then ask students to pick five of the questions and copy them into their notes, including their answers. One of the five should be a calculation question and the students should be encouraged to show their working to arrive at the answer.
- Demonstrate how to complete a calculation from a titration by building up the steps on the board. Encourage students to generate a flow chart to help them answer this type of question, showing, in the correct order, all the steps that need to be followed.
- Then give the students a few calculation questions to answer using their flow chart to help them.

Plenaries

Bingo – Make bingo cards with six answers on the card, including key words (e.g. moles), units (e.g. grams) and numbers to answer simple calculations. Then ask questions. The students cross off an answer if it appears on their bingo card. The first student to call 'house' (with all the answers) is the winner and could get a prize. *(5 minutes)*

Calculations – Give the students an A4 whiteboard (or laminated paper), washable pen and eraser. Show a calculation question on the board, then students try to answer the question, write the answer on the board and hold it up to you for instant feedback. Encourage students to report the answers including the units. Support students by encouraging them to play the game in pairs and wait until other students have answered the question before writing their own. Extend students by using questions that require unit conversion. *(10 minutes)*

Support

- Give students the lines of working out for each question, but in the wrong order, or incomplete. Allow the students to cut and stick them in the correct order and fill in the missing information.

Extend

- Encourage students to tackle titration questions from AS-level papers.

Further teaching suggestions

Question writing

- Ask students to write their own titration examination question-and-mark scheme.

Question slideshow

- Generate multiple-choice questions, one per PowerPoint slide in a single slideshow. Play the question slideshow and allow small groups to go to the interactive whiteboard to try to answer the sequence of questions. Time each group and give the group that correctly answers the questions fastest a small prize.

Mind map

- Give the students the key terms, e.g. mass and concentration, on separate cards. Split the class into small groups and give each team some sticky-tac, sugar paper, pens and a pack of cards. Ask the students to generate a mind map of these points by choosing two cards, one to be the first word of the sentence and the other to be the last word. Draw an arrow between the cards to show the flow of the sentence and write the middle words of the sentence on the arrow. Each word can have as many arrows going into or coming out of it as is needed. Encourage the students to use all the key words at least once.

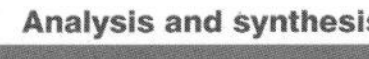

Analysis and synthesis

C3 4.4 Titration calculations

Learning objectives

- How can we calculate concentrations from reacting volumes?
- How can we calculate the amount of acid or alkali needed in a neutralisation reaction?

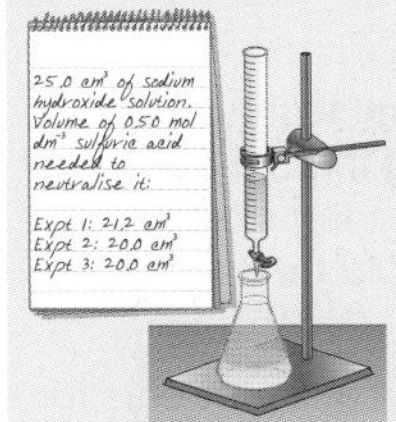

Figure 1 From results like these we can work out the concentration of the unknown solution – in this case the sodium hydroxide solution

Higher

Calculating concentrations

The concentration of a solute in a solution is the number of moles of solute dissolved in one cubic decimetre of solution. We write these units as **moles per decimetre cubed** or **mol/dm³** for short. So if we know the mass of solute dissolved in a certain volume of solution, we can work out its concentration.

As an example, imagine that we make a solution of sodium hydroxide in water. We dissolve exactly 40 g of sodium hydroxide to make exactly 1 dm³ of solution. We know how to work out the mass of 1 mole of sodium hydroxide (NaOH). We add up the relative atomic masses of sodium, oxygen and hydrogen:

$$23\ (\mathrm{Na}) + 16\ (\mathrm{O}) + 1\ (\mathrm{H}) = 40\,\mathrm{g} = \text{mass of 1 mole of NaOH}$$

Therefore we know that the solution contains 1 mole of sodium hydroxide in 1 dm³ of solution. So the concentration of sodium hydroxide in the solution is 1 mol/dm³. See Worked example 1.

Maths skills

Worked example 1

But what if we use 40 g of sodium hydroxide to make 500 cm³ of solution instead of 1 dm³? (Remember that 1 dm³ = 1000 cm³.)

Solution

To find the concentration of the solution we must work out how much sodium hydroxide there would be if we had 1000 cm³ (1 dm³) of the solution.

40 g of NaOH are dissolved in 500 cm³ of solution, so

$\frac{40}{500}$ g of NaOH would be dissolved in 1 cm³ of solution, and

$\frac{40}{500} \times 1000\,\mathrm{g} = \mathbf{80\,g}$ of NaOH would be dissolved in 1000 cm³ of solution.

The mass of 1 mole of NaOH is 40 g, so 80 g of NaOH is 80 ÷ 40 moles = **2 moles**.

2 moles of NaOH are dissolved in 1 dm³ of solution. So the concentration of NaOH in the solution is **2 mol/dm³**.

Sometimes we know the concentration of a solution and need to work out the mass of solute in a certain volume. See Worked example 2.

Titration calculations

In a titration we always have one solution with a concentration which we know accurately. We can put this in the burette. Then we can place the other solution, with an unknown concentration, in a conical flask. We do this using a pipette. This ensures we know the volume of this solution accurately. The result from the titration is used to calculate the number of moles of the substance in the solution in the conical flask. See Worked example 3.

236

Titration calculations

Higher

Maths skills

Worked example 2

What mass of potassium sulfate, K_2SO_4, is there in 250 cm³ of a 1 mol/dm³ solution?

Solution

In 1 dm³ of solution there would be 1 mole of K_2SO_4

The mass of 1 mole of K_2SO_4 is $(2 \times 39) + 32 + (4 \times 16)\,\mathrm{g} = 174\,\mathrm{g}$, so

in 1000 cm³ of solution there would be 174 g of K_2SO_4, and

in 1 cm³ of solution there are $\frac{174}{1000}$ g of K_2SO_4

So in 250 cm³ of solution there are $\frac{174}{1000} \times 250$ g of K_2SO_4 = **43.5 g** of K_2SO_4

There is **43.5 g of K_2SO_4** in 250 cm³ of 1 mol/dm³ potassium sulfate solution.

Maths skills

Worked example 3

A student put 25.0 cm³ of sodium hydroxide solution of unknown concentration into a conical flask using a pipette. The sodium hydroxide reacted with exactly 20.0 cm³ of 0.50 mol/dm³ sulfuric acid added from a burette. What was the concentration of the sodium hydroxide solution?

Solution

The equation for this reaction is:

$$2NaOH(aq) + H_2SO_4(aq) \rightarrow Na_2SO_4(aq) + 2H_2O(l)$$

This equation tells us that 2 moles of NaOH reacts with 1 mole of H_2SO_4.

The concentration of the H_2SO_4 is 0.50 mol/dm³, so

0.50 moles of H_2SO_4 are dissolved in 1000 cm³ of acid, and

$\frac{0.50}{1000}$ moles of H_2SO_4 are dissolved in 1 cm³ of acid, therefore

$\frac{0.50}{1000} \times 20.0$ moles of H_2SO_4 are dissolved in 20.0 cm³ of acid.

So there are 0.010 moles of H_2SO_4 dissolved in 20.0 cm³ of acid.

The equation for the reaction tells us that 0.010 moles of H_2SO_4 will react with exactly 2 × 0.010 moles of NaOH. This means that there must have been 0.020 moles of NaOH in the 25.0 cm³ of solution in the conical flask. To calculate the concentration of NaOH in the solution in the flask we need to calculate the number of moles of NaOH in 1 dm³ (1000 cm³) of solution.

0.020 moles of NaOH are dissolved in 25.0 cm³ of solution, so

$\frac{0.020}{25}$ moles of NaOH are dissolved in 1 cm³ of solution, and there will be

$\frac{0.020}{25} \times 1000 = 0.80$ moles of NaOH in 1000 cm³ solution.

The concentration of the sodium hydroxide solution is 0.80 mol/dm³.

Summary questions

1. In a titration, a 25.0 cm³ sample of nitric acid (HNO_3) reacted exactly with 20.0 cm³ of 0.40 mol/dm³ sodium hydroxide solution.
   - **a** Write down a balanced symbol equation for this reaction.
   - **b** Calculate the number of moles of sodium hydroxide added.
   - **c** Write down the number of moles of HNO_3 in the acid.
   - **d** Calculate the concentration of the nitric acid.

Key points

- To calculate the concentration of a solution, given the mass of solute in a certain volume:
  1. Calculate the mass (in grams) of solute in 1 cm³ of solution.
  2. Calculate the mass (in grams) of solute in 1000 cm³ of solution.
  3. Convert the mass (in grams) to moles.
- To calculate the mass of solute in a certain volume of solution of known concentration:
  1. Calculate the mass (in grams) of the solute there is in 1 dm³ (1000 cm³) of solution.
  2. Calculate the mass (in grams) of solute in 1 cm³ of solution.
  3. Calculate the mass (in grams) of solute there is in the given volume of the solution.

237

Summary answers

1 **a** $HNO_3(aq) + NaOH(aq) \rightarrow NaNO_3(aq) + H_2O(l)$

b 0.008 moles

c 0.008 moles

d 0.32 mol/dm³

C3 4.5 Chemical analysis

Learning objectives

Students should learn:

- how chemical analysis is used
- how results of chemical analysis can be interpreted.

Learning outcomes

Most students should be able to:

- list some uses of chemical analysis
- interpret analyses by matching against known data.

Some students should also be able to:

- explain how results of chemical analysis can be interpreted in a variety of contexts.

Specification link-up: Chemistry C3.4

- Interpret and evaluate the results of analyses carried out to identify elements and compounds for forensic, health or environmental purposes. *[C3.4]*

Lesson structure

Starters

Video – Find a short video of science being used in a TV programme, e.g. go on to www.youtube.com and search for *CSI*, forensics, pathology, environmental health. Ask students to suggest what field of science we are looking at today. [analytical science] *(5 minutes)*

Think – Ask students to think about the following two words: quantitative and qualitative. They should define each of these words. Support students by giving them the definitions and they have to choose which key word matches which definition. Extend students by asking them to explain the differences and similarities of quantitative and qualitative results. *(10 minutes)*

Main

- Split the class into two groups. Ask each group to work on one activity from the Student Book, using secondary sources of information to answer all of the questions. Give students 20 min to become 'experts' on their case study. Then ask students to work in pairs. One member of the pair should be from the DNA analysis group and the second from the river pollution group. The pairs should then work together with the 'expert' teaching their partner about their case study. Stress to students that they will not be asked to recall the case studies in the Student Book but they can expect to be given different scenarios to interpret in the examination.
- Ask students to make a poster to summarise what electrophoresis is. Students should be encouraged to use colours and could summarise the information in diagrams as well as text.

Plenaries

Have a think! – Ask students to suggest other applications of analytical techniques, e.g. quality control in food manufacture. Support students by showing them a collage of pictures made from other applications of analytical science. Extend students by asking them to use their knowledge to suggest what 'wet' chemistry tests could be used. *(5 minutes)*

Sentences – Choose a key word from the topic, e.g. 'qualitative'. Then ask students for a number from 1 to 19 and count down to the appropriate line in the Student Book text. Ask another student to pick a number from 1 to the maximum number of words in a sentence, and count to this word. Write the randomly chosen word onto the board, with the key word. Students should then try to write a sentence that is scientifically and grammatically correct and contains both words. Select a few students to share their sentences with the rest of the class. Repeat the process with other key words. *(10 minutes)*

Support

- Provide students with an outline of the poster about electrophoresis with some information already on it, such as 'fill in the missing words in the prose', diagrams that need annotating and titles, etc.

Extend

- Ask students to contrast the benefits and drawbacks of qualitative and quantitative analytical chemistry.

Further teaching suggestions

Forensic science video

- The website www.teachers.tv has a 20-minute video about forensic science called *Crime Scene Investigation*, which could be used.

Electrophoresis demonstration

- Some schools have an electrophoresis kit, which could be demonstrated to students.

Extracting DNA

- Students can extract the DNA from kiwi fruit. Go to www.sep.org.uk and search for forensic chemistry.

Analysis alphabet

- Ask students to complete an alphabet based on analysis. For example, 'F is for forensic science.' Students should use their Student Books and work in small groups. They will need to be creative to obtain 26 different sentences.

Revision cards

- Supply five A6 revision cards to each student. Each card should represent a different instrumental analysis technique. On the front, the student should write five questions. Encourage students to try to write in an examination style. Then, on the reverse of the card, students should write a full answer.

How Science Works — Analysis and synthesis

C3 4.5 Chemical analysis

Learning objectives

- How is chemical analysis used?
- How can we interpret the results?

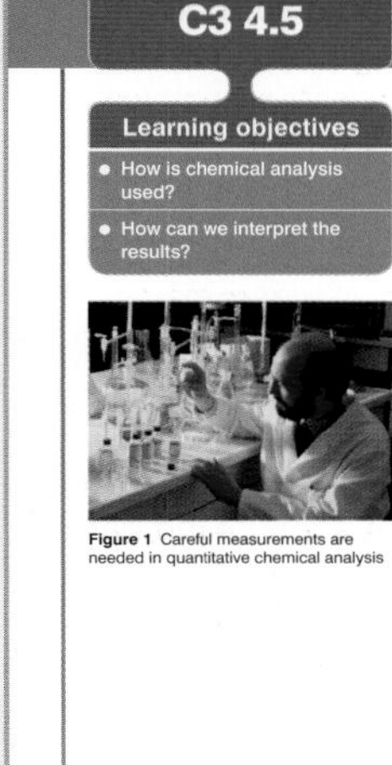

Figure 1 Careful measurements are needed in quantitative chemical analysis

As you saw in C2 3.8 Instrumental analysis, chemists have instruments to help them analyse unknown substances. They can also use traditional chemical analysis such as the tests you have seen in this chapter. These are sometimes known as 'wet' chemistry. The tests in C3 4.1 Tests for negative ions and C3 4.2 Tests for positive ions can tell us which ions are in solution (called **qualitative** testing). The titrations on C3 4.3 Titrations and C3 4.4 Titration calculations can tell us how much of the ions are present (called **quantitative** testing).

Whether instrumental or 'wet' chemistry is used, the accuracy of the data collected depends on the expertise of the tester.

a What do we call testing that relies on observations – qualitative or quantitative?

b What do we call testing that relies on measurements – qualitative or quantitative?

Analysis in forensic science

Forensic chemistry uses both qualitative and quantitative analysis. Chemists in forensic labs help solve crimes by analysing:

- drugs
- paints
- remnants of explosives
- fire debris
- gunshot residues
- fibres
- soil samples
- toxic chemicals (used in chemical weapons)
- biological toxins (used in biological weapons).

Figure 2 Analysing a DNA sample by gel electrophoresis

A technique called **gel electrophoresis** is used to analyse DNA. Analysing DNA using this technique produces a plate which carries a series of bands, according to the composition of the DNA. The bands are unique to each individual (except identical twins).

Activity

DNA analysis

Forensic scientists analysed DNA (the genetic fingerprint) from three suspects in a criminal investigation using this method. They compared their results with the DNA from a sample found at the scene of a crime. The following results were obtained:

a How is evidence like this used to decide whether someone accused of a crime is guilty or not?

b According to the results, which of the suspects A, B or C could have been present at the scene of the crime? Explain your answer.

c It might not be a good idea to rely on this kind of evidence alone when deciding if someone is guilty of a crime or not. Why not?

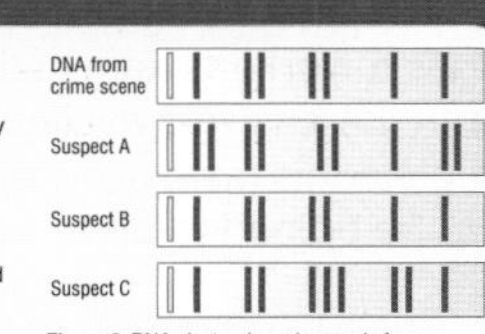

Figure 3 DNA electrophoresis sample from a crime scene and samples from three suspects. The police have a national DNA database on computer to match against samples taken at crime scenes.

Analysis in pollution control

Environmental scientists need to monitor cases of environmental pollution. For example, if a river gets polluted, they will test the water and trace the origin of any pollutants.

Activity

River pollution

The residents of a village on the banks of a river are worried about pollution. A new factory making batteries has been built upstream of the village. Cadmium, a toxic heavy metal, is used in the factory. Some people in the village enjoy fishing. They believe that there are fewer fish to catch since the factory opened six months ago. The manager of the factory promised to get his chemists to carry out some tests.

They collected water from the sites A and B shown in Figure 4. Then they analysed the water in the lab.

Two months later they reassured the residents that cadmium levels were safe.

a Some people in the village wanted the Environment Agency to carry out their own tests for cadmium. Why?

b When questioned, the factory manager admitted the instrument used to analyse the water samples was 20 years old. Why did this worry the concerned villagers?

c Give two ways in which the factory could have made sure the data collected were more repeatable, reproducible and valid?

- **i** How could the factory argue that levels of cadmium were safe?
- **ii** How could the villagers argue that they were not convinced it was safe even if the data were collected properly?

Figure 4 Testing river water

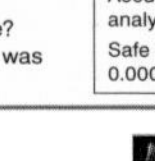

TECHNICAL INFORMATION
Cadmium levels found in a water sample taken at 6.00 am on 5th January at each site:
Site A = 0.000004 g/litre
Site B = 0.000002 g/litre
Accuracy of method used to analyse water samples + or −50%.
Safe level of cadmium = 0.000002 g/dm^3

Analysis in medicine

Another use of genetic fingerprinting is in the treatment of leukaemia (a blood disease). Bone marrow is transplanted from a healthy donor to the patient. After the operation, samples of blood from the patient and donor are analysed for their DNA. Doctors are looking for a match between the bands on the two electrophoresis plates. If they are the same, then the patient's blood cells, and the DNA in them, have come from the transplanted bone marrow. This means that the transplant has been successful.

Doctors can also study metal ion concentrations of a few parts per billion in patients. For example, they can look for cobalt and chromium ions in the blood of patients with hip replacements. A concentration of metal ions above 7 parts per billion can indicate that the joint will fail.

Figure 5 Doctors can look for metal ions in blood as a sign that an artificial hip joint is in danger of wearing out

Summary questions

1 Describe two ways in which the analysis of blood can help the work of doctors in a hospital.

2 The maximum safe level of cadmium for freshwater fish is 0.000002 g/dm^3 What is the safe level of cadmium in mol/dm^3? (A_r value of Cd = 112) [H]

Key points

- Scientists working in environmental monitoring, medicine and forensic science all need to analyse substances.
- The results of their analysis are often matched against existing databases to identify substances (or suspects in the case of forensics).

Answers to in-text questions

a Qualitative.

b Quantitative.

Summary answers

1 For example, to test if a bone-marrow transplant has been successful; to test if an artificial hip joint is in danger of failing; or any other sensible answer.

2 1.79×10^{-8} mol/dm^3 (or 1.8×10^{-8} to two significant figures).

C3 4.6 Chemical equilibrium

Learning objectives

Students should learn:

- what is meant by 'equilibrium' **[HT only]**
- how to change the amount of product in a reversible reaction. **[HT only]**

Learning outcomes

Most students should be able to:

- state a definition of a reversible reaction **[HT only]**
- recognise a reversible reaction **[HT only]**
- explain how to change the amount of product in a reversible reaction. **[HT only]**

Specification link-up: Chemistry C3.5

- When a reversible reaction occurs in a closed system, equilibrium is reached when the reactions occur at exactly the same rate in each direction. *[C3.5.1 c)]* **[HT only]**
- The relative amounts of all the reacting substances at equilibrium depend on the conditions of the reaction. *[C3.5.1 d)]* **[HT only]**

Lesson structure

Starters

Questions – Give each student an a mini-whiteboard (or laminated sheet of paper), a whiteboard pen and an eraser. Then ask them the series of questions below. Students should note down their answers and show you for immediate assessment. These questions remind the students about C2 4.8 Energy and reversible reactions. If students are unsure of the answer, they could refer to the Student Book or wait for other students to hold up their answers and then use these responses to inform their answer.

- What is the symbol to show a reversible reaction? [$\rightleftharpoons$]
- Give an example of a reversible reaction. [the Haber process, hydration of dehydrated copper sulfate, thermal decomposition of ammonium chloride]
- What does exothermic mean? [gives out energy]
- What does endothermic mean? [takes in energy] *(5 minutes)*

Equations – Give students a selection of equations to complete. You could support students by giving them the missing words to fit in. Extend students by asking them to write balanced symbol equations for the reactions. For example:

- nitrogen + [hydrogen] $\rightleftharpoons$ ammonia
- sulfur oxide + oxygen $\rightleftharpoons$ [sulfur trioxide]
- iodine monochloride + [chlorine] $\rightleftharpoons$ iodine trichloride

(10 minutes)

Main

- Students often struggle to understand what a *dynamic* equilibrium is. Place a number of polystyrene balls in a tray and ask for two student helpers. Scatter a few of the balls around the tray. Ask one student to take balls out as another student puts balls in. Ask students how the number of balls in the tray changes – it does not: the number in the tray stays the same. However, the balls are individually in the tray some of the time and out of it the rest of the time. This is a model of a dynamic equilibrium.
- Ask students to imagine that they have been commissioned by a revision website to create a GCSE webpage to help teach about chemical equilibrium. Show students a selection of websites such as www.bbc.co.uk (search for 'revision'). Ask them to discuss in groups what they like and dislike about the sites. Explain that they need to have a webpage that includes a worked examination question and an extra question for the reader to attempt, with the answers available by hyperlink. Students could work in small teams to complete this task, allowing them to divide up the task as they please. The best ones could be uploaded on to the virtual learning environment for students to use.

Plenaries

Text (Txt) – Give students an outline of a mobile phone and ask them to summarise the learning objectives into a text message. *(5 minutes)*

Assessment for Learning (AfL) – Give students a selection of fictitious candidate answers for an examination question. Ask students to work in pairs to mark the question and give feedback to the candidate. Support students by giving them the mark scheme. Extend students by not giving them the mark scheme and instead comparing their marking with that of the 'examiners'. *(10 minutes)*

Support

- Give students a template in Publisher with prompt questions to help them complete the webpage.

Extend

- Ask students to define a closed system [matter can't move freely into or out of the system]. (An open system is therefore one in which matter and energy can move in and out freely. An isolated system doesn't allow matter or energy into or out of the system.)

Further teaching suggestions

Making models

- You may wish to demonstrate a model of dynamic equilibrium with two measuring cylinders that can hold the same volume. Both should have an open-ended glass tube but with different diameters. Put 25 cm^3 of water into one cylinder. Transfer water from one cylinder to the other using a finger over the end of each tube in turn. (Keep the tubes in the same cylinder) until the level in each cylinder does not change any more. Using this idea of a model being a simpler way to understand an observation, encourage students to make their own models of a dynamic equilibrium and share them with the rest of the class.

Group activity

- Split the class into small groups and give each team a different reversible reaction. Ask students to work together to suggest how to change the yield of the reaction.

Reversible reactions

- You may wish to demonstrate equilibrium reactions to the students, for example, adding acid and then alkali to bromine water. Alkali removes the yellow colour and acid restores it. Students can investigate the effect of adding acid and alkali to potassium chromate(VI). See www.practicalchemistry.org (search for 'equilibrium').

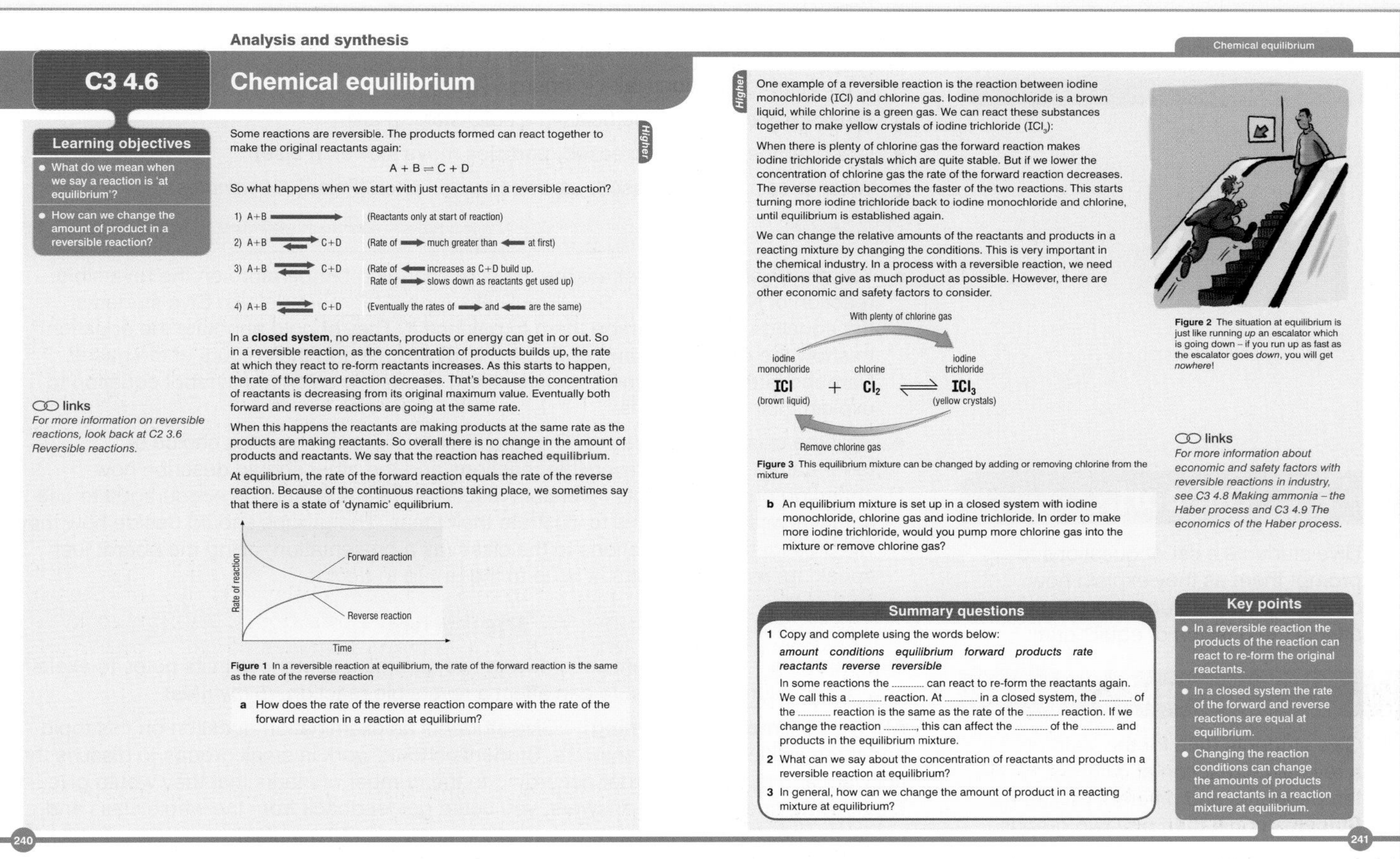

Analysis and synthesis

C3 4.6 Chemical equilibrium

Learning objectives

- What do we mean when we say a reaction is 'at equilibrium'?
- How can we change the amount of product in a reversible reaction?

Higher

Some reactions are reversible. The products formed can react together to make the original reactants again:

$$A + B \rightleftharpoons C + D$$

So what happens when we start with just reactants in a reversible reaction?

1) A+B → (Reactants only at start of reaction)

2) A+B ⇌ C+D (Rate of → much greater than ← at first)

3) A+B ⇌ C+D (Rate of ← increases as C+D build up. Rate of → slows down as reactants get used up)

4) A+B ⇌ C+D (Eventually the rates of → and ← are the same)

In a **closed system**, no reactants, products or energy can get in or out. So in a reversible reaction, as the concentration of products builds up, the rate at which they react to re-form reactants increases. As this starts to happen, the rate of the forward reaction decreases. That's because the concentration of reactants is decreasing from its original maximum value. Eventually both forward and reverse reactions are going at the same rate.

When this happens the reactants are making products at the same rate as the products are making reactants. So overall there is no change in the amount of products and reactants. We say that the reaction has reached **equilibrium**.

At equilibrium, the rate of the forward reaction equals the rate of the reverse reaction. Because of the continuous reactions taking place, we sometimes say that there is a state of 'dynamic' equilibrium.

Figure 1 In a reversible reaction at equilibrium, the rate of the forward reaction is the same as the rate of the reverse reaction

a How does the rate of the reverse reaction compare with the rate of the forward reaction in a reaction at equilibrium?

links
For more information on reversible reactions, look back at C2 3.6 Reversible reactions.

240

Chemical equilibrium

Higher

One example of a reversible reaction is the reaction between iodine monochloride (ICl) and chlorine gas. Iodine monochloride is a brown liquid, while chlorine is a green gas. We can react these substances together to make yellow crystals of iodine trichloride (ICl_3):

When there is plenty of chlorine gas the forward reaction makes iodine trichloride crystals which are quite stable. But if we lower the concentration of chlorine gas the rate of the forward reaction decreases. The reverse reaction becomes the faster of the two reactions. This starts turning more iodine trichloride back to iodine monochloride and chlorine, until equilibrium is established again.

We can change the relative amounts of the reactants and products in a reacting mixture by changing the conditions. This is very important in the chemical industry. In a process with a reversible reaction, we need conditions that give as much product as possible. However, there are other economic and safety factors to consider.

Figure 3 This equilibrium mixture can be changed by adding or removing chlorine from the mixture

b An equilibrium mixture is set up in a closed system with iodine monochloride, chlorine gas and iodine trichloride. In order to make more iodine trichloride, would you pump more chlorine gas into the mixture or remove chlorine gas?

Figure 2 The situation at equilibrium is just like running *up* an escalator which is going down – if you run up as fast as the escalator goes *down*, you will get *nowhere*!

links
For more information about economic and safety factors with reversible reactions in industry, see C3 4.8 Making ammonia – the Haber process and C3 4.9 The economics of the Haber process.

Summary questions

1 Copy and complete using the words below:
amount conditions equilibrium forward products rate reactants reverse reversible
In some reactions the .... can react to re-form the reactants again. We call this a .... reaction. At .... in a closed system, the .... of the .... reaction is the same as the rate of the .... reaction. If we change the reaction ...., this can affect the .... of the .... and products in the equilibrium mixture.

2 What can we say about the concentration of reactants and products in a reversible reaction at equilibrium?

3 In general, how can we change the amount of product in a reacting mixture at equilibrium?

Key points

- In a reversible reaction the products of the reaction can react to re-form the original reactants.
- In a closed system the rate of the forward and reverse reactions are equal at equilibrium.
- Changing the reaction conditions can change the amounts of products and reactants in a reaction mixture at equilibrium.

241

Answers to in-text questions

a The two rates are equal.

b Pump more chlorine gas in.

Summary answers

1 products, reversible, equilibrium, rate, forward (reverse), reverse (forward), conditions, amount, reactants

2 They remain constant.

3 Change the conditions or add reactants and/or remove products.

C3 4.7 Altering conditions

Learning objectives

Students should learn:

- how changing pressure can affect reversible reactions that involve gases **[HT only]**
- how changing the temperature affects reversible reactions. **[HT only]**

Learning outcomes

Most students should be able to:

- state the effect of changing the pressure on an equilibrium mixture involving gases **[HT only]**
- state the effect on an equilibrium mixture of changing the temperature, given the energy change of the forward reaction. **[HT only]**

Some students should also be able to:

- explain the effects of changing temperature and pressure in a given reversible reaction. **[HT only]**

Specification link-up: Chemistry C3.4

- If the temperature is raised, the yield from the endothermic reaction increases and the yield from the exothermic reaction decreases. *[C3.5.1 e)]* **[HT only]**
- If the temperature is lowered, the yield from the endothermic reaction decreases and the yield from the exothermic reaction increases. *[C3.5.1 f)]* **[HT only]**
- In gaseous reactions, an increase in pressure will favour the reaction that produces the least number of molecules as shown by the symbol equation for that reaction. *[C3.5.1 g)]* **[HT only]**

Lesson structure

Starters

Particle model – Ask students to think back to the particle model of matter, which they met in KS3. Ask them to use a labelled diagram of particles in a gas state to explain gas pressure. You could support students by supplying them with a diagram and labels to add. Extend students by asking them how pressure could be increased. *(5 minutes)*

True or false? – Create two A4 sheets of paper, one with 'True' written on and the other with 'False'. Then secure these at each side of the classroom. Read out the statements in the list below. Students should then stand by the poster that shows their response. If they are unsure, they could stand in the centre or move more towards one side than the other to show how sure they are of their answer.

- Exothermic reactions give out energy. [True]
- Endothermic reactions take in energy. [True]
- Pressure is caused by particles in a solid state. [False]
- When temperature is increased, particles move slower. [False]
- When pressure is increased, particles move faster. [False] *(10 minutes)*

Main

- You may wish to demonstrate the effect of changing temperature on the reversible reaction of $N_2O_4 \rightleftharpoons 2NO_2$. See 'Practical support' for more details. Give students the equation and encourage them to balance it. They should annotate the equation to show the exothermic and endothermic directions and the colours of the gases. Demonstrate the experiment and ask them to use the balanced symbol equation to explain their observations.
- Split the class into two teams. One group should explain how the choice of temperature can affect reversible reactions and the other should describe how changing pressure affects reversible reactions. Teams should be encouraged to use actual reversible reactions to illustrate their ideas. The groups should decide how they will present their explanations to the class (as a presentation; using the board; just speaking; as a poster, etc). It is up to the group to decide.

Plenaries

Explanations – Each student should write two summary sentences in its notes to explain how temperature and pressure can affect a reversible reaction. *(5 minutes)*

AfL (Assessment for Learning) – Give students an examination question on this topic and three fictitious student answers. Students should work in small groups to discuss the answers and put them in order according to the number of marks that they would give. Then, using questions-and-answers, you could get feedback from the whole class and reveal the positive and negative points of the answers. *(10 minutes)*

Support

- Give students a set of questions to prompt them as they explain how changing temperature and pressure affects the position of equilibrium.

Extend

- Give students data for the yield of a reversible reaction at different temperatures and different pressures. Students could then plot two graphs to show the trend in yield as these variables change. Some students may plot the data on one graph with two *x*-axis scales. They could then be encouraged to use the graphs to conclude how temperature and pressure affect this specific reaction. You could encourage these students to explain the trends using information from the Student Book.

Practical support

Demonstration: the effect of temperature and pressure on equilibrium

Equipment and materials required

Plastic 50 cm^3 syringe with Luer lock and caps half-filled with dry NO_2, 3 × sealed boiling tubes with equilibrium mixture of N_2O_4 and NO_2, ice bath, warm water bath, boiling tube holder, fume cupboard, eye protection.

Details

Put a sealed boiling tube in warm water, a second in ice water and the third in the boiling tube holder. Contrast the colour change. Push the gas syringe to show how changing the pressure affects the concentration of gases and therefore the colour.

Safety: See CLEAPSS Hazcard 68 – N_2O_4 and NO_2 are very toxic. Great care needs to be taken with this demonstration. Preparation of these gases uses lead salts which are also toxic. Preparation and the demonstration must be done in a working fume cupboard. Waste must be collected by a specialist collector. Further information and safety advice can be found at www.practicalchemistry.org. Glass syringes could be used instead of boiling tubes.

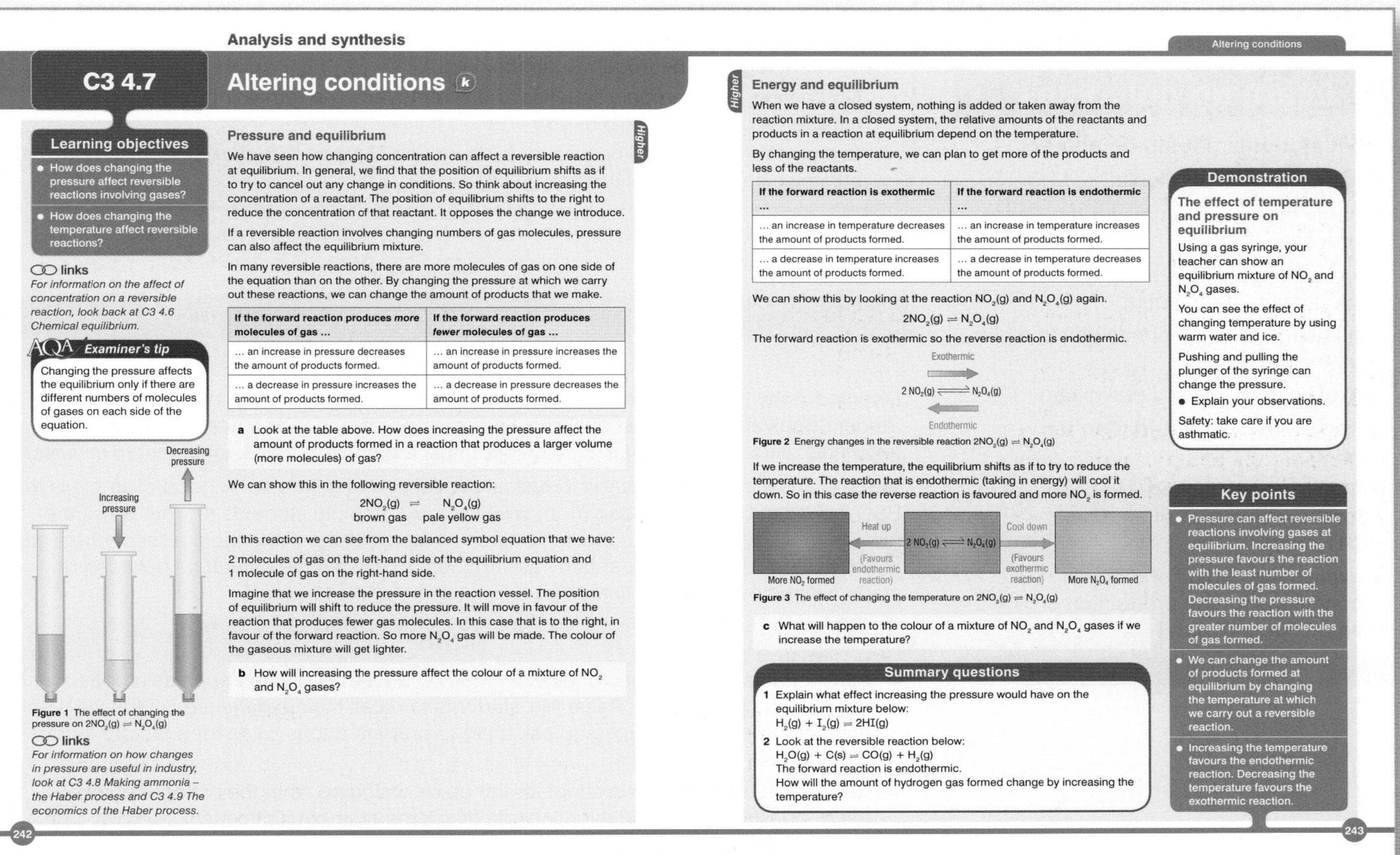

Analysis and synthesis

C3 4.7 Altering conditions

Learning objectives

- How does changing the pressure affect reversible reactions involving gases?
- How does changing the temperature affect reversible reactions?

Pressure and equilibrium

We have seen how changing concentration can affect a reversible reaction at equilibrium. In general, we find that the position of equilibrium shifts as if to try to cancel out any change in conditions. So think about increasing the concentration of a reactant. The position of equilibrium shifts to the right to reduce the concentration of that reactant. It opposes the change we introduce.

If a reversible reaction involves changing numbers of gas molecules, pressure can also affect the equilibrium mixture.

In many reversible reactions, there are more molecules of gas on one side of the equation than on the other. By changing the pressure at which we carry out these reactions, we can change the amount of products that we make.

If the forward reaction produces more molecules of gas ...	If the forward reaction produces fewer molecules of gas ...
... an increase in pressure decreases the amount of products formed.	... an increase in pressure increases the amount of products formed.
... a decrease in pressure increases the amount of products formed.	... a decrease in pressure decreases the amount of products formed.

a Look at the table above. How does increasing the pressure affect the amount of products formed in a reaction that produces a larger volume (more molecules) of gas?

We can show this in the following reversible reaction:

$$2NO_2(g) \rightleftharpoons N_2O_4(g)$$

brown gas — pale yellow gas

In this reaction we can see from the balanced symbol equation that we have:

2 molecules of gas on the left-hand side of the equilibrium equation and 1 molecule of gas on the right-hand side.

Imagine that we increase the pressure in the reaction vessel. The position of equilibrium will shift to reduce the pressure. It will move in favour of the reaction that produces fewer gas molecules. In this case that is to the right, in favour of the forward reaction. So more N_2O_4 gas will be made. The colour of the gaseous mixture will get lighter.

b How will increasing the pressure affect the colour of a mixture of NO_2 and N_2O_4 gases?

links: For information on the affect of concentration on a reversible reaction, look back at C3 4.6 Chemical equilibrium.

AQA Examiner's tip: Changing the pressure affects the equilibrium only if there are different numbers of molecules of gases on each side of the equation.

Decreasing pressure
Increasing pressure

Figure 1 The effect of changing the pressure on $2NO_2(g) \rightleftharpoons N_2O_4(g)$

links: For information on how changes in pressure are useful in industry, look at C3 4.8 Making ammonia – the Haber process and C3 4.9 The economics of the Haber process.

Energy and equilibrium (Higher)

When we have a closed system, nothing is added or taken away from the reaction mixture. In a closed system, the relative amounts of the reactants and products in a reaction at equilibrium depend on the temperature.

By changing the temperature, we can plan to get more of the products and less of the reactants.

If the forward reaction is exothermic ...	If the forward reaction is endothermic ...
... an increase in temperature decreases the amount of products formed.	... an increase in temperature increases the amount of products formed.
... a decrease in temperature increases the amount of products formed.	... a decrease in temperature decreases the amount of products formed.

We can show this by looking at the reaction $NO_2(g)$ and $N_2O_4(g)$ again.

$$2NO_2(g) \rightleftharpoons N_2O_4(g)$$

The forward reaction is exothermic so the reverse reaction is endothermic.

Exothermic
$2NO_2(g) \rightleftharpoons N_2O_4(g)$
Endothermic

Figure 2 Energy changes in the reversible reaction $2NO_2(g) \rightleftharpoons N_2O_4(g)$

If we increase the temperature, the equilibrium shifts as if to try to reduce the temperature. The reaction that is endothermic (taking in energy) will cool it down. So in this case the reverse reaction is favoured and more NO_2 is formed.

More NO_2 formed — Heat up (Favours endothermic reaction) — $2NO_2(g) \rightleftharpoons N_2O_4(g)$ — Cool down (Favours exothermic reaction) — More N_2O_4 formed

Figure 3 The effect of changing the temperature on $2NO_2(g) \rightleftharpoons N_2O_4(g)$

c What will happen to the colour of a mixture of NO_2 and N_2O_4 gases if we increase the temperature?

Summary questions

1 Explain what effect increasing the pressure would have on the equilibrium mixture below:
$H_2(g) + I_2(g) \rightleftharpoons 2HI(g)$

2 Look at the reversible reaction below:
$H_2O(g) + C(s) \rightleftharpoons CO(g) + H_2(g)$
The forward reaction is endothermic.
How will the amount of hydrogen gas formed change by increasing the temperature?

Demonstration

The effect of temperature and pressure on equilibrium

Using a gas syringe, your teacher can show an equilibrium mixture of NO_2 and N_2O_4 gases.

You can see the effect of changing temperature by using warm water and ice.

Pushing and pulling the plunger of the syringe can change the pressure.

- Explain your observations.

Safety: take care if you are asthmatic.

Key points

- Pressure can affect reversible reactions involving gases at equilibrium. Increasing the pressure favours the reaction with the least number of molecules of gas formed. Decreasing the pressure favours the reaction with the greater number of molecules of gas formed.
- We can change the amount of products formed at equilibrium by changing the temperature at which we carry out a reversible reaction.
- Increasing the temperature favours the endothermic reaction. Decreasing the temperature favours the exothermic reaction.

Further teaching suggestions

Simulation

- There are a number of simulations available for the reversible reactions. These allow students to change the conditions to investigate the effect on yield. One such simulation could be projected on to an interactive whiteboard. Teams could then come to the board and change the temperature and pressure separately in order to obtain their own set of results to plot the graphs.

Rate versus yield

- A temperature of 0 °C would achieve a better percentage yield in some reversible reactions where the forward direction is exothermic. However, this temperature would never be used. Ask students to explain why it would not be used by industry. In their answer they need to refer to the collision theory.

Word search

- Create a word search involving key words from this section. However, do not share the words with the students: they should refer to the key points section to decide which words they can find. Support some students by giving them the number of words to be found in the search, along with their first letter.

Answers to in-text questions

a It decreases the amount of product formed.

b It will turn a lighter brown colour.

c It will turn a darker brown colour.

Summary answers

1 No effect because there are the same number of molecules (or moles) of gas on either side of the equation.

2 The amount of hydrogen will increase.

C3 4.8 Making ammonia – the Haber process

Learning objectives

Students should learn:

- that ammonia is an important chemical
- how ammonia is made
- how ammonia can be made without wasting raw materials.

Learning outcomes

Most students should be able to:

- state a use for ammonia
- name the raw materials for the Haber process
- write the word equation for the production of ammonia
- quote the reaction conditions to make ammonia
- explain how waste is minimised.

Some students should also be able to:

- construct a balanced-symbol equation for the production of ammonia **[HT only]**
- explain the choice of reaction conditions for the production of ammonia. **[HT only]**

Specification link-up: Chemistry C3.5

- The raw materials for the Haber process are nitrogen and hydrogen. Nitrogen is obtained from the air and hydrogen may be obtained from natural gas or other sources. *[C3.5.1 a)]*
- The purified gases are passed over a catalyst of iron at a high temperature (about 450 °C) and a high pressure (about 200 atmospheres). Some of the hydrogen and nitrogen reacts to form ammonia. The reaction is reversible so ammonia breaks down again into nitrogen and hydrogen:

  nitrogen + hydrogen $\rightleftharpoons$ ammonia

  On cooling, the ammonia liquefies and is removed. The remaining hydrogen and nitrogen are recycled. *[C3.5.1 b)]*
- Evaluate the conditions necessary in an industrial process to maximise yield and minimise environmental impact. *[C3.5]*

Lesson structure

Starters

Directed Activity Relating to Text (DART) – Ask the students to study the topic in the Student Book and make a list of all the elements and compounds that are mentioned. [elements: nitrogen, hydrogen, iron; compounds: ammonia, nitrates, methane] *(5 minutes)*

Video – Show a video about the industrial production of ammonia. A good example is the RSC *Industrial Chemistry* video (from www.rsc.org). Ask the students to write down the raw materials and reaction conditions. Support students by giving them a list of chemicals to choose from. Extend students by asking them to explain the difference between raw materials and reactants. *(10 minutes)*

Main

- Ask students whether they have ever been on a tour of a factory such as a pottery or a car-manufacturing plant. Ask a few students to recall their experiences of the visit. Split the class into small groups and ask them to prepare a tour guide for a factory tour of an ammonia plant.
- Encourage them to write a script for the guide including what they should point out to the guests. They could even include diagrams or images of certain sections. The students could extend this activity by making a visitor guidebook.
- Students could perform their own audio revision piece for the Haber process. They could record it using digital technology and this could be put on to the school website to help others with revision.
- The Haber process in an important industrial reaction met at many levels in chemistry. Other commercially available scientific posters could be shown to students to give them ideas about how to approach designing an effective poster. Encourage them to make a poster to highlight the important aspects of the Haber process, including a balanced symbol equation.
- Some students could discuss in detail the sometimes conflicting factors affecting rate of reaction, yield, safety and economics of the Haber process, although this is expanded in C3 4.9 The economics of the Haber process.

Plenaries

Mnemonic – Ask students to summarise the Haber process and then create a mnemonic to help them remember it. You could encourage some students to share theirs with the rest of the class. *(5 minutes)*

Questions and answers – Give each student a slip of paper (about A5 in size). Ask them to write a question and its answer on the paper about the Haber process and reversible reactions. Then ask them to cut each answer free from the question and give them to you. Now give a question and answer to each student but they should not match. Read the first question out, then the student with the correct answer should read it out. Then read that student's question and so on around the room. Support students by asking them to work in small groups and give them questions that they must answer. Extend students by asking them to include balanced symbol equations in their answers. *(10 minutes)*

Support

- You could provide the students with a simple script for the factory tour. However, the paragraphs could be in the wrong order. Students should then order the information and copy it out into their own work.

Extend

- Ask students to discover what Le Châtelier's principle is and how it can be used to maximise yield from the Haber process.

Further teaching suggestions

Research ideas

- Before the Haber process was invented, ammonia had already been discovered and was being used for a variety of things including fertilisers. Ask the students to research where this ammonia came from [urine].
- Ammonia is an important chemical: ask students to find three uses for ammonia.
- There are a number of computer simulations and models for the Haber process available. In some of them, it is possible for students to vary conditions and view the effect that it has on yield.

Answers to in-text questions

a From nitrates in the soil, absorbed through their roots.

b Nitrogen and hydrogen.

c The gases are cooled down and ammonia is liquefied/condensed and run off.

d The unreacted nitrogen and hydrogen are recycled into the reaction vessel.

Analysis and synthesis

C3 4.8

Making ammonia – the Haber process

Learning objectives

- Why is ammonia important?
- How do we make ammonia?
- How can we make ammonia without wasting raw materials?

We need plants for food, and as a way of maintaining oxygen in the air. Plants need nitrogen to grow. Although nitrogen gas makes up about 80 per cent of the air, most plants cannot use it directly.

Instead, plants absorb soluble nitrates from the soil through their roots. When we harvest crops the nitrogen in plants is lost. That's because the plants do not die and decompose to replace the nitrogen in the soil. So farmers need to put nitrogen back into the soil for the next year's crops.

Nowadays we usually do this by adding nitrate fertilisers to the soil. We make these fertilisers using a process invented by a German chemist called Fritz Haber.

a Where do plants get their nitrogen from?

Figure 1 Plants are surrounded by nitrogen in the air. They cannot use this nitrogen, and rely on soluble nitrates in the soil instead. We supply these by spreading fertiliser on the soil.

The Haber process

The Haber process provides us with a way of turning the nitrogen in the air into ammonia. We can use ammonia in many different ways. One of the most important of these is to make fertilisers.

The raw materials for making ammonia are:

- nitrogen from the air
- hydrogen which we mainly get from natural gas (containing methane, CH_4).

The nitrogen and hydrogen are purified. Then they are passed over an iron catalyst at high temperatures (about 450 °C) and pressures (about 200 atmospheres). The product of this reversible reaction is ammonia.

b What are the two gases needed to make ammonia?

1 Gas stream containing hydrogen and nitrogen

2 The nitrogen/hydrogen mixture is compressed to a pressure of 200 atm and heated to 450 °C

3 Reaction vessel containing iron catalyst

4 The mixture of gases emerging from the reactor is cooled. Ammonia liquefies and is separated.

5 Unreacted nitrogen and hydrogen are returned to the reaction vessel via the compressor

Cooling chamber

Figure 2 The Haber process

The reaction used in the Haber process is reversible. This means that the ammonia breaks down again into nitrogen and hydrogen. We remove the ammonia by cooling the gases so that the ammonia liquefies. It can then be separated from the unreacted nitrogen and hydrogen that remain as gases. The unreacted nitrogen and hydrogen are recycled back into the reaction mixture. They then have a chance to react again.

$$N_2 + 3H_2 \xrightleftharpoons[]{450\,°C \text{ and } 200\text{ atm}} 2NH_3$$

nitrogen   hydrogen   ammonia

By removing the ammonia that forms we can reduce the rate of the reverse reaction. This helps to stop the ammonia that is formed from breaking down into nitrogen and hydrogen.

We carry out the Haber process in carefully chosen conditions. These are decided to give a reasonable yield of ammonia as quickly as possible.

c How is ammonia removed from the reaction mixture?
d How do we make sure the reactants are not wasted?

Summary questions

1 Copy and complete using the words and numbers below:

air fertilisers gas 450 hydrogen iron liquefies nitrogen 200 cooling

Ammonia is an important chemical used for making ………… . The raw materials to make it are ………… (from the …………) and ………… (mainly from natural …………). These are reacted at about ………… °C and ………… atmospheres pressure using an ………… catalyst. Ammonia is removed from the reaction mixture by ………… it. This ………… the ammonia gas.

2 Draw a flowchart to show how the Haber process is used to make ammonia.

links
For more information about the choice of conditions, see C3 4.9 The economics of the Haber process.

Did you know …?

Fritz Haber's work on the Haber process took place just before the First World War. When the war started, he was involved in the first chemical weapons used in warfare. His wife, also a chemist, opposed his work on chemical weapons. She committed suicide before the war ended.

Key points

- Ammonia is an important chemical for making other products, including fertilisers.
- Ammonia is made from nitrogen and hydrogen in the Haber process.
- We carry out the Haber process under conditions which are chosen to give a reasonable yield of ammonia as quickly as possible.
- Any unreacted nitrogen and hydrogen are recycled in the Haber process.

244 245

Summary answers

1 fertilisers, nitrogen, air, hydrogen, gas, 450, 200, iron, cooling, liquefies

2

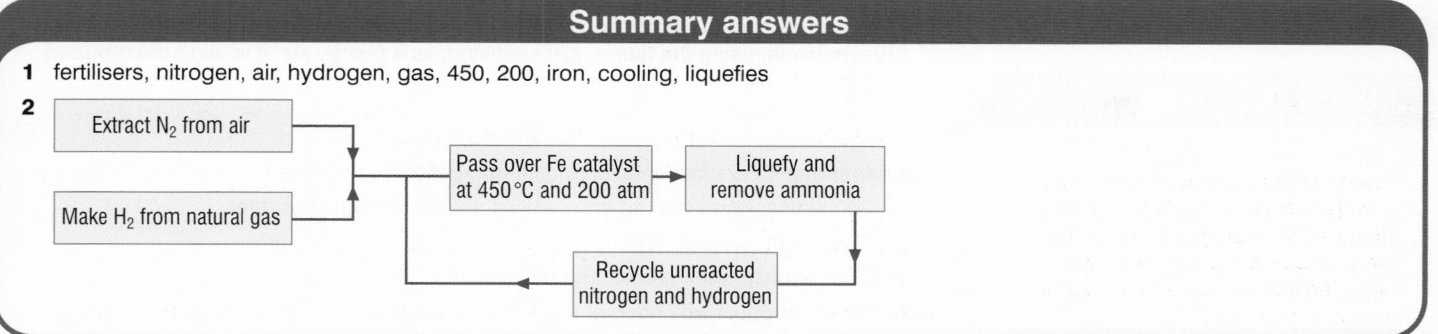

C3 4.9 The economics of the Haber process

Learning objectives

Students should learn:

- why a temperature of 450 °C is used for the Haber process **[HT only]**
- why a pressure of about 200 atmospheres is used for the Haber process. **[HT only]**

Learning outcomes

Most students should be able to:

- state the operating temperature and pressure used in the Haber process.

Some students should also be able to:

- explain the effects of changing temperature and pressure in a given reversible reaction **[HT only]**
- justify the choice of conditions in the Haber process. **[HT only]**

Specification link-up: Chemistry C3.5

- These factors, together with reaction rates, are important when determining the optimum conditions in industrial processes, including the Haber process. *[C3.5.1 h)]* **[HT only]**
- Describe and evaluate the effects of changing the conditions of temperature and pressure on a given reaction or process. *[C3.5]* **[HT only]**
- Evaluate the conditions necessary in an industrial process to maximise yield and minimise environmental impact. *[C3.5]*
- Evaluate the conditions used in industrial processes in terms of energy requirements. *[C3.5]*

Lesson structure

Starters

Flow chart – Students have already encountered the Haber process. To refresh their memory, give the students an unfinished flow chart of the industrial process for them to complete. *(5 minutes)*

True or false – Create two A4 sheets of paper, one with 'True' written on and the other with 'False'. Then secure these at each side of the classroom. Read out the following statements and students should then stand by the poster that shows their response. If they are unsure, they could stand in the centre or move more towards one side than the other to show how sure they are of their answer.

- The Haber process makes ammonia. [True]
- The Haber process is a reversible reaction. [True]
- The reactants for the Haber process are natural gas and air. [False]
- The Haber process is carried out at room temperature and pressure. [False]
- There are more moles of gas on the products' side than the reactants' side in the Haber process. [False]

Support students by allowing them to work in pairs to decide if the statements are true or false. Extend students by asking them to correct the false statements. *(10 minutes)*

Main

- Give the students data for the yield of the Haber process at different temperatures and different pressures. Students could then plot two graphs to show the trend in yield as these variables change. You could then encourage them to use the graphs to make conclusions about how changing temperature or pressure affects this reaction. Students could then try to explain these trends using information from the Student Book.
- Split the class into two teams. One group should explain the choice of temperature in the industrial Haber process and the other should explain the choice of pressure. The groups should decide how they will present their explanations to the class – as a presentation, using the board, just speaking, as a poster, etc. It is up to the group to decide.

Plenaries

Explanations – Give students three different reaction conditions for an ammonia factory:

0 °C, 200 atmospheres and an iron catalyst. [high yield but slow rate]

450 °C, 200 atmospheres and no catalyst. [high yield but slow rate]

450 °C, 20 atmospheres and an iron catalyst. [low yield, fast rate]

Ask them to suggest why none of these sets of conditions is optimum. *(5 minutes)*

AfL (Assessment for Learning) – Give students an examination question on this topic and allow them to answer it. Students should then use the mark scheme and assess their own work. You could use either Foundation or Higher Tier, depending on the ability of the students in the class. *(10 minutes)*

Support

- Students may need support to interpret graphs that show how yield is affected as temperature or pressure changes. These students could be given a set of questions that lead them through the interpretation of the graph.

Extend

- Show students both pressure and temperature data on the same yield graph for the Haber process. This requires much higher graphical skills to interpret it.

Further teaching suggestions

Video

- *Industrial Chemistry* (from www.rsc.org), which includes the Haber process could be shown to remind students of the reaction.

Simulation

- There are a number of simulations available for the Haber process. These allow students to change the conditions to investigate the affect on yield. One such simulation could be projected on to an interactive whiteboard. Teams could then come to the board and change the temperature and pressure separately in order to get their own set of results to plot the graphs.

Analysis and synthesis

C3 4.9

The economics of the Haber process

Learning objectives

- Why do we use a pressure of about 200 atmospheres for the Haber process?
- Why do we use a temperature of 450°C for the Haber process?

The effect of pressure

Higher

We have just seen how ammonia is made in the Haber process. Nitrogen and hydrogen react to make ammonia in a reversible reaction:

$$N_2 + 3H_2 \rightleftharpoons 2NH_3$$

As the balanced equation above shows, there are 4 molecules of gas on the left-hand side of the equation ($N_2 + 3H_2$). But on the right-hand side there are only 2 molecules of gas ($2NH_3$). This means that the volume of the reactants is much greater than the volume of the products. So an increase in pressure will tend to shift the position of equilibrium to the right, producing more ammonia.

a How does the volume of the products in the Haber process compare with the volume of the reactants?

To get the maximum possible yield of ammonia, we need to make the pressure as high as possible. But very high pressures need lots of energy to compress the gases. Very high pressures also need expensive reaction vessels and pipes. They have to be strong enough to withstand very high pressures. Otherwise there is always the danger of an explosion. To avoid the higher costs of building a stronger chemical plant, the Haber process uses a pressure of 200 atmospheres. This is a good compromise – it gives a lower yield than it would with even higher pressures but it does reduce the expense.

Figure 1 It is very expensive to build chemical plants that operate at high pressures

AQA Examiner's tip

Remember that the yield is the amount of product obtained compared with (that is, divided by) the amount we would obtain if all of the reactants were used up to make product.

The effect of temperature

The effect of temperature on the Haber process is more complicated than the effect of pressure. The forward reaction is exothermic. So if the temperature is low, this would increase the amount of ammonia in the reaction mixture at equilibrium.

Higher

But at a low temperature, the rate of the reaction would be very slow. That's because the particles would collide less often and would have less energy. To make ammonia commercially we must get the reaction to go quickly. We don't want to waste time waiting for the ammonia to be produced.

To do this we need another compromise. A reasonably high temperature is used to get the reaction going at a reasonable rate, even though this reduces the yield of ammonia. Look at Figure 2.

Figure 2 The conditions for the Haber process are a compromise between getting a reasonable yield of ammonia and getting the reaction to take place at a fast enough rate

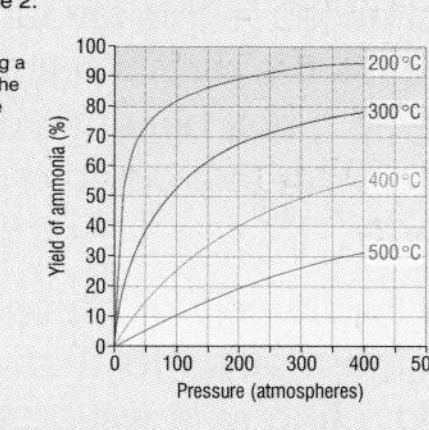

A lower temperature can also reduce the effectiveness of the iron catalyst. See below.

The effect of a catalyst

We also use an iron catalyst to speed up the reaction. A catalyst speeds up the rate of both the forward and reverse reactions by the same amount. Therefore it does not affect the actual yield of ammonia but we do get it formed more quickly.

Did you know ... ?

The Haber process is sometimes called the Haber–Bosch process, since Fritz Haber found out how to make ammonia from nitrogen and hydrogen but Carl Bosch carried out the work to find the best conditions for the reaction. Bosch and his team carried out 6500 experiments to find the best catalyst for the reaction.

Summary questions

1 Copy and complete using the words below:

decreases exothermic fewer increasing left pressure released

The Haber process is .... so energy is .... during the reaction. This means that .... the temperature .... the yield of ammonia. Increasing the .... will increase the yield of ammonia, because there are .... molecules of gas on the right-hand side of the equation than on the ....-hand side.

2 Look at Figure 2.
- **a** What is the approximate yield of ammonia at a temperature of 500°C and 400 atmospheres pressure?
- **b** What is the approximate yield of ammonia at a temperature of 500°C and 100 atmospheres pressure?
- **c** What is the approximate yield of ammonia at a temperature of 200°C and 400 atmospheres pressure?
- **d** What is the approximate yield of ammonia at a temperature of 200°C and 100 atmospheres pressure?
- **e** Why is the Haber process carried out at around 200 atmospheres and 450°C?

Key points

- The Haber process uses a pressure of around 200 atmospheres to increase the amount of ammonia produced.
- Although higher pressures would produce more ammonia, they would make the chemical plant too expensive to build and run.
- A temperature of about 450°C is used for the reaction. Although lower temperatures would increase the yield of ammonia, it would be produced too slowly.

Answers to in-text questions

a The volume of products is smaller.

Summary answers

1 exothermic, released, increasing, decreases, pressure, fewer, left

2
- **a** 32%
- **b** 10%
- **c** 95%
- **d** 82%
- **e** This combines optimum conditions for rate of reaction, amount of ammonia at equilibrium and cost.

Summary answers

1 a no reaction/dissolves
  b white precipitate forms
  c white precipitate forms
  d red flame (brick red)
  e sodium iodide
  f copper carbonate

2 potassium sulfate

3 a $NaOH(aq) + HCl(aq) \rightarrow NaCl(aq) + H_2O(l)$
  b i 0.01 mol ii 0.01 mol
  c i $0.4\,mol/dm^3$ ii $16\,g/dm^3$

4 a $X(g) + 2Y(g) \rightleftharpoons 2Z(g)$
  b i increase it ii increase it
  c i decrease it ii increase it

5 The yield of ammonia at equilibrium decreases if we increase the temperature but a moderately high temperature of 450 °C is chosen to increase the rate at which ammonia is formed. The high pressure of 200 atmospheres is used because this favours formation of ammonia (and increases the rate of reaction) but even higher pressures are not used because of safety factors and the high cost of a plant that could resist extremely high pressures.

6 a Use the pH meter because it produces a continuous variable, which is more powerful than the categoric variable of a colour change. It will produce greater accuracy than using a colour chart comparison.
  b As little as can be reasonably measured. This produces more plotting points on the graph and therefore a more accurate graph.
  c To ensure that the pH represents the whole of the contents of the flask. It increases the accuracy. This is a potential area for human error and therefore random errors. Mixing ensures an even distribution of all types of particles throughout the mixture.
  d To increase the accuracy of your method [accept improve reliability].
  e As a line graph of the mean pH against the volume of alkali added.
  f You should have a pH of 7 when the same volume of acid has been added, as there is alkali in the flask.

Kerboodle resources

Resources available for this chapter on Kerboodle are:

- Chapter map: Analysis and synthesis
- Practical: Identifying positive ions (C3 4.1)
- Support: What compound is it? (C3 4.1 and 4.2)
- How Science Works: Time to titrate (C3 4.4)
- Maths skills: Titrations (C3 4.4)
- Simulation: The effect of temperature and pressure on the position of equilibrium (C3 4.7)
- Extension: Balancing act (C3 4.7)
- How Science Works: The Haber process (C3 4.9)
- Interactive activity: Analysis and synthesis
- Revision podcast: The Haber process
- Test yourself: Analysis and synthesis
- On your marks: Titrations
- Examination-style questions: Analysis and synthesis
- Answers to examination-style questions: Analysis and synthesis

Summary questions

1 Copy and complete the table.

Add dilute acid	Add sodium hydroxide solution	Add dilute nitric acid followed by silver nitrate solution	Flame test	Substance
Nothing observed	Nothing observed	Pale yellow precipitate	Yellow	e
Fizzing – gas turns limewater cloudy	Pale blue precipitate formed	Not needed	Nothing observed	f
a	b	c	d	Calcium chloride

2 An unknown compound gave the following positive tests when analysed:

- The Bunsen flame turned lilac in a flame test.
- When dissolved in dilute hydrochloric acid and barium chloride solution added, a white precipitate was formed.

Name the unknown compound.

3 A student put $25.0\,cm^3$ of sodium hydroxide solution of an unknown concentration into a conical flask using a pipette. The sodium hydroxide reacted with exactly $20.0\,cm^3$ of $0.50\,mol/dm^3$ hydrochloric acid added from a burette.

a Write the balanced symbol equation for this reaction.
b i How many moles of hydrochloric acid are there in $20.0\,cm^3$ of $0.50\,mol/dm^3$ hydrochloric acid?
  ii How many moles of sodium hydroxide will this react with?
c What is the concentration of the sodium hydroxide solution:
  i in moles per dm^3?
  ii in grams per dm^3?
  (A_r values: Na = 23; O = 16; H = 1) [H]

4 A chemical reaction can make product Z from reactants X and Y. Under the reaction conditions, X, Y and Z are gases.

X, Y and Z react in the proportions 1 : 2 : 2. The reaction is carried out at 250 °C and 100 atmospheres. The reaction is reversible, and it is exothermic in the forward direction.

a Write a balanced symbol equation for this (reversible) reaction.
b How would increasing the pressure affect:
  i the amount of Z formed
  ii the rate at which Z is formed?
c How would increasing the temperature affect:
  i the amount of Z formed
  ii the rate at which Z is formed?

5 Explain why the conditions in the Haber process (200 atmospheres pressure and a temperature of 450 °C) ar described as a 'compromise'.

6 You have been asked to design an investigation to find out the changes in pH as you add an acid to an alkali. You have a burette into which you have been told to ad some dilute hydrochloric acid. The flask will contain yo sodium hydroxide solution.

You have to add the acid to the alkali and get a reading of the pH as more and more acid is added.

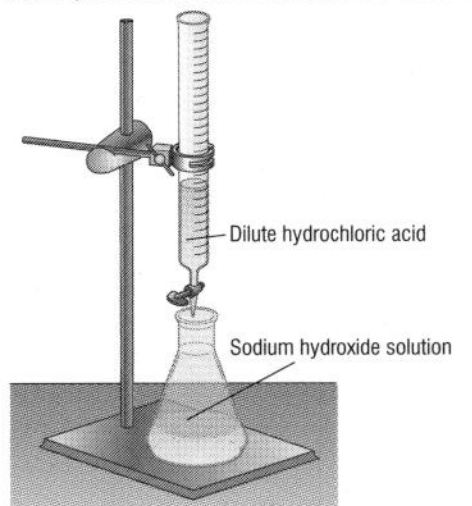

a You have a choice of either using an indicator and recording the colour changes or a pH meter and recording the pH. Which would you choose? Explain your choice.
b How much acid would you add at a time?
  Explain your answer.
c Why would you stir the flask before checking the pH?
d Why would it be important to repeat the titration at leas one more time?
e How would you present your results?
f If you knew that both the acid and the alkali were of the same concentration, how might you check the accura of your results?

Practical suggestions

Practicals	AQA	k		
Flame tests – spray solution into flame or use wooden splints soaked in solutions overnight or use nichrome wire loops.	✓	✓	✓	
Try tests using mixtures of two salts, e.g. flame tests on solutions containing pairs of the listed ions.	✓		✓	
Fe^{2+} with sodium hydroxide solution – note that the initial colour is quickly oxidised.	✓	✓	✓	
React carbonates with acid and test the gas for CO_2 using a drop of limewater on a glass rod.	✓	✓	✓	
Distinguishing between the halide ions using silver nitrate solution.	✓			
Identifying unknown single salts using the tests in the content.	✓		✓	
Plan a suitable order of tests to use on a solution that contains an unknown single salt.	✓		✓	
Strong acid/strong alkali titrations (HCl/NaOH) to find unknown concentration (using indicators and pH sensors to determine titration endpoints).	✓		✓	

QA Examination-style questions

'he label is from a packet of low sodium salt.

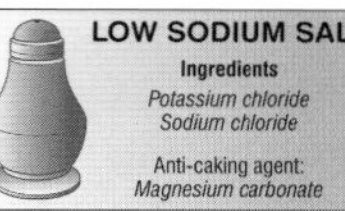
LOW SODIUM SALT
Ingredients
Potassium chloride
Sodium chloride
Anti-caking agent:
Magnesium carbonate

a A student tested some low sodium salt to show that it contains carbonate ions. Describe and give the result of a test for carbonate ions. (2)

The student identified chloride ions using acidified silver nitrate solution.

i Name the acid the student should use to acidify the solution. (1)

ii State what you would **see** when acidified silver nitrate solution is added to a solution of low sodium salt. (1)

Flame tests can be used to identify potassium ions and sodium ions.

i Describe how to do a flame test on low sodium salt. (2)

ii Suggest why it is difficult to identify **both** of these ions in low sodium salt using a flame test. (1)

AQA, 2009

'he diagram shows the main stages in the Haber process.

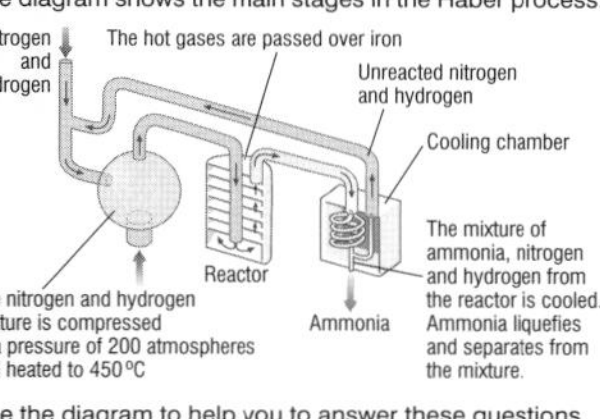
Nitrogen and hydrogen
The hot gases are passed over iron
Unreacted nitrogen and hydrogen
Cooling chamber
The mixture of ammonia, nitrogen and hydrogen from the reactor is cooled. Ammonia liquefies and separates from the mixture.
Reactor
Ammonia
The nitrogen and hydrogen mixture is compressed to a pressure of 200 atmospheres and heated to 450 °C

se the diagram to help you to answer these questions.

Complete the word equation for the reaction that takes place in the reactor.

nitrogen + ... ⇌ .... (1)

What does the symbol ⇌ mean? (1)

What is the purpose of the iron in the reactor? (1)

What is done with the unreacted nitrogen and hydrogen? (1)

AQA, 2008

3 Sulfuric acid and potassium hydroxide react to produce sodium sulfate and water.

A student was given 200 cm^3 of potassium hydroxide solution and 200 cm^3 of a solution of sulfuric acid. The student was asked to do a titration to find the volumes of the solutions that react to produce a neutral solution. The student was given some methyl orange indicator, which is yellow in alkaline solutions and red in acidic solutions.

a *In this question you will be assessed on using good English, organising information clearly and using specialist terms where appropriate.*

Describe how the student should do a titration to find the volume of the sulfuric acid that reacts with 25.0 cm^3 of the potassium hydroxide solution. Include in your answer the names of any apparatus that the student should use. (6)

b The student found that 20.0 cm^3 of sulfuric acid reacted with 25.0 cm^3 of the potassium hydroxide solution.

The equation for this reaction is:

$$H_2SO_4 + 2KOH \rightarrow K_2SO_4 + 2H_2O$$

i The potassium hydroxide solution had a concentration of 5.60 g/dm^3. Calculate the concentration of the potassium hydroxide solution in moles/dm^3.
(Relative atomic masses: H = 1, O = 16, K = 39) [H] (2)

ii Calculate the concentration of the sulfuric acid in moles/dm^3. [H] (3)

4 Hydrogen for use in the Haber process can be produced by reacting methane with steam. The methane and steam mixture is passed over a nickel catalyst at 800 °C and a pressure of 30 atmospheres. The equation for this reaction is:

$$CH_4(g) + H_2O(g) \rightleftharpoons CO(g) + 3H_2(g)$$

The forward reaction is endothermic.

a i What conditions of temperature would increase the yield of products for the forward reaction at equilibrium? Explain your answer. (2)

ii Suggest why a temperature of 800 °C, not higher or lower, is used in the industrial process. [H] (2)

b i What conditions of pressure would increase the yield of products from the forward reaction at equilibrium? Explain your answer. (2)

ii Suggest **one** reason why a pressure of 30 atmospheres is used in the industrial process. [H] (1)

Practicals	AQA	k	📖	⚙
Demonstration of the effect of adding acid and then alkali to bromine water to show the effect of changing conditions on equilibrium.	✓		✓	
Investigation of the effect of adding acid and then alkali to a solution of potassium chromate.	✓		✓	
Modelling dynamic equilibrium with two 25 cm^3 measuring cylinders, each with an open-ended glass tube but with different diameters. Put 25 cm^3 of water into one cylinder. Transfer water from one cylinder to the other using a finger over the end of each tube in turn (keep the tubes in the same cylinder) until the level in each cylinder does not change any more.	✓		✓	
Demonstration of effect of temperature and pressure on equilibrium using 50 cm^3 of NO_2/N_2O_4 in a gas syringe.	✓		✓	

Examination-style answers

1 a Add hydrochloric acid or any named acid. Carbon dioxide or bubbles or fizzes or gas turns limewater cloudy. *(2 marks)*

b i Nitric acid. *(1 mark)*

ii (White) precipitate/solid. *(1 mark)*

c i Use a blue/colourless Bunsen burner flame, add substance to flame e.g. by using a (platinum/nichrome) wire or by spraying solution. *(2 marks)*

ii Yellow or sodium colour masks/obscures lilac or potassium colour. *(1 mark)*

2 a nitrogen + hydrogen ⇌ ammonia *(1 mark)*

b Reversible (reaction). *(1 mark)*

c Catalyst or to speed up reaction. *(1 mark)*

d Recycled. *(1 mark)*

3 a There is a clear and detailed scientific description of how to do a titration. The answer shows almost faultless spelling, punctuation and grammar. It is coherent and in an organised, logical sequence. It contains a range of appropriate and relevant specialist terms used accurately. *(5–6 marks)*

There is a scientific description of how to do a titration. There are some errors in spelling, punctuation and grammar. The answer has some structure and organisation. The use of specialist terms has been attempted but not always accurately. *(3–4 marks)*

There is a brief description of how to do a titration. The spelling, punctuation and grammar are very weak. The answer is poorly organised with almost no specialist terms and/or their use demonstrating a general lack of understanding of their meaning. *(1–2 marks)*

No relevant content. *(0 marks)*

Examples of chemistry points made in the response:

- rinse burette with (sulfuric) acid
- put (sulfuric) acid into burette
- rinse pipette with potassium hydroxide solution
- use pipette to measure 25.0 cm^3 of potassium hydroxide solution
- put (25.0 cm^3) potassium hydroxide solution into a (conical) flask
- add indicator
- record volume of acid in burette
- add acid to alkali
- a small amount at a time (near end point)
- until indicator changes colour
- record final volume/volume added
- repeat (until consistent results).

b i 0.1 (moles per dm^3) gains 2 marks. If incorrect, then correct working gains 1 mark, e.g. 39 + 16 + 1 or (1 mole) KOH = 56 (g). *(2 marks)*

ii 0.0625 (moles per dm^3) gains 3 marks. If incorrect, then 0.125 or (25 × 0.1)/(20 × 2) gains 2 marks. (25 × 0.1)/20 gains 1 mark. Allow error carried forward using answer from **3bi**. *(3 marks)*

4 a i High temperature; forward reaction is endothermic or forward reaction takes in energy (allow heat). *(2 marks)*

ii Not higher because energy costs would be high(er) or apparatus would be more expensive or apparatus would need to withstand higher temperatures; not lower because rate of reaction would be slow(er) or catalyst would not work. *(2 marks)*

b i Low pressure; forward reaction produces more molecules (of gas) or more (gas) molecules on right side of equation. *(2 marks)*

ii Some pressure is needed to move gases through apparatus or to increase the rate of reaction. *(1 mark)*

C3 5.1 Structure of alcohols, carboxylic acids and esters

Learning objectives

Students should learn:

- the names and formulae of some simple alcohols, carboxylic acids and esters
- how to show the structures of alcohols, carboxylic acids and esters.

Learning outcomes

Most students should be able to:

- name an example of an alcohol, carboxylic acid and an ester
- recognise the formula of an alcohol, carboxylic acid and ester.

Some students should also be able to:

- explain the difference between an alcohol, a carboxylic acid and an ester
- draw the structure of an alcohol, a carboxylic acid and an ester.

Specification link-up: Chemistry C3.6

- Alcohols contain the functional group —OH. Methanol, ethanol and propanol are the first three members of a homologous series of alcohols. *[C3.6.1 a)]*
- Ethanoic acid is a member of the carboxylic acids, which have the functional group —COOH. *[C3.6.2 a)]*
- … Esters have the functional group —COO—… *[C3.6.3 a)]*
- Represent the structures of alcohols in the following forms:

```
    H   H
    |   |
H — C — C — O — H
    |   |
    H   H
```

CH_3CH_2OH *[C3.6]*

- Represent the structures of carboxylic acids in the following forms:

```
    H
    |
H — C — C = O
    |   |
    H   O — H
```

CH_3COOH *[C3.6]*

Answers to in-text questions

a CH_3OH

b Take the name of the alkane with the same number of carbon atoms; remove the last letter, 'e'; add the ending 'oic' (or remove 'ane' and add 'anoic').

c HCOOH

Support

- You can create a 'cut-and-stick' activity about the different organic chemicals that students need to study. Students could create a table in which the column headings are 'alcohol', 'carboxylic acid' and 'ester'. They could then cut and stick information and examples into the correct columns.

Extend

- Ask students to name simple alcohols and carboxylic acids. The prefix is the same as those used for alkanes and alkenes, however the ending is specific to the homologous series.

Lesson structure

Starters

Link-up – Show students a molecular model of an alkane, its displayed formula and molecular formula. Ask students to suggest what the link is between all of the images [they are all the same compound]. *(5 minutes)*

Definitions – Ask students to think back to the work they did in C1 Chapters 4, 5 and 6. Ask them to define the following key terms: organic compound, molecule, hydrocarbon, alkane, alkene. Support students by encouraging them to work in pairs and use their previous work and the Student Book to help them. Extend students by asking them to define a homologous series [chemical family] and give an example [e.g. alkane, alkene, alcohol]. *(10 minutes)*

Main

- Give students an A4 sheet of paper and ask them to fold it into equal thirds. At the top of each third they should write the name of one homologous series from: alcohol, carboxylic acid and ester. On one side they should state the characteristics of that homologous series, using secondary sources such as the Student Book and the internet to help them. On the reverse they should draw displayed formulae of some examples of compounds that would be in that group.
- Students could then use the school's marking policy to review each other's work, marking and making suggestions for improvements. Students could then work on the advice of their peers to improve their work before handing it to the teacher for marking.

Plenaries

Classification – Give students a set of three cards: a red card with the word 'carboxylic acid' on it, a blue card with 'ester' and a white card with 'alcohol'. Show students different chemical names, formulae and diagrams of organic molecules. Students have to hold the correct card up to correctly identify which homologous series it belongs to. Support students by using simple organic compounds that they have met during the lesson. They can then look back at their notes to help them. Extend students by using unfamiliar organic compounds. *(5 minutes)*

Circle the group – On the interactive whiteboard, show students a selection of different complex organic molecules, which contain ester, alcohol and carboxylic acid functional groups. Ask for volunteers to circle and label the different functional groups. *(5 minutes)*

Further teaching suggestions

Group activity

- Split the class into three groups. Each group should research one homologous series from alcohols, carboxylic acids and esters. Once the research has been completed, the students work in groups of three. Each group should have a researcher from the alcohol, carboxylic acid and ester teams. The students in each group should then work together to summarise their findings in not more than an A5 sheet of paper.

Molecular models

- Students can use molecular model kits to make different examples of chemicals and classify which homologous series they belong to.

Text summary

- Students could be encouraged to translate the key points into text language. The more artistic students could then draw a mobile phone with the display showing the points.

Organic chemistry

C3 5.1 Structures of alcohols, carboxylic acids and esters

Learning objectives

- What are the names and formulae of some simple alcohols, carboxylic acids and esters?
- How can we show the structures of alcohols, carboxylic acids and esters?

The substances that form the basis of all living things, including ourselves, are organic compounds. Organic molecules all contain carbon atoms. They tend to form the 'backbone' of organic molecules.

You have already met some organic compounds in C1 Chapters 4, 5 and 6. You will probably recall how we draw molecules of alkanes and alkenes.

Methane Ethane Propane Ethene Propene (Double bond)

Figure 1 The structures of the smallest alkane and alkene molecules. The lines between atoms represent covalent bonds. Notice the double bonds in ethene and propene. This type of 2D drawing to show the structure of a molecule is called a 'displayed formula'.

Both alkanes and alkenes are made of only carbon and hydrogen atoms. However, there are many more 'families' of organic compounds that also contain a few other types of atom.

In this chapter we will learn about some organic compounds made up of carbon, hydrogen and oxygen. The three 'families' we will look at are called alcohols, carboxylic acids and esters.

Did you know ...?

Not all carbon compounds are classified as organic compounds. The oxides of carbon and compounds containing carbonate and hydrogencarbonate ions are classed as inorganic compounds.

Alcohols

You have also met a member of the alcohol 'family' before. This was ethanol, which can be used as a biofuel. So what is an alcohol?

Imagine removing a H atom from an alkane molecule and replacing it with an O–H group. This would give us an alcohol molecule. The –OH group of atoms is an example of a **functional group**.

A functional group gives a 'family' of organic compounds their characteristic reactions. A 'family' of compounds with the same functional group is called a **homologous series**.

You need to know the first three members of the homologous series of alcohols:

Methanol Ethanol Propanol

Figure 2 The structures and names of the first three alcohol molecules. We name alcohols from the alkane with the same number of carbon atoms. Just take the 'e' from the end of the alkane's name and replace it with 'ol'.

If you were asked for the formula of ethanol, you might count the atoms in a molecule and write C_2H_6O. This is correct but chemists can give more information in a structural formula. This doesn't show all the bonds, as in a displayed formula, but shows what is bonded to each carbon atom. So for ethanol a chemist will often show its formula as CH_3CH_2OH (often shortened to C_2H_5OH). Can you see how this relates to the structure of ethanol shown in Figure 2?

a What is the structural formula of methanol?

links
For more information on the use of ethanol as a biofuel, see C1 4.5 Alternative fuels.

links
For information about alcohols, see C3 5.2 Properties and uses of alcohols.

AQA Examiner's tip

If you draw displayed formulae, make sure you show all the bonds and all the atoms.

Carboxylic acids

You will certainly know of one carboxylic acid. Ethanoic acid is the main acid in vinegar. All carboxylic acids contain the –COOH functional group. Look at the first three members of the homologous series of carboxylic acids:

Methanoic acid Ethanoic acid Propanoic acid

Figure 3 The structures and names of the first three carboxylic acid molecules

We show the structural formula of each of the carboxylic acids in Figure 3 as HCOOH, CH_3COOH and CH_3CH_2COOH (or C_2H_5COOH).

b What is the rule for naming a carboxylic acid?
c What is the structural formula of methanoic acid?

Figure 4 Vinegar contains less than 10 per cent ethanoic acid but the acid provides the characteristic sharp taste and smell

Esters

Esters are closely related to carboxylic acids. If we replace the H atom in the –COOH group by a hydrocarbon group we get an ester. Here is the ester called ethyl ethanoate:

Ethyl ethanoate

Figure 5 The structure of the ester ethyl ethanoate

An ester's structural formula always has the –COO– functional group in it. The structural formula of ethyl ethanoate is $CH_3COOCH_2CH_3$ (or $CH_3COOC_2H_5$).

links
For information about the homologous series of carboxylic acids and esters, see C3 5.3 Carboxylic acids and esters.

Summary questions

1 Copy and complete using the words and symbols below:
esters –COOH functional homologous –OH
A group of atoms in an organic molecule, such as –OH, is called a .... group. All alcohols contain the .... group, the carboxylic acids contain the .... group, and the .... contain the –COO– group. Each of these 'families' of organic compounds are called .... series.

2 **a** Which homologous series do the following compounds belong to?
  i CH_3COOCH_3
  ii $CH_3CH_2CH_2CH_2OH$
  iii HCOOH
b Name the following compounds:
  i $CH_3CH_2CH_2OH$
  ii $CH_3COOC_2H_5$
  iii HCOOH
c Draw a displayed formula showing all the bonds in butanoic acid, which contains 4 carbon atoms.

Key points

- The homologous series of alcohols contain the –OH functional group.
- The homologous series of carboxylic acids contain the –COOH functional group.
- The homologous series of esters contain the –COO– functional group.

Summary answers

1 functional, –OH, –COOH, esters, homologous

2 **a** i esters ii alcohols iii carboxylic acids
b i propanol/propan-1-ol ii ethyl ethanoate iii methanoic acid
c

```
     H   H   H
     |   |   |       O
H — C — C — C — C  //
     |   |   |      \
     H   H   H       O — H
```

C3 5.2 Properties and uses of alcohols

Learning objectives

Students should learn:

- the properties of alcohols
- the main uses of alcohols
- the products of the full oxidation of ethanol.

Learning outcomes

Most students should be able to:

- describe the properties of alcohols (including their reaction with sodium and their oxidation to carboxylic acids)
- state the main uses of alcohols
- write a word equation for the combustion of an alcohol.

Some students should also be able to:

- construct a balanced symbol equation for the combustion of an alcohol. **[HT only]**

Answers to in-text questions

a Ethanol.

b An oxidising agent.

Support

- Give students information about the combustion of the three alcohols. They could then cut-and-stick this information to create their results table.

Extend

- Students could use a data logger to monitor the pH of a solution of ethanol as detailed in the 'Practical support'. Ask students to print out the data as a graph and add a curve of best fit. Encourage students to annotate the graph to explain the shape of the curve of best fit using oxidation of ethanol in their answers.

Specification link-up: Chemistry C3.6

- Methanol, ethanol and propanol:
  - dissolve in water to form a neutral solution
  - react with sodium to produce hydrogen
  - burn in air
  - are used as a fuels and solvents, and ethanol is the main alcohol in alcoholic drinks. *[C3.6.1 b)]*
- Ethanol can be oxidised to ethanoic acid, either by chemical oxidising agents or by microbial action. Ethanoic acid is the main acid in vinegar. *[C3.6.1 c)]*

Lesson structure

Starters

Connection – Show students an image of a fermentation vessel, a bottle of beer, gasohol station and a bottle of methylated spirit. Ask students to say how all the images are similar [they all contain alcohol]. Support students by encouraging them to think back to the work that they completed in C1 Chaper 5.5 Ethanol. Extend students by asking them to find out more about the constituents of methylated spirits. *(5 minutes)*

Remember – Ask students to recall the products of complete combustion of a hydrocarbon. Encourage students to write a general equation for combustion of a saturated hydrocarbon [hydrocarbon + oxygen → carbon dioxide + water]. You could extend Higher Tier students by encouraging them to write a general symbol equation for this reaction [$C_xH_{2x+2} + \frac{1}{2}(3x + 1)O_2 \rightarrow xCO_2 + (x + 1)H_2O$]. Give students some examples of hydrocarbons and ask them to use their general equation to write specific equations for the combustion of these saturated hydrocarbons. *(10 minutes)*

Main

- Allow students to set up a solution of ethanol (less than 15 per cent by volume) and monitor its pH over a number of days (see 'Practical support' for more details). Once students have set the experiment up encourage them to make a prediction using scientific ideas in their work.
- Allow students to observe the combustion of different alcohols (see 'Practical support' for more details). Encourage students to draw the displayed formula of each alcohol, and write an equation for the reaction that is occurring. Their observations could be recorded in an appropriate results table.
- Ask students to create a spider diagram to show the different reactions of ethanol. Students should have the word ethanol in the centre of the page and then add information about the reactions that it undergoes. Higher Tier students could include a balanced symbol equation for the combustion reaction.

Plenaries

Finish the equation – Ask students to finish the following equations:

ethanol + [oxygen] → carbon dioxide + [water]

methanol + oxygen → [carbon dioxide] + [water]

Support students by giving them the missing words. Extend students by asking them to write balanced symbol equations for each reaction and to classify the reactions [oxidation or combustion]. *(5 minutes)*

Flow chart – Ask students to make a flow chart to explain how ethanol is oxidised when it is open to the air. Support students by giving them the outline of the flow chart. *(10 minutes)*

Summary answers

1 neutral, hydrogen, carbon, water, carboxylic, ethanoic, microbes

2 Alcoholic drinks (ethanol), solvents and fuels.

3 Place methanol, ethanol and propanol in separate spirit burners and weigh them. Heat a known volume of water for a set time and record the rise in temperature. Reweigh the spirit burners and remaining alcohols. Use $Q = mc\Delta T$ to work out the energy released by each alcohol and then scale up/down to get energy released per gram of each alcohol burned.

Practical support

Comparing the reactions of methanol, ethanol and propanol

Equipment and materials required

Three spirit burners, methanol, ethanol, propanol, sodium metal kept under oil, dry boiling tube, boiling-tube holder, tweezers, white tile, filter paper, scalpel, eye protection and safety screen.

Details

Allow students to ignite each of the spirit burners and observe the flame produced.

As a demonstration, wearing eye protection, add 10 cm^3 of ethanol to a dry Petri dish. Cut a piece of sodium (maximum 2 mm square) and wipe off the excess oil with filter paper. Using the tweezers, drop the sodium metal into the ethanol and observe. Ensure all of the sodium has reacted before disposing of the solution formed.

Safety: CLEAPSS Hazcard 88 Sodium – highly flammable; CLEAPSS Hazcard 40B Methanol – highly flammable/toxic; CLEAPSS Hazcard 40A Ethanol – harmful/highly flammable; CLEAPSS Hazcard 84A Propanol – highly flammable/irritant. For spirit burners see CLEAPSS Guide L195 'Safe chemicals, safer reactions'.

Oxidation by microbes in air

Equipment and materials required

Ethanol, universal indicator, 100 cm^3 conical flask, 100 cm^3 measuring cylinder, 10 cm^3 measuring cylinder

Details

Measure 50 cm^3 of tap water to a conical flask. Add a few drops of universal indicator and then add 5 cm^3 of ethanol. Leave the conical flask on a window sill and observe the change in pH over the course of several weeks, swirling the flask occasionally. Keep an identical stoppered flask to compare it with. This experiment works better with beer that has had carbon dioxide removed (shaking or boiling) because flies carrying the microbes are attracted by the smell.

Safety: CLEAPSS Hazcard 40A Ethanol – harmful/highly flammable; CLEAPSS Hazcard 32 Universal indicator – highly flammable/irritant.

Organic chemistry

C3 5.2 Properties and uses of alcohols

Learning objectives

- What are the properties of alcohols?
- What are the main uses of alcohols?
- What is produced when ethanol is fully oxidised?

Alcohols, especially ethanol, are commonly used in everyday products. We have already seen how ethanol is the main alcohol we refer to in alcoholic drinks. This is made by fermenting sugars from plant material. It is becoming an important alternative fuel to petrol and diesel. We also saw how ethanol can be made from ethene and steam in industry.

a What is the main alcohol in alcoholic drinks?

Alcohols dissolve many of the same substances as water. In fact the alcohols with smaller molecules mix very well with water, giving neutral solutions. The alcohols can also dissolve many other organic compounds. This property makes them useful as solvents. For example, we can remove ink stains from permanent marker pens using methylated spirits.

Methylated spirits ('meths') is mainly ethanol but has the more toxic methanol mixed with it. It also has a purple dye added and other substances to make it unpleasant to drink. Alcohols are also used as solvents in products such as perfumes, aftershaves and mouthwashes.

links

For more information on the production of ethanol and its use as a fuel, look back to C1 5.5 Ethanol.

links

For information on the dangers of drinking ethanol, see C3 5.4 Organic issues.

Figure 1 Alcohols are used as solvents in perfumes

Did you know ...?

Ethanol is the main solvent in many perfumes but a key ingredient in some perfumes is octanol. It evaporates more slowly and so holds the perfume on the skin for longer.

Reactions of alcohols

Practical/Demonstration

Comparing the reactions of methanol, ethanol and propanol

a Ignite and observe the flame in three spirit burners – one containing methanol, one ethanol and the other propanol.
- Compare the three combustion reactions.

b Watch your teacher add a small piece of sodium metal to each of the alcohols.
- Compare the reactions. Which gas is given off?

c Watch your teacher boil some of each alcohol with acidified potassium dichromate(VI) solution.
- What do you see happen in each reaction?

Combustion

The use of ethanol (and also methanol) as fuels shows that the alcohols are flammable. Ethanol is used in spirit burners. It burns with a 'clean' blue flame:

$$\text{ethanol} + \text{oxygen} \rightarrow \text{carbon dioxide} + \text{water}$$
$$C_2H_5OH + 3O_2 \rightarrow 2CO_2 + 3H_2O$$

252

Properties and uses of alcohols

Reaction with sodium

The alcohols react in a similar way to water when sodium is added. The sodium fizzes, giving off hydrogen gas, and dissolves away to form a solution. Their reactions are not as vigorous as the reaction we observe with water.

Oxidation

Combustion is one way to oxidise an alcohol. However, when we use chemical oxidising agents, such as potassium dichromate(VI), we get different products. An alcohol is oxidised to a carboxylic acid when boiled with acidified potassium dichromate(VI) solution.

b Is potassium dichromate(VI) a reducing agent or an oxidising agent?

So ethanol can be oxidised to ethanoic acid. The same reaction takes place if ethanol is left exposed to air. Microbes in the air produce ethanoic acid from the ethanol. That's why bottles of beer or wine taste and smell like vinegar when they are left open.

Figure 2 Alcohols are flammable. They produce carbon dioxide and water in their combustion reactions.

Practical

Oxidation by microbes in air

Add 5 cm^3 of ethanol to 50 cm^3 of water in a conical flask, mix and test the pH of the solution formed.
Then mix 5 cm^3 of ethanol with 50 cm^3 of water in a second conical flask but this time seal the flask with a stopper.

Leave both solutions for a few weeks, swirling occasionally.
- What happens to the pH of the solutions?
- Explain your observations.

Summary questions

1 Copy and complete using the words below:

hydrogen ethanoic water carbon carboxylic neutral microbes

Methanol, ethanol and propanol all dissolve in water to form a .... solution. The alcohols react with sodium metal, with the sodium fizzing to produce .... gas. When they burn in air, they form .... dioxide and .... as the products of combustion. They are also oxidised by acidified potassium dichromate(VI) solution to form .... acids.

The ethanol in alcoholic drinks is oxidised to .... acid by .... if left exposed to the air.

2 List the main uses of alcohols.

3 Plan an investigation to see which alcohol – methanol, ethanol or propanol – releases most energy per gram when it burns.

Key points

- Alcohols are used as solvents and fuels, and ethanol is the main alcohol in alcoholic drinks.
- Alcohols burn in air, forming carbon dioxide and water.
- With sodium metal, alcohols react to form a solution, and hydrogen gas is given off.
- Ethanol can be oxidised to ethanoic acid, either by chemical oxidising agents or by the action of microbes. Ethanoic acid is the main acid in vinegar.

253

Further teaching suggestions

Fermentation question

- You may wish to ask students to explain why 'air locks' are not needed in breweries when the fermentation is going rapidly (many breweries work with open fermenters) but are needed once fermentation slows down.

Advert

- You could ask students to create an advert to explain why methylated spirits must not be consumed.

Original questions

- Students could generate three questions to which each key point would be an answer.

Comparing ethanol and water

- Ask students to use their knowledge of the properties of water and ethanol along with secondary sources of information such as data books to contrast the properties of water and ethanol. Students could compare melting point, boiling point, chemical reactions, appearance and density. Students can complete a number of tests on alcohols to find out their properties. See www.chemistryteachers.org and search for 'properties of alcohols'.

Oxidation of ethanol

- You may wish to demonstrate the oxidation of ethanol using potassium dichromate. See CLEAPSS Hazcard 40A.

C3 5.3 Carboxylic acids and esters

Learning objectives

Students should learn:

- how to recognise carboxylic acids from their properties
- the uses of carboxylic acids and esters
- why carboxylic acids are described as weak acids **[HT only]**
- how esters can be made.

Learning outcomes

Most students should be able to:

- describe the properties and an example of a carboxylic acid
- state a use of a carboxylic acid and an ester
- describe how to make an ester.

Some students should also be able to:

- explain the difference between strong and weak acids in terms of ionisation **[HT only]**
- explain why carboxylic acids are categorised as weak acids. **[HT only]**

Answers to in-text questions

a Carbon dioxide.

b Ethyl ethanoate.

c They evaporate easily as they have low boiling points.

Support

- Students may find it difficult to remember what a carboxylic acid is and how this relates to an ester. Support students by giving them an incomplete summary about carboxylic acids and esters. You could then give them the missing information to complete the summary. Students could then refer to this when tackling other activities.

Extend

- Give students the names of a selection of esters. They should then write a word equation for the formation of that ester.

Specification link-up: Chemistry C3.6

- Carboxylic acids:
  - dissolve in water to produce acidic solutions
  - react with carbonates to produce carbon dioxide
  - react with alcohols in the presence of an acid catalyst to produce esters
  - do not ionise completely when dissolved in water and so are weak acids **[HT only]**
  - aqueous solutions of weak acids have a higher pH value than aqueous solutions of strong acids with the same concentration *[C3.6.2 b)]* **[HT only]**
- Ethyl ethanoate is the ester produced from ethanol and ethanoic acid. Esters have the functional group —COO—. They are volatile compounds with distinctive smells and are used as flavourings and perfumes. *[C3.6.3 a)]*

Controlled Assessment: C4.1 Plan practical ways to develop and test candiates' own scientific ideas. *[C4.1.1 c)]*; C4.3 Collect primary and secondary data. *[C4.3.2 a)]*

Lesson structure

Starters

Group – Give students a written selection of 10 different compounds, all of which should be esters, carboxylic acids or alcohols. Some of the compounds should be given as a molecular formula, some as a structural formula and some as a displayed formula. Ask the students to group the compounds by the chemical family (homologous series) they belong to. Support students by giving them a brief definition of each homologous series. Extend students by asking them to try to name the compounds. *(5 minutes)*

Explain – Ask students to explain how they would find out which of two acids (of equal concentration) is strong and which is weak [pH scale/common reactions]. Give students time to think about the question individually. Then split the class into four groups. Ask each group to decide on the method and feed back to the class. *(10 minutes)* **[HT only]**

Main

- Allow students to compare ethanoic acid (a weak acid) with hydrochloric acid (a strong acid). Extend students by asking them to write word equations for the reactions with sodium carbonate.
- Encourage students to reflect on their method and results. Students should be able to suggest that their conclusions about the relative strength of the acids are valid only if the concentration of the two acids is the same. They should also appreciate that using a pH sensor rather than universal indicator gives improved accuracy when finding pH values.
- Carefully explain the difference between strong and weak acids as opposed to concentrated and dilute solutions of acids. Ask how they can have a solution of a strong acid (high degree of ionisation/dissociation) but with a relatively high pH value, e.g. 5 [very dilute solution]. **[HT only]**
- Create a card sort made up of 30 questions and their answers on separate cards. The questions should range from easy questions such as 'Which functional group is in an alcohol?' to examination-style questions including equations and summary questions. Students should match the questions with their answers. This activity can be differentiated by giving fewer questions to some students. Students could also work in teams.
- Then ask students to pick five of the questions, together with their answers, and record them in their notes.

Plenaries

Bingo – Make bingo cards with six answers on the card, including key words – e.g. ester, alcohol, carboxylic acid, organic chemical. Then ask questions. The students cross off the answer if it appears on their bingo card. The first student to call 'house' (with all the answers) is the winner and could get a prize. *(5 minutes)*

Prediction – Ask students to predict how ethanoic acid and hydrochloric acid with the same concentration would react with magnesium. [Both acids would react with the metal to produce hydrogen gas and a metal salt; however, the rate of reaction with hydrochloric acid would be greater.] Support students by encouraging them to work in pairs and reflect on the practical in which they reacted sodium carbonate with the two acids. You may wish to demonstrate the reaction of magnesium with hydrochloric acid to act as memory aid. Extend students by asking them to write equations for these reactions. *(10 minutes)*

Practical support

Comparing ethanoic acid and hydrochloric acid

Equipment and materials required

Two dropping pipettes, spatula, 10 cm^3 measuring cylinder, 2 × boiling tubes, test-tube rack, universal indicator or pH sensor with data logger or pH meter, 2 mol/dm^3 hydrochloric acid (irritant), 2 mol/dm^3 ethanoic acid (irritant), sodium carbonate powder (irritant), eye protection.

Details

Measure 5 cm^3 of hydrochloric acid into a boiling tube and record the pH. Add half a spatula of sodium carbonate powder and observe. Repeat using ethanoic acid. It is worth noting that weak acids are Higher Tier only but reaction with carbonates is Foundation Tier.

Safety: CLEAPSS Hazcard 47A Hydrochloric acid – corrosive; CLEAPSS Hazcard 38A Ethanoic acid – corrosive; CLEAPSS Hazcad 95A Sodium carbonate – irritant.

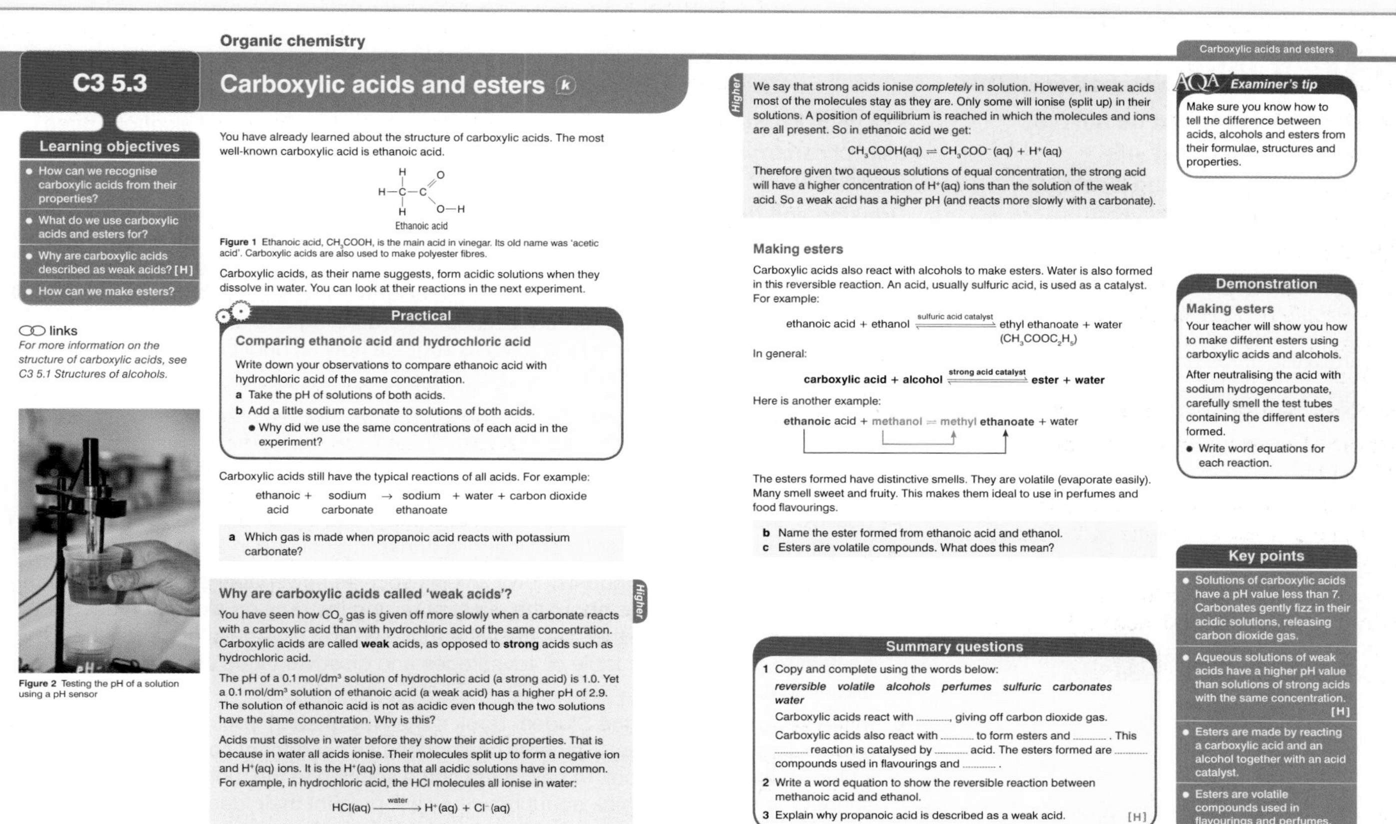

Organic chemistry

C3 5.3 Carboxylic acids and esters

Learning objectives
- How can we recognise carboxylic acids from their properties?
- What do we use carboxylic acids and esters for?
- Why are carboxylic acids described as weak acids? [H]
- How can we make esters?

links
For more information on the structure of carboxylic acids, see C3 5.1 Structures of alcohols.

Figure 2 Testing the pH of a solution using a pH sensor

You have already learned about the structure of carboxylic acids. The most well-known carboxylic acid is ethanoic acid.

Ethanoic acid

Figure 1 Ethanoic acid, CH_3COOH, is the main acid in vinegar. Its old name was 'acetic acid'. Carboxylic acids are also used to make polyester fibres.

Carboxylic acids, as their name suggests, form acidic solutions when they dissolve in water. You can look at their reactions in the next experiment.

Practical

Comparing ethanoic acid and hydrochloric acid

Write down your observations to compare ethanoic acid with hydrochloric acid of the same concentration.
a Take the pH of solutions of both acids.
b Add a little sodium carbonate to solutions of both acids.
- Why did we use the same concentrations of each acid in the experiment?

Carboxylic acids still have the typical reactions of all acids. For example:

ethanoic acid + sodium carbonate → sodium ethanoate + water + carbon dioxide

a Which gas is made when propanoic acid reacts with potassium carbonate?

Why are carboxylic acids called 'weak acids'? (Higher)

You have seen how CO_2 gas is given off more slowly when a carbonate reacts with a carboxylic acid than with hydrochloric acid of the same concentration. Carboxylic acids are called **weak** acids, as opposed to **strong** acids such as hydrochloric acid.

The pH of a 0.1 mol/dm^3 solution of hydrochloric acid (a strong acid) is 1.0. Yet a 0.1 mol/dm^3 solution of ethanoic acid (a weak acid) has a higher pH of 2.9. The solution of ethanoic acid is not as acidic even though the two solutions have the same concentration. Why is this?

Acids must dissolve in water before they show their acidic properties. That is because in water all acids ionise. Their molecules split up to form a negative ion and $H^+(aq)$ ions. It is the $H^+(aq)$ ions that all acidic solutions have in common. For example, in hydrochloric acid, the HCl molecules all ionise in water:

$$HCl(aq) \xrightarrow{water} H^+(aq) + Cl^-(aq)$$

We say that strong acids ionise *completely* in solution. However, in weak acids most of the molecules stay as they are. Only some will ionise (split up) in their solutions. A position of equilibrium is reached in which the molecules and ions are all present. So in ethanoic acid we get:

$$CH_3COOH(aq) \rightleftharpoons CH_3COO^-(aq) + H^+(aq)$$

Therefore given two aqueous solutions of equal concentration, the strong acid will have a higher concentration of $H^+(aq)$ ions than the solution of the weak acid. So a weak acid has a higher pH (and reacts more slowly with a carbonate).

Making esters

Carboxylic acids also react with alcohols to make esters. Water is also formed in this reversible reaction. An acid, usually sulfuric acid, is used as a catalyst. For example:

ethanoic acid + ethanol $\xrightleftharpoons{\text{sulfuric acid catalyst}}$ ethyl ethanoate + water ($CH_3COOC_2H_5$)

In general:

carboxylic acid + alcohol $\xrightleftharpoons{\text{strong acid catalyst}}$ **ester + water**

Here is another example:

ethanoic acid + methanol ⇌ methyl ethanoate + water

The esters formed have distinctive smells. They are volatile (evaporate easily). Many smell sweet and fruity. This makes them ideal to use in perfumes and food flavourings.

b Name the ester formed from ethanoic acid and ethanol.
c Esters are volatile compounds. What does this mean?

Summary questions

1 Copy and complete using the words below:
reversible volatile alcohols perfumes sulfuric carbonates water
Carboxylic acids react with .... giving off carbon dioxide gas.
Carboxylic acids also react with .... to form esters and .... This .... reaction is catalysed by .... acid. The esters formed are .... compounds used in flavourings and ....
2 Write a word equation to show the reversible reaction between methanoic acid and ethanol.
3 Explain why propanoic acid is described as a weak acid. [H]

AQA Examiner's tip

Make sure you know how to tell the difference between acids, alcohols and esters from their formulae, structures and properties.

Demonstration

Making esters

Your teacher will show you how to make different esters using carboxylic acids and alcohols.

After neutralising the acid with sodium hydrogencarbonate, carefully smell the test tubes containing the different esters formed.
- Write word equations for each reaction.

Key points
- Solutions of carboxylic acids have a pH value less than 7. Carbonates gently fizz in their acidic solutions, releasing carbon dioxide gas.
- Aqueous solutions of weak acids have a higher pH value than solutions of strong acids with the same concentration. [H]
- Esters are made by reacting a carboxylic acid and an alcohol together with an acid catalyst.
- Esters are volatile compounds used in flavourings and perfumes.

254 255

Further teaching suggestions

Ester samples

- You may have a selection of esters available in school. Put a small amount of an ester into a 100 cm^3 beaker and dip cotton wool buds in. Then allow students to smell different samples of esters. You can also add a few drops of the ester to water and this can help students to smell the ester more effectively. You could extend students by asking them to suggest the uses of esters, e.g. food additives.

Ester synthesis

- You might wish to allow students to make an ester. You could even use a microscale method, which would reduce the amounts of chemicals used and the hazards associated with them. Go to www.practicalchemistry.org and search for 'esters'.

Acids in rain

- Students could be encouraged to find out which acid is naturally in rain (carbonic acid – a weak acid) and how it is formed. They could then go on to discover/revise how acid rain differs [contains strong acids such as nitric and sulfuric acid] and how it affects the environment.

Synoptic

- You may wish to give students samples of ethanol, ethanoic acid and ethyl ethanoate and ask them to use simple tests to find out which chemical is which.

Summary answers

1 carbonates, alcohols, water, reversible, sulfuric, volatile, perfumes

2 methanoic acid + ethanol ⇌ ethyl methanoate + water

3 Strong acids will ionise completely in water but weak acids, such as propanoic acid, only ionise partially. There is a large proportion of the propanoic acid molecules present in its solution.

C3 5.4 Organic issues

Learning objectives

Students should learn:

- that there are advantages and disadvantages to using alcohols, carboxylic acids and esters.

Learning outcomes

Most students should be able to:

- recall the advantages and disadvantages of using alcohols, carboxylic acids and esters.

Some students should also be able to:

- evaluate in detail the advantages and disadvantages for using alcohols, carboxylic acids and esters.

Specification link-up: Chemistry C3.6

- Evaluate the social and economic advantages and disadvantages of the uses of alcohols, carboxylic acids and esters. *[C3.6]*

Lesson structure

Starters

5, 4, 3, 2, 1 – Ask students to give five uses of alcohols, four names of alcohols, three uses of carboxylic acids, two names of carboxylic acids and one use of an ester. *(5 minutes)*

Drinking and driving – Show students some government material against drinking and driving. Ask students to make a list of three reasons why people do drink alcohol and three disadvantages to drinking alcohol. Encourage volunteers to read out their reasons. Support students by asking them to work in pairs. You could supply them with a number of statements, which they then order as reasons for drinking and disadvantages of drinking. Extend students by asking them to suggest how alcohol can negatively affect a person's health, finance and social life. *(10 minutes)*

Main

- In the first activity in the Student Book ('Raising the cost of alcoholic drinks?') encourage students to think about how they could discover reliable data about the positive and negative effects of the cost of alcohol and the effect on society. Research the move to increase the price of alcoholic drinks in Scotland in 2010.
- You may wish to show students news articles such as newspaper clippings and archive articles on news websites about the effect of raising the duty on alcohol. Students could then consider whether the articles are biased and use the information they contain to generate their table of advantages and disadvantages of raising the price of alcoholic drinks.
- In the second activity in the Student Book ('The way ahead') you may wish to secure a room-swap to the library. Students could then use secondary sources of information including the Student Book to help them find a solution to the fuel crisis. You could encourage students to think about how they will present their ideas to the class – a podcast, a presentation, a poster, a text message, etc. Each group could then share its work with the rest of the class. At the end of the presentations, students could vote for the idea they think is the most creative, the most scientific and the most likely to be adopted by government.

Plenaries

Advantage or disadvantage? – Ask students to put their thumb up if the statement is an advantage or down if the statement is a disadvantage.

- In the UK most of the cost of alcoholic drinks is tax. [advantage or disadvantage, depending on viewpoint]
- People can be dependent on alcohol. [disadvantage]
- Alcohol can help relieve stress. [advantage]
- Alcohol has been linked to high blood pressure. [disadvantage]
- Alcohol has been linked to heart disease. [disadvantage]
- New farming land can be created by deforestation. [disadvantage]
- Biodiesel increases crop prices. [disadvantage] *(5 minutes)*

AfL (Assessment for Learning) – Give the students an examination question and time their attempt. Give the students one minute per mark. After the time is up, show the students the mark scheme and allow them to mark their own work. Ask students to reflect on their answers and suggest how they could improve their examination technique, e.g. answer more briefly. The appropriate tier of entry could be chosen based on the ability of the students in the class. *(10 minutes)*

Support

- Give students a writing frame to list the advantages and disadvantages of using alcohols. They could be given a list of statements that they could sort into the correct part of the writing frame. To support them further, allow them to contrast the benefits and drawbacks of the uses.

Extend

- Ask students to group the advantages and disadvantages of the uses of these organic chemicals as social, economic or environmental.

Further teaching suggestions

Debate

- You could have a debate on the motion 'The sale of cheap alcoholic drinks in supermarkets should be banned.' Split the class into two groups, one arguing for and the other against. Allow students to prepare their arguments and then hold the debate.

Research and presentation

- Split the class into three groups. Each group should be given one of the organic chemical families that it has met, esters, alcohols and carboxylic acids, to evaluate their uses. Each group should use secondary sources of information including the Student Book to evaluate their chemical family's use. They should also find out the advantages and disadvantages of some uses. They should consider where they are getting their information from and assess whether the source is reliable and credible. Each group should then decide whether it is for or against the uses. Then encourage each group to report its thoughts back to the whole class, including a conclusion where it states whether they are for or against the use of this family of organic chemicals.

How Science Works

Organic chemistry

C3 5.4

Organic issues

Learning objectives

- What are the issues involved in the use of alcohols, carboxylic acids and esters?

Figure 1 Addiction to alcohol can cause many problems for the individuals themselves and society as a whole

links

For more information on the use of ethanol as a solvent in methylated spirits, see C3 5.2 Properties and uses of alcohols.

Did you know ... ?

When European winemakers produce more wine than they can sell, it is distilled to make ethanol for industrial purposes (not for alcoholic drinks).

Ethanol in drinks

An alcoholic drink, such as one glass of red wine, can help people relax. It can help relieve the stress for some people after a hard day at work. However, too many people are drinking more than the maximum recommended amount of alcohol. This puts their health at risk. Alcohol has been associated with high blood pressure and heart disease. Excess alcohol can also damage the liver. In extreme cases, a liver transplant is the only way to avoid death.

Alcohol is a socially acceptable drug. Like other drugs, some people become dependent on it. Alcoholics are addicted to ethanol. Many ruin their lives because their behaviour changes as a result of drinking alcohol.

Ethanol is used in methylated spirits as a solvent. Some desperate addicts will drink this because it contains a lot of ethanol and it's cheap. They will do this even though it has had toxic methanol added to it. It also contains emetics (substances that make you vomit) and foul-tasting chemicals.

Drinking 'meths' causes liver failure, blindness and an early death. Other chemicals are also added to methylated spirits to make it more difficult to distil off the ethanol. By including chemicals with similar boiling points, people can't separate off the ethanol for drinking.

a Why is it very difficult to distil ethanol from methylated spirits?

Alcoholic drinks are more expensive than methylated spirits because they have tax added on. The government can use the income generated for many good causes. However, we should weigh this against the costs of dealing with:

- the health problems
- days lost at work
- policing antisocial behaviour.

Figure 2 Binge drinking can cause violent behaviour. In severe cases it leads to alcoholic poisoning. Doctors are reporting an increasing number of alcohol-related health problems in 20- to 30-year-olds. These problems used to be seen mainly in middle-aged alcoholics.

Activity

Raising the cost of alcoholic drinks?

Some people are campaigning against the sale of cheap alcoholic drinks in supermarkets. They argue that it makes alcohol more easily available, especially to young adults. Cheap promotions in bars are also criticised for encouraging excessive drinking.

- Make a table with points for and against the banning of cheap alcoholic drinks in supermarkets and 'happy hours' in bars.

Ethanol and esters as biofuels

You have seen that ethanol can be used as a biofuel. It is made by fermenting sugars from crops. We also looked at biodiesel. This is made from plant oils which are esters. In processing these esters, the oils are broken down into long-chain carboxylic acids. They are then reacted with methanol or ethanol (in the presence of a catalyst) to make the esters used as biodiesel.

b Which alcohol is used to make an ethyl ester used as biodiesel?

However, the land used for biofuel crops could be used for food crops. With an ever-increasing world population, feeding ourselves will become more of an issue. We will need more land for farming – both for fuels and food.

This new farming land is often made by cutting down and burning tropical rainforests. This destroys habitats of wildlife and contributes towards increasing the percentage of carbon dioxide in the atmosphere. Yet alternatives to crude oil are needed urgently, so what can we do?

Activity

The way ahead

Work as a small group to make a list of ideas for a new government to help deal with the fuel crisis facing us.

Present your ideas to the rest of the class and arrive at a single list together.

Summary questions

1 a What are the social affects of drinking alcohol and driving?
  b Some people would prefer it to be illegal to have any trace of alcohol in the blood at all when driving. At present you must have more than 80 mg of alcohol per 100 cm^3 of blood in order to be prosecuted. What might be a difficulty with a limit of 0 mg of ethanol?

2 If more and more farmers grow crops for biofuels, what could happen to the price of crop-based foods such as bread? How might this become a cyclical problem which gets worse and then better at regular intervals?

Organic issues

links

For more information on using ethanol as a biofuel and vegetable oils for making biodiesel, look back to C1 4.5 Alternative fuels.

Did you know ... ?

You can buy this machine to make your own biodiesel at home. It can produce up to 50 litres a day from a mixture of used or new cooking oils and methanol.

Figure 3 A domestic biodiesel generator

Key points

- Alcohols, carboxylic acids and esters have many uses which benefit society.
- Some of these substances, such as ethanol and solvents, can be abused.
- In future, the use of biofuels, such as ethanol and esters, could help society as crude oil supplies run out.
- Future uses of biofuels might conflict with the need to feed the world.

256 257

Answers to in-text questions

a Methylated spirits contains other substances with similar boiling points to ethanol.

b Ethanol.

Summary answers

1 a An increased chance of having fatal road accidents, causing suffering for many people associated with the victims of the crime. The offenders, if they survive, can face a prison sentence and often lose their jobs.

b Some everyday products use a small amount of ethanol as a solvent, e.g. mouthwash, so innocent drivers who have not been drinking could face prosecution if the limit was zero.

2 The price will go up as demand outstrips supply. As the price of food crops increases, farmers might well switch back from fuel crops to food crops. The price of fuel crops then increases, and so on.

Summary answers

1. **a** A
   b C
   c esters
   d B
   e A = $CH_3CH_2CH_2COOH$ C = $CH_3CH_2CH_2CH_2OH$
2. **a** Sodium sinks, fizzes, getting smaller and smaller until it has all dissolved away.
   b Hydrogen.
3. **a**

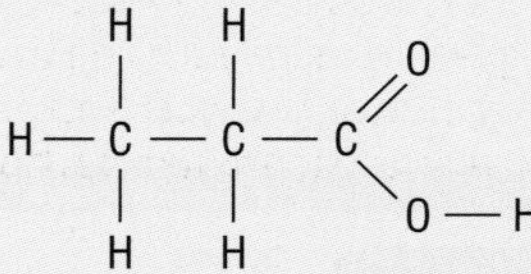

   b Pass the gas through limewater, which should turn cloudy/milky.
4. **a** Ethyl ethanoate.
   b ethanoic acid + ethanol $\xrightleftharpoons{\text{sulfuric acid}}$ ethyl ethanoate + water
   c It has a boiling point that is relatively low/not much above room temperature.
   d Methanoic acid and propanol.
5. **a** For example, alcoholic drinks/solvents/fuels/making esters.
   b For example, vinegar/making esters.
   c For example, perfumes/flavourings.
6. For example, add sodium carbonate – only ethanoic acid would fizz (giving off carbon dioxide); then add sodium to the other two remaining liquids – only ethanol would fizz (giving off hydrogen); the remaining liquid must be ethyl ethanoate (smell to check).
7. **a** $CH_3CH_2CH_2OH$
   b **i** The blue cobalt chloride paper turns pink/condensation is seen inside the U-tube.
   ii The limewater turns cloudy/milky.
   c When propanol burns, carbon dioxide and water are formed as products.
   d propanol + oxygen → carbon dioxide + water
   e $C_3H_7OH + 4\frac{1}{2}O_2 \rightarrow 3CO_2 + 4H_2O$
   or $2C_3H_7OH + 9O_2 \rightarrow 6CO_2 + 8H_2O$
8. **a** The pH of the propanoic acid solution will be higher than the pH of the dilute nitric acid.
   b Nitric acid molecules ionise completely when added to water but in an aqueous solution of propanoic acid only a small proportion of molecules ionise. This means the solution of nitric acid has a higher concentration of hydrogen ions, $H^+(aq)$.

Summary questions

1 Look at the three organic molecules A, B and C below:

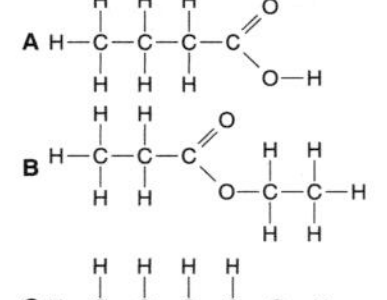

C $H-C(H)(H)-C(H)(H)-C(H)(H)-C(H)(H)-O-H$

Answer these questions about **A**, **B** and **C**:

a Which one is a carboxylic acid?

b Which one is an alcohol?

c Which homologous series of organic compounds does B belong to?

d Which of the compounds can be represented as $CH_3CH_2COOCH_2CH_3$?

e Using a structural formula as shown in part **d**, give the structural formulae of the other two compounds.

2 a Describe what you would see happen if a small piece of sodium metal was dropped into a beaker containing some ethanol.

b Name the gas given off in the reaction described in part **a**.

3 a Draw the displayed formula of propanoic acid, showing all the atoms and bonds.

b Some calcium carbonate powder is dropped into a test tube of a solution of propanoic acid. How would you positively identify the gas given off?

4 a What is the name of this compound?

b Write a word equation showing how this compound can be made. Include the catalyst.

c The compound shown is volatile. What can you say about its boiling point?

d Name the carboxylic acid and the alcohol we would use to make propyl methanoate.

5 State one use of:

a alcohols

b carboxylic acids

c esters.

6 Describe how could you distinguish between sample[s of] ethanol, ethanoic acid and ethyl ethanoate using sim[ple] tests.

7 Propanol was burned in a spirit burner and the produ[cts] of combustion were tested as shown below:

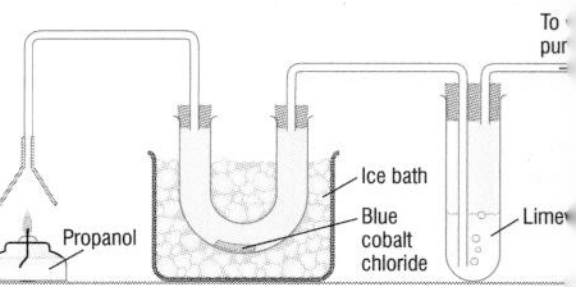

a In the style of Question 1 **d**, write the structural for[mula] of propanol.

b After the propanol has been burning for a while, wh[at] happens in:

i the U-tube

ii the boiling tube containing limewater?

c What does this experiment show?

d Write a word equation to show the combustion of propanol.

e Write a balanced symbol equation to show the combustion of propanol.

8 You are given a 0.1 mol/dm³ solution of nitric acid, an[d a] solution of propanoic acid with the same concentrati[on].

a What can you predict about the pH of the two solutions?

b Explain your answer to part **a**.

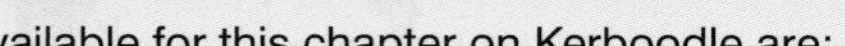

Kerboodle resources

Resources available for this chapter on Kerboodle are:

- Chapter map: Organic chemistry
- Practical: Comparing the reactions of methanol, ethanol and propanol (C3 5.2)
- Support: Organic molecules (C3 5.1–5.3)
- Extension: Identifying organic molecules (C3 5.1–5.3)
- How Science Works: Drinking and driving (C3 5.4)
- Viewpoints: Thinking about drinking (C3 5.4)
- WebQuest: Is it time alcohol was banned? (C3 5.4)
- Interactive activity: Properties of alcohols, carboxylic acids and esters
- Revision podcast: Organic families
- Test yourself: Organic chemistry
- On your marks: Alcohols
- Examination-style questions: Organic chemistry
- Answers to examination-style questions: Organic chemistry

QA Examination-style questions

1 The displayed formula of a compound is

```
   H   H
   |   |
H—C—C—O—H
   |   |
   H   H
```

Choose an answer from the lists to complete each sentence.

a This compound is called .... (1)

methanol ethanol propanol

b The structural formula of this compound is .... (1)

CH_3CH_2OH CH_3COOH CH_3COOCH_3

c The functional group in this compound is .... (1)

–COOH –OH –COO–

2 The structural formulae of three substances A, B and C are shown.

CH_3CH_2COOH (A) $CH_3COOCH_2CH_3$ (B) $CH_3CH_2CH_2OH$ (C)

a Match each substance with the group of compounds to which it belongs.

i alcohol
ii carboxylic acid
iii ester (1)

b Which substance is called ethyl ethanoate? (1)

c Which substance has the formula $C_3H_6O_2$? (1)

3 A technician found three bottles of liquids on a shelf. The labels had fallen off. The labels were ethanol, ethanoic acid and ethyl ethanoate. The technician wrote A, B and C on the three bottles and tested the liquids. The technician's results are shown in the table.

Test	Results		
	A	B	C
Add a few drops of liquid to 2 cm³ of water and add Universal Indicator.	red	green	green
Add a few drops of sodium carbonate solution to 2 cm³ of liquid.	fizzed	no reaction	no reaction
Add a small piece of sodium to 2 cm³ of liquid.	not tested	fizzed	no reaction

Match the labels with the liquids A, B and C, giving reasons for your conclusions. (4)

4 Methylated spirits is a mixture of ethanol with 5 to 10% methanol. Methanol is much more toxic than ethanol. Methylated spirits is used as a fuel for spirit burners and as a solvent for cleaning and decorating.
The equation for the complete combustion of ethanol is:

$CH_3CH_2OH + 3O_2 \rightarrow 2CO_2 + 3H_2O$

a i Write a word equation for the complete combustion of methanol. (1)

ii Suggest **two** reasons why methylated spirits is suitable for use as a fuel in spirit burners used as camping stoves. (2)

b i What is the structural formula of methanol? (1)

ii Explain why methanol mixes completely with ethanol. (2)

c Methylated spirits can be used to clean glass windows and mirrors. Suggest **two** properties that make it suitable for cleaning glass. (2)

d Methylated spirits containing 90 to 95% ethanol costs much less than vodka containing 40% ethanol. This is because alcohol sold in the UK for drinking is taxed at a high rate by the UK government.

i Suggest **one** reason why the UK government tax alcohol for drinking at a high rate. (1)

ii Why is methylated spirits not taxed at the same rate? (1)

5 a The flavour and smell of wine that is kept in bottles for some time may improve.

One of the reactions that may take place is:

$CH_3COOH + CH_3CH_2OH \rightarrow CH_3COOCH_2CH_3 + H_2O$

Explain why this reaction would improve the smell of the wine. (2)

b Beer contains about 5% ethanol. Explain why beer that is left exposed to the air for several hours tastes sour. (3)

6 Sulfuric acid ionises completely when dissolved in water.

a Balance the equation that shows what happens to sulfuric acid in water.

$H_2SO_4(aq) \rightarrow H^+(aq) + SO_4^{2-}(aq)$ [H] (1)

b Write a balanced symbol equation to show what happens when ethanoic acid, CH_3COOH, is dissolved in water. [H] (2)

c You have been given two solutions of acids with the same concentration. One is ethanoic acid, the other is sulfuric acid. Describe and give the results of a test that you could use to show which acid is in each solution. Explain the results of your test. [H] (5)

Practical suggestions

Practicals	AQA	k		
Investigation of the reactions of ethanol.	✓		✓	
Comparison of properties of ethanol with water.	✓		✓	
Oxidation of ethanol using aqueous potassium dichromate.	✓		✓	
Design and carry out an investigation of the oxidation of dilute solutions of ethanol (e.g. wine or beer) by exposing to the air for several days.	✓		✓	
Comparison of the reactions of methanol, ethanol and propanol.	✓		✓	
Investigation of the reactions of ethanoic acid.	✓		✓	
Distinguishing between samples of ethanol, ethanoic acid and ethyl ethanoate using simple chemical tests.	✓	✓	✓	
Preparation of ethyl ethanoate using ethanol and ethanoic acid with sulfuric acid as a catalyst. Recognise the ester by smell after neutralising the acid with sodium hydrogencarbonate.	✓		✓	
Add drops of esters to water to smell more effectively.	✓		✓	

Examination-style answers

1 a ethanol *(1 mark)*

b CH_3CH_2OH *(1 mark)*

c –OH *(1 mark)*

2 a **i** C **ii** A **iii** B *(1 mark)*

b B *(1 mark)*

c A *(1 mark)*

3 A is ethanoic acid, B is ethanol, C is ethyl ethanoate gains one mark, plus any three from: A turns UI red or is acid (in solution); A fizzed or reacts or produces gas with sodium carbonate solution; B turns UI green or is neutral (in solution); B fizzed or reacts or produces gas/hydrogen with sodium; C is neutral and does not react with either sodium carbonate or sodium. *(4 marks)*

4 a **i** methanol + oxygen → carbon dioxide + water *(1 mark)*

ii Any **two** from: burns cleanly; leaves no residue; can be carried in bottles or without special/heavy equipment; produces a lot of energy (for its mass/volume); easy to ignite. *(2 marks)*

b **i** CH_3OH *(1 mark)*

ii They mix because they are chemically similar or have similar properties, as they are in the same group/series of compounds or have similar formulae or the same functional group or are both small covalent molecules. *(2 marks)*

c Any **two** from: (good) solvent or dissolves grease; evaporates easily/low boiling point; leaves no residue; does not react with/attack glass (allow neutral solution). *(2 marks)*

d **i** Any **one** from: to raise money/revenue for the government; to discourage (excessive) drinking (of alcohol). *(1 mark)*

ii Any **one** from: it is toxic/poisonous; it is not suitable/intended for drinking. *(1 mark)*

5 a Any **two** from: ester produced or $CH_3COOCH_2CH_3$ is an ester; esters have a good/pleasant smell; CH_3COOH removed or amount reduced; CH_3COOH has a poor smell or smells of vinegar. *(2 marks)*

b Ethanol is oxidised or ethanoic acid is formed; by oxygen from the air or by microbes; and ethanoic acid tastes sour or like vinegar. *(3 marks)*

6 a $H_2SO_4\ (aq) \rightarrow 2\ H^+\ (aq) + SO_4^{2-}\ (aq)$ *(1 mark)*

b $CH_3COOH\ (aq) \rightarrow CH_3COO^-\ (aq) + H^+\ (aq)$ one mark for CH_3COO^- (aq) and one mark for H^+ (aq) *(2 marks)*

c One mark for test – e.g. add universal indicator, add sodium/calcium/named carbonate. One mark for result – e.g. turns red/orange; fizzes/effervesces/produces gas/carbon dioxide; one mark for difference in result e.g. sulfuric – red and ethanoic – orange or faster reaction with sulfuric; one mark for sulfuric is strong acid; ethanoic is weak; one mark for ethanoic does not ionise completely or ethanoic has higher pH. *(5 marks)*

Examination-style answers

1 a i To remove solids/undissolved substances. *(1 mark)*

ii To kill microorganisms/microbes/harmful bacteria *(1 mark)*

b To monitor/check the levels of any of the following: dissolved solids or chlorine or pH or microorganisms *(1 mark)*

c i Calcium or magnesium or Ca^{2+} or Mg^{2+} *(1 mark)*

ii When heated temporary hard water produces scale or a solid deposit; this is deposited on pipes and heating elements and so reduces the efficiency of heaters or increases energy costs or blocks pipes or damages boilers. (Accept HT answers that explain in terms of hydrogencarbonate ions/HCO_3^{2-} being decomposed by heating; calcium/magnesium carbonate is insoluble/deposited/precipitated. *(3 marks)*

iii Good for bones or teeth, reduces heart disease. *(1 mark)*

2 a Products are at a lower level than reactants or products have less energy than reactants. *(1 mark)*

b Diagram with line drawn from reactants to products that peaks below original line. *(1 mark)*

c No effect. *(1 mark)*

d 4359 kJ per mole gains 3 marks. 4359 (no units) gains 2 marks. If incorrect, 5 × (C—H) + (C—O) + (O—H) + 3 × (O=O) gains 1 mark, 5 × 413 + 336 + 464 + 1494 gains 1 mark. *(3 marks)*

e 6004 (kJ per mole) gains 2 marks. 4 × (C=O) + 6 × (O—H) or 4 × 805 + 6 × 464 gains 1 mark. *(2 marks)*

f 1645 (kJ per mole) or error carried forward from answers to **d** and **e**. *(1 mark)*

g 35800/35760 or answer to **f** divided by 46 correctly evaluated gains 2 marks. If incorrect, 46 or answer to **f** divided by incorrectly calculated M_r gains 1 mark. *(2 marks)*

3 One mark for each correct part of names (allow formulae of compounds or ions): A is magnesium sulfate, B is potassium bromide, C is lithium chloride, D is iron(II) sulfate. *(8 marks)*

4 a One mark each for: X is ethyl ethanoate, Y is ethanoic acid, Z is ethanol. *(3 marks)*

b One mark each for: X is an ester, Y is a carboxylic acid, Z is an alcohol. *(3 marks)*

c One mark each for *either* use universal indicator or full-range indicator or pH indicator or pH meter/probe; Y turns red or has low(er) pH; Z turns green or has pH7; *or* use any named carbonate or hydrogencarbonate; Y fizzes or produces bubbles/carbon dioxide; Z has no reaction. (Do not accept reaction with sodium.) *(3 marks)*

5 There is a clear and detailed scientific description of the reasons why ammonia plants operate using compromise conditions of 200 atmospheres and 450 °C. The answer shows almost faultless spelling, punctuation and grammar. It is coherent and in an organised, logical sequence. It contains a range of appropriate and relevant specialist terms used accurately. *(5–6 marks)*

There is a scientific description of the reasons why ammonia plants operate using conditions of 200 atmospheres and 450 °C. There are some errors in spelling, punctuation and grammar. The answer has some structure and organisation. The use of specialist terms has been attempted but not always accurately. *(3–4 marks)*

There is a brief description of the reasons why ammonia plants operate at 200 atmospheres and/or 450 °C. The spelling, punctuation and grammar are very weak. The answer is poorly organised with almost no specialist terms and/or their use demonstrating a general lack of understanding of their meaning. *(1–2 marks)*

No relevant content. *(0 marks)*

Examples of chemistry points made in the response:

- Effect of pressure: high pressure increases yield; because fewer (gaseous) product molecules (Le Châtelier); high pressure increases rate of reaction.
- Effect of temperature: low temperature increases yield; because exothermic reaction (Le Châtelier); at low temperature rate is slow; at low temperature catalyst does not work.
- Conditions are compromised/optimum, give reasonable yield at reasonable rate and cost: 200 atmospheres is the highest pressure that is economical; high pressure increases costs of plant/energy; high pressure increases safety risks; 450 °C because of conflict between rate and yield; 450 °C gives a reasonably fast rate of reaction; higher temperatures would increase the rate but would decrease yield; above 450 °C yield becomes too low; below 450 °C the rate becomes too slow, although percentage yield/conversion is low; unreacted gases are recycled and not wasted.

Examination-style questions

1 The diagram shows a simplified flow diagram of a water treatment works which supplies drinking water.

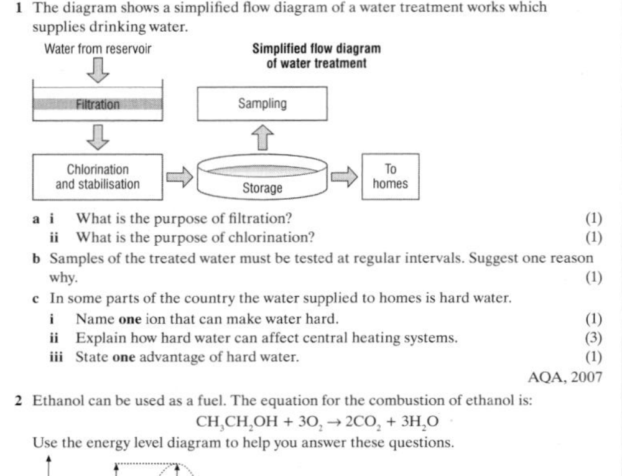

a i What is the purpose of filtration? (1)
ii What is the purpose of chlorination? (1)
b Samples of the treated water must be tested at regular intervals. Suggest one reason why. (1)
c In some parts of the country the water supplied to homes is hard water.
i Name **one** ion that can make water hard. (1)
ii Explain how hard water can affect central heating systems. (3)
iii State **one** advantage of hard water. (1)

AQA, 2007

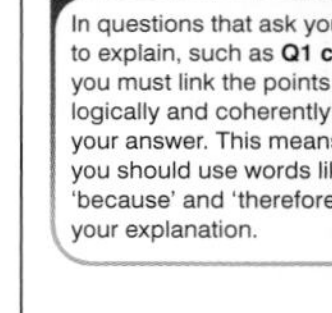

AQA Examiner's tip
In questions that ask you to explain, such as Q1 c i you must link the points logically and coherently in your answer. This means you should use words like 'because' and 'therefore' your explanation.

2 Ethanol can be used as a fuel. The equation for the combustion of ethanol is:

$$CH_3CH_2OH + 3O_2 \rightarrow 2CO_2 + 3H_2O$$

Use the energy level diagram to help you answer these questions.

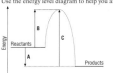

Energy, Reactants, Products, A, B, C

a How can you tell from the diagram that this reaction is exothermic? (1)
b Copy the diagram and draw a line to show what happens if a catalyst is used. (1)
c What is the effect of a catalyst on the overall energy change for the reaction? (1)

The student decided to calculate the energy change using bond energies. He wrote this equation to help.

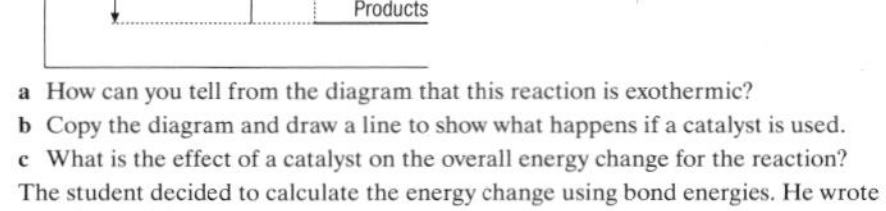

H–C(H)(H)–C(H)(H)–O–H + 3[O=O] → 2[O=C=O] + 3[H–O–H]

He looked up the following bond energies.

Bond	Bond energy in kJ per mole	Bond	Bond energy in kJ per mole
C–H	413	O–H	464
C–C	347	O=O	498
C–O	336	C=O	805

d Calculate the energy needed to break the bonds in the reactants. [H] (3)
e Calculate the energy released when the bonds in the products are formed. [H] (2)
f Calculate the energy change for the reaction. [H] (1)
g Calculate the energy change in joules per g of ethanol. [H] (2)

260

3 Some chemical tests were done on solutions of four substances, A, B, C and D. The table shows the tests and the results of the tests.

Substance	Flame test colour	Sodium hydroxide solution added	Nitric acid and silver nitrate solution added	Hydrochloric acid and barium chloride added
A	No colour	White precipitate that dissolves in excess sodium hydroxide	No reaction	White precipitate
B	Lilac	No reaction	Cream precipitate	No reaction
C	Crimson	No reaction	White precipitate	No reaction
D	No colour	Green precipitate	No reaction	White precipitate

Name the four substances A, B, C and D. (8)

When compound X with the formula $CH_3COOCH_2CH_3$ is heated with dilute hydrochloric acid it reacts with water. Two compounds Y and Z are produced. The equation for the reaction is:

$$\underset{\mathbf{X}}{CH_3COOCH_2CH_3} + H_2O \rightarrow \underset{\mathbf{Y}}{CH_3COOH} + \underset{\mathbf{Z}}{CH_3CH_2OH}$$

a Name compounds X, Y and Z. (3)

b For each of X, Y and Z, name the group of organic compounds to which it belongs. (3)

c Y and Z can be separated by distillation. They are both colourless liquids. Describe a simple test that you could use to distinguish between Y and Z. Give the results of the test for both Y and Z. (3)

In this question you will be assessed for using good English, organising information clearly and using specialist terms where appropriate.

Ammonia is produced by the Haber process. In the process nitrogen and hydrogen are mixed. The pressure is increased to about 200 atmospheres. The gases are passed over an iron catalyst at about 450 °C. The equation for the reaction is:

$$N_2(g) + 3H_2(g) \rightleftharpoons 2NH_3(g)$$

The reaction between nitrogen and hydrogen is reversible. This affects the amount of ammonia that it is possible to obtain from the process. The graph shows how the pressure and temperature affect the percentage of ammonia that can be produced.

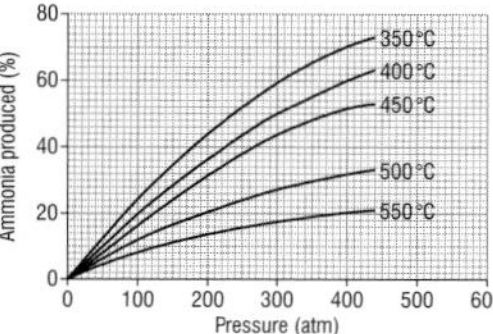

Use this information, together with your knowledge of the process, to explain why many industrial ammonia plants operate at 200 atmospheres and 450 °C. [H] (6)

AQA, 2006

AQA Examiner's tip

Many candidates find questions like **Q3** about chemical tests difficult because they do not know the tests and their results. Your teacher should give you some practice in doing the tests in the lab. Write out a checklist of the reagents used for the tests and their results and learn them thoroughly.

AQA Examiner's tip

You should be familiar with the formulae of organic compounds and the functional groups named in the specification and should be able to identify the type of organic compound from its structural formula.

Notes